TIME SERIES
Data Analysis and Theory

EXPANDED EDITION

HOLDEN-DAY SERIES IN TIME SERIES ANALYSIS

Enders A. Robinson, Editor

Box and Jenkins: *Time Series Analysis: Forecasting and Control, Rev. Ed.*
Jenkins and Watts: *Spectral Analysis and Its Applications*
Parzen: *Stochastic Processes*
Robinson: *Multichannel Time Series Analysis with Digital Computer Programs*
Robinson and Silvia: *Digital Foundations of Time Series Analysis: Volume I — The Box-Jenkins Approach*
Robinson and Silvia: *Digital Foundations of Time Series Analysis: Volume II — Wave-Equation Space-Time Processing*
Robinson and Silvia: *Digital Signal Processing and Time Series Analysis*
Rozanov: *Stationary Random Processes*

TIME SERIES

Data Analysis and Theory

EXPANDED EDITION

DAVID R. BRILLINGER

The University of California, Berkeley

HOLDEN-DAY, INC.

San Francisco

Dusseldorf Johannesburg London

Panama Singapore Sydney

Figure 1-1-3 has been provided by the courtesy of E. W. Carpenter, Figure 8 in "Explosion Seismology," *Science*, **147**: 363-373, 22 January 1965. Copyright 1965 by the American Association for the Advancement of Science.

TIME SERIES
Data Analysis and Theory
Expanded Edition

500 Sansome Street, San Francisco, California 94111

Library of Congress Catalog Card Number: 80-84117
ISBN: 0-8162-1150-7
1234567890 0123456789
Printed in the United States of America

To My Family

PREFACE TO THE EXPANDED EDITION

The 1975 edition of *Time Series: Data Analysis and Theory* has been expanded to include the survey paper "Fourier Analysis of Stationary Processes." The intention of the first edition was to develop the many important properties and uses of the discrete Fourier transforms of the observed values of time series. The Addendum indicates the extension of the results to continuous series, spatial series, point processes and random Schwartz distributions. Extensions to higher-order spectra and nonlinear systems are also suggested.

The Preface to the 1975 edition promised a Volume Two devoted to the aforementioned extensions. The author found that there was so much existing material, and developments were taking place so rapidly in those areas, that whole volumes could be devoted to each. He chose to concentrate on research, rather than exposition.

From the letters that he has received the author is convinced that his intentions with the first edition have been successfully realized. He thanks those who wrote for doing so.

D. R. B.

Berkeley, California
August, 1980

PREFACE TO THE FIRST EDITION

The initial basis of this work was a series of lectures that I presented to the members of Department 1215 of Bell Telephone Laboratories, Murray Hill, New Jersey, during the summer of 1967. Ram Gnanadesikan of that Department encouraged me to write the lectures up in a formal manner. Many of the worked examples that are included were prepared that summer at the Laboratories using their GE 645 computer and associated graphical devices.

The lectures were given again, in a more elementary and heuristic manner, to graduate students in Statistics at the University of California, Berkeley, during the Winter and Spring Quarters of 1968 and later to graduate students in Statistics and Econometrics at the London School of Economics during the Lent Term, 1969. The final manuscript was completed in mid 1972. It is hoped that the references provided are near complete for the years before then.

I feel that the book will prove useful as a text for graduate level courses in time series analysis and also as a reference book for research workers interested in the frequency analysis of time series. Throughout, I have tried to set down precise definitions and assumptions whenever possible. This undertaking has the advantage of providing a firm foundation from which to reach for real-world applications. The results presented are generally far from the best possible; however, they have the advantage of flowing from a single important mixing condition that is set down early and gives continuity to the book.

Because exact results are simply not available, many of the theorems of the work are asymptotic in nature. The applied worker need not be put off by this. These theorems have been set down in the spirit that the indicated asymptotic moments and distributions may provide reasonable approximations to the desired finite sample results. Unfortunately not too much work has gone into checking the accuracy of the asymptotic results, but some references are given.

The reader will note that the various statistics presented are immediate functions of the discrete Fourier transforms of the observed values of the time series. Perhaps this is what characterizes the work of this book. The discrete Fourier transform is given such prominence because it has important empirical and mathematical properties. Also, following the work of Cooley and Tukey (1965), it may be computed rapidly. The definitions, procedures, techniques, and statistics discussed are, in many cases, simple extensions of existing multiple regression and multivariate analysis techniques. This pleasant state of affairs is indicative of the widely pervasive nature of the important statistical and data analytic procedures.

The work is split into two volumes. This volume is, in general, devoted to aspects of the linear analysis of stationary vector-valued time series. Volume Two, still in preparation, is concerned with nonlinear analysis and the extension of the results of this volume to stationary vector-valued continuous series, spatial series, and vector-valued point processes.

Dr. Colin Mallows of Bell Telephone Laboratories provided the author with detailed comments on a draft of this volume. Professor Ingram Olkin of Stanford University also commented on the earlier chapters of that draft. Mr. Jostein Lillestöl read through the galleys. Their suggestions were most helpful.

I learned time series analysis from John W. Tukey. I thank him now for all the help and encouragement he has provided.

D.R.B.

Berkeley, California
June, 1974

CONTENTS

3 Analytic Properties of Fourier Transforms and Complex Matrices 49

4 Stochastic Properties of Finite Fourier Transforms 88

5 The Estimation of Power Spectra 116

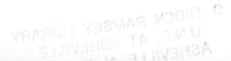

1

THE NATURE OF TIME SERIES AND THEIR FREQUENCY ANALYSIS

1.1 INTRODUCTION

In this work we will be concerned with the examination of r vector-valued functions

$$\begin{bmatrix} X_1(t) \\ \cdot \\ \cdot \\ \cdot \\ X_r(t) \end{bmatrix} \tag{1.1.1}$$

where $X_j(t)$, $j = 1, \ldots, r$ is real-valued and t takes on the values $0, \pm 1, \pm 2, \ldots$. Such an entity of measurements will be referred to as an **r vector-valued time series.** The index t will often refer to the time of recording of the measurements.

An example of a vector-valued time series is the collection of mean monthly temperatures recorded at scattered locations. Figure 1.1.1 gives such a series for the locations listed in Table 1.1.1. Figure 1.1.2 indicates the positions of these locations. Such data may be found in World Weather Records (1965). This series was provided by J. M. Craddock, Meteorological Office, Bracknell. Another example of a vector-valued time series is the set of signals recorded by an array of seismometers in the aftermath of an earthquake or nuclear explosion. These signals are discussed in Keen et al (1965) and Carpenter (1965). Figure 1.1.3 presents an example of such a record.

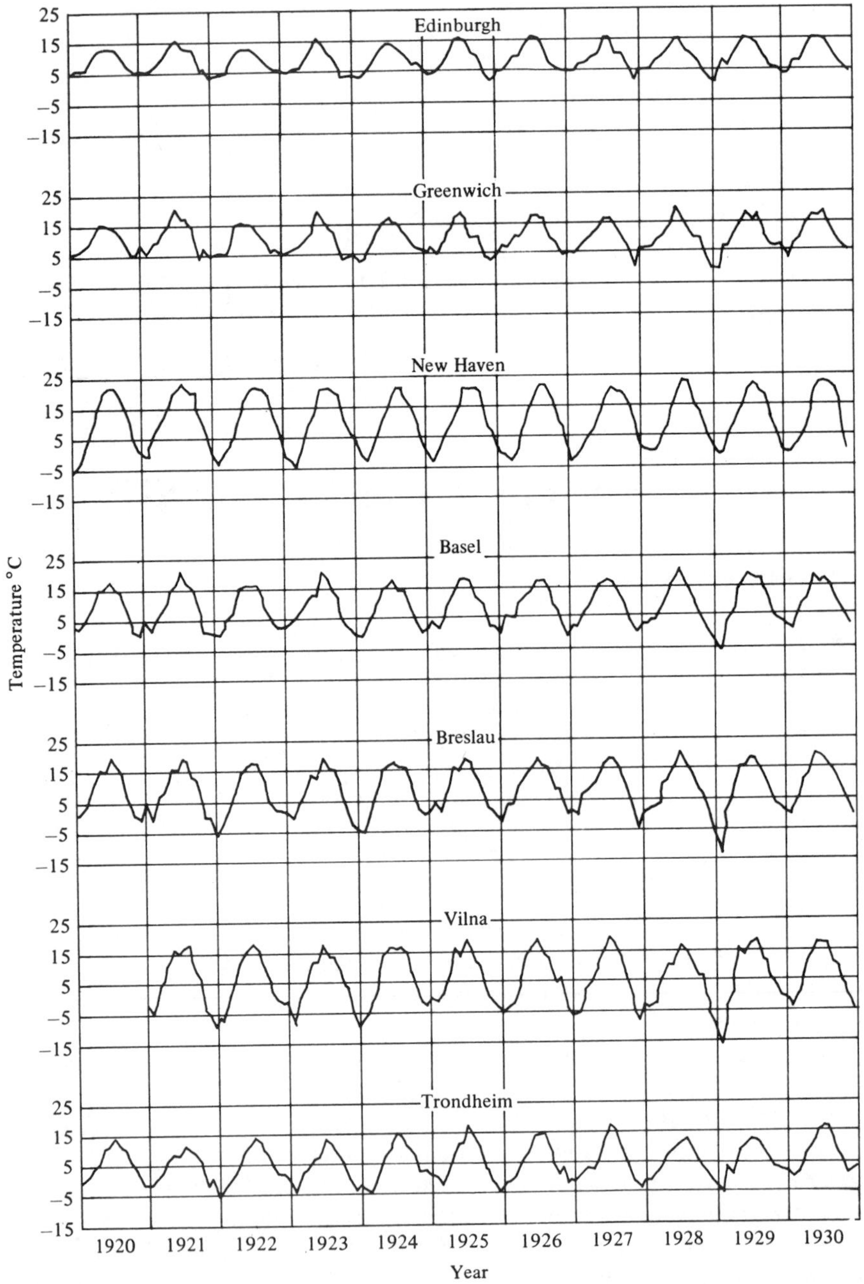

Figure 1.1.1 Monthly mean temperatures in °C at 14 stations for the years 1920–1930.

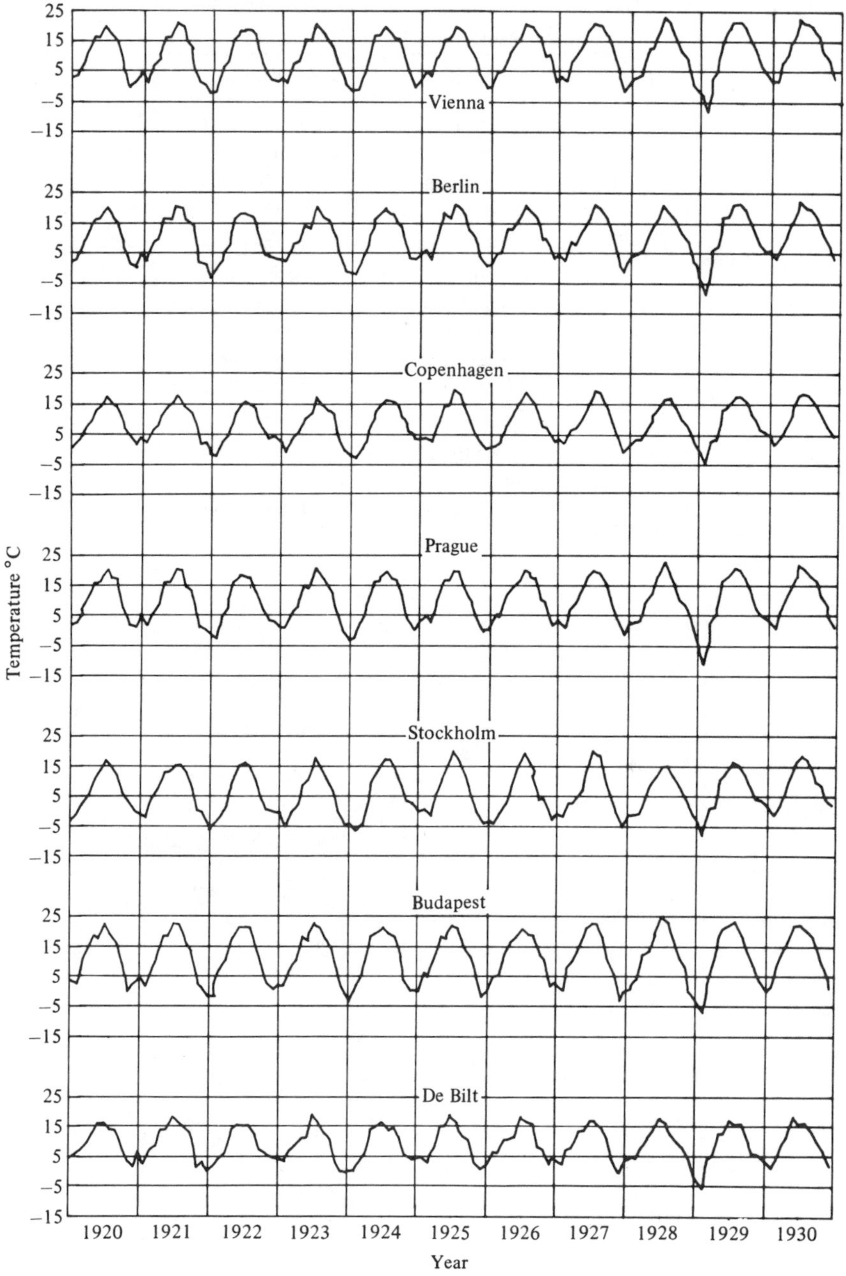
Vienna
Berlin
Copenhagen
Prague
Stockholm
Budapest
De Bilt
Temperature °C
25
15
5
−5
−15
1920
1921
1922
1923
1924
1925
1926
1927
1928
1929
1930
Year

Table 1.1.1 Stations and Time Periods of Temperature Data Used in Worked Examples

Index	*City*	*Period Available*
1	Vienna	1780–1950
2	Berlin	1769–1950
3	Copenhagen	1798–1950
4	Prague	1775–1939
5	Stockholm	1756–1960
6	Budapest	1780–1947
7	DeBilt	1711–1960
8	Edinburgh	1764–1959
9	Greenwich	1763–1962
10	New Haven	1780–1950
11	Basel	1755–1957
12	Breslau	1792–1950
13	Vilna	1781–1938
14	Trondheim	1761–1946

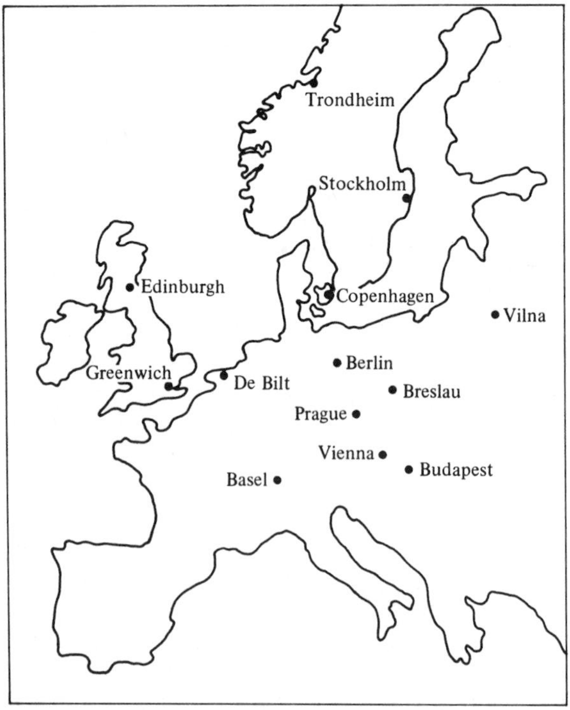

Figure 1.1.2 Locations of the temperature stations (except New Haven, U.S.A.).

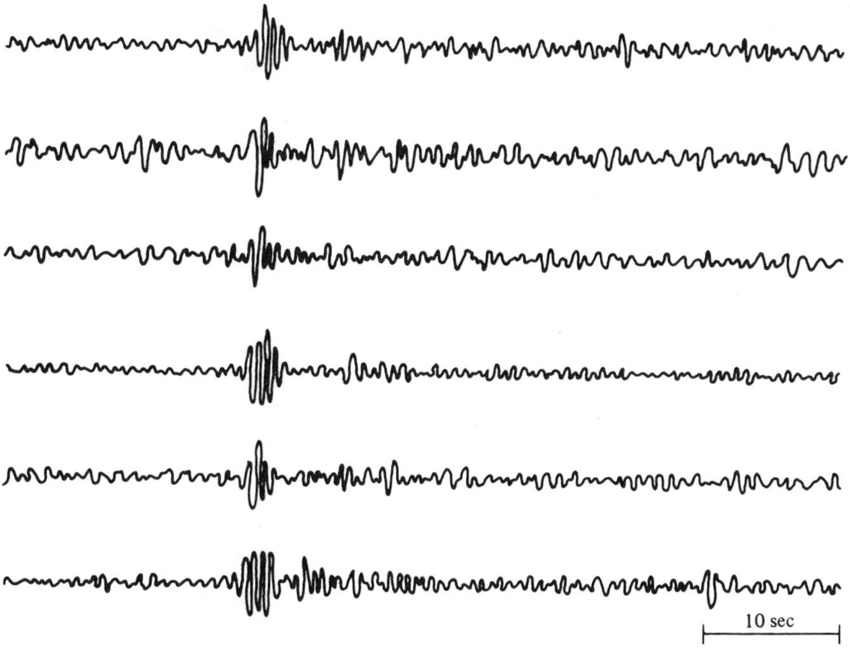

Figure 1.1.3 Signals recorded by an array of seismometers at the time of an event.

These examples are taken from the physical sciences; however, the social sciences also lead to the consideration of vector-valued time series. Figure 1.1.4 is a plot of exports from the United Kingdom separated by destination during the period 1958–1968. The techniques discussed in this work will sometimes be useful in the analysis of such a series although the results obtained are not generally conclusive due to a scarcity of data and departure from assumptions.

An inspection of the figures suggests that the individual component series are quite strongly interrelated. Much of our concern in this work will center on examining interrelations of component series. In addition there are situations in which we are interested in a single series on its own. For example, Singleton and Poulter (1967) were concerned with the call of a male killer whale and Godfrey (1965) was concerned with the quantity of cash held within the Federal Reserve System for the purpose of meeting interbank check-handling obligations each month. Figure 1.1.5 is a graph of the annual mean sunspot numbers for the period 1760–1965; see Waldmeir (1961). This series has often been considered by statisticians; see Yule (1927), Whittle (1954), Brillinger and Rosenblatt (1967b). Generally speaking it will be enough to consider single component series as particular cases of vector-valued series corresponding to $r = 1$. However, it is typically much more informative if we carry out a vector analysis, and it is wise to search out series related to any single series and to include them in the analysis.

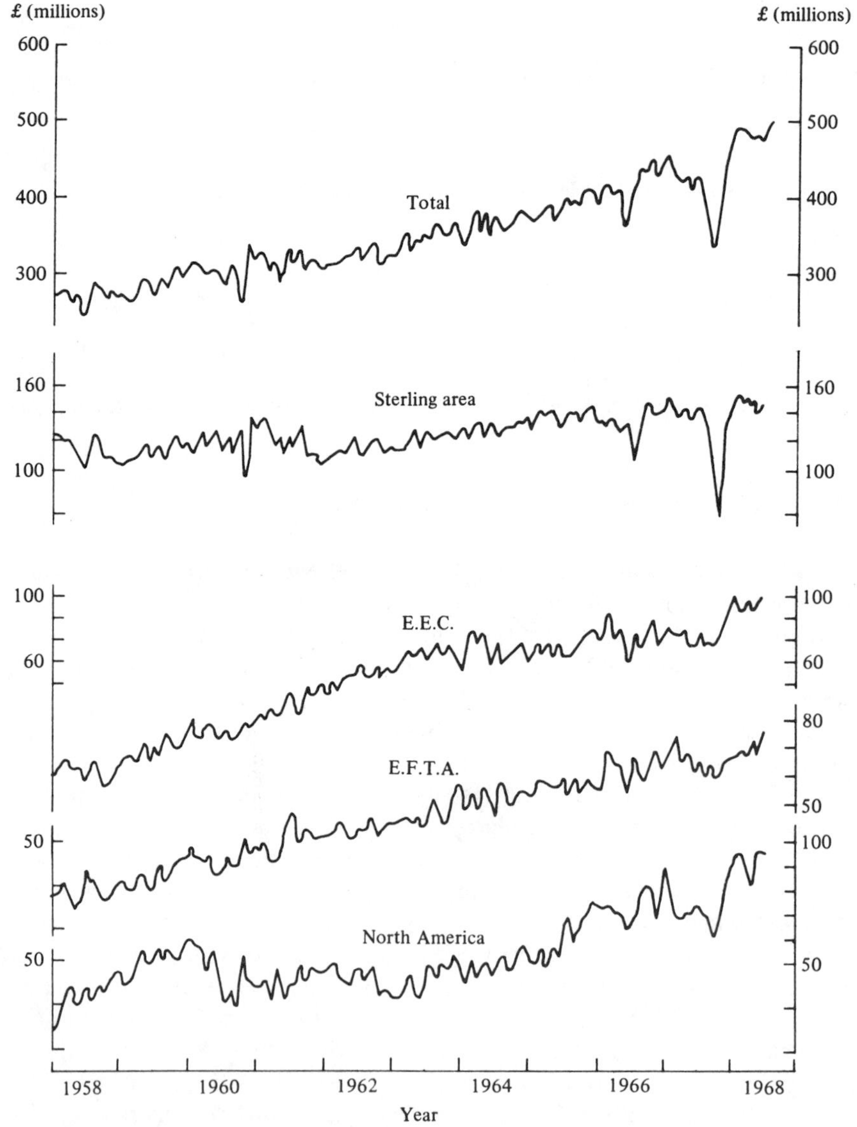

Figure 1.1.4 Value of United Kingdom exports by destination for 1958–1968.

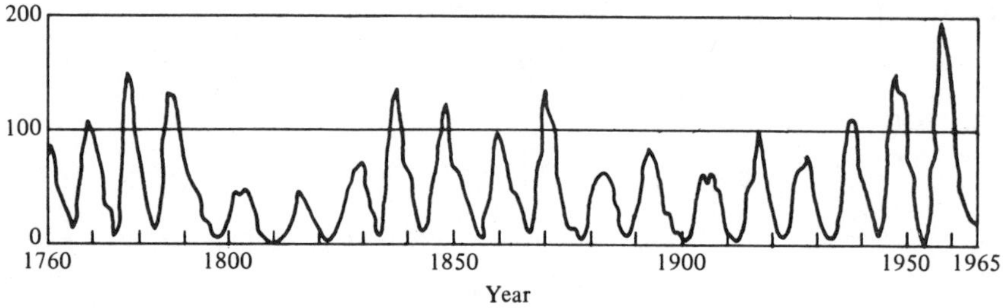

Figure 1.1.5 Annual mean sunspot numbers for 1760–1965.

1.2 A REASON FOR HARMONIC ANALYSIS

The principal mathematical methodology we will employ in our analysis of time series is harmonic analysis. This is because of our decision to restrict consideration to series resulting from experiments not tied to a specific time origin or, in other words, experiments invariant with respect to translations of time. This implies, for example, that the proportion of the values $X(t)$, $t > u$, falling in some interval I, should be approximately the same as the proportion of the values $X(t)$, $t > u + v$, falling in I for all v.

The typical physical experiment appears to possess, in large part, this sort of time invariance. Whether a physicist commenced to measure the force of gravity one day or the next does not seem to matter for most purposes. A cursory examination of the series of the previous section suggests: the temperature series of Figure 1.1.1 are reasonably stable in time; portions of the seismic series appear stable; the export series do not appear stationary; and the sunspot series appear possibly so. The behavior of the export series is typical of that of many socioeconomic series. Since people learn from the past and hence alter their behavior, series relating to them are not generally time invariant. Later we will discuss methods that may allow removing a stationary component from a nonstationary series; however, the techniques of this work are principally directed toward the analysis of series stable in time.

The requirement of elementary behavior under translations in time has certain analytic implications. Let $f(t)$ be a real or complex-valued function defined for $t = 0, \pm 1, \ldots$. If we require

$$f(t + u) = f(t) \qquad \text{for } t, u = 0, \pm 1, \ldots, \tag{1.2.1}$$

then clearly $f(t)$ is constant. We must therefore be less stringent than expression (1.2.1) in searching for functions behaving simply under time translations. Let us require instead

$$f(t + u) = C_u f(t) \qquad \text{for } t, u = 0, \pm 1, \ldots \text{ with } C_1 \neq 0. \tag{1.2.2}$$

Setting $u = 1$ and proceeding recursively gives

$$f(t + 1) = C_1 f(t) = C_1^2 f(t - 1) = \cdots = C_1^{t+1} f(0) \tag{1.2.3}$$

if $t \geqslant 0$ or in the case $t \leqslant 0$

$$f(t) = C_1^{-1} f(t + 1) = C_1^{-2} f(t + 2) = \cdots = C_1^t f(0). \tag{1.2.4}$$

In either case, if we write $C_1 = \exp\{\alpha\}$, α real or complex, then we see that the general solution of expression (1.2.2) may be written

$$f(t) = f(0) \exp\{\alpha t\} \tag{1.2.5}$$

and that $C_u = \exp\{\alpha u\}$. The bounded solutions of expression (1.2.2) are seen to occur for $\alpha = i\lambda$, λ real, where $i = \sqrt{-1}$. In summary, if we look for functions behaving simply with respect to time translation, then we are led to the sinusoids $\exp\{i\lambda t\}$, λ real; the parameter λ is called the **frequency** of the sinusoid. If in fact

$$f(t) = \sum_j c_j \exp\{i\lambda_j t\}, \tag{1.2.6}$$

then

$$\begin{aligned} f(t + u) &= \sum_j c_j \exp\{i\lambda_j u\} \exp\{i\lambda_j t\} \\ &= \sum_j C_j \exp\{i\lambda_j t\}, \end{aligned} \tag{1.2.7}$$

with $C_j = c_j \exp\{i\lambda_j u\}$. In other words, if a function of interest is a sum of cosinusoids, then its behavior under translations is also easily described. We have, therefore, in the case of experiments leading to results that are deterministic functions, been led to functions that can be developed in the manner of (1.2.6). The study of such functions is the concern of harmonic or Fourier analysis; see Bochner (1959), Zygmund (1959), Hewitt and Ross (1963), Wiener (1933), Edwards (1967).

In Section 2.7 we will see that an important class of operations on time series, filters, is also most easily described and investigated through harmonic analysis.

With experiments that result in random or stochastic functions, $X(t)$, time invariance leads us to investigate the class of experiments such that $\{X(t_1),\ldots,X(t_k)\}$ has the same probability structure as $\{X(t_1 + u),\ldots,X(t_k + u)\}$ for all u and $t_1, \ldots, t_k$. The results of such experiments are called **stationary stochastic processes**; see Doob (1953), Wold (1938), and Khintchine (1934).

1.3 MIXING

A second important requirement that we will place upon the time series that we consider is that they have a short span of dependence. That is, the measurements $X(t)$ and $X(s)$ are becoming unrelated or statistically independent of each other as $t - s \rightarrow \infty$.

This requirement will later be set down in a formal manner with Assumptions 2.6.1 and 2.6.2(l). It allows us to define relevant population parameters and implies that various estimates of interest are asymptotically Gaussian in the manner of the central limit theorem.

Many series that are reasonably stationary appear to satisfy this sort of requirement; possibly because as time progresses they are subjected to random shocks, unrelated to what has gone before, and these random shocks eventually form the prime content of the series.

A requirement that a time series have a weak memory is generally referred to as a **mixing assumption**; see Rosenblatt (1956b).

1.4 HISTORICAL DEVELOPMENT

The basic tool that we will employ, in the analysis of time series, is the finite Fourier transform of an observed section of the series.

The taking of the Fourier transform of an empirical function was proposed as a means of searching for hidden periodicities in Stokes (1879). Schuster (1894), (1897), (1900), (1906a), (1906b), in order to avoid the annoyance of considering relative phases, proposed the consideration of the modulus-squared of the finite Fourier transform. He called this statistic the **periodogram.** His motivation was also the search for hidden periodicities.

The consideration of the periodogram for general stationary processes was initiated by Slutsky (1929, 1934). He developed many of the statistical properties of the periodogram under a normal assumption and a mixing assumption. Concurrently Wiener (1930) was proposing a very general form of harmonic analysis for time series and beginning a study of vector processes.

The use of harmonic analysis as a tool for the search of hidden periodicities was eventually replaced by its much more important use for inquiring into relations between series; see Wiener (1949) and Press and Tukey (1956). An important statistic in this case is the **cross-periodogram,** a product of the finite Fourier transforms of two series. It is inherent in Wiener (1930) and Goodman (1957); the term cross-periodogram appears in Whittle (1953).

The periodogram and cross-periodogram are second-order statistics and thus are especially important in the consideration of Gaussian processes. Higher order analogs are required for the consideration of various aspects of non-Gaussian series. The **third-order periodogram,** a product of three finite Fourier transforms, appears in Rosenblatt and Van Ness (1965), and the **kth order periodogram,** a product of k finite Fourier transforms, in Brillinger and Rosenblatt (1967a, b).

The instability of periodogram-type statistics is immediately apparent when they are calculated from empirical functions; see Kendall (1946), Wold (1965), and Chapter 5 of this text. This instability led Daniell (1946) to propose a numerical smoothing of the periodogram which has now become basic to most forms of frequency analysis.

Papers and books, historically important in the development of the mathematical foundations of the harmonic analysis of time series, include: Slutsky (1929), Wiener (1930), Khintchine (1934), Wold (1938), Kolmogorov (1941a, b), Cramér (1942), Blanc-Lapiere and Fortet (1953), and Grenander (1951a).

Papers and books, historically important in the development of the empirical harmonic analysis of time series, include: Schuster (1894, 1898), Tukey (1949), Bartlett (1948), Blackman and Tukey (1958), Grenander and Rosenblatt (1957), Bartlett (1966), Hannan (1960), Stumpff (1937), and Chapman and Bartels (1951).

Wold (1965) is a bibliography of papers on time series analysis. Burkhardt (1904) and Wiener (1938) supply a summary of the very early work. Simpson (1966) and Robinson (1967) provide many computer programs useful in analyzing time series.

1.5 THE USES OF THE FREQUENCY ANALYSIS

This section contains a brief survey of some of the fields in which spectral analysis has been employed. There are three principal reasons for using spectral analysis in the cases to be presented (i) to provide useful descriptive statistics, (ii) as a diagnostic tool to indicate which further analyses might be relevant, and (iii) to check postulated theoretical models. Generally, the success experienced with the technique seems to vary directly with the length of series available for analysis.

Physics If the spectral analysis of time series is viewed as the study of the individual frequency components of some time series of interest, then the first serious application of this technique may be regarded as having occurred in 1664 when Newton broke sunlight into its component parts by passing it through a prism. From this experiment has grown the subject of spectroscopy (Meggers (1946), McGucken (1970), and Kuhn (1962)), in which there is investigation of the distribution of the energy of a radiation field as a function of frequency. (This function will later be called a power spectrum.) Physicists have applied spectroscopy to identifying chemical elements, to determine the direction and rate of movement of celestial bodies, and to testing general relativity. The spectrum is an important parameter in the description of color; Wright (1958).

The frequency analysis of light is discussed in detail in Born and Wolfe (1959); see also Schuster (1904), Wiener (1953), Jennison (1961), and Sears (1949).

Power spectra have been used frequently in the fields of turbulence and fluid mechanics; see Meecham and Siegel (1964), Kampé de Fériet (1954), Hopf (1952), Burgers (1948), Friedlander and Topper (1961), and Batchelor (1960). Here one typically sets up a model leading to a theoretical power spectrum and checks it empirically. Early references are given in Wiener (1930).

Electrical Engineering Electrical engineers have long been concerned with the problem of measuring the power in various frequency bands of some electromagnetic signal of interest. For example, see Pupin (1894), Wegel and Moore (1924), and Van der Pol (1930). Later, the invention of radar gave stimulus to the problem of signal detection, and frequency analysis proved a useful tool in its investigation; see Wiener (1949), Lee and Wiesner (1950), and Solodovnikov (1960). Frequency analysis is now firmly involved in the areas of coding, information theory, and communications; see Gabor (1946), Middleton (1960), and Pinsker (1964). In many of these problems, Maxwell's equations lead to an underlying model of some use.

Acoustics Frequency analysis has proved itself important in the field of acoustics. Here the power spectrum has generally played the role of a descriptive statistic. For example, see Crandall and Sacia (1924), Beranek (1954), and Majewski and Hollien (1967). An important device in this connection is the sound spectrograph which permits the display of time-dependent spectra; see Fehr and McGahan (1967). Another interesting device is described in Noll (1964).

Geophysics Tukey (1965a) has given a detailed description and bibliography of the uses of frequency analysis in geophysics; see also Tukey (1965b), Kinosita (1964), Sato (1964), Smith et al (1967), Labrouste (1934), Munk and MacDonald (1960), *Ocean Wave Spectra* (1963), Haubrich and MacKenzie (1965), and various authors (1966). A recent dramatic example involves the investigation of the structure of the moon by the frequency analysis of seismic signals, resulting from man-made impacts on the moon; see Latham et al (1970).

Other Engineering Harmonic analysis has been employed in many areas of engineering other than electrical: for example, in aeronautical engineering, Press and Tukey (1956), Takeda (1964); in naval engineering, Yamanouchi (1961), Kawashima (1964); in hydraulics, Nakamura and Murakami (1964); and in mechanical engineering, Nakamura (1964), Kaneshige (1964), Crandall (1958), Crandall (1963). Civil engineers find spectral techniques useful in understanding the responses of buildings to earthquakes.

Medicine A variety of medical data is collected in the form of time series; for example, electroencephalograms and electrocardiograms. References to the frequency analysis of such data include: Alberts et al (1965), Bertrand and Lacape (1943), Gibbs and Grass (1947), Suhara and Suzuki (1964), and Yuzuriha (1960). The correlation analysis of EEG's is discussed in Barlow (1967); see also Wiener (1957, 1958).

Economics Two books, Granger (1964) and Fishman (1969), have appeared on the application of frequency analysis to economic time series. Other references include: Beveridge (1921), Beveridge (1922), Nerlove (1964), Cootner (1964), Fishman and Kiviat (1967), Burley (1969), and Brillinger and Hatanaka (1970). Bispectral analysis is employed in Godfrey (1965).

Biology Frequency analysis has been used to investigate the circadian rhythm present in the behavior of certain plants and animals; for example, see Aschoff (1965), Chance et al (1967), Richter (1967). Frequency analysis is also useful in constructing models for human hearing; see Mathews (1963).

Psychology A frequency analysis of data, resulting from psychological tests, is carried out in Abelson (1953).

Numerical Analysis Spectral analysis has been used to investigate the independence properties of pseudorandom numbers generated by various recursive schemes; see Jagerman (1963) and Coveyou and MacPherson (1967).

1.6 INFERENCE ON TIME SERIES

The purpose of this section is to record the following fact that the reader will soon note for himself in proceeding through this work: the theory and techniques employed in the discussion of time series statistics are entirely elementary. The basic means of constructing estimates is the method of moments. Asymptotic theory is heavily relied upon to provide justifications. Much of what is presented is a second-order theory and is therefore most suitable for Gaussian processes. Sufficient statistics, maximum likelihood statistics, and other important concepts of statistical inference are only barely mentioned.

A few attempts have been made to bring the concepts and methods of current statistical theory to bear on stationary time series; see Bartlett (1966), Grenander (1950), Slepian (1954), and Whittle (1952). Likelihood ratios have been considered in Striebel (1959), Parzen (1963), and Gikman and Skorokhod (1966). General frameworks for time series analysis have been described in Rao (1963), Stigum (1967), and Rao (1966); see also Hájek (1962), Whittle (1961), and Arato (1961).

It should be pointed out that historically there have been two rather distinct approaches to the analysis of time series: the frequency or harmonic approach and the time domain approach. This work is concerned with the former, while the latter is exemplified by the work of Mann and Wald (1943), Quenouille (1957), Durbin (1960), Whittle (1963), Box and Jenkins (1970). The differences between these two analyses is discussed in Wold (1963). With the appearance of the Fast Fourier Algorithm, however, it may be more efficient to carry out computations in the frequency domain even when the time domain approach is adopted; see Section 3.6, for example.

1.7 EXERCISES

1.7.1 If $f(\cdot)$ is complex valued and $f(t_1 + u_1, \ldots, t_k + u_k) = C_{u_1 \ldots u_k} f(t_1, \ldots, t_k)$ for $t_j, u_j = 0, \pm 1, \pm 2, \ldots, j = 1, \ldots, k$, prove that $f(t_1, \ldots, t_k) = f(0, \ldots, 0) \exp\{\Sigma \alpha_j t_j\}$ for some $\alpha_1, \ldots, \alpha_k$. See Aczel (1969).

1.7.2 If $f(t)$ is complex valued, continuous, and $f(t + u) = C_u f(t)$ for $-\infty < t$, $u < \infty$, prove that $f(t) = f(0) \exp\{\alpha t\}$ for some α.

1.7.3 If $\mathbf{f}(t)$ is r vector valued, with complex components, and $\mathbf{f}(t + u) = \mathbf{C}_u \mathbf{f}(t)$ for $t, u = 0, \pm 1, \pm 2, \ldots$ and $\mathbf{C}_u$ an $r \times r$ matrix function, prove that $\mathbf{f}(t) = \mathbf{C}_1{}^t\, \mathbf{f}(0)$ if $\mathrm{Det}\{\mathbf{f}(0), \ldots, \mathbf{f}(r-1)\} \neq 0$, where

$$\mathbf{C}_u = \exp\{\mathbf{A}u\}$$
$$\mathbf{A} = \ln \mathbf{C}_1.$$

See Doeblin (1938) and Kirchener (1967).

1.7.4 Let $W(\alpha)$, $-\infty < \alpha < \infty$ be an absolutely integrable function satisfying

$$\int_{-\infty}^{\infty} W(\alpha) d\alpha = 1.$$

Let $f(\alpha)$, $-\infty < \alpha < \infty$ be a bounded function continuous at $\alpha = \lambda$. Show that $\varepsilon^{-1} \int W[\varepsilon^{-1}(\lambda - \alpha)] d\alpha = 1$ and

$$\lim_{\varepsilon \to 0} \varepsilon^{-1} \int_{-\infty}^{\infty} f(\alpha) W[\varepsilon^{-1}(\lambda - \alpha)] d\alpha = f(\lambda).$$

1.7.5 Prove that for $\lambda \neq 0, \pm 2\pi, \ldots,$

$$\text{(a)} \quad \sum_{j=1}^{T} \sin j\lambda = \frac{\cos \frac{\lambda}{2} - \cos (T + \frac{1}{2})\lambda}{2 \sin \frac{\lambda}{2}}$$

$$\text{(b)} \quad \tfrac{1}{2} + \sum_{j=1}^{T} \cos j\lambda = \frac{\sin (T + \frac{1}{2})\lambda}{2 \sin \frac{\lambda}{2}}$$

(c) $$\sum_{j=-T}^{T} \exp\{-i\lambda j\} = \frac{\sin (T+\frac{1}{2})\lambda}{\sin \frac{\lambda}{2}}$$

(d) $$\int_{-\pi}^{\pi} \frac{\sin (T+\frac{1}{2})\lambda}{\sin \frac{\lambda}{2}} d\lambda = 2\pi.$$

1.7.6 Let $X_1, \ldots, X_r$ be independent random variables with $EX_j = \mu_j$ and var $X_j = \sigma_j^2$. Consider linear combinations $Y = \sum_j a_j X_j$, $\sum_j a_j = 1$. We have $EY = \sum_j a_j \mu_j$. Prove that var Y is minimized by the choice $a_J = \sigma_J^{-2}/\sum_j \sigma_j^{-2}$, $J = 1, \ldots, r$.

1.7.7 Prove that $\sum_{t=0}^{T-1} \exp\{i(2\pi ts)/T\} = T$ if $s = 0, \pm T, \pm 2T, \ldots$ and $= 0$ for other integral values of s.

1.7.8 If X is a real-valued random variable with finite second moment and θ is real valued, prove that $E(X-\theta)^2 = \text{var } X + (EX - \theta)^2$.

1.7.9 Let l denote the space of two-sided sequences $x = \{x_t, t = 0, \pm 1, \pm 2, \ldots\}$. Let $\mathcal{A}$ denote an operation on l that is linear, $[\mathcal{A}(\alpha x + \beta y) = \alpha\mathcal{A}x + \beta\mathcal{A}y$ for α, β scalars and $x, y \in l$,] and time invariant, $[\mathcal{A}y = Y$ if $\mathcal{A}x = X$, $y_t = x_{t+u}$, $Y_t = X_{t+u}$ for some $u = 0, \pm 1, \pm 2, \ldots$]. Prove that there exists a function $A(\lambda)$ such that $(\mathcal{A}x)_t = A(\lambda)x_t$ if $x_t = \exp\{i\lambda t\}$.

1.7.10 Consider a sequence $c_0, c_1, c_2, \ldots$, its partial sums $S_T = \sum_{t=0}^{T} c_t$, and the Cesàro means

$$\sigma_T = \frac{S_0 + \cdots + S_T}{T+1} = \sum_{t=0}^{T}\left(1 - \frac{t}{T+1}\right)c_t.$$

If $S_T \to S$, prove that $\sigma_T \to S$ (as $T \to \infty$); see Knopp (1948).

1.7.11 Let $\begin{bmatrix}\mathbf{X}\\ Y\end{bmatrix}$ be a vector-valued random variable with Y real-valued and $EY^2 < \infty$. Prove that $\phi(\mathbf{X})$ with $E\phi(\mathbf{X})^2 < \infty$ that minimizes $E[Y - \phi(\mathbf{X})]^2$ is given by $\phi(\mathbf{X}) = E\{Y \mid \mathbf{X}\}$.

1.7.12 Show that for $n = 1, 2, \ldots$

$$\sum_{u=-n+1}^{n-1}\left(1 - \frac{|u|}{n}\right)\exp\{-i\lambda u\} = \frac{1}{n}\left[\frac{\sin \frac{n\lambda}{2}}{\sin \frac{\lambda}{2}}\right]^2$$

and from this show

$$\int_{-\pi}^{\pi}\left[\frac{\sin \frac{n\lambda}{2}}{\sin \frac{\lambda}{2}}\right]^2 d\lambda = 2\pi n.$$

1.7.13 Show that the identity

$$\sum_{k=m}^{n} u_k v_k = \sum_{k=m}^{n-1} U_k(v_k - v_{k+1}) - U_{m-1}v_m + U_n v_n$$

holds, where $0 \leqslant m \leqslant n$, $U_k = u_0 + \cdots + u_k$, $(k \geqslant 0)$, $U_{-1} = 0$. (Abel's transformation)

1.7.14 (a) Let $f(x)$, $0 \leqslant x \leqslant 1$, be integrable and have an integrable derivative $f^{(1)}(x)$. Show that

$$\frac{1}{n}\sum_{j=0}^{n} f\left(\frac{j}{n}\right) = \int_0^1 f(x)dx + \frac{1}{2n}(f(0) + f(1)) + \frac{1}{n}\int_0^1 \left(nx - [nx] - \frac{1}{2}\right) f^{(1)}(x)dx$$

with $[y]$ denoting the integral part of y.

(b) Let $f^{(k)}(x)$, $k = 0, 1, 2, \ldots$ denote the kth derivative of $f(x)$. Suppose $f^{(k)}(x)$, $0 \leqslant x \leqslant 1$, is integrable for $k = 0, 1, 2, \ldots, K$. Show that

$$\frac{1}{n}\sum_{j=0}^{n} f\left(\frac{j}{n}\right) = \int_0^1 f(x)dx + \frac{1}{n}f(0) + \sum_{k=1}^{K} (-1)^k B_k(0) n^{-k} \times (f^{(k-1)}(1) - f^{(k-1)}(0)) + (-1)^{K+1}\int_0^1 B_K(nx - [nx]) f^{(K)}(x)dx$$

where $B_k(y)$ denotes the kth Bernoulli polynomial. (Euler-MacLaurin)

2

FOUNDATIONS

2.1 INTRODUCTION

In this chapter we present portions of both the stochastic and deterministic approaches to the foundations of time series analysis. The assumptions made in either approach will be seen to lead to the definition of similar parameters of interest, and implications for practice are generally the same. In fact it will be shown that the two approaches are equivalent in a certain sense. An important part of this chapter will be to develop the invariance properties of the parameters of interest for a class of transformations of the series called filters. Proofs of the theorems and lemmas are given at the end of the book.

The notation that will be adopted throughout this text includes bold face letters **A, B** which denote **matrices.** If a matrix **A** has entries A_{jk} we sometimes indicate it by $[A_{jk}]$. Given an $r \times s$ matrix **A,** its $s \times r$ **transpose** is denoted by $\mathbf{A}^\tau$, and the matrix whose entries are the complex conjugates of those of **A** is denoted by $\overline{\mathbf{A}}$. Det **A** denotes the **determinant** of the square matrix **A;** the **trace of A,** is indicated by tr **A.** $|\mathbf{A}|$ denotes the sum of the absolute values of the entries of **A,** and **I,** the **identity** matrix. An ***r* vector** is an $r \times 1$ matrix.

We denote the **expected value** of a random variable X by EX generally, and sometimes, by ave X. This will reduce the possibility of confusion in certain expressions. We denote the variance of X by var X. If (X,Y) is a bivariate random variable, we denote the covariance of X with Y by cov $\{X,Y\}$. We signify the correlation of X with Y by cor $\{X,Y\}$.

If z is a complex number, we indicate its real part by Re z and its imaginary part by Im z. We therefore have the representation

$$z = \text{Re } z + i \text{ Im } z. \tag{2.1.1}$$

We denote the modulus of z, $[(\text{Re } z)^2 + (\text{Im } z)^2]^{1/2}$ by $|z|$ and its argument, $\tan^{-1} \{\text{Im } z/\text{Re } z\}$, by arg z. If x and y are real numbers, we will write

$$x \equiv y \pmod{\alpha} \tag{2.1.2}$$

when the difference $x - y$ is an integral multiple of α.

The following functions will prove useful in our work: the **Kronecker delta**

$$\begin{aligned} \delta\{\alpha\} &= 1 \qquad \text{if } \alpha = 0 \\ &= 0 \qquad \text{otherwise} \end{aligned} \tag{2.1.3}$$

and the **Kronecker comb**

$$\begin{aligned} \eta\{\alpha\} &= 1 \qquad \text{if } \alpha \equiv 0 \pmod{2\pi} \\ &= 0 \qquad \text{otherwise.} \end{aligned} \tag{2.1.4}$$

Likewise the following generalized functions will be useful: the **Dirac delta function**, $\delta(\alpha)$, $-\infty < \alpha < \infty$, with the property

$$\int_{-\infty}^{\infty} f(\alpha)\delta(\alpha)\, d\alpha = f(0) \tag{2.1.5}$$

for all functions $f(\alpha)$ continuous at 0, and the **Dirac comb**

$$\eta(\alpha) = \sum_{j=-\infty}^{\infty} \delta(\alpha - 2\pi j) \tag{2.1.6}$$

for $-\infty < \alpha < \infty$ with the property

$$\int_{-\infty}^{\infty} f(\alpha)\eta(\alpha)\, d\alpha = \sum_{j=-\infty}^{\infty} f(2\pi j) \tag{2.1.7}$$

for all suitable functions $f(\alpha)$. These last functions are discussed in Lighthill (1958), Papoulis (1962), and Edwards (1967). Exercise 1.7.4 suggests that $\varepsilon^{-1}W(\varepsilon^{-1}\alpha)$, for small ε, provides an approximate Dirac delta function.

2.2 STOCHASTICS

On occasion it may make sense to think of a particular r vector-valued time series $\mathbf{X}(t)$ as being a member of an ensemble of vector time series which are generated by some random scheme. We can denote such an ensemble by

$\{\mathbf{X}(t,\theta);\ \theta \in \Theta$ and $t = 0, \pm 1, \pm 2, \ldots,\}$ where θ denotes a random variable taking values in Θ. If $\mathbf{X}(t,\theta)$ is a measurable function of θ, then $\mathbf{X}(t,\theta)$ is a random variable and we can talk of its **finite dimensional distributions** given by relations such as

$$F_{a_1, \ldots, a_k}(x_1, \ldots, x_k; t_1, \ldots, t_k) = \text{Prob}\,\{X_{a_1}(t_1,\theta) \leqslant x_1, \ldots, X_{a_k}(t_k,\theta) \leqslant x_k\}$$
$$a_1, \ldots, a_k = 1, \ldots, r,\ k = 1, 2, \ldots \quad (2.2.1)$$

and we can consider functionals such as

$$\text{ave}\, X_a(t,\theta) = \int x\, dF_a(x;t) = c_a(t), \quad (2.2.2)$$

$$\text{var}\, X_a(t,\theta) = \int [x - c_a(t)]^2 dF_a(x;t) = c_{aa}(t,t), \quad (2.2.3)$$

and

$$\text{cov}\,\{X_a(t_1,\theta), X_b(t_2,\theta)\} = \iint [x_1 - c_a(t_1)][x_2 - c_b(t_2)]dF_{ab}(x_1,x_2;\, t_1,t_2) = c_{ab}(t_1,t_2) \qquad \text{for } a, b = 1, \ldots, r \quad (2.2.4)$$

if the integrals involved exist. Once a θ has been generated (in accordance with its probability distribution), the function $\mathbf{X}(t,\theta)$, with θ fixed, will be described as a **realization, trajectory,** or **sample path** of the time series.

Since there will generally be no need to include θ specifically as an argument in $\mathbf{X}(t,\theta)$, we will henceforth denote $\mathbf{X}(t,\theta)$ by $\mathbf{X}(t)$. $\mathbf{X}(t)$ will be called a **time series, stochastic process,** or **random function.**

The interested reader may refer to Cramér and Leadbetter (1967), Yaglom (1962), or Doob (1953) for more details of the probabilistic foundations of time series. Function $c_a(t)$, defined in (2.2.2), is called the **mean function** of the time series $X_a(t)$. Function $c_{aa}(t_1,t_2)$, as derived from (2.2.4), is called the **(auto) covariance function** of $X_a(t)$, and $c_{ab}(t_1,t_2)$, defined in (2.2.4), is called the **cross-covariance function** of $X_a(t)$ with $X_b(t)$. $c_a(t)$ will exist if and only if ave $|X_a(t)| < \infty$. By the Schwarz inequality we have

$$|c_{ab}(t_1,t_2)|^2 \leqslant c_{aa}(t_1)c_{bb}(t_2), \quad (2.2.5)$$

and $c_{ab}(t_1,t_2)$ will exist if $c_{aa}(t_1,t_1),\ c_{bb}(t_2,t_2) < \infty$.

$$\rho_{aa}(t_1,t_2) = c_{aa}(t_1,t_2)/\{c_{aa}(t_1,t_1)c_{aa}(t_2,t_2)\}^{1/2}$$

is called the **(auto) correlation function** of $X_a(t)$ and

$$\rho_{ab}(t_1,t_2) = c_{ab}(t_1,t_2)/\{c_{aa}(t_1,t_1)c_{bb}(t_2,t_2)\}^{1/2}$$

is called the **cross-correlation function** of $X_a(t_1)$ with $X_b(t_2)$.

We will say that the series $X_a(t)$ and $X_b(t)$ are **orthogonal** if $c_{ab}(t_1,t_2) = 0$ for all t_1, t_2.

2.3 CUMULANTS

Consider for the present an r variate random variable $(Y_1, \ldots, Y_r)$ with ave $|Y_j|^r < \infty, j = 1, \ldots, r$, where the Y_j are real or complex.

Definition 2.3.1 The rth order joint cumulant, cum $(Y_1, \ldots, Y_r)$, of $(Y_1, \ldots, Y_r)$ is given by

$$\text{cum}\,(Y_1, \ldots, Y_r) = \sum (-1)^{p-1}(p-1)!\Big(\text{ave} \prod_{j \in \nu_1} Y_j\Big) \cdots \Big(\text{ave} \prod_{j \in \nu_p} Y_j\Big) \tag{2.3.1}$$

where the summation extends over all partitions $(\nu_1, \ldots, \nu_p), p = 1, \ldots, r$, of $(1, \ldots, r)$.

An important special case of this definition occurs when $Y_j = Y, j = 1, \ldots, r$. The definition gives then the cumulant of order r of a univariate random variable.

Theorem 2.3.1 cum $(Y_1, \ldots, Y_r)$ is given by the coefficient of $(i)^r t_1 \cdots t_r$ in the Taylor series expansion of log (ave exp $i \sum_{j=1}^r Y_j t_j$) about the origin.

This last is sometimes taken as the definition of cum $(Y_1, \ldots, Y_r)$. Properties of cum $(Y_1, \ldots, Y_r)$ include:

(i) $\text{cum}\,(a_1 Y_1, \ldots, a_r Y_r) = a_1 \cdots a_r \;\text{cum}\,(Y_1, \ldots, Y_r)$ for $a_1, \ldots, a_r$ constant

(ii) $\text{cum}\,(Y_1, \ldots, Y_r)$ is symmetric in its arguments

(iii) if any group of the Y's are independent of the remaining Y's, then $\text{cum}\,(Y_1, \ldots, Y_r) = 0$

(iv) for the random variable $(Z_1, Y_1, \ldots, Y_r)$, $\text{cum}\,(Y_1 + Z_1, Y_2, \ldots, Y_r) = \text{cum}\,(Y_1, Y_2, \ldots, Y_r) + \text{cum}\,(Z_1, Y_2, \ldots, Y_r)$

(v) for μ constant and $r = 2, 3, \ldots$

$$\text{cum}\,(Y_1 + \mu, Y_2, \ldots, Y_r) = \text{cum}\,(Y_1, Y_2, \ldots, Y_r) \tag{2.3.2}$$

(vi) if the random variables $(Y_1, \ldots, Y_r)$ and $(Z_1, \ldots, Z_r)$ are independent, then

$$\text{cum}\,(Y_1 + Z_1, \ldots, Y_r + Z_r) = \text{cum}\,(Y_1, \ldots, Y_r) + \text{cum}\,(Z_1, \ldots, Z_r) \tag{2.3.3}$$

(vii) $\text{cum}\, Y_j = EY_j$ for $j = 1, \ldots, r$

(viii) $\text{cum}\,(Y_j, \bar{Y}_j) = \text{var}\, Y_j$ for $j = 1, \ldots, r$

(ix) $\text{cum}\,(Y_j, \bar{Y}_k) = \text{cov}(Y_j, Y_k)$ for $j, k = 1, \ldots, r$.

Cumulants will provide us with a means of defining parameters of interest, with useful measures of the joint statistical dependence of random variables (see (iii) above) and with a convenient tool for proving theorems. Cumulants have also been called **semi-invariants** and are discussed in Dressel (1940), Kendall and Stuart (1958), and Leonov and Shiryaev (1959).

A standard normal variate has characteristic function $\exp\{-t^2/2\}$. It follows from the theorem therefore that its cumulants of order greater than 2 are 0. Also, from (iii), all the joint cumulants of a collection of independent variates will be 0. Now a general multivariate normal is defined to be a vector of linear combinations of independent normal variates. It now follows from (i) and (vi) that all the cumulants of order greater than 2 are 0 for a multivariate normal.

We will have frequent occasion to discuss the joint cumulants of polynomial functions of random variables. Before presenting expressions for the joint cumulants of such variates, we introduce some terminology due to Leonov and Shiryaev (1959). Consider a (not necessarily rectangular) two-way table

$$\begin{array}{ccc} (1,1) & \cdots & (1,J_1) \\ \cdot & & \cdot \\ \cdot & & \cdot \\ \cdot & & \cdot \\ (I,1) & \cdots & (I,J_I) \end{array} \tag{2.3.4}$$

and a partition $P_1 \cup P_2 \cup \cdots \cup P_M$ of its entries. We shall say that sets $P_{m'}, P_{m''}$, of the partition, **hook** if there exist $(i_1,j_1) \in P_{m'}$ and $(i_2,j_2) \in P_{m''}$ such that $i_1 = i_2$. We shall say that the sets $P_{m'}$ and $P_{m''}$ **communicate** if there exists a sequence of sets $P_{m_1} = P_{m'}, P_{m_2}, \ldots, P_{m_N} = P_{m''}$ such that P_{m_n} and $P_{m_{n+1}}$ hook for $n = 1, 2, \ldots, N-1$. A partition is said to be **indecomposable** if all sets communicate. If the rows of Table 2.3.4 are denoted $R_1, \ldots, R_I$, then a partition $P_1 \cdots P_M$ is indecomposable if and only if there exist no sets $P_{m_1}, \ldots, P_{m_N}$, $(N < M)$, and rows $R_{i_1}, \ldots, R_{i_0}$, $(0 < I)$, with

$$P_{m_1} \cup \cdots \cup P_{m_N} = R_{i_1} \cup \cdots \cup R_{i_0}. \tag{2.3.5}$$

The next lemma indicates a result relating to indecomposable partitions.

Lemma 2.3.1 Consider a partition $P_1 \cdots P_M$, $M > 1$, of Table 2.3.4. Given elements $r_{ij}, s_m; j = 1, \ldots, J_i; i = 1, \ldots, I; m = 1, \ldots, M$; define the function $\phi(r_{ij}) = s_m$ if $(i,j) \in P_m$. The partition is indecomposable if and only if the $\phi(r_{ij_1}) - \phi(r_{ij_2})$; $1 \leqslant j_1, j_2 \leqslant J_i$; $i = 1, \ldots, I$ generate all the elements of the set $\{s_m - s_{m'}; 1 \leqslant m, m' \leqslant M\}$ by additions and subtractions. Alternately, given elements t_i, $i = 1, \ldots, I$ define the function

$\psi(r_{ij}) = t_i; j = 1, \ldots, J_i; i = 1, \ldots, I$. The partition is indecomposable if and only if the $\psi(r_{ij}) - \psi(r_{i'j'})$; $(i,j), (i',j') \in P_m$; $m = 1, \ldots, M$ generate all the elements of the set $\{t_i - t_{i'};\ 1 \leqslant i,i' \leqslant I\}$ by addition and subtraction.

We remark that the set $\{t_i - t_{i'};\ 1 \leqslant i,i' \leqslant I\}$ is generated by $I - 1$ independent differences, such as $t_1 - t_I, \ldots, t_{I-1} - t_I$. It follows that when the partition is indecomposable, we may find $I - 1$ independent differences among the $\psi(r_{ij}) - \psi(r_{i'j'})$; $(i,j), (i',j') \in P_m$; $m = 1, \ldots, M$.

Theorem 2.3.2 Consider a two-way array of random variables X_{ij}; $j = 1, \ldots, J_i$; $i = 1, \ldots, I$. Consider the I random variables

$$Y_i = \prod_{j=1}^{J_i} X_{ij}, \qquad i = 1, \ldots, I. \tag{2.3.6}$$

The joint cumulant cum $(Y_1, \ldots, Y_I)$ is then given by

$$\sum_{\mathbf{v}} \operatorname{cum}(X_{ij};\ ij \in \nu_1) \cdots \operatorname{cum}(X_{ij};\ ij \in \nu_p) \tag{2.3.7}$$

where the summation is over all indecomposable partitions $\mathbf{v} = \nu_1 \cup \cdots \cup \nu_p$ of the Table 2.3.4.

This theorem is a particular case of a result of work done by Leonov and Shiryaev (1959).

We briefly mention an example of the use of this theorem. Let $(X_1, \ldots, X_4)$ be a 4-variate normal random variable. Its cumulants of order greater than 2 will be 0. Suppose we wish $\operatorname{cov}\{X_1X_2, X_3X_4\}$. Following the details of Theorem 2.3.2 we see that

$$\begin{aligned} \operatorname{cov}\{X_1X_2, X_3X_4\} &= \operatorname{cov}\{X_1, X_3) \operatorname{cov}\{X_2, X_4\} \\ &\quad + \operatorname{cov}\{X_1, X_4\} \operatorname{cov}\{X_2, X_3\}. \end{aligned} \tag{2.3.8}$$

This is a case of a result of Isserlis (1918).

We end this section with a definition extending that of the mean function and autocovariance function given in Section 2.2. Given the r vector-valued time series $\mathbf{X}(t)$, $t = 0, \pm 1, \ldots$ with components $X_a(t)$, $a = 1, \ldots, r$, and $E|X_a(t)|^k < \infty$, we define

$$c_{a_1,\ldots,a_k}(t_1, \ldots, t_k) = \operatorname{cum}\{X_{a_1}(t_1), \ldots, X_{a_k}(t_k)\} = c_{X_{a_1},\ldots,X_{a_k}}(t_1, \ldots, t_k) \tag{2.3.9}$$

for $a_1, \ldots, a_k = 1, \ldots, r$ and $t_1, \ldots, t_k = 0, \pm 1, \ldots$. Such a function will be called a **joint cumulant function of order k** of the series $\mathbf{X}(t)$, $t = 0, \pm 1, \ldots$.

2.4 STATIONARITY

An r vector-valued time series $\mathbf{X}(t)$, $t = 0, \pm 1, \ldots$ is called **strictly stationary** when the whole family of its finite dimensional distributions is invariant under a common translation of the time arguments or, when the joint distribution of $X_{a_1}(t_1 + t), \ldots, X_{a_k}(t_k + t)$ does not depend on t for $t, t_1, \ldots, t_k = 0, \pm 1, \ldots$ and $a_1, \ldots, a_k = 1, \ldots, r$, $k = 1, 2, \ldots$.

Examples of strictly stationary series include a series of independent identically distributed r vector-valued variates, $\boldsymbol{\varepsilon}(t)$, $t = 0, \pm 1, \ldots$ and a series that is a deterministic function of such variates as

$$\mathbf{X}(t) = \mathbf{f}[\boldsymbol{\varepsilon}(t), \boldsymbol{\varepsilon}(t-1), \boldsymbol{\varepsilon}(t+1), \ldots] \qquad \text{for } t = 0, \pm 1, \ldots. \tag{2.4.1}$$

More examples of strictly stationary series will be given later.

In this section, and throughout this text, the time domain of the series is assumed to be $t = 0, \pm 1, \ldots$. We remark that if I is any finite stretch of integers, then a series $\mathbf{X}(t)$, $t \in I$, that is relatively stationary over I, may be extended to be strictly stationary over all the integers. (The stationary extension of series defined and relatively stationary over an interval is considered in Parthasarathy and Varadhan (1964).) The important thing, from the standpoint of practice, is that the series be approximately stationary over the time period of observation.

An r vector-valued series $\mathbf{X}(t)$, $t = 0, \pm 1, \ldots$ is called **second-order stationary** or **wide-sense stationary** if

$$\begin{aligned} c_a(t) &= EX_a(t) = c_a, \\ c_{ab}(t+u,t) &= \operatorname{cov}\{X_a(t+u), X_b(t)\} = c_{ab}(u) \\ &\text{for } t, u = 0, \pm 1, \ldots \quad \text{and } a, b = 1, \ldots, r. \end{aligned}$$

We note that a strictly stationary series with finite second-order moments is second-order stationary.

On occasion we write the covariance function, of a second-order stationary series, in an unsymmetric form as

$$\begin{aligned} c_{ab}(t) &= c_{ab}(t,0) \\ &= c_{ab}(t+u,u) \qquad u = 0, \pm 1, \pm 2, \ldots. \end{aligned} \tag{2.4.2}$$

We indicate the $r \times r$ matrix-valued function with entries $c_{ab}(u)$ by $\mathbf{c}_{XX}(u)$ and refer to it as the **autocovariance function** of the series $\mathbf{X}(t)$, $t = 0, \pm 1, \ldots$. If we extend the definition of cov to vector-valued random variables $\mathbf{X}, \mathbf{Y}$ by writing

$$\operatorname{cov}\{\mathbf{X}, \mathbf{Y}\} = E\{[\mathbf{X} - E(\mathbf{X})]\overline{[\mathbf{Y} - E(\mathbf{Y})]}^{\tau}\}, \tag{2.4.3}$$

then we may define the autocovariance function of the series $\mathbf{X}(t)$ by

$$\mathbf{c}_{XX}(u) = \operatorname{cov}\{\mathbf{X}(t+u), \mathbf{X}(t)\} \tag{2.4.4}$$

for $t, u = 0, \pm 1, \ldots$ in the second-order stationary case.

If the vector-valued series $\mathbf{X}(t)$, $t = 0, \pm 1, \ldots$ is strictly stationary with $E|X_j(t)|^k < \infty$, $j = 1, \ldots, r$, then

$$c_{a_1,\ldots,a_k}(t_1 + u, \ldots, t_k + u) = c_{a_1,\ldots,a_k}(t_1, \ldots, t_k) \tag{2.4.5}$$

for $t_1, \ldots, t_k$, $u = 0, \pm 1, \ldots$. In this case we will sometimes use the asymmetric notation

$$c_{a_1,\ldots,a_k}(t_1, \ldots, t_{k-1}) = c_{a_1,\ldots,a_k}(t_1, \ldots, t_{k-1}, 0) \tag{2.4.6}$$

to remove the redundancy. This assumption of finite moments need not cause concern, for in practice all series available for analysis appear to be strictly bounded, $|X_j(t)| < C$, $j = 1, \ldots, r$ for some finite C and so all moments exist.

2.5 SECOND-ORDER SPECTRA

Suppose that the series $\mathbf{X}(t)$, $t = 0, \pm 1, \ldots$ is stationary and that, following the discussion of Section 1.3, its span of dependence is small in the sense that $X_a(t)$ and $X_b(t + u)$ are becoming increasingly less dependent as $|u| \to \infty$ for $a, b = 1, \ldots, r$. It is then reasonable to postulate that

$$\sum_{u=-\infty}^{\infty} |c_{ab}(u)| < \infty \qquad \text{for } a, b = 1, \ldots, r. \tag{2.5.1}$$

In this case we define the **second-order spectrum** of the series $X_a(t)$ with the series $X_b(t)$ by

$$f_{ab}(\lambda) = (2\pi)^{-1} \sum_{u=-\infty}^{\infty} c_{ab}(u) \exp\{-i\lambda u\}$$
$$\text{for } -\infty < \lambda < \infty,\ a, b = 1, \ldots, r. \tag{2.5.2}$$

Under the condition (2.5.1), $f_{ab}(\lambda)$ is bounded and uniformly continuous. The fact that the components of $\mathbf{X}(t)$ are real-valued implies that

$$f_{ab}(\lambda) = \bar{f}_{ab}(-\lambda) = f_{ba}(-\lambda) = \bar{f}_{ba}(\lambda). \tag{2.5.3}$$

Also an examination of expression (2.5.2) shows that $f_{ab}(\lambda)$ has period 2π with respect to λ.

The real-valued parameter λ appearing in (2.5.2) is called the **radian** or **angular frequency per unit time** or more briefly the **frequency.** If $b = a$, then $f_{aa}(\lambda)$ is called the **power spectrum** of the series $X_a(t)$ at frequency λ. If $b \neq a$, then $f_{ab}(\lambda)$ is called the **cross-spectrum** of the series $X_a(t)$ with the series $X_b(t)$

at frequency λ. We note that if $X_a(t) = X_b(t)$, $t = 0, \pm 1, \ldots$ with probability 1, then $f_{ab}(\lambda)$, the cross-spectrum, is in fact the power spectrum $f_{aa}(\lambda)$. Re $f_{ab}(\lambda)$ is called the **co-spectrum** and Im $f_{ab}(\lambda)$ is called the **quadrature spectrum.** $\phi_{ab}(\lambda) = \arg f_{ab}(\lambda)$ is called the **phase spectrum,** while $|f_{ab}(\lambda)|$ is called the **amplitude spectrum.**

Suppose that the autocovariance functions $c_{ab}(u)$, $u = 0, \pm 1, \ldots$ are collected together into the matrix-valued function $\mathbf{c}_{XX}(u)$, $u = 0, \pm 1, \ldots$ having $c_{ab}(u)$ as the entry in the ath row and bth column. Suppose likewise that the second-order spectra, $f_{ab}(\lambda)$, $-\infty < \lambda < \infty$, are collected together into the matrix-valued function $\mathbf{f}_{XX}(\lambda)$, $-\infty < \lambda < \infty$, having $f_{ab}(\lambda)$ as the entry in the ath row and bth column. Then the definition (2.5.2) may be written

$$\mathbf{f}_{XX}(\lambda) = (2\pi)^{-1} \sum_{u=-\infty}^{\infty} \mathbf{c}_{XX}(u) \exp\{-i\lambda u\} \quad \text{for } -\infty < \lambda < \infty. \tag{2.5.4}$$

The $r \times r$ matrix-valued function, $\mathbf{f}_{XX}(\lambda)$, $-\infty < \lambda < \infty$, is called the **spectral density matrix** of the series $\mathbf{X}(t)$, $t = 0, \pm 1, \ldots$. Under the condition (2.5.1), the relation (2.5.4) may be inverted to obtain the representation

$$\mathbf{c}_{XX}(u) = \int_{-\pi}^{\pi} \exp\{iu\lambda\} \mathbf{f}_{XX}(\lambda) d\lambda \qquad \text{for } u = 0, \pm 1, \ldots. \tag{2.5.5}$$

In Theorem 2.5.1 we shall see that the matrix $\mathbf{f}_{XX}(\lambda)$ is Hermitian, non-negative definite, that is, $\overline{\mathbf{f}_{XX}(\lambda)} = \mathbf{f}_{XX}(\lambda)^\tau$ and $\boldsymbol{\alpha}^\tau \mathbf{f}_{XX}(\lambda) \bar{\boldsymbol{\alpha}} \geq 0$ for all r vectors $\boldsymbol{\alpha}$ with complex entries.

Theorem 2.5.1 Let $\mathbf{X}(t)$, $t = 0, \pm 1, \ldots$ be a vector-valued series with autocovariance function $\mathbf{c}_{XX}(u) = \operatorname{cov}\{\mathbf{X}(t+u), \mathbf{X}(t)\}$, $t, u = 0, \pm 1, \ldots$ satisfying

$$\sum_{u=-\infty}^{\infty} |\mathbf{c}_{XX}(u)| < \infty. \tag{2.5.6}$$

Then the spectral density matrix

$$\mathbf{f}_{XX}(\lambda) = (2\pi)^{-1} \sum_{u=-\infty}^{\infty} \mathbf{c}_{XX}(u) \exp\{-i\lambda u\} \tag{2.5.7}$$

is Hermitian, non-negative definite.

In the case $r = 1$, this implies that the power spectrum is real and non-negative.

In the light of this theorem and the symmetry and periodicity properties indicated above, a power spectrum may be displayed as a non-negative function on the interval $[0,\pi]$. We will discuss the properties of power spectra in detail in Chapter 5.

In the case that the vector-valued series $\mathbf{X}(t)$, $t = 0, \pm 1, \ldots$ has finite second-order moments, but does not necessarily satisfy some mixing condition of the character of (2.5.1), we can still obtain a spectral representation of the nature of (2.5.5). Specifically we have the following:

Theorem 2.5.2 Let $\mathbf{X}(t)$, $t = 0, \pm 1, \ldots$ be a vector-valued series that is second-order stationary with finite autocovariance function $\mathbf{c}_{XX}(u) = \operatorname{cov}\{\mathbf{X}(t+u), \mathbf{X}(t)\}$, for $t, u = 0, \pm 1, \ldots$. Then there exists an $r \times r$ matrix-valued function $\mathbf{F}_{XX}(\lambda)$, $-\pi < \lambda \leqslant \pi$, whose entries are of bounded variation and whose increments are non-negative definite, such that

$$\mathbf{c}_{XX}(u) = \int_{-\pi}^{\pi} \exp\{iu\lambda\} d\mathbf{F}_{XX}(\lambda) \qquad \text{for } u = 0, \pm 1, \ldots. \tag{2.5.8}$$

This function is given by

$$\mathbf{F}_{XX}(\lambda) = \lim_{T \to \infty} (2\pi)^{-1} \sum_{u=-T}^{T} \mathbf{c}_{XX}(u)[\exp\{-i\lambda u\} - 1]/(-iu) \qquad \text{for } -\pi < \lambda \leqslant \pi. \tag{2.5.9}$$

The function $\mathbf{F}_{XX}(\lambda)$ is called the **spectral measure** of the series $\mathbf{X}(t)$, $t = 0, \pm 1, \ldots$. In the case that (2.5.1) holds, it is given by

$$\mathbf{F}_{XX}(\lambda) = \int_0^{\lambda} \mathbf{f}_{XX}(\alpha) d\alpha \qquad \text{for } -\pi < \lambda \leqslant \pi. \tag{2.5.10}$$

The representation (2.5.8) was obtained by Herglotz (1911) in the real-valued case and by Cramér (1942) in the vector-valued case.

2.6 CUMULANT SPECTRA OF ORDER k

Suppose that the series $\mathbf{X}(t)$, $t = 0, \pm 1, \ldots$ is stationary and that its span of dependence is small enough that

$$\sum_{u_1, \ldots, u_{k-1} = -\infty}^{\infty} |c_{a_1, \ldots, a_k}(u_1, \ldots, u_{k-1})| < \infty. \tag{2.6.1}$$

In this case, we define the **kth order cumulant spectrum,** $f_{a_1, \ldots, a_k}(\lambda_1, \ldots, \lambda_{k-1}) \equiv f_{X_{a_1}, \ldots, X_{a_k}}(\lambda_1, \ldots, \lambda_{k-1})$ by

$$f_{a_1, \ldots, a_k}(\lambda_1, \ldots, \lambda_{k-1}) = (2\pi)^{-k+1} \sum_{u_1, \ldots, u_{k-1} = -\infty}^{\infty} c_{a_1, \ldots, a_k}(u_1, \ldots, u_{k-1}) \exp\left\{-i \sum_{j=1}^{k-1} u_j \lambda_j\right\} \tag{2.6.2}$$

for $-\infty < \lambda_j < \infty$, $a_1, \ldots, a_k = 1, \ldots, r$, $k = 2, 3, \ldots$. We will extend the definition (2.6.2) to the case $k = 1$ by setting $f_a = c_a = EX_a(t)$,

$a = 1, \ldots, r$. We will sometimes add a symbolic argument λ_k to the function of (2.6.2) writing $f_{a_1,\ldots,a_k}(\lambda_1, \ldots, \lambda_k)$ in order to maintain symmetry. λ_k may be taken to be related to the other λ_j by $\sum_1^k \lambda_j \equiv 0 \pmod{2\pi}$.

We note that $f_{a_1,\ldots,a_k}(\lambda_1, \ldots, \lambda_k)$ is generally complex-valued. It is also bounded and uniformly continuous in the manifold $\sum_1^k \lambda_j \equiv 0 \pmod{2\pi}$. We have the inverse relation

$$c_{a_1,\ldots,a_k}(u_1, \ldots, u_{k-1}) = \int_{-\pi}^{\pi} \cdots \int_{-\pi}^{\pi} \exp\left\{i \sum_1^{k-1} \lambda_j u_j\right\} f_{a_1,\ldots,a_k}(\lambda_1, \ldots, \lambda_{k-1}) d\lambda_1 \cdots d\lambda_{k-1} \quad (2.6.3)$$

and in symmetric form

$$c_{a_1,\ldots,a_k}(u_1, \ldots, u_k) = \int_{-\pi}^{\pi} \cdots \int_{-\pi}^{\pi} \exp\left\{i \sum_1^{k} \lambda_j u_j\right\} f_{a_1,\ldots,a_k}(\lambda_1, \ldots, \lambda_k) \times \eta(\lambda_1 + \cdots + \lambda_k) d\lambda_1 \cdots d\lambda_k \quad (2.6.4)$$

where

$$\eta(\lambda) = \sum_{j=-\infty}^{\infty} \delta(\lambda + 2\pi j) \quad (2.6.5)$$

is the Dirac comb of (2.1.6).

We will frequently assume that our series satisfy

Assumption 2.6.1 $\mathbf{X}(t)$ is a strictly stationary r vector-valued series with components $X_j(t)$, $j = 1, \ldots, r$ all of whose moments exist, and satisfying (2.6.1) for $a_1, \ldots, a_k = 1, \ldots, r$ and $k = 2, 3, \ldots$.

We note that all cumulant spectra, of all orders, exist for series satisfying Assumption 2.6.1. In the case of a Gaussian process, it amounts to nothing more than $\Sigma\, |c_{ab}(u)| < \infty$, $a, b = 1, \ldots, r$.

Cumulant spectra are defined and discussed in Shiryaev (1960), Leonov (1964), Brillinger (1965) and Brillinger and Rosenblatt (1967a, b). The idea of carrying out a Fourier analysis of the higher moments of a time series occurs in Blanc-Lapierre and Fortet (1953).

The third-order spectrum of a single series has been called the **bispectrum;** see Tukey (1959) and Hasselman, Munk and MacDonald (1963). The fourth-order spectrum has been called the **trispectrum.**

On occasion we will find the following assumption useful.

Assumption 2.6.2(l) Given the r-vector stationary process $\mathbf{X}(t)$ with components $X_j(t)$, $j = 1, \ldots, r$, there is an $l \geqslant 0$ with

$$\sum_{v_1,\ldots,v_{k-1}=-\infty}^{\infty} \{1 + |v_j|^l\} |c_{a_1,\ldots,a_k}(v_1, \ldots, v_{k-1})| < \infty \quad (2.6.6)$$

for $j = 1, \ldots, k - 1$ and any k tuple $a_1, \ldots, a_k$ when $k = 2, 3, \ldots$.

This assumption implies, for $l > 0$, that well-separated (in time) values of the process are even less dependent than implied by Assumption 2.6.1, the extent of dependence depending directly on l. Equation (2.6.6) implies that $f_{a_1,\ldots,a_k}(\lambda_1, \ldots, \lambda_k)$ has bounded and uniformly continuous derivatives of order $\leqslant l$.

If instead of expressions (2.6.1) or (2.6.6) we assume only ave $|X_a(t)|^k < \infty$, $a = 1, \ldots, r$, then the $f_{a_1,\ldots,a_k}(\lambda_1, \ldots, \lambda_k)$ appearing in (2.6.4) are Schwartz distributions of order $\leqslant 2$. These distributions, or generalized functions, are found in Schwartz (1957, 1959). In the case $k = 2$, Theorem 2.5.2 shows they are measures.

Several times in later chapters we will require a stronger assumption than the commonly used Assumption 2.6.1. It is the following:

Assumption 2.6.3 The r vector-valued series $\mathbf{X}(t)$, $t = 0, \pm 1, \ldots$ satisfies Assumption 2.6.1. Also if

$$C_k = \sup_{a_1,\ldots,a_k} \sum_{v_1,\ldots,v_{k-1}} |c_{a_1,\ldots,a_k}(v_1, \ldots, v_{k-1})|, \tag{2.6.7}$$

then

$$\sum_k C_k z^k/k! < \infty \tag{2.6.8}$$

for z in a neighborhood of 0.

This assumption will allow us to obtain probability 1 bounds for various statistics of interest. If $\mathbf{X}(t)$, $t = 0, \pm 1, \ldots$ is Gaussian, all that is required is that the covariance function be summable. Exercise 2.13.36 indicates the form of the assumption for another example of interest.

2.7 FILTERS

In the analysis of time series we often have occasion to apply some manipulatory operation. An important class of operations consists of those that are linear and time invariant. Specifically, consider an operation whose domain consists of r vector-valued series $\mathbf{X}(t)$, $t = 0, \pm 1, \ldots$ and whose range consists of s vector-valued series $\mathbf{Y}(t)$, $t = 0, \pm 1, \ldots$. We write

$$\mathbf{Y}(t) = \mathcal{A}[\mathbf{X}](t) \tag{2.7.1}$$

to indicate the action of the operation. The operation is **linear** if for series $\mathbf{X}_1(t)$, $\mathbf{X}_2(t)$, $t = 0, \pm 1, \ldots$ in its domain and for constants α_1, α_2 we have

$$\mathcal{A}[\alpha_1\mathbf{X}_1 + \alpha_2\mathbf{X}_2](t) = \alpha_1\mathcal{A}[\mathbf{X}_1](t) + \alpha_2\mathcal{A}[\mathbf{X}_2](t). \tag{2.7.2}$$

Next for given u let $T^u\mathbf{X}(t)$, $t = 0, \pm 1, \ldots$ denote the series $\mathbf{X}(t + u)$, $t = 0, \pm 1, \ldots$. The operation $\mathcal{A}$ is **time invariant** if

$$\mathcal{A}[T^u\mathbf{X}](t) = \mathcal{A}[\mathbf{X}](t + u) \qquad \text{for } t, u = 0, \pm 1, \ldots. \tag{2.7.3}$$

We may now set down the definition: an operation $\mathcal{A}$ carrying r vector-valued series into s vector-valued series and possessing the properties (2.7.2) and (2.7.3) is called an $s \times r$ **linear filter.**

The domain of an $s \times r$ linear filter may include $r \times r$ matrix-valued functions $\mathbf{U}(t)$, $t = 0, \pm 1, \ldots$. Denote the columns of $\mathbf{U}(t)$ by $\mathbf{U}_j(t)$, $j = 1, \ldots, r$ and we then define

$$\mathcal{A}[\mathbf{U}](t) = [\mathcal{A}[\mathbf{U}_1](t) \cdots \mathcal{A}[\mathbf{U}_r](t)]. \tag{2.7.4}$$

The range of this extended operation is seen to consist of $s \times r$ matrix-valued functions.

An important property of filters is that they transform cosinusoids into cosinusoids. In particular we have

Lemma 2.7.1 Let $\mathcal{A}$ be a linear time invariant operation whose domain includes the $r \times r$ matrix-valued series

$$\mathbf{e}(t) = \exp\{i\lambda t\}\mathbf{I} \tag{2.7.5}$$

$t = 0, \pm 1, \ldots; -\infty < \lambda < \infty$ where $\mathbf{I}$ is the $r \times r$ identity matrix. Then there is an $s \times r$ matrix $\mathbf{A}(\lambda)$ such that

$$\mathcal{A}[\mathbf{e}](t) = \exp\{i\lambda t\}\mathbf{A}(\lambda). \tag{2.7.6}$$

In other words a linear time invariant operation carries complex exponentials of frequency λ over into complex exponentials of the same frequency λ. The function $\mathbf{A}(\lambda)$ is called the **transfer function** of the operation. We see that $\mathbf{A}(\lambda + 2\pi) = \mathbf{A}(\lambda)$.

An important class of $s \times r$ linear filters takes the form

$$\begin{aligned} \mathbf{Y}(t) &= \sum_{u=-\infty}^{\infty} \mathbf{a}(t - u)\mathbf{X}(u) \\ &= \sum_{u=-\infty}^{\infty} \mathbf{a}(u)\mathbf{X}(t - u) \end{aligned} \tag{2.7.7}$$

$t = 0, \pm 1, \ldots$, where $\mathbf{X}(t)$ is an r vector-valued series, $\mathbf{Y}(t)$ is an s vector-valued series, and $\mathbf{a}(u)$, $u = 0, \pm 1, \ldots$ is a sequence of $s \times r$ matrices satisfying

$$\sum_{u=-\infty}^{\infty} |\mathbf{a}(u)| < \infty. \tag{2.7.8}$$

We call such a filter an $s \times r$ **summable filter** and denote it by $\{\mathbf{a}(u)\}$. The transfer function of the filter (2.7.7) is seen to be given by

$$\mathbf{A}(\lambda) = \sum_{u=-\infty}^{\infty} \mathbf{a}(u) \exp\{-i\lambda u\} \quad \text{for } -\infty < \lambda < \infty. \tag{2.7.9}$$

It is a uniformly continuous function of λ in view of (2.7.8). The function $\mathbf{a}(u)$, $u = 0, \pm 1, \ldots$ is called the **impulse response** of the filter in view of the fact that if the domain of the filter is extended to include $r \times r$ matrix valued series and we take the input series to be the impulse

$$\begin{aligned} \mathbf{X}(t) &= \mathbf{I} \qquad \text{for } t = 0 \\ &= 0 \qquad \text{for } t \neq 0 \end{aligned} \tag{2.7.10}$$

then the output series is $\mathbf{a}(t)$, $t = 0, \pm 1, \ldots$.

An $s \times r$ filter $\{\mathbf{a}(u)\}$ is said to be **realizable** if $\mathbf{a}(u) = \mathbf{0}$ for $u = -1, -2, -3, \ldots$. From (2.7.7) we see that such a filter has the form

$$\mathbf{Y}(t) = \sum_{u=0}^{\infty} \mathbf{a}(u)\mathbf{X}(t - u) \tag{2.7.11}$$

and so $\mathbf{Y}(t)$ only involves the values of the $\mathbf{X}$ series for present and past times. In this case the domain of $\mathbf{A}(\lambda)$ may be extended to be the region $-\infty < \operatorname{Re} \lambda < \infty$, $\operatorname{Im} \lambda \geqslant 0$.

On occasion we may wish to apply a succession of filters to the same series. In this connection we have

Lemma 2.7.2 If $\{\mathbf{a}_1(t)\}$ and $\{\mathbf{a}_2(t)\}$ are $s \times r$ summable filters with transfer functions $\mathbf{A}_1(\lambda)$, $\mathbf{A}_2(\lambda)$, respectively, then $\{\mathbf{a}_1(t) + \mathbf{a}_2(t)\}$ is an $s \times r$ summable filter with transfer function $\mathbf{A}_1(\lambda) + \mathbf{A}_2(\lambda)$.

If $\{\mathbf{b}_1(t)\}$ is an $r \times q$ summable filter with transfer function $\mathbf{B}_1(\lambda)$ and $\{\mathbf{b}_2(t)\}$ is an $s \times r$ summable filter with transfer function $\mathbf{B}_2(\lambda)$, then $\{\mathbf{b}_2 * \mathbf{b}_1(t)\}$, the filter resulting from applying first $\{\mathbf{b}_1(t)\}$ followed by $\{\mathbf{b}_2(t)\}$, is a $s \times q$ summable filter with transfer function $\mathbf{B}_2(\lambda)\mathbf{B}_1(\lambda)$.

The second half of this lemma demonstrates the advantage of considering transfer functions as well as the time domain coefficients of a filter. The convolution expression

$$\mathbf{b}_2 * \mathbf{b}_1(t) = \sum_u \mathbf{b}_2(t - u)\mathbf{b}_1(u) \tag{2.7.12}$$

takes the form of a multiplication in the frequency domain.

Let $\{\mathbf{a}(t)\}$ be an $r \times r$ summable filter. If an $r \times r$ filter $\{\mathbf{b}(t)\}$ exists such that

$$\begin{aligned} \mathbf{b} * \mathbf{a}(t) &= \mathbf{I} \qquad \text{for } t = 0 \\ &= 0 \qquad \text{for } t \neq 0, \end{aligned} \tag{2.7.13}$$

then $\{\mathbf{a}(t)\}$ is said to be **nonsingular.** The filter $\{\mathbf{b}(t)\}$ is called the **inverse** of $\{\mathbf{a}(t)\}$. It exists if the matrix $\mathbf{A}(\lambda)$ is nonsingular for $-\infty < \lambda < \infty$; its transfer function is $\mathbf{A}(\lambda)^{-1}$.

On occasion we will refer to an ***l* summable filter.** This is a summable filter satisfying the condition

$$\sum_{u=-\infty}^{\infty} [1 + |u|^l]|\mathbf{a}(u)| < \infty \qquad \text{for some } l > 0. \tag{2.7.14}$$

Two examples of l summable filters follow. The operation indicated by

$$Y(t) = (2M + 1)^{-1} \sum_{u=-M}^{M} X(t + u) \tag{2.7.15}$$

is an l summable filter, for all l, with coefficients

$$\begin{aligned} a(u) &= (2M + 1)^{-1} \qquad && u = 0, \pm 1, \ldots, \pm M \\ &= 0 && \text{otherwise} \end{aligned} \tag{2.7.16}$$

and transfer function

$$A(\lambda) = (2M + 1)^{-1} \frac{\sin (2M + 1)\frac{\lambda}{2}}{\sin \frac{\lambda}{2}} \qquad \text{for } -\infty < \lambda < \infty. \tag{2.7.17}$$

We will see the shape of this transfer function in Section 3.2. For M not too small, $A(\lambda)$ is a function with its mass concentrated in the neighborhood of frequencies $\lambda \equiv 0 \pmod{2\pi}$. The general effect of this filter will be to smooth functions to which it is applied.

Likewise the operation indicated by

$$Y(t) = X(t) - X(t - 1) \tag{2.7.18}$$

is an l summable filter, for all l, with coefficients

$$\begin{aligned} a(u) &= 1 \qquad && \text{for } u = 0 \\ &= -1 && \text{for } u = 1 \\ &= 0 && \text{otherwise} \end{aligned} \tag{2.7.19}$$

and transfer function

$$A(\lambda) = 2i \exp\left\{\frac{-i\lambda}{2}\right\} \sin \frac{\lambda}{2}. \tag{2.7.20}$$

This transfer function has most of its mass in the neighborhood of frequencies $\lambda = \pm\pi, \pm 3\pi, \ldots$. The effect of this filter will be to remove the slowly varying part of a function and retain the rapidly varying part.

We will often be applying filters to stochastic series. In this connection we have

Lemma 2.7.3 If $\mathbf{X}(t)$ is a stationary r vector-valued series with $E|\mathbf{X}(t)| < \infty$, and $\{\mathbf{a}(t)\}$ is an $s \times r$ summable filter, then

$$\mathbf{Y}(t) = \sum_{u=-\infty}^{\infty} \mathbf{a}(t-u)\mathbf{X}(u) \tag{2.7.21}$$

$t = 0, \pm 1, \ldots$ exists with probability 1 and is an s vector-valued stationary series. If $E|\mathbf{X}(t)|^k < \infty$, $k > 0$, then $E|\mathbf{Y}(t)|^k < \infty$.

An important use of this lemma is in the derivation of additional stationary time series from stationary time series already under discussion. For example, if $\boldsymbol{\varepsilon}(t)$ is a sequence of independent identically distributed r vector variates and $\{\mathbf{a}(t)\}$ is an $s \times r$ filter, then the s vector-valued series

$$\mathbf{X}(t) = \sum_{u=-\infty}^{\infty} \mathbf{a}(t-u)\boldsymbol{\varepsilon}(u) \tag{2.7.22}$$

is a strictly stationary series. It is called a **linear process.**

Sometimes we will want to deal with a linear time invariant operation whose transfer function $\mathbf{A}(\lambda)$ is not necessarily the Fourier transform of an absolutely summable sequence. In the case that

$$\int_{-\pi}^{\pi} \mathbf{A}(\lambda)\overline{\mathbf{A}(\lambda)}^{\tau} d\lambda < \infty \tag{2.7.23}$$

it is possible to define the output of such a filter as a limit in mean square. Specifically we have

Theorem 2.7.1 Let $\mathbf{X}(t)$, $t = 0, \pm 1, \ldots$ be an r vector-valued series with absolutely summable autocovariance function. Let $\mathbf{A}(\lambda)$ be an $s \times r$ matrix-valued function satisfying (2.7.23). Set

$$\mathbf{a}(u) = (2\pi)^{-1} \int_{-\pi}^{\pi} \mathbf{A}(\lambda) \exp\{iu\lambda\} d\lambda \tag{2.7.24}$$

$u = 0, \pm 1, \ldots$. Then

$$\mathbf{Y}(t) = \underset{T\to\infty}{\text{l.i.m}} \sum_{u=-T}^{T} \mathbf{a}(t-u)\mathbf{X}(u) \tag{2.7.25}$$

exists for $t = 0, \pm 1, \ldots$.

Results of this character are discussed in Rosenberg (1964) for the case in which the conditions of Theorem 2.5.2 are satisfied plus

$$\int \mathbf{A}(\lambda) d\mathbf{F}_{XX}(\lambda)\overline{\mathbf{A}(\lambda)}^{\tau} < \infty.$$

Two 1×1 filters satisfying (2.7.23) will be of particular importance in our work. A 1×1 filter $\{a(u)\}$ is said to be a **band-pass filter, centered at the frequency λ_0 and with band-width** 2Δ if its transfer function has the form

$$\begin{aligned} A(\lambda) &= 1 \qquad \text{for } |\lambda \pm \lambda_0| \leqslant \Delta \\ &= 0 \qquad \text{otherwise} \end{aligned} \tag{2.7.26}$$

in the domain $-\pi < \lambda < \pi$. Typically Δ is small. If $\lambda_0 = 0$, the filter is called a **low-pass filter.** In the case that

$$X(t) = \sum_{j=1}^{k} R_j \cos(\lambda_j t + \phi_j) \tag{2.7.27}$$

for constants R_j, ϕ_j, k and the transfer function $A(\lambda)$ is given by (2.7.26), we see that the filtered series is given by

$$\sum_j R_j \cos(\lambda_j t + \phi_j) \tag{2.7.28}$$

with the summation extending over j such that $|\lambda_j \pm \lambda_0| \leqslant \Delta$. In other words, components whose frequencies are near λ_0 remain unaffected, whereas other components are removed.

A second useful 1×1 filter is the **Hilbert transform.** Its transfer function is purely imaginary and given by $-i \operatorname{sgn} \lambda$, that is

$$\begin{aligned} A(\lambda) &= -i \qquad \text{for } 0 < \lambda < \pi \\ &= \;\;0 \qquad \text{for } \lambda = 0 \\ &= \;\;i \qquad \text{for } -\pi < \lambda < 0. \end{aligned} \tag{2.7.29}$$

If the series $X(t)$, $t = 0, \pm 1, \ldots$ is given by (2.7.27), then the series resulting from the application of the filter with transfer function (2.7.29) is

$$\sum_{j=1}^{k} R_j \sin(\lambda_j t + \phi_j). \tag{2.7.30}$$

The series that is the Hilbert transform of a series $X(t)$ will be denoted $X^H(t)$, $t = 0, \pm 1, \ldots$.

Lemma 2.7.4 indicates how the procedure of **complex demodulation** (see Tukey (1961)) may be used to obtain a band-pass filter centered at a general frequency λ_0 and the corresponding Hilbert transform from a low-pass filter.

In complex demodulation we first form the pair of real-valued series

$$\begin{aligned} Y_1(t) &= \cos \lambda_0 t \, X(t) \\ Y_2(t) &= \sin \lambda_0 t \, X(t) \end{aligned} \tag{2.7.31}$$

for $t = 0, \pm 1, \ldots$ and then the pair of series

$$\begin{aligned} W_1(t) &= \sum_u a(t-u) Y_1(u) \\ W_2(t) &= \sum_u a(t-u) Y_2(u) \end{aligned} \tag{2.7.32}$$

where $\{a(t)\}$ is a low-pass filter. The series, $W_1(t)$, $W_2(t)$ $-\infty < t < \infty$, are called the **complex demodulates** of the series $X(t)$, $-\infty < t < \infty$. Because $\{a(t)\}$ is a low-pass filter, they will typically be substantially smoother than the series $X(t)$, $-\infty < t < \infty$. If we further form the series

$$\begin{aligned} V_1(t) &= \cos \lambda_0 t\, W_1(t) + \sin \lambda_0 t\, W_2(t) \\ V_2(t) &= \sin \lambda_0 t\, W_1(t) - \cos \lambda_0 t\, W_2(t) \end{aligned} \tag{2.7.33}$$

for $-\infty < t < \infty$, then the following lemma shows that the series $V_1(t)$ is essentially a band-pass filtered version of the series $X(t)$, while the series $V_2(t)$ is essentially a band-pass filtered version of the series $X^H(t)$.

Lemma 2.7.4 Let $\{a(t)\}$ be a filter with transfer function $A(\lambda)$, $-\infty < \lambda < \infty$. The operation carrying the series $X(t)$, $-\infty < t < \infty$, into the series $V_1(t)$ of (2.7.33) is linear and time invariant with transfer function

$$\frac{A(\lambda - \lambda_0) + A(\lambda + \lambda_0)}{2}. \tag{2.7.34}$$

The operation carrying the series $X(t)$ into $V_2(t)$ of (2.7.33) is linear and time invariant with transfer function

$$\frac{A(\lambda - \lambda_0) - A(\lambda + \lambda_0)}{2i}. \tag{2.7.35}$$

In the case that $A(\lambda)$ is given by

$$\begin{aligned} A(\lambda) &= 1 \qquad \text{for } |\lambda| \leqslant \Delta \\ &= 0 \qquad \text{otherwise} \end{aligned} \tag{2.7.36}$$

for $-\pi < \lambda < \pi$ and Δ small, functions (2.7.34) and (2.7.35) are seen to have the forms

$$\begin{aligned} &\tfrac{1}{2} \qquad \text{for } |\lambda \pm \lambda_0| \leqslant \Delta \\ &0 \qquad \text{otherwise} \end{aligned} \tag{2.7.37}$$

$-\pi < \lambda, \lambda_0 < \pi$ and

$$\begin{aligned} &-\tfrac{1}{2}i \qquad \text{for } |\lambda - \lambda_0| \leqslant \Delta \\ &\tfrac{1}{2}i \qquad \text{for } |\lambda + \lambda_0| \leqslant \Delta \\ &0 \qquad \text{otherwise} \end{aligned} \tag{2.7.38}$$

for $-\pi < \lambda, \lambda_0 < \pi$.

Bunimovitch (1949), Oswald (1956), Dugundji (1958), and Deutsch (1962) discuss the interpretation and use of the output of such filters.

2.8 INVARIANCE PROPERTIES OF CUMULANT SPECTRA

The principal parameters involved in our discussion of the frequency analysis of stationary time series are the cumulant spectra. At the same time we will often be applying filters to the series or it will be the case that some filtering operation has already been applied. It is therefore important that we understand the effect of a filter on the cumulant spectra of stationary series. The effect is of an elementary algebraic nature.

Theorem 2.8.1 Let $\mathbf{X}(t)$ be an r vector series satisfying Assumption 2.6.1 and $\mathbf{Y}(t) = \sum_u \mathbf{a}(t-u)\mathbf{X}(u)$, where $\{\mathbf{a}(t)\}$ is an $s \times r$ summable filter. $\mathbf{Y}(t)$ satisfies Assumption 2.6.1. Its cumulant spectra

$$g_{b_1,\ldots,b_k}(\lambda_1, \ldots, \lambda_k),\ b_1, \ldots, b_k = 1, \ldots, s;\ k = 2, 3, \ldots$$

are given by

$$g_{b_1,\ldots,b_k}(\lambda_1, \ldots, \lambda_k) = \sum_{j_1=1}^{r} \cdots \sum_{j_k=1}^{r} A_{b_1 j_1}(\lambda_1)\cdots A_{b_k j_k}(\lambda_k) f_{j_1,\ldots,j_k}(\lambda_1, \ldots, \lambda_k). \tag{2.8.1}$$

Some cases of this theorem are of particular importance.

Example 2.8.1 Let $X(t)$ and $Y(t)$ be real-valued with power spectra $f_{XX}(\lambda)$, $f_{YY}(\lambda)$, respectively; then

$$f_{YY}(\lambda) = |A(\lambda)|^2 f_{XX}(\lambda) \tag{2.8.2}$$

where $A(\lambda)$ is the transfer function of the filter.

Example 2.8.2 Let $\mathbf{f}_{XX}(\lambda)$ and $\mathbf{f}_{YY}(\lambda)$ signify the $r \times r$ and $s \times s$ matrices of second-order spectra of $\mathbf{X}(t)$ and $\mathbf{Y}(t)$, respectively. Then

$$\mathbf{f}_{YY}(\lambda) = \mathbf{A}(\lambda)\,\mathbf{f}_{XX}(\lambda)\overline{\mathbf{A}(\lambda)}^{\tau}. \tag{2.8.3}$$

If $s = 1$, then the power spectrum of $Y(t)$ is given by

$$\sum_{j=1}^{r} \sum_{k=1}^{r} A_j(\lambda)\bar{A}_k(\lambda) f_{jk}(\lambda). \tag{2.8.4}$$

As power spectra are non-negative, we may conclude from (2.8.4) that

$$\sum_{j=1}^{r} \sum_{k=1}^{r} A_j \bar{A}_k f_{jk}(\lambda) \geqslant 0 \tag{2.8.5}$$

for all (complex) $A_1, \ldots, A_r$ and so obtain the result of Theorem 2.5.1 — that the matrix $\mathbf{f}_{XX}(\lambda)$ is non-negative definite — from the case $r = 1$.

Example 2.8.3 If $\mathbf{X}(t)$, $\mathbf{Y}(t)$, $t = 0, \pm 1, \ldots$ are both r vector-valued with $\mathbf{Y}$ related to $\mathbf{X}$ through

$$Y_j(t) = \sum_u b_j(t - u)X_j(u) \qquad j = 1, \ldots, r, \tag{2.8.6}$$

then the cumulant spectra of $\mathbf{Y}(t)$ are given by

$$B_{a_1}(\lambda_1)\cdots B_{a_k}(\lambda_k)f_{a_1,\ldots,a_k}(\lambda_1, \ldots, \lambda_k) \tag{2.8.7}$$

where $B_j(\lambda)$ denotes the transfer function of the filter $\{b_j(u)\}$.

Later we will see that Examples 2.8.1 and 2.8.3 provide convenient means of interpreting the power spectrum, cross-spectrum, and higher order cumulant spectra.

2.9 EXAMPLES OF STATIONARY TIME SERIES

The definition of, and several elementary examples of, a stationary time series was presented in Section 2.4. As stationary series are the basic entities of our analysis, it is of value to have as many examples as possible.

Example 2.9.1 (A Pure Noise Series) Let $\boldsymbol{\varepsilon}(t)$, $t = 0, \pm 1, \ldots$ be a sequence of independent, identically distributed r vector-valued random variables. Such a series clearly forms a stationary time series.

Example 2.9.2 (Linear Process) Let $\boldsymbol{\varepsilon}(t)$, $t = 0, \pm 1, \ldots$ be the r vector-valued pure noise series of the previous example. Let

$$\mathbf{X}(t) = \sum_u \mathbf{a}(t - u)\boldsymbol{\varepsilon}(u) \tag{2.9.1}$$

where $\{\mathbf{a}(u)\}$ is an $s \times r$ summable filter. Following Lemma 2.7.3, this series is a stationary s vector-valued series.

If only a finite number of the $\mathbf{a}(u)$ in expression (2.9.1) are nonzero, then the series $\mathbf{X}(t)$ is referred to as a **moving average process.** If $\mathbf{a}(0)$, $\mathbf{a}(m) \neq \mathbf{0}$ and $\mathbf{a}(u) = \mathbf{0}$ for $u > m$ and $u < 0$, the process is said to be of order m.

Example 2.9.3 (Cosinusoid) Suppose that $\mathbf{X}(t)$ is an r vector-valued series with components

$$X_j(t) = R_j \cos(\omega_j t + \phi_j) \qquad \text{for } j = 1, \ldots, r, \tag{2.9.2}$$

where $R_1, \ldots, R_r$ are constant, $\phi_1, \ldots, \phi_{r-1}$ are uniform on $(-\pi,\pi)$, and $\phi_1 + \cdots + \phi_r = 0$. This series is stationary, because if any finite collection of values is considered and then the time points are all shifted by t, their structure is unchanged.

Example 2.9.4 (*Stationary Gaussian Series*) An r vector time series $\mathbf{X}(t)$, $t = 0, \pm 1, \pm 2, \ldots$ is a Gaussian series if all of its finite dimensional distributions are multivariate Gaussian (normal). If $E\mathbf{X}(t) = \boldsymbol{\mu}$ and $E\mathbf{X}(t)\mathbf{X}(u)^{\tau} = \mathbf{R}(t - u)$ for all t, u, then $\mathbf{X}(t)$ is stationary in this case, for the series is determined by its first- and second-order moment properties.

We note that if $\mathbf{X}(t)$ is a stationary r vector Gaussian series, then

$$\mathbf{Y}(t) = \sum_{u} \mathbf{a}(t - u)\mathbf{X}(u) \tag{2.9.3}$$

for an $s \times r$ filter $\{\mathbf{a}(t)\}$ is a stationary s vector Gaussian series.

Extensive discussions of stationary Gaussian series are found in Blanc-Lapierre and Fortet (1965), Loève (1963), and Cramér and Leadbetter (1967).

Example 2.9.5 (*Stationary Markov Processes*) An r vector time series $\mathbf{X}(t)$, $t = 0, \pm 1, \pm 2, \ldots$ is said to be an **r vector Markov process** if the conditional probability

$$\text{Prob}\{\mathbf{X}(t) \leqslant \mathbf{X} \mid \mathbf{X}(s_1) = \mathbf{x}_1, \ldots, \mathbf{X}(s_n) = \mathbf{x}_n, \mathbf{X}(s) = \mathbf{x}\} \tag{2.9.4}$$

(for any $s_1 < s_2 < \cdots < s_n < s < t$) is equal to the conditional probability

$$\text{Prob}\{\mathbf{X}(t) \leqslant \mathbf{X} \mid \mathbf{X}(s) = \mathbf{x}\} = P(s,\mathbf{x},t,\mathbf{X}) \qquad s < t. \tag{2.9.5}$$

The function $P(s,\mathbf{x},t,\mathbf{X})$ is called the **transition probability function.** It and an **initial probability** $\text{Prob}\{\mathbf{X}(0) \leqslant \mathbf{x}_0\}$, completely determine the probability law of the process. Extensive discussions of Markov processes and in particular stationary Markov processes may be found in Doob (1953), Dynkin (1960), Loève (1963), and Feller (1966).

A particularly important example is that of the Gaussian stationary Markov process. In the real-valued case, its autocorrelation function takes a simple form.

Lemma 2.9.1 If $X(t)$, $t = 0, \pm 1, \pm 2, \ldots$ is a nondegenerate real-valued, Gaussian, stationary Markov process, then its autocovariance function is given by $c_{XX}(0)\rho^{|u|}$, for some ρ, $-1 < \rho < 1$.

Another class of examples of real-valued stationary Markov processes is given in Wong (1963). Bernstein (1938) considers the generation of Markov processes as solutions of stochastic difference and differential equations.

An example of a stationary Markov r vector process is provided by $\mathbf{X}(t)$, the solution of

$$\mathbf{X}(t) = \mathbf{a}\mathbf{X}(t - 1) + \boldsymbol{\varepsilon}(t) \tag{2.9.6}$$

where $\boldsymbol{\varepsilon}(t)$ is an r vector pure noise series and $\mathbf{a}$ an $r \times r$ matrix with all eigenvalues less than 1 in absolute value.

Example 2.9.6 (*Autoregressive Schemes*) Equation (2.9.6) leads us to consider r vector processes $\mathbf{X}(t)$ that are generated by schemes of the form

$$\mathbf{X}(t) + \mathbf{a}(1)\mathbf{X}(t-1) + \cdots + \mathbf{a}(m)\mathbf{X}(t-m) = \boldsymbol{\varepsilon}(t) \qquad (2.9.7)$$

where $\boldsymbol{\varepsilon}(t)$ is an r vector pure noise series and $\mathbf{a}(1), \ldots, \mathbf{a}(m)$ are $r \times r$ matrices. If the roots of Det $\mathbf{A}(z) = 0$ lie outside the unit circle where

$$\mathbf{A}(z) = \mathbf{I} + \mathbf{a}(1)z + \cdots + \mathbf{a}(m)z^m \qquad (2.9.8)$$

it can be shown (Section 3.8) that (2.9.7) has a stationary solution. Such an $\mathbf{X}(t)$ is referred to as an r vector-valued **autoregressive process of order** m.

Example 2.9.7 (*Mixing Moving Average and Autoregressive Process*) On occasion we combine the moving average and autoregressive schemes. Consider the r vector-valued process $\mathbf{X}(t)$ satisfying

$$\mathbf{X}(t) + \mathbf{a}(1)\mathbf{X}(t-1) + \cdots + \mathbf{a}(m)\mathbf{X}(t-m) = \boldsymbol{\varepsilon}(t) + \mathbf{b}(1)\boldsymbol{\varepsilon}(t-1) + \cdots + \mathbf{b}(n)\boldsymbol{\varepsilon}(t-n) \qquad (2.9.9)$$

where $\boldsymbol{\varepsilon}(t)$ is an s vector-valued pure noise series, $\mathbf{a}(j)$, $j = 1, \ldots, m$ are $r \times r$ matrices, and $\mathbf{b}(k)$, $k = 1, \ldots, n$ are $r \times s$ matrices. If a stationary $\mathbf{X}(t)$, satisfying expression (2.9.9), exists it is referred to as a **mixed moving average autoregressive process of order** (m,n).

If the roots of

$$\text{Det}[\mathbf{I} + \mathbf{a}(1)z + \cdots + \mathbf{a}(m)z^m] = 0 \qquad (2.9.10)$$

lie outside the unit circle, then an $\mathbf{X}(t)$ satisfying (2.9.9) is in fact a linear process

$$\mathbf{X}(t) = \sum_{u=0}^{\infty} \mathbf{c}(t-u)\boldsymbol{\varepsilon}(u) \qquad (2.9.11)$$

where $\mathbf{C}(\lambda) = \mathbf{A}(\lambda)^{-1}\mathbf{B}(\lambda)$; see Section 3.8.

Example 2.9.8 (*Functions of Stationary Series*) If we have a stationary series (such as a pure noise series) already at hand and we form time invariant measurable functions of that series then we have generated another stationary series. For example, suppose $\mathbf{X}(t)$ is a stationary series and $\mathbf{Y}(t) = \sum_u \mathbf{a}(t-u)\mathbf{X}(u)$ for some $s \times r$ filter $\{\mathbf{a}(u)\}$. We have seen, (Lemma 2.7.1), that under regularity conditions $\mathbf{Y}(t)$ is also stationary. Alternatively we can form a $\mathbf{Y}(t)$ through nonlinear functions as by

$$\mathbf{Y}(t) = \mathbf{f}[\mathbf{X}(t), \mathbf{X}(t-1)] \qquad (2.9.12)$$

for some measurable $\mathbf{f}[\mathbf{x}_1,\mathbf{x}_2]$; see Rosenblatt (1964). In fact, in a real sense, all stationary functions are of the form of expression (2.9.12), $\mathbf{f}$ possibly having an infinite number of arguments. Any stationary time series, defined on a probability space, can be put in the form

$$\mathbf{Y}(t) = \mathbf{f}(U^t\theta) \tag{2.9.13}$$

where U is a transformation that preserves probabilities and θ lies in the probability space; see Doob (1953) p. 509. We can often take θ in the unit interval; see Choksi (1966).

Unfortunately relations such as (2.9.12) and (2.9.13) generally are not easy to work with. Consequently investigators (Wiener (1958), Balakrishnan (1964), Shiryaev (1960), McShane (1963), and Meecham and Siegel (1964)) have turned to series generated by nonlinear relations of the form

$$\begin{aligned} Y(t) = \sum_{u_1} a_1(t - u_1)X(u_1) + \sum_{u_1}\sum_{u_2} a_2(t - u_1, t - u_2)X(u_1)X(u_2) \\ + \sum_{u_1}\sum_{u_2}\sum_{u_3} a_3(t - u_1, t - u_2, t - u_3)X(u_1)X(u_2)X(u_3) + \cdots \end{aligned} \tag{2.9.14}$$

in the hope of obtaining more reasonable results. Nisio (1960, 1961) has investigated $Y(t)$ of the above form for the case in which $X(t)$ is a pure noise Gaussian series. Meecham (1969) is concerned with the case where $Y(t)$ is nearly Gaussian.

We will refer to expansions of the form of expression (2.9.14) as **Volterra functional expansions**; see Volterra (1959) and Brillinger (1970a).

In connection with $Y(t)$ generated by expression (2.9.14), we have

Theorem 2.9.1 If the series $X(t)$, $t = 0, \pm 1, \ldots$ satisfies Assumption 2.6.1 and

$$Y(t) = \sum_{J=0}^{L} \sum_{u_1,\ldots,u_J} a_J(t - u_1, \ldots, t - u_J)X(u_1)\cdots X(u_J) \tag{2.9.15}$$

with the a_J absolutely summable and $L < \infty$, then the series $Y(t)$, $t = 0, \pm 1, \ldots$ also satisfies Assumption 2.6.1.

We see, for example, that the series $X(t)^J$, $t = 0, \pm 1, \ldots$ satisfies Assumption 2.6.1 when the series $X(t)$ does. The theorem generalizes to r vector-valued series and in that case provides an extension of Lemma 2.7.3.

Example 2.9.9 (Solutions of Stochastic Difference and Differential Equations) We note that literature is developing on stationary processes that satisfy random difference and differential equations; see, for example, Kampé de Fériet (1965), Ito and Nisio (1964), and Mortensen (1969).

In certain cases (see Ito and Nisio (1964)) the solution of a stochastic equation may be expressed in the form (2.9.14).

Example 2.9.10 (*Solutions of Volterra Functional Relations*) On occasion, we may be given $Y(t)$ and wish to define $X(t)$ as a series satisfying expression (2.9.14). This provides a model for frequency demultiplication and the appearance of lower order harmonics.

2.10 EXAMPLES OF CUMULANT SPECTRA

In this section we present a number of examples of cumulant spectra of order k for a number of r vector-valued stationary time series of interest.

Example 2.10.1 (*A Pure Noise Series*) Suppose that $\boldsymbol{\varepsilon}(t)$ is an r vector pure noise series with components $\varepsilon_a(t), a - 1, \ldots, r$. Let

$$K_{a_1,\ldots,a_k} = \operatorname{cum}\{\varepsilon_{a_1}(t), \ldots, \varepsilon_{a_k}(t)\}$$

exist; then $c_{a_1,\ldots,a_k}(u_1, \ldots, u_{k-1}) = K_{a_1,\ldots,a_k}\delta\{u_1\}\cdots\delta\{u_{k-1}\}$, where $\delta\{x\}$ is the Kronecker delta. We see directly that

$$f_{a_1,\ldots,a_k}(\lambda_1, \ldots, \lambda_k) = (2\pi)^{-k+1}K_{a_1,\ldots,a_k}. \tag{2.10.1}$$

Example 2.10.2 (*A Linear Process*) Suppose that

$$\mathbf{X}(t) = \sum_{u=-\infty}^{\infty} \mathbf{a}(t - u)\boldsymbol{\varepsilon}(u) \tag{2.10.2}$$

where $\{\mathbf{a}(t)\}$ is an $s \times r$ filter and $\boldsymbol{\varepsilon}(t)$ an r vector pure noise series. From Theorem 2.8.1 we have

$$f_{a_1,\ldots,a_k}(\lambda_1, \ldots, \lambda_k) = (2\pi)^{-k+1} \sum_{j_1=1}^{r} \cdots \sum_{j_k=1}^{r} A_{a_1j_1}(\lambda_1)\cdots A_{a_kj_k}(\lambda_k)K_{j_1,\ldots,j_k}. \tag{2.10.3}$$

The result of this example may be combined with that of the previous example to obtain the spectra of moving average and autoregressive processes.

Example 2.10.2 (*Stationary Gaussian Series*) The characteristic function of a multivariate Gaussian variable, with mean vector $\boldsymbol{\mu}$ and variance-covariance matrix $\boldsymbol{\Sigma}$, is given by

$$\exp\{i\boldsymbol{\mu}^\tau\mathbf{t} - \tfrac{1}{2}\mathbf{t}^\tau\boldsymbol{\Sigma}\mathbf{t}\}. \tag{2.10.4}$$

We see from this that all cumulant functions of order greater than 2 must vanish for a Gaussian series and therefore all cumulant spectra of order greater than 2 also vanish for such a series.

We see that cumulant spectra of order greater than 2, in some sense, measure the non-normality of a series.

Example 2.10.3 (Cosinusoids) Suppose that $\mathbf{X}(t)$ is an r vector process with components $X_a(t) = R_a \cos(\omega_a t + \varphi_a)$, $a = 1, \ldots, r$, where R_a is constant, $\omega_1 + \cdots + \omega_r \equiv 0 \pmod{2\pi}$, and $\varphi_1, \ldots, \varphi_{r-1}$ independent and uniform on $(-\pi,\pi]$ while $\varphi_1 + \cdots + \varphi_r \equiv 0 \pmod{2\pi}$. $\mathbf{X}(t)$ is stationary. We note that the members of any proper subset of $\varphi_1, \ldots, \varphi_r$ are independent of each other and so joint cumulants involving such proper subsets vanish. Therefore

$$\begin{aligned}\operatorname{cum}\{X_1(t_1), \ldots, X_r(t_r)\} &= \operatorname{ave}\{X_1(t_1) \times \cdots \times X_r(t_r)\} \\ &= R_1 \cdots R_r \cos(\omega_1 t_1 + \cdots + \omega_r t_r) 2^{-r+1}. \qquad (2.10.5)\end{aligned}$$

This is a function of $t_1 - t_r, \ldots, t_{r-1} - t_r$ as $\omega_1 + \cdots + \omega_r \equiv 0 \pmod{2\pi}$. We have

$$c_{1\ldots r}(u_1, \ldots, u_{r-1}) = R_1 \cdots R_r \cos(\omega_1 u_1 + \cdots + \omega_{r-1} u_{r-1}) 2^{-r+1} \qquad (2.10.6)$$

and so

$$\begin{aligned}f_{1\ldots r}(\lambda_1, \ldots, \lambda_r) = \frac{1}{2^r} R_1 \cdots R_r [\eta(\lambda_1 + \omega_1) \cdots \eta(\lambda_{r-1} + \omega_{r-1}) + \eta(\lambda_1 - \omega_1) \\ \cdots \eta(\lambda_{r-1} - \omega_{r-1})]. \qquad (2.10.7)\end{aligned}$$

$\eta(\lambda)$ was defined by (2.1.6), see also Exercise 2.13.33.

In the case that $r = 1$, the power spectrum of the series $X(t) = R \cos(\omega t + \phi)$, is seen to be

$$f_{XX}(\lambda) = \tfrac{1}{4} R^2 [\eta(\lambda - \omega) + \eta(\lambda + \omega)]. \qquad (2.10.8)$$

It has peaks at the frequencies $\lambda \equiv \pm\omega \pmod{2\pi}$. This provides one of the reasons for calling λ the frequency. We see that $\omega/(2\pi)$ is the number of complete cycles the cosinusoid $\cos(\omega t + \phi)$ passes through when t increases by one unit. For this reason $\lambda/(2\pi)$ is called the **frequency in cycles per unit time.** Its reciprocal, $2\pi/\lambda$, is called the **period.** λ itself is the **angular frequency** in radians per unit time.

Example 2.10.4 (Volterra Functional Expansions) We return to Example 2.9.8 and have

Theorem 2.10.1 Let $Y(t)$, $t = 0, \pm 1, \ldots$ be given by (2.9.15) where $\sum |a_J(u_1, \ldots, u_J)| < \infty$, and

$$A_J(\lambda_1, \ldots, \lambda_J) = \sum_{u_1} \cdots \sum_{u_J} a_J(u_1, \ldots, u_J) \exp\{-i(\lambda_1 u_1 + \cdots + \lambda_J u_J)\}$$
$$\text{for } J = 1, \ldots, L. \qquad (2.10.9)$$

Then the Ith order cumulant spectrum of the series $Y(t)$, $t = 0, \pm 1, \ldots$ is given by

$$\eta(\lambda_1 + \cdots + \lambda_I) f_{Y \ldots Y}(\lambda_1, \ldots, \lambda_I) d\lambda_1 \cdots d\lambda_I$$
$$= \sum_{M \geq 1} \sum_{J_1} \cdots \sum_{J_I} \int \cdots \int \eta\Big(\sum_1^{J_1} \alpha_{1j} - \lambda_1\Big) \cdots A_1(\alpha_{1j}; j = 1, \ldots, J_1)$$
$$\cdots \eta\Big(\sum \alpha_{ij}; (i,j) \in P_1\Big) \cdots \eta\Big(\sum \alpha_{ij}; (i,j) \in P_M\Big) f_{X \ldots X}(\alpha_{ij}; (i,j) \in P_1) \cdots$$
$$f_{X \ldots X}(\alpha_{ij}; (i,j) \in P_M) d\alpha_{11} \cdots d\alpha_{IJ_I}, \quad (2.10.10)$$

where the outer sum is over all the indecomposable partitions $\{P_1, \ldots, P_M\}$, $M = 1, 2, \ldots$ of Table 2.3.4.

We have used the symmetrical notation for the cumulant spectra in Equation (2.10.10). Theorem 2 in Shiryaev (1960) provides a related result.

2.11 THE FUNCTIONAL AND STOCHASTIC APPROACHES TO TIME SERIES ANALYSIS

Currently two different approaches are adopted by workers in time series: the **stochastic** approach and the **functional** approach. The former, generally adopted by probabilists and statisticians (Doob, (1953) and Cramér and Leadbetter (1967)), is that described in Section 2.2. A given time series is regarded as being selected stochastically from an ensemble of possible series. We have a set Θ of r vector functions $\boldsymbol{\theta}(t)$. After defining a probability measure on Θ, we obtain a random function $\mathbf{X}(t,\theta)$, whose samples are the given functions $\theta(t)$. Alternatively given $\mathbf{X}(t)$, we can set up an index $\theta = \mathbf{X}(\cdot)$ and take Θ to be the set of all θ. We then may set $\mathbf{X}(t,\theta) = \mathbf{X}(t,\mathbf{X}(\cdot))$. In any case we find ourselves dealing with measure theory and probability spaces.

In the second approach, a given r vector time series is interpreted as a mathematical function and the basic ensemble of time functions takes the form $\{\mathbf{X}(t,v) = \mathbf{X}(t + v) \mid v = 0, \pm 1, \pm 2, \ldots\}$, where $\mathbf{X}(t)$ is the given r vector function. This approach is taken in Wiener (1930), for example, and is called **generalized harmonic analysis.**

The distinction, from the point of view of the theoretician, is the different mathematical tools required and the different limiting processes involved.

Suppose that $\mathbf{X}(t)$ has components $X_a(t)$, $a = 1, \ldots, r$. In the functional approach we assume that limits of the form

$$\lim_{S,T \to \infty} \frac{\sum_{t=-S}^{T-1} X_a(t)}{S + T} = m_a \quad (2.11.1)$$

$$\lim_{S,T \to \infty} \frac{\sum_{t=-S}^{T-1} X_a(t) X_b(t + u)}{S + T} = m_{ab}(u) \quad (2.11.2)$$

exist. A form of stationarity obtains as

$$\lim_{S,T\to\infty} \frac{\sum_{t=-S}^{T-1} X_a(t+v)}{S+T} = m_a \tag{2.11.3}$$

$$\lim_{S,T\to\infty} \frac{\sum_{t=-S}^{T-1} X_a(t+v)X_b(t+v+u)}{S+T} = m_{ab}(u) \tag{2.11.4}$$

independently of v for $v = 0, \pm 1, \pm 2, \ldots$. We now define a cross-covariance function by

$$c_{ab}(u) = m_{ab}(u) - m_a m_b. \tag{2.11.5}$$

If

$$\sum_{u=-\infty}^{\infty} |c_{ab}(u)| < \infty, \tag{2.11.6}$$

we can define a second-order spectrum $f_{ab}(\lambda)$ as in Section 2.5.

Suppose that the functions $X_a(t)$, $a = 1, \ldots, r$ are such that

(i) for given real $x_1, \ldots, x_k$ and $t_1, \ldots, t_k$ the proportions, $F_{a_1,\ldots,a_k}^{(T)}(x_1, \ldots, x_k; t_1, \ldots, t_k)$, of t's in the interval $[-S,T)$ such that

$$X_{a_1}(t+t_1) \leqslant x_1, \ldots, X_{a_k}(t+t_k) \leqslant x_k \tag{2.11.7}$$

tends to a limit $F_{a_1,\ldots,a_k}(x_1, \ldots, x_k; t_1, \ldots, t_k)$ (at points of continuity of this function) as $S, T \to \infty$ and

(ii) a compactness assumption such as

$$(S+T)^{-1} \sum_{t=-S}^{T-1} |X_j(t)|^u < M \tag{2.11.8}$$

is satisfied for all S, T and some $u > 0$.

In this case the $F_{a_1,\ldots,a_k}(x_1, \ldots, x_k; t_1, \ldots, t_k)$ provide a consistent and symmetric family of finite dimensional distributions and so can be associated with some stochastic process by the Kolmogorov extension theorem; see Doob (1953). The limit in (i) depends only on the differences $t_1 - t_k, \ldots, t_{k-1} - t_k$ and so the associated process is strictly stationary. If in (ii) we have $u \geqslant k$ and $\tilde{\mathbf{X}}(t)$ is the associated stationary process, then

$$E\tilde{X}_{a_1}(t_1)\cdots\tilde{X}_{a_{k-1}}(t_{k-1})\tilde{X}_{a_k}(0) = \lim_{S,T\to\infty} (S+T)^{-1} \sum_{t=-S}^{T-1} X_{a_1}(t+t_1)\cdots X_{a_{k-1}}(t+t_{k-1})X_{a_k}(t) \tag{2.11.9}$$

and the association makes sense. $\tilde{\mathbf{X}}(t)$ will satisfy Assumption 2.6.1 if the cumulant-type functions derived from $\mathbf{X}(t)$ satisfy (2.6.1).

In other words, if the function (of the functional approach) satisfies certain regularity conditions, then there is a strictly stationary process whose analysis is equivalent.

Conversely: if $\mathbf{X}(t)$ is **ergodic** (metrically transitive), then with probability 1 any sample path satisfies the required limiting properties and can be taken as the basis for a functional approach.[1]

In conclusion, we have

Theorem 2.11.1 If an r vector function satisfies (i) and (ii) above, then a stationary stochastic process can be associated with it having the same limiting properties. Alternatively, if a stationary process is ergodic, then with probability 1 any of its sample paths can be taken as the basis for a functional approach.

These two approaches are directly comparable to the two approaches to statistics through **kollectivs** (Von Mises (1964)) and measurable functions (Doob (1953)); see also Von Mises and Doob (1941).

The condition that $\mathbf{X}(t)$ be ergodic is not overly restrictive for our purposes since it is ergodic when it satisfies Assumption 2.6.1 and is determined by its moments; Leonov (1960). We note that a general stationary process is a mixture of ergodic processes (Rozanov (1967)), and the associated process obtained by the above procedure will correspond to some component of the mixture. The limits in (i) will exist with probability 1; however, they will generally be random variables.

Wold (1948) discusses relations between the functional and stochastic approaches in the case of second-order moments.

We note that the limits required in expressions (2.11.1) and (2.11.2) follow under certain conditions from the existence of the limits in (i); see Wintner (1932).

We will return to a discussion of the functional approach to time series analysis in Section 3.9.

2.12 TRENDS

One simple form of departure from the assumption of stationarity is that the series $X(t)$, $t = 0, \pm 1, \ldots$ has the form

$$X(t) = m(t) + \varepsilon(t) \qquad t = 0, \pm 1, \ldots \tag{2.12.1}$$

[1] $\mathbf{X}(t)$ is ergodic if for any real-valued $f[\mathbf{x}]$ with ave $|f[\mathbf{X}(t)]| < \infty$, with probability 1,

$$\lim_{T \to \infty} T^{-1} \sum_{t=0}^{T-1} f[\mathbf{X}(t)] = \text{ave}\, f[\mathbf{X}(t)].$$

See Cramér and Leadbetter (1967), Wiener et al (1967), Halmos (1956), Billingsley (1965), and Hopf (1937).

where the series $\varepsilon(t)$, $t = 0, \pm 1, \ldots$ is stationary, while $m(t)$, $t = 0, \pm 1, \ldots$ is a nonconstant deterministic function. If, in addition, $m(t)$ does not satisfy conditions of the character of those of Section 2.11, then a harmonic analysis of $X(t)$ is not directly available. Our method of analysis of such series will be to try to isolate the effects of $m(t)$ and $\varepsilon(t)$ for separate analysis.

If the function $m(t)$, $t = 0, \pm 1, \ldots$ varies slowly, it will be referred to as a **trend.** Many series occurring in practice appear to possess such a trend component. The series of United Kingdom exports graphed in Figure 1.1.4 appear to have this characteristic. In Section 5.11 we will discuss the estimation of trend functions of simple form.

2.13 EXERCISES

2.13.1 Let $X(t) = \cos(\lambda t + \theta)$, where θ has a uniform distribution on $(-\pi,\pi]$. Determine the finite dimensional distributions of the process, the mean function, $c_X(t)$, and the autocovariance function $c_{XX}(t_1,t_2)$.

2.13.2 If $(Y_1, \ldots, Y_r)$ is an r variate chance quantity for which cum $(Y_{j_1}, \ldots, Y_{j_s})$ exists $j_1, \ldots, j_s = 1, \ldots, r$ and $Z_k = \sum_j a_{kj} Y_j$, $k = 1, \ldots, s$, prove that $\text{cum}(Z_{k_1}, \ldots, Z_{k_s}) = \sum_{j_1} \cdots \sum_{j_s} a_{k_1 j_1} \cdots a_{k_s j_s} \text{cum}(Y_{j_1}, \ldots, Y_{j_s})$, $k_1, \ldots, k_s = 1, \ldots, s$.

2.13.3 Denote cum ($Y_1[m_1$ times$], \ldots, Y_r[m_r$ times$]$) and cum ($Z_1[n_1$ times$], \ldots, Z_s[n_s$ times$]$) by $K_{m_1 \ldots m_r}(\mathbf{Y})$ and $K_{n_1 \ldots n_s}(\mathbf{Z})$, respectively and let $K^{[m]}(\mathbf{Y})$ and $K^{[n]}(\mathbf{Z})$, $m = m_1 + \cdots + m_r$, $n = n_1 + \cdots + n_s$ denote the vectors with these components. Denote the transformation of 2.13.2 by $\mathbf{Z} = \mathbf{AY}$, where A is an $s \times r$ matrix. Prove that $K^{[n]}(\mathbf{Z}) = \mathbf{A}^{[n]} K^{[n]}(\mathbf{Y})$, where $\mathbf{A}^{[n]}$ is the nth symmetric Kronecker power of $\mathbf{A}$; see Hua (1963) pp. 10, 100.

2.13.4 Determine the transfer function of the filter of (2.9.6).

2.13.5 Show that the power spectrum of the (wide sense stationary) series $X(t) = R\cos(\omega t + \phi)$ where R is a constant, ω is a random variable with continuous density function $f(\omega)$ and ϕ is an independent uniform variate on $(-\pi,\pi]$ is given by

$$R^2 \sum_{j=-\infty}^{\infty} [f(\lambda + 2\pi j) + f(-\lambda + 2\pi j)]/4.$$

2.13.6 Prove that the transfer function of the $r \times r$ filter indicated by

$$\mathbf{Y}(t) = (2N + 1)^{-1} \sum_{u=-N}^{N} \mathbf{X}(t - u)$$

has off-diagonal elements 0 and diagonal elements $[\sin(2N + 1)\lambda/2]/[(2N + 1)\sin \lambda/2]$.

2.13.7 If $Y_1(t) = \sum_u a_{11}(t - u)X_1(u)$, $Y_2(t) = \sum_u a_{22}(t - u)X_2(u)$, where $\{X_1(t), X_2(t)\}$ satisfies Assumption 2.6.1. Suppose the transfer functions

$A_{11}(\lambda)$, $A_{22}(\lambda)$ are not 0. Denote the second-order spectra of $\{X_1(t), X_2(t)\}$ by $f_{jk}(\lambda)$ and those of $\{Y_1(t), Y_2(t)\}$ by $g_{jk}(\lambda)$, $j, k = 1, 2$. Prove that

$$\frac{|f_{12}(\lambda)|^2}{[f_{11}(\lambda) f_{22}(\lambda)]} = \frac{|g_{12}(\lambda)|^2}{[g_{12}(\lambda) g_{22}(\lambda)]}.$$

2.13.8 Prove that $\delta(x) = \lim_{n\to\infty} nW(nx)$ where $\int |W(x)|dx < \infty$ and $\int W(x)dx = 1$.

2.13.9 If $\mathbf{X}(t)$, $\mathbf{Y}(t)$ are statistically independent r vector series with cumulant spectra $\mathbf{f_a}(\boldsymbol{\lambda}) = f_{a_1,\ldots,a_k}(\lambda_1, \ldots, \lambda_k)$ $\mathbf{g_a}(\boldsymbol{\lambda}) = g_{a_1,\ldots,a_k}(\lambda_1, \ldots, \lambda_k)$, respectively, prove that the cumulant spectra of $\mathbf{X}(t) + \mathbf{Y}(t)$ are given by $\mathbf{f_a}(\boldsymbol{\lambda}) + \mathbf{g_a}(\boldsymbol{\lambda})$.

2.13.10 If $X(t)$ and $a(t)$ are real-valued, $Y(t) = \sum_u a(t-u)X(u)$ and $X(t)$ has cumulant spectra $f_{X,\ldots,X}(\lambda_1, \ldots, \lambda_k)$, prove that the cumulant spectra of $Y(t)$ are given by $A(\lambda_1)\cdots A(\lambda_k) f_{X,\ldots,X}(\lambda_1, \ldots, \lambda_k)$.

2.13.11 Prove that $\overline{f}_{a_1,\ldots,a_k}(\lambda_1, \ldots, \lambda_k) = f_{a_1,\ldots,a_k}(-\lambda_1, \ldots, -\lambda_k)$ for an r vector series with real-valued components.

2.13.12 If $\mathbf{X}(t)$ is a stationary Gaussian Markov r vector process with

$$E(\mathbf{X}(u) - \boldsymbol{\mu})(\mathbf{X}(0) - \boldsymbol{\mu})^\tau = \mathbf{c}_{XX}(u),\ \boldsymbol{\mu} = E\mathbf{X}(u),$$

prove that

$$\mathbf{c}_{XX}(u) = [\mathbf{c}_{XX}(1)(\mathbf{c}_{XX}(0))^{-1}]^u\, \mathbf{c}_{XX}(0),\ u \geqslant 0$$

and $\mathbf{c}_{XX}(u) = (\mathbf{c}_{XX}(-u))^\tau$, $u < 0$.

2.13.13 Prove that the power spectrum of a real-valued stationary Gaussian Markov process has the form $\sigma^2/(1 + \rho^2 - 2\rho \cos \lambda)2\pi$, $-\pi < \lambda \leqslant \pi$, $-1 < \rho < 1$.

2.13.14 Give an example to indicate that $\tilde{\mathbf{X}}(t)$ of Section 2.11 is not necessarily ergodic.

2.13.15 Let $\mathbf{X}^{(N)}(t)$, $t = 0, \pm 1, \ldots$; $N = 1, 2, \ldots$ be a sequence of series satisfying Assumption 2.6.1. Suppose

$$|\text{cum}\{X_{a_1}{}^{(N)}(t + u_1), \ldots, X_{a_{k-1}}^{(N)}(t + u_{k-1}), X_{a_k}{}^{(N)}(t)\}| < g_{a_1\ldots a_k}(u_1, \ldots, u_{k-1})$$

for $t, u_1, \ldots, u_{k-1} = 0, \pm 1, \ldots$; $N = 1, 2, \ldots$ where

$$\sum_{u_1,\ldots,u_{k-1}} g_{a_1\ldots a_k}(u_1, \ldots, u_{k-1}) < \infty.$$

Suppose, as $N \to \infty$, all the finite dimensional distributions of the process $\mathbf{X}^{(N)}(t)$, $t = 0, \pm 1, \ldots$ tend in distribution to those of a process $\mathbf{X}(t)$, $t = 0, \pm 1, \ldots$. Show that

$$\sum_{u_1,\ldots,u_{k-1}} |\text{cum}\{X_{a_1}(t + u_1), \ldots, X_{a_{k-1}}(t + u_{k-1}), X_{a_k}(t)\}| < \infty.$$

2.13.16 Show that the transfer function of the filter

$$Y(t) = X(t) - 2 \cos \omega X(t-1) + X(t-2)$$

vanishes at $\lambda = \pm\omega$. Discuss the effect of this filter on the series

$$X(t) = R \cos(\omega t + \phi) + \varepsilon(t) \qquad t = 0, \pm 1, \ldots.$$

2.13.17 Let $X(t) = 1$ for $(2j-1)^2 \leqslant t \leqslant (2j)^2$ and

$$-(2j)^2 \leqslant t \leqslant -(2j-1)^2 \qquad j = 1, 2, \ldots.$$

Let $X(t) = -1$ for

$$(2j)^2 < t < (2j+1)^2 \quad \text{and} \quad -(2j+1)^2 < t < -(2j)^2 \qquad j = 0, 1, 2, \ldots.$$

Prove that $X(t)$ satisfies the conditions of Section 2.11 and determine the associated stochastic process.

2.13.18 Let $X(t) = R\cos(\omega t + \varphi)$ where R, ω, and φ are constants. Prove that $X(t)$ satisfies the conditions of Section 2.11 and determine the associated stochastic process.

2.13.19 Let $X(t)$, $t = 0, \pm 1, \ldots$ and $Y(t)$, $t = 0, \pm 1, \ldots$ be independent series with mean 0 and power spectra $f_{XX}(\lambda)$, $f_{YY}(\lambda)$ respectively. Show that the power spectrum of the series $X(t)Y(t)$, $t = 0, \pm 1, \ldots$ is

$$\int_{-\pi}^{\pi} f_{XX}(\lambda - \alpha) f_{YY}(\alpha) d\alpha.$$

2.13.20 Let $X(t)$, $t = 0, \pm 1, \ldots$ be a Gaussian series with mean 0 and power spectrum $f_{XX}(\lambda)$. Show that the power spectrum of the series $X(t)^2$, $t = 0, \pm 1, \ldots$ is

$$2\int_{-\pi}^{\pi} f_{XX}(\lambda - \alpha) f_{XX}(\alpha) d\alpha.$$

2.13.21 If $X(t)$ is a real-valued series satisfying Assumption 2.6.1, prove, directly, that $[X(t)]^2$ also satisfies Assumption 2.6.1 and determine its cumulant spectra.

2.13.22 If $\mathbf{X}(t)$ satisfies Assumption 2.6.2(l), $\mathbf{Y}(t) = \sum_u \mathbf{a}(t-u)\mathbf{X}(u)$ for $\mathbf{a}(u)$ an $s \times r$ filter with $\Sigma\, |u|^l |a_{jk}(u)| < \infty, j = 1, \ldots, s, k = 1, \ldots, r$ for some l, then $\mathbf{Y}(t)$ satisfies Assumption 2.6.2(l).

2.13.23 An $s \times r$ filter $\mathbf{a}(u)$ is said to have **rank** t if $\mathbf{A}(\lambda)$ has rank t for each λ. Prove that in this case $\mathbf{a}(u)$ is equivalent in effect to applying first a $t \times r$ filter then an $s \times t$ filter.

2.13.24 If $\mathbf{X}(t) = \sum_{u=0}^{\infty} \mathbf{a}(t-u)\boldsymbol{\varepsilon}(u)$ with $\boldsymbol{\varepsilon}(t)$ an r vector pure noise series and $\{\mathbf{a}(u)\}$ an $s \times r$ summable filter, prove that $\mathbf{f}_{XX}(\lambda)$ may be written in the form $\Phi(e^{i\lambda})\, \overline{\Phi(e^{i\lambda})}^{\tau}$ where $\Phi(z)$ is an $s \times r$ matrix valued function with components analytic in the disc $|z| \leqslant 1$.

2.13.25 If $X(t) = \sum_{u=0}^{\infty} a(t-u)\varepsilon(u)$ with $\varepsilon(u)$ a real-valued pure noise series and $\Sigma a(u)^2 < \infty$, prove that the kth order cumulant spectrum, $f_{X\ldots X}(\lambda_1, \ldots, \lambda_k)$ has the form $\Phi(e^{i\lambda_1})\cdots\Phi(e^{i\lambda_k})$, $\lambda_1 + \cdots + \lambda_k \equiv 0 (\text{mod } 2\pi)$, with $\Phi(z)$ analytic in the disc $|z| \leqslant 1$.

2.13.26 If $\mathbf{X}(t)$ is a moving average process of order m, prove that $\mathbf{c}_{XX}(u) = \mathbf{0}$ for $|u| > m$.

2.13.27 If we adopt the functional approach to time series analysis, demonstrate that $Y(t) = \sum_u a(t-u)X(u)$, $\sum_{-\infty}^{\infty} |a(u)| < \infty$ defines a filter. Indicate the relation between the spectra of $Y(t)$ and those of $X(t)$.

2.13.28 Show that $V_1(t)$, $V_2(t)$ of (2.7.33) come from $X(t)$ through filters with coefficients $\{a(u) \cos \lambda_0 u\}$, $\{a(u) \sin \lambda_0 u\}$, respectively.

2.13.29 Prove that $\delta(ax) = |a|^{-1}\delta(x)$.

2.13.30 Let $\mathbf{X}(t)$ be a stationary r vector valued series with $X_j(t) = \rho_j X_j(t-1) + \varepsilon_j(t)$, $|\rho_j| < 1$, $j = 1, \ldots, r$ where $\boldsymbol{\varepsilon}(t)$ is an r vector pure noise series. Prove that

$$c_{a_1,\ldots,a_k}(t_1, \ldots, t_k) = K_{a_1,\ldots,a_k}(\prod_{j\in\mathbf{a}} \rho_j^{t_j-\tau})/(-\prod_{j\in\mathbf{a}} \rho_j^{\tau})$$

where $\tau = \min(t_1, \ldots, t_k)$, $\mathbf{a} = (a_1, \ldots, a_k)$ and $K_{a_1,\ldots,a_k} = \operatorname{cum}\{\varepsilon_{a_1}(t), \ldots, \varepsilon_{a_k}(t)\}$.

2.13.31 Let $\Phi(T)$, $T = 1, 2, \ldots$ be a sequence of positive numbers with the properties; $\Phi(T) \to \infty$ and $\Phi(T+1)/\Phi(T) \to 1$ as $T \to \infty$. Let $\mathbf{X}(t)$, $t = 0, \pm 1, \ldots$ be an r vector-valued function with the property

$$\lim_{T\to\infty} \Phi(T)^{-1} \sum_{t=0}^{T-|u|} \mathbf{X}(t+u)\mathbf{X}(t)^\tau = \mathbf{m}_{XX}(u)$$

for $u = 0, \pm 1, \ldots$. Show that there exists an $r \times r$ matrix-valued function $\mathbf{G}_{XX}(\lambda)$, $-\pi < \lambda \leqslant \pi$, such that

$$\mathbf{m}_{XX}(u) = \int_{-\pi}^{\pi} \exp\{iu\lambda\} d\mathbf{G}_{XX}(\lambda) \qquad \text{for } u = 0, \pm 1, \ldots.$$

Hint: Define $\mathbf{I}_{XX}^{(T)}(\lambda)$ as in the proof of Theorem 2.5.2. See Bochner (1959) p. 329, and Grenander (1954). Prove $\int_{-\pi}^{\pi} \mathbf{A}(\alpha)\mathbf{I}_{XX}^{(T)}(\alpha)d\alpha \to \int_{-\pi}^{\pi} \mathbf{A}(\alpha)d\mathbf{G}_{XX}(\alpha)$ for $\mathbf{A}(\alpha)$ continuous on $[-\pi,\pi]$ also.

2.13.32 Let $\mathbf{X}(t)$, $t = 0, \pm 1, \ldots$ be a vector-valued stationary series with cumulant spectra $f_{a_1,\ldots,a_k}(\lambda_1, \ldots, \lambda_k)$. Evaluate the cumulant spectra of the time-reversed series $\mathbf{X}(-t)$, $t = 0, \pm 1, \ldots$.

2.13.33 Show that

$$(2\pi)^{-1} \sum_{u=-\infty}^{\infty} \exp\{-i\lambda u\} = \sum_{j=-\infty}^{\infty} \delta(\lambda - 2\pi j) = \eta(\lambda)$$

for $-\infty < \lambda < \infty$. (The Poisson Summation Formula, Edwards (1967))

2.13.34 Show that the function $c_{XX}(u) = 1$, for $|u| \leqslant m$ and $c_{XX}(u) = 0$ otherwise, cannot be an autocovariance function.

2.13.35 Let $X_j(t)$, $t = 0, \pm 1, \ldots$; $j = 0, \ldots, J-1$ be J independent realizations of a stationary process. Let

$$Y(sJ + j) = X_j(s) \qquad \text{for } j = 0, \ldots, J-1;\ s = 0, \pm 1, \ldots.$$

Show that $Y(t)$, $t = 0, \pm 1, \ldots$ is a stationary series. Show that its power spectrum is $f_{XX}(\lambda J)$.

2.13.36 In the case that $\mathbf{X}(t)$, $t = 0, \pm 1, \ldots$ is a linear process with

$$\mathbf{X}(t) = \sum_u \mathbf{a}(t-u)\boldsymbol{\varepsilon}(u),$$

$$\sum |\mathbf{a}(u)| < \infty, \text{ and } K_{a_1 \ldots a_k} = \operatorname{cum}\{\varepsilon_{a_1}(0), \ldots, \varepsilon_{a_k}(0)\},$$

show that Assumption 2.6.3 is satisfied provided for z in a neighborhood of 0

$$\sum_k \sup_a |K_{a_1 \ldots a_k}| z^k / k! < \infty.$$

2.13.37 A filter is called **stable** if it carries a bounded input series over into a bounded output series. Show that a summable filter is stable.

2.13.38 Let $x(t)$, $t = 0, \pm 1, \ldots$ be an autoregressive process of order 1. Let $\varepsilon(t)$, $t = 0, \pm 1, \ldots$ be a pure noise series. Let $X(t) = x(t) + \varepsilon(t)$. Show that the series $X(t)$, $t = 0, \pm 1, \ldots$ is a mixed moving average autoregressive process of order (1,1).

2.13.39 State and prove an extension of Theorem 2.9.1 in which both the series $\mathbf{X}(t)$ and $\mathbf{Y}(t)$ are vector-valued.

3

ANALYTIC PROPERTIES OF FOURIER TRANSFORMS AND COMPLEX MATRICES

3.1 INTRODUCTION

The principal analytic tool that we will employ with time series is the Fourier transform. In this chapter we present those portions of Fourier analysis that will be required for our discussion. All the functions considered in this chapter will be fixed, rather than stochastic. Stochastic properties of Fourier transforms will be considered in the next chapter.

Among the topics discussed here are the following: the degree of approximation of a function by the partial sums of its Fourier series; the improvement of this approximation by the insertion of convergence factors; the Fourier transform of a finite set of values; the rapid numerical evaluation of Fourier transforms; the spectrum of a matrix and its relation to the approximation of one matrix by another of reduced rank; mathematical properties of functions of Fourier transforms; and finally the spectral or harmonic representation of functions possessing a generalized harmonic analysis. We begin by considering the Fourier series of a given function $A(\lambda)$.

3.2 FOURIER SERIES

Let $A(\lambda)$, $-\infty < \lambda < \infty$, be a complex-valued function of period 2π such that

$$\int_{-\pi}^{\pi} |A(\lambda)| d\lambda < \infty. \tag{3.2.1}$$

The Fourier coefficients of $A(\lambda)$ are given by

$$a(u) = (2\pi)^{-1} \int_{-\pi}^{\pi} \exp\{iu\lambda\} A(\lambda) d\lambda \qquad u = 0, \pm 1, \ldots \tag{3.2.2}$$

and the Fourier series of $A(\lambda)$ is then given by

$$\sum_{u=-\infty}^{\infty} \exp\{-i\lambda u\} a(u). \tag{3.2.3}$$

There is extensive literature concerning Fourier series and Fourier coefficients; see Zygmund (1959) and Edwards (1967), for example. Much of this literature is concerned with the behavior of the partial sums

$$A^{(n)}(\lambda) = \sum_{u=-n}^{n} \exp\{-i\lambda u\} a(u) \qquad n = 0, 1, 2, \ldots. \tag{3.2.4}$$

In this work we will have frequent occasion to examine the nearness of the $A^{(n)}(\lambda)$ to $A(\lambda)$ for large n. Begin by noting, from expression (3.2.2) and Exercise 1.7.5, that

$$A^{(n)}(\lambda) = \int_{-\pi}^{\pi} \frac{\sin (n + \frac{1}{2})\alpha}{2\pi \sin \frac{1}{2}\alpha} A(\lambda - \alpha) d\alpha. \tag{3.2.5}$$

The function

$$D_n(\alpha) = \frac{\sin (n + \frac{1}{2})\alpha}{2\pi \sin \frac{1}{2}\alpha} \qquad 0 \leqslant \alpha \leqslant \pi \tag{3.2.6}$$

is plotted in Figure 3.2.1 for the values $n = 1, 3, 5, 10$. We note that it fluctuates in sign and that it is concentrated in the neighborhood of $\alpha = 0$, becoming more concentrated as n increases. Also

$$\int_{-\pi}^{\pi} \frac{\sin (n + \frac{1}{2})\alpha}{2\pi \sin \frac{1}{2}\alpha} d\alpha = 1 \tag{3.2.7}$$

from Exercise 1.7.5. In consequence of these properties of the function (3.2.6) we see that $A^{(n)}(\lambda)$ is a weighted average of the function $A(\lambda - \alpha)$, with weight concentrated in the neighborhood of $\alpha = 0$. We would expect $A^{(n)}(\lambda)$ to be near $A(\lambda)$ for large n, if the function $A(\alpha)$ is not too irregular. In fact we can show that $A^{(n)}(\lambda)$ tends to $A(\lambda)$ as $n \to \infty$ if, for example, $A(\alpha)$ is of bounded variation; see Edwards (1967) p. 150.

Under supplementary regularity conditions we can measure the rapidity of approach of $A^{(n)}(\lambda)$ to $A(\lambda)$ as $n \to \infty$. Suppose

$$\sum_{u=-\infty}^{\infty} |u|^k |a(u)| < \infty \qquad \text{for some } k \geqslant 0. \tag{3.2.8}$$

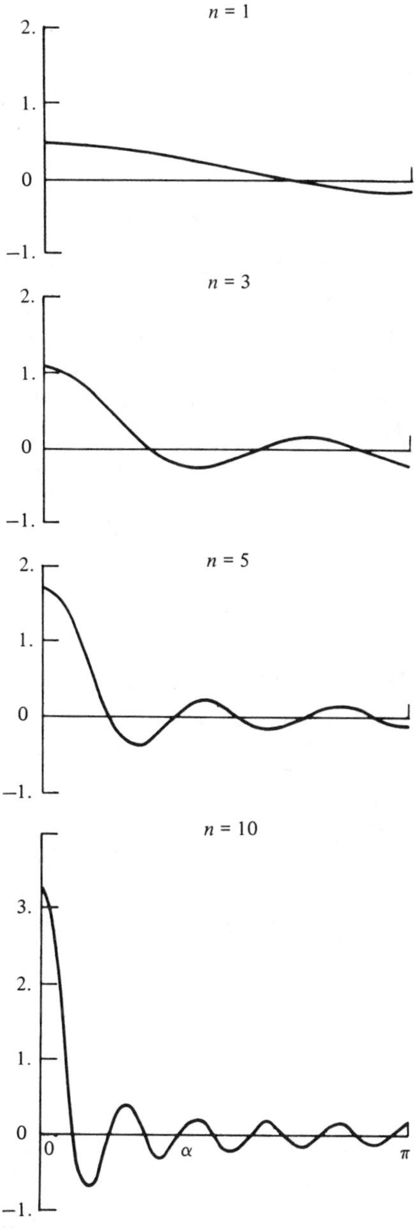

Figure 3.2.1 Plot of $D_n(\alpha) = \sin(n + \frac{1}{2})\alpha/2\pi \sin \frac{1}{2}\alpha$.

This condition is tied up with the degree of smoothness of $A(\alpha)$. Under it $A(\alpha)$ has bounded continuous derivatives of order $\leqslant k$. We have therefore

$$A(\lambda) - A^{(n)}(\lambda) = \sum_{|u|>n} \exp\{-i\lambda u\} a(u) \tag{3.2.9}$$

and so

$$|A(\lambda) - A^{(n)}(\lambda)| \leqslant \sum_{|u|>n} |a(u)|$$
$$\leqslant n^{-k} \sum_{|u|>n} |u|^k |a(u)| = o(n^{-k}).^1 \tag{3.2.10}$$

In summary, the degree of approximation of $A(\lambda)$ by $A^{(n)}(\lambda)$ is intimately related to the smoothness of $A(\lambda)$.

We warn the reader that $A^{(n)}(\lambda)$ need not necessarily approach $A(\lambda)$ as $n \to \infty$ even in the case that $A(\lambda)$ is a bounded continuous function of λ; see Edwards (1967) p. 150, for example. However, the relationship between the two functions is well illustrated by (3.2.5). The behavior of $A(\lambda) - A^{(n)}(\lambda)$ is especially disturbed in the neighborhood of discontinuities of $A(\lambda)$. Gibbs' phenomenon involving the nondiminishing overshooting of a functional value can occur; see Hamming (1962) p. 295 or Edwards (1967) p. 172.

3.3 CONVERGENCE FACTORS

Fejér (1900, 1904) recognized that the partial sums of a Fourier series might be poor approximations of a function of interest even if the function were continuous. He therefore proposed that instead of the partial sum (3.2.4) we consider the sum

$$\sum_{u=-n+1}^{n-1} \left(1 - \frac{|u|}{n}\right) a(u) \exp\{-i\lambda u\}. \tag{3.3.1}$$

Using expression (3.2.2) and Exercise 1.7.12 we see that (3.3.1) may be written

$$\int_{-\pi}^{\pi} \frac{1}{2\pi n} \left[\frac{\sin n\alpha/2}{\sin \alpha/2}\right]^2 A(\lambda - \alpha) d\alpha. \tag{3.3.2}$$

The function

$$\frac{1}{2\pi n} \left[\frac{\sin n\alpha/2}{\sin \alpha/2}\right]^2 \qquad 0 \leqslant \alpha \leqslant \pi \tag{3.3.3}$$

[1] We will make use of the Landau o, O notations writing $\alpha_n = o(\beta_n)$ when $\alpha_n/\beta_n \to 0$ as $n \to \infty$ and writing $\alpha_n = O(\beta_n)$ when $|\alpha_n/\beta_n|$ is bounded for sufficiently large n.

is plotted in Figure 3.3.1 for $n = 2, 4, 6, 11$. It is seen to be non-negative, concentrated in the neighborhood of $\alpha = 0$ and, following Exercise 1.7.12, such that

$$\int_{-\pi}^{\pi} \frac{1}{2\pi n} \left[\frac{\sin n\alpha/2}{\sin \alpha/2}\right]^2 d\alpha = 1. \tag{3.3.4}$$

It is blunter than the function (3.2.6) of the previous section and has fewer ripples. This greater regularity leads to the convergence of (3.3.2) to $A(\lambda)$ in the case that $A(\alpha)$ is a continuous function in contrast to the behavior of (3.2.5); see Edwards (1967) p. 87. The insertion of the factors $1 - |u|/n$ in expression (3.3.1) has expanded the class of functions that may be reasonably represented by trigonometric series.

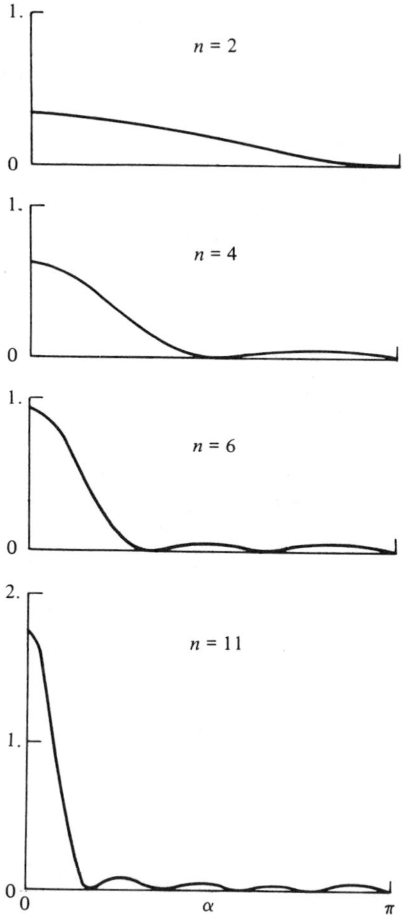

Figure 3.3.1 Plot of $\sin^2 \frac{1}{2}n\alpha/2\pi n \sin^2 \frac{1}{2}\alpha$.

In general we may consider expressions of the form

$$\sum_u \exp\{-i\lambda u\} h\left(\frac{u}{n}\right) a(u) \tag{3.3.5}$$

for some function $h(x)$ with $h(0) = 1$ and $h(x) = 0$ for $|x| > 1$. The multiplier, $h(u/n)$, appearing in (3.3.5) is called a **convergence factor**; see, for example, Moore (1966). If we set

$$H^{(n)}(\lambda) = \frac{1}{2\pi} \sum_u h\left(\frac{u}{n}\right) \exp\{-iu\lambda\}, \tag{3.3.6}$$

(3.3.5) may be written

$$\int_{-\pi}^{\pi} H^{(n)}(\alpha) A(\lambda - \alpha) d\alpha, \tag{3.3.7}$$

indicating that (3.3.5) is a weighted average of the function of interest.

A wide variety of convergence factors $h(u/n)$ have been proposed. Some of these are listed in Table 3.3.1 along with their associated $H^{(n)}(\lambda)$. The typical shape of $h(u/n)$ involves a maximum of 1 at $u = 0$, followed by a steady decrease to 0 as $|u|$ increases to n. Convergence factors have also been called **data windows** and **tapers**; see Tukey (1967).

The typical form of $H^{(n)}(\lambda)$ is that of a blocky weight function in the neighborhood of 0 that becomes more concentrated as $n \to \infty$. In fact it follows from (3.3.6) that

$$\int_{-\pi}^{\pi} H^{(n)}(\alpha) d\alpha = 1 \tag{3.3.8}$$

as we would expect. An examination of expression (3.3.7) suggests that for some purposes we may wish to choose $H^{(n)}(\lambda)$ to be non-negative. The second and third entries of Table 3.3.1 possess this property. The function $H^{(n)}(\lambda)$ has been called a **frequency window** and a **kernel.** From (3.3.7) we see that the nearness of (3.3.5) to $A(\lambda)$ relates to the degree of concentration of the function $H^{(n)}(\alpha)$ about $\alpha = 0$. Various measures of this concentration, or bandwidth, have been proposed. Press and Tukey (1956) suggested the half-power width given by $\alpha_L - \alpha_U$, where α_L and α_U are the first positive and negative α such that $H^{(n)}(\alpha) = H^{(n)}(0)/2$. Grenander (1951) suggested the measure

$$\left[\int_{-\pi}^{\pi} \alpha^2 H^{(n)}(\alpha) d\alpha\right]^{1/2}. \tag{3.3.9}$$

This is the mean-squared error about 0 if $H^{(n)}(\alpha)$ is considered as a probability distribution on $(-\pi,\pi)$. Parzen (1961) has suggested the measure

$$\frac{1}{H^{(n)}(0)} = \frac{2\pi}{\sum h(u/n)}. \tag{3.3.10}$$

Table 3.3.1 Some Particular Convergence Factors

$h(u/n),\ 0 \leqslant \lvert u\rvert \leqslant n$	$H^{(n)}(\lambda),\ -\pi < \lambda < \pi$	*Authors*
1	$D_n(\lambda) = \dfrac{\sin(n+\frac{1}{2})\lambda}{2\pi \sin\frac{1}{2}\lambda}$	Dirichlet [Edwards (1967)]
$1 - \dfrac{\lvert u\rvert}{n}$	$\dfrac{1}{2\pi n}\left[\dfrac{\sin n\lambda/2}{\sin\lambda/2}\right]^2$	Fejér, Bartlett [Edwards (1967), Parzen (1963)]
$1 - \dfrac{6u^2}{n^2}\left(1 - \dfrac{\lvert u\rvert}{n}\right)\ 0 \leqslant \lvert u\rvert \leqslant \dfrac{n}{2}$ $2\left(1 - \dfrac{\lvert u\rvert}{n}\right)^3\ \ \dfrac{n}{2} \leqslant \lvert u\rvert \leqslant n$	$\dfrac{2+\cos\lambda}{4\pi n^3}\left[\dfrac{\sin n\lambda/4}{\sin\lambda/4}\right]^4$	de la Vallé-Poussin, Jackson, Parzen [Akhiezer (1956), Parzen (1961)]
$\dfrac{1}{2}\left(1 + \cos\dfrac{\pi u}{n}\right)$	$\frac{1}{2}D_n(\lambda) + \frac{1}{4}D_n\left(\lambda - \dfrac{\pi}{n}\right)$ $+ \frac{1}{4}D_n\left(\lambda + \dfrac{\pi}{n}\right)$	Hamming, Tukey [Blackman and Tukey (1958)]
$\left(1 - \dfrac{\lvert u\rvert}{n}\right)\cos\dfrac{\pi u}{n}$ $+ \dfrac{1}{\pi}\sin\dfrac{\pi\lvert u\rvert}{n}$	$\doteq n2\pi\dfrac{(1+\cos n\lambda)}{(n^2\lambda^2 - \pi^2)^2}$	Bohman [Bohman (1960)]
$\rho^{\lvert u\rvert/n}\quad 0 \leqslant \lvert u\rvert \leqslant n$ $0 < \rho < 1$	$\doteq \dfrac{1}{2\pi}\dfrac{1 - \rho^{2/n}}{1 - 2\rho^{1/n}\cos\lambda + \rho^{2/n}}$	Poisson [Edwards (1967)]
$\dfrac{\sin\dfrac{u}{n}}{\dfrac{u}{n}}$	$\doteq \dfrac{n}{2}\quad \lvert\lambda\rvert \leqslant \dfrac{1}{n}$ $\doteq 0$ otherwise	Riemann, Lanczos [Edwards (1967), Lanczos (1956)]
$\exp\left\{-\dfrac{u^2}{2n^2}\right\}$	$\doteq \dfrac{n}{\sqrt{2\pi}}\exp\left\{-\dfrac{n^2\lambda^2}{2}\right\}$	Gauss Weierstrass [Akhiezer (1956)]
$\dfrac{1}{(1+u^2/n^2)}$	$\doteq \pi n \exp\{-n\lvert\lambda\rvert\}$	Cauchy, Abel, Poisson [Akhiezer (1956)]
$1 - \dfrac{u^2}{n^2}$	$D_n(\lambda) + \dfrac{1}{n^2}\dfrac{d^2D_n(\lambda)}{d\lambda^2}$	Riesz, Bochner Parzen [Bochner (1936), Parzen (1961)]
$1\quad 0 \leqslant \lvert u\rvert \leqslant fn$ $\dfrac{1}{2}\left(1 + \cos\pi\left[\dfrac{\lvert u\rvert}{n} - f\right]/(1-f)\right)$ $fn \leqslant \lvert u\rvert \leqslant n$ where $0 < f < 1$		Tukey [Tukey (1967)]

This is the width of the rectangle of the same maximum height and area as $H^{(n)}(\alpha)$.

A measure that is particularly easy to handle is

$$\begin{aligned}\beta_n{}^H &= \left[\int_{-\pi}^{\pi} (1 - \cos \alpha) H^{(n)}(\alpha) d\alpha\right]^{1/2} \\ &= \left[\int_{-\pi}^{\pi} 2 \sin^2 \frac{\alpha}{2}\, H^{(n)}(\alpha) d\alpha\right]^{1/2} \\ &= \left[1 - \left[h\left(\frac{1}{n}\right) + h\left(-\frac{1}{n}\right)\right]/2\right]^{1/2}. \end{aligned} \tag{3.3.11}$$

Its properties include: if $h(u)$ has second derivative $h''(0)$ at $u = 0$, then

$$\begin{aligned}\beta_n{}^H &\backsim \frac{1}{n}\left[-\frac{h''(0)}{2}\right]^{1/2} \\ &\backsim \frac{1}{n}[\tfrac{1}{2} \textstyle\int \alpha^2 H(\alpha) d\alpha]^{1/2}, \end{aligned} \tag{3.3.12}$$

showing a connection with Grenander's measure (3.3.9). Alternately if the kernel being employed is the convolution of kernels $G^{(n)}(\alpha)$, $H^{(n)}(\alpha)$, then we can show that

$$\beta_n^{G*H} \backsim [(\beta_n{}^G)^2 + (\beta_n{}^H)^2]^{1/2} \tag{3.3.13}$$

for large n. Finally, if

$$h_q = \lim_{u \to 0} \frac{1 - h(u)}{|u|^q} \tag{3.3.14}$$

exists for some $q > 0$, as Parzen (1961) assumes, then

$$\beta_n{}^H \backsim [h_q n^{-q}]^{1/2}. \tag{3.3.15}$$

Table 3.3.2 gives the values of $\beta_n{}^H$ and $1/H^{(n)}(0)$ for the kernels of Table 3.3.1. The entries of this table give an indication of the relative asymptotic concentration of the various kernels.

The following theorem gives an alternate means of examining the asymptotic degree of approximation.

Theorem 3.3.1 Suppose $A(\lambda)$ has bounded derivatives of order $\leqslant P$. Suppose

$$H(\alpha) = (2\pi)^{-1} \int h(x) \exp\{-i\alpha x\} dx \tag{3.3.16}$$

with

$$\int |\alpha|^P |H(\alpha)| d\alpha \leqslant K \tag{3.3.17}$$

Table 3.3.2 Bandwidths of the Kernels

Kernel	$\beta_n{}^H$	$1/H^{(n)}(0)$
Dirichlet	0	π/n
Fejér	$1/\sqrt{n}$	$2\pi/n$
de la Vallé-Poussin	$\sqrt{6}/n$	$4\pi/3n$
Hamming	$\pi/2n$	$2\pi/n$
Bohman	$\pi/\sqrt{2}n$	$\pi^3/4n$
Poisson	$\sqrt{\log 1/\rho}/\sqrt{n}$	$\pi(\log 1/\rho)/n$
Riemann	$1/\sqrt{6}n$	$2/n$
Gauss	$1/\sqrt{2}n$	$\sqrt{2\pi}/n$
Cauchy	$1/n$	$1/\pi n$
Riesz	$1/n$	$3\pi/2n$
Tukey	0	$2\pi/(n + fn)$

for some finite K, then

$$\sum_u \exp\{-i\lambda u\}\, h\left(\frac{u}{n}\right) a(u) = \int_{-\infty}^{\infty} nH(n\alpha)A(\lambda - \alpha)d\alpha$$
$$= A(\lambda) + \sum_{p=1}^{P-1} \frac{1}{p!}\frac{1}{n^p}\left\{\int_{-\infty}^{\infty} \alpha^p H(\alpha)d\alpha\right\}\frac{d^p A(\lambda)}{d\lambda^p} + O(n^{-P}). \tag{3.3.18}$$

Expression (3.3.11) gives a useful indication of the manner in which the nearness of (3.3.5) to $A(\lambda)$ depends on the convergence factors employed. If possible it should be arranged that

$$\int \alpha^p H(\alpha)d\alpha = i^{-p}\left.\frac{d^p h(x)}{dx^p}\right|_{x = 0} \tag{3.3.19}$$

be 0 for $p = 1, 2, \ldots$. If $h(x) = h(-x)$, then this is the case for odd values of p. The requirement for even p is equivalent to requiring that $h(x)$ be very flat near $x = 0$. The last function of Table 3.3.1 is notable in this respect.

In fact, the optimum $h(u/n)$ will depend on the particular $A(\lambda)$ of interest. A considerable mathematical theory has been developed concerning the best approximation of functions by trigonometric polynomials; see Akhiezer (1956) or Timan (1963), for example. Bohman (1960) and Akaike (1968) were concerned with the development of convergence factors appropriate for a broad class of functions; see also Timan (1962), Shapiro (1969), Hoff (1970), Butzer and Nessel (1971).

Wilkins (1948) indicates asymptotic expansions of the form of (3.3.18) that are valid under less restrictive conditions.

As an application of the discussion of this section we now turn to the problem of filter design. Suppose that we wish to determine time domain coefficients $a(u)$, $u = 0, \pm 1, \ldots$ of a filter with prespecified transfer function $A(\lambda)$. The relation between $a(u)$ and $A(\lambda)$ is given by the expressions

$$a(u) = (2\pi)^{-1} \int_{-\pi}^{\pi} \exp\{iu\lambda\} A(\lambda) d\lambda \tag{3.3.20}$$

$$A(\lambda) = \sum_{u=-\infty}^{\infty} \exp\{-i\lambda u\} a(u). \tag{3.3.21}$$

The filter has the form

$$Y(t) = \sum_{u=-\infty}^{\infty} a(t-u) X(u) \tag{3.3.22}$$

if $X(t)$, $t = 0, \pm 1, \ldots$ is the initial series. Generally $a(u)$ does not vanish for large $|u|$ and only a finite stretch of the $X(t)$ series is available. These facts lead to difficulty in applying (3.3.22). We can consider the problem of determining a finite length filter with transfer function near $A(\lambda)$. This may be formalized as the problem of determining multipliers $h(u/n)$ so that

$$\sum_{u=-n}^{n} \exp\{-i\lambda u\} h\left(\frac{u}{n}\right) a(u) \tag{3.3.23}$$

is near $A(\lambda)$. This is the problem discussed above.

Suppose that we wish to approximate a low-pass filter with cut-off frequency, $\Omega < \pi$; that is, the desired transfer function is

$$\begin{aligned} A(\lambda) &= 1 \qquad \text{for } |\lambda| \leqslant \Omega \\ &= 0 \qquad \text{otherwise} \end{aligned} \tag{3.3.24}$$

$-\pi < \lambda < \pi$. This filter has coefficients

$$\begin{aligned} a(u) = \frac{1}{2\pi} \int_{-\Omega}^{\Omega} \exp\{i\lambda u\} d\lambda &= \frac{\Omega}{\pi} \qquad \text{for } u = 0 \\ &= \frac{\sin \Omega u}{u\pi} \qquad \text{for } u \neq 0. \end{aligned} \tag{3.3.25}$$

The preceding discussion leads us to consider a filter of, for example, the form

$$Y(t) = \sum_{u=-n}^{n} h\left(\frac{u}{n}\right) \frac{\sin \Omega u}{u\pi} X(t-u) \tag{3.3.26}$$

for some convergence factors $h(u/n)$.

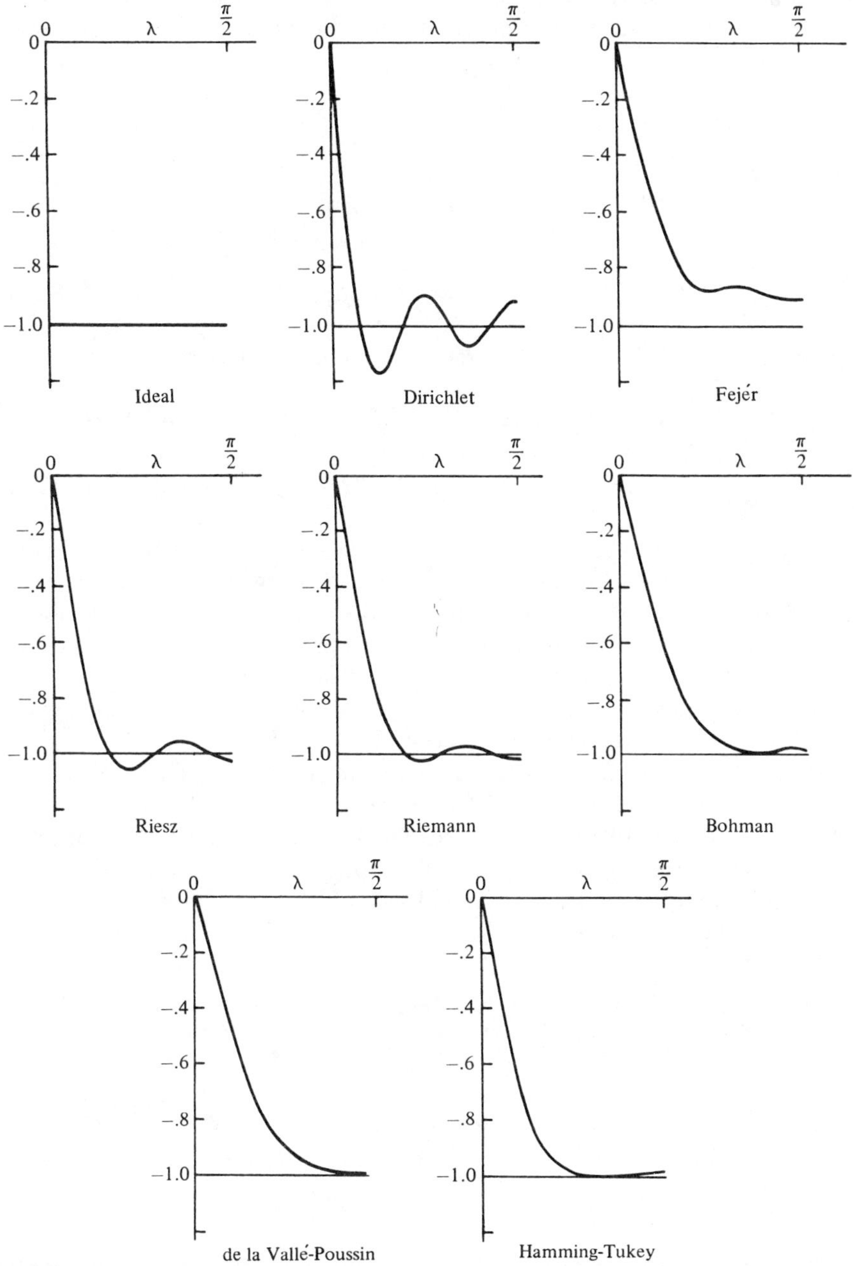

Figure 3.3.2 Transfer function of ideal Hilbert transform and approximations with various factors, $n = 7$.

Suppose alternately that we would like to realize numerically the Hilbert transform introduced in Section 2.7. Its transfer function is

$$\begin{aligned} A(\lambda) &= -i \qquad 0 < \lambda < \pi \\ &= 0 \qquad \lambda = 0 \\ &= i \qquad -\pi < \lambda < 0. \end{aligned} \tag{3.3.27}$$

The filter coefficients are therefore

$$\begin{aligned} a(u) &= 0 \qquad u \text{ even} \\ &= \frac{2}{\pi u} \qquad u \text{ odd.} \end{aligned} \tag{3.3.28}$$

Suppose n is odd. We are led therefore to consider filters of the form

$$Y(t) = \frac{2}{\pi}\left\{h\left(\frac{1}{n}\right)[X(t-1) - X(t+1)] + \frac{1}{3}h\left(\frac{3}{n}\right)[X(t-3) - X(t+3)] + \cdots + \frac{1}{n}h(1)[X(t-n) - X(t+n)]\right\}. \tag{3.3.29}$$

Figure 3.3.2 indicates the imaginary part of the ideal $A(\lambda)$ of (3.3.27) for $0 < \lambda < \pi/2$ and the imaginary part of the $A(\lambda)$ achieved by (3.3.29) with $n = 7$ for a variety of the convergence factors of Table 3.3.1. Because of the symmetries of the functions involved we need present only the functions for this restricted frequency range. The importance of inserting convergence factors is well demonstrated by these diagrams. We also see the manner in which different convergence factors can affect the result.

References concerned with the design of digital filters include: Kuo and Kaiser (1966), Wood (1968) and No. 3 of the IEEE Trans. Audio Electro. (1968). Goodman (1960) was concerned with the numerical realization of a Hilbert transform and of band-pass filters. In Section 3.6 we will discuss a means of rapidly evaluating the filtered series $Y(t)$. Parzen (1963) discusses a variety of the topics of this section.

3.4 FINITE FOURIER TRANSFORMS AND THEIR PROPERTIES

Given the sequence $a(u)$, $u = 0, \pm 1, \ldots$ our work in the previous sections has led us to consider expressions of the form

$$\sum_{u=-n}^{n} \exp\{-i\lambda u\} a(u). \tag{3.4.1}$$

For fixed n, such an expression is called a **finite Fourier transform** of the sequence $a(u)$, $u = 0, \pm 1, \ldots, \pm n$. Such transforms will constitute the essential statistics of our analysis of time series.

Before proceeding further, it is worthwhile to alter our notation slightly and to consider the general case of a vector-valued sequence. Specifically we consider an r vector-valued sequence $\mathbf{X}(0), \mathbf{X}(1), \ldots, \mathbf{X}(T-1)$, whose domain is $0, 1, \ldots, T-1$ rather than $-n, -n+1, \ldots, -1, 0, 1, \ldots, n$. We define the finite Fourier transform of this sequence to be

$$\mathbf{d}_X{}^{(T)}(\lambda) = \sum_{t=0}^{T-1} \mathbf{X}(t) \exp\{-i\lambda t\} \qquad -\infty < \lambda < \infty. \tag{3.4.2}$$

In the case that $T = 2n + 1$, n being an integer, we may write

$$\mathbf{d}_X{}^{(T)}(\lambda) = \exp\{-i\lambda n\} \sum_{u=-n}^{n} \mathbf{X}(u+n) \exp\{-i\lambda u\}$$

and thus we see that the only essential difference between definitions of the form (3.4.1) and the form (3.4.2) is a multiplier of modulus 1. Which definition is more convenient depends on the situation being discussed.

Among the properties of the definition (3.4.2) we note

$$\mathbf{d}_X{}^{(T)}(\lambda + 2\pi) = \mathbf{d}_X{}^{(T)}(\lambda). \tag{3.4.3}$$

Also, if the components of $\mathbf{X}(t)$ are real-valued, then

$$\overline{\mathbf{d}_X{}^{(T)}(\lambda)} = \mathbf{d}_X{}^{(T)}(-\lambda). \tag{3.4.4}$$

These two properties imply that, in the case of real-valued components, the principal domain of $\mathbf{d}_X{}^{(T)}(\lambda)$ may be taken to be $0 \leqslant \lambda \leqslant \pi$. Continuing, we note that if $\mathbf{X}(t), \mathbf{Y}(t), t = 0, \ldots, T-1$ are given and if α and β are scalars, then

$$\mathbf{d}_{\alpha X+\beta Y}^{(T)}(\lambda) = \alpha \mathbf{d}_X{}^{(T)}(\lambda) + \beta \mathbf{d}_Y{}^{(T)}(\lambda). \tag{3.4.5}$$

On occasion we may wish to relate the finite Fourier transform of the convolution of two sequences to the Fourier transforms of the two sequences themselves. We have

Lemma 3.4.1 Let $\mathbf{X}(t), t = 0, \pm 1, \ldots$ be r vector-valued and uniformly bounded. Let $\mathbf{a}(t), t = 0, \pm 1, \ldots$ be $s \times r$ matrix-valued and such that

$$\sum_{u=-\infty}^{\infty} \{1 + |u|\} \, |\mathbf{a}(u)| < \infty. \tag{3.4.6}$$

Set

$$\mathbf{Y}(t) = \sum_{u} \mathbf{a}(t-u)\mathbf{X}(u). \tag{3.4.7}$$

Then there is a finite K such that

$$|\mathbf{d}_Y{}^{(T)}(\lambda) - \mathbf{A}(\lambda)\mathbf{d}_X{}^{(T)}(\lambda)| < K \qquad -\infty < \lambda < \infty \tag{3.4.8}$$

where

$$\mathbf{A}(\lambda) = \sum_{u=-\infty}^{\infty} \mathbf{a}(u) \exp\{-i\lambda u\}. \tag{3.4.9}$$

We see that the finite Fourier transform of a filtered series is approximately the product of the transfer function of the filter and the finite Fourier transform of the series. This result will later provide us with a useful means of realizing the digital filtering of a series of interest. See also Lemma 6.3.1.

We now indicate a few examples of finite Fourier transforms. For these cases it is simplest to take the symmetric definition

$$\sum_{t=-n}^{n} X(t) \exp\{-i\lambda t\}. \tag{3.4.10}$$

Example 1 (*Constant*) Suppose $X(t) = 1$, $t = 0, \pm 1, \ldots$, then expression (3.4.10) equals

$$\sum_{t=-n}^{n} \exp\{-i\lambda t\} = \frac{\sin (n + \frac{1}{2})\lambda}{\sin \frac{1}{2}\lambda} = 2\pi D_n(\lambda). \tag{3.4.11}$$

This function was plotted in Figure 3.2.1 for $0 < \lambda < \pi$. Notice that it has peaks at $\lambda = 0, \pm 2\pi, \ldots$.

Example 2 (*Cosinusoid*) Suppose $X(t) = \exp\{i\omega t\}$, $t = 0, \pm 1, \ldots$ with ω real-valued, then (3.4.10) equals

$$\sum_{t=-n}^{n} \exp\{-i(\lambda - \omega)t\} = \frac{\sin (n + \frac{1}{2})(\lambda - \omega)}{\sin \frac{1}{2}(\lambda - \omega)} = 2\pi D_n(\lambda - \omega). \tag{3.4.12}$$

This is the transform of Example 1 translated along by ω units. It has peaks at $\lambda = \omega, \omega \pm 2\pi, \ldots$.

Example 3 (*Trigonometric Polynomial*) Suppose $X(t) = \sum_k \rho_k \exp\{i\omega_k t\}$. Clearly, from what has gone before, (3.4.10) equals

$$\sum_k \rho_k \frac{\sin (n + \frac{1}{2})(\lambda - \omega_k)}{\sin \frac{1}{2}(\lambda - \omega_k)}, \tag{3.4.13}$$

an expression with large amplitude at $\lambda = \omega_k \pm 2\pi l$, $l = 0, \pm 1, \ldots$.

Example 4 (*Monomials*) Suppose $X(t) = t^k$, $t = 0, \pm 1, \ldots$, k a positive integer. Expression (3.4.10) becomes

$$\sum_{t=-n}^{n} t^k \exp\{-i\lambda t\} = (i)^{-k} \frac{\partial^k}{\partial \lambda^k}\left[\frac{\sin (n + \frac{1}{2})\lambda}{\sin \frac{1}{2}\lambda}\right]. \tag{3.4.14}$$

This transform behaves like the derivatives of the transform of Example 1. Notice that it is concentrated in the neighborhood of $\lambda = 0, \pm 2\pi, \ldots$ for large n.

A polynomial $\sum_k \alpha_k t^k$ will behave as a linear combination of functions of the form (3.4.14).

Example 5 (*Monomial Amplitude Cosinusoid*) Suppose $X(t) = t^k \exp\{i\omega t\}$, then (3.4.10) is

$$\sum_{t=-n}^{n} t^k \exp\{-i(\lambda - \omega)t\} = i^{-k} \frac{\partial^k}{\partial \lambda^k}\left[\frac{\sin (n + \frac{1}{2})(\lambda - \omega)}{\sin \frac{1}{2}(\lambda - \omega)}\right]. \quad (3.4.15)$$

This is the function of Example 4 translated along by ω frequency units.

The general nature of these results is the following: the Fourier transform of a function $X(t)$ is concentrated in amplitude near $\lambda = 0, \pm 2\pi, \ldots$ if $X(t)$ is constant or slowly changing with t. It is concentrated near $\lambda = \omega$, $\omega \pm 2\pi, \ldots$ if $X(t)$ is a cosinusoid of frequency ω or is a cosinusoid of frequency ω multiplied by a polynomial in t.

The transform (3.4.2) may be inverted by the integration

$$\mathbf{X}(t) = (2\pi)^{-1} \int_0^{2\pi} \exp\{it\lambda\} \mathbf{d}_X^{(T)}(\lambda) d\lambda \qquad t = 0, \ldots, T-1. \quad (3.4.16)$$

Alternatively it is seen to be inverted by the sum

$$\mathbf{X}(t) = T^{-1} \sum_{s=0}^{T-1} \exp\left\{i\frac{2\pi st}{T}\right\} \mathbf{d}_X^{(T)}\left(\frac{2\pi s}{T}\right) \qquad t = 0, \ldots, T-1. \quad (3.4.17)$$

The Tr vectors $\mathbf{d}_X^{(T)}(2\pi s/T)$, $s = 0, \ldots, T-1$, are sometimes referred to as the **discrete Fourier transform** of $\mathbf{X}(t)$, $t = 0, \ldots, T-1$. We will discuss its numerical evaluation and properties in the next two sections.

The discrete Fourier transform may be written in matrix form. Let $\mathfrak{X}$ denote the $r \times T$ matrix whose columns are $\mathbf{X}(0), \ldots, \mathbf{X}(T-1)$ successively. Let $\mathfrak{D}$ denote the $r \times T$ matrix whose columns are $\mathbf{d}_X^{(T)}(2\pi s/T)$, $s = 0, \ldots, T-1$. Let $\mathfrak{F}$ denote the $T \times T$ matrix with $\exp\{-i(2\pi st/T)\}$ in row $s+1$ and column $t+1$ for $s,t = 0, \ldots, T-1$. Then we see that we have

$$\mathfrak{D} = \mathfrak{X}\mathfrak{F}. \quad (3.4.18)$$

The cases $T = 1, 2, 3, 4$ are seen to correspond to the respective matrices

$$[1] \tag{3.4.19}$$

$$\begin{bmatrix} 1 & 1 \\ 1 & -1 \end{bmatrix} \tag{3.4.20}$$

$$\begin{bmatrix} 1 & 1 & 1 \\ 1 & -\frac{1}{2} - i\frac{\sqrt{3}}{2} & -\frac{1}{2} + i\frac{\sqrt{3}}{2} \\ 1 & -\frac{1}{2} + i\frac{\sqrt{3}}{2} & -\frac{1}{2} - i\frac{\sqrt{3}}{2} \end{bmatrix} \tag{3.4.21}$$

$$\begin{bmatrix} 1 & 1 & 1 & 1 \\ 1 & -i & -1 & i \\ 1 & -1 & 1 & -1 \\ 1 & i & -1 & -i \end{bmatrix}. \tag{3.4.22}$$

General discussions of discrete and finite Fourier transforms are given in Stumpff (1937), Whittaker and Robinson (1944), Schoenberg (1950), Cooley, Lewis, and Welch (1967). Further properties are given in the exercises at the end of this chapter.

3.5 THE FAST FOURIER TRANSFORM

In this book the discrete Fourier transform will be the basic entity from which statistics of interest will be formed. It is therefore important to be able to calculate readily the discrete Fourier transform of a given set of numbers $X(t)$, $0 \leqslant t \leqslant T - 1$.

We have

$$d_X^{(T)}\left(\frac{2\pi j}{T}\right) = \sum_{t=0}^{T-1} \exp\left\{\frac{-i2\pi jt}{T}\right\} X(t) \qquad j = 0, \ldots, T-1 \tag{3.5.1}$$

and note that T^2 complex multiplications are required if we calculate the discrete Fourier transform directly from its definition. If T is composite (the product of several integers), then elementary procedures to reduce the required number of multiplications have often been employed; see Cooley et al (1967). Recently formal algorithms, which reduce the required number of multiplications to what must be a near minimum, have appeared; see Good (1958), Cooley and Tukey (1965), Gentleman and Sande (1966), Cooley et al (1967), Bergland (1967), Bingham et al (1967) and Brigham and Morrow (1967). For a formulation in terms of a composition series of a finite group see Posner (1968) and Cairns (1971).

We now indicate the form of these Fast Fourier Transform Algorithms beginning with two elementary cases. The underlying idea is to reduce the calculation of the discrete Fourier transform, of a long stretch of data, to the calculation of successive Fourier transforms of shorter sets of data. We begin with

Theorem 3.5.1 Let $T = T_1T_2$, where T_1 and T_2 are integers; then

$$d_X^{(T)}(2\pi T^{-1}[j_1T_2 + j_2]) = \sum_{t_1=0}^{T_1-1} \exp\{-i2\pi T_1^{-1}j_1t_1\} \exp\{-i2\pi T^{-1}j_2t_1\} \times \sum_{t_2=0}^{T_2-1} \exp\{-i2\pi T_2^{-1}j_2t_2\}X(t_1 + t_2T_1)$$

$$0 \leqslant j_1 \leqslant T_1 - 1, \qquad 0 \leqslant j_2 \leqslant T_2 - 1. \quad (3.5.2)$$

We note that $j_1T_2 + j_2$ runs through all integers j, $0 \leqslant j \leqslant T - 1$ for $0 \leqslant j_1 \leqslant T_1 - 1$ and $0 \leqslant j_2 \leqslant T_2 - 1$. We note that $(T_1 + T_2)T_1T_2$ complex multiplications are required in (3.5.2) to perform discrete Fourier transforms of orders T_1 and T_2. Certain additional operations will be required to insert the terms $\exp\{-i2\pi T^{-1}j_2t_1\}$.

A different algorithm is provided by the following theorem in which we let $\hat{X}(t)$ denote the period T extension of $X(0), \ldots, X(T - 1)$.

Theorem 3.5.2 Let $T = T_1T_2$, where T_1 and T_2 are relatively prime integers; then for $j \equiv j_1(\text{mod } T_1), j \equiv j_2(\text{mod } T_2), 0 \leqslant j_1 \leqslant T_1 - 1, 0 \leqslant j_2 \leqslant T_2 - 1$

$$d_X^{(T)}(2\pi T^{-1}j) = \sum_{t_1=0}^{T_1-1} \sum_{t_2=0}^{T_2-1} \exp\{-i2\pi(T_1^{-1}t_1j_1 + T_2^{-1}t_2j_2)\}\hat{X}(t_1T_2 + t_2T_1). \quad (3.5.3)$$

The number of complex multiplications required is again $(T_1 + T_2)T_1T_2$. In this case we must determine, for each j, j_1 and j_2 above and use this information to select the appropriate Fourier coefficient. Notice that the $\exp\{-i2\pi T^{-1}j_2t_1\}$ terms of (3.5.2) are absent and that the result is symmetric in T_1 and T_2. Good (1971) contrasts the two Fast Fourier Algorithms.

When we turn to the case in which $T = T_1 \ldots T_k$, for general k, with $T_1, \ldots, T_k$ integers, the extension of Theorem 3.5.1 is apparent. In (3.5.2), T_2 is now composite and so the inner Fourier transform, with respect to t_2, may be written in iterated form (in the form of (3.5.2) itself). Continuing in this way it is seen that the $d_X^{(T)}(2\pi T^{-1}j), j = 0, \ldots, T - 1$ may be derived by k successive discrete Fourier transforms of orders $T_1, \ldots, T_k$ in turn. The number of complex multiplications required is $(T_1 + \cdots + T_k)T$. Specific formulas for this case may be found in Bingham et al (1967).

The generalization of Theorem 3.5.2 is as follows:

Theorem 3.5.3 Let $T = T_1 \cdots T_k$, where $T_1, \ldots, T_k$ are relatively prime in pairs. Let $j \equiv j_l \pmod{T_l}$, $0 \leqslant j_l \leqslant T_l - 1$, $l = 1, \ldots, k$; then

$$d_X^{(T)}(2\pi T^{-1}j) = \sum_{t_1=0}^{T_1-1} \cdots \sum_{t_k=0}^{T_k-1} \exp\{-i2\pi(T_1^{-1}t_1 j_1 + \cdots + T_k^{-1}t_k j_k)\}$$
$$\times \hat{X}(t_1 T\, T_1^{-1} + \cdots + t_k T\, T_k^{-1}) \qquad j = 0, \ldots, T-1. \quad (3.5.4)$$

($\hat{X}(t)$ is here the periodic extension, with period T, of $X(t)$.)

By way of explanation of this result, we note that the numbers $t_1 T/T_1 + \cdots + t_k T/T_k$, when reduced mod T, run through all integers t, $0 \leqslant t \leqslant T - 1$ for $0 \leqslant t_1 \leqslant T_1 - 1, \ldots, 0 \leqslant t_k \leqslant T_k - 1$. For each j we must determine the $j_1, \ldots, j_k$ above, and select the appropriate Fourier coefficient from those that have been calculated. This may be done by setting up a table of the residues of j, $0 \leqslant j \leqslant T - 1$.

The number of complex multiplications indicated in Theorem 3.5.3 is also $(T_1 + \cdots + T_k)T$. We see that we will obtain the greatest saving if the T_j are small. If $T = 2^n$, we see that essentially $2T \log_2 T$ multiplications are needed. At the end of Section 3.4, we gave the discrete Fourier transform for the cases $T = 1, 2, 3, 4$. Examination of the results shows that fewer than the indicated number of operations may be required, the cases $T = 4$ and $T = 8$ being particularly important. Additional gains can be achieved by taking note of the real nature of the $X(t)$ or by transforming more than one series; see Cooley et al (1967) and Exercise 3.10.30.

It often occurs that T is not highly composite and one is not interested in the values of $d_X^{(T)}(\lambda)$ at frequencies of the form $2\pi j/T, j = 0, \ldots, T - 1$. If this is so, we can add $S - T$ zeros to the $X(t)$ values, choosing $S > T$ to be highly composite. The transform $d_X^{(T)}(\lambda)$ is now obtained for $\lambda = 2\pi j/S$, $j = 0, 1, \ldots, S - 1$.

Quite clearly we can combine the technique of Theorem 3.5.3, where the factors of T are relatively prime, with the previously indicated procedure for dealing with general factors. The number of extra multiplications by cosinusoids may be reduced in this way. See Hamming (1962), p. 74, for the case $T = 12$. A FORTRAN program for the mixed radix Fast Fourier Transform may be found in Singleton (1969).

In conclusion we remark that the Fast Fourier Transform is primarily an efficient numerical algorithm. Its use or nonuse does not affect the basis of statistical inference. Its effect has been to radically alter the calculations of empirical time series analysis.

3.6 APPLICATIONS OF DISCRETE FOURIER TRANSFORMS

Suppose the values $X(t)$, $Y(t)$, $t = 0, \ldots, T-1$ are available. We will sometimes require the convolution

$$\sum_{\substack{t \\ 0 \leq t+u \leq T-1}} X(t+u)Y(t) \qquad u = 0, \pm 1, \ldots. \tag{3.6.1}$$

If

$$d_X^{(T)}(\lambda) = \sum_{t=0}^{T-1} X(t) \exp\{-i\lambda t\}$$
$$d_Y^{(T)}(\lambda) = \sum_{t=0}^{T-1} Y(t) \exp\{-i\lambda t\}, \tag{3.6.2}$$

then we quickly see that the convolution (3.6.1) is the coefficient of $\exp\{-i\lambda u\}$ in the trigonometric polynomial $d_X^{(T)}(\lambda)\overline{d_Y^{(T)}(\lambda)}$. It is therefore given by

$$(2\pi)^{-1} \int_0^{2\pi} d_X^{(T)}(\lambda)\overline{d_Y^{(T)}(\lambda)} \exp\{iu\lambda\}\, d\lambda \qquad \text{for } u = 0, \pm 1, \ldots. \tag{3.6.3}$$

This occurrence suggests that we may be able to compute (3.6.1) by means of a discrete Fourier transform and so take advantage of the Fast Fourier Transform Algorithm. In fact we have

Lemma 3.6.1 Given $X(t)$, $Y(t)$, $t = 0, \ldots, T-1$ and an integer $S > T$, the convolution (3.6.1) is given by

$$S^{-1} \sum_{s=0}^{S-1} d_X^{(T)}\left(\frac{2\pi s}{S}\right)\overline{d_Y^{(T)}\left(\frac{2\pi s}{S}\right)} \exp\left\{i\frac{2\pi s u}{S}\right\}$$
$$\text{for } u = 0, \pm 1, \ldots, \pm(S-T). \tag{3.6.4}$$

In general (3.6.4) equals

$$\sum_{\substack{t \\ 0 \leq s+u \leq T-1}} X(s+u)Y(t)\eta\left\{\frac{2\pi(s-t)}{S}\right\}. \tag{3.6.5}$$

We may obtain the desired values of the convolution from (3.6.4) by taking S large enough. If S is taken to be highly composite then the discrete Fourier transforms required in the direct evaluation of (3.6.4) may be rapidly calculated by means of the Fast Fourier Transform Algorithm of the previous section. Consequently the convolution (3.6.1) may well be more rapidly computed by this procedure rather than by using its definition (3.6.1) directly. This fact was noted by Sande; see Gentleman and Sande (1966), and also Stockham (1966). From (3.6.5) we see that for $S - T < |u| \leqslant T - 1$, expression (3.6.4) gives (3.6.1) plus some additional terms. For moderate values of $|u|$ it will approximately equal (3.6.1). It can be obtained for all u by taking $S \geqslant 2T$.

One situation in which one might require the convolution (3.6.1) is in the estimation of the moment function $m_{12}(u) = E[X_1(t+u)\, X_2(t)]$ for some stationary bivariate series. An unbiased estimate of $m_{12}(u)$ is provided by

$$(T - |u|)^{-1} \sum_{\substack{t \\ 0 \leq t+u \leq T-1}} X_1(t+u)X_2(t) \tag{3.6.6}$$

an expression of the form of (3.6.1). Exercise 3.10.7 indicates how the result of Lemma 3.6.1 might be modified to construct an estimate of $c_{12}(u) = \operatorname{cov}\{X_1(t+u), X_2(t)\}$.

Another situation in which the result of Lemma 3.6.1 proves useful is in the calculation of the filtered values of Section 3.3:

$$Y(t) = \sum_{u=0}^{T-1} a(t-u)X(u) \qquad \text{for } t = 0, \pm 1, \ldots \tag{3.6.7}$$

given the values $X(t)$, $t = 0, \ldots, T-1$. Suppose the transfer function of the filter $\{a(u)\}$ is $A(\lambda)$. Then Lemmas 3.4.1 and 3.6.1 suggest that we form

$$S^{-1} \sum_{s=0}^{S-1} d_X{}^{(T)}\left(\frac{2\pi s}{S}\right) A\left(\frac{2\pi s}{S}\right) \exp\left\{i\frac{2\pi st}{S}\right\} \qquad \text{for } t = 0, \pm 1, \ldots. \tag{3.6.8}$$

These values should be near the desired filtered values. In fact by direct substitution we see that (3.6.8) equals

$$\sum_{u=0}^{T-1} \left[\sum_{l=-\infty}^{\infty} a(t-u+lS)\right] X(u) \tag{3.6.9}$$

and so if $a(u)$ falls off rapidly as $|u| \to \infty$ and $0 \leqslant t \leqslant T-1$ expression (3.6.8) should be near (3.6.7). If S is taken to be highly composite then the calculations indicated in (3.6.8) may be reduced by means of the Fast Fourier Transform. We might introduce convergence factors.

We remark that Lemma 3.6.1 has the following extension:

Lemma 3.6.2 Given $X_j(t)$, $t = 0, \ldots, T-1$, $j = 1, \ldots, r$ and an integer $S > T$ the expression

$$\sum_{\substack{t \\ 0 \leq t+u_j \leq T-1}} X_1(t+u_1) \cdots X_{r-1}(t+u_{r-1})X_r(t) \tag{3.6.10}$$

equals

$$S^{-r+1} \sum_{s_1=0}^{S-1} \cdots \sum_{s_{r-1}=0}^{S-1} d_1{}^{(T)}\left(\frac{2\pi s_1}{S}\right) \cdots d_{r-1}^{(T)}\left(\frac{2\pi s_{r-1}}{S}\right) \overline{d_r{}^{(T)}\left(\frac{2\pi[s_1 + \cdots + s_{r-1}]}{S}\right)}$$
$$\times \exp\left\{i\frac{2\pi[s_1u_1 + \cdots + s_{r-1}u_{r-1}]}{S}\right\}$$
$$u_j = 0, \pm 1, \ldots, \pm(S-T), j = 1, \ldots, r-1. \tag{3.6.11}$$

We conclude this section by indicating some uses of the finite Fourier transform. Suppose

$$2\pi X(t) = R\cos(\omega t + \phi) \qquad \text{with } -\pi < \omega < \pi; \tag{3.6.12}$$

then

$$\sum_{t=-n}^{n} X(t)\exp\{-i\lambda t\} = \tfrac{1}{2}Re^{i\phi}D_n(\lambda - \omega) + \tfrac{1}{2}Re^{-i\phi}D_n(\lambda + \omega) \tag{3.6.13}$$

following Example 2 of Section 3.4. By inspection we see that the amplitude of expression (3.6.13) is large for λ near $\pm\omega$ and not otherwise, $-\pi < \lambda < \pi$. In consequence the finite Fourier transform (3.6.13) should prove useful in detecting the frequency of a cosinusoid of unknown frequency. This use was proposed in Stokes (1879).

We remark that if $X(t)$ contains two unknown frequencies, say

$$2\pi X(t) = R_1\cos(\omega_1 t + \phi_1) + R_2\cos(\omega_2 t + \phi_2), \tag{3.6.14}$$

then we may have difficulty resolving ω_1 and ω_2 if they are close to one another for

$$\begin{aligned}\sum_{t=-n}^{n} X(t)\exp\{-i\lambda t\} = {}& \tfrac{1}{2}R_1\exp\{i\phi_1\}D_n(\lambda - \omega_1)\\ &+ \tfrac{1}{2}R_2\exp\{i\phi_2\}D_n(\lambda - \omega_2)\\ &+ \tfrac{1}{2}R_1\exp\{-i\phi_1\}D_n(\lambda + \omega_1)\\ &+ \tfrac{1}{2}R_2\exp\{-i\phi_2\}D_n(\lambda + \omega_2).\end{aligned} \tag{3.6.15}$$

This function will not have obvious peaks in amplitude at $\lambda = \pm\omega_1, \pm\omega_2$ if ω_1 and ω_2 are so close together that the ripples of the D_n functions interfere with one another. This difficulty may be reduced by tapering the $X(t)$ series prior to forming the Fourier transform. Specifically consider

$$\begin{aligned}\sum_{t=-n}^{n} h\left(\frac{t}{n}\right)X(t)\exp\{-i\lambda t\} = {}& \tfrac{1}{2}R_1\exp\{i\phi_1\}H^{(n)}(\lambda - \omega_1)\\ &+ \tfrac{1}{2}R_2\exp\{i\phi_2\}H^{(n)}(\lambda - \omega_2)\\ &+ \tfrac{1}{2}R_1\exp\{-i\phi_1\}H^{(n)}(\lambda + \omega_1)\\ &+ \tfrac{1}{2}R_2\exp\{-i\phi_2\}H^{(n)}(\lambda + \omega_2)\end{aligned} \tag{3.6.16}$$

in the case of (3.6.14) where we have made use of (3.3.6). If the convergence factors $h(u/n)$ are selected so that $H^{(n)}(\lambda)$ is concentrated in some interval, say $|\lambda| < \Delta/n$, then the amplitude of (3.6.16) should have obvious peaks if $|\omega_1 - \omega_2| > 2\Delta/n$.

Other uses of the finite Fourier transform include: the evaluation of the latent values of a matrix of interest, see Lanczos (1955); the estimation of the mixing distribution of a compound distribution, see Medgyessy (1961); and the determination of the cumulative distribution function of a random variable from the characteristic function, see Bohman (1960).

3.7 COMPLEX MATRICES AND THEIR EXTREMAL VALUES

We turn to a consideration of matrices whose entries are complex numbers and remark that the spectral density matrix introduced in Section 2.5 is an example of such a matrix. Begin with several definitions. If $\mathbf{Z} = [Z_{jk}]$ is a $J \times K$ matrix with the complex number Z_{jk} in the jth row and kth column, then we define $\overline{\mathbf{Z}} = [\overline{Z}_{jk}]$ to be the matrix whose entries are the complex conjugates of the entries of $\mathbf{Z}$. Let $\mathbf{Z}^\tau = [Z_{kj}]$ denote the **transpose** of $\mathbf{Z}$. We then say that $\mathbf{Z}$ is **Hermitian** if $\overline{\mathbf{Z}}^\tau = \mathbf{Z}$. If $\mathbf{Z}$ is $J \times J$ Hermitian then we say that $\mathbf{Z}$ is **non-negative definite** if

$$\sum_{j,k=1}^{J} \alpha_j \bar{\alpha}_k Z_{jk} \geqslant 0 \tag{3.7.1}$$

for all complex scalars α_j, $j = 1, \ldots, J$. A square matrix $\mathbf{Z}$ is **unitary** if $\mathbf{Z}^{-1} = \overline{\mathbf{Z}}^\tau$ or equivalently $\mathbf{Z}\overline{\mathbf{Z}}^\tau = \mathbf{I}$ with $\mathbf{I}$ the identity matrix. The complex number μ is called a **latent value** or **latent root** of the $J \times J$ matrix $\mathbf{Z}$ if

$$\text{Det}(\mathbf{Z} - \mu\mathbf{I}) = 0 \tag{3.7.2}$$

where $\mathbf{I}$ is the identity of the same dimension as $\mathbf{Z}$. Because $\text{Det}(\mathbf{Z} - \mu\mathbf{I})$ is a polynomial of order J in μ, the equation (3.7.2) has at most J distinct roots. It is a classic result (MacDuffee (1946)) that corresponding to any latent value μ there is always a J vector $\boldsymbol{\alpha}$ such that

$$\mathbf{Z}\boldsymbol{\alpha} = \mu\boldsymbol{\alpha}. \tag{3.7.3}$$

Such an $\boldsymbol{\alpha}$ is called a **latent vector** of $\mathbf{Z}$. If $\mathbf{Z}$ is Hermitian, then its latent values are real-valued; see MacDuffee (1946). We denote the jth largest of these by μ_j or $\mu_j(\mathbf{Z})$ for $j = 1, \ldots, J$. The corresponding latent vector is denoted by $\boldsymbol{\alpha}_j$ or $\boldsymbol{\alpha}_j(\mathbf{Z})$. The collection of latent values of a square matrix is called its **spectrum.** We will shortly discuss the connection between this spectrum and the previously defined second-order spectrum of a stationary series.

Given a matrix $\mathbf{Z}$, we note that the matrices $\mathbf{Z}\overline{\mathbf{Z}}^\tau$, $\overline{\mathbf{Z}}^\tau\mathbf{Z}$ are always Hermitian and non-negative definite. Also, following Theorem 2.5.1, we note that if $\mathbf{X}(t)$, $t = 0, \pm 1, \ldots$ is an r vector-valued stationary series with absolutely summable covariance function, then $\mathbf{f}_{XX}(\lambda)$, its spectral density matrix, is Hermitian and non-negative definite. We remark that if

$$\mathfrak{F} = [\exp\{-i2\pi jk/T\}] \tag{3.7.4}$$

is the matrix of the discrete Fourier transform considered in Section 3.4, then the matrix $T^{-1/2}\mathfrak{F}$ is unitary. Its latent values are given in Exercise 3.10.12.

It is sometimes useful to be able to reduce computations involving complex matrices to computations involving only real matrices. Lemma 3.7.1 below gives an important isomorphism between complex matrices and real matrices. We first set down the notation; if $\mathbf{Z} = [Z_{jk}]$ with $Z_{jk} = \operatorname{Re} Z_{jk} + i \operatorname{Im} Z_{jk}$, then

$$\begin{aligned} \operatorname{Re} \mathbf{Z} &= [\operatorname{Re} Z_{jk}] \\ \operatorname{Im} \mathbf{Z} &= [\operatorname{Im} Z_{jk}]. \end{aligned} \tag{3.7.5}$$

Lemma 3.7.1 To any $J \times K$ matrix $\mathbf{Z}$ with complex entries there corresponds a $(2J) \times (2K)$ matrix $\mathbf{Z}^R$ with real entries such that

(i) if $\mathbf{Z} = \mathbf{X} + \mathbf{Y}$, then $\mathbf{Z}^R = \mathbf{X}^R + \mathbf{Y}^R$
(ii) if $\mathbf{Z} = \mathbf{X}\mathbf{Y}$, then $\mathbf{Z}^R = \mathbf{X}^R\mathbf{Y}^R$
(iii) if $\mathbf{Y} = \mathbf{Z}^{-1}$, then $\mathbf{Y}^R = (\mathbf{Z}^R)^{-1}$
(iv) $\operatorname{Det} \mathbf{Z}^R = |\operatorname{Det} \mathbf{Z}|^2$
(v) if $\mathbf{Z}$ is Hermitian, then $\mathbf{Z}^R$ is symmetric
(vi) if $\mathbf{Z}$ is unitary, then $\mathbf{Z}^R$ is orthogonal
(vii) if the latent values and vectors of $\mathbf{Z}$ are μ_j, $\boldsymbol{\alpha}_j$, $j = 1, \ldots, J$, then those of $\mathbf{Z}^R$ are, respectively,

$$\mu_j, \begin{bmatrix} \operatorname{Re} \boldsymbol{\alpha}_j \\ \operatorname{Im} \boldsymbol{\alpha}_j \end{bmatrix}; \qquad \mu_j, \begin{bmatrix} -\operatorname{Im} \boldsymbol{\alpha}_j \\ \operatorname{Re} \boldsymbol{\alpha}_j \end{bmatrix} \qquad j = 1, \ldots, J, \tag{3.7.6}$$

providing the dimensions of the matrices appearing throughout the lemma are appropriate.

In fact the correspondence of this lemma may be taken to be

$$\mathbf{Z}^R = \begin{bmatrix} \operatorname{Re} \mathbf{Z} & \operatorname{Im} \mathbf{Z} \\ -\operatorname{Im} \mathbf{Z} & \operatorname{Re} \mathbf{Z} \end{bmatrix}. \tag{3.7.7}$$

It is discussed in Wedderburn (1934), Lanczos (1956), Bellman (1960), Brenner (1961), Good (1963), and Goodman (1963). The correspondence is exceedingly useful for carrying out numerical computations involving matrices with complex-valued entries. However, Ehrlich (1970) suggests that we should stick to complex arithmetic when convenient.

Latent vectors and values are important in the construction of representations of matrices by more elementary matrices. In the case of a Hermitian matrix we have

Theorem 3.7.1 If $\mathbf{H}$ is a $J \times J$ Hermitian matrix, then

$$\mathbf{H} = \sum_{j=1}^{J} \mu_j \mathbf{U}_j \overline{\mathbf{U}}_j^{\tau} \tag{3.7.8}$$

where μ_j is the jth latent value of $\mathbf{H}$ and $\mathbf{U}_j$ is the corresponding latent vector.

The theorem has the following:

Corollary 3.7.1 If $\mathbf{H}$ is $J \times J$ Hermitian, then it may be written $\mathbf{U}\mathbf{M}\overline{\mathbf{U}}^\tau$ where $\mathbf{M} = \text{diag}\{\mu_j; j = 1, \ldots, J\}$ and $\mathbf{U} = [\mathbf{U}_1 \cdots \mathbf{U}_J]$ is unitary. Also if $\mathbf{H}$ is non-negative definite, then $\mu_j \geqslant 0, j = 1, \ldots, J$.

This theorem is sometimes known as the **Spectral Theorem.** In the case of matrices of arbitrary dimension we have

Theorem 3.7.2 If $\mathbf{Z}$ is $J \times K$, then

$$\mathbf{Z} = \sum_{j \leq J,K} \mu_j \mathbf{U}_j \overline{\mathbf{V}}_j{}^\tau \tag{3.7.9}$$

where $\mu_j{}^2$ is the jth latent value of $\mathbf{Z}\overline{\mathbf{Z}}^\tau$ (or $\overline{\mathbf{Z}}^\tau\mathbf{Z}$), $\mathbf{U}_j$ is the jth latent vector of $\mathbf{Z}\overline{\mathbf{Z}}^\tau$ and $\mathbf{V}_j$ is the jth latent vector of $\overline{\mathbf{Z}}^\tau\mathbf{Z}$ and it is understood $\mu_j \geqslant 0$.

The theorem has the following:

Corollary 3.7.2 If $\mathbf{Z}$ is $J \times K$, then it may be written $\mathbf{U}\mathbf{M}\overline{\mathbf{V}}^\tau$ where the $J \times K$ $\mathbf{M} = \text{diag}\{\mu_j;\ j = 1, \ldots, J\}$, the $J \times J$ $\mathbf{U} = [\mathbf{U}_1 \cdots \mathbf{U}_J]$ is unitary and the $K \times K$ $\mathbf{V} = [\mathbf{V}_1 \cdots \mathbf{V}_K]$ is also unitary.

This theorem is given in Autonne (1915). Structure theorems for matrices are discussed in Wedderburn (1934) and Hua (1963); see also Schwerdtfeger (1960). The representation $\mathbf{Z} = \mathbf{U}\mathbf{M}\overline{\mathbf{U}}^\tau$ is called the singular value decomposition of $\mathbf{Z}$. A computer program for it is given in Businger and Golub (1969).

An important class of matrices, in the subject of time series analysis, is the class of finite Toeplitz matrices. We say that a matrix $\mathbf{C} = [C_{jk}]$ is **finite Toeplitz** if C_{jk} depends only on $j - k$, that is, $C_{jk} = c(j - k)$ for some function $c(.)$. These matrices are discussed in Widom (1965) where other references may be found. Finite Toeplitz matrices are important in time series analysis for the following reason; if $X(t)$, $t = 0, \pm 1, \ldots$ is a real-valued stationary series with autocovariance function $c_{XX}(u)$, $u = 0, \pm 1, \ldots$, then the covariance matrix of the stretch $X(t)$, $t = 0, \ldots, T - 1$ is a finite Toeplitz matrix with $c_{XX}(j - k)$ in the jth row and kth column.

We will sometimes be interested in the latent roots and vectors of the covariance matrix of $X(t)$, $t = 0, \ldots, T - 1$ for a stationary $X(t)$. Various approximate results are available concerning these in the case of large T. Before indicating certain of these we first introduce an important class of finite Toeplitz matrices. A square matrix $\mathbf{Z} = [Z_{jk}]$ is said to be a **circulant** of order T if $Z_{jk} = z(k - j)$ for some function $z(.)$ of period T, that is,

$$\mathbf{Z} = \begin{bmatrix} z(0) & z(1) & \cdots & z(T-1) \\ z(T-1) & z(0) & \cdots & z(T-2) \\ z(T-2) & z(T-1) & \cdots & z(T-3) \\ \cdot & \cdot & \cdots & \cdot \\ \cdot & \cdot & \cdots & \cdot \\ \cdot & \cdot & \cdots & \cdot \\ z(1) & z(2) & \cdots & z(0) \end{bmatrix}. \tag{3.7.10}$$

In connection with the latent values and vectors of a circulant we have

Theorem 3.7.3 Let $\mathbf{Z} = [z(k - j)]$ be a $T \times T$ circulant matrix, then its latent values are given by

$$\sum_{j=0}^{T-1} z(j) \exp\{-i2\pi jk/T\} \qquad k = 0, \ldots, T-1 \tag{3.7.11}$$

and the corresponding latent vectors by

$$T^{-1/2} [\exp\{-i2\pi jk/T\}; j = 0, \ldots, T-1] \qquad k = 0, \ldots, T-1 \tag{3.7.12}$$

respectively.

The latent values are seen to provide the discrete Fourier transform of the sequence $z(t)$, $t = 0, \ldots, T - 1$. The matrix of latent vectors is proportional to the matrix $\mathfrak{F}$ of Section 3.4. Theorem 3.7.3 may be found in Aitken (1954), Schoenberg (1950), Hamburger and Grimshaw (1951) p. 94, Good (1950), and Whittle (1951).

Let us return to the discussion of a general square finite Toeplitz matrix $\mathbf{C} = [c(j - k)]$, $j, k = 1, \ldots, T$. Consider the related circulant matrix $\mathbf{Z}$ whose kth entry in the first row is $c(1 - k) + c(1 - k + T)$, where we consider $c(T) = 0$. Following Theorem 3.7.3 the latent values of $\mathbf{Z}$ are

$$\begin{aligned} \mu_k(\mathbf{Z}) &= \sum_{j=0}^{T-1} [c(-j) + c(-j + T)] \exp\{-i2\pi jk/T\} \\ &= \sum_{u=-T+1}^{T-1} c(u) \exp\{i2\pi uk/T\} \qquad k = 0, \ldots, T-1 \end{aligned} \tag{3.7.13}$$

giving a discrete Fourier transform of the $c(u)$, $u = 0, \pm 1, \ldots, \pm(T - 1)$. Let $\mathfrak{F}_T$ denote the $T \times T$ matrix whose columns are the vectors (3.7.12). Let $\mathbf{M}_T$ denote the diagonal matrix with corresponding entries $\mu_k(\mathbf{Z})$, then

$$\mathbf{Z} = \mathfrak{F}_T \mathbf{M}_T \overline{\mathfrak{F}}_T^{\tau} \tag{3.7.14}$$

and we may consider approximating $\mathbf{C}$ by $\mathfrak{F}_T\mathbf{M}_T\bar{\mathfrak{F}}^\tau = \mathbf{Z}$. We have

$$\|\mathbf{C} - \mathbf{Z}\|^2 = \sum_{j,k} |C_{jk} - Z_{jk}|^2$$
$$= \sum_{u=-T+1}^{T-1} |u||c(u)|^2 \tag{3.7.15}$$

giving us a bound on the difference between $\mathbf{C}$ and $\mathbf{Z}$. This bound may be used to place bounds on the differences between the latent roots and vectors of $\mathbf{C}$ and $\mathbf{Z}$. For example the Wielandt-Hoffman Theorem (Wilkinson (1965)) indicates that there is an ordering $\mu_{i_1}(\mathbf{C}), \ldots, \mu_{i_T}(\mathbf{C})$ of the latent roots $\mu_1(\mathbf{C}), \ldots, \mu_T(\mathbf{C})$ of $\mathbf{C}$ such that

$$\sum_{k=1}^{T} |\mu_{i_k}(\mathbf{C}) - \sum_{u=-T+1}^{T-1} c(u) \exp\{i2\pi ku/T\}|^2 \leqslant \sum_{u=-T+1}^{T-1} |u|\ |c(u)|^2. \tag{3.7.16}$$

If

$$\sum_{u=-\infty}^{\infty} |u|\ |c(u)|^2 < \infty, \tag{3.7.17}$$

then the latent roots of $\mathbf{C}$ are tending to be distributed like the values of the discrete Fourier transform of $c(u)$, $u = 0, \pm 1, \ldots, \pm(T-1)$ as $T \to \infty$. A variety of results of this nature may be found in Grenander and Szego (1958); see also Exercise 3.10.14. This sort of result indicates a connection between the power spectrum of a stationary time series (defined as the Fourier transform of its autocovariance function) and the spectrum (defined to be the collection of latent values) of the covariance matrix of long stretches of the series. We return to this in Section 4.7.

Results concerning the difference between the latent vectors of $\mathbf{C}$ and those of $\mathbf{Z}$ may be found in Gavurin (1957) and Davis and Kahan (1969). We remark that the above discussion may be extended to the case of vector-valued time series and block Toeplitz matrices; see Exercise 3.10.15.

The representation (3.7.9) is important in the approximation of a matrix by another matrix of reduced rank. We have the following:

Theorem 3.7.4 Let $\mathbf{Z}$ be $J \times K$. Among $J \times K$ matrices $\mathbf{A}$ of rank $L \leqslant J, K$

$$\mu_j([\mathbf{Z} - \mathbf{A}]\overline{[\mathbf{Z} - \mathbf{A}]}^\tau) \tag{3.7.18}$$

is minimized by

$$\mathbf{A} = \sum_{j=1}^{L} \mu_j \mathbf{U}_j \bar{\mathbf{V}}_j^\tau \tag{3.7.19}$$

where μ_j, $\mathbf{U}_j$, $\mathbf{V}_j$ are given in Theorem 3.7.2. The minimum achieved is μ_{j+L}^2.

We see that we construct **A** from the terms in (3.7.9) corresponding to the L largest μ_j; see Okamoto (1969) for the case of real symmetric **Z** and **A.**

Corollary 3.7.4 The above choice of **A** also minimizes

$$\|\mathbf{Z} - \mathbf{A}\|^2 = \sum_{j=1}^{J} \sum_{k=1}^{K} |Z_{jk} - A_{jk}|^2 \tag{3.7.20}$$

for **A** of rank $L \leqslant J, K$. The minimum achieved is

$$\sum_{j>L} \mu_j^2. \tag{3.7.21}$$

Results of the form of this corollary are given in Eckart and Young (1936), Kramer and Mathews (1956), and Rao (1965) for the case of real **Z, A.**

3.8 FUNCTIONS OF FOURIER TRANSFORMS

Let $\mathbf{X}(t)$, $t = 0, \pm 1, \ldots$ be a vector-valued time series of interest. In order to discuss the statistical properties of certain series resulting from the application of operators to the series $\mathbf{X}(t)$, we must now develop several analytic results concerning functions of Fourier transforms. We begin with the following:

Definition 3.8.1 Let C denote the space of complex numbers. A complex-valued function $f(\mathbf{z})$ defined for $\mathbf{z} = (z_1, \ldots, z_n) \in D$, an open subset of C^n, is holomorphic in D if each point $\mathbf{w} = (w_1, \ldots, w_n) \in D$ is contained in an open neighborhood U such that $f(\mathbf{z})$ has a convergent power series expansion

$$f(\mathbf{z}) = \sum_{k_1,\ldots,k_n=0}^{\infty} a_{k_1\ldots k_n}(z_1 - w_1)^{k_1} \cdots (z_n - w_n)^{k_n} \tag{3.8.1}$$

for all $\mathbf{z} \in U$.

A result that is sometimes useful in determining holomorphic functions is provided by

Theorem 3.8.1 Suppose $F_j(y_1, \ldots, y_m; z_1, \ldots, z_n)$, $j = 1, \ldots, m$ are holomorphic functions of $m + n$ variables in a neighborhood of $(u_1, \ldots, u_m; v_1, \ldots, v_n) \in C^{m+n}$. If $F_j(u_1, \ldots, u_m; v_1, \ldots, v_n) = 0$, $j = 1, \ldots, m$, while the determinant of the Jacobian matrix

$$\frac{\partial(F_1, \ldots, F_m)}{\partial(y_1, \ldots, y_m)} \tag{3.8.2}$$

is nonzero at $(u_1, \ldots, u_m; v_1, \ldots, v_n)$, then the equations

$$F_j(y_1, \ldots, y_m; z_1, \ldots, z_n) = 0 \qquad j = 1, \ldots, m \tag{3.8.3}$$

have a unique solution $y_j = y_j(z_1, \ldots, z_n), j = 1, \ldots, m$ which is holomorphic in a neighborhood of $(v_1, \ldots, v_n)$.

This theorem may be found in Bochner and Martin (1948) p. 39. It implies, for example, that the zeros of a polynomial are holomorphic functions of the coefficients of the polynomial in a region where the polynomial has distinct roots. It implies a fortiori that the latent values of a matrix are holomorphic functions of the elements of the matrix in a region of distinct latent values; see Exercise 3.10.19.

Let $V_+(l)$, $l \geq 0$, denote the space of functions $z(\lambda)$, $-\infty < \lambda < \infty$, that are Fourier transforms of the form

$$z(\lambda) = \sum_{u=0}^{\infty} a(u) \exp\{-iu\lambda\} \tag{3.8.4}$$

with the $a(u)$ real-valued and satisfying

$$\sum_{u=0}^{\infty} [1 + |u|^l]|a(u)| < \infty. \tag{3.8.5}$$

Under the condition (3.8.5) the domain of $z(\lambda)$ may be extended to consist of complex λ with $-\infty < \text{Re}\,\lambda < \infty$, $\text{Im}\,\lambda \geq 0$. We then have

Theorem 3.8.2 If $z_j(\lambda)$ belongs to $V_+(l)$, $j = 1, \ldots, n$ and $f(z_1, \ldots, z_n)$ is a holomorphic function in a neighborhood of the range of values $\{z_1(\lambda), \ldots, z_n(\lambda)\}$; $-\infty < \text{Re}\,\lambda < \infty$, $\text{Im}\,\lambda \geq 0$, then $f(z_1(\lambda), \ldots, z_n(\lambda))$ also belongs to $V_+(l)$.

This theorem may be deduced from results in Gelfand et al (1964). The first theorems of this nature were given by Wiener (1933) and Levy (1933).

As an example of the use of this theorem consider the following: let $\{\mathbf{a}(u)\}$, $u = 0, 1, 2, \ldots$ be an $r \times r$ realizable l summable filter with transfer function $\mathbf{A}(\lambda)$ satisfying $\text{Det}\,\mathbf{A}(\lambda) \neq 0$, $-\infty < \text{Re}\,\lambda < \infty$, $\text{Im}\,\lambda \geq 0$. This last condition implies that the entries of $\mathbf{A}(\lambda)^{-1}$ are holomorphic functions of the entries of $\mathbf{A}(\lambda)$ in a neighborhood of the range of $\mathbf{A}(\lambda)$; see Exercise 3.10.37. An application of Theorem 3.8.2 indicates that the entries of $\mathbf{B}(\lambda) = \mathbf{A}(\lambda)^{-1}$ are in $V_+(l)$ and so $\mathbf{B}(\lambda)$ is the transfer function of an $r \times r$ realizable l summable filter $\{\mathbf{b}(u)\}$, $u = 0, 1, 2, \ldots$. In particular we see that if $\mathbf{X}(t)$, $t = 0, \pm 1, \ldots$ is a stationary r vector-valued series with $E|\mathbf{X}(t)| < \infty$, then the relation

$$\mathbf{Y}(t) = \sum_{u=0}^{\infty} \mathbf{a}(u)\mathbf{X}(t-u) \tag{3.8.6}$$

may, with probability 1, be inverted to give

$$\mathbf{X}(t) = \sum_{u=0}^{\infty} \mathbf{b}(u)\mathbf{Y}(t-u) \tag{3.8.7}$$

for some $\mathbf{b}(u)$, $u = 0, 1, 2, \ldots$ with

$$\sum_{u=0}^{\infty} [1 + |u|^l]|\mathbf{b}(u)| < \infty. \tag{3.8.8}$$

We remark that the condition $\text{Det } \mathbf{A}(\lambda) \neq 0$, $-\infty < \text{Re } \lambda < \infty$, $\text{Im } \lambda \geqslant 0$ is equivalent to the condition

$$\text{Det}\left[\sum_{u=0}^{\infty} \mathbf{a}(u)z^u\right] \tag{3.8.9}$$

and has no roots in the unit disc $|z| \leqslant 1$. In the case that $\mathbf{Y}(t) = \boldsymbol{\varepsilon}(t)$, a pure noise series with finite mean, the above reasoning indicates that if

$$\text{Det}[\mathbf{I} + \mathbf{a}(1)z + \cdots + \mathbf{a}(m)z^m] \tag{3.8.10}$$

has no roots in the unit disc, then the autoregressive scheme

$$\mathbf{X}(t) + \mathbf{a}(1)\mathbf{X}(t-1) + \cdots + \mathbf{a}(m)\mathbf{X}(t-m) = \boldsymbol{\varepsilon}(t) \tag{3.8.11}$$

has, with probability 1, a stationary solution of the form

$$\mathbf{X}(t) = \sum_{u=0}^{\infty} \mathbf{b}(t-u)\boldsymbol{\varepsilon}(u) \tag{3.8.12}$$

with

$$\sum_{u=0}^{\infty} [1 + |u|^l]|\mathbf{b}(u)| < \infty \tag{3.8.13}$$

for all $l \geqslant 0$.

An alternate set of results of the above nature is sometimes useful. We set down

Definition 3.8.2 A complex-valued function $f(\mathbf{z})$ defined for $\mathbf{z} = (z_1, \ldots, z_n) \in D$, an open subset of C^n, is real holomorphic in D if each point $\mathbf{w} = (w_1, \ldots, w_n) \in D$ is contained in an open neighborhood U such that $f(\mathbf{z})$ has a convergent power series expansion

$$f(\mathbf{z}) = \sum_{k_1,\ldots,k_n=0}^{\infty} \sum_{j_1,\ldots,j_n=0}^{\infty} a_{k_1\ldots k_n;j_1\ldots j_n}(z_1 - w_1)^{k_1}\overline{(z_1 - w_1)}^{j_1} \cdots (z_n - w_n)^{k_n}\overline{(z_n - w_n)}^{j_n} \tag{3.8.14}$$

for all $\mathbf{z} \in U$.

We next introduce $V(l)$, $l \geqslant 0$, the space of functions $z(\lambda)$, $-\infty < \lambda < \infty$, that are Fourier transforms of the form

$$z(\lambda) = \sum_{u=-\infty}^{\infty} a(u) \exp\{-iu\lambda\} \tag{3.8.15}$$

with the $a(u)$ real-valued and satisfying

$$\sum_{u=-\infty}^{\infty} [1 + |u|^l]|a(u)| < \infty. \tag{3.8.16}$$

We then have the following:

Theorem 3.8.3 If $z_j(\lambda)$ belongs to $V(l)$, $j = 1, \ldots, n$ and $f(z_1, \ldots, z_n)$ is a real holomorphic function in a neighborhood of the range of values $\{z_1(\lambda), \ldots, z_n(\lambda);\ -\infty < \lambda < \infty\}$, then $f(z_1(\lambda), \ldots, z_n(\lambda))$ also belongs to $V(l)$.

This theorem again follows from the work of Gelfand et al (1964). Comparing this theorem with Theorem 3.8.2, we note that the required domain of regularity of $f(\cdot)$ is smaller here and its values are allowed to be more general.

As an application of this theorem: let $\{\mathbf{a}(u)\}$ $u = 0, \pm 1, \pm 2, \ldots$ be an $r \times r$ l summable filter with transfer function $\mathbf{A}(\lambda)$ satisfying Det $\mathbf{A}(\lambda) \neq 0$, $-\infty < \lambda < \infty$. Then there exists an l summable filter $\{\mathbf{b}(u)\} u = 0, \pm 1, \ldots$ with transfer function $\mathbf{B}(\lambda) = \mathbf{A}(\lambda)^{-1}$. Or with the same notation, there exists an l summable filter $\{\mathbf{c}(u)\} u = 0, \pm 1, \ldots$ with transfer function $\mathbf{C}(\lambda) = (\mathbf{A}(\lambda)\overline{\mathbf{A}(\lambda)})^{-1}$.

As an example of the joint use of Theorems 3.8.2 and 3.8.3 we mention the following result useful in the linear prediction of real-valued stationary series.

Theorem 3.8.4 Let $X(t)$, $t = 0, \pm 1, \ldots$ be a real-valued series with mean 0 and $\operatorname{cov}\{X(t+u), X(t)\} = c_{XX}(u)$, $t, u = 0, \pm 1, \ldots$. Suppose

$$\sum_{u=-\infty}^{\infty} [1 + |u|^l]|c_{XX}(u)| < \infty \qquad \text{for some } l \geqslant 0. \tag{3.8.17}$$

Suppose $f_{XX}(\lambda) \neq 0$, $-\infty < \lambda < \infty$. Then we may write

$$X(t) = \sum_{u=0}^{\infty} b(u)\, \varepsilon(t-u) \qquad \text{for } t = 0, \pm 1, \ldots \tag{3.8.18}$$

where the series

$$\varepsilon(t) = \sum_{u=0}^{\infty} a(u) X(t-u) \tag{3.8.19}$$

has mean 0 and autocovariance function $c_{\varepsilon\varepsilon}(u) = \delta\{u\}$. The coefficients satisfy

$$\Sigma\, [1 + |u|^l]|a(u)|,\ \Sigma\, [1 + |u|^l]|b(u)| < \infty. \tag{3.8.20}$$

The $\{a(u)\}$, $\{b(u)\}$ required here are determined somewhat indirectly. If

$$A(\lambda) = \sum_{u=0}^{\infty} a(u) \exp\{-i\lambda u\}$$

$$B(\lambda) = \sum_{u=0}^{\infty} b(u) \exp\{-i\lambda u\}, \tag{3.8.21}$$

then we see that it is necessary to have

$$B(\lambda) = A(\lambda)^{-1} \tag{3.8.22}$$

$$f_{XX}(\lambda) = |B(\lambda)|^2 = |A(\lambda)|^{-2} \tag{3.8.23}$$

$$\log f_{XX}(\lambda) = \log B(\lambda) + \overline{\log B(\lambda)}. \tag{3.8.24}$$

As (3.8.17) holds and $f_{XX}(\lambda)$ does not vanish, we may write

$$\log f_{XX}(\lambda) = \sum_{u=-\infty}^{\infty} g(u) \exp\{-i\lambda u\} \tag{3.8.25}$$

with

$$g(u) = (2\pi)^{-1} \int_{-\pi}^{\pi} \exp\{i\lambda u\} \log f_{XX}(\lambda) d\lambda \tag{3.8.26}$$

and

$$\sum_{u=-\infty}^{\infty} [1 + |u|^l]|g(u)| < \infty \tag{3.8.27}$$

following Theorem 3.8.3. Expression (3.8.24) suggests defining

$$B(\lambda) = \exp\left\{\tfrac{1}{2}g(0) + \sum_{u=1}^{\infty} g(u) \exp\{-i\lambda u\}\right\}. \tag{3.8.28}$$

The corresponding $\{a(u)\}$, $\{b(u)\}$ satisfy expression (3.8.20) following Theorem 3.8.2.

Theorems 3.8.2 and 3.8.3 have previously been used in a time series context in Hannan (1963). Arens and Calderon (1955) and Gelfand et al (1964) are general references to the theorems. Baxter (1963) develops an inequality, using these procedures, that may be useful in bounding the error of finite approximations to certain Fourier transforms.

3.9 SPECTRAL REPRESENTATIONS IN THE FUNCTIONAL APPROACH TO TIME SERIES

In Section 2.7 we saw that the effect of linear time invariant operations on a time series $\mathbf{X}(t)$, $t = 0, \pm 1, \ldots$ was easily illustrated if the series could be written as a sum of cosinusoids, that is, if for example

$$\mathbf{X}(t) = \sum_j \exp\{i\lambda_j t\}\mathbf{z}(j), \tag{3.9.1}$$

the $\mathbf{z}(j)$ being r vectors. In this section we consider representations of a series $\mathbf{X}(t)$ that have the nature of expression (3.9.1), but apply to a broader class of time series: such representations will be called **spectral representations.** They have the general form

$$\mathbf{X}(t) \backsim \int_{-\pi}^{\pi} \exp\{i\lambda t\} d\mathbf{Z}_X(\lambda) \qquad t = 0, \pm 1, \ldots \tag{3.9.2}$$

for some r vector-valued $\mathbf{Z}_X(\lambda)$. We begin with

Theorem 3.9.1 Let $\mathbf{X}(t)$, $t = 0, \pm 1, \ldots$ be an r vector-valued function such that

$$\lim_{S\to\infty} (2S+1)^{-1} \sum_{s=-S}^{S} \mathbf{X}(t+u+s)\mathbf{X}(t+s)^\tau = \mathbf{m}_{XX}(u) \tag{3.9.3}$$

exists for $t, u = 0, \pm 1, \ldots$. Then the following limit exists,

$$\mathbf{G}_{XX}(\lambda) = \lim_{U\to\infty} (2\pi)^{-1} \sum_{u=-U}^{U} \mathbf{m}_{XX}(u)[\exp\{-iu\lambda\} - 1]/(-iu) \qquad -\pi < \lambda \leqslant \pi. \tag{3.9.4}$$

Also there exists an r vector-valued $\mathbf{Z}_X(\lambda;s)$, $-\pi < \lambda \leqslant \pi$, $s = 0, \pm 1, \ldots$ such that

$$\mathbf{X}(t+s) \backsim \int_{-\pi}^{\pi} \exp\{i\lambda t\} d\mathbf{Z}_X(\lambda;s) \qquad s, t = 0, \pm 1, \ldots \tag{3.9.5}$$

in the sense that

$$\lim_{S\to\infty} (2S+1)^{-1} \sum_{s=-S}^{S} \left\|\mathbf{X}(t+s) - \int_{-\pi}^{\pi} \exp\{i\lambda t\} d\mathbf{Z}_X(\lambda;s)\right\|^2 = 0 \qquad t = 0, \pm 1, \ldots. \tag{3.9.6}$$

$\mathbf{Z}_X(\lambda;s)$ also satisfies

$$\lim_{T\to\infty} \lim_{S\to\infty} (2S+1)^{-1} \sum_{s=-S}^{S} \left\|\mathbf{Z}_X(\lambda;s) - (2\pi)^{-1} \sum_{t=-T}^{T} \mathbf{X}(t+s)[\exp\{-it\lambda\} - 1]/(-it)\right\|^2 = 0 \qquad -\pi < \lambda \leqslant \pi \tag{3.9.7}$$

and

$$\lim_{S\to\infty} (2S+1)^{-1} \sum_{s=-S}^{S} \mathbf{Z}_X(\lambda;s)\overline{\mathbf{Z}_X(\mu;s)}^{\tau} = \mathbf{G}_{XX}(\min\{\lambda, \mu\})$$

$$0 \leqslant \lambda, \mu \leqslant \pi. \quad (3.9.8)$$

The matrix $\mathbf{G}_{XX}(\lambda)$ of (3.9.4) may be seen to be bounded, non-negative definite, nondecreasing as a function of λ, $0 \leqslant \lambda \leqslant \pi$, and such that $\mathbf{G}_{XX}(-\lambda) = \mathbf{G}_{XX}(\lambda)^{\tau}$. Exercise 2.13.31 indicates a related result.

Expression (3.9.5) provides a representation for $\mathbf{X}(t+s)$ as a sum of cosinusoids of differing phases and amplitudes. Suppose that $\{\mathbf{a}(u)\}$, $u = 0, \pm 1, \ldots$ is a filter whose coefficients vanish for sufficiently large $|u|$. Let $\mathbf{A}(\lambda)$ denote the transfer function of this filter. Then if we set

$$\mathbf{Y}(t) = \sum_u \mathbf{a}(t-u)\mathbf{X}(u) \qquad t = 0, \pm 1, \ldots, \quad (3.9.9)$$

we see that the filtered series has the representation

$$\mathbf{Y}(t+s) \backsim \int_{-\pi}^{\pi} \exp\{i\lambda t\}\mathbf{A}(\lambda)d\mathbf{Z}_X(\lambda;s) \qquad s, t = 0, \pm 1, \ldots. \quad (3.9.10)$$

The cosinusoids making up $\mathbf{X}(t+s)$ have become multiplied by the transfer function of the filter.

A version of Theorem 3.9.1 is given in Bass (1962a,b); however, the theorem itself follows from a representation theorem of Wold (1948).

An alternate form of spectral representation was given by Wiener (1930) and a discrete vector-valued version of his result is provided by

Theorem 3.9.2 Let $\mathbf{X}(t)$, $t = 0, \pm 1, \ldots$ be an r vector-valued function such that

$$\sum_{t=-\infty}^{\infty} (1+t^2)^{-1} \|\mathbf{X}(t)\|^2 < \infty \quad (3.9.11)$$

then there exists an r vector-valued $\mathbf{Z}_X(\lambda)$, $-\pi < \lambda \leqslant \pi$, with $\mathbf{Z}_X(\pi) - \mathbf{Z}_X(-\pi) = \mathbf{X}(0)$, such that

$$\mathbf{X}(t) \backsim \int_{-\pi}^{\pi} \exp\{i\lambda t\} d\mathbf{Z}_X(\lambda) \qquad t = 0, \pm 1, \ldots. \quad (3.9.12)$$

Expression (3.9.12) holds in the sense of the formal integration by parts

$$\mathbf{X}(t) = e^{i\pi t}\mathbf{Z}_X(\pi) - e^{-i\pi t}\mathbf{Z}_X(-\pi) + it\int_{-\pi}^{\pi} \exp\{i\lambda t\}\mathbf{Z}_X(\lambda)d\lambda. \quad (3.9.13)$$

The function $\mathbf{Z}_X(\lambda)$ satisfies

$$\lim_{T\to\infty} \int_{-\pi}^{\pi} \|\mathbf{Z}_X(\lambda) - \frac{\lambda}{2\pi}\mathbf{X}(0) - \frac{1}{2\pi}\sum_{1\leq|t|\leq T} \mathbf{X}(t)\exp\{-i\lambda t\}/(-it)\|^2 d\lambda = 0. \tag{3.9.14}$$

If $\mathbf{X}(t)$ also satisfies expression (3.9.3) and $\mathbf{G}_{XX}(\lambda)$ is given by (3.9.4), then

$$\lim_{\varepsilon\to 0} \frac{1}{2\varepsilon}\int_0^{\lambda} [\mathbf{Z}_X(\alpha+\varepsilon) - \mathbf{Z}_X(\alpha-\varepsilon)]\overline{[\mathbf{Z}_X(\alpha+\varepsilon) - \mathbf{Z}_X(\alpha-\varepsilon)]}^{\tau} d\alpha = \mathbf{G}_{XX}(\lambda) \tag{3.9.15}$$

at points of continuity of $\mathbf{G}_{XX}(\lambda)$, $0 \leqslant \lambda \leqslant \pi$.

A theorem of Wiener (1933), p.138, applies to show that expression (3.9.11) holds if

$$\lim_{T\to\infty} \sup\, (2T+1)^{-1} \sum_{t=-T}^{T} \|\mathbf{X}(t)\|^2 < \infty. \tag{3.9.16}$$

Expression (3.9.12) may clearly be used to illustrate the effect of linear filters on the series $\mathbf{X}(t)$.

Yet another means of obtaining a spectral representation for a fixed series $\mathbf{X}(t)$, $t = 0, \pm 1, \ldots$ is to make use of the theory of Schwartz distributions; see Schwartz (1957, 1959) and Edwards (1967) Chap. 12. We will obtain a spectral representation for a stochastic series in Section 4.6. Bertrandias (1960, 1961) also considers the case of fixed series as does Heninger (1970).

3.10 EXERCISES

3.10.1 Suppose $A(\lambda) = 1$ for $|\lambda \pm \omega| < \Delta$ with Δ small and $A(\lambda) = 0$ otherwise for $-\pi < \lambda < \pi$. Show that

$$\begin{aligned} a(u) &= 2\Delta/\pi & u &= 0 \\ &= 2\cos\omega u \sin\Delta u/(\pi u) & u &\neq 0. \end{aligned}$$

3.10.2 Let $A(\lambda)$ denote the transfer function of a filter. Show that the filter leaves polynomials of degree k invariant if and only if $A(0) = 1$, $A^{(j)}(0) = 0$, $1 \leqslant j \leqslant k$. (Here $A^{(j)}(\lambda)$ denotes the jth derivative.) See Schoenberg (1946) and Brillinger (1965a).

3.10.3 If

$$H(\alpha) = (2\pi)^{-1}\int_{-1}^{1} h(x)\exp\{-i\alpha x\}dx$$

with $|H(\alpha)| \leqslant K(1 + |\alpha|)^{-2}$, show that $H^{(n)}(\lambda)$ of (3.3.6) is given by

$$n\sum_{j=-\infty}^{\infty} H(n[\lambda - 2\pi j]) \qquad \text{for } -\infty < \lambda < \infty.$$

3.10.4 If $\mathfrak{F}$ denotes the matrix with $\exp\{-i2\pi(j-1)(k-1)/T\}$ in row j, column k, $1 \leqslant j, k \leqslant T$, show that $\mathfrak{F}\bar{\mathfrak{F}}^\tau = T\mathbf{I}$ and $\mathfrak{F}^4 = T^2\mathbf{I}$.

3.10.5 If $D_n(\lambda)$ is given by expression (3.2.6), prove that $(2n+1)^{-1}D_n(\lambda)$ tends to $\eta\{\lambda\}/2\pi$ as $n \to \infty$.

3.10.6 Prove that expression (3.4.14) tends to $2\pi(-i)^k \dfrac{d^k\eta(\lambda)}{d\lambda^k}$ as $n \to \infty$.

3.10.7 Let $c_X{}^{(T)}$, $c_Y{}^{(T)}$ denote the means of the values $X(t)$, $Y(t)$, $t = 0, \ldots, T-1$. Show that

$$\sum_{\substack{t \\ 0 \leq t+u \leq T-1}} [X(t+u) - c_X{}^{(T)}][Y(t) - c_Y{}^{(T)}]$$

is given by

$$S^{-1} \sum_{s=1}^{S-1} d_X{}^{(T)}\left(\frac{2\pi s}{S}\right) \overline{d_Y{}^{(T)}\left(\frac{2\pi s}{S}\right)} \exp\{i2\pi su/S\}$$

for $u = 0, \pm 1, \ldots, \pm(S-T)$.

3.10.8 Let $\hat{X}_j(t)$, $t = 0, \pm 1, \ldots$ denote the period T extension of $X_j(t)$, $t = 0, \ldots, T-1$ for $j = 1, \ldots, r$. Show that the expression

$$\sum_{t=0}^{T-1} \hat{X}_1(t+u_1)\cdots\hat{X}_{r-1}(t+u_{r-1})\hat{X}_r(t)$$
$$= T^{-r+1} \sum_{s_1=0}^{T-1} \cdots \sum_{s_{r-1}=0}^{T-1} d_1{}^{(T)}\left(\frac{2\pi s_1}{T}\right) \cdots$$
$$d_{r-1}^{(T)}\left(\frac{2\pi s_{r-1}}{T}\right) \overline{d_r{}^{(T)}\left(\frac{2\pi[s_1 + \cdots + s_{r-1}]}{T}\right)} \exp\left\{i\frac{2\pi[s_1u_1 + \cdots + s_{r-1}u_{r-1}]}{T}\right\}$$

for $u_1, \ldots, u_{r-1} = 0, \pm 1, \ldots$.

3.10.9 Let

$$\mathbf{Y}(t) = \sum_{u=0}^{T-1} \mathbf{a}(t-u)\hat{\mathbf{X}}(u) \qquad \text{for } t = 0, \pm 1, \ldots.$$

Show that $\mathbf{d}_Y{}^{(T)}(\lambda) = \mathbf{A}(\lambda)\mathbf{d}_X{}^{(T)}(\lambda)$, $-\infty < \lambda < \infty$.

3.10.10 Let $n^{(T)}(u_1, \ldots, u_{r-1})$ denote expression (3.6.10). Show that

$$\sum_{u_1=-T+1}^{T-1} \cdots \sum_{u_{r-1}=-T+1}^{T-1} n^{(T)}(u_1, \ldots, u_{r-1}) = \left[\sum_{t=0}^{T-1} X_1(t)\right] \cdots \left[\sum_{t=0}^{T-1} X_r(t)\right]$$
$$= T^{-r+1} \sum_{s_1=0}^{T-1} \cdots \sum_{s_{r-1}=0}^{T-1} d_1{}^{(T)}\left(\frac{2\pi s_1}{T}\right) \cdots$$
$$d_{r-1}^{(T)}\left(\frac{2\pi s_{r-1}}{T}\right) \overline{d_r{}^{(T)}\left(\frac{2\pi[s_1 + \cdots + s_{r-1}]}{T}\right)}.$$

3.10.11 If $\mathbf{W} = \mathbf{Z}^{-1}$, show that

$$\operatorname{Re} \mathbf{W} = \{\operatorname{Re} \mathbf{Z} + (\operatorname{Im} \mathbf{Z})(\operatorname{Re} \mathbf{Z})^{-1}(\operatorname{Im} \mathbf{Z})\}^{-1}$$
$$\operatorname{Im} \mathbf{W} = -(\operatorname{Re} \mathbf{W})(\operatorname{Im} \mathbf{Z})(\operatorname{Re} \mathbf{Z})^{-1}.$$

3.10.12 Let $\mathfrak{F}$ denote the matrix of Exercise 3.10.4. Show that its latent values are $T^{1/2}$, $-iT^{1/2}$, $-T^{1/2}$, $iT^{1/2}$ with multiplicities $[T/4]+1$, $[(T+1)/4]$, $[(T+2)/4]$, $[(T+3)/4]-1$ respectively. (Here $[N]$ denotes the integral part of N.) See Lewis (1939).

3.10.13 If the Hermitian matrix $\mathbf{Z}$ has latent values $\mu_1, \ldots, \mu_r$ and corresponding latent vectors $\mathbf{U}_1, \ldots, \mathbf{U}_r$, prove that the matrix $\mathbf{Z} - \mu_1 \mathbf{U}_1 \overline{\mathbf{U}}_1^\tau$ has latent values $0, \mu_2, \ldots, \mu_r$ and latent vectors $\mathbf{U}_1, \ldots, \mathbf{U}_r$. Show how this result may be used to reduce the calculations required in determining the latent values and vectors of $\mathbf{Z}$ from those of $\mathbf{Z}^R$.

3.10.14 Use the inequality (3.7.16) to prove the following theorem; let

$$f(\lambda) = \sum_{u=-\infty}^{\infty} c(u) \exp\{i\lambda u\}$$

$-\infty < \lambda < \infty$ with $\sum_u |u|\ |c(u)|^2 < \infty$. Let $\mathbf{C}^{(T)} = [c(j-k)]$, $j, k = 1, \ldots, T$. If $F[.]$ is a function with a uniformly bounded derivative on the range of $f(\lambda)$, $-\infty < \lambda < \infty$, then

$$\lim_{T\to\infty} \frac{F[\mu_1(\mathbf{C}^{(T)})] + \cdots + F[\mu_T(\mathbf{C}^{(T)})]}{T} = (2\pi)^{-1} \int_0^{2\pi} F[f(\lambda)]d\lambda.$$

Theorems of this sort are given in Grenander and Szegö (1958).

3.10.15 A $(Tr) \times (Tr)$ matrix $\mathbf{Z}$ is said to be a **block circulant** if it is made up of $r \times r$ matrices $\mathbf{Z}_{jk} = \mathbf{z}(k-j)$ for some $r \times r$ matrix-valued function $\mathbf{z}(.)$ of period T. Prove that the latent values of $\mathbf{Z}$ are given by the latent values of

$$\sum_{j=0}^{T-1} \mathbf{z}(j) \exp\{-i2\pi jk/T\} \qquad k = 0, \ldots, T-1 \qquad (*)$$

and the corresponding latent vectors by

$$[\exp\{-i2\pi jk/T\}\mathbf{u}_{lk}; j = 0, \ldots, T-1]$$
$$k = 0, \ldots, T-1,\ l = 0, \ldots, r-1$$

where $\mathbf{u}_{lk}$ are the latent vectors of (*); see Friedman (1961). Indicate how this result may be used to determine the inverse of a block circulant matrix.

3.10.16 Let $\mathbf{Z}$ be a $J \times J$ Hermitian matrix. Show that

$$\mu_j(\mathbf{Z}) = \inf_{\mathbf{D}} \sup_{\mathbf{Dx}=0} \frac{\bar{\mathbf{x}}^\tau \mathbf{Z} \mathbf{x}}{\bar{\mathbf{x}}^\tau \mathbf{x}}$$

for $\mathbf{x}$ a J vector and $\mathbf{D}$ a matrix of rank $\leqslant j-1$ that has J rows. This is the Courant-Fischer Theorem and may be found in Bellman (1960).

3.10.17 If the pure noise series $\varepsilon(t)$, $t = 0, \pm 1, \ldots$ has moments of all orders, prove that the autoregressive scheme (3.8.11) has, with probability 1, a solution $\mathbf{X}(t)$ satisfying Assumption 2.6.1 provided that the polynomial (3.8.10) has no roots in the unit disc.

3.10.18 If $\mathbf{A}$, $\mathbf{B}$ are $r \times r$ complex matrices and $\mathbf{F}(\mathbf{B};\mathbf{A}) = \mathbf{BA}$, prove that the determinant of the Jacobian $\partial\mathbf{F}/\partial\mathbf{B}$ is given by $(\text{Det } \mathbf{A})^r$; see Deemer and Olkin (1951) and Khatri (1965a).

3.10.19 Let $\mathbf{Z}$ be an $r \times r$ complex matrix with distinct latent values μ_j, $j = 1, \ldots, r$. Prove that the μ_j are holomorphic functions of the entries of $\mathbf{Z}$. *Hint:* Note that the μ_j are the solutions of the equation $\text{Det}(\mathbf{Z} - \mu\mathbf{I}) = 0$ and use Theorem 3.8.1; see Portmann (1960).

3.10.20 Let $\mathbf{Z}_0$ be an $r \times r$ complex matrix with distinct latent values. Show that there exists a nonsingular $\mathbf{Q}$ whose entries are holomorphic functions of the entries of $\mathbf{Z}$ for all $\mathbf{Z}$ in a neighborhood of $\mathbf{Z}_0$ and such that $\mathbf{Q}^{-1}\mathbf{Z}\mathbf{Q}$ is a diagonal matrix in the neighborhood; see Portmann (1960).

3.10.21 If $\mathbf{Z}_0$, $\mathbf{Z}$ of Exercise 3.10.20 are Hermitian, then the columns of $\mathbf{Q}$ are orthogonal. Conclude that a unitary matrix, $\mathbf{U}$, whose entries are real holomorphic functions of the entries of $\mathbf{Z}$ may be determined so that $\bar{\mathbf{U}}^{\tau}\mathbf{Z}\mathbf{U}$ is a diagonal matrix.

3.10.22 If $\{\mathbf{a}(u)\}$, $u = 0, \pm 1, \ldots$ is an $r \times r$ realizable filter and $\{\mathbf{b}(u)\}$ its inverse exists, prove that the $\mathbf{b}(u)$, $u = 0, 1, \ldots$ are given by: $\mathbf{a}(0)\mathbf{b}(0) = \mathbf{I}$, $\mathbf{a}(0)\mathbf{b}(1) + \mathbf{a}(1)\mathbf{b}(0) = \mathbf{0}$, $\mathbf{a}(0)\mathbf{b}(2) + \mathbf{a}(1)\mathbf{b}(1) + \mathbf{a}(2)\mathbf{b}(0) = \mathbf{0}, \ldots$.

3.10.23 Prove Exercise 2.13.22 using the results of Section 3.8.

3.10.24 Let $\rho(S)$ be a monotonically increasing function such that $\lim_{S\to\infty} \rho(S+1)/\rho(S) = 1$. Let $X(t)$, $t = 0, \pm 1, \ldots$ be a function such that

$$\lim_{S\to\infty} [\rho(S)]^{-1} \sum_{s=-S}^{S} X(t+u+s)X(t+s)$$

exists for $t, u = 0, \pm 1, \ldots$. Indicate the form that Theorem 3.9.1 takes for such an $X(t)$.

3.10.25 Adopt the notation of Theorem 3.9.1. If the moments $m_{a_1 \ldots a_k}(u_1, \ldots, u_{k-1})$ of expression (2.11.9) exist and are given by the Fourier-Stieltjes transforms of the functions $M_{a_1 \ldots a_k}(\lambda_1, \ldots, \lambda_{k-1})$, $-\pi < \lambda_j \leqslant \pi$, prove that

$$\lim_{S\to\infty} (2S+1)^{-1} \sum_{s=-S}^{S} dZ_{a_1}(\lambda_1;s) \cdots dZ_{a_k}(\lambda_k;s)$$
$$= \eta(\lambda_1 + \cdots + \lambda_k)dM_{a_1 \ldots a_k}(\lambda_1, \ldots, \lambda_{k-1})d\lambda_k.$$

3.10.26 Let $\mathbf{Z}$ be a $J \times J$ Hermitian matrix with ordered latent vectors $\mathbf{x}_1, \ldots, \mathbf{x}_J$. Show that

$$\mu_j(\mathbf{Z}) = \max_{\mathbf{x}} \frac{\bar{\mathbf{x}}^{\tau}\mathbf{Z}\mathbf{x}}{\bar{\mathbf{x}}^{\tau}\mathbf{x}}$$

where the maximum is over $\mathbf{x}$ orthogonal to $\mathbf{x}_1, \ldots, \mathbf{x}_{j-1}$. Equality occurs for $\mathbf{x} = \mathbf{x}_j$.

3.10.27 Let **A** be an $r \times r$ Hermitian matrix with latent roots and vectors μ_j, $\mathbf{V}_j, j = 1, \ldots, r$. Given ϕ mapping the real line into itself, the $r \times r$ matrix-valued function $\phi(\mathbf{A})$ is defined by

$$\phi(\mathbf{A}) = \sum_j \phi(\mu_j)\mathbf{X}_j\overline{\mathbf{X}}_j^{\tau}.$$

Show that $\phi(\mathbf{A})^R = \phi(\mathbf{A}^R)$.

3.10.28 Show that there exist constants K, L such that

$$\int_0^{2\pi} |\Delta^{(T)}(\alpha)|d\alpha < K \log T, \quad \int_0^{2\pi} |\Delta^{(T)}(\alpha)|^p d\alpha < LT^{p-1}$$

for $p, T > 1$ where $\Delta^{(T)}(\alpha) = \sum_{t=0}^{T-1} \exp\{-i\alpha t\}$.

3.10.29 Suppose the conditions of Theorem 3.3.1 are satisfied and in addition the Pth derivative of $A(\alpha)$ is continuous at $\alpha = \lambda$. Show that the last expression of (3.3.18) may be replaced by

$$A(\lambda) + \sum_{p=1}^{P} \frac{1}{p!} n^{-p}\{\textstyle\int \alpha^p H(\alpha)d\alpha\} \frac{d^p A(\lambda)}{d\lambda^p} + o(n^{-P}).$$

3.10.30 Let real-valued data $X(t)$, $Y(t)$, $t = 0, \ldots, T-1$ be given. Set $Z(t) = X(t) + iY(t)$. Show that

$$\begin{aligned}
\operatorname{Re} d_X^{(T)}(\lambda) &= \{\operatorname{Re} d_Z^{(T)}(\lambda) + \operatorname{Re} d_Z^{(T)}(-\lambda)\}/2 \\
\operatorname{Im} d_X^{(T)}(\lambda) &= \{\operatorname{Im} d_Z^{(T)}(\lambda) - \operatorname{Im} d_Z^{(T)}(-\lambda)\}/2 \\
\operatorname{Re} d_Y^{(T)}(\lambda) &= \{\operatorname{Im} d_Z^{(T)}(\lambda) + \operatorname{Im} d_Z^{(T)}(-\lambda)\}/2 \\
\operatorname{Im} d_Y^{(T)}(\lambda) &= \{-\operatorname{Re} d_Z^{(T)}(\lambda) + \operatorname{Re} d_Z^{(T)}(-\lambda)\}/2.
\end{aligned}$$

This exercise indicates how the Fourier transforms of two real-valued sets of data may be found with one application of a Fourier transform to a complex-valued set of data; Bingham et al (1967).

3.10.31 Prove that for S an integer

$$\sum_{v=-\infty}^{\infty} a(Sv) = S^{-1} \sum_{s=0}^{S-1} A(2\pi s/S)$$

when $a(u)$, $A(\lambda)$ are related as in (3.2.2).

3.10.32 If $\boldsymbol{\alpha}$ is an r vector and **Z** is an $r \times r$ Hermitian matrix, show that

$$\overline{\boldsymbol{\alpha}}^{\tau}\mathbf{A}\boldsymbol{\alpha} = \boldsymbol{\beta}^{\tau}\mathbf{A}^R\boldsymbol{\beta}$$

where

$$\boldsymbol{\beta} = \begin{bmatrix} \operatorname{Re} \boldsymbol{\alpha} \\ \operatorname{Im} \boldsymbol{\alpha} \end{bmatrix}.$$

3.10.33 With the notation of Corollary 3.7.2, set $\mathbf{M}^+ = \operatorname{diag}\{\mu_j{}^+, j = 1, \ldots, J\}$ where $\mu^+ = 1/\mu$ if $\mu \neq 0$, $\mu^+ = 0$ if $\mu = 0$. Then the $K \times J$ matrix $\mathbf{Z}^+ = \mathbf{V}\mathbf{M}^+\overline{\mathbf{U}}^\tau$ is called the **generalized inverse** of $\mathbf{Z}$. Show that
(a) $\mathbf{Z}\mathbf{Z}^+\mathbf{Z} = \mathbf{Z}$
(b) $\mathbf{Z}^+\mathbf{Z}\mathbf{Z}^+ = \mathbf{Z}^+$
(c) $\overline{(\mathbf{Z}\mathbf{Z}^+)}^\tau = \mathbf{Z}\mathbf{Z}^+$
(d) $\overline{(\mathbf{Z}^+\mathbf{Z})}^\tau = \mathbf{Z}^+\mathbf{Z}$.

3.10.34 Show for $S \geqslant T$ that

(a) $d_X{}^{(T)}(\lambda) = S^{-1} \sum_{s=0}^{S-1} d_X{}^{(T)}\left(\frac{2\pi s}{S}\right)\Delta^{(T)}\left(\lambda - \frac{2\pi s}{S}\right)$

(b) $d_X{}^{(T)}(\lambda) = (2\pi)^{-1} \int_0^{2\pi} d_X{}^{(T)}(\alpha)\Delta^{(T)}(\lambda - \alpha)d\alpha$.

3.10.35 If $A^{(n)}(\lambda)$ is given by expression (3.2.4), show for $m \geqslant n$ that

(a) $A^{(n)}(\lambda) = (2m + 1)^{-1}2\pi \sum_{s=0}^{2m-1} A^{(n)}\left(\frac{2\pi s}{2m + 1}\right)D_n\left(\lambda - \frac{2\pi s}{2m + 1}\right)$

(b) $A^{(n)}(\lambda) = \int_0^{2\pi} A^{(n)}(\alpha)D_n(\lambda - \alpha)d\alpha$.

3.10.36 Use the singular value decomposition to show that a $J \times K$ matrix $\mathbf{A}$ of rank L may be written $\mathbf{A} = \mathbf{B}\mathbf{C}$, where $\mathbf{B}$ is $J \times L$ and $\mathbf{C}$ is $L \times K$.

3.10.37 Let $\mathbf{Z}_0$ be an $r \times r$ matrix with Det $\mathbf{Z}_0 \neq 0$. Show that the entries of $\mathbf{Z}^{-1}$ are holomorphic functions of $\mathbf{Z}$ in a neighborhood of $\mathbf{Z}_0$.

4

STOCHASTIC PROPERTIES OF FINITE FOURIER TRANSFORMS

4.1 INTRODUCTION

Consider an r vector-valued sequence $\mathbf{X}(t)$, $t = 0, \pm 1, \ldots$. In the previous chapter we considered various properties of the finite Fourier transform

$$\mathbf{d}_X^{(T)}(\lambda) = \sum_{t=0}^{T-1} \mathbf{X}(t) \exp\{-i\lambda t\} \qquad -\infty < \lambda < \infty \tag{4.1.1}$$

in the case that $\mathbf{X}(t)$ was a fixed, nonstochastic function. In this chapter we present a variety of properties of $\mathbf{d}_X^{(T)}(\lambda)$, if $\mathbf{X}(t)$, $t = 0, \pm 1, \ldots$ is a stationary time series. We will also consider asymptotic distributions, probability 1 bounds, behavior under convolution as well as develop the Cramér representation of $\mathbf{X}(t)$.

In previous chapters we have seen that Fourier transforms possess a wealth of valuable mathematical properties. For example, in Chapter 3 we saw that the discrete Fourier transform has the important numerical property of being rapidly computable by the Fast Fourier Transform Algorithm, while in this chapter we will see that it has useful and elementary statistical properties. For all of the reasons previously given, the Fourier transform is an obvious entity on which to base an analysis of a time series of interest.

However, before developing stochastic properties of the transform (4.1.1) we first define two types of complex-valued random variables. These variables will prove important in our development of the distributions of various time series statistics.

4.2 THE COMPLEX NORMAL DISTRIBUTION

If $\mathbf{X}$ is an r vector-valued random variable having real-valued components and having a multivariate normal distribution with mean $\boldsymbol{\mu}_X$ and covariance matrix $\boldsymbol{\Sigma}_{XX}$, write: $\mathbf{X}$ is $N_r(\boldsymbol{\mu}_X,\boldsymbol{\Sigma}_{XX})$. Throughout this text we will often have to consider r vector-valued random variables $\mathbf{X}$ whose individual components are complex-valued. If, for such an $\mathbf{X}$, the $2r$ vector-valued variate with real components

$$\begin{bmatrix} \operatorname{Re} \mathbf{X} \\ \operatorname{Im} \mathbf{X} \end{bmatrix} \tag{4.2.1}$$

is distributed as

$$N_{2r}\left(\begin{bmatrix} \operatorname{Re} \boldsymbol{\mu}_X \\ \operatorname{Im} \boldsymbol{\mu}_X \end{bmatrix}; \tfrac{1}{2}\begin{bmatrix} \operatorname{Re} \boldsymbol{\Sigma}_{XX} & -\operatorname{Im} \boldsymbol{\Sigma}_{XX} \\ \operatorname{Im} \boldsymbol{\Sigma}_{XX} & \operatorname{Re} \boldsymbol{\Sigma}_{XX} \end{bmatrix}\right) \tag{4.2.2}$$

for some r vector $\boldsymbol{\mu}_X$ and $r \times r$ Hermitian non-negative definite $\boldsymbol{\Sigma}_{XX}$, we will write: $\mathbf{X}$ is $N_r^C(\boldsymbol{\mu}_X,\boldsymbol{\Sigma}_{XX})$. Then $\mathbf{X}$ is **complex multivariate normal** with mean $\boldsymbol{\mu}_X$ and covariance matrix $\boldsymbol{\Sigma}_{XX}$, which leads us to

$$E\mathbf{X} = \boldsymbol{\mu}_X, \tag{4.2.3}$$

$$E\{(\mathbf{X} - \boldsymbol{\mu}_X)\overline{(\mathbf{X} - \boldsymbol{\mu}_X)}^\tau\} = \boldsymbol{\Sigma}_{XX}, \tag{4.2.4}$$

and

$$E\{(\mathbf{X} - \boldsymbol{\mu}_X)(\mathbf{X} - \boldsymbol{\mu}_X)^\tau\} = \mathbf{0}. \tag{4.2.5}$$

We remark that within the class of complex vector-valued random variables whose real and imaginary parts have a joint multivariate normal distribution, the complex multivariate normals have the property that if (4.2.4) is diagonal, then the components of $\mathbf{X}$ are statistically independent; see Exercise 4.8.1. Various properties of the complex multivariate normal are given in Wooding (1956), Goodman (1963), James (1964), and in Exercises 4.8.1 to 4.8.3. We mention the properties: if $\boldsymbol{\Sigma}_{XX}$ is nonsingular, then the probability element of $\mathbf{X}$ is given by

$$\pi^{-r}(\operatorname{Det} \boldsymbol{\Sigma}_{XX})^{-1} \exp\{-\overline{(\mathbf{X} - \boldsymbol{\mu}_X)}^\tau\boldsymbol{\Sigma}_{XX}^{-1}(\mathbf{X} - \boldsymbol{\mu}_X)\} \prod_j (d \operatorname{Re} X_j)(d \operatorname{Im} X_j) \tag{4.2.6}$$

for $-\infty < \operatorname{Re} X_j, \operatorname{Im} X_j < \infty$. And in the case $r = 1$ if X is $N_1^C(\mu_X,\sigma_{XX})$, then $\operatorname{Re} X$ and $\operatorname{Im} X$ are independent $N_1(\operatorname{Re} \mu_X,\sigma_{XX}/2)$ and $N_1(\operatorname{Im} \mu_X,\sigma_{XX}/2)$, respectively.

Turning to a different class of variates, suppose $\mathbf{X}_1, \ldots, \mathbf{X}_n$ are independent $N_r(\mathbf{0},\boldsymbol{\Sigma}_{XX})$ variates. Then the $r \times r$ matrix-valued random variable

$$\mathbf{W} = \sum_{j=1}^{n} \mathbf{X}_j\mathbf{X}_j^\tau \tag{4.2.7}$$

is said to have a **Wishart distribution** of dimension r and degrees of freedom n. We write: $\mathbf{W}$ is $W_r(n,\mathbf{\Sigma}_{XX})$. If on the other hand $\mathbf{X}_1, \ldots, \mathbf{X}_n$ are independent $N_r^C(\mathbf{0},\mathbf{\Sigma}_{XX})$ variates, then the $r \times r$ matrix-valued random variable

$$\mathbf{W} = \sum_{j=1}^{n} \mathbf{X}_j \overline{\mathbf{X}}_j^{\tau} \tag{4.2.8}$$

is said to have a **complex Wishart distribution** of dimension r and degrees of freedom n. In this case we write: $\mathbf{X}$ is $W_r^C(n,\mathbf{\Sigma}_{XX})$. The complex Wishart distribution was introduced in Goodman (1963). Various of its properties are given in Exercises 4.8.4 to 4.8.8 and in Srivastava (1965), Gupta (1965), Kabe (1966, 1968), Saxena (1969), and Miller (1968, 1969). Its density function may be seen to be given by

$$\left[\pi^{r(r-1)/2} \prod_{j=1}^{n} \Gamma(n-j+1)\right]^{-1} (\text{Det } \mathbf{\Sigma}_{XX})^{-n} (\text{Det } \mathbf{W})^{n-r} \times \exp\{-\text{tr } \mathbf{\Sigma}_{XX}^{-1}\mathbf{W}\} \tag{4.2.9}$$

for $n \geqslant r$ and $\mathbf{W} \geqslant \mathbf{0}$. Other properties include:

$$\overline{\mathbf{W}} = \mathbf{W}^{\tau}, \tag{4.2.10}$$

$$E\mathbf{W} = n\mathbf{\Sigma}_{XX}, \tag{4.2.11}$$

and

$$\begin{aligned} \text{cov}\{W_{jk}, W_{lm}\} &= E\{(W_{jk} - n\Sigma_{jk})\overline{(W_{lm} - n\Sigma_{lm})}\} \\ &= n\,\Sigma_{jl}\,\overline{\Sigma}_{km}. \end{aligned} \tag{4.2.12}$$

The complex Wishart distribution will be useful in the development of approximations to the distributions of estimates of spectral density matrices.

In later sections of this text, we will require the concept of a sequence of variates being **asymptotically normal.** We will say that the r vector-valued sequence ζ_T, $T = 1, 2, \ldots$ is asymptotically $N_r(\mathbf{y}_T,\mathbf{\Sigma}_T)$ if the sequence $\mathbf{\Sigma}_T^{-1/2}(\zeta_T - \mathbf{y}_T)$ tends, in distribution, to $N_r(\mathbf{0},\mathbf{I})$. We will also say that the r vector-valued sequence ζ_T, $T = 1, 2, \ldots$ is asymptotically $N_r^C(\mathbf{y}_T,\mathbf{\Sigma}_T)$ if the sequence $\mathbf{\Sigma}_T^{-1/2}(\zeta_T - \mathbf{y}_T)$ tends, in distribution, to $N_r^C(\mathbf{0},\mathbf{I})$.

4.3 STOCHASTIC PROPERTIES OF THE FINITE FOURIER TRANSFORM

Consider the r vector-valued stationary series $\mathbf{X}(t)$, $t = 0, \pm 1, \ldots$. In this section we will develop asymptotic expressions for the cumulants of the finite Fourier transform of an observed stretch of the series. In Section 3.3 we saw that certain benefits could result from the insertion of convergence factors into the direct definition of the finite Fourier transform. Now let us begin by inserting convergence factors here and then deducing the results for the simple Fourier transform as a particular case. We begin with

Assumption 4.3.1 $h(u), -\infty < u < \infty$ is bounded, is of bounded variation, and vanishes for $|u| > 1$.

Suppose $h_a(u)$ satisfies this assumption for $a = 1, \ldots, r$. The finite Fourier transform we consider is defined by

$$\mathbf{d}_X^{(T)}(\lambda) = [\sum_t h_a(t/T)X_a(t)\exp\{-i\lambda t\}]$$
$$= [d_a^{(T)}(\lambda)] \qquad \text{for } -\infty < \lambda < \infty, a = 1, \ldots, r. \quad (4.3.1)$$

In the present context we will refer to the function $h_a(t/T)$ as a **taper** or **data window.** The transform involves at most the values $\mathbf{X}(t)$, $t = 0, \pm 1, \ldots, \pm(T-1)$ of the series. If $h_a(u) = 0$ for $u < 0$, then it involves only the values $\mathbf{X}(t)$, $t = 0, \ldots, T-1$. This means that the asymptotic results we develop apply to either one-sided or two-sided statistics. If a segment of the series is missing, within the time period of observation, then the data available may be handled directly by taking $h(t/T)$ to vanish throughout the missing segment. If the component series are observed over different time intervals, this is handled by having the $h_a(t/T)$ nonzero over different time intervals.

Set

$$H_{a_1 \ldots a_k}^{(T)}(\lambda) = \sum_t \left[\prod_1^k h_{a_j}(t/T)\right] \exp\{-i\lambda t\}$$
$$\text{for } -\infty < \lambda < \infty \text{ and } a_1, \ldots, a_k = 1, \ldots, r. \quad (4.3.2)$$

If

$$H_{a_1 \ldots a_k}(\lambda) = \int \left[\prod_1^k h_{a_j}(t)\right] \exp\{-i\lambda t\}\,dt \quad (4.3.3)$$

and if it is possible to apply the Poisson summation formula (Edwards (1967) p. 173), then we may write

$$H_{a_1 \ldots a_k}^{(T)}(\lambda) = T \sum_{l=-\infty}^{\infty} H_{a_1 \ldots a_k}(T[\lambda + 2\pi l]). \quad (4.3.4)$$

The discussion of convergence factors in Section 3.3 suggests that $H_{a_1 \ldots a_k}(\lambda)$ will have substantial magnitude only for λ near 0. This implies that the function (4.3.2) will have substantial magnitude only for λ near some multiple of 2π.

We repeat the definition

$$c_{a_1\ldots a_k}(u_1, \ldots, u_{k-1}) = \operatorname{cum}\{X_{a_1}(t+u_1), \ldots, X_{a_{k-1}}(t+u_{k-1}), X_{a_k}(t)\}, \tag{4.3.5}$$

and if

$$\sum_{u_1} \cdots \sum_{u_{k-1}} |c_{a_1\ldots a_k}(u_1, \ldots, u_{k-1})| < \infty, \tag{4.3.6}$$

we also repeat the definition

$$f_{a_1\ldots a_k}(\lambda_1, \ldots, \lambda_{k-1}) = (2\pi)^{-k+1} \sum_{u_1} \cdots \sum_{u_{k-1}} \exp\{-i(\lambda_1 u_1 + \cdots + \lambda_{k-1}u_{k-1})\} \times c_{a_1\ldots a_k}(u_1, \ldots, u_{k-1}). \tag{4.3.7}$$

Now we have

Theorem 4.3.1 Let $\mathbf{X}(t)$, $t = 0, \pm 1, \ldots$ be a stationary r vector-valued series satisfying (4.3.6). Suppose $h_a(u)$, $-\infty < u < \infty$, satisfies Assumption 4.3.1 for $a = 1, \ldots, r$. Then

$$\operatorname{cum}\{d_{a_1}^{(T)}(\lambda_1), \ldots, d_{a_k}^{(T)}(\lambda_k)\} = (2\pi)^{k-1} H_{a_1\ldots a_k}^{(T)}\left(\sum_1^k \lambda_j\right) f_{a_1\ldots a_k}(\lambda_1, \ldots, \lambda_{k-1}) + o(T). \tag{4.3.8}$$

The error term is uniform in $\lambda_1, \ldots, \lambda_k$.

If $\lambda_1 + \cdots + \lambda_k \equiv 0 \pmod{2\pi}$, then

$$\operatorname{cum}\{d_{a_1}^{(T)}(\lambda_1), \ldots, d_{a_k}^{(T)}(\lambda_k)\} \sim (2\pi)^{k-1} T H_{a_1\ldots a_k}(0) f_{a_1\ldots a_k}(\lambda_1, \ldots, \lambda_{k-1}). \tag{4.3.9}$$

If $\lambda_1 + \cdots + \lambda_k \not\equiv 0 \pmod{2\pi}$, then the cumulant will be of reduced order. Expression (4.3.9) suggests that we can base an estimate of the cumulant spectrum (4.3.7) on the $d_{a_1}^{(T)}(\lambda_1), \ldots, d_{a_k}^{(T)}(\lambda_k)$ with $\lambda_1 + \cdots + \lambda_k \equiv 0 \pmod{2\pi}$.

There are circumstances in which the error term of (4.3.8) is of smaller order of magnitude than $o(T)$. Suppose, in place of (4.3.6), we have

$$\sum_{u_1} \cdots \sum_{u_{k-1}} [1 + |u_j|] |c_{a_1\ldots a_k}(u_1, \ldots, u_{k-1})| < \infty \qquad j = 1, \ldots, k-1, \tag{4.3.10}$$

then we can prove

Theorem 4.3.2 Let $X(t)$, $t = 0, \pm 1, \ldots$ be a stationary r vector-valued series satisfying (4.3.10). Suppose $h_a(u)$, $-\infty < u < \infty$, satisfies Assumption 4.3.1 for $a = 1, \ldots, r$. Then

$$\operatorname{cum}\{d_{a_1}^{(T)}(\lambda_1), \ldots, d_{a_k}^{(T)}(\lambda_k)\}$$
$$= (2\pi)^{k-1} H_{a_1 \ldots a_k}^{(T)}\left(\sum_1^k \lambda_j\right) f_{a_1 \ldots a_k}(\lambda_1, \ldots, \lambda_{k-1}) + O(1). \quad (4.3.11)$$

The error term is uniform in $\lambda_1, \ldots, \lambda_k$.

Qualitatively the results of Theorem 4.3.1 are the same as those of Theorem 4.3.2. However, this theorem suggests to us that decreasing the span of dependence of series, as is the effect of expression (4.3.10) over (4.3.6), reduces the size of the asymptotic error term. Exercise 4.8.14 indicates that the error term may be further reduced by choosing the $h_a(u)$ to have Fourier transforms rapidly falling off to 0 as $|\lambda|$ increases.

The convergence factor

$$\begin{aligned} h(u) &= 1 \qquad \text{for } 0 \leqslant u < 1 \\ &= 0 \qquad \text{otherwise} \end{aligned} \quad (4.3.12)$$

is of special interest. In this case the Fourier transform is

$$\mathbf{d}_X^{(T)}(\lambda) = \sum_{t=0}^{T-1} \mathbf{X}(t) \exp\{-i\lambda t\} \qquad \text{for } -\infty < \lambda < \infty. \quad (4.3.13)$$

Also, from expression (4.3.2)

$$\begin{aligned} H_{a_1 \ldots a_k}^{(T)}(\lambda) &= \sum_{t=0}^{T-1} \exp\{-i\lambda t\} \\ &= \Delta^{(T)}(\lambda). \end{aligned} \quad (4.3.14)$$

The function $\Delta^{(T)}(\lambda)$ has the properties: $\Delta^{(T)}(\lambda) = T$ for $\lambda \equiv 0 \pmod{2\pi}$, $\Delta^{(T)}(2\pi s/T) = 0$ for s an integer with $s \not\equiv 0 \pmod{T}$. Also $|\Delta^{(T)}(\lambda)| \leqslant 1/|\sin \frac{1}{2}\lambda|$ and so $\Delta^{(T)}(\lambda)$ is of reduced magnitude for λ not near a multiple of 2π. Expression (4.3.11) here takes the form

$$\operatorname{cum}\{d_{a_1}^{(T)}(\lambda_1), \ldots, d_{a_k}^{(T)}(\lambda_k)\}$$
$$= (2\pi)^{k-1} \Delta^{(T)}\left(\sum_1^k \lambda_j\right) f_{a_1 \ldots a_k}(\lambda_1, \ldots, \lambda_{k-1}) + O(1). \quad (4.3.15)$$

This joint cumulant has substantial magnitude for $\lambda_1 + \cdots + \lambda_k$ near some multiple of 2π. Note that the first term on the right side of expression (4.3.15) vanishes for $\lambda_j = 2\pi s_j/T$, s_j an integer, if $s_1 + \cdots + s_k \not\equiv 0 \pmod{T}$.

Expression (4.3.15) was developed in Brillinger and Rosenblatt (1967a); other references to this type of material include: Davis (1953), Root and Pitcher (1955), and Kawata (1960, 1966). Exercise 4.8.21 suggests that, on occasion, it may be more efficient to carry out the tapering through computations in the frequency domain.

4.4 ASYMPTOTIC DISTRIBUTION OF THE FINITE FOURIER TRANSFORM

In the previous section we developed asymptotic expressions for the joint cumulants of the finite Fourier transforms of a stationary time series. In this section we use these expressions to develop the limiting distribution of the transform. We set $\mathbf{c}_X = E\mathbf{X}(t)$ and have

Theorem 4.4.1 Let $\mathbf{X}(t)$, $t = 0, \pm 1, \ldots$ be an r vector-valued series satisfying Assumption 2.6.1. Let $s_j(T)$ be an integer with $\lambda_j(T) = 2\pi s_j(T)/T \to \lambda_j$ as $T \to \infty$ for $j = 1, \ldots, J$. Suppose $2\lambda_j(T), \lambda_j(T) \pm \lambda_k(T) \not\equiv 0 \pmod{2\pi}$ for $1 \leq j < k \leq J$. Let

$$\mathbf{d}_X^{(T)}(\lambda) = \sum_{t=0}^{T-1} \mathbf{X}(t) \exp\{-i\lambda t\} \qquad \text{for } -\infty < \lambda < \infty. \tag{4.4.1}$$

Then $\mathbf{d}_X^{(T)}(\lambda_j(T))$, $j = 1, \ldots, J$ are asymptotically independent $N_r^C(\mathbf{0}, 2\pi T\mathbf{f}_{XX}(\lambda_j))$ variates respectively. Also if $\lambda = 0, \pm 2\pi, \ldots$, $\mathbf{d}_X^{(T)}(\lambda)$ is asymptotically $N_r(T\mathbf{c}_X, 2\pi T\mathbf{f}_{XX}(\lambda))$ independently of the previous variates and if $\lambda = \pm\pi, \pm 3\pi, \ldots$, $\mathbf{d}_X^{(T)}(\lambda)$ is asymptotically $N_r(\mathbf{0}, 2\pi T\,\mathbf{f}_{XX}(\lambda))$ independently of the previous variates.

In the case $\lambda = 0$,

$$\mathbf{d}_X^{(T)}(0) = \sum_{t=0}^{T-1} \mathbf{X}(t) \tag{4.4.2}$$

and the theorem is seen to provide a central limit theorem for the series $\mathbf{X}(t)$. Other central limit theorems for stationary series are given in Rosenblatt (1956, 1961), Leonov and Shiryaev (1960), Iosifescu and Theodorescu (1969) p. 22, and Philipp (1969). The asymptotic normality of Fourier coefficients themselves is investigated in Kawata (1965, 1966).

If the conditions of the theorem are satisfied and $\lambda_j = \lambda$, $j = 1, \ldots, J$, then we see that the $\mathbf{d}_X^{(T)}(\lambda_j(T))$, $j = 1, \ldots, J$ are approximately a sample of size J from $N_r^C(\mathbf{0}, 2\pi T\mathbf{f}_{XX}(\lambda))$. This last remark will prove useful later in the development of estimates of $\mathbf{f}_{XX}(\lambda)$ and in the suggesting of approximate distributions for a variety of statistics of interest.

If the series $\mathbf{X}(t)$, $t = 0, \pm 1, \ldots$ is tapered prior to evaluating its finite Fourier transform, then an alternate form of central limit theorem is available. It is

Theorem 4.4.2 Let $\mathbf{X}(t)$, $t = 0, \pm 1, \ldots$ be an r vector-valued series satisfying Assumption 2.6.1. Suppose $2\lambda_j$, $\lambda_j \pm \lambda_k \not\equiv 0 \pmod{2\pi}$ for $1 \leqslant j < k \leqslant J$. Let

$$d_a^{(T)}(\lambda) = \sum_{t=0}^{T-1} h_a\left(\frac{t}{T}\right) X_a(t) \exp\{-i\lambda t\} \tag{4.4.3}$$

where $h_a(t)$ satisfies Assumption 4.3.1, $a = 1, \ldots, r$. Then the $\mathbf{d}_X^{(T)}(\lambda_j)$, $\lambda_j \not\equiv 0 \pmod{2\pi}$, $j = 1, \ldots, J$ are asymptotically independent $N_r^C(\mathbf{0}, 2\pi T[H_{ab}(0) f_{ab}(\lambda_j)])$ variates. Also if $\lambda = 0, \pm 2\pi, \ldots$, $\mathbf{d}_X^{(T)}(\lambda)$ is asymptotically $N_r(T[c_a H_a(0)], 2\pi T[H_{ab}(0) f_{ab}(\lambda)])$ independently of the previous variates and if $\lambda = \pm\pi, \pm 3\pi, \ldots$, $\mathbf{d}_X^{(T)}(\lambda)$ is asymptotically $N_r(\mathbf{0}, 2\pi T[H_{ab}(0) f_{ab}(\lambda)])$ independently of the previous variates.

If the same taper $h(t)$ is applied to each of the components of $\mathbf{X}(t)$, then we see that the asymptotic covariance matrix of $\mathbf{d}_X^{(T)}(\lambda)$ has the form

$$2\pi T \int h(t)^2 dt \; \mathbf{f}_{XX}(\lambda) \qquad \text{for } -\infty < \lambda < \infty. \tag{4.4.4}$$

Under additional regularity conditions on the $h_a(t)$, $a = 1, \ldots, r$, we can obtain a theorem pertaining to sequences $\lambda_j(T)$ of frequencies tending to limits λ_j, $j = 1, \ldots, J$; see Brillinger (1970) and Exercise 4.8.20. The corresponding $\mathbf{d}_X^{(T)}(\lambda_j(T))$ will be asymptotically independent provided the $\lambda_j(T)$, $\lambda_k(T)$ are not too near each other, (mod 2π), for $1 \leqslant j < k \leqslant J$. Exercise 4.8.23 gives the asymptotic behavior of Fourier transforms based on disjoint stretches of data.

Suppose that $X(t)$, $t = 0, \pm 1, \ldots$ is a real-valued stationary series whose power spectrum $f_{XX}(\lambda)$ is near constant, equal $\sigma^2/(2\pi)$ say, $-\infty < \lambda < \infty$. From Theorem 4.4.1 we might expect the values $d_X^{(T)}(2\pi s/T)$, $s = 1, \ldots, (T-1)/2$ to be approximately independent $N_1^C(0, T\sigma^2)$ variates and *a fortiori* the values Re $d_X^{(T)}(2\pi s/T)$, Im $d_X^{(T)}(2\pi s/T)$, $s = 1, \ldots, (T-1)/2$ to be approximately independent $N_1(0, T\sigma^2/2)$ variates. We turn to a partial empirical examination of this conclusion.

Consider the series $V(t)$, $t = 0, 1, \ldots$ of mean monthly temperatures in Vienna for the period 1780–1950; this series, partially plotted in Figure 1.1.1, has a strong yearly periodic component. In an attempt to obtain a series with near constant power spectrum, we have reduced this periodic component by subtracting from each monthly value the average of the values for the same month across the whole stretch of data. Specifically we have formed the series

$$X(j + 12k) = V(j + 12k) - \sum_k V(j + 12k)/\sum_k 1 \tag{4.4.5}$$

for $j = 0, \ldots, 11$ and $k = 0, 1, \ldots$. We then evaluated the Fourier transform $d_X^{(T)}(2\pi s/T)$, $s = 1, \ldots, (T - 1)/2$ taking $T = 2048 = 2^{11}$ so that the Fast Fourier Transform Algorithm could be used.

Figures 4.4.1 and 4.4.2 are normal probability plots of the values

$$\text{Re } d_X^{(T)}\left(\frac{2\pi s}{T}\right) \qquad s = 1, \ldots, 1000$$

$$\text{Im } d_X^{(T)}\left(\frac{2\pi s}{T}\right) \qquad s = 1, \ldots, 1000, \tag{4.4.6}$$

respectively. The construction of such plots is described in Chernoff and Lieberman (1954). The estimated power spectrum of this series, given in Section 7.8, falls off slowly as λ increases and is approximately constant. If each of the variates has the same marginal normal distribution, the values should lie near straight lines. The plots obtained are essentially straight lines, with slight tailing off at the ends, suggesting that the conclusions of Theorem 4.4.1 are reasonable, at least for this series of values.

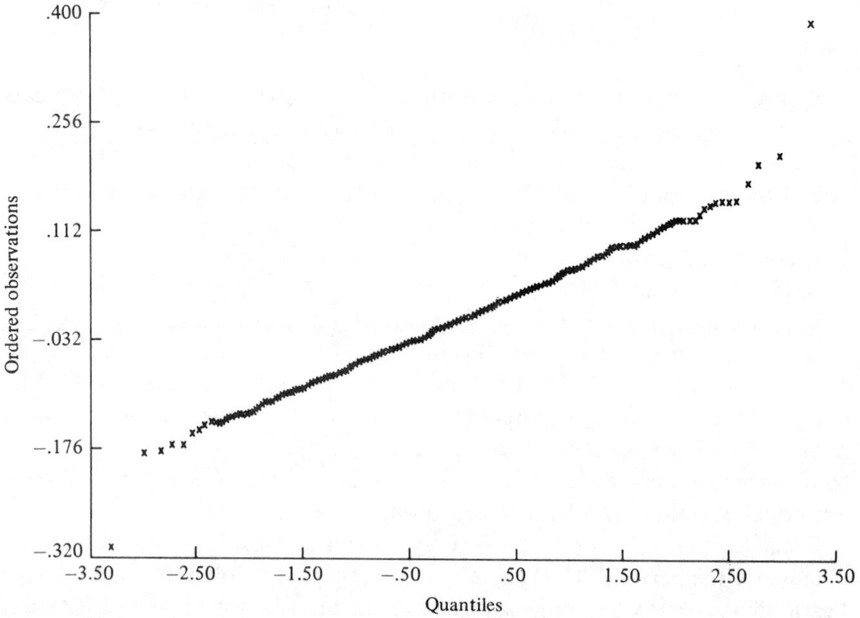

Figure 4.4.1 Normal probability plot of real part of discrete Fourier transform of seasonally adjusted Vienna mean monthly temperatures 1780–1950.

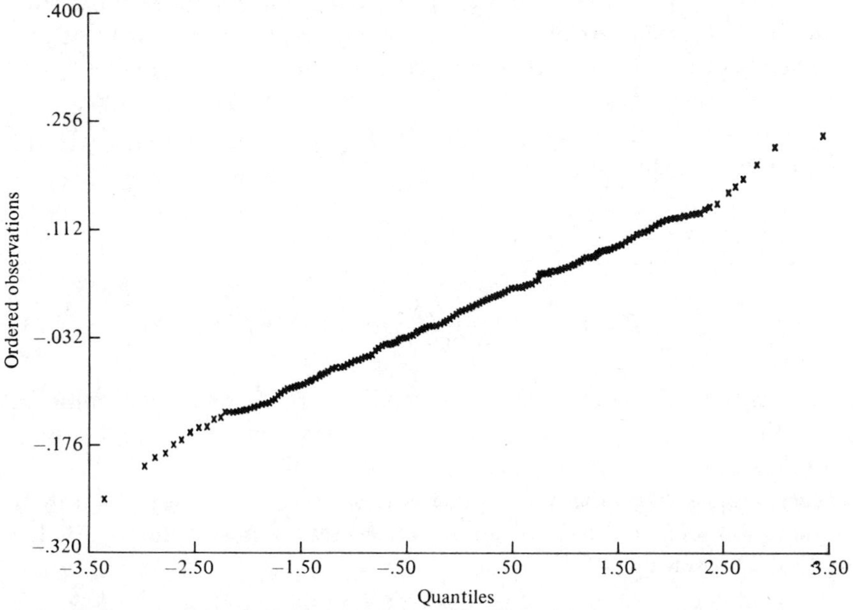

Figure 4.4.2 Normal probability plot of imaginary part of discrete Fourier transform of seasonally adjusted Vienna mean monthly temperatures 1780–1950.

The theorems in this section may provide a justification for the remark, often made in the communications theory literature, that the output of a narrow band-pass filter is approximately Gaussian; see Rosenblatt (1961). Consider the following narrow band-pass transfer function centered at λ_0

$$\begin{aligned} A(\lambda) &= 1 \qquad \text{for } |\lambda \pm \lambda_0| < \frac{\pi}{T} \qquad -\pi < \lambda < \pi \\ &= 0 \qquad \text{otherwise.} \end{aligned} \tag{4.4.7}$$

If a series $X(t)$, $t = 0, \pm 1, \ldots$ is taken as input to this filter, then expression (3.6.8), of the previous chapter, indicates that the output series of the filter will be approximately

$$T^{-1}\, d_X^{(T)}\left(\frac{2\pi s}{T}\right) \exp\left\{i\frac{2\pi st}{T}\right\} + T^{-1}\, d_X^{(T)}\left(-\frac{2\pi s}{T}\right) \exp\left\{-i\frac{2\pi st}{T}\right\}$$
$$t = 0, \pm 1, \ldots. \tag{4.4.8}$$

Here s is the integral part of $T\lambda_0/(2\pi)$ and so $2\pi s/T \doteq \lambda_0$. Theorem 4.4.1 now suggests that the variate (4.4.8) is asymptotically $N(0,4\pi T^{-1}f_{XX}(\lambda_0))$ in the case $\lambda_0 \not\equiv 0 \pmod{\pi}$. Related references to this result are Leonov and Shiryaev (1960), Picinbono (1959), and Rosenblatt (1956c).

Exercise 4.8.23 contains the useful result that finite Fourier transforms based on successive stretches of data are asymptotically independent and identically distributed in certain circumstances.

4.5 PROBABILITY 1 BOUNDS

It is sometimes useful to have a bound on the fluctuations of a finite Fourier transform

$$d_X^{(T)}(\lambda) = \sum_t h\left(\frac{t}{T}\right) X(t) \exp\{-i\lambda t\} \tag{4.5.1}$$

as a function of frequency λ and sample size T. In this connection we mention

Theorem 4.5.1 Let the real-valued series $X(t)$, $t = 0, \pm 1, \ldots$ satisfy Assumption 2.6.3 and have mean 0. Let $h(t)$ satisfy Assumption 4.3.1. Let $d_X^{(T)}(\lambda)$ be given by (4.5.1). Then

$$\overline{\lim_{T\to\infty}} \sup_\lambda |d_X^{(T)}(\lambda)|/(T \log T)^{1/2} \leq 2\{2\pi \int h(t)^2 dt \sup_\lambda f_{XX}(\lambda)\}^{1/2} \tag{4.5.2}$$

with probability 1.

This means that for $K > 1$, there is probability 1 that only finitely many of the events

$$\sup_\lambda |d_X^{(T)}(\lambda)| > K(T \log T)^{1/2}\, 2\{2\pi \int h(t)^2 dt \sup_\lambda f_{XX}(\lambda)\}^{1/2} \qquad T = 1, 2, \ldots \tag{4.5.3}$$

occur.

We see from expression (4.5.2) that under the indicated conditions, the Fourier transform has a rate of growth at most of order $(T \log T)^{1/2}$. If $X(t)$ is bounded by a constant, M say, then we have the elementary inequality

$$\sup_\lambda |d_X^{(T)}(\lambda)| \leq MT \tag{4.5.4}$$

giving a growth rate of order T. On the other hand, if we consider $|d_X^{(T)}(\lambda)|$ for a single frequency λ, we are in the realm of the law of the iterated logarithm; see Maruyama (1949), Parthasarathy (1960), Philipp (1967), Iosifescu (1968), and Iosifescu and Theodorescu (1969). This law leads to a rate of growth of order $(T \log \log T)^{1/2}$; other results of the nature of (4.5.2) are given in Salem and Zygmund (1956), Whittle (1959), and Kahane (1968).

An immediate implication of Theorem 4.5.1 is that, under the stated regularity conditions,

$$\sup_{\lambda} |d_X^{(T)}(\lambda)|/T \to 0 \tag{4.5.5}$$

with probability 1 as $T \to \infty$. In particular, taking $\lambda = 0$ we see

$$\sum_{t=0}^{T-1} X(t)/T \to 0 \tag{4.5.6}$$

with probability 1 as $T \to \infty$ — this last is the strong law of large numbers. Results similar to this are given in Wiener and Wintner (1941).

Turning to the development of a different class of asymptotic results, suppose the s vector-valued series $\mathbf{Y}(t)$, $t = 0, \pm 1, \ldots$ is a filtered version of $\mathbf{X}(t)$, say

$$\mathbf{Y}(t) = \sum_{u=-\infty}^{\infty} \mathbf{a}(t-u)\mathbf{X}(u) \tag{4.5.7}$$

for some $s \times r$ matrix-valued filter $\{\mathbf{a}(u)\}$. On occasion we will be interested in relating the finite Fourier transform of $\mathbf{Y}(t)$ to that of $\mathbf{X}(t)$. Lemma 3.4.1 indicates that if $\mathbf{X}(t)$, $t = 0, \pm 1, \ldots$ is bounded and

$$\sum_{u=-\infty}^{\infty} (1 + |u|)|\mathbf{a}(u)| < \infty \tag{4.5.8}$$

then there is a finite K such that

$$\sup_{\lambda} |\mathbf{d}_Y^{(T)}(\lambda) - \mathbf{A}(\lambda)\mathbf{d}_X^{(T)}(\lambda)| < K, \tag{4.5.9}$$

where

$$\begin{bmatrix} \mathbf{d}_X^{(T)}(\lambda) \\ \mathbf{d}_Y^{(T)}(\lambda) \end{bmatrix} = \sum_{t=0}^{T-1} \begin{bmatrix} \mathbf{X}(t) \\ \mathbf{Y}(t) \end{bmatrix} \exp\{-i\lambda t\}, \tag{4.5.10}$$

and

$$\mathbf{A}(\lambda) = \sum_{u=-\infty}^{\infty} \mathbf{a}(u) \exp\{-i\lambda u\}. \tag{4.5.11}$$

In the case that $\mathbf{X}(t)$, $t = 0, \pm 1, \ldots$ is an r vector-valued stochastic series we have

Theorem 4.5.2 Let the r vector-valued $\mathbf{X}(t)$, $t = 0, \pm 1, \ldots$ satisfy Assumption 2.6.3 and have mean $\mathbf{0}$. Let $\mathbf{Y}(t)$ be given by (4.5.7) where $\{\mathbf{a}(u)\}$ satisfies condition (4.5.8), then there is a finite L such that

$$\overline{\lim_{T\to\infty}} \sup_{\lambda} |\mathbf{d}_Y^{(T)}(\lambda) - \mathbf{A}(\lambda)\mathbf{d}_X^{(T)}(\lambda)|/(\log T)^{1/2} < L \tag{4.5.12}$$

with probability 1.

Expression (4.5.12) indicates a possible rate of growth for

$$\sup_{\lambda} |\mathbf{d}_Y^{(T)}(\lambda) - \mathbf{A}(\lambda)\mathbf{d}_X^{(T)}(\lambda)|$$

to be of order $(\log T)^{1/2}$. In Theorem 4.5.3 we will see that this rate of growth may be reduced to the order $T^{-1/2}(\log T)^{1/2}$ if the series are tapered prior to evaluating the Fourier transform.

Theorem 4.5.3 Let the r vector-valued $\mathbf{X}(t)$, $t = 0, \pm 1, \ldots$ satisfy Assumption 2.6.3 and have mean **0**. Let $\mathbf{Y}(t)$ be given by (4.5.7) where $\{\mathbf{a}(u)\}$ satisfies (4.5.8). Let

$$\begin{bmatrix} \mathbf{d}_X^{(T)}(\lambda) \\ \mathbf{d}_Y^{(T)}(\lambda) \end{bmatrix} = \sum_t h\left(\frac{t}{T}\right) \begin{bmatrix} \mathbf{X}(t) \\ \mathbf{Y}(t) \end{bmatrix} \exp\{-i\lambda t\} \tag{4.5.13}$$

where $h(u) = 0$ for $|u| \geqslant 1$ and has a uniformly bounded derivative. Then there is a finite L such that

$$\overline{\lim_{T\to\infty}} \sup_{\lambda} |\mathbf{d}_Y^{(T)}(\lambda) - \mathbf{A}(\lambda)\mathbf{d}_X^{(T)}(\lambda)| T^{1/2}/(\log T)^{1/2} < L \tag{4.5.14}$$

with probability 1.

In the case that $\mathbf{X}(t)$, $t = 0, \pm 1, \ldots$ is a series of independent variates, $\mathbf{Y}(t)$ given by (4.5.7) is a linear process. Expressions (4.5.12) and (4.5.14) suggest how we can learn about the sampling properties of the Fourier transform of a linear process from the sampling properties of the Fourier transform of a series of independent variates. This simplification was adopted by Bartlett (1966) Section 9.2.

On certain occasions it may be of interest to have a cruder bound on the growth of $\sup |d_X^{(T)}(\lambda)|$ when the series $X(t)$, $t = 0, \pm 1, \ldots$ satisfies the weaker Assumption 2.6.1.

Theorem 4.5.4 Let the real-valued series $X(t)$, $t = 0, \pm 1, \ldots$ satisfy Assumption 2.6.1 and have mean 0. Let $h(t)$ satisfy Assumption 4.3.1 and let $d_X^{(T)}(\lambda)$ be given by (4.5.1). Then for any $\varepsilon > 0$,

$$T^{-1/2-\varepsilon} \sup |d_X^{(T)}(\lambda)| \to 0 \tag{4.5.15}$$

with probability 1 as $T \to \infty$.

4.6 THE CRAMÉR REPRESENTATION

In Section 3.9 we developed two spectral representations for series involved in the functional approach to time series while in this section we indicate a spectral representation in the stochastic approach. The representation is due to Cramér (1942).

Suppose $\mathbf{X}(t)$, $t = 0, \pm 1, \ldots$ is an r vector-valued series. Consider the tapering function

$$h(u) = 1 \quad \text{for } |u| \leqslant 1$$
$$= 0 \quad \text{otherwise} \tag{4.6.1}$$

giving the finite Fourier transform

$$\mathbf{d}_X^{(T)}(\lambda) = \sum_{t=-T}^{T} \mathbf{X}(t) \exp\{-i\lambda t\}. \tag{4.6.2}$$

This transform will provide the basis for the representation. Set

$$2\pi\, \mathbf{Z}_X^{(T)}(\lambda) = \int_0^\lambda \mathbf{d}_X^{(T)}(\alpha) d\alpha. \tag{4.6.3}$$

We see

$$2\pi\, \mathbf{Z}_X^{(T)}(\lambda) = \sum_{t=-T}^{T} \mathbf{X}(t)[1 - \exp\{-i\lambda t\}]/(-it) \tag{4.6.4}$$

if we understand

$$[1 - \exp\{-i\lambda t\}]/(-it) = \lambda \qquad \text{for } t = 0. \tag{4.6.5}$$

Define

$$\eta(\lambda) = \sum_{j=-\infty}^{\infty} \delta(\lambda + 2\pi j) \tag{4.6.6}$$

to be the period 2π extension of the Dirac delta function. We may now state

Theorem 4.6.1 Let $\mathbf{X}(t)$, $t = 0, \pm 1, \ldots$ satisfy Assumption 2.6.1. Let $\mathbf{Z}_X^{(T)}(\lambda)$, $-\infty < \lambda < \infty$, be given by (4.6.4). Then there exists $\mathbf{Z}_X(\lambda)$, $-\infty < \lambda < \infty$, such that $\mathbf{Z}_X^{(T)}(\lambda)$ tends to $\mathbf{Z}_X(\lambda)$ in mean of order ν, for any $\nu > 0$. Also $\mathbf{Z}_X(\lambda + 2\pi) = \mathbf{Z}_X(\lambda)$, $\overline{\mathbf{Z}_X(\lambda)} = \mathbf{Z}_X(-\lambda)$ and

$$\operatorname{cum}\{Z_{a_1}(\lambda_1), \ldots, Z_{a_k}(\lambda_k)\}$$
$$= \int_0^{\lambda_1} \cdots \int_0^{\lambda_k} \eta\left(\sum_1^k \alpha_j\right) f_{a_1 \ldots a_k}(\alpha_1, \ldots, \alpha_{k-1}) d\alpha_1 \cdots d\alpha_k \tag{4.6.7}$$

for $a_1, \ldots, a_k = 1, \ldots, r$; $k = 2, 3, \ldots$.

We may rewrite (4.6.7) in differential notation as

$$\operatorname{cum}\{dZ_{a_1}(\lambda_1), \ldots, dZ_{a_k}(\lambda_k)\} = \eta\left(\sum_1^k \lambda_j\right) f_{a_1 \ldots a_k}(\lambda_1, \ldots, \lambda_{k-1}) d\lambda_1 \cdots d\lambda_k. \tag{4.6.8}$$

Expression (4.6.8) indicates that

$$\operatorname{cov}\{d\mathbf{Z}_X(\lambda), d\mathbf{Z}_X(\mu)\} = \eta(\lambda - \mu)\, \mathbf{f}_{XX}(\lambda) d\lambda d\mu \tag{4.6.9}$$

where $\mathbf{f}_{XX}(\lambda)$ denotes the spectral density matrix of the series $\mathbf{X}(t)$. The increments of $\mathbf{Z}_X(\lambda)$ are orthogonal unless $\lambda \equiv \mu \pmod{2\pi}$. Also joint cumulants of increments are negligible unless $\sum_1^k \lambda_j \equiv 0 \pmod{2\pi}$. The increments of $\mathbf{Z}_X(\lambda)$ mimic the behavior of $\mathbf{d}_X^{(T)}(\lambda)$ as given in Section 4.3.

In Theorem 4.6.2 we will need to consider a stochastic integral of the form

$$\int_0^{2\pi} \phi(\lambda) d\mathbf{Z}_X(\lambda). \tag{4.6.10}$$

If

$$\int_0^{2\pi} \phi(\lambda)\mathbf{f}_{XX}(\lambda)\overline{\phi(\lambda)}^\tau d\lambda < \infty, \tag{4.6.11}$$

this integral exists when defined as

$$\underset{N\to\infty}{\text{l.i.m}} \frac{2\pi}{N} \sum_{n=0}^{N-1} \phi\left(\frac{2\pi n}{N}\right)\left[\mathbf{Z}_X\left(\frac{2\pi(n+1)}{N}\right) - \mathbf{Z}_X\left(\frac{2\pi n}{N}\right)\right]. \tag{4.6.12}$$

See Cramér and Leadbetter (1967) Section 5.3. We may now state the **Cramér representation** of the series $\mathbf{X}(t)$, $t = 0, \pm 1, \ldots$.

Theorem 4.6.2 Under the conditions of Theorem 4.6.1

$$\mathbf{X}(t) = \int_0^{2\pi} \exp\{i\lambda t\} d\mathbf{Z}_X(\lambda) \qquad t = 0, \pm 1, \ldots \tag{4.6.13}$$

with probability 1, where $\mathbf{Z}_X(\lambda)$ satisfies the properties indicated in Theorem 4.6.1.

It is sometimes convenient to rewrite the representation (4.6.13) in a form involving variates with real-valued components. To this end set

$$\begin{aligned} \mathbf{U}_X(\lambda) &= \text{Re}\, \mathbf{Z}_X(\lambda) \\ \mathbf{V}_X(\lambda) &= -\text{Im}\, \mathbf{Z}_X(\lambda). \end{aligned} \tag{4.6.14}$$

These satisfy

$$\begin{aligned} \mathbf{U}_X(\lambda + 2\pi) &= \mathbf{U}_X(\lambda) \\ \mathbf{V}_X(\lambda + 2\pi) &= \mathbf{V}_X(\lambda), \end{aligned} \tag{4.6.15}$$

and

$$\begin{aligned} \mathbf{U}_X(-\lambda) &= \mathbf{U}_X(\lambda) \\ \mathbf{V}_X(-\lambda) &= -\mathbf{V}_X(\lambda). \end{aligned} \tag{4.6.16}$$

If we make the substitutions

$$\begin{aligned} \mathbf{U}_X(\lambda) &= [\mathbf{Z}_X(\lambda) + \mathbf{Z}_X(-\lambda)]/2 \\ \mathbf{V}_X(\mu) &= -[\mathbf{Z}_X(\mu) - \mathbf{Z}_X(-\mu)]/(2i), \end{aligned} \tag{4.6.17}$$

then from expression (4.6.8) we see that

$$\operatorname{cum}\{dU_{a_1}(\lambda_1), \ldots, dU_{a_k}(\lambda_k), dV_{b_1}(\mu_1), \ldots, dV_{b_l}(\mu_l)\}$$
$$= i^l 2^{-k-l} \sum_{\varepsilon_1} \cdots \sum_{\varepsilon_k} \sum_{\gamma_1} \cdots \sum_{\gamma_l} \eta(\varepsilon_1\lambda_1 + \cdots + \varepsilon_k\lambda_k + \gamma_1\mu_1 + \cdots + \gamma_l\mu_l)\gamma_1 \cdots \gamma_l$$
$$\times f_{a_1 \ldots a_k b_1 \ldots b_l}(\varepsilon_1\lambda_1, \ldots, \varepsilon_k\lambda_k, \gamma_1\mu_1, \ldots, \gamma_{l-1}\mu_{l-1}) d\lambda_1 \cdots d\lambda_k d\mu_1 \cdots d\mu_l \tag{4.6.18}$$

where the summations extend over $\varepsilon, \gamma = \pm 1$. In the case $k + \gamma = 2$, these relations give

$$\operatorname{cov}\{d\mathbf{U}_X(\lambda), d\mathbf{U}_X(\mu)\} = \tfrac{1}{2}\{\eta(\lambda - \mu) + \eta(\lambda + \mu)\} \operatorname{Re} \mathbf{f}_{XX}(\lambda) d\lambda d\mu, \tag{4.6.19}$$

$$\operatorname{cov}\{d\mathbf{U}_X(\lambda), d\mathbf{V}_X(\mu)\} = \tfrac{1}{2}\{\eta(\lambda - \mu) - \eta(\lambda + \mu)\} \operatorname{Im} \mathbf{f}_{XX}(\lambda) d\lambda d\mu, \tag{4.6.20}$$

$$\operatorname{cov}\{d\mathbf{V}_X(\lambda), d\mathbf{V}_X(\mu)\} = \tfrac{1}{2}\{\eta(\lambda - \mu) - \eta(\lambda + \mu)\} \operatorname{Re} \mathbf{f}_{XX}(\lambda) d\lambda d\mu. \tag{4.6.21}$$

The Cramér representation (4.6.13) takes the form

$$\mathbf{X}(t) = \int_0^{2\pi} [\cos \lambda t \, d\mathbf{U}_X(\lambda) + \sin \lambda t \, d\mathbf{V}_X(\lambda)] \qquad t = 0, \pm 1, \ldots \tag{4.6.22}$$

in these new terms.

The Cramér representation is especially useful for indicating the effect of operations on series of interest. For example, consider the filtered series

$$\mathbf{Y}(t) = \sum_u \mathbf{a}(t - u)\mathbf{X}(u) \qquad t = 0, \pm 1, \ldots \tag{4.6.23}$$

where the series $\mathbf{X}(t)$ has Cramér representation (4.6.13). If

$$\mathbf{A}(\lambda) = \sum_u \mathbf{a}(u) \exp\{-i\lambda u\} \qquad -\infty < \lambda < \infty, \tag{4.6.24}$$

with

$$\int_0^{2\pi} \mathbf{A}(\lambda)\mathbf{f}_{XX}(\lambda)\overline{\mathbf{A}(\lambda)}^\tau d\lambda < \infty, \tag{4.6.25}$$

then

$$\mathbf{Y}(t) = \int_0^{2\pi} \exp\{i\lambda t\}\mathbf{A}(\lambda) d\mathbf{Z}_X(\lambda) \qquad t = 0, \pm 1, \ldots. \tag{4.6.26}$$

In differential notation the latter may be written

$$d\mathbf{Z}_Y(\lambda) = \mathbf{A}(\lambda) d\mathbf{Z}_X(\lambda) \qquad -\infty < \lambda < \infty. \tag{4.6.27}$$

As an example of an application of (4.6.27) we remark that it, together with (4.6.9), gives the direct relation

$$\mathbf{f}_{YY}(\lambda) = \mathbf{A}(\lambda)\mathbf{f}_{XX}(\lambda)\overline{\mathbf{A}(\lambda)}^{\tau} \tag{4.6.28}$$

of Section 2.8.

Suppose the filter is a band-pass filter with transfer function, $-\pi < \lambda \leqslant \pi$,

$$\begin{aligned} A(\lambda) &= 1 \qquad \text{for } |\lambda \pm \omega| < \Delta \\ &= 0 \qquad \text{otherwise} \end{aligned} \tag{4.6.29}$$

applied to each coordinate of the series $\mathbf{X}(t)$, $t = 0, \pm 1, \ldots$.

Suppose, as we may, that the Cramér representation of $\mathbf{X}(t)$ is written

$$\mathbf{X}(t) = \int_{-\pi}^{\pi} \exp\{i\lambda t\} d\mathbf{Z}_X(\lambda). \tag{4.6.30}$$

Then the band-pass filtered series may be written

$$\begin{aligned} \mathbf{Y}(t) &= \left\{ \int_{-\omega-\Delta}^{-\omega+\Delta} + \int_{\omega-\Delta}^{\omega+\Delta} \right\} \exp\{i\lambda t\} d\mathbf{Z}_X(\lambda) \\ &\doteq \exp\{i\omega t\} d\mathbf{Z}_X(\omega) + \exp\{-i\omega t\} d\mathbf{Z}_X(-\omega) \\ &\doteq 2[\cos \omega t \, d\mathbf{U}_X(\omega) + \sin \omega t \, d\mathbf{V}_X(\omega)] \end{aligned} \tag{4.6.31}$$

for small Δ. The effect of band-pass filtering is seen to be the lifting, from the Cramér representation, of cosinusoids of frequency near $\pm\omega$. For small Δ this series, $\mathbf{Y}(t)$, is sometimes called **the component of frequency** ω of $\mathbf{X}(t)$ and is denoted by $\mathbf{X}(t,\omega)$, suppressing the dependence on Δ. By considering a bank of exhaustive and mutually exclusive band-pass filters with transfer functions such as

$$\begin{aligned} A_j(\lambda) &= 1 \qquad \text{for } |\lambda \pm 2j\Delta| < \Delta \\ &= 0 \qquad \text{otherwise} \end{aligned} \tag{4.6.32}$$

$j = 0, 1, \ldots, J$ where $(2J + 1)\Delta = \pi$, we see that a series $\mathbf{X}(t)$, $t = 0, \pm 1, \ldots$ may be thought of as the sum of its individual frequency components,

$$\mathbf{X}(t) = \sum_{j=0}^{J} \mathbf{X}(t,2j\Delta) \qquad t = 0, \pm 1, \ldots. \tag{4.6.33}$$

We will see, later in this work, that many useful statistical procedures have the character of elementary procedures applied to the separate frequency components of a series of interest.

Let us next consider the effect of forming the Hilbert transform of each component of $\mathbf{X}(t)$, $t = 0, \pm 1, \ldots$. The transfer function of the Hilbert transform is

$$A(\lambda) = -i \operatorname{sgn} \lambda \qquad \text{for } -\pi < \lambda \leqslant \pi. \tag{4.6.34}$$

If we write the Cramér representation of $\mathbf{X}(t)$ in the form

$$\mathbf{X}(t) = \int_{-\pi}^{\pi} [\cos \lambda t \, d\mathbf{U}_X(\lambda) + \sin \lambda t \, d\mathbf{V}_X(\lambda)], \tag{4.6.35}$$

then we quickly see

$$\mathbf{X}(t)^H = \int_{-\pi}^{\pi} [\sin \lambda t \, d\mathbf{U}_X(\lambda) - \cos \lambda t \, d\mathbf{V}_X(\lambda)]. \tag{4.6.36}$$

The cosinusoids of the representation have been shifted through a phase angle of $\pi/2$. In the case of $\mathbf{X}(t,\omega)$, the component of frequency ω in the series $\mathbf{X}(t)$, we see from (4.6.36) that

$$\mathbf{X}(t,\omega)^H \doteq 2[\sin \omega t \, d\mathbf{U}_X(\omega) - \cos \omega t \, d\mathbf{V}_X(\omega)] \tag{4.6.37}$$

and so

$$\mathbf{X}(t,\omega) + i\mathbf{X}(t,\omega)^H \doteq 2 \exp\{i\omega t\} d\mathbf{Z}_X(\omega), \tag{4.6.38}$$

for example. Function (4.6.38) provides us with another interpretation of the differential $d\mathbf{Z}_X(\omega)$, appearing in the Cramér representation.

Next let us consider the covariance matrix of the $2r$ vector-valued series

$$\begin{bmatrix} \mathbf{X}(t,\omega) \\ \mathbf{X}(t,\omega)^H \end{bmatrix}. \tag{4.6.39}$$

Elementary calculations show that it is given by

$$\begin{bmatrix} \operatorname{Re} \mathbf{f}_{XX}(\omega) & \operatorname{Im} \mathbf{f}_{XX}(\omega) \\ -\operatorname{Im} \mathbf{f}_{XX}(\omega) & \operatorname{Re} \mathbf{f}_{XX}(\omega) \end{bmatrix} 4\Delta = \mathbf{f}_{XX}(\omega)^R 4\Delta \tag{4.6.40}$$

in the case $\omega \not\equiv 0 \pmod{\pi}$ and by

$$\begin{bmatrix} \operatorname{Re} \mathbf{f}_{XX}(\omega) & \operatorname{Im} \mathbf{f}_{XX}(\omega) \\ -\operatorname{Im} \mathbf{f}_{XX}(\omega) & \operatorname{Re} \mathbf{f}_{XX}(\omega) \end{bmatrix} 2\Delta = \mathbf{f}_{XX}(\omega)^R 2\Delta \tag{4.6.41}$$

in the case $\omega \equiv 0 \pmod{\pi}$. These results provide us with a useful interpretation of the real and imaginary parts of the spectral density matrix of a series of interest.

As another example of the use of the Cramér representation let us see what form a finite Fourier transform takes in terms of it. Suppose

$$\mathbf{d}_X^{(T)}(\lambda) = \sum_t h\left(\frac{t}{T}\right) \mathbf{X}(t) \exp\{-i\lambda t\} \tag{4.6.42}$$

for some tapering function $h(u)$. By direct substitution we see that

$$\mathbf{d}_X^{(T)}(\lambda) = \int_0^{2\pi} H^{(T)}(\lambda - \alpha) d\mathbf{Z}_X(\alpha) \tag{4.6.43}$$

where

$$H^{(T)}(\lambda) = \sum_t h\left(\frac{t}{T}\right) \exp\{-i\lambda t\}. \tag{4.6.44}$$

From what we have seen in previous discussions of tapering, for large values of T, the function $H^{(T)}(\lambda - \alpha)$ is concentrated in the neighborhood of $\lambda \equiv \alpha \pmod{2\pi}$. Therefore, from (4.6.43), for large values of T, $\mathbf{d}_X^{(T)}(\lambda)$ is essentially getting at $d\mathbf{Z}_X(\lambda)$. As a final remark we mention that (4.6.8) and (4.6.43) imply

$$\begin{aligned} \operatorname{cum}\{d_{a_1}{}^{(T)}(\lambda_1), \ldots, d_{a_k}{}^{(T)}(\lambda_k)\} \\ = \int_0^{2\pi} \cdots \int_0^{2\pi} H^{(T)}(\lambda_1 - \alpha_1) \cdots H^{(T)}(\lambda_k - \alpha_k)\eta(\alpha_1 + \cdots + \alpha_k) \\ \times f_{a_1 \ldots a_k}(\alpha_1, \ldots, \alpha_{k-1}) d\alpha_1 \cdots d\alpha_k \end{aligned} \tag{4.6.45}$$

exactly. The latter may usefully be compared with the asymptotic expression (4.3.8).

In fact Cramér (1942) developed the representation (4.6.13) under the conditions of Theorem 2.5.2. In this more general case, the function $\mathbf{Z}_X(\lambda)$ satisfies

$$\operatorname{cov}\{d\mathbf{Z}_X(\lambda), d\mathbf{Z}_X(\mu)\} = \eta(\lambda - \mu) d\mathbf{F}_{XX}(\lambda) d\mu \tag{4.6.46}$$

where $\mathbf{F}_{XX}(\lambda)$ denotes the $r \times r$ matrix-valued function whose existence was indicated in Theorem 2.5.2. The integral representation now holds in an integral in mean square sense only; the proof of Theorem 3.9.1 may be modified to provide this result.

4.7 PRINCIPAL COMPONENT ANALYSIS AND ITS RELATION TO THE CRAMÉR REPRESENTATION

Let $\mathbf{Y}$ be a J vector-valued random variable with covariance matrix $\mathbf{\Sigma}_{YY}$. If the components of $\mathbf{Y}$ are intercorrelated and $J > 2$ or 3, then it is often difficult to understand the essential statistical nature of $\mathbf{Y}$. Consider, therefore, the problem of obtaining a variate $\boldsymbol{\zeta}$ more elementary than $\mathbf{Y}$, yet containing most of the statistical information in $\mathbf{Y}$. We will require $\boldsymbol{\zeta}$ to have the form

$$\boldsymbol{\zeta} = \mathbf{A}\mathbf{Y} \tag{4.7.1}$$

for some $K \times J$ matrix $\mathbf{A}$ with $K < J$. And we will formalize the requirement that $\boldsymbol{\zeta}$ contains much of the statistical information in $\mathbf{Y}$ by requiring that it minimize

$$\min_{\mathbf{B},\mathbf{C}} E\{(\mathbf{Y} - \mathbf{B} - \mathbf{C}\boldsymbol{\zeta})^{\tau}(\mathbf{Y} - \mathbf{B} - \mathbf{C}\boldsymbol{\zeta})\} \qquad \mathbf{B}\ J \times 1,\ \mathbf{C}\ J \times K. \tag{4.7.2}$$

We have

Theorem 4.7.1 Let $\mathbf{Y}$ be a J vector-valued variate with covariance matrix $\boldsymbol{\Sigma}_{YY}$. The $K \times J$ matrix $\mathbf{A}$ that minimizes (4.7.2) for $\boldsymbol{\zeta}$ of the form (4.7.1) is given by

$$\mathbf{A} = \begin{bmatrix} \mathbf{U}_1^{\tau} \\ \cdot \\ \cdot \\ \cdot \\ \mathbf{U}_K^{\tau} \end{bmatrix} \tag{4.7.3}$$

where $\mathbf{U}_j$ is the jth latent vector of $\boldsymbol{\Sigma}_{YY}, j = 1, \ldots, J$. The minimum achieved is

$$\sum_{j>K} \mu_j \tag{4.7.4}$$

where μ_j is the jth latent root of $\boldsymbol{\Sigma}_{YY}$. The extremal $\mathbf{B}$, $\mathbf{C}$ are given by

$$\mathbf{C} = \mathbf{A}^{\tau}$$
$$\mathbf{B} = E\{\mathbf{Y}\} - \mathbf{C}\mathbf{A}E\{\mathbf{Y}\}. \tag{4.7.5}$$

The individual components of $\boldsymbol{\zeta}$ are called the **principal components** of $\mathbf{Y}$. They are seen to have the form $\zeta_j = \mathbf{U}_j^{\tau}\mathbf{Y}$ and to satisfy

$$\operatorname{var} \zeta_j = \mu_j$$
$$\operatorname{cov}\{\zeta_j, \zeta_k\} = 0 \qquad \text{if } j \neq k. \tag{4.7.6}$$

The variate $\boldsymbol{\zeta}$, therefore, has a more elementary statistical structure than $\mathbf{Y}$.

The theorem has led us to consider approximating the J vector-valued $\mathbf{Y}$ by

$$\mathbf{B} + [\mathbf{U}_1 \cdots \mathbf{U}_K]\boldsymbol{\zeta} \tag{4.7.7}$$

and its jth component by

$$B_j + \sum_k U_{jk}\zeta_k. \tag{4.7.8}$$

The error caused by replacing $\mathbf{Y}$ by (4.7.7) is seen from (4.7.4) to depend on the magnitude of the latent roots with $j > K$. If $K = J$ then the error is 0 and expression (4.7.8) is seen to provide a representation for $\mathbf{Y}$ in terms of uncorrelated variates ζ_j.

Principal components will be discussed in greater detail in Chapter 9. They were introduced by Hotelling (1933). Theorem 4.7.1 is essentially due to Kramer and Mathews (1956) and Rao (1964, 1965).

We now turn to the case in which the variate $\mathbf{Y}$ refers to a stretch of values $X(t)$, $t = -T, \ldots, T$ of some real-valued stationary time series. In this case

$$\mathbf{Y} = \begin{bmatrix} X(-T) \\ \cdot \\ \cdot \\ \cdot \\ X(0) \\ \cdot \\ \cdot \\ \cdot \\ X(T) \end{bmatrix}. \tag{4.7.9}$$

Suppose $X(t)$, $t = 0, \pm 1, \ldots$ has autocovariance function $c_{XX}(u)$, $u = 0, \pm 1, \ldots$, so

$$\mathbf{\Sigma}_{YY} = \begin{bmatrix} c_{XX}(0) & c_{XX}(1) & \cdots & c_{XX}(2T) \\ c_{XX}(-1) & \cdot & & \cdot \\ \cdot & & & \cdot \\ \cdot & & & \cdot \\ \cdot & & & \cdot \\ c_{XX}(-2T) & \cdot & \cdots & c_{XX}(0) \end{bmatrix}. \tag{4.7.10}$$

Following the discussion above, the principal components of the variate (4.7.9) will be based upon the latent vectors of the matrix (4.7.10). This matrix is finite Toeplitz and so, from Section 3.7, its latent values and vectors are approximately

$$\sum_{t=-2T}^{2T} c_{XX}(t) \exp\{i2\pi st\}/(2T+1) \tag{4.7.11}$$

and

$$(2T+1)^{-1/2}[\exp\{-i2\pi st/(2T+1)\}; t = -T, \ldots, T] \tag{4.7.12}$$

respectively $s = -T, \ldots, T$. The principal components of (4.7.9) are therefore approximately

$$(2T+1)^{-1/2} \sum_{t=-T}^{T} \exp\{-i2\pi st/(2T+1)\} X(t) \qquad s = -T, \ldots, T. \tag{4.7.13}$$

If we refer back to expression (4.6.2), then we see that (4.7.13) is $d_X^{(T)}(2\pi s/(2T+1))$, the finite Fourier transform on which the Cramér representation was based and which we have proposed be taken as a basic statistic in computations with an observed stretch of series.

Following Theorem 4.7.1 we are led to approximate $X(t)$ by

$$(2T+1)^{-1}\sum_s d_X^{(T)}(2\pi s/(2T+1))\exp\{i2\pi st/(2T+1)\} \qquad t=-T,\ldots,T \tag{4.7.14}$$

where the summation in expression (4.7.14) is over s corresponding to the K largest values of (4.7.11). If we take $K = 2T + 1$ and let $T \to \infty$, then we would expect the value (4.7.14) to be very near $X(t)$. In fact, (4.7.14) tends to

$$\int_0^{2\pi} \exp\{i\lambda t\}dZ_X(\lambda), \tag{4.7.15}$$

if

$$(2\pi)^{-1}\int_0^{\lambda} d_X^{(T)}(\alpha)d\alpha \to Z_X(\lambda) \tag{4.7.16}$$

in some sense. Expression (4.7.15) is seen to be the Cramér representation of $X(t)$. The Cramér representation therefore results from a limit of a principal component analysis of $X(t)$, $t = 0, \pm 1, \ldots$.

Craddock (1965) carried out an empirical principal component analysis of a covariance matrix resulting from a stretch of time series values. The principal components he obtained have the cosinusoidal nature of (4.7.13).

The collection of latent values of a matrix is sometimes referred to as its spectrum, which in the case of matrix (4.7.10) are seen, from (4.7.11), to equal approximately $2\pi f_{XX}(2\pi s/(2T + 1))$, $s = -T, \ldots, T$, where $f_{XX}(\lambda)$ is the power spectrum of the series $X(t)$, $t = 0, \pm 1, \ldots$. We have therefore been led to an immediate relation between two different sorts of spectra.

4.8 EXERCISES

4.8.1 Let $\mathbf{Y} = \mathbf{U} + i\mathbf{V}$ where $\mathbf{U}$ and $\mathbf{V}$ are jointly multivariate normal. Let $E\mathbf{Y} = \boldsymbol{\mu}$ and $E(\mathbf{Y} - \boldsymbol{\mu})(\mathbf{Y} - \boldsymbol{\mu})^\tau = \mathbf{0}$. Prove that the individual components of $\mathbf{Y}$ are statistically independent if $E(\mathbf{Y} - \boldsymbol{\mu})\overline{(\mathbf{Y} - \boldsymbol{\mu})}^\tau$ is diagonal.

4.8.2 If $\mathbf{Y}$ is $N_r^C(\mathbf{0},\boldsymbol{\Sigma})$, prove that $\mathbf{AY}$ is $N_s^C(\mathbf{0},\mathbf{A}\boldsymbol{\Sigma}\overline{\mathbf{A}}^\tau)$ for any $s \times r$ matrix $\mathbf{A}$. Conclude that if the entries of $\mathbf{Y}$ are independent $N_1^C(0,\sigma^2)$ variates and if $\mathbf{A}$ is $r \times r$ unitary, then the entries of $\mathbf{AY}$ are also independent $N_1^C(0,\sigma^2)$. Also conclude that the marginal distributions of a multivariate complex normal are complex normal.

4.8.3 If $\mathbf{X}$ is $N_r^C(\boldsymbol{\mu},\boldsymbol{\Sigma})$ and $\operatorname{Im}\boldsymbol{\Sigma} = \mathbf{0}$, show that $\operatorname{Re}\mathbf{X}$ and $\operatorname{Im}\mathbf{X}$ are statistically independent.

4.8.4 If $\mathbf{W}$ is distributed as $W_r^C(n,\boldsymbol{\Sigma})$ prove that W_{jj} is distributed as $\Sigma_{jj}\chi^2_{2n}/2$.

4.8.5 If $\mathbf{W}$ is distributed as $W_r^C(n,\boldsymbol{\Sigma})$, prove that $E\mathbf{W} = n\boldsymbol{\Sigma}$. Also prove that $E(W_{jk} - n\Sigma_{jk})(W_{lm} - n\Sigma_{lm}) = n\Sigma_{jm}\Sigma_{lk}$.

4.8.6 If $\mathbf{W}$ is distributed as $W_r^C(n,\mathbf{\Sigma})$, prove that $\mathbf{\Sigma}^{-1/2}\mathbf{W}\mathbf{\Sigma}^{-1/2}$ is distributed as $W_r^C(n,\mathbf{I})$ if $\mathbf{\Sigma}$ is nonsingular.

4.8.7 Let $\mathbf{Y}$ be distributed as $N_n^C(\boldsymbol{\mu},\sigma^2\mathbf{I})$ and let

$$\overline{\mathbf{Y}}^\tau\mathbf{Y} = \overline{\mathbf{Y}}^\tau\mathbf{A}_1\mathbf{Y} + \cdots + \overline{\mathbf{Y}}^\tau\mathbf{A}_K\mathbf{Y}$$

where $\mathbf{A}_k$ is Hermitian of rank n_k. A necessary and sufficient condition for the forms $\overline{\mathbf{Y}}^\tau\mathbf{A}_k\mathbf{Y}$ to be distributed independently with $\overline{\mathbf{Y}}^\tau\mathbf{A}_k\mathbf{Y}$ distributed as (noncentral chi-squared)$\sigma^2\chi_{2n_k}^2(\overline{\boldsymbol{\mu}}^\tau\boldsymbol{\mu}/\sigma^2)/2$ is that $n_1 + \cdots + n_K = n$; see Brillinger (1973).

4.8.8 Let $\mathbf{W}$ be distributed as $W_{r+s}^C(n,\mathbf{\Sigma})$. Let it be partitioned into

$$\mathbf{W} = \begin{bmatrix} \mathbf{W}_{11} & \mathbf{W}_{12} \\ \mathbf{W}_{21} & \mathbf{W}_{22} \end{bmatrix}$$

with $\mathbf{W}_{11}$ $r \times r$ and $\mathbf{W}_{22}$ $s \times s$. Suppose that $\mathbf{\Sigma}$ is similarly partitioned. Prove that $\mathbf{W}_{22} - \mathbf{W}_{21}\mathbf{W}_{11}^{-1}\mathbf{W}_{12}$ is distributed as

$$W_s^C(n - r, \mathbf{\Sigma}_{22} - \mathbf{\Sigma}_{21}\mathbf{\Sigma}_{11}^{-1}\mathbf{\Sigma}_{12}).$$

If $\mathbf{\Sigma}_{12} = \mathbf{0}$, prove that $\mathbf{W}_{21}\mathbf{W}_{11}^{-1}\mathbf{W}_{12}$ is distributed as $W_s^C(r,\mathbf{\Sigma}_{22})$ and is independent of $\mathbf{W}_{22} - \mathbf{W}_{21}\mathbf{W}_{11}^{-1}\mathbf{W}_{12}$.

4.8.9 Let $\mathbf{Y}$ be $N_r^C(\mathbf{0},\mathbf{\Sigma})$. Prove that

$$EY_{a_1}\cdots Y_{a_j}\overline{Y}_{b_1}\cdots\overline{Y}_{b_k} = 0 \qquad \text{if } j \neq k,$$

and equals

$$\begin{vmatrix} \Sigma_{a_1b_1} & \cdots & \Sigma_{a_1b_j} \\ \cdot & & \cdot \\ \cdot & & \cdot \\ \cdot & & \cdot \\ \Sigma_{a_jb_1} & \cdots & \Sigma_{a_jb_j} \end{vmatrix}^{+} \qquad (*)$$

if $j = k$ where (*) denotes a $j \times j$ permanent; see Goodman and Dubman (1969).

4.8.10 Let $\mathbf{X}(t)$, $t = 0, \pm 1, \ldots$ be a stationary process with finite moments and satisfying $\mathbf{X}(t + T) = \mathbf{X}(t)$, $t = 0, \pm 1, \ldots$ for some positive integer T. (Such an $\mathbf{X}(t)$ is called a **circular process**.) Prove that

$$\operatorname{cum}\{d_{a_1}^{(T)}(2\pi s_1/T), \ldots, d_{a_k}^{(T)}(2\pi s_k/T)\} = 0$$

for $s_1, \ldots, s_k$ integers with $s_1 + \cdots + s_k \not\equiv 0 \pmod{2\pi}$.

4.8.11. If $\mathbf{X}(t)$, $t = 0, \pm 1, \ldots$ is a stationary Gaussian series, prove that for $k > 2$

$$\operatorname{cum}\{d_{a_1}^{(T)}(\lambda_1), \ldots, d_{a_k}^{(T)}(\lambda_k)\} = 0.$$

4.8.12 Let $\mathbf{X}(t)$, $t = 0, \pm 1, \ldots$ be an r vector-valued pure noise series with

$$\operatorname{cum}\{X_{a_1}(t), \ldots, X_{a_k}(t)\} = \kappa_{a_1 \ldots a_k}.$$

If $d_a^{(T)}[\lambda]$ is given by expression (4.3.13), prove that

$$\text{cum}\{d_{a_1}^{(T)}(\lambda_1), \ldots, d_{a_k}^{(T)}(\lambda_k)\} = \Delta^{(T)}\left(\sum_1^k \lambda_j\right)\kappa_{a_1\ldots a_k}.$$

4.8.13 Let $\mathbf{X}(t)$, $t = 0, \pm 1, \ldots$ be an r vector-valued stationary series satisfying expression (4.3.6). Let $T = \min_j T_j$. If $d_a^{(T)}(\lambda)$ is given by expression (4.3.13) show that

$$\text{cum}\{d_{a_1}^{(T_1)}(\lambda_1), \ldots, d_{a_k}^{(T_k)}(\lambda_k)\}$$
$$= (2\pi)^{k-1}\Delta^{(T)}\left(\sum_1^k \lambda_j\right) f_{a_1\ldots a_k}(\lambda_1, \ldots, \lambda_{k-1}) + o(T).$$

4.8.14 Suppose the conditions of Theorem 4.3.2 are satisfied. Suppose that $H_a(\lambda)$ of (4.3.3) satisfies

$$|H_a(\lambda)| < K(1 + |\lambda|)^{-\nu}$$

for some finite K where $\nu > 2$, $a = 1, \ldots, r$. Then prove that

$$\text{cum}\{d_{a_1}^{(T)}(\lambda_1), \ldots, d_{a_k}^{(T)}(\lambda_k)\} = (2\pi)^{k-1}H_{a_1\ldots a_k}^{(T)}\left(\sum_1^k \lambda_j\right) f_{a_1\ldots a_k}(\lambda_1, \ldots, \lambda_{k-1})$$
$$+ O\left[\min_l \left(1 + T\left|\sum_1^k \lambda_j - 2\pi l\right|\right)^{-\nu}\right].$$

4.8.15 Let $\mathbf{X}(t)$, $t = 0, \pm 1, \ldots$ be an r vector-valued series. Suppose the stretch of values $\mathbf{X}(t)$, $t = 0, \ldots, T - 1$ is given. Prove that $\mathbf{d}_X^{(T)}(2\pi s/T)$, $s = 0, \ldots, T/2$ is a sufficient statistic.

4.8.16 Under the conditions of Theorem 4.6.1, prove that $\mathbf{Z}_X(\lambda)$ is continuous in mean of order ν for any $\nu > 0$.

4.8.17 Let $\mathbf{Y}$ be a J vector-valued random variable with covariance matrix $\mathbf{\Sigma}_{YY}$. Determine the linear combination $\boldsymbol{\alpha}^\tau\mathbf{Y}$, with $\boldsymbol{\alpha}^\tau\boldsymbol{\alpha} = 1$, that has maximum variance.

4.8.18 Making use of Exercise 3.10.15, generalize the discussion of Section 4.7 to the case of vector-valued series.

4.8.19 Under the conditions of Theorem 4.4.2, prove that if $\lambda \not\equiv 0 \pmod{\pi}$, then $\arg\{d_a^{(T)}(\lambda)\}$ tends in distribution to a uniform variate on the interval $(0, 2\pi)$ as $T \to \infty$. What is the distribution if $\lambda \equiv 0 \pmod{\pi}$?

4.8.20 Suppose the conditions of Theorem 4.4.2 are satisfied. Suppose $H_a(\lambda)$ of (4.3.3) satisfies

$$|H_a(\lambda)| < K(1 + |\lambda|)^{-\nu}$$

for some finite K where $\nu > 2$, $a = 1, \ldots, r$. Suppose $\lambda_j(T) \to \lambda_j$ as $T \to \infty$ with $\min_l T|\lambda_j(T) - 2\pi l|$, $\min_l T|\lambda_j(T) \pm \lambda_k(T) - 2\pi l| \to \infty$ as $T \to \infty$, $1 \leqslant j < k \leqslant J$. Prove that the conclusions of Theorem 4.4.2 apply to $\mathbf{d}_X^{(T)}(\lambda_j(T))$, $j = 1, \ldots, J$.

4.8.21 Let $d_X^{(T)}(\lambda) = \sum_{t=0}^{T-1} X(t) \exp\{-i\lambda t\}$. Show that

$$\sum_t h_a\left(\frac{t}{T}\right)X(t) \exp\{-i\lambda t\} = T^{-1} \sum_{s=0}^{T-1} H_a^{(T)}\left(\frac{2\pi s}{T} - \lambda\right) d_X^{(T)}\left(\frac{2\pi s}{T}\right),$$

where $H_a^{(T)}(\lambda)$ is given by (4.3.2).

4.8.22 Let $X(t)$, $t = 0, \ldots, T-1$ be a sequence of independent normal variates with mean 0 and variance σ^2. Let

$$d_X^{(T)}(\lambda) = \sum_{t=0}^{T-1} X(t) \exp\{-i\lambda t\}.$$

(a) Show that $d_X^{(T)}(2\pi s/T)$ is distributed as $N_1^C(0,T\sigma^2)$ for s an integer with $2\pi s/T \not\equiv 0 \pmod{\pi}$.

(b) Indicate the distribution of

$$I_{XX}^{(T)}\left(\frac{2\pi s}{T}\right) = (2\pi T)^{-1}\left|d_X^{(T)}\left(\frac{2\pi s}{T}\right)\right|^2.$$

(c) Indicate the distribution of $\arg\left\{d_X^{(T)}\left(\frac{2\pi s}{T}\right)\right\}$.

4.8.23 Let $\mathbf{X}(t)$, $t = 0, \pm 1, \ldots$ be an r vector-valued series satisfying Assumption 2.6.1. Let $h_a(u)$, $-\infty < u < \infty$, satisfy Assumption 4.3.1. Let

$$d_a^{(V)}(\lambda,l) = \sum_{v=0}^{V-1} h_a\left(\frac{v}{V}\right) X_a(v + lV) \exp\{-i\lambda(v + lV)\}$$

for $-\infty < \lambda < \infty$; $l = 0, \ldots, L-1$; $a = 1, \ldots, r$. Show that $\mathbf{d}_X^{(V)}(\lambda,l) = [d_a^{(V)}(\lambda,l)]$, $l = 0, \ldots, L-1$ are asymptotically independent

$$N_r^C(\mathbf{0},2\pi V[H_{ab}(0) f_{ab}(\lambda)])$$

variates if $\lambda \not\equiv 0 \pmod{\pi}$ and asymptotically $N_r(\mathbf{0},2\pi V[H_{ab}(0) f_{ab}(\lambda)])$ variates if $\lambda = \pm\pi, \pm 3\pi, \ldots$, as $V \to \infty$. *Hint:* This result follows directly from Theorem 4.4.2 with $\mathbf{X}(t)$ and the $h_a(u)$ suitably redefined.

4.8.24 If $\mathbf{X}$ is $N_r^C(\mathbf{0},\mathbf{\Sigma})$ and $\mathbf{A}$ is $r \times r$ Hermitian, show that $\overline{\mathbf{X}}^\tau\mathbf{A}\mathbf{X}$ is distributed as $\sum_{j=1}^{r} \mu_j(u_j^2 + v_j^2)$ where $\mu_1, \ldots, \mu_r$ are the latent values of

$$\mathbf{\Sigma}^{1/2}\mathbf{A}\mathbf{\Sigma}^{1/2}$$

and $u_1, \ldots, u_r, v_1, \ldots, v_r$ are independent $N(0,1)$ variates.

4.8.25 Let $\mathbf{X}_1, \ldots, \mathbf{X}_n$ be independent $N_r^C(\boldsymbol{\mu},\mathbf{\Sigma})$ variates. Show that $\hat{\boldsymbol{\mu}} = \sum_j \mathbf{X}_j/n$ and $\hat{\mathbf{\Sigma}} = \sum_j(\mathbf{X}_j - \hat{\boldsymbol{\mu}}_j)\overline{(\mathbf{X}_j - \hat{\boldsymbol{\mu}}_j)}^\tau/n$ are the maximum likelihood estimates of $\boldsymbol{\mu}$ and $\mathbf{\Sigma}$; see Giri (1965).

4.8.26 Show that a $W_r^C(n,\mathbf{\Sigma})$ variate with $\operatorname{Im} \mathbf{\Sigma} = \mathbf{0}$ may be represented as $\frac{1}{2}\{\mathbf{W}_{11} + \mathbf{W}_{22} + i(\mathbf{W}_{12} - \mathbf{W}_{21})\}$ where

$$\begin{bmatrix} \mathbf{W}_{11} & \mathbf{W}_{12} \\ \mathbf{W}_{21} & \mathbf{W}_{22} \end{bmatrix} \text{ is } W_{2r}\left(n, \begin{bmatrix} \mathbf{\Sigma} & \mathbf{0} \\ \mathbf{0} & \mathbf{\Sigma} \end{bmatrix}\right).$$

Conclude that the real part of such a complex Wishart is distributed as $\frac{1}{2}W_2(2n,\mathbf{\Sigma})$.

4.8.27 Use the density function (4.2.9) to show that $\mathbf{W} = W_r^C(n,\mathbf{I})$ may be represented as $(\mathbf{X} + i\mathbf{Y})\overline{(\mathbf{X} + i\mathbf{Y})}^\tau$ where $\mathbf{X}$, $\mathbf{Y}$ are lower triangular, X_{jk}, Y_{jk}, $1 \leqslant k < j \leqslant r$ are independent $N_1(0,1)$ variates and X_{jj}^2, Y_{jj}^2 are independent χ_{n-j+1}^2.

4.8.28 Under the conditions of Exercise 4.8.25, show that $\mathfrak{J}^2 = \bar{\hat{\mathbf{u}}}^\tau\hat{\mathbf{\Sigma}}^{-1}\hat{\mathbf{u}}$ is distributed as $\chi'^2_{2r}(2n\bar{\mathbf{u}}^\tau\mathbf{\Sigma}^{-1}\mathbf{u})\chi^2_{2(n-r)}$ where $\chi'^2_{2r}(\delta)$ denotes a noncentral chi-square variate with $2r$ degrees of freedom and noncentrality parameter δ and $\chi^2_{2(n-r)}$ denotes an independent central chi-squared with $2(n - r)$ degrees of freedom; see Giri (1965).

4.8.29 Under the conditions of Theorem 4.4.2, show that the asymptotic distribution of $\mathbf{d}_X^{(T)}(\lambda)$ is unaffected by the omission of any finite number of the $\mathbf{X}(t)$.

4.8.30 Let W be distributed as $W_r^C(n,\mathbf{\Sigma})$. Show that

$$\text{cum}\{W_{a_1b_1}, \ldots, W_{a_kb_k}\} = n\sum_P \{\Sigma_{a_1bP_1}\cdots\Sigma_{a_kbP_k}\}$$

where the summation is over all permutations, P, of the set $\{1, 2, \ldots, k\}$ with the property that P leaves no proper subset of $\{1, 2, \ldots, k\}$ invariant. Show that the number of such permutations is $(k - 1)!$

4.8.31 Let W be distributed as $W_r(n,\mathbf{\Sigma})$. Show that

$$\text{cum}\{W_{a_1b_1}, \ldots, W_{a_kb_k}\}$$
$$= n\sum_P \{\Sigma_{a_1bP_1}\cdots\Sigma_{a_kbP_k} + \text{the } 2^{k-1} - 1 \text{ similar terms obtained by making}$$
$$\text{all possible interchanges } a_j \leftrightarrow b_j \text{ for } j = 1, \ldots, k - 1\}$$

where P is as in the previous exercise. Show that the number of terms summed in all is $2^{k-1}(k - 1)!$

4.8.32 Let $\mathbf{W}$ be distributed as $W_2\left(n, \begin{bmatrix} 1 & \rho \\ \rho & 1 \end{bmatrix}\right)$. Show that the density function of $x = W_{12}$ is given by

$$\frac{|x|^{(n-1)/2}}{\sqrt{\pi}\Gamma(n/2)2^{(n-1)/2}\sqrt{1 - \rho^2}} \exp\{\rho x/(1 - \rho^2)\} K_{(n-1)/2}(|x|/(1 - \rho^2))$$

where K_ν is the modified Bessel function of the second kind and order ν; see Pearson et al (1929) and Wishart and Bartlett (1932).

4.8.33 Let $\mathbf{W}$ be distributed as $W_2^C\left(n, \begin{bmatrix} 1 & \rho \\ \bar{\rho} & 1 \end{bmatrix}\right)$.

(a) Show that the density of $x = W_{12}$ (with respect to Re x, Im x) is given by

$$\frac{|x|^{n-1}}{\pi\Gamma(n)2^n\sqrt{1 - |\rho|^2}} \exp \text{Re}\{x\bar{\rho}/(1 - |\rho|^2)\} K_{n-1}(|x|/(1 - |\rho|^2)).$$

(b) Show that the density of $y = \text{Re } W_{12}$ is given by

$$\frac{|y|^{n-1/2}}{\sqrt{\pi}\Gamma(n)2^{n-1/2}(\sqrt{1-\beta^2})^{n-1/2}} \times \exp\{\alpha y/(1-|\rho|^2)\}K_{n-1/2}(|y|\sqrt{1-\beta^2}/(1-|\rho|^2))$$

where $\alpha = \text{Re } \rho$, $\beta = \text{Im } \rho$.

(c) Show that the density of $z = \text{Im } W_{12}$ is given by

$$\frac{|z|^{n-1/2}}{\sqrt{\pi}\Gamma(n)2^{n-1/2}(\sqrt{1-\alpha^2})^{n-1/2}} \times \exp\{\beta z/(1-|\rho|^2)\}K_{n-1/2}(|z|\sqrt{1-\alpha^2}/(1-|\rho|^2)).$$

(d) Show that the density of $\phi = \arg W_{12}$ is given by

$$\frac{(1-|\rho|^2)^n}{\pi\Gamma(n)} \sum_{k=0}^{\infty} \frac{2^{k-1}|\rho|^k\Gamma(n+k/2)\Gamma(1+k/2)}{\Gamma(k+1)} \cos^k(\phi-\phi_0)$$

where $\phi_0 = \arg \rho$.

(e) Show that the density of $w = |W_{12}|$ is given by

$$\frac{w^n}{\Gamma(n)2^{n-1}} I_0(w|\rho|/(1-|\rho|^2))K_{n-1}(w/(1-|\rho|^2))$$

where I_0 is the modified Bessel function of the first kind and order 0.

All of the densities of this exercise were derived in Goodman (1957).

4.8.34 Let $X(t)$, $t = 0, \pm 1, \ldots$ be a stationary Gaussian series with mean 0 and power spectrum $f_{XX}(\lambda)$, $-\infty < \lambda < \infty$. Show that the Cramér representation may be written

$$X(t) = \int_{-\pi}^{\pi} \exp\{i\lambda t\}\sqrt{f_{XX}(\lambda)}dB(\lambda) \qquad t = 0, \pm 1, \ldots$$

where $B(\lambda)$, $0 \leqslant \lambda \leqslant \pi$, is a complex Brownian motion process satisfying $\text{cov}\{B(\lambda), B(\mu)\} = \min\{\lambda, \mu\}$ and $B(-\lambda) = \overline{B(\lambda)}$.

4.8.35 Suppose the series $Y(t)$, $t = 0, \pm 1, \ldots$ is given by (2.9.15). Show that it has the form

$$Y(t) = \sum_{J=0}^{L} \int \cdots \int \exp\{i(\lambda_1 + \cdots + \lambda_J)t\}A_J(\lambda_1, \ldots, \lambda_J)dZ_X(\lambda_1)\cdots dZ_X(\lambda_J)$$

in terms of the Cramér representation of the series $X(t)$, $t = 0, \pm 1, \ldots$.

4.8.36 (a) Let $\mathbf{W}$ be $W_r(n, \mathbf{\Sigma})$ and $\boldsymbol{\alpha}, \boldsymbol{\beta}, \boldsymbol{\gamma}, \boldsymbol{\delta}$ be r vectors. Show that

$$\text{cov}\{\boldsymbol{\alpha}^\tau\mathbf{W}\boldsymbol{\beta}, \boldsymbol{\gamma}^\tau\mathbf{W}\boldsymbol{\delta}\} = n(\boldsymbol{\alpha}^\tau\mathbf{\Sigma}\boldsymbol{\gamma})(\boldsymbol{\beta}^\tau\mathbf{\Sigma}\boldsymbol{\delta}) + n(\boldsymbol{\alpha}^\tau\mathbf{\Sigma}\boldsymbol{\delta})(\boldsymbol{\beta}^\tau\mathbf{\Sigma}\boldsymbol{\gamma}).$$

(b) Let $\mathbf{W}$ be $W_r^C(n, \mathbf{\Sigma})$ and $\boldsymbol{\alpha}, \boldsymbol{\beta}, \boldsymbol{\gamma}, \boldsymbol{\delta}$ complex r vectors. Show that

$$\text{cov}\{\overline{\boldsymbol{\alpha}}^\tau\mathbf{W}\boldsymbol{\beta}, \overline{\boldsymbol{\gamma}}^\tau\mathbf{W}\boldsymbol{\delta}\} = n(\overline{\boldsymbol{\alpha}}^\tau\mathbf{\Sigma}\boldsymbol{\gamma})(\overline{\boldsymbol{\delta}}^\tau\mathbf{\Sigma}\boldsymbol{\beta}).$$

4.8.37 Let $X(t)$, $t = 0, \pm 1, \ldots$ be a 0 mean, real-valued series satisfying Assumption 2.6.1. Let u be a non-negative integer. Then Theorem 2.9.1 indicates that the series $Y(t) = X(t + u)X(t)$ also satisfies Assumption 2.6.1. Use this and Theorem 4.4.1 to show that

$$m_{XX}^{(T)}(u) = T^{-1} \sum_{t=0}^{T-1-u} X(t + u)X(t)$$

is asymptotically normal with mean $c_{XX}(u)$ and variance

$$\frac{2\pi}{T}\left[\int_0^{2\pi} (1 + \cos 2u\alpha)\, f_{XX}(\alpha)^2 d\alpha \right.$$
$$\left. + \int_0^{2\pi}\int_0^{2\pi} \exp\{iu(\alpha - \beta)\}\, f_{XXXX}(\alpha,-\alpha,\beta)d\alpha d\beta\right].$$

5

THE ESTIMATION OF POWER SPECTRA

5.1 POWER SPECTRA AND THEIR INTERPRETATION

Let $X(t)$, $t = 0, \pm 1, \ldots$ be a real-valued time series with mean function

$$EX(t) = c_X \qquad t = 0, \pm 1, \ldots \tag{5.1.1}$$

and autocovariance function

$$\operatorname{cov}\{X(t+u), X(t)\} = c_{XX}(u) \qquad t, u = 0, \pm 1, \ldots. \tag{5.1.2}$$

Suppose that the autocovariance function satisfies

$$\sum_{u=-\infty}^{\infty} |c_{XX}(u)| < \infty, \tag{5.1.3}$$

then the **power spectrum** of the series $X(t)$, $t = 0, \pm 1, \ldots$ at frequency λ is defined to be the Fourier transform

$$f_{XX}(\lambda) = (2\pi)^{-1} \sum_{u=-\infty}^{\infty} \exp\{-i\lambda u\} c_{XX}(u) \qquad \text{for } -\infty < \lambda < \infty. \tag{5.1.4}$$

We have seen in Section 2.5 that this power spectrum is non-negative, even, and of period 2π with respect to λ. This evenness and periodicity means that we may take the interval $[0, \pi]$ as the fundamental domain of definition of $f_{XX}(\lambda)$ if we wish.

Under the condition (5.1.3), $f_{XX}(\lambda)$ is a bounded uniformly continuous function. Also the relation (5.1.4) may be inverted and the autocovariance function $c_{XX}(u)$ expressed as

$$c_{XX}(u) = \int_{-\pi}^{\pi} \exp\{i\alpha u\} f_{XX}(\alpha) d\alpha \qquad \text{for } u = 0, \pm 1, \ldots. \tag{5.1.5}$$

In particular setting $u = 0$ gives

$$\text{var } X(t) = \int_{-\pi}^{\pi} f_{XX}(\alpha) d\alpha. \tag{5.1.6}$$

In Sections 2.8 and 4.6 we saw that the power spectrum transforms in an elementary manner if the series is filtered in a linear time invariant way. Specifically, suppose the series $Y(t)$, $t = 0, \pm 1, \ldots$ results when the series $X(t)$, $t = 0, \pm 1, \ldots$ is passed through a filter having transfer function $A(\lambda)$, $-\infty < \lambda < \infty$. Then, from Example 2.8.1, the power spectrum of the series $Y(t)$, $t = 0, \pm 1, \ldots$ is given by

$$f_{YY}(\lambda) = |A(\lambda)|^2 f_{XX}(\lambda) \qquad \text{for } -\infty < \lambda < \infty. \tag{5.1.7}$$

We may use expressions (5.1.6) and (5.1.7) to see that

$$\text{var } Y(t) = \int_{-\pi}^{\pi} |A(\alpha)|^2 f_{XX}(\alpha) d\alpha. \tag{5.1.8}$$

Expression (5.1.8) suggests one possible means of interpreting the power spectrum. Suppose we take for $-\pi < \alpha \leqslant \pi$ and Δ small

$$\begin{aligned} A(\alpha) &= (4\Delta)^{-1/2} && \text{if } |\alpha \pm \lambda| < \Delta \\ &= 0 && \text{otherwise} \end{aligned} \tag{5.1.9}$$

and then extend $A(\alpha)$ outside of the interval $(-\pi,\pi]$ periodically. This transfer function corresponds to a filter proportional to a band-pass filter; see Section 2.7. The output series $Y(t)$, $t = 0, \pm 1, \ldots$ is therefore proportional to $X(t,\lambda)$, the component of frequency λ in the series $X(t)$, $t = 0, \pm 1, \ldots$; see Section 4.6. From expressions (5.1.8) and (5.1.9) we now see that

$$\text{var } Y(t) \doteq f_{XX}(\lambda) \qquad \text{for } t = 0, \pm 1, \ldots. \tag{5.1.10}$$

This means that $f_{XX}(\lambda)$ may be interpreted as proportional to the variance of $X(t,\lambda)$, the component of frequency λ in the series $X(t)$, $t = 0, \pm 1, \ldots$. Incidentally, we remark that

$$EY(t) = A(0)c_X. \tag{5.1.11}$$

This equals 0 if λ is farther than Δ from $0, \pm 2\pi, \ldots$, and so

$$EY(t)^2 \doteq f_{XX}(\lambda) \qquad \text{if } \lambda \neq 0, \pm 2\pi, \ldots \text{ for } t = 0, \pm 1, \ldots. \tag{5.1.12}$$

Now if $Y(t)$ is taken to be the voltage applied across the terminals of the simple electric circuit of Figure 5.1.1 containing a resistance of $R = 1$ ohm, then the instantaneous power dissipated is $Y(t)^2$. An examination of (5.1.12) now indicates that $f_{XX}(\lambda)$ may be interpreted as the expected amount of power dissipated in a certain electric circuit by the component in $X(t)$ of frequency λ. This example is the reason that $f_{XX}(\lambda)$ is often referred to as a "power" spectrum.

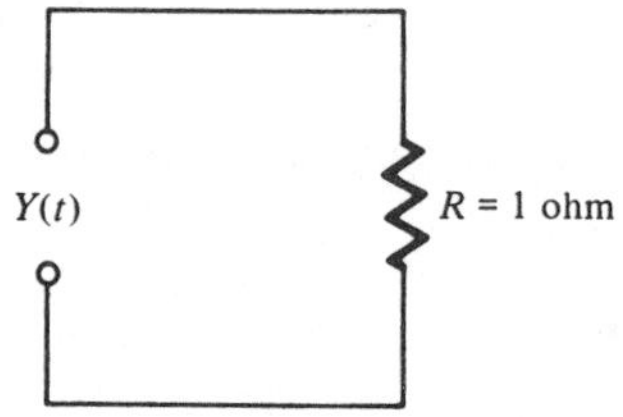

Figure 5.1.1 An elementary electric circuit with voltage $Y(t)$ applied at time t.

Roberts and Bishop (1965) have discussed a simple vibratory system for illustrating the value of the power spectrum. It consists of a cylindrical brass tube with a jet of air blown across an open end; this may be thought of as $X(t)$. The output signal is the pressure at the closed end of the tube, while the transfer function of this system is sketched in Figure 5.1.2. The peaks in the figure occur at frequencies

$$\lambda_n = \frac{(2n - 1)c}{4l} \tag{5.1.13}$$

where l = length of the tube, c = velocity of sound, and $n = 1, 2, \ldots$. The output of this system will have pressure proportional to

$$\sum_n f_{XX}(\lambda_n) \tag{5.1.14}$$

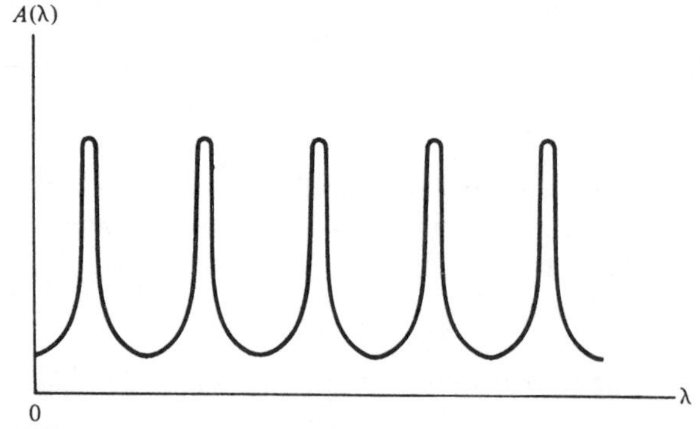

Figure 5.1.2 Approximate form of transfer function of system consisting of brass tube with air blown across one end.

where the λ_n are given by (5.1.13). A microphone at the closed end allows one to hear the output.

We conclude this section by presenting some examples of autocovariance functions and the corresponding power spectra, which are given in Figure 5.1.3. For example, if $c_{XX}(u)$ is concentrated near $u = 0$, then $f_{XX}(\lambda)$ is near constant. If $c_{XX}(u)$ falls off slowly as u increases, then $f_{XX}(\lambda)$ is concentrated near $\lambda = 0, \pm 2\pi, \ldots$. If $c_{XX}(u)$ oscillates about 0 as u increases, then $f_{XX}(\lambda)$ has substantial mass away from $\lambda \equiv 0 \pmod{2\pi}$.

We now turn to the development of an estimate of $f_{XX}(\lambda)$, $-\infty < \lambda < \infty$, and a variety of statistical properties of the proposed estimate. For additional discussion the reader may wish to consult certain of the following

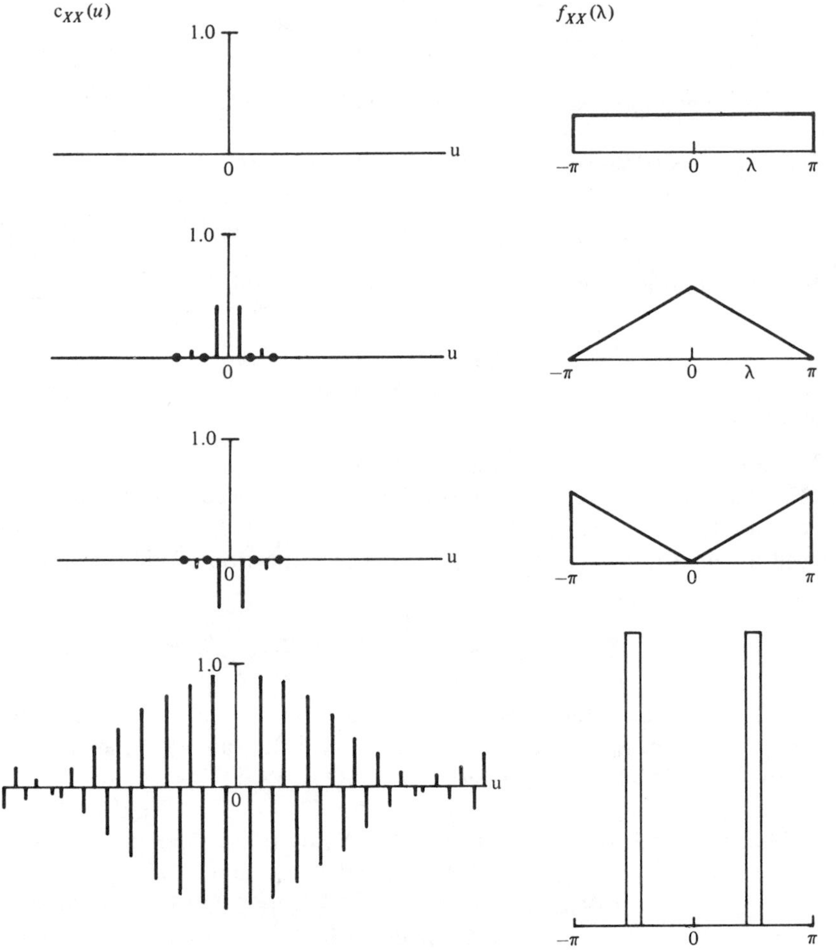

Figure 5.1.3 Selected autocovariances, $c_{XX}(u)$, and associated power spectra, $f_{XX}(\lambda)$.

review papers concerning the estimation of power spectra: Tukey (1959a,b), Jenkins (1961), Parzen (1961), Priestley (1962a), Bingham et al (1967), and Cooley et al (1970).

5.2 THE PERIODOGRAM

Suppose $X(t)$, $t = 0, \pm 1, \ldots$ is a stationary series with mean function c_X and power spectrum $f_{XX}(\lambda)$, $-\infty < \lambda < \infty$. Suppose also that the values $X(0), \ldots, X(T-1)$ are available and we are interested in estimating $f_{XX}(\lambda)$. Then we first compute the finite Fourier transform

$$d_X^{(T)}(\lambda) = \sum_{t=0}^{T-1} \exp\{-i\lambda t\} X(t). \tag{5.2.1}$$

Following Theorem 4.4.2, this variate is asymptotically

$$\begin{aligned} &N_1^C(0, 2\pi T f_{XX}(\lambda)) && \text{if } \lambda \not\equiv 0 \pmod{\pi} \\ &N_1(Tc_X, 2\pi T f_{XX}(\lambda)) && \text{if } \lambda = 0, \pm 2\pi, \ldots \\ &N_1(0, 2\pi T f_{XX}(\lambda)) && \text{if } \lambda = \pm\pi, \pm 3\pi, \ldots. \end{aligned} \tag{5.2.2}$$

These distributions suggest a consideration of the statistic

$$\begin{aligned} I_{XX}^{(T)}(\lambda) &= (2\pi T)^{-1} |d_X^{(T)}(\lambda)|^2 \\ &= (2\pi T)^{-1} \left| \sum_{t=0}^{T-1} \exp\{-i\lambda t\} X(t) \right|^2 \end{aligned} \tag{5.2.3}$$

as an estimate of $f_{XX}(\lambda)$ in the case $\lambda \not\equiv 0 \pmod{2\pi}$.

The statistic $I_{XX}^{(T)}(\lambda)$ of (5.2.3) is called the **second-order periodogram,** or more briefly periodogram, of the values $X(0), \ldots, X(T-1)$. It was introduced by Schuster (1898) as a tool for the identification of hidden periodicities because in the case

$$X(t) = \sum_j \rho_j \cos(\omega_j t + \phi_j) \qquad t = 0, \ldots, T-1, \tag{5.2.4}$$

$I_{XX}^{(T)}(\lambda)$ has peaks at the frequencies $\lambda \doteq \pm\omega_j \pmod{2\pi}$.

We note that $I_{XX}^{(T)}(\lambda)$, given by (5.2.3), has the same symmetry, non-negativity, and periodicity properties as $f_{XX}(\lambda)$.

Figure 5.2.1 is a plot of monthly rainfall in England for the period 1920 to 1930; the finite Fourier transform, $d_X^{(T)}(\lambda)$, of values for 1780–1960 was calculated using the Fast Fourier Transform Algorithm. The periodogram $I_{XX}^{(T)}(\lambda)$ was then calculated and is given as Figure 5.2.2. It is seen to be a rather irregular function of λ. This irregularity is also apparent in Figures

5.2.3 and 5.2.4 which are the lower and upper 100 periodogram ordinates of the series of mean monthly sunspot numbers (see Figure 1.1.5 for mean annual numbers). Other examples of periodograms are given in Wold (1965). In each of these examples $I_{XX}^{(T)}(\lambda)$ is a very irregular function of λ despite the fact that $f_{XX}(\lambda)$ is probably a regular function of λ. It appears that $I_{XX}^{(T)}(\lambda)$ is an inefficient estimate of $f_{XX}(\lambda)$ and so we turn to a consideration of alternate estimates. First, we will present some theorems relating to the statistical behavior of $I_{XX}^{(T)}(\lambda)$ in an attempt to understand the source of the irregularity and so construct better estimates.

First consider the expected value of the periodogram. We have

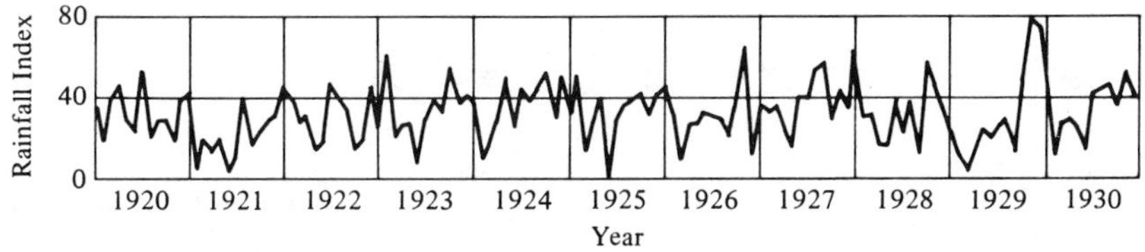

Figure 5.2.1. Composite index of rainfall for England and Wales for the years 1920–1930.

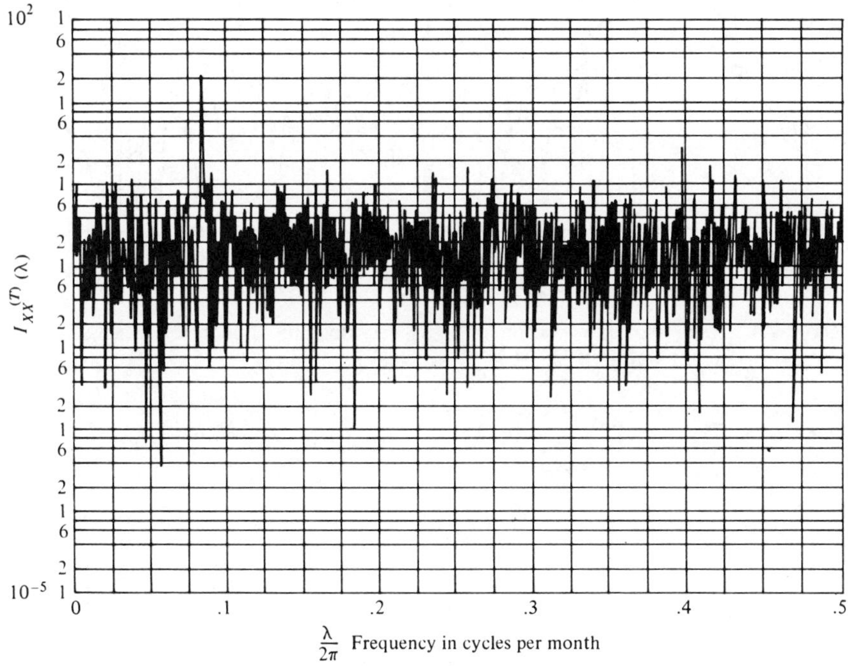

Figure 5.2.2 Periodogram of composite rainfall series of England and Wales for the years 1789–1959. (Logarithmic plot)

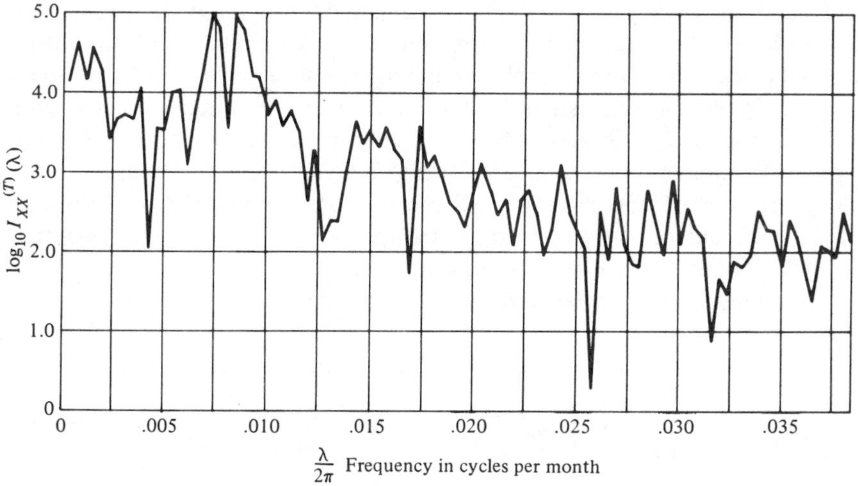

Figure 5.2.3 Low frequency portion of $\log_{10}$ periodogram of monthly mean sunspot numbers for the years 1750–1965.

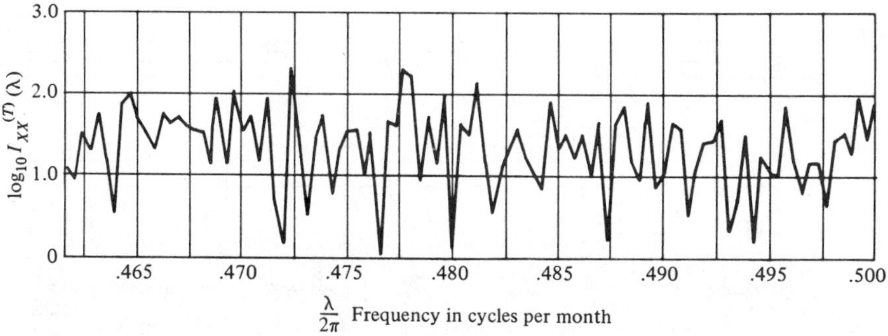

Figure 5.2.4 High frequency portion of $\log_{10}$ periodogram of monthly mean sunspot numbers for the years 1750–1965.

Theorem 5.2.1 Let $X(t)$, $t = 0, \pm 1, \ldots$ be a time series with $EX(t) = c_X$, $\operatorname{cov}\{X(t+u), X(t)\} = c_{XX}(u)$, $t, u = 0, \pm 1, \ldots$. Suppose

$$\sum_u |c_{XX}(u)| < \infty, \tag{5.2.5}$$

then

$$EI_{XX}^{(T)}(\lambda) = (2\pi T)^{-1} \int_{-\pi}^{\pi} \left[\frac{\sin T(\lambda - \alpha)/2}{\sin (\lambda - \alpha)/2}\right]^2 f_{XX}(\alpha) d\alpha + (2\pi T)^{-1} \left[\frac{\sin T\lambda/2}{\sin \lambda/2}\right]^2 c_X{}^2 \quad \text{for } -\infty < \lambda < \infty. \tag{5.2.6}$$

In the case that $\lambda \not\equiv 0 \pmod{2\pi}$, the final term in (5.2.6) is reduced in size and we see that $EI_{XX}^{(T)}(\lambda)$ is essentially a weighted average of the power spectrum of interest with weight concentrated in the neighborhood of λ. In the limit we have

Corollary 5.2.1 Under the conditions of the theorem, $I_{XX}^{(T)}(\lambda)$ is an asymptotically unbiased estimate of $f_{XX}(\lambda)$ if $\lambda \not\equiv 0 \pmod{2\pi}$.

The next theorem gives a bound for the asymptotic bias of $I_{XX}^{(T)}(\lambda)$.

Theorem 5.2.2 Under the conditions of Theorem 5.2.1 and if

$$\sum_u |u|\,|c_{XX}(u)| < \infty, \tag{5.2.7}$$

we have

$$EI_{XX}^{(T)}(\lambda) = f_{XX}(\lambda) + (2\pi T)^{-1}\left[\frac{\sin T\lambda/2}{\sin \lambda/2}\right]^2 c_X^2 + O(T^{-1}). \tag{5.2.8}$$

The $O(T^{-1})$ term is uniform in λ.

We remark that in the case $\lambda = 2\pi s/T$, s an integer $\not\equiv 0 \pmod T$, the second term on the right side of expressions (5.2.6) and (5.2.8) drops out leading to useful simple results. Now a consideration of $I_{XX}^{(T)}(\lambda)$ only for frequencies of the form $2\pi s/T$, s an integer $\not\equiv 0 \pmod T$, amounts to a consideration of $I^{(T)}_{X-c_X^{(T)},X-c_X^{(T)}}(\lambda)$, the periodogram of the sample values $X(t)$, $t = 0, \ldots, T-1$ with mean

$$c_X^{(T)} = T^{-1}\sum_{t=0}^{T-1} X(t) \tag{5.2.9}$$

removed, because we have the identity

$$d^{(T)}_{X-c_X^{(T)}}\left(\frac{2\pi s}{T}\right) = d_X^{(T)}\left(\frac{2\pi s}{T}\right) \tag{5.2.10}$$

for $s \not\equiv 0 \pmod T$. In view of the fact that the basic definition of a power spectrum is based on covariances and so is mean invariant, the restricted consideration of $I_{XX}^{(T)}(2\pi s/T)$, s an integer $\not\equiv 0 \pmod T$, seems reasonable. We will return to this case in Theorem 5.2.4 below.

We have seen in Sections 3.3 and 4.6 that advantages result from tapering observed values prior to computing their Fourier transform. We now turn to the construction of a modified periodogram that is appropriate for a series of tapered values. Suppose that we have formed

$$d_X^{(T)}(\lambda) = \sum_t h\left(\frac{t}{T}\right)X(t)\exp\{-i\lambda t\} \tag{5.2.11}$$

for some taper $h(u)$ satisfying Assumption 4.3.1. Then Theorem 4.4.2 suggests that the distribution of $d_X^{(T)}(\lambda)$ may be approximated by

$$N_1^C(0,2\pi T\{\int h(t)^2dt\}f_{XX}(\lambda)) \tag{5.2.12}$$

in the case $\lambda \not\equiv 0 \pmod{\pi}$. This suggests that we might consider the statistic

$$\begin{aligned} I_{XX}^{(T)}(\lambda) &= \left(2\pi \sum_t h\left(\frac{t}{T}\right)^2\right)^{-1} |d_X^{(T)}(\lambda)|^2 \\ &= \left(2\pi \sum_t h\left(\frac{t}{T}\right)^2\right)^{-1} |\sum_t h\left(\frac{t}{T}\right)X(t) \exp\{-i\lambda t\}|^2 \end{aligned} \tag{5.2.13}$$

as an estimate of $f_{XX}(\lambda)$ in the tapered case.

We have replaced $T \int h(t)^2dt$ by the sum of the squares of the taper coefficients as this is easily computed. Suppose we set

$$H(\lambda) = \int h(u) \exp\{-i\lambda u\} du \tag{5.2.14}$$

and

$$H^{(T)}(\lambda) = \sum_t h\left(\frac{t}{T}\right) \exp\{-i\lambda t\}. \tag{5.2.15}$$

If it is possible to apply the Poisson summation formula, then these two are connected by

$$H^{(T)}(\lambda) = T \sum_{l=-\infty}^{\infty} H(T[\lambda + 2\pi l]) \tag{5.2.16}$$

and $H^{(T)}(\lambda)$ is seen to have substantial magnitude for large T only if $\lambda \doteq 0 \pmod{2\pi}$. This observation will help us in interpreting expression (5.2.17). We can now state

Theorem 5.2.3 Let $X(t)$, $t = 0, \pm 1, \ldots$ be a real-valued series satisfying the conditions of Theorem 5.2.1. Let $h(u)$ satisfy Assumption 4.3.1. Let $I_{XX}^{(T)}(\lambda)$ be given by (5.2.13). Then

$$\begin{aligned} EI_{XX}^{(T)}(\lambda) &= \left(\int_{-\pi}^{\pi} |H^{(T)}(\alpha)|^2 d\alpha\right)^{-1} \int_{-\pi}^{\pi} |H^{(T)}(\alpha)|^2 f_{XX}(\lambda - \alpha) d\alpha \\ &+ \left(\int_{-\pi}^{\pi} |H^{(T)}(\alpha)|^2 d\alpha\right)^{-1} |H^{(T)}(\lambda)|^2 c_X{}^2 \quad \text{for } -\infty < \lambda < \infty. \end{aligned} \tag{5.2.17}$$

In the case that $\lambda \not\equiv 0 \pmod{2\pi}$, the final term in (5.2.17) will be of reduced magnitude. The first term on the right side of (5.2.17) is seen to be a weighted average of the power spectrum of interest with weight concentrated in the neighborhood of λ and relative weight determined by the taper. This expression is usefully compared with expression (5.2.6) corresponding to the nontapered case. If $f_{XX}(\alpha)$ has a substantial peak for α in the neighborhood of λ, then the expected value of $I_{XX}^{(T)}(\lambda)$, given by (5.2.6) or (5.2.17), can differ quite substantially from $f_{XX}(\lambda)$. The advantage of employing a taper is now apparent. It can be taken to have a shape to reduce the effect of neighboring peaks.

Continuing our investigation of the statistical properties of the periodogram as an estimate of the power spectrum, we find Theorem 5.2.4 describes the covariance structure of $I_{XX}^{(T)}(\lambda)$ in the nontapered case and when it is of the special form $2\pi s/T$, s an integer.

Theorem 5.2.4 Let $X(t)$, $t = 0, \pm 1, \ldots$ be a real-valued series satisfying Assumption 2.6.2(1). Let $I_{XX}^{(T)}(\lambda)$ be given by expression (5.2.3). Let r, s be integers with r, s, $r \pm s \not\equiv 0 \pmod T$. Let $\mu = 2\pi r/T$, $\lambda = 2\pi s/T$. Then

$$\operatorname{var} I_{XX}^{(T)}(\lambda) = f_{XX}(\lambda)^2 + O(T^{-1}) \tag{5.2.18}$$

$$\operatorname{cov}\{I_{XX}^{(T)}(\lambda), I_{XX}^{(T)}(\mu)\} = O(T^{-1}). \tag{5.2.19}$$

The $O(T^{-1})$ terms are uniform in λ, μ of the indicated form.

In connection with the conditions of this theorem, we remark that $I_{XX}^{(T)}(2\pi r/T) = I_{XX}^{(T)}(2\pi s/T)$ if $r + s$ or $r - s \equiv 0 \pmod T$ so the estimates are then identical.

This theorem has a crucial implication for statistical practice. It suggests that no matter how large T is taken, the variance of $I_{XX}^{(T)}(\lambda)$ will tend to remain at the level $f_{XX}(\lambda)^2$. If an estimate with a variance smaller than this is desired, it is not to be obtained by simply increasing the sample length and continuing to use the periodogram. The theorem also suggests a reason for the irregularity of Figures 5.2.2 to 5.2.4 — namely, adjacent periodogram ordinates are seen to have small covariance relative to their variances. In fact we will see in Theorem 5.2.6 that distinct periodogram ordinates are asymptotically independent.

Theorem 5.2.5 describes the asymptotic covariance structure of the periodogram when λ is not necessarily of the form $2\pi s/T$.

Theorem 5.2.5 Let $X(t)$, $t = 0, \pm 1, \ldots$ be a real-valued series satisfying Assumption 2.6.2(1). Let $I_{XX}^{(T)}(\lambda)$ be given by expression (5.2.3). Suppose $\lambda, \mu \not\equiv 0 \pmod{2\pi}$. Then

$$\operatorname{cov}\{I_{XX}^{(T)}(\lambda), I_{XX}^{(T)}(\mu)\} = \left\{\left[\frac{\sin T(\lambda+\mu)/2}{T\sin(\lambda+\mu)/2}\right]^2 + \left[\frac{\sin T(\lambda-\mu)/2}{T\sin(\lambda-\mu)/2}\right]^2\right\} f_{XX}(\lambda)^2 + O(T^{-1}) \tag{5.2.20}$$

Given $\varepsilon > 0$, the $O(T^{-1})$ term is uniform in λ, μ deviating from all multiples of 2π by at least ε.

We remark that expression (5.2.20) is more informative than (5.2.19) in that it indicates the transition of $\operatorname{cov}\{I_{XX}^{(T)}(\lambda), I_{XX}^{(T)}(\mu)\}$ into var $I_{XX}^{(T)}(\lambda)$ as $\mu \to \lambda$. It also suggests the reason for the reduced covariance in the case that λ, μ have the particular forms $2\pi s/T$, $2\pi r/T$ with s, r as integers.

We now complete our investigation of the elementary asymptotic properties of the periodogram by indicating its asymptotic distribution under regularity conditions. Theorem 4.4.1 indicated the asymptotic normality of $d_X^{(T)}(\lambda)$ for λ of the form $2\pi s/T$, s an integer. An immediate application of this theorem gives

Theorem 5.2.6 Let $X(t)$, $t = 0, \pm 1, \ldots$ be a real-valued series satisfying Assumption 2.6.1. Let $s_j(T)$ be an integer with $\lambda_j(T) = 2\pi s_j(T)/T$ tending to λ_j as $T \to \infty$ for $j = 1, \ldots, J$. Suppose $2\lambda_j(T)$, $\lambda_j(T) \pm \lambda_k(T) \not\equiv 0 \pmod{2\pi}$ for $1 \leqslant j < k \leqslant J$ and $T = 1, 2, \ldots$. Let

$$I_{XX}^{(T)}(\lambda) = (2\pi T)^{-1} \left|\sum_{t=0}^{T-1} X(t) \exp\{-i\lambda t\}\right|^2 \tag{5.2.21}$$

for $-\infty < \lambda < \infty$. Then $I_{XX}^{(T)}(\lambda_j(T))$, $j = 1, \ldots, J$ are asymptotically independent $f_{XX}(\lambda_j)\chi_2^2/2$ variates. Also if $\lambda = \pm\pi, \pm 3\pi, \ldots$, $I_{XX}^{(T)}(\lambda)$ is asymptotically $f_{XX}(\lambda)\chi_1^2$ independently of the previous variates.

In Theorem 5.2.6 χ_ν^2 denotes a chi-squared variate with ν degrees of freedom. The particular case of $\chi_2^2/2$ is an exponential variate with mean 1.

A practical implication of the theorem is that it may prove reasonable to approximate the distribution of a periodogram ordinate, $I_{XX}^{(T)}(\lambda)$, by a multiple of a χ_2^2 variate. Some empirical evidence for this assertion is provided by Figure 5.2.5 which is a two degree of freedom chi-squared probability plot of the values $I_{XX}^{(T)}(2\pi s/T)$, $s = T/4, \ldots, T/2$, for the series of mean monthly sunspot numbers. We have chosen these particular values of s because Figures 5.2.4 and 5.4.3 suggest that $f_{XX}(\lambda)$ is approximately constant for the corresponding frequency interval. If the values graphed in a two degree of freedom chi-squared probability plot actually have a distribution that is a multiple of χ_2^2, then the points plotted should tend to fall along a straight line. There is substantial evidence of this happening in Figure 5.2.5. Such plots are described in Wilk et al (1962).

Theorem 5.2.6 reinforces the suggestion, made in the discussion of Theorem 5.2.4, that the periodogram might prove an ineffective estimate of the power spectrum. For large T its distribution is approximately that of a multiple of a chi-squared variate with two degrees of freedom and hence is very unstable. In Section 5.4 we will turn to the problem of constructing estimates that are reasonably stable.

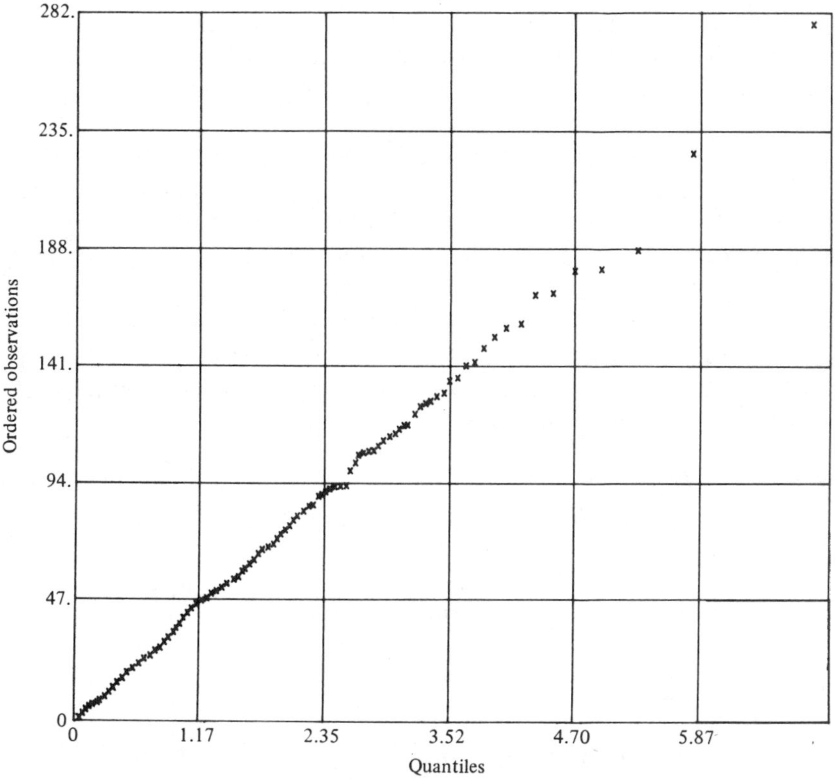

Figure 5.2.5 χ_2^2 probability plot of the upper 500 periodogram ordinates of monthly mean sunspot numbers for the years 1750–1965.

The mean and variance of the asymptotic distribution of $I_{XX}^{(T)}(2\pi s/T)$ are seen to be consistent with the large sample mean and variance of $I_{XX}^{(T)}(2\pi s/T)$ given by expressions (5.2.8) and (5.2.18), respectively.

Theorem 5.2.6 does not describe the asymptotic distribution of $I_{XX}^{(T)}(\lambda)$ when $\lambda \equiv 0 \pmod{2\pi}$. Theorem 4.4.1 indicates that the asymptotic distribution is $f_{XX}(\lambda)\chi_1^2$ when $EX(t) = c_X = 0$. In the case that $c_X \neq 0$, Theorem 4.4.1 suggests approximating the large sample distribution by $f_{XX}(\lambda)\chi'^2_1$ where χ'^2_1 denotes a noncentral chi-squared variate with one degree of freedom and noncentrality parameter $|c_X|\sqrt{T/(2\pi f_{XX}(\lambda))}$.

Turning to the tapered case we have

Theorem 5.2.7 Let $X(t)$, $t = 0, \pm 1, \ldots$ be a real-valued series satisfying Assumption 2.6.1. Suppose $2\lambda_j$, $\lambda_j \pm \lambda_k \not\equiv 0 \pmod{2\pi}$ for $1 \leqslant j < k \leqslant J$. Let $h(u)$ satisfy Assumption 4.3.1. Let

$$I_{XX}^{(T)}(\lambda) = \left(2\pi \sum_t h\left(\frac{t}{T}\right)^2\right)^{-1} \left|\sum_t h\left(\frac{t}{T}\right) X(t) \exp\{-i\lambda t\}\right|^2 \quad (5.2.22)$$

for $-\infty < \lambda < \infty$. Then $I_{XX}^{(T)}(\lambda_j)$, $j = 1, \ldots, J$ are asymptotically independent $f_{XX}(\lambda_j)\chi_2^2/2$ variates. Also if $\lambda = \pm\pi, \pm 3\pi, \ldots$, $I_{XX}^{(T)}(\lambda)$ is asymptotically $f_{XX}(\lambda)\chi_1^2$, independently of the previous variates.

With the definition and limiting procedure adopted, the limiting distribution of $I_{XX}^{(T)}(\lambda)$ is the same whether or not the series has been tapered. The hope is, however, that in large samples the tapered estimate will have less bias. A result of extending Theorem 5.2.5 to tapered values in the case of 0 mean is

Theorem 5.2.8 Let $X(t)$, $t = 0, \pm 1, \ldots$ be a real-valued series satisfying Assumption 2.6.2(1) and having mean 0. Let $h(u)$ satisfy Assumption 4.3.1 and let $I_{XX}^{(T)}(\lambda)$ be given by (5.2.22). Then

$$\operatorname{cov}\{I_{XX}^{(T)}(\lambda), I_{XX}^{(T)}(\mu)\} = |H_2^{(T)}(0)|^{-2}\{|H_2^{(T)}(\lambda - \mu)|^2 + |H_2^{(T)}(\lambda + \mu)|^2\} f_{XX}(\lambda)^2 + O(T^{-1}) \tag{5.2.23}$$

for $-\infty < \lambda, \mu < \infty$. The error term is uniform in λ, μ.

Here

$$H_2^{(T)}(\lambda) = \sum h\left(\frac{t}{T}\right)^2 \exp\{-i\lambda t\}. \tag{5.2.24}$$

The extent of dependence of $I_{XX}^{(T)}(\lambda)$ and $I_{XX}^{(T)}(\mu)$ is seen to fall off as the function $H_2^{(T)}$ falls off.

Bartlett (1950, 1966) developed expressions for the mean and covariance of the periodogram under regularity conditions; he also suggested approximating its distribution by a multiple of a chi-squared with two degrees of freedom. Other references to the material of this section include: Slutsky (1934), Grenander and Rosenblatt (1957), Kawata (1959), Hannan (1960), Akaike (1962b), Walker (1965), and Olshen (1967).

5.3 FURTHER ASPECTS OF THE PERIODOGRAM

The power spectrum, $f_{XX}(\lambda)$, of the series $X(t)$, $t = 0, \pm 1, \ldots$ was defined by

$$f_{XX}(\lambda) = (2\pi)^{-1} \sum_{u=-\infty}^{\infty} \exp\{-i\lambda u\} c_{XX}(u) \tag{5.3.1}$$

where $c_{XX}(u)$, $u = 0, \pm 1, \ldots$ was the autocovariance function of the series. This suggests an alternate means of estimating $f_{XX}(\lambda)$. We could first estimate $c_{XX}(u)$ by an expression of the form

$$c_{XX}^{(T)}(u) = T^{-1} \sum_{0 \leq t, t+u \leq T-1} (X(t+u) - c_X^{(T)})(X(t) - c_X^{(T)})$$
$$u = 0, \pm 1, \ldots \quad (5.3.2)$$

where

$$c_X^{(T)} = T^{-1} \sum_{t=0}^{T-1} X(t) \quad (5.3.3)$$

and then, taking note of (5.3.1), estimate $f_{XX}(\lambda)$ by

$$(2\pi)^{-1} \sum_u \exp\{-i\lambda u\} c_{XX}^{(T)}(u). \quad (5.3.4)$$

If we substitute expression (5.3.2) into (5.3.4), we see that this estimate takes the form

$$(2\pi T)^{-1} \sum_{s=0}^{T-1} \sum_{t=0}^{T-1} \exp\{is\lambda - it\lambda\}(X(s) - c_X^{(T)})(X(t) - c_X^{(T)})$$
$$= (2\pi T)^{-1} \left| \sum_{t=0}^{T-1} \exp\{-it\lambda\}(X(t) - c_X^{(T)})\right|^2 = I_{X-c_X^{(T)}, X-c_X^{(T)}}^{(T)}(\lambda), \quad (5.3.5)$$

that is, the periodogram of the deviations of the observed values from their mean. We noted in the discussion of Theorem 5.2.2 that

$$I_{X-c_X^{(T)}, X-c_X^{(T)}}^{(T)}(\lambda) = I_{XX}^{(T)}(\lambda) \quad (5.3.6)$$

for λ of the form $2\pi s/T$, $s \not\equiv 0 \pmod{T}$ and so Theorems 5.2.4 and 5.2.6 in fact relate to estimates of the form (5.3.5).

In the tapered case where

$$d_X^{(T)}(\lambda) = \sum_t h(t/T)X(t) \exp\{-i\lambda t\} \quad (5.3.7)$$

we see directly that

$$E d_X^{(T)}(\lambda) = c_X \sum_t h(t/T) \exp\{-i\lambda t\}$$
$$= c_X H^{(T)}(\lambda) \quad (5.3.8)$$

suggesting the consideration of the 0 mean statistic

$$d_X^{(T)}(\lambda) - c_X^{(T)} H^{(T)}(\lambda) = d_X^{(T)}(\lambda) - d_X^{(T)}(0) H^{(T)}(\lambda)/H^{(T)}(0) \quad (5.3.9)$$

where

$$c_X^{(T)} = \sum_t h(t/T)X(t) / \sum_t h(t/T) = d_X^{(T)}(0)/H^{(T)}(0). \quad (5.3.10)$$

We have therefore been led to consider the Fourier transform

$$d_{X-c_X^{(T)}}^{(T)}(\lambda) = d_X^{(T)}(\lambda) - d_X^{(T)}(0)H^{(T)}(\lambda)/H^{(T)}(0) \tag{5.3.11}$$

based on mean-corrected values. We remark that, in terms of the Cramér representation of Section 4.6, this last may be written

$$d_{X-c_X^{(T)}}^{(T)}(\lambda) = \int_{-\pi}^{\pi} [H^{(T)}(\lambda - \alpha) - H^{(T)}(\lambda)H^{(T)}(-\alpha)/H^{(T)}(0)]dZ_X(\alpha) \tag{5.3.12}$$

showing the reduction of frequency components for λ near $0, \pm 2\pi, \ldots$. In the light of this discussion it now seems appropriate to base spectral estimates on the modified periodogram

$$I_{X-c_X^{(T)},X-c_X^{(T)}}^{(T)}(\lambda) = (2\pi \sum_t h(t/T)^2)^{-1}|d_{X-c_X^{(T)}}^{(T)}(\lambda)|^2 \tag{5.3.13}$$

in the tapered case. The expected value of this statistic is indicated in Exercise 5.13.22.

Turning to another aspect of the periodogram, we have seen that the periodogram ordinates $I_{XX}^{(T)}(\lambda_j)$, $j = 1, \ldots, J$, are asymptotically independent for distinct λ_j, $j = 1, \ldots, J$. In Theorem 5.3.1 we will see that periodogram ordinates of the same frequency, but based on different stretches of data, are also asymptotically independent.

Theorem 5.3.1 Let $X(t)$, $t = 0, \pm 1, \ldots$ be a real-valued series satisfying Assumption 2.6.1. Let $h(u)$ satisfy Assumption 4.3.1 and vanish for $u < 0$. Let

$$I_{XX}^{(V)}(\lambda,l) = \left(2\pi \sum_v h\left(\frac{v}{V}\right)^2\right)^{-1} |\sum_{v=0}^{V-1} h\left(\frac{v}{V}\right)X(v + lV)\exp\{-i\lambda(v + lV)\}|^2 \tag{5.3.14}$$

for $-\infty < \lambda < \infty$, $l = 0, \ldots, L - 1$. Then, as $V \to \infty$, $I_{XX}^{(V)}(\lambda,l)$, $l = 0, \ldots, L - 1$, are asymptotically independent $f_{XX}(\lambda)\chi_2^2/2$ variates if $\lambda \not\equiv 0 \pmod \pi$ and asymptotically independent $f_{XX}(\lambda)\chi_1^2$ variates if $\lambda = \pm\pi, \pm 3\pi, \ldots$.

This result will suggest a useful means of constructing spectral estimates later. It is interesting to note that we can obtain asymptotically independent periodogram values either by splitting the data into separate segments, as we do here, or by evaluating them at neighboring frequencies, as in Theorem 5.2.7.

We conclude this section by indicating several probability 1 results relating to the periodogram. We begin by giving an almost sure bound for $I_{XX}^{(T)}(\lambda)$ as a function of λ and T.

Theorem 5.3.2 Let $X(t)$, $t = 0, \pm 1, \ldots$ be a real-valued series satisfying Assumption 2.6.3 and having mean 0. Let $h(u)$ satisfy Assumption 4.3.1. Let

$$I_{XX}^{(T)}(\lambda) = \left(2\pi \sum_t h\left(\frac{t}{T}\right)^2\right)^{-1} \left|\sum_t h\left(\frac{t}{T}\right) X(t) \exp\{-i\lambda t\}\right|^2. \quad (5.3.15)$$

Then

$$\overline{\lim_{T\to\infty}} \sup_\lambda I_{XX}^{(T)}(\lambda)/\log T \leqslant 2 \sup_\lambda f_{XX}(\lambda) \quad (5.3.16)$$

with probability 1.

In words, the rate of growth of the periodogram is at most of order $\log T$, uniformly in λ, under the indicated conditions. A practical implication of this is that the maximum deviation of $I_{XX}^{(T)}(\lambda)$ from $f_{XX}(\lambda)$ as a function of λ becomes arbitrarily large as $T \to \infty$. This is yet another indication of the fact that $I_{XX}^{(T)}(\lambda)$ is often an inappropriate estimate of $f_{XX}(\lambda)$.

We now briefly investigate the effect of a linear time invariant operation on the periodogram. Suppose

$$Y(t) = \sum_u a(t-u)X(u) \quad (5.3.17)$$

for some filter $\{a(u)\}$ satisfying

$$\sum_u (1 + |u|)|a(u)| < \infty \quad (5.3.18)$$

and having transfer function $A(\lambda)$. Theorem 4.5.2 indicated that under regularity conditions

$$\sup_\lambda |d_Y^{(T)}(\lambda) - A(\lambda)d_X^{(T)}(\lambda)| = O(\sqrt{\log T}) \quad (5.3.19)$$

almost surely. Elementary algebra then indicates that

$$I_{YY}^{(T)}(\lambda) = |A(\lambda)|^2 I_{XX}^{(T)}(\lambda) + O(T^{-1/2}\log T) \quad (5.3.20)$$

with probability 1, the error term being uniform in λ. In words, the effect of filtering on a periodogram is, approximately, multiplication by the modulus squared of the transfer function of the filter. This parallels the effect of filtering on the power spectrum as given in expression (5.1.7).

5.4 THE SMOOTHED PERIODOGRAM

In this section we make our first serious proposal for an estimate of the power spectrum. The discussion following Theorem 5.2.4 indicated that a critical disadvantage of the periodogram as an estimate of the power spectrum, $f_{XX}(\lambda)$, was that its variance was approximately $f_{XX}(\lambda)^2$, under reasonable regularity conditions, even when based on a lengthy stretch of data. On many occasions we require an estimate of greater precision than this and feel that it must exist. In fact, Theorem 5.2.6 suggests a means of constructing an improved estimate.

Suppose $s(T)$ is an integer with $2\pi s(T)/T$ near λ. Then Theorem 5.2.6 indicates that the $(2m + 1)$ adjacent periodogram ordinates $I_{XX}^{(T)}(2\pi[s(T) + j]/T), j = 0, \pm 1, \ldots, \pm m$ are approximately independent $f_{XX}(\lambda)\chi_2^2/2$ variates, if $2[s(T) + j] \not\equiv 0 \pmod{T}, j = 0, \pm 1, \ldots, \pm m$. These values may therefore provide $(2m + 1)$ approximately independent estimates of $f_{XX}(\lambda)$, which suggests an estimate having the form

$$f_{XX}^{(T)}(\lambda) = (2m + 1)^{-1} \sum_{j=-m}^{m} I_{XX}^{(T)}\left(\frac{2\pi[s(T) + j]}{T}\right) \quad \text{if } \lambda \not\equiv 0 \pmod{\pi}, \tag{5.4.1}$$

that is, a simple average of the periodogram ordinates in the neighborhood of λ. A further examination of Theorem 5.2.6 suggests the consideration of the estimate

$$\begin{aligned} f_{XX}^{(T)}(\lambda) &= (2m)^{-1}\left(\sum_{j=-m}^{-1} + \sum_{j=1}^{m}\right) I_{XX}^{(T)}\left(\lambda + \frac{2\pi j}{T}\right) \\ &= m^{-1} \sum_{j=1}^{m} I_{XX}^{(T)}\left(\lambda + \frac{2\pi j}{T}\right) \end{aligned} \tag{5.4.2}$$

if $\lambda = 0, \pm 2\pi, \pm 4\pi, \ldots$ or if $\lambda = \pm\pi, \pm 3\pi, \ldots$ and T is even, and

$$\begin{aligned} f_{XX}^{(T)}(\lambda) &= (2m)^{-1} \sum_{j=1}^{m} \left\{ I_{XX}^{(T)}\left(\lambda - \frac{\pi}{T} + \frac{2\pi j}{T}\right) + I_{XX}^{(T)}\left(\lambda + \frac{\pi}{T} - \frac{2\pi j}{T}\right)\right\} \\ &= m^{-1} \sum_{j=1}^{m} I_{XX}^{(T)}\left(\lambda - \frac{\pi}{T} + \frac{2\pi j}{T}\right) \end{aligned} \tag{5.4.3}$$

if $\lambda = \pm\pi, \pm 3\pi, \ldots$ and T is odd.

The estimate given by expressions (5.4.1) to (5.4.3) is seen to have the same non-negativity, periodicity, and symmetry properties as $f_{XX}(\lambda)$ itself. It is based on the values $d_X^{(T)}(2\pi s/T)$, $s = 0, \ldots, T - 1$ and so may be rapidly computed by the Fast Fourier Transform Algorithm if T happens to be highly composite. We will investigate its statistical properties shortly.

In preparation for Theorem 5.4.1 set

$$F_T(\lambda) = (2\pi T)^{-1}\left[\frac{\sin T\lambda/2}{\sin \lambda/2}\right]^2, \tag{5.4.4}$$

the Fejér kernel of Section 3.3. Then set

$$A_T^m(\lambda) = (2m + 1)^{-1} \sum_{j=-m}^{m} F_T\left(\lambda - \frac{2\pi j}{T}\right) \tag{5.4.5}$$

and set

$$B_T^m(\lambda) = (2m)^{-1}\left(\sum_{j=-m}^{-1} + \sum_{j=1}^{m}\right)F_T\left(\lambda - \frac{2\pi j}{T}\right)$$

$$C_T^m(\lambda) = (2m)^{-1}\sum_{j=1}^{m}\left\{F_T\left(\lambda - \frac{\pi}{T} + \frac{2\pi j}{T}\right) + F_T\left(\lambda + \frac{\pi}{T} - \frac{2\pi j}{T}\right)\right\}$$

$$\text{for } -\infty < \lambda < \infty. \quad (5.4.6)$$

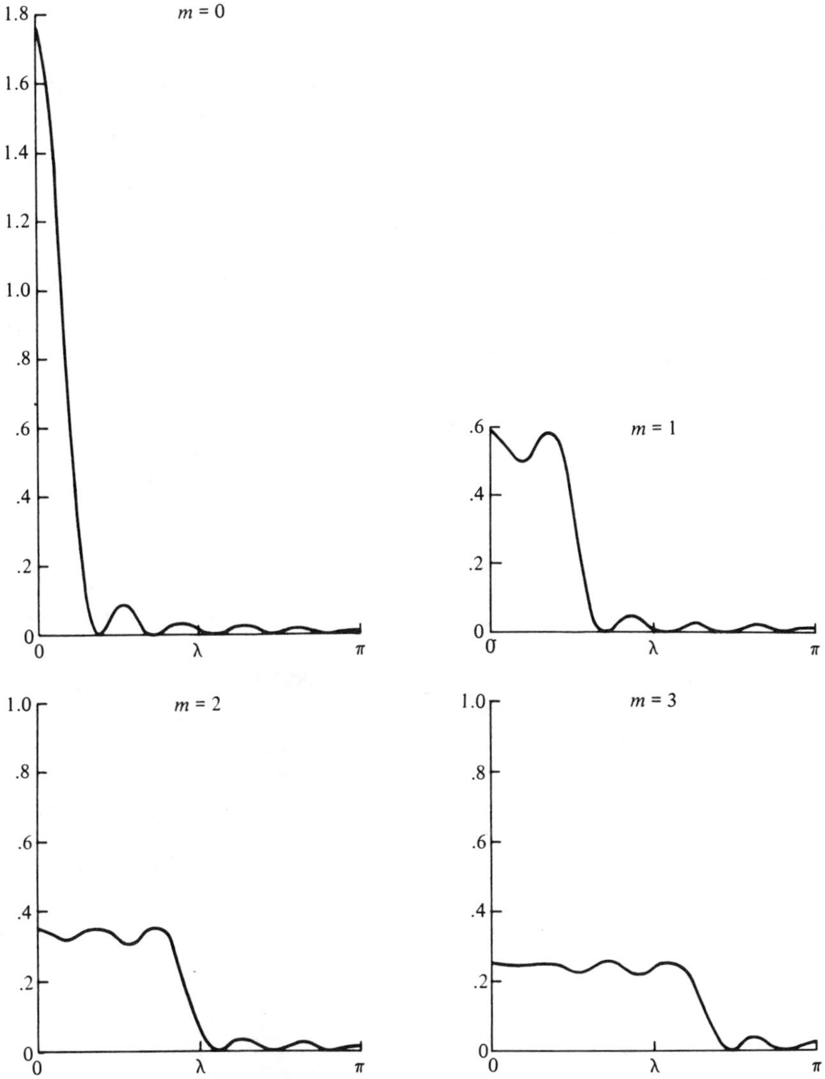

Figure 5.4.1 Plot of the kernel $A_T^m(\lambda)$ for $T = 11$ and $m = 0, 1, 2, 3$.

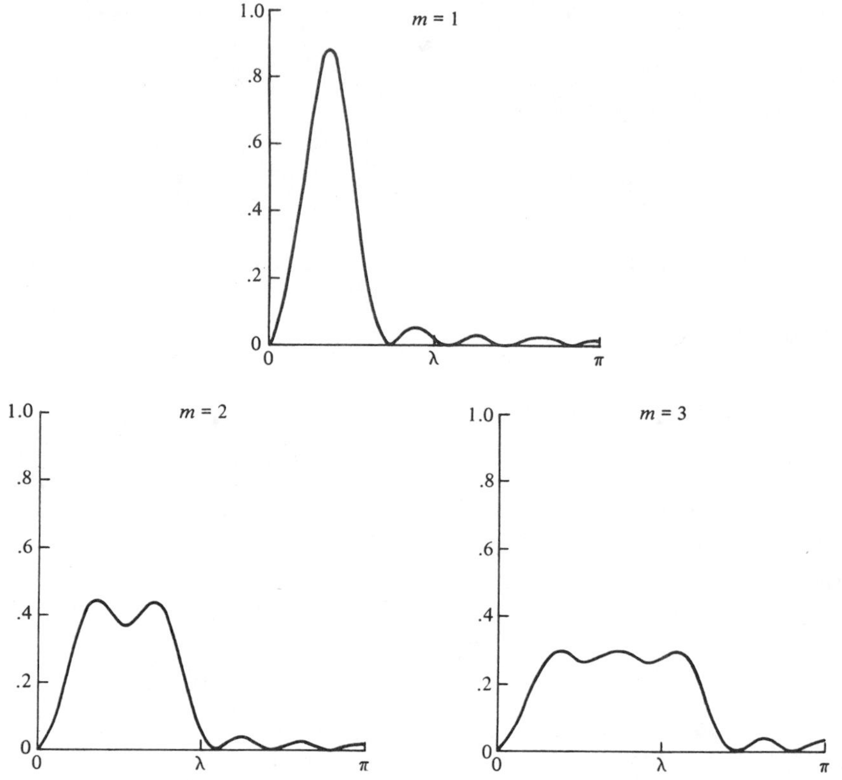

Figure 5.4.2 Plot of the kernel $B_T^m(\lambda)$ for $T = 11$ and $m = 1, 2, 3$.

Taking note of the properties of $F_T(\lambda)$, indicated in Section 3.3, we see that $A_T^m(\lambda)$, $B_T^m(\lambda)$, and $C_T^m(\lambda)$ are non-negative, have unit integral over the interval $(-\pi,\pi)$ and have period 2π. They are concentrated principally in the interval $(-2\pi m/T, 2\pi m/T)$ for $-\pi < \lambda < \pi$. $A_T^m(\lambda)$ is plotted in Figure 5.4.1 for the values $T = 11$, $m = 0, 1, 2, 3$. It is seen to have an approximate rectangular shape as was to be expected from the definition (5.4.5). $B_T^m(\lambda)$ is plotted in Figure 5.4.2 for the values $T = 11$, $m = 1, 2, 3$. It is seen to have a shape similar to that of $A_T^m(\lambda)$ except that in the immediate neighborhood of the origin it is near 0.

Turning to an investigation of the expected value of $f_{XX}^{(T)}(\lambda)$ we have

Theorem 5.4.1 Let $X(t)$, $t = 0, \pm 1, \ldots$ be a real-valued series with $EX(t) = c_X$ and $\text{cov}\{X(t+u), X(t)\} = c_{XX}(u)$ for $t, u = 0, \pm 1, \ldots$. Suppose

$$\sum_{u=-\infty}^{\infty} |c_{XX}(u)| < \infty. \tag{5.4.7}$$

Let $f_{XX}^{(T)}(\lambda)$ be given by (5.4.1) to (5.4.3). Then

$$
\begin{aligned}
Ef_{XX}^{(T)}(\lambda) &= \int_{-\pi}^{\pi} A_T^m(\alpha) f_{XX}\left(\frac{2\pi s(T)}{T} - \alpha\right) d\alpha \qquad \text{if } \lambda \not\equiv 0 \ (\text{mod } 2\pi) \\
&= \int_{-\pi}^{\pi} B_T^m(\alpha) f_{XX}(\lambda - \alpha) d\alpha \qquad \begin{array}{c}\text{if } \lambda \equiv 0 \ (\text{mod } 2\pi) \\ \text{or if } \lambda = \pm\pi, \pm 3\pi, \ldots \\ \text{and } T \text{ is even}\end{array} \\
&= \int_{-\pi}^{\pi} C_T^m(\alpha) f_{XX}(\lambda - \alpha) d\alpha \qquad \text{if } \lambda = \pm\pi, \pm 3\pi, \ldots \text{ and } T \text{ is odd.}
\end{aligned}
\tag{5.4.8}
$$

The expected value of $f_{XX}^{(T)}(\lambda)$ is a weighted average of the power spectrum of interest, $f_{XX}(\alpha)$, with weight concentrated in a band of width $4\pi m/T$ about λ in the case $\lambda \not\equiv 0$ (mod 2π). In the case $\lambda \equiv 0$ (mod 2π), $Ef_{XX}^{(T)}(\lambda)$ remains a weighted average of $f_{XX}(\alpha)$ with weight concentrated in the neighborhood of λ with the difference that values of $f_{XX}(\alpha)$ in the immediate neighborhood of 0 are partially excluded. The latter is a reflection of the difficulty resulting from not knowing $EX(t)$. If m is not too large compared to T and $f_{XX}(\alpha)$ is smooth, then $Ef_{XX}^{(T)}(\lambda)$ can be expected to be near $f_{XX}(\lambda)$ in both cases. A comparison of expressions (5.2.6) and (5.4.8) suggests that the bias of $f_{XX}^{(T)}(\lambda)$ will generally be greater than that of $I_{XX}^{(T)}(\lambda)$ as the integral extends over a greater essential range in the former case. We will make detailed remarks concerning the question of bias later.

The theorem has the following:

Corollary 5.4.1 Suppose in addition that $\lambda - 2\pi s(T)/T = O(T^{-1})$, m is constant with respect to T and

$$\sum_u |u| \, |c_{XX}(u)| < \infty, \tag{5.4.9}$$

then

$$Ef_{XX}^{(T)}(\lambda) = f_{XX}(\lambda) + O(T^{-1}) \qquad \text{for } -\infty < \lambda < \infty. \tag{5.4.10}$$

In the limit, $f_{XX}^{(T)}(\lambda)$ is an asymptotically unbiased estimate of $f_{XX}(\lambda)$.

In summary, with regard to its first moment, $f_{XX}^{(T)}(\lambda)$ seems a reasonable estimate of $f_{XX}(\lambda)$ provided that m is not too large with respect to T. The estimate seems reasonable in the case $\lambda \equiv 0$ (mod 2π) even if $EX(t)$ is unknown. Turning to the second-order moment structure of this estimate we have

Theorem 5.4.2 Let $X(t)$, $t = 0, \pm 1, \ldots$ be a real-valued series satisfying Assumption 2.6.2(1). Let $f_{XX}^{(T)}(\lambda)$ be given by (5.4.1) to (5.4.3) with

$\lambda - 2\pi s(T)/T = O(T^{-1})$. Suppose $\lambda \pm \mu \not\equiv 0 \pmod{2\pi}$ and that m does not depend on T. Then

$$\begin{aligned} \operatorname{var} f_{XX}^{(T)}(\lambda) &= \frac{f_{XX}(\lambda)^2}{2m+1} + O(T^{-1}) \qquad \text{if } \lambda \not\equiv 0 \pmod{\pi} \\ &= \frac{f_{XX}(\lambda)^2}{m} + O(T^{-1}) \qquad \text{if } \lambda \equiv 0 \pmod{\pi}. \end{aligned} \tag{5.4.11}$$

Also

$$\operatorname{cov}\{f_{XX}^{(T)}(\lambda), f_{XX}^{(T)}(\mu)\} = O(T^{-1}). \tag{5.4.12}$$

In the case $\lambda \not\equiv 0 \pmod{\pi}$, the effect of averaging $2m + 1$ adjacent periodogram ordinates has been to produce an estimate whose asymptotic variance is $1/(2m + 1)$ times that of the periodogram. Therefore, contemplate choosing a value of m so large that an acceptable level of stability in the estimate is achieved. However, following the discussion of Theorem 5.4.1, note that the bias of the estimate $f_{XX}^{(T)}(\lambda)$ may well increase as m is increased and thus some compromise value for m will have to be selected.

The variance of $f_{XX}^{(T)}(\lambda)$ in the case of $\lambda \equiv 0 \pmod{\pi}$ is seen to be approximately double that in the $\lambda \not\equiv 0 \pmod{\pi}$ case. This reflects the fact that the estimate in the former case is based approximately on half as many independent statistics. The asymptotic distribution of $f_{XX}^{(T)}(\lambda)$ under certain regularity conditions is indicated in the following:

Theorem 5.4.3 Let $X(t)$, $t = 0, \pm 1, \ldots$ be a real-valued series satisfying Assumption 2.6.1. Let $f_{XX}^{(T)}(\lambda)$ be given by (5.4.1) to (5.4.3) with $2\pi s(T)/T \to \lambda$ as $T \to \infty$. Suppose $\lambda_j \pm \lambda_k \not\equiv 0 \pmod{2\pi}$ for $1 \leqslant j < k \leqslant J$. Then $f_{XX}^{(T)}(\lambda_1), \ldots, f_{XX}^{(T)}(\lambda_J)$ are asymptotically independent with $f_{XX}^{(T)}(\lambda)$ asymptotically $f_{XX}(\lambda)\chi^2_{4m+2}/(4m + 2)$ if $\lambda \not\equiv 0 \pmod{\pi}$, asymptotically $f_{XX}(\lambda)\chi_{2m}{}^2/(2m)$ if $\lambda \equiv 0 \pmod{\pi}$.

This theorem will prove especially useful when it comes time to suggest approximate confidence limits for $f_{XX}(\lambda)$.

Figure 5.4.3 presents the logarithm base 10 of $f_{XX}^{(T)}(\lambda)$, given by (5.4.1) to (5.4.3), for the series of monthly sunspot numbers whose periodogram was given in Figures 5.2.3 and 5.2.4. The statistic $f_{XX}^{(T)}(\lambda)$ is calculated for $0 \leqslant \lambda \leqslant \pi$, $m = 2, 5, 10, 20$ and the growing stability of the estimate as m increases is immediately apparent. The figures suggest that $f_{XX}(\lambda)$ has a lot of mass in the neighborhood of 0. This in turn suggests that neighboring values of the series tend to cluster together. An examination of the series itself (Figure 1.1.5) confirms this remark. The periodogram and the plots corresponding to $m = 2, 5, 10$ suggest a possible peak in the spectrum in the neighborhood of the frequency $.015\pi$. This frequency corresponds to the

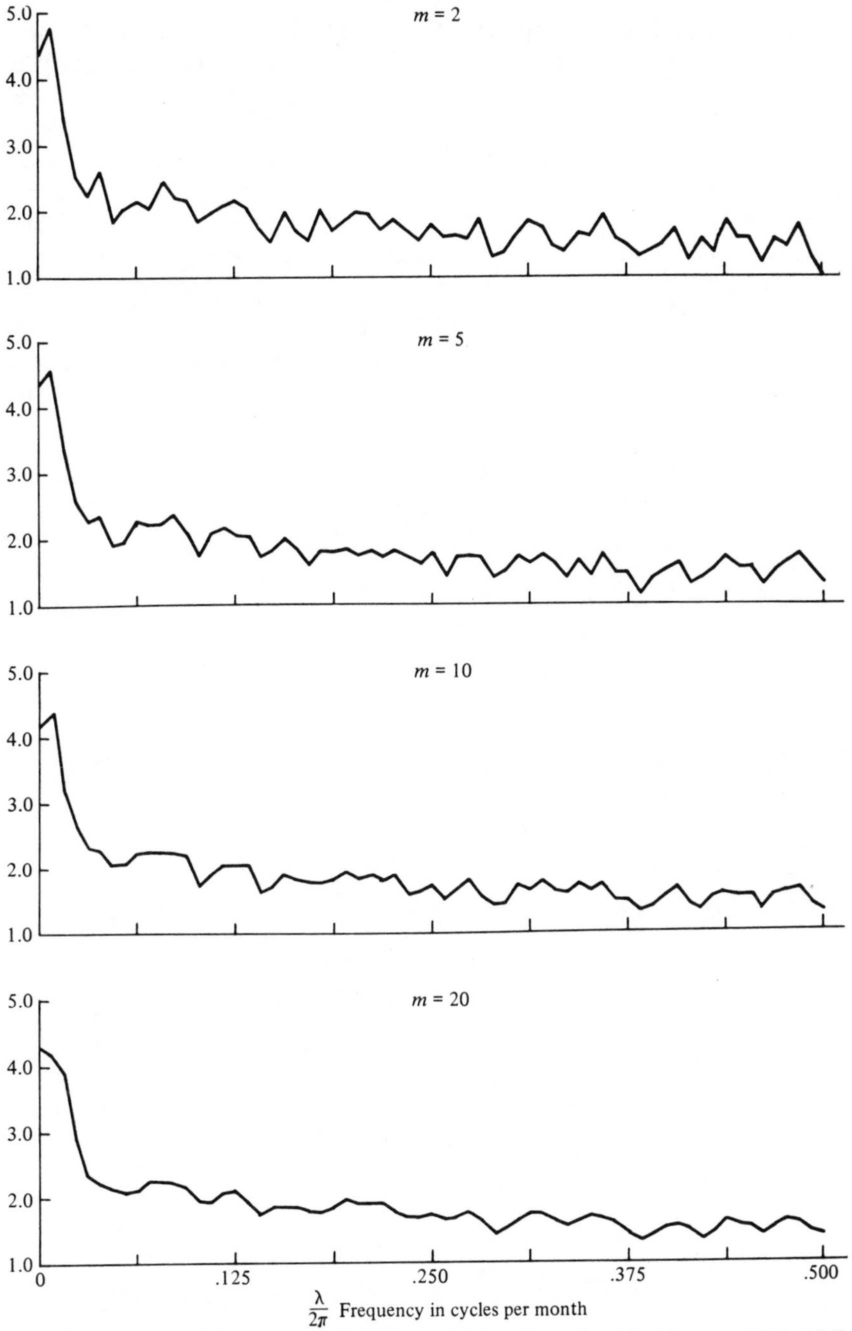

Figure 5.4.3 $\log_{10} f_{XX}^{(T)}(\lambda)$ for monthly mean sunspot numbers for the years 1750–1965 with $2m + 1$ periodogram ordinates averaged.

eleven-year solar cycle suggested by Schwabe in 1843; see Newton (1958). This peak has disappeared in the case $m = 20$ indicating that the bias of the estimate has become appreciable. Because this peak is of special interest, we have plotted $f_{XX}^{(T)}(\lambda)$ in the case $m = 2$ in an expanded scale in Figure 5.4.4. In this figure there is an indication of a peak near the frequency $.030\pi$ that is the first harmonic of $.015\pi$.

Figures 5.4.5 to 5.4.8 present the spectral estimate $f_{XX}^{(T)}(\lambda)$ for the series of mean monthly rainfall whose periodogram was given as Figure 5.2.2. The statistic is calculated for $m = 2, 5, 7, 10$. Once again the increasing stability of the estimate as m increases is apparent. The substantial peak

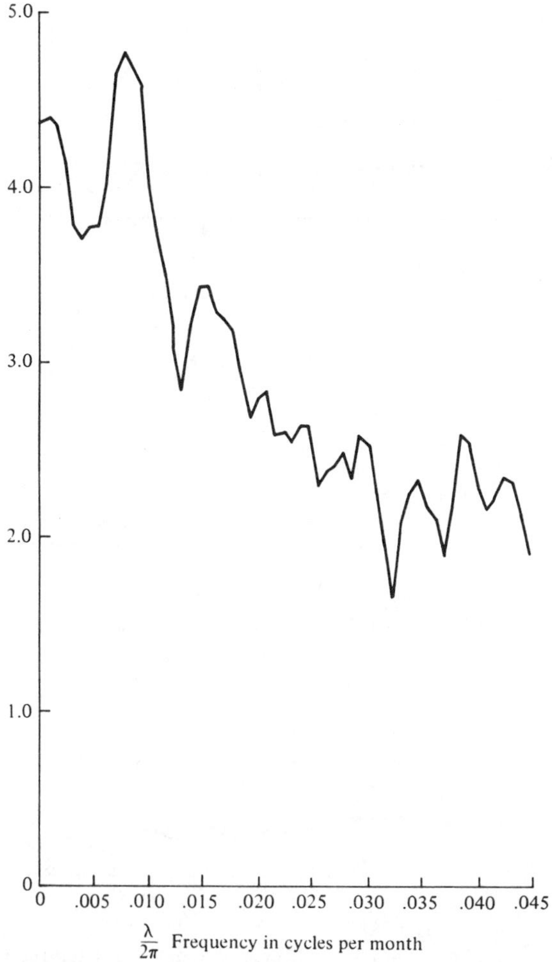

Figure 5.4.4 Low frequency portion of $\log_{10} f_{XX}^{(T)}(\lambda)$ for monthly mean sunspot numbers for the years 1750–1965 with five periodogram ordinates averaged.

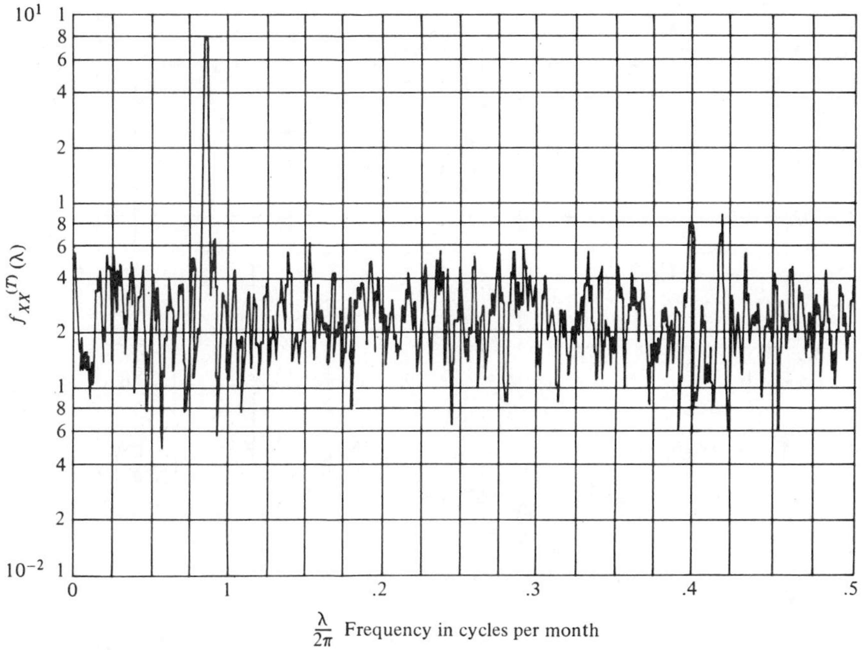

Figure 5.4.5 $f_{XX}^{(T)}(\lambda)$ of composite rainfall series of England and Wales for the years 1789–1959 with five periodogram ordinates averaged. (Logarithmic plot.)

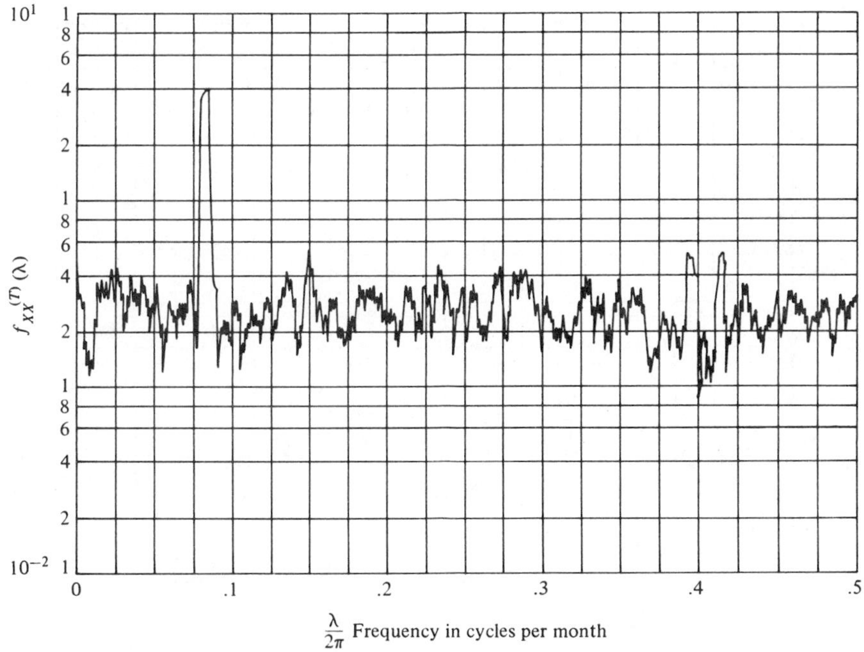

Figure 5.4.6 $f_{XX}^{(T)}(\lambda)$ of composite rainfall series of England and Wales for the years 1789–1959 with eleven periodogram ordinates averaged. (Logarithmic plot.)

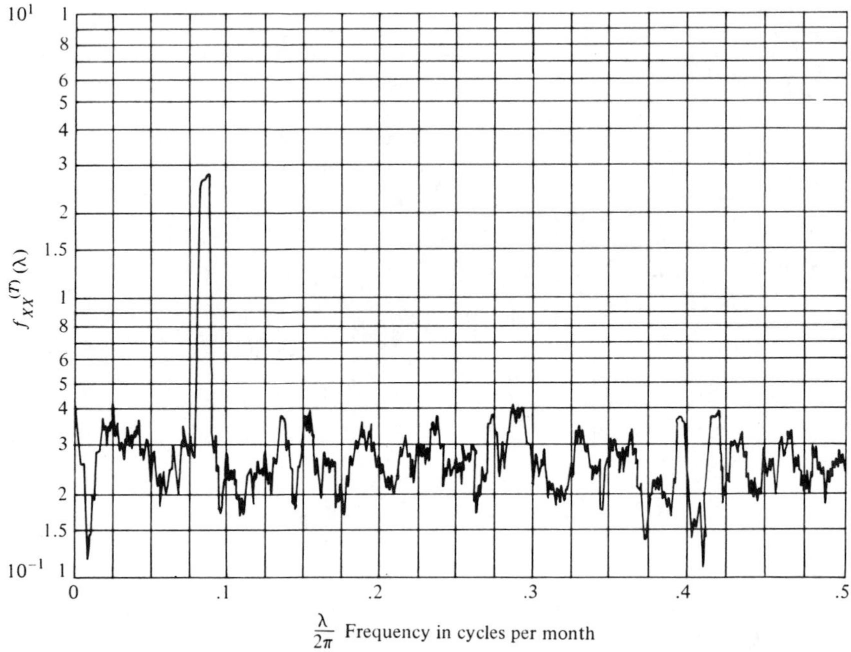

Figure 5.4.7 $f_{XX}^{(T)}(\lambda)$ of composite rainfall series of England and Wales for the years 1789–1959 with fifteen periodogram ordinates averaged. (Logarithmic plot.)

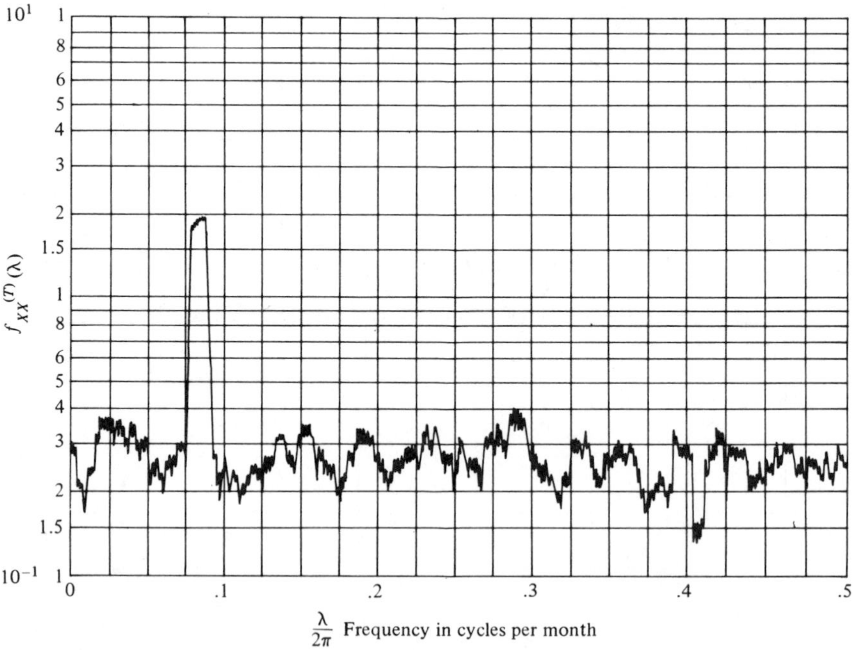

Figure 5.4.8 $f_{XX}^{(T)}(\lambda)$ of composite rainfall series of England and Wales for the years 1789–1959 with twenty-one periodogram ordinates averaged. (Logarithmic plot.)

in the figures occurs at a frequency of one cycle per year as would be expected in the light of the seasonal nature of the series. For other values of λ, $f_{XX}^{(T)}(\lambda)$ is near constant suggesting that the series is made up approximately of an annual component superimposed on a pure noise series.

Figure 5.4.9 presents some empirical evidence related to the validity of Theorem 5.4.3. It is a χ_{30}^2 probability plot of the values $f_{XX}^{(T)}(2\pi s/T)$, $T = 2592$, $s = T/4, \ldots, (T/2) - 1$, for the series of monthly sunspots. $f_{XX}^{(T)}(\lambda)$ has been formed by smoothing 15 adjacent periodogram ordinates. If $f_{XX}(\lambda)$ is near constant for $\pi/2 < \lambda < \pi$, as the estimated spectra suggest, and it is reasonable to approximate the distribution of $f_{XX}^{(T)}(\lambda)$ by a multiple of χ_{30}^2, as Theorem 5.4.3 suggests, then the plotted values should tend to fall along a straight line. In fact the bulk of the points plotted in Figure 5.4.9 appear to do this. However, there is definite curvature for the rightmost points. The direction of this curvature suggests that the actual distribution may have a shorter right-hand tail than that of a multiple of χ_{30}^2.

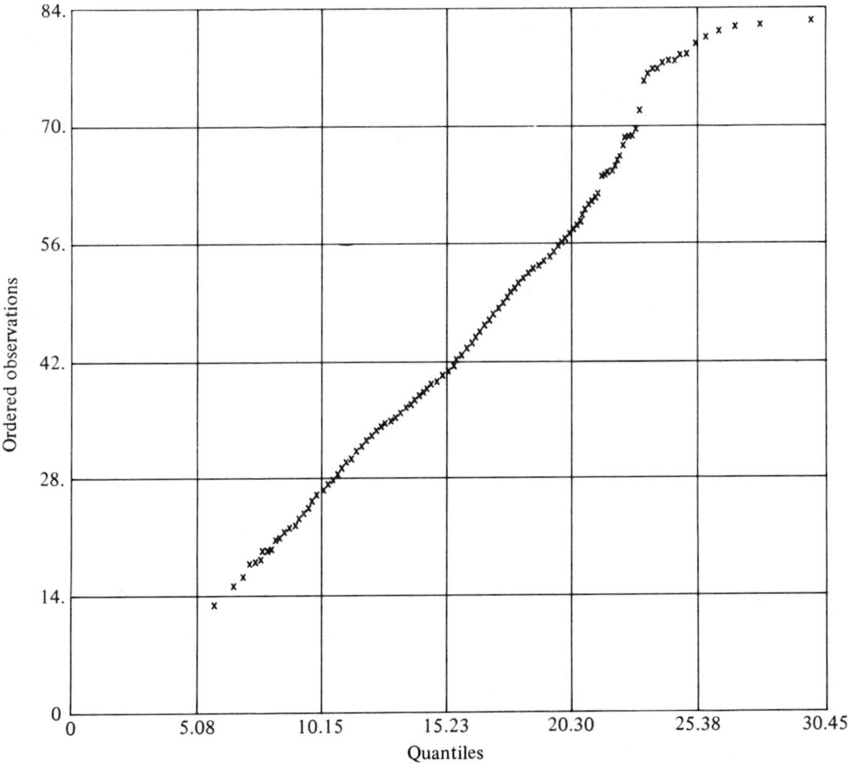

Figure 5.4.9 χ_{30}^2 probability plot of the upper 500 power spectrum estimates, when fifteen periodograms are averaged, of monthly mean sunspot numbers for the years 1750–1965.

It is important to note how informative it is to have calculated $f_{XX}^{(T)}(\lambda)$ not just for a single value of m, but rather for a succession of values. The figures for small values of m help in locating any nearly periodic components and their frequencies, while the figures for large values of m give exceedingly smooth curves that could prove useful in model fitting. In the case that the values $I_{XX}^{(T)}(2\pi s/T)$, $s = 1, 2, \ldots$ are available (perhaps calculated using the Fast Fourier Transform), it is an elementary matter to prepare estimates for a succession of values of m.

The suggestion that an improved spectral estimate might be obtained by smoothing the periodogram was made by Daniell (1946); see also Bartlett (1948b, 1966), the paper by Jones (1965), and the letter by Tick (1966). Bartlett (1950) made use of the χ^2 distribution for smoothed periodogram estimates.

5.5 A GENERAL CLASS OF SPECTRAL ESTIMATES

The spectral estimate of the previous section weights all periodogram ordinates in the neighborhood of λ equally. If $f_{XX}(\alpha)$ is near constant for α near λ, then this is undoubtedly a reasonable procedure; however, if $f_{XX}(\alpha)$ varies to an extent, then it is perhaps more reasonable to weight periodogram ordinates in the immediate neighborhood of λ more heavily than those at a distance. We proceed to construct an estimate that allows differential weighting.

Let W_j, $j = 0, \pm 1, \ldots, \pm m$ be weights satisfying

$$\sum_{j=-m}^{m} W_j = 1. \tag{5.5.1}$$

Let $s(T)$ be an integer such that $2\pi s(T)/T$ is near λ and $2[s(T) + j] \not\equiv 0 \pmod{T}$, $j = 0, \pm 1, \ldots, \pm m$. Consider the estimate

$$f_{XX}^{(T)}(\lambda) = \sum_{j=-m}^{m} W_j I_{XX}^{(T)}\left(\frac{2\pi[s(T) + j]}{T}\right) \qquad \text{if } \lambda \not\equiv 0 \pmod{\pi} \tag{5.5.2}$$

$$= \sum_{j=1}^{m} W_j I_{XX}^{(T)}\left(\lambda + \frac{2\pi j}{T}\right) \Big/ \left[\sum_{1}^{m} W_j\right] \qquad \begin{array}{l}\text{if } \lambda \equiv 0 \pmod{2\pi}\\ \text{or if } \lambda = \pm\pi, \pm 3\pi, \ldots\\ \text{and } T \text{ is even}\end{array} \tag{5.5.3}$$

$$= \sum_{j=1}^{m} W_j I_{XX}^{(T)}\left(\lambda + \frac{\pi}{T} - \frac{2\pi j}{T}\right) \Big/ \left[\sum_{1}^{m} W_j\right] \qquad \begin{array}{l}\text{if } \lambda = \pm\pi, \pm 3\pi, \ldots\\ \text{and } T \text{ is odd}\end{array} \tag{5.5.4}$$

where

$$I_{XX}^{(T)}(\lambda) = (2\pi T)^{-1} \left| \sum_{t=0}^{T-1} X(t) \exp\{-i\lambda t\} \right|^2 \qquad \text{for } -\infty < \lambda < \infty. \tag{5.5.5}$$

In order to discuss the expected value of this estimate we must define certain functions. Set

$$F_T(\lambda) = (2\pi T)^{-1}\left[\frac{\sin T\lambda/2}{\sin \lambda/2}\right]^2 \qquad \text{for } -\infty < \lambda < \infty. \tag{5.5.6}$$

Set

$$A_T(\lambda) = \sum_{j=-m}^{m} W_j F_T\left(\lambda - \frac{2\pi j}{T}\right), \tag{5.5.7}$$

$$B_T(\lambda) = \sum_{j=1}^{m} W_j F_T\left(\lambda - \frac{2\pi j}{T}\right)\Big/\left[\sum_1^m W_j\right], \tag{5.5.8}$$

$$C_T(\lambda) = \sum_{j=1}^{m} W_j F_T\left(\lambda - \frac{\pi}{T} + \frac{2\pi j}{T}\right)\Big/\left[\sum_1^m W_j\right]. \tag{5.5.9}$$

Because of the shape of $F_T(\lambda)$, both $A_T(\lambda)$ and $B_T(\lambda)$ will be weight functions principally concentrated in the interval $(-2\pi m/T, 2\pi m/T)$ for $-\pi < \lambda < \pi$. $B_T(\lambda)$ will differ from $A_T(\lambda)$ in having negligible mass for $-2\pi/T < \lambda < 2\pi/T$. In the case of equal weights, $A_T(\lambda)$ and $B_T(\lambda)$ are rectangular in shape. In the general case, the shape of $A_T(\lambda)$ will mimic that of $W_j, j = -m, \ldots, 0, \ldots, m$.

Turning to an investigation of the properties of this estimate we begin with

Theorem 5.5.1 Let $X(t)$, $t = 0, \pm 1, \ldots$ be a real-valued series with $EX(t) = c_X$ and $\operatorname{cov}\{X(t+u), X(t)\} = c_{XX}(u)$ for $t, u = 0, \pm 1, \ldots$. Suppose

$$\sum_{u=-\infty}^{\infty} |c_{XX}(u)| < \infty. \tag{5.5.10}$$

Let $f_{XX}^{(T)}(\lambda)$ be given by (5.5.2) to (5.5.4), then

$$Ef_{XX}^{(T)}(\lambda) = \int_{-\pi}^{\pi} A_T(\alpha) f_{XX}\left(\frac{2\pi s(T)}{T} - \alpha\right)d\alpha \qquad \text{if } \lambda \not\equiv 0 \pmod{\pi} \tag{5.5.11}$$

$$= \int_{-\pi}^{\pi} B_T(\alpha) f_{XX}(-\alpha)d\alpha \qquad \begin{array}{l}\text{if } \lambda \equiv 0 \pmod{2\pi} \\ \text{or if } \lambda = \pm\pi, \pm 3\pi, \ldots \text{ and } T \text{ is even}\end{array} \tag{5.5.12}$$

$$= \int_{-\pi}^{\pi} C_T(\alpha) f_{XX}(\lambda - \alpha)d\alpha \qquad \text{if } \lambda = \pm\pi, \pm 3\pi, \ldots \text{ and } T \text{ is odd.} \tag{5.5.13}$$

The expected value of the estimate (5.5.2) to (5.5.4) differs from that of the estimate of Section 5.4 in the nature of the weighted average of the power spectrum $f_{XX}(\alpha)$. Because we can affect the character of the weighted average by the choice of the W_j, we may well be able to produce an estimate with less bias than that of Section 5.4 in the case that $f_{XX}(\alpha)$ varies in the neighborhood of λ.

Corollary 5.5.1 Suppose in addition to the assumptions of the theorem, $\lambda - 2\pi s(T)/T = O(T^{-1})$ and

$$\sum_u |u|\,|c_{XX}(u)| < \infty, \tag{5.5.14}$$

then

$$Ef_{XX}^{(T)}(\lambda) = f_{XX}(\lambda) + O(T^{-1}) \qquad \text{for } -\infty < \lambda < \infty. \tag{5.5.15}$$

In the limit, $f_{XX}^{(T)}(\lambda)$ is an asymptotically unbiased estimate of $f_{XX}(\lambda)$.

Turning to the second-order moment structure, we have

Theorem 5.5.2 Let $X(t)$, $t = 0, \pm 1, \ldots$ be a real-valued series satisfying Assumption 2.6.2(1). Let $f_{XX}^{(T)}(\lambda)$ be given by (5.5.2) to (5.5.4) with $\lambda - 2\pi s(T)/T = O(T^{-1})$. Suppose $\lambda \pm \mu \not\equiv 0 \pmod{2\pi}$, then

$$\operatorname{var} f_{XX}^{(T)}(\lambda) = f_{XX}(\lambda)^2 \sum_{j=-m}^{m} W_j^2 + O(T^{-1}) \qquad \text{if } \lambda \not\equiv 0 \pmod{\pi} \tag{5.5.16}$$

$$= f_{XX}(\lambda)^2 \sum_{j=1}^{m} W_j^2 \Big/ \left[\sum_{1}^{m} W_j\right]^2 \qquad \text{if } \lambda \equiv 0 \pmod{\pi}. \tag{5.5.17}$$

Also

$$\operatorname{cov}\{f_{XX}^{(T)}(\lambda), f_{XX}^{(T)}(\mu)\} = O(T^{-1}). \tag{5.5.18}$$

The variance of the estimate is seen to be proportional to $\sum_j W_j^2$ for large T. We remark that

$$\sum W_k^2 = \sum_k \left(W_k - \frac{\sum W_j}{2m+1}\right)^2 + \frac{(\sum W_j)^2}{2m+1} \tag{5.5.19}$$

and so $\sum_j W_j^2$ is minimized, subject to $\sum_j W_j = 1$, by setting

$$W_j = \frac{1}{2m+1} \qquad j = 0, \pm 1, \ldots, \pm m. \tag{5.5.20}$$

It follows that the large sample variance of $f_{XX}^{(T)}(\lambda)$ is minimized by taking it to be the estimate of Section 5.4. Following the discussion after Theorem

5.5.1, it may well be the case that the estimate of Section 5.4 has greater bias than the estimate (5.5.2) involving well-chosen W_j.

Turning to an investigation of limiting distributions we have

Theorem 5.5.3 Let $X(t)$, $t = 0, \pm 1, \ldots$ be a real-valued series satisfying Assumption 2.6.1. Let $f_{XX}^{(T)}(\lambda)$ given by (5.5.2) to (5.5.4) with $2\pi s(T)/T \rightarrow \lambda$ as $T \rightarrow \infty$. Suppose $\lambda_j \pm \lambda_k \not\equiv 0 \pmod{2\pi}$ for $1 \leqslant j < k \leqslant J$. Then $f_{XX}^{(T)}(\lambda_1), \ldots, f_{XX}^{(T)}(\lambda_J)$ are asymptotically independent with $f_{XX}^{(T)}(\lambda)$ asymptotically

$$f_{XX}(\lambda) \sum_{j=-m}^{m} W_j \chi_2^2(j)/2 \qquad \text{if } \lambda \not\equiv 0 \pmod{\pi} \tag{5.5.21}$$

and

$$f_{XX}(\lambda)\left[\sum_{j=1}^{m} W_j \chi_2^2(j)/2\right] \Big/ \left[\sum_{1}^{m} W_j\right]^2 \qquad \text{if } \lambda \equiv 0 \pmod{\pi}. \tag{5.5.22}$$

The different chi-squared variates appearing are statistically independent.

The asymptotic distribution of $f_{XX}^{(T)}(\lambda)$ is seen to be that of a weighted combination of independent chi-squared variates. It may prove difficult to use this as an approximating distribution in practice; however, a standard statistical procedure (see Satterthwaite (1941) and Box (1954)) is to approximate the distribution of such a variate by a multiple, $\theta\chi_\nu^2$, of a chi-squared whose mean and degrees of freedom are determined by equating first- and second-order moments. Here we are led to set

$$\theta\nu = \sum_{j=-m}^{m} W_j = 1 \tag{5.5.23}$$

$$\theta^2 2\nu = \sum_{j=-m}^{m} W_j^2 \tag{5.5.24}$$

or

$$\nu = \frac{2}{\sum_{j=-m}^{m} W_j^2} \tag{5.5.25}$$

and

$$\theta = \frac{1}{\nu}. \tag{5.5.26}$$

In the case $W_j = 1/(2m + 1)$, this leads us back to the approximation suggested by Theorem 5.4.3. The approximation of the distribution of power spectral estimates by a multiple of a chi-squared was suggested by Tukey (1949). Other approximations are considered in Freiberger and Grenander (1959), Slepian (1958), and Grenander et al (1959).

In this section we have obtained a flexible estimate of the power spectrum by introducing a variable weighting scheme for periodogram ordinates. We have considered asymptotic properties of the estimate under a limiting procedure involving the weighting of a constant number, $2m + 1$, of periodogram ordinates as $T \to \infty$. For some purposes this procedure may suggest valid large sample approximations, for other purposes it might be better to allow m to increase with T. We turn to an investigation of this alternate limiting procedure in the next section, which will lead us to estimates that are asymptotically normal and consistent.

It is possible to employ different weights W_j or different m in separate intervals of the frequency domain if the character of $f_{XX}(\lambda)$ differs in those intervals.

5.6 A CLASS OF CONSISTENT ESTIMATES

The class of estimates we consider in this section has the form

$$f_{XX}^{(T)}(\lambda) = \frac{2\pi}{T}\sum_{s=1}^{T-1} W^{(T)}\left(\lambda - \frac{2\pi s}{T}\right) I_{XX}^{(T)}\left(\frac{2\pi s}{T}\right) \qquad -\infty < \lambda < \infty \tag{5.6.1}$$

where $W^{(T)}(\alpha)$, $-\infty < \alpha < \infty$, $T = 1, 2, \ldots$ is a family of weight functions of period 2π whose mass is arranged so that estimate (5.6.1) essentially involves a weighting of $2m_T + 1$ periodogram ordinates in the neighborhood of λ. In order to obtain an estimate of diminishing variance as $T \to \infty$, we will therefore require $m_T \to \infty$ in contrast to the constant m of Section 5.5. Also the range of frequencies involved in the estimate (5.6.1) is $2\pi(2m_T + 1)/T$, and so, in order to obtain an asymptotically unbiased estimate, we will require $m_T/T \to 0$ as $T \to \infty$. The estimate inherits the smoothness properties of $W^{(T)}(\alpha)$.

A convenient manner in which to construct the weight function $W^{(T)}$, appearing in the estimate (5.6.1) and having the properties referred to, is to consider a sequence of scale parameters B_T, $T = 1, 2, \ldots$ with the properties $B_T > 0$, $B_T \to 0$, $B_T T \to \infty$ as $T \to \infty$ and to set

$$W^{(T)}(\alpha) = \sum_{j=-\infty}^{\infty} B_T^{-1} W(B_T^{-1}[\alpha + 2\pi j]) \qquad -\infty < \alpha < \infty, \tag{5.6.2}$$

where $W(\beta)$, $-\infty < \beta < \infty$, is a fixed function satisfying

Assumption 5.6.1 $W(\beta)$, $-\infty < \beta < \infty$, is real-valued, even, of bounded variation

$$\int_{-\infty}^{\infty} W(\beta)d\beta = 1 \tag{5.6.3}$$

and

$$\int |W(\beta)|d\beta < \infty. \tag{5.6.4}$$

If we choose $W(\beta)$ to be 0 for $|\beta| > 2\pi$, then we see that the estimate (5.6.1) involves the weighting of the $2B_TT + 1$ periodogram ordinates whose frequencies fall in the interval $(\lambda - 2\pi B_T, \lambda + 2\pi B_T)$. In terms of the introduction to this section, the identification $m_T = B_TT$ is made.

Because $W^{(T)}(\alpha)$ has period 2π, the same will be true of $f_{XX}^{(T)}(\lambda)$. Likewise, because $W^{(T)}(-\alpha) = W^{(T)}(\alpha)$, we will have $f_{XX}^{(T)}(-\lambda) = f_{XX}^{(T)}(\lambda)$. The estimate (5.6.1) is not necessarily non-negative under Assumption 5.6.1; however if, in addition, we assume $W(\beta) \geqslant 0$, then it will be the case that $f_{XX}^{(T)}(\lambda) \geqslant 0$. Because of (5.6.3), $\int_0^{2\pi} W^{(T)}(\alpha)d\alpha = 1$.

In view of (5.6.2) we may set down the following alternate expressions for $f_{XX}^{(T)}(\lambda)$,

$$\begin{aligned} f_{XX}^{(T)}(\lambda) &= \sum_{s \not\equiv 0 (\text{mod } T)} B_T^{-1} W\left(B_T^{-1}\left[\lambda - \frac{2\pi s}{T}\right]\right) I_{XX}^{(T)}\left(\frac{2\pi s}{T}\right) \\ &= \sum_{s=-\infty}^{\infty} B_T^{-1} W\left(B_T^{-1}\left[\lambda - \frac{2\pi s}{T}\right]\right) I_{X-c_X^{(T)},X-c_X^{(T)}}^{(T)}\left(\frac{2\pi s}{T}\right) \end{aligned} \tag{5.6.5}$$

that help to explain its character. For large T, in view of (5.6.3), the sum of the weights appearing in (5.6.1) should be near 1. The worker may wish to alter (5.6.1) so the sum of the weights is exactly 1. This alteration will have no effect on the asymptotic expressions given below.

Turning to a large sample investigation of the mean of $f_{XX}^{(T)}(\lambda)$ we have

Theorem 5.6.1 Let $X(t)$, $t = 0, \pm 1, \ldots$ be a real-valued series with $EX(t) = c_X$ and $\text{cov}\{X(t + u), X(t)\} = c_{XX}(u)$ for $t, u = 0, \pm 1, \ldots$. Suppose

$$\sum_{u=-\infty}^{\infty} |u|\, |c_{XX}(u)| < \infty. \tag{5.6.6}$$

Let $f_{XX}^{(T)}(\lambda)$ be given by (5.6.1) where $W(\beta)$ satisfies Assumption 5.6.1. Then

$$\begin{aligned} Ef_{XX}^{(T)}(\lambda) &= \frac{2\pi}{T}\sum_{s=1}^{T-1} W^{(T)}\left(\lambda - \frac{2\pi s}{T}\right) f_{XX}\left(\frac{2\pi s}{T}\right) + O(T^{-1}) \\ &= \int_{-\infty}^{\infty} W(\beta) f_{XX}(\lambda - B_T\beta)d\beta + O(B_T^{-1}T^{-1}) \end{aligned}$$

$$\text{for } -\infty < \lambda < \infty. \tag{5.6.7}$$

The error terms are uniform in λ.

The expected value of $f_{XX}^{(T)}(\lambda)$ is seen to be a weighted average of the function $f_{XX}(\alpha)$, $-\infty < \alpha < \infty$, with weight concentrated in an interval containing λ and of length proportional to B_T. We now have

Corollary 5.6.1 Under the conditions of the theorem and if $B_T \to 0$ as $T \to \infty$, $f_{XX}^{(T)}(\lambda)$ is an asymptotically unbiased estimate of $f_{XX}(\lambda)$, that is,

$$\lim_{T\to\infty} Ef_{XX}^{(T)}(\lambda) = f_{XX}(\lambda) \qquad \text{for } -\infty < \lambda < \infty. \tag{5.6.8}$$

The property of being asymptotically unbiased was also possessed by the estimate of Section 5.5. Turning to second-order large sample properties of the estimate we have

Theorem 5.6.2 Let $X(t)$, $t = 0, \pm 1, \ldots$ be a real-valued series satisfying Assumption 2.6.2(1). Let $f_{XX}^{(T)}(\lambda)$ be given by (5.6.1) where $W(\beta)$ satisfies Assumption 5.6.1. Then

$$\begin{aligned}
&\operatorname{cov}\{f_{XX}^{(T)}(\lambda), f_{XX}^{(T)}(\mu)\} \\
&= \left(\frac{2\pi}{T}\right)^2 \sum_{s=1}^{T-1} W^{(T)}\left(\lambda - \frac{2\pi s}{T}\right) W^{(T)}\left(\mu - \frac{2\pi s}{T}\right) f_{XX}\left(\frac{2\pi s}{T}\right)^2 \\
&\quad + \left(\frac{2\pi}{T}\right)^2 \sum_{s=1}^{T-1} W^{(T)}\left(\lambda - \frac{2\pi s}{T}\right) W^{(T)}\left(\mu + \frac{2\pi s}{T}\right) f_{XX}\left(\frac{2\pi s}{T}\right)^2 + O(T^{-1}) \\
&= 2\pi T^{-1} \int_0^{2\pi} W^{(T)}(\lambda - \alpha) W^{(T)}(\mu - \alpha) f_{XX}(\alpha)^2 d\alpha \\
&\quad + 2\pi T^{-1} \int_0^{2\pi} W^{(T)}(\lambda - \alpha) W^{(T)}(\mu + \alpha) f_{XX}(\alpha)^2 d\alpha + O(B_T^{-2}T^{-2}) \\
&\quad + O(T^{-1}) \qquad \text{for } -\infty < \lambda, \mu < \infty.
\end{aligned} \tag{5.6.9}$$

In Corollary 5.6.2 below we make use of the function

$$\begin{aligned}
\eta\{\lambda\} &= 1 \qquad \text{if } \lambda \equiv 0 \pmod{2\pi} \\
&= 0 \qquad \text{otherwise.}
\end{aligned} \tag{5.6.10}$$

This is a periodic extension of the Kronecker delta function

$$\begin{aligned}
\delta\{\lambda\} &= 1 \qquad \text{if } \lambda = 0 \\
&= 0 \qquad \text{otherwise.}
\end{aligned} \tag{5.6.11}$$

Corollary 5.6.2 Under the conditions of Theorem 5.6.2 and if $B_T T \to \infty$ as $T \to \infty$

$$\begin{aligned}
&\lim_{T\to\infty} B_T T \operatorname{cov}\{f_{XX}^{(T)}(\lambda), f_{XX}^{(T)}(\mu)\} \\
&\qquad = 2\pi[\eta\{\lambda - \mu\} + \eta\{\lambda + \mu\}] f_{XX}(\lambda)^2 \int_{-\infty}^{\infty} W(\beta)^2 d\beta.
\end{aligned} \tag{5.6.12}$$

In the case of $\mu = \lambda$, this corollary indicates

$$\begin{aligned} \operatorname{var} f_{XX}^{(T)}(\lambda) &\backsim B_T^{-1}T^{-1}2\pi f_{XX}(\lambda)^2 \int_{-\infty}^{\infty} W(\beta)^2 d\beta \qquad \text{if } \lambda \not\equiv 0 \pmod{\pi} \\ &\backsim B_T^{-1}T^{-1}4\pi f_{XX}(\lambda)^2 \int_{-\infty}^{\infty} W(\beta)^2 d\beta \qquad \text{if } \lambda \equiv 0 \pmod{\pi}. \end{aligned} \tag{5.6.13}$$

In either case the variance of $f_{XX}^{(T)}(\lambda)$ is tending to 0 as $B_T T \to \infty$. In Corollary 5.6.1 we saw that $Ef_{XX}^{(T)}(\lambda) \to f_{XX}(\lambda)$ as $T \to \infty$ if $B_T \to 0$. Therefore the estimate (5.6.1) has the property

$$\lim_{T\to\infty} E|f_{XX}^{(T)}(\lambda) - f_{XX}(\lambda)|^2 = 0 \tag{5.6.14}$$

under the conditions of Theorem 5.6.2 and if $B_T \to 0$, $B_T T \to \infty$ as $T \to \infty$. Such an estimate is called **consistent in mean square.**

Notice that in expression (5.6.13) we have a doubling of variance at $\lambda = 0, \pm\pi, \pm 2\pi, \ldots$. Expression (5.6.9) is much more informative in this connection. It indicates that the transition between the usual asymptotic behavior and that at $\lambda = 0, \pm\pi, \pm 2\pi, \ldots$ takes place in intervals about these points whose length is of the order of magnitude of B_T.

We see from (5.6.12), that $f_{XX}^{(T)}(\lambda), f_{XX}^{(T)}(\mu)$ are asymptotically uncorrelated as $T \to \infty$, provided $\lambda - \mu, \lambda + \mu \not\equiv 0 \pmod{2\pi}$. Turning to the asymptotic distribution of $f_{XX}^{(T)}(\lambda)$, we have

Theorem 5.6.3 Let $X(t)$, $t = 0, \pm 1, \ldots$ be a real-valued series satisfying Assumption 2.6.1. Let $f_{XX}^{(T)}(\lambda)$ be given by (5.6.1) with $W(\beta)$ satisfying Assumption 5.6.1. Suppose $f_{XX}(\lambda_j) \neq 0, j = 1, \ldots, J$. Then $f_{XX}^{(T)}(\lambda_1), \ldots, f_{XX}^{(T)}(\lambda_J)$ are asymptotically normal with covariance structure given by (5.6.12) as $T \to \infty$ with $B_T T \to \infty$, $B_T \to 0$.

The estimate considered in Section 5.4 had an asymptotic distribution proportional to chi-squared under the assumption that we were smoothing a fixed number of periodogram ordinates. Here the number of periodogram ordinates being smoothed is increasing to ∞ with T and so it is not surprising that an asymptotic normal distribution results. One interesting implication of the theorem is that $f_{XX}^{(T)}(\lambda)$ and $f_{XX}^{(T)}(\mu)$ are asymptotically independent if $\lambda \pm \mu \not\equiv 0 \pmod{2\pi}$. The theorem has the following:

Corollary 5.6.3 Under the conditions of Theorem 5.6.3 and if $f_{XX}(\lambda) \neq 0$, $\log_{10} f_{XX}^{(T)}(\lambda)$ is asymptotically normal with

$$\begin{aligned} \overrightarrow{\operatorname{var}} \log_{10} f_{XX}^{(T)}(\lambda) &\backsim B_T^{-1}T^{-1}(\log_{10} e)^2\, 2\pi \int W(\beta)^2 d\beta \qquad \text{if } \lambda \not\equiv 0 \pmod{\pi} \\ &\backsim B_T^{-1}T^{-1}(\log_{10} e)^2\, 4\pi \int W(\beta)^2 d\beta \qquad \text{if } \lambda \equiv 0 \pmod{\pi}. \end{aligned} \tag{5.6.15}$$

This corollary suggests that the variance of $\log f_{XX}^{(T)}(\lambda)$ may not depend too strongly on the magnitude of $f_{XX}(\lambda)$, nor generally on λ, for large T. Therefore, it is probably more sensible to plot the statistic $\log f_{XX}^{(T)}(\lambda)$, rather than $f_{XX}^{(T)}(\lambda)$ itself. In fact this has been the standard engineering practice and is what is done for the various estimated spectra of this chapter.

Consistent estimates of the power spectrum were obtained by Grenander and Rosenblatt (1957) and Parzen (1957, 1958). The asymptotic mean and variance were considered by these authors and by Blackman and Tukey (1958). Asymptotic normality has been demonstrated by Rosenblatt (1959), Brillinger (1965b, 1968), Brillinger and Rosenblatt (1967a), Hannan (1970), and Anderson (1971) under various conditions. Jones (1962a) is also of interest.

In the case that the data have been tapered prior to forming a power spectral estimate, Theorem 5.6.3 takes the form

Theorem 5.6.4 Let $X(t)$, $t = 0, \pm 1, \ldots$ be a real-valued series satisfying Assumption 2.6.1. Let $h(t)$, $-\infty < t < \infty$, be a taper satisfying Assumption 4.3.1. Let $W(\alpha)$, $-\infty < \alpha < \infty$, satisfy Assumption 5.6.1. Set

$$I_{X-c_X^{(T)},X-c_X^{(T)}}^{(T)}(\lambda) = \left(2\pi \sum_t h\left(\frac{t}{T}\right)^2\right)^{-1} \left|\sum_t h\left(\frac{t}{T}\right)(X(t) - c_X^{(T)}) \exp\{-i\lambda t\}\right|^2 \tag{5.6.16}$$

where

$$c_X^{(T)} = \sum_t h\left(\frac{t}{T}\right)X(t) \Big/ \left[\sum_t h\left(\frac{t}{T}\right)\right]. \tag{5.6.17}$$

Set

$$f_{XX}^{(T)}(\lambda) = \frac{2\pi}{T} \sum_s W^{(T)}\left(\lambda - \frac{2\pi s}{T}\right) I_{X-c_X^{(T)},X-c_X^{(T)}}^{(T)}\left(\frac{2\pi s}{T}\right). \tag{5.6.18}$$

Let $B_T \to 0$, $B_T T \to \infty$ as $T \to \infty$. Then $f_{XX}^{(T)}(\lambda_1), \ldots, f_{XX}^{(T)}(\lambda_J)$ are asymptotically jointly normal with

$$\lim_{T\to\infty} E f_{XX}^{(T)}(\lambda) = f_{XX}(\lambda) \tag{5.6.19}$$

and

$$\lim_{T\to\infty} B_T T \operatorname{cov}\{f_{XX}^{(T)}(\lambda), f_{XX}^{(T)}(\mu)\}$$
$$= 2\pi[\eta\{\lambda - \mu\} + \eta\{\lambda + \mu\}] \int h(t)^4 dt \left[\int h(t)^2 dt\right]^{-2} \int W(\alpha)^2 d\alpha\, f_{XX}(\lambda)^2 \tag{5.6.20}$$

for $-\infty < \lambda, \mu < \infty$.

By comparison with expression (5.6.12) the limiting variance of the tapered estimate is seen to differ from that of the untapered estimate by the factor

$$\int_{-1}^{1} h(t)^4 dt \Big/ \left[\int_{-1}^{1} h(t)^2 dt \right]^2. \tag{5.6.21}$$

By Schwarz's inequality this factor is $\geqslant 1$. In the case where we employ a cosine taper extending over the first and last 10 percent of the data, its value is 1.116. It is hoped that in many situations the bias of the tapered estimate will be reduced so substantially as to more than compensate for this increase in variance. Table 3.3.1 gives some useful tapers.

5.7 CONFIDENCE INTERVALS

In order to communicate an indication of the possible nearness of an estimate to a parameter, it is often desirable to provide a confidence interval for the parameter based on the estimate. The asymptotic distributions determined in the previous sections for the various spectral estimates may be used in this connection. We first set down some notation. Let $z(\alpha)$, $\chi_\nu^2(\alpha)$ denote numbers such that

$$\text{Prob}\,[z < z(\alpha)] = \alpha \tag{5.7.1}$$

and

$$\text{Prob}\,[\chi_\nu^2 < \chi_\nu^2(\alpha)] = \alpha \tag{5.7.2}$$

where z is a standard normal variate and χ_ν^2 a chi-squared variate with ν degrees of freedom.

Consider first the estimate of Section 5.4,

$$f_{XX}^{(T)}(\lambda) = (2m+1)^{-1} \sum_{j=-m}^{m} I_{XX}^{(T)}\left(\frac{2\pi[s(T)+j]}{T}\right) \tag{5.7.3}$$

for $2\pi s(T)/T$ near $\lambda \not\equiv 0 \pmod{\pi}$. Theorem 5.4.3 suggests approximating its distribution by $f_{XX}(\lambda)\chi_{4m+2}^2/(4m+2)$. This leads to the following 100γ percent confidence interval for $f_{XX}(\lambda)$

$$\frac{(4m+2)f_{XX}^{(T)}(\lambda)}{\chi_{4m+2}^2\left(\frac{1+\gamma}{2}\right)} < f_{XX}(\lambda) < \frac{(4m+2)f_{XX}^{(T)}(\lambda)}{\chi_{4m+2}^2\left(\frac{1-\gamma}{2}\right)}. \tag{5.7.4}$$

If we take logarithms, this interval becomes

$$\log f_{XX}^{(T)}(\lambda) - \log\left\{\chi_{4m+2}^2\left(\frac{1+\gamma}{2}\right)/(4m+2)\right\} < \log f_{XX}(\lambda)$$
$$< \log f_{XX}^{(T)}(\lambda) - \log\left\{\chi_{4m+2}^2\left(\frac{1-\gamma}{2}\right)/(4m+2)\right\}. \tag{5.7.5}$$

The degrees of freedom and multipliers of chi-squared will be altered in the case $\lambda \equiv 0 \pmod{\pi}$ in accordance with the details of Theorem 5.4.3.

In Figure 5.7.1 we have set 95 percent limits around the estimate, corresponding to $m = 2$, of Figure 5.4.4. We have inserted these limits in two manners. In the upper half of Figure 5.7.1 we have proceeded in accordance with expression (5.7.5). In the lower half, we have set the limits around a strongly smoothed spectral estimate; this procedure has the advantage of causing certain peaks to stand out.

In Section 5.5, we considered the estimate

$$\sum_{j=-m}^{m} W_j I_{XX}^{(T)}\left(\frac{2\pi[s(T)+j]}{T}\right) \qquad \lambda \not\equiv 0 \pmod{\pi}, \tag{5.7.6}$$

involving a variable weighting of periodogram ordinates. Its asymptotic distribution was found to be that of a weighted sum of exponential variates. This last is generally not a convenient distribution to work with; however, in the discussion of Theorem 5.5.3 it was suggested it be approximated by $f_{XX}(\lambda)\chi_\nu^2/\nu$ where

$$\nu = \frac{2}{\sum_{j=-m}^{m} W_j^2} \tag{5.7.7}$$

in the case $\lambda \not\equiv 0 \pmod{\pi}$. Taking this value of ν, we are led to the following 100γ percent confidence interval for $\log f_{XX}(\lambda)$,

$$\log f_{XX}^{(T)}(\lambda) - \log\left\{\chi_\nu^2\left(\frac{1+\gamma}{2}\right)/\nu\right\} < \log f_{XX}(\lambda)$$
$$< \log f_{XX}^{(T)}(\lambda) - \log\left\{\chi_\nu^2\left(\frac{1-\gamma}{2}\right)/\nu\right\}. \tag{5.7.8}$$

If $W_j = 1/(2m+1)$, $j = 0, \pm 1, \ldots, \pm m$, then the interval (5.7.8) is the same as the interval (5.7.5).

If ν is large, then $\log_{10}\{\chi_\nu^2/\nu\}$ is approximately normal with mean 0 and variance $2(.4343)^2/\nu$. The interval (5.7.8) is therefore approximately

$$\log f_{XX}^{(T)}(\lambda) - z\left(\frac{1+\gamma}{2}\right)(.4343)\sqrt{\sum_{j=-m}^{m} W_j^2} < \log f_{XX}(\lambda)$$
$$< \log f_{XX}^{(T)}(\lambda) - z\left(\frac{1-\gamma}{2}\right)(.4343)\sqrt{\sum_{j=-m}^{m} W_j^2}. \tag{5.7.9}$$

Interval (5.7.9) leads us directly into the approximation suggested by the results of Section 5.6. The estimate considered there had the form

$$f_{XX}^{(T)}(\lambda) = \frac{2\pi}{T}\sum_{s=1}^{T-1} W^{(T)}\left(\lambda - \frac{2\pi s}{T}\right) I_{XX}^{(T)}\left(\frac{2\pi s}{T}\right). \tag{5.7.10}$$

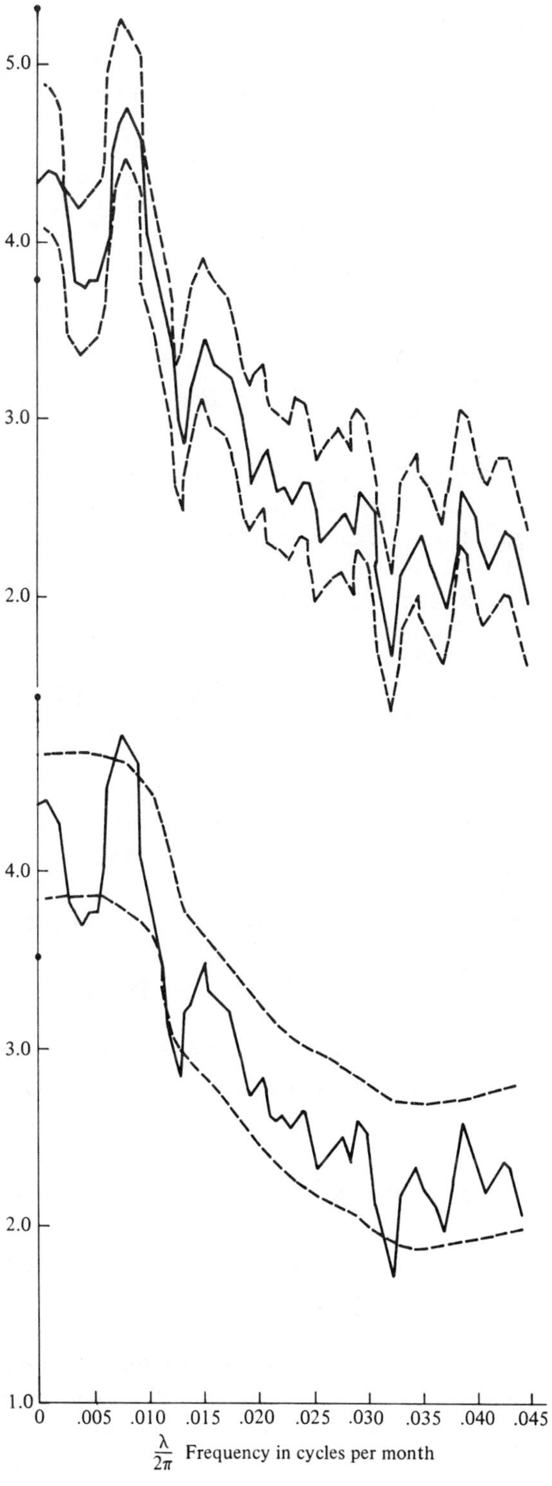

Figure 5.7.1 Two manners of setting 95 percent confidence limits about the power spectrum estimate of Figure 5.4.4.

Corollary 5.6.3 leads us to set down the following 100γ percent confidence interval for $\log_{10} f_{XX}(\lambda)$,

$$\log_{10} f_{XX}^{(T)}(\lambda) - z\left(\frac{1+\gamma}{2}\right)(.4343)\sqrt{\frac{2\pi \int W(\beta)^2 d\beta}{B_T T}} < \log_{10} f_{XX}(\lambda)$$
$$< \log_{10} f_{XX}^{(T)}(\lambda) + z\left(\frac{1+\gamma}{2}\right)(.4343)\sqrt{\frac{2\pi \int W(\beta)^2 d\beta}{B_T T}}. \quad (5.7.11)$$

Because

$$\sum_{-m}^{m} W_j^2 = \sum_s \left[\frac{2\pi}{T} W^{(T)}\left(\lambda - \frac{2\pi s}{T}\right)\right]^2 \doteq \frac{2\pi}{B_T T}\int W(\beta)^2 d\beta, \quad (5.7.12)$$

the interval (5.7.11) is in essential agreement with the interval (5.7.9). The intervals (5.7.9) and (5.7.11) are relevant to the case $\lambda \not\equiv 0 \pmod{\pi}$. If $\lambda \equiv 0 \pmod{\pi}$, then the variance of the estimate is approximately doubled, indicating we should broaden the intervals by a factor of $\sqrt{2}$.

In the case that we believe $f_{XX}(\alpha)$ to be a very smooth function in some interval about λ, an *ad hoc* procedure is also available. We may estimate the variance of $f_{XX}^{(T)}(\lambda)$ from the variation of $f_{XX}^{(T)}(\alpha)$, in the neighborhood of λ. For example, this might prove a reasonable procedure for the frequencies $\pi/2 < \lambda < \pi$ in the case of the series of monthly mean sunspot numbers analyzed previously.

The confidence intervals constructed in this section apply to the spectral estimate at a single frequency λ. A proportion $1 - \gamma$ of the values may be expected to fall outside the limits. On occasion we may wish a confidence region valid for the whole frequency range. Woodroofe and Van Ness (1967) determined the asymptotic distribution of the variate

$$\sup_{0 < n < N_T} \left| f_{XX}^{(T)}\left(\frac{\pi n}{N_T}\right) - f_{XX}\left(\frac{\pi n}{N_T}\right)\right| \quad (5.7.13)$$

where $N_T \to \infty$ as $T \to \infty$. An approximate confidence region for $f_{XX}(\lambda)$, $0 < \lambda < \pi$, might be determined from this asymptotic distribution.

5.8 BIAS AND PREFILTERING

In this section we will carry out a more detailed analysis of the bias of the proposed estimates of a power spectrum. We will indicate how an elementary operation, called **prefiltering**, can often be used to reduce this bias. We begin by considering the periodogram of a series of tapered values. For convenience, assume $EX(t) = 0$, although the general conclusions reached will be relevant to the nonzero mean case as well.

Let

$$d_X^{(T)}(\lambda) = \sum_t h\left(\frac{t}{T}\right)X(t)\exp\{-i\lambda t\} \qquad \text{for } -\infty < \lambda < \infty, \quad (5.8.1)$$

where $h(u)$ is a tapering function vanishing for $u < 0$, $u > 1$. The periodogram here is taken to be

$$I_{XX}^{(T)}(\lambda) = \left(2\pi \sum_t h\left(\frac{t}{T}\right)^2\right)^{-1} |d_X^{(T)}(\lambda)|^2. \quad (5.8.2)$$

If we define the kernel

$$K^{(T)}(\alpha) = |H^{(T)}(\alpha)|^2 \Big/ \int_{-\pi}^{\pi} |H^{(T)}(\alpha)|^2 d\alpha \quad (5.8.3)$$

where

$$H^{(T)}(\alpha) = \sum_t h\left(\frac{t}{T}\right)\exp\{-it\alpha\}, \quad (5.8.4)$$

then Theorem 5.2.3 indicates that

$$EI_{XX}^{(T)}(\lambda) = \int_{-\pi}^{\pi} K^{(T)}(\alpha) f_{XX}(\lambda - \alpha) d\alpha. \quad (5.8.5)$$

We proceed to examine this expected value in greater detail. Set

$$k^{(T)}(u) = \int_{-\pi}^{\pi} K^{(T)}(\alpha)\exp\{-iu\alpha\}d\alpha = \sum_t h\left(\frac{t}{T}\right)h\left(\frac{t+u}{T}\right) \Big/ \sum_t h\left(\frac{t}{T}\right)^2. \quad (5.8.6)$$

As we might expect from Section 3.3, we have

Theorem 5.8.1 Let $X(t)$, $t = 0, \pm 1, \ldots$ be a real-valued series with 0 mean and autocovariance function satisfying

$$\sum |u|^P |c_{XX}(u)| < \infty \qquad \text{for some } P \geq 1. \quad (5.8.7)$$

Let the tapering function $h(u)$ be such that $k^{(T)}(u)$ given by (5.8.6) satisfies

$$k^{(T)}(u) = 1 + k_1\frac{u}{T} + k_2\frac{u^2}{T^2} + \cdots + k_{P-1}\frac{u^{P-1}}{T^{P-1}} + o\left(\frac{|u|^P}{T^P}\right) + O(T^{-P}) \qquad \text{for } |u| \leq T. \quad (5.8.8)$$

Let $I_{XX}^{(T)}(\lambda)$ be given by (5.8.2), then

$$EI_{XX}^{(T)}(\lambda) = \int_{-\pi}^{\pi} K^{(T)}(\alpha) f_{XX}(\lambda - \alpha) d\alpha$$
$$= f_{XX}(\lambda) + \sum_{p=1}^{P-1} i^p T^{-p} k_p f_{XX}^{\{p\}}(\lambda) + O(T^{-P}) \quad (5.8.9)$$

where $f_{XX}^{\{p\}}(\lambda)$ is the pth derivative of $f_{XX}(\lambda)$. The error term is uniform in λ.

From its definition, $k^{(T)}(u) = k^{(T)}(-u)$ and so the k_p in (5.8.8) are 0 for odd p. The dominant bias term appearing in (5.8.9) is therefore

$$-T^{-2}k_2 \frac{d^2 f_{XX}(\lambda)}{d\lambda^2}. \tag{5.8.10}$$

This term is seen to depend on both the kernel employed and the spectrum being estimated. We will want to chose a taper so that $|k_2|$ is small. In fact, if we use the definition (3.3.11) of bandwidth, then the bandwidth of the kernel $K^{(T)}(\alpha)$ is $\sqrt{|k_2|}/T$ also implying the desirability of small $|k_2|$. The bandwidth is an important parameter in determining the extent of bias. In real terms, the student will have difficulty in distinguishing (or resolving) peaks in the spectrum closer than $\sqrt{|k_2|}/T$ apart. This was apparent to an extent from Theorem 5.2.8, which indicated that the statistics $I_{XX}^{(T)}(\lambda)$ and $I_{XX}^{(T)}(\mu)$ were highly dependent for λ near μ. Expressions (5.8.9) and (5.8.10) do indicate that the bias will be reduced in the case that $f_{XX}(\alpha)$ is near constant in a neighborhood of λ; this remark will prove the basis for the operation of prefiltering to be discussed later.

Suppose next that the estimate

$$f_{XX}^{(T)}(\lambda) = \sum_{j=-m}^{m} W_j I_{XX}^{(T)}\left(\frac{2\pi[s(T)+j]}{T}\right) \tag{5.8.11}$$

with $2\pi s(T)/T$ near λ and

$$\sum_{j=-m}^{m} W_j = 1 \tag{5.8.12}$$

is considered. Because

$$Ef_{XX}^{(T)}(\lambda) = \sum_{j=-m}^{m} W_j E I_{XX}^{(T)}\left(\frac{2\pi[s(T)+j]}{T}\right) \tag{5.8.13}$$

the remarks following Theorem 5.8.1 are again relevant and imply that the bias of (5.8.11) will be reduced in the case that k_2 is small or $f_{XX}(\alpha)$ is near constant. An alternate way to look at this is to note, from (5.8.5), that

$$Ef_{XX}^{(T)}(\lambda) = \int_{-\pi}^{\pi} \sum_j W_j K^{(T)}\left(\alpha - \frac{2\pi j}{T}\right) f_{XX}\left(\frac{2\pi s(T)}{T} - \alpha\right) d\alpha \tag{5.8.14}$$

where the kernel

$$\sum_j W_j K^{(T)}\left(\alpha - \frac{2\pi j}{T}\right) \tag{5.8.15}$$

appearing in (5.8.14) has the shape of a function taking the value W_j for α near $2\pi j/T, j = 0, \pm 1, \ldots, \pm m$. In crude terms, this kernel extends over an interval m times broader than that of $K^{(T)}(\alpha)$ and so if $f_{XX}(\alpha)$ is not constant, the bias of (5.8.11) may be expected to be greater than that of $I_{XX}^{(T)}(\lambda)$. It will generally be difficult to resolve peaks in $f_{XX}(\lambda)$ nearer than $m\sqrt{k_2}/T$ with the statistic (5.8.11). The smoothing with weights W_j has caused a loss in resolution of the estimate $I_{XX}^{(T)}(\lambda)$. It must be remembered, however, that the smoothing was introduced to increase the stability of the estimate and it is hoped that the smoothed estimate will be better in some overall sense.

We now turn to a more detailed investigation of the consistent estimate introduced in Section 5.6. This estimate is given by

$$f_{XX}^{(T)}(\lambda) = \frac{2\pi}{T}\sum_{s=1}^{T-1} W^{(T)}\left(\lambda - \frac{2\pi s}{T}\right) I_{XX}^{(T)}\left(\frac{2\pi s}{T}\right) \tag{5.8.16}$$

with $I_{XX}^{(T)}(\lambda)$ given by (5.8.2) and $W^{(T)}(\alpha)$ given by (5.6.2).

Theorem 5.8.2 Let $X(t)$, $t = 0, \pm 1, \ldots$ be a real-valued series with $EX(t) = 0$ and autocovariance function satisfying

$$\sum_{u=-\infty}^{\infty} |u|^P \, |c_{XX}(u)| < \infty \tag{5.8.17}$$

for some $P \geqslant 1$. Let the tapering function $h(u)$ be such that $k^{(T)}(u)$ of (5.8.6) satisfies (5.8.8) for $|u| \leqslant T$. Let $f_{XX}^{(T)}(\lambda)$ be given by (5.8.16) where $W(\alpha)$ satisfies Assumption 5.6.1. Then

$$\begin{aligned} Ef_{XX}^{(T)}(\lambda) &= \frac{2\pi}{T}\sum_{s=1}^{T-1} W^{(T)}\left(\lambda - \frac{2\pi s}{T}\right) f_{XX}\left(\frac{2\pi s}{T}\right) \\ &\quad + \frac{2\pi}{T}\sum_{p=1}^{P-1} T^{-p} k_p i^p \sum_{s=1}^{T-1} W^{(T)}\left(\lambda - \frac{2\pi s}{T}\right) f_{XX}^{\{p\}}\left(\frac{2\pi s}{T}\right) + O(T^{-P}) \\ &= \int_{-\pi}^{\pi} W^{(T)}(\alpha) f_{XX}(\lambda - \alpha)\,d\alpha + O(B_T^{-1}T^{-1}) \\ &\qquad \text{for } -\infty < \lambda < \infty. \end{aligned} \tag{5.8.18}$$

The error terms are uniform in λ.

From expression (5.8.18) we see that advantages accrue from tapering in this case as well. Expression (5.8.18) indicates that the expected value is given, approximately, by a weighted average with kernel $W^{(T)}(\alpha)$ of the power spectrum of interest. The bandwidth of this kernel is

$$\left\{\int_{-\pi}^{\pi} [1 - \cos\alpha] W^{(T)}(\alpha)\,d\alpha\right\}^{1/2} \backsim B_T\left\{\frac{1}{2}\int_{-\infty}^{\infty} \alpha^2 W(\alpha)\,d\alpha\right\}^{1/2} \tag{5.8.19}$$

and so is of order $O(B_T)$.

In Corollary 5.8.2 we set

$$W_p = \int_{-\infty}^{\infty} \beta^p W(\beta)\,d\beta. \tag{5.8.20}$$

Corollary 5.8.2 Suppose in addition to the conditions of Theorem 5.8.2

$$\int_{-\infty}^{\infty} |\beta|^P |W(\beta)| d\beta < \infty, \tag{5.8.21}$$

then

$$Ef_{XX}^{(T)}(\lambda) = f_{XX}(\lambda) + \sum_{p=1}^{P-1} \frac{(-B_T)^p}{p!} W_p f_{XX}^{\{p\}}(\lambda) + O(B_T^P) + O(B_T^{-1}T^{-1})$$
$$\text{for } -\infty < \lambda < \infty. \tag{5.8.22}$$

Because $W(\beta) = W(-\beta)$, the terms in (5.8.22) with p odd drop out. We see that the bias, up to order B_T^{P-1}, may be eliminated by selecting a $W(\beta)$ such that $W_p = 0$ for $p = 1, \ldots, P-1$. Clearly such a $W(\beta)$ must take on negative values somewhere leading to complications in some situations. If $P = 3$, then (5.8.22) becomes

$$Ef_{XX}^{(T)}(\lambda) = f_{XX}(\lambda) + \tfrac{1}{2} B_T^2 W_2 \frac{d^2 f_{XX}(\lambda)}{d\lambda^2} + O(B_T^3) + O(B_T^{-1}T^{-1}). \tag{5.8.23}$$

Now from expression (5.8.19) the bandwidth of the kernel $W^{(T)}(\alpha)$ is essentially $B_T\sqrt{W_2/2}$ and once again the bias is seen to depend directly on both the bandwidth of the kernel and the smoothness of $f_{XX}(\alpha)$ for α near λ.

The discussion of Section 3.3 gives some help with the question of which kernel $W^{(T)}(\alpha)$ to employ in the smoothing of the periodogram. Luckily this question can be made academic in large part by a judicious filtering of the data prior to estimating the power spectrum. We have seen that $Ef_{XX}^{(T)}(\lambda)$ is essentially given by

$$\int_{-\pi}^{\pi} W^{(T)}(\lambda - \alpha) f_{XX}(\alpha) d\alpha. \tag{5.8.24}$$

Now if $f_{XX}(\alpha)$ is constant, f_{XX}, then (5.8.24) equals f_{XX} exactly. This suggests that the nearer $f_{XX}(\alpha)$ is to being constant, the smaller the bias. Suppose that the series $X(t)$, $t = 0, \pm 1, \ldots$ is passed through a filter with transfer function $A(\lambda)$. Denote the filtered series by $Y(t)$, $t = 0, \pm 1, \ldots$. From Example 2.8.1, the power spectrum of this series is given by

$$f_{YY}(\lambda) = |A(\lambda)|^2 f_{XX}(\lambda) \tag{5.8.25}$$

with inverse relation

$$f_{XX}(\lambda) = |A(\lambda)|^{-2} f_{YY}(\lambda) \qquad \text{if } A(\lambda) \neq 0. \tag{5.8.26}$$

Let $f_{YY}^{(T)}(\lambda)$ be an estimate of the power spectrum of the series $Y(t)$. Relation (5.8.26) suggests the consideration of the statistic

$$|A(\lambda)|^{-2} f_{YY}^{(T)}(\lambda) \tag{5.8.27}$$

as an estimate of $f_{XX}(\lambda)$. Following the discussion above the expected value of this estimate essentially equals

$$|A(\lambda)|^{-2}\int_{-\pi}^{\pi} W^{(T)}(\lambda-\alpha)f_{YY}(\alpha)d\alpha$$
$$= |A(\lambda)|^{-2}\int_{-\pi}^{\pi} W^{(T)}(\lambda-\alpha)|A(\alpha)|^2 f_{XX}(\alpha)d\alpha. \quad (5.8.28)$$

Had $A(\lambda)$ been chosen so that $|A(\alpha)|^2 f_{XX}(\alpha)$ were constant, then (5.8.28) would equal $f_{XX}(\lambda)$ exactly. This result suggests that in a case where $f_{XX}(\alpha)$ is not near constant, we should attempt to find a filter, with transfer function $A(\lambda)$, such that the filtered series $Y(t)$ has near constant power spectrum; then we should estimate this near constant power spectrum from a stretch of the series $Y(t)$; and finally, we take $|A(\lambda)|^{-2} f_{YY}{}^{(T)}(\lambda)$ as an estimate of $f_{XX}(\lambda)$. This procedure is called **spectral estimation by prefiltering or prewhitening;** it was proposed in Press and Tukey (1956). Typically the filter has been determined by *ad hoc* methods; however, one general procedure has been proposed by Parzen and Tukey. It is to determine the filter by fitting an autoregressive scheme to the data. Specifically, for some m, determine $a^{(T)}(1), \ldots, a^{(T)}(m)$ to minimize

$$\sum_{t=m}^{T-1} [X(t) - a^{(T)}(1)X(t-1) - \cdots - a^{(T)}(m)X(t-m)]^2 \quad (5.8.29)$$

then form the filtered series

$$Y(t) = X(t) - a^{(T)}(1)X(t-1) - \cdots - a^{(T)}(m)X(t-m) \qquad \text{for } t = m, \ldots, T-1 \quad (5.8.30)$$

and proceed as above. In the case that the series $X(t)$ is approximately an autoregressive scheme of order m, see Section 2.9; this must be a near optimum procedure. It seems to work well in other cases also.

A procedure of similar character, but not requiring any filtering of the data, is if the series $Y(t)$ were obtained from the series $X(t)$ by filtering with transfer function $A(\lambda)$, then following (5.3.20) we have

$$I_{YY}{}^{(T)}(\lambda) \doteq |A(\lambda)|^2 I_{XX}{}^{(T)}(\lambda) \quad (5.8.31)$$

and so

$$f_{YY}{}^{(T)}(\lambda) = \frac{2\pi}{T}\sum_{s=1}^{T-1} W^{(T)}\left(\lambda - \frac{2\pi s}{T}\right) I_{YY}{}^{(T)}\left(\frac{2\pi s}{T}\right)$$
$$= \frac{2\pi}{T}\sum_{s=1}^{T-1} W^{(T)}\left(\lambda - \frac{2\pi s}{T}\right)\left|A\left(\frac{2\pi s}{T}\right)\right|^2 I_{XX}{}^{(T)}\left(\frac{2\pi s}{T}\right). \quad (5.8.32)$$

The discussion above now suggests the following estimate of $f_{XX}(\lambda)$,

$$|A(\lambda)|^{-2}\frac{2\pi}{T}\sum_{s=1}^{T-1} W^{(T)}\left(\lambda-\frac{2\pi s}{T}\right)\left|A\left(\frac{2\pi s}{T}\right)\right|^2 I_{XX}^{(T)}\left(\frac{2\pi s}{T}\right) \qquad (5.8.33)$$

where the function $A(\alpha)$ has been chosen in the hope that $|A(\alpha)|^2 f_{XX}(\alpha)$ is near constant. This estimate is based directly on the discrete Fourier transform of the values $X(t)$, $t = 0, \ldots, T-1$ and is seen to involve the smoothing of weighted periodogram ordinates. In an extreme situation where $f_{XX}(\alpha)$ appears to have a high peak near $2\pi S/T \neq \lambda$, we may wish to take $A(2\pi S/T) = 0$. The sum in (5.8.33) now excludes the periodogram ordinate $I_{XX}^{(T)}(2\pi S/T)$ altogether. We remark that the ordinate $I_{XX}^{(T)}(0)$ is already missing from the estimate (5.8.16). Following the discussion of Theorem 5.2.2, this is equivalent to forming the periodogram of the mean adjusted values

$$X(t) - c_X^{(T)} \qquad \text{for } t = 0, \ldots, T-1 \qquad (5.8.34)$$

where

$$c_X^{(T)} = T^{-1}\sum_{t=0}^{T-1} X(t). \qquad (5.8.35)$$

A similar situation holds if the ordinate $I_{XX}^{(T)}(2\pi S/T)$ is dropped from the estimate. Since the values $d_X^{(T)}(2\pi s/T)$, $s = 0, \ldots, S-1, S+1, \ldots, T/2$ are unaffected by whether or not a multiple of the series $\exp\{\pm i2\pi St/T\}$, $t = 0, \ldots, T-1$ is subtracted, dropping $I_{XX}^{(T)}(2\pi S/T)$ is equivalent to forming the periodogram of the values $X(t)$, $t = 0, \ldots, T-1$ with best fitting sinusoid of frequency $2\pi S/T$ removed. The idea of avoiding certain frequencies in the smoothing of the periodogram appears in Priestley (1962b), Bartlett (1967) and Brillinger and Rosenblatt (1967b).

Akaike (1962a) discusses certain aspects of prefiltering. We sometimes have a good understanding of the character of the filter function, $A(\lambda)$, used in a prefiltering and so are content to examine the estimated spectrum of the filtered series $Y(t)$ and not bother to divide it by $|A(\lambda)|^2$.

5.9 ALTERNATE ESTIMATES

Up until this point, the spectral estimates discussed have had the character of a weighted average of periodogram values at the particular frequencies $2\pi s/T$, $s = 0, \ldots, T-1$. This estimate is useful because these particular periodogram values may be rapidly calculated using the Fast Fourier Transform Algorithm of Section 3.5 if T is highly composite and in addition their joint large sample statistical behavior is elementary; see Theorems 4.4.1 and 5.2.6. In this section we turn to the consideration of certain other estimates.

The estimate considered in Section 5.6 has the specific form

$$\frac{2\pi}{T}\sum_{s=1}^{T-1} W^{(T)}\left(\lambda - \frac{2\pi s}{T}\right) I^{(T)}_{X-c_X^{(T)},X-c_X^{(T)}}\left(\frac{2\pi s}{T}\right) \qquad \text{for } -\infty < \lambda < \infty. \tag{5.9.1}$$

If the discrete average in (5.9.1) is replaced by a continuous one, this estimate becomes

$$\int_0^{2\pi} W^{(T)}(\lambda - \alpha) I^{(T)}_{X-c_X^{(T)},X-c_X^{(T)}}(\alpha)d\alpha$$
$$= B_T^{-1}\int_{-\infty}^{\infty} W(B_T^{-1}\alpha) I^{(T)}_{X-c_X^{(T)},X-c_X^{(T)}}(\lambda - \alpha)d\alpha. \tag{5.9.2}$$

Now

$$I^{(T)}_{X-c_X^{(T)},X-c_X^{(T)}}(\alpha) = (2\pi)^{-1}\sum_{u=-T+1}^{T-1} c_{XX}{}^{(T)}(u)\exp\{-i\lambda u\} \tag{5.9.3}$$

where

$$c_{XX}{}^{(T)}(u) = T^{-1}\sum_{0\leq t,t+u\leq T-1}[X(t+u) - c_X{}^{(T)}][X(t) - c_X{}^{(T)}]. \tag{5.9.4}$$

If this is substituted into (5.9.2), then that estimate takes the form

$$(2\pi)^{-1}\sum_{u=-T+1}^{T-1} w(B_T u)c_{XX}{}^{(T)}(u)\exp\{-i\lambda u\} \tag{5.9.5}$$

where

$$w(u) = \int_{-\infty}^{\infty} W(\alpha)\exp\{iu\alpha\}d\alpha. \tag{5.9.6}$$

The estimate (5.9.5) is of the general form investigated by Grenander (1951a), Grenander and Rosenblatt (1957), and Parzen (1957); it contains as particular cases the early estimates of Bartlett (1948b), Hamming and Tukey (1949), and Bartlett (1950). Estimate (5.9.5) was generally employed until the Fast Fourier Transform Algorithm came into common use.

In fact the estimates (5.9.1) and (5.9.2) are very much of the same character as well as nearly equal. For example, Exercise 5.13.15 shows that (5.9.5) may be written as the following discrete average of periodogram values

$$\frac{2\pi}{S}\sum_{s=1}^{S-1} W^{(T)}\left(\lambda - \frac{2\pi s}{S}\right) I_{XX}{}^{(T)}\left(\frac{2\pi s}{S}\right) \tag{5.9.7}$$

for any integer $S \geqslant 2T - 1$; see also Parzen (1957). The expression (5.9.7) requires twice as many periodogram values as (5.9.1). In the case that S is highly composite it may be rapidly computed by the Fast Fourier Transform of the series

$$X'(t) = X(t) \qquad t = 0, \ldots, T-1$$
$$= 0 \qquad t = T, \ldots, S-1 \tag{5.9.8}$$

or by computing $c_{XX}^{(T)}(u)$, $u = 0, \pm 1, \ldots$ using a Fast Fourier Transform as described in Exercise 3.10.7, and then evaluating expression (5.9.5), again using a Fast Fourier Transform. In the reverse direction, Exercise 5.13.15 shows that the estimate (5.9.1) may be written as the following continuous average of periodogram values

$$\int_0^{2\pi} \left\{ \frac{2\pi}{T} \sum_{s=1}^{T-1} W^{(T)}\left(\lambda - \frac{2\pi s}{T}\right) D_{T-1}\left(\frac{2\pi s}{T} - \alpha\right) \right\} I_{X-c_X^{(T)}, X-c_X^{(T)}}^{(T)}(\alpha) d\alpha \tag{5.9.9}$$

where

$$D_{T-1}(\alpha) = \frac{\sin (T - \frac{1}{2})\alpha}{2\pi \sin \frac{1}{2}\alpha}. \tag{5.9.10}$$

A uniform bound for the difference between the two estimates is provided by

Theorem 5.9.1 Let $W(\alpha)$, $-\infty < \alpha < \infty$, satisfy Assumption 5.6.1 and have a bounded derivative. Then

$$\left| \frac{2\pi}{T} \sum_{s=1}^{T-1} W^{(T)}\left(\lambda - \frac{2\pi s}{T}\right) I_{XX}^{(T)}\left(\frac{2\pi s}{T}\right) - \int_{-\pi}^{\pi} W^{(T)}(\lambda - \alpha) I_{X-c_X^{(T)}, X-c_X^{(T)}}^{(T)}(\alpha) d\alpha \right|$$
$$\leqslant L B_T^{-1} T^{-1} (B_T^{-1} + \log T) \int_0^{2\pi} I_{X-c_X^{(T)}, X-c_X^{(T)}}^{(T)}(\alpha) d\alpha$$
$$\leqslant L B_T^{-1} T^{-1} (B_T^{-1} + \log T) c_{XX}^{(T)}(0) \tag{5.9.11}$$

for some finite L and $-\infty < \lambda < \infty$.

It is seen that, in the case that B_T does not tend to 0 too quickly, the asymptotic behavior of the two estimates is essentially identical.

The discussion of the interpretation of power spectra given in Section 5.1 suggests a spectral estimate. Specifically, let $A(\alpha)$ denote the transfer function of a band-pass filter with the properties

$$A(\alpha) \doteq 0 \qquad \text{for } |\alpha \pm \lambda| > \Delta \tag{5.9.12}$$

$-\pi < \alpha, \lambda < \pi$, Δ small, and

$$\int_{-\pi}^{\pi} |A(\alpha)|^2 d\alpha = 1. \tag{5.9.13}$$

The construction of filters with such properties was discussed in Sections 2.7, 3.3, and 3.6. If $X(t,\lambda)$, $t = 0, \pm 1, \ldots$ denotes the output series of such a filter, then

$$EX(t,\lambda)^2 = \int_{-\pi}^{\pi} |A(\alpha)|^2 f_{XX}(\alpha)d\alpha + |A(0)|^2 c_X{}^2$$
$$\doteq f_{XX}(\lambda) \qquad \text{if } \lambda \not\equiv 0 \pmod{2\pi}. \tag{5.9.14}$$

This suggests the consideration of the estimate

$$T^{-1} \sum_{t=0}^{T-1} X(t,\lambda)^2 \tag{5.9.15}$$

in the case $\lambda \not\equiv 0 \pmod{2\pi}$. In fact, it appears that this last is the first spectral estimate used in practice; see Pupin (1894), Wegel and Moore (1924), Blanc-Lapierre and Fortet (1953). It is the one generally employed in real-time or analog situations. Turning to a discussion of its character, we begin by supposing that

$$A(\alpha) \doteq \sqrt{\frac{T}{4\pi(2m+1)}} \qquad \text{for } |\alpha \pm \lambda| \leqslant \frac{2\pi}{T}(m + \tfrac{1}{2})$$
$$\doteq 0 \qquad \text{otherwise} \qquad -\pi < \alpha < \pi. \tag{5.9.16}$$

If $d_{X(\cdot,\lambda)}^{(T)}(2\pi s/T)$ denotes the discrete Fourier transform of the filtered values $X(t,\lambda)$, $t = 0, \ldots, T-1$ and $2\pi s(T)/T \doteq \lambda$, then

$$d_{X(\cdot,\lambda)}^{(T)}\left(\frac{2\pi[s(T)+s]}{T}\right) \doteq \sqrt{\frac{T}{4\pi(2m+1)}}\, d_X{}^{(T)}\left(\frac{2\pi[s(T)+s]}{T}\right)$$
$$d_{X(\cdot,\lambda)}^{(T)}\left(\frac{2\pi[T-s(T)-s]}{T}\right) \doteq \sqrt{\frac{T}{4\pi(2m+1)}}\, d_X{}^{(T)}\left(\frac{2\pi[T-s(T)-s]}{T}\right)$$
$$\text{for } s = 0, \pm 1, \ldots, \pm m \tag{5.9.17}$$

and approximately equals 0 otherwise. Using Parseval's formula

$$T^{-1} \sum_{t=0}^{T-1} X(t,\lambda)^2 = T^{-2} \sum_{s} \left| d_{X(\cdot,\lambda)}^{(T)}\left(\frac{2\pi[s(T)+s]}{T}\right)\right|^2 \tag{5.9.18}$$

and so is approximately equal to

$$(2m+1)^{-1} \sum_{s=-m}^{m} I_{XX}{}^{(T)}\left(\frac{2\pi[s(T)+s]}{T}\right). \tag{5.9.19}$$

The estimate (5.9.15) therefore has similar form to estimate (5.4.1).

In Theorem 5.3.1 we saw that periodogram ordinates of the same frequency, $\lambda \not\equiv 0 \pmod{\pi}$, but based on different stretches of data were asymptotically independent $f_{XX}(\lambda)\chi_2{}^2/2$ variates. This result suggests that we construct a spectral estimate by averaging the periodograms of different stretches of data. In fact we have

Theorem 5.9.2 Let $X(t)$, $t = 0, \pm 1, \ldots$ be a real-valued series satisfying Assumption 2.6.1. Let

$$I_{XX}^{(V)}(\lambda,l) = (2\pi V)^{-1}\left|\sum_{v=0}^{V-1} X(v + lV) \exp\{-i\lambda(v + lV)\}\right|^2 \tag{5.9.20}$$

for $-\infty < \lambda < \infty$, $l = 0, \ldots, L - 1$. Let

$$f_{XX}^{(T)}(\lambda) = L^{-1} \sum_{l=0}^{L-1} I_{XX}^{(V)}(\lambda,l) \tag{5.9.21}$$

where $T = LV$. Then $f_{XX}^{(T)}(\lambda)$ is asymptotically $f_{XX}(\lambda)\chi_{2L}^2/(2L)$ if $\lambda \not\equiv 0 \pmod{\pi}$ and asymptotically $f_{XX}(\lambda)\chi_L^2/L$ if $\lambda = \pm\pi, \pm 3\pi, \ldots$ as $V \to \infty$.

Bartlett (1948b, 1950) proposed the estimate (5.9.21); it is also discussed in Welch (1967) and Cooley, Lewis, and Welch (1970). This estimate has the advantage of requiring fewer calculations than other estimates, especially when V is highly composite. In addition it allows us to examine the assumption of stationarity. Welch (1967) proposes the use of periodograms based on overlapping stretches of data. Akcasu (1961) and Welch (1961) considered spectral estimates based on the Fourier transform of the data. The result of this theorem may be used to construct approximate confidence limits for $f_{XX}(\lambda)$, if we think of the $I_{XX}^{(V)}(\lambda,l)$, $l = 0, \ldots, L - 1$ as L independent estimates of $f_{XX}(\lambda)$.

In the previous section it was suggested that an autoregressive scheme be fitted to the data in the course of estimating a power spectrum. Parzen (1964) suggested that we estimate the spectrum of the residual series $Y(t)$ for a succession of values m and when that estimate becomes nearly flat we take

$$|A^{(T)}(\lambda)|^{-2}\, T^{-1} \sum_{t=0}^{T-1} Y(t)^2 \tag{5.9.22}$$

as the estimate of $f_{XX}(\lambda)$, where $A^{(T)}(\lambda)$ is the transfer function of the filter carrying the series over to the residual series. This procedure is clearly related to prefiltering. Certain of its statistical properties are considered in Kromer (1969), Akaike (1969a), and Section 8.10.

In the course of the work of this chapter we have seen the important manner in which the band-width parameter m, or B_T, affects the statistical behavior of the estimate. In fact, if we carried out some prewhitening of the series, the shape of the weight function appearing in the estimate appears unimportant. What is important is its band-width. We have expected the student to determine m, or B_T, from the desired statistical stability. If the desired stability was not clear, a succession of band-widths were to be employed. Leppink (1970) proposes that we estimate B_T from the data and indicates an estimate; see also Picklands (1970). Daniels (1962) and Akaike (1968b) suggest procedures for modifying the estimate.

In the case that $X(t)$, $t = 0, \pm 1, \ldots$ is a 0 mean Gaussian series, an estimate of $f_{XX}(\lambda)$, based solely on the values

$$Y(t) = \operatorname{sgn} X(t) \qquad t = 0, \ldots, T-1 \tag{5.9.23}$$

(where $\operatorname{sgn} X = 1$ if $X > 0$, $\operatorname{sgn} X = -1$ if $X < 0$), was proposed by Goldstein; see Rodemich (1966), and discussed in Hinich (1967), McNeil (1967), and Brillinger (1968). Rodemich (1966) also considered the problem of constructing estimates of $f_{XX}(\lambda)$ from the values of $X(t)$ grouped in a general way.

Estimates have been constructed by Jones (1962b) and Parzen (1963a) for the case in which certain values $X(t)$, $t = 0, \ldots, T-1$ are missing in a systematic manner. Brillinger (1972) considers estimation for the case in which the values $X(\tau_1), \ldots, X(\tau_n)$ are available $\tau_1, \ldots, \tau_n$ being the times of events of some point process. Akaike (1960) examines the effect of observing $X(t)$ for t near the values $0, 1, \ldots, T-1$ rather than exactly at these values; this has been called **jittered sampling.**

Pisarenko (1972) has proposed a flexible class of nonlinear estimates. Let the data be split into L segments. Let $c_{XX}^{(T)}(u,l)$, $u = 0, \pm 1, \ldots$; $l = 0, \ldots, L-1$ denote the autocovariance estimate of segment l. Let $\mu_j^{(T)}$, $\mathbf{U}_j^{(T)}$, $j = 1, \ldots, J$ denote the latent roots and vectors of $[L^{-1} \sum_l c_{XX}^{(T)}(j-k,l);$ $j, k = 1, \ldots, J]$. Pisarenko suggests the following estimate of $f_{XX}(\lambda)$,

$$h\left[\sum_{k=1}^{J} H(\mu_k^{(T)})(2\pi J)^{-1}\left|\sum_{j=1}^{J} U_{jk}^{(T)} \exp\{-i\lambda j\}\right|^2\right] \tag{5.9.24}$$

where $H(x)$, $0 < x < \infty$, is a strictly monotonic function with inverse $h(.)$. He was motivated by the definition 3.10.27 of a function of a matrix. In the case $H(x) = x$, the estimate (5.9.24) may be written

$$L^{-1}\sum_{l=0}^{L-1}(2\pi J)^{-1}\sum_{u=-J}^{J}\left(1 - \frac{|u|}{J}\right)c_{XX}^{(T)}(u,l)\exp\{-iu\lambda\}. \tag{5.9.25}$$

This is essentially the estimate (5.9.21) if $J = V$. In the case $H(x) = x^{-1}$, the estimate takes the form

$$\left[(2\pi J)^{-1}\sum_{j,k=1}^{J} C_{jk}^{(T)} \exp\{-i\lambda(j-k)\}\right]^{-1} \tag{5.9.26}$$

with $[C_{jk}^{(T)}]$ the inverse of the matrix whose latent values were computed. The estimate (5.9.26) was suggested by Capon (1969) as having high resolution. Pisarenko (1972) argues that if $J, L \to \infty$ as $T \to \infty$, and if the series is normal, then the estimate (5.9.24) will be asymptotically normal with variance

$$L^{-1} f_{XX}(\lambda)^2 [1 + \eta\{2\lambda\}]. \tag{5.9.27}$$

Capon and Goodman (1970) suggest approximating the distribution of (5.9.26) by $f_{XX}(\lambda)\chi^2_{2L-2J+1}/2L$ if $\lambda \not\equiv 0 \pmod{\pi}$ and by $f_{XX}(\lambda)\chi^2_{L-J+1}/L$ if $\lambda = \pm\pi, \pm 3\pi, \ldots$.

Sometimes we are interested in fitting a parametric model for the power spectrum. A useful general means of doing this was proposed in Whittle (1951, 1952a, 1961). Some particular models are considered in Box and Jenkins (1970).

5.10 ESTIMATING THE SPECTRAL MEASURE AND AUTOCOVARIANCE FUNCTION

Let $X(t)$, $t = 0, \pm 1, \ldots$ denote a real-valued series with autocovariance function $c_{XX}(u)$, $u = 0, \pm 1, \ldots$ and spectral density $f_{XX}(\lambda)$, $-\infty < \lambda < \infty$. There are a variety of situations in which we would like to estimate the spectral measure

$$F_{XX}(\lambda) = \int_0^\lambda f_{XX}(\alpha)d\alpha \tag{5.10.1}$$

introduced in Section 2.5. There are also situations in which we would like to estimate the autocovariance function

$$c_{XX}(u) = \int_0^{2\pi} \exp\{iu\alpha\} f_{XX}(\alpha)d\alpha \tag{5.10.2}$$

itself, and situations in which we would like to estimate a broad-band spectral average of the form

$$\int_0^{2\pi} W(\lambda - \alpha) f_{XX}(\alpha)d\alpha, \tag{5.10.3}$$

$W(\alpha)$ being a weight function of period 2π concentrated near $\alpha \equiv 0 \pmod{2\pi}$.

The parameters (5.10.1), (5.10.2), and (5.10.3) are all seen to be particular cases of the general form

$$J(A) = \int_0^{2\pi} A(\alpha) f_{XX}(\alpha)d\alpha \qquad \text{for some function } A(\alpha),\ 0 \leqslant \alpha < 2\pi. \tag{5.10.4}$$

For this reason we turn to a brief investigation of estimates of the parameter (5.10.4) for given $A(\alpha)$. This problem was considered by Parzen (1957).

As a first estimate we consider the statistic

$$J^{(T)}(A) = \frac{2\pi}{T}\sum_{s=1}^{T-1} A\left(\frac{2\pi s}{T}\right) I_{XX}^{(T)}\left(\frac{2\pi s}{T}\right) \tag{5.10.5}$$

where $I_{XX}^{(T)}(\lambda)$, $-\infty < \lambda < \infty$, is the periodogram of a stretch of values $X(t)$, $t = 0, \ldots, T-1$. Taking a discrete average at the points $2\pi s/T$ allows a possible use of the Fast Fourier Transform Algorithm in the course of the calculations.

Setting

$$\begin{aligned} A(\alpha) &= 1 \qquad 0 \leqslant \alpha \leqslant \lambda \\ &= 0 \qquad \text{otherwise} \end{aligned} \tag{5.10.6}$$

amounts to proposing

$$F_{XX}^{(T)}(\lambda) = \frac{2\pi}{T}\sum_{0<\frac{2\pi s}{T}\leq\lambda} I_{XX}^{(T)}\left(\frac{2\pi s}{T}\right) \tag{5.10.7}$$

as an estimate of $F_{XX}(\lambda)$. Taking

$$A(\alpha) = \exp\{iu\alpha\} \qquad 0 \leqslant \alpha < 2\pi, \tag{5.10.8}$$

we see from Exercise 3.10.8 that we are proposing the circular autocovariance function

$$\hat{c}_{XX}^{(T)}(u) = T^{-1}\sum_{t=0}^{T-1} [\hat{X}(t+u) - c_X^{(T)}][\hat{X}(t) - c_X^{(T)}] \tag{5.10.9}$$

as an estimate of $c_{XX}(u)$. (Here $\hat{X}(t)$, $t = 0, \pm 1, \ldots$ is the period T extension of $X(t)$, $t = 0, \ldots, T-1$.) Taking

$$A(\alpha) = W(\lambda - \alpha) \qquad 0 \leqslant \alpha \leqslant 2\pi, \tag{5.10.10}$$

leads to our considering a spectral estimate of the form

$$f_{XX}^{(T)}(\lambda) = \frac{2\pi}{T}\sum_{s=1}^{T-1} W\left(\lambda - \frac{2\pi s}{T}\right) I_{XX}^{(T)}\left(\frac{2\pi s}{T}\right). \tag{5.10.11}$$

The statistic of Exercise 5.13.31 is sometimes used to test the hypothesis that a stationary Gaussian series has power spectrum $f_{XX}(\lambda)$. We have

Theorem 5.10.1 Let $X(t)$, $t = 0, \pm 1, \ldots$ be a real-valued series satisfying Assumption 2.6.2(1). Let $A_j(\alpha)$, $0 \leqslant \alpha < 2\pi$, be bounded and of bounded variation for $j = 1, \ldots, J$. Then

$$EJ^{(T)}(A_j) = \frac{2\pi}{T} \sum_{s=1}^{T-1} A_j\left(\frac{2\pi s}{T}\right) f_{XX}\left(\frac{2\pi s}{T}\right) + O(T^{-1})$$
$$= \int_0^{2\pi} A_j(\alpha) f_{XX}(\alpha) d\alpha + O(T^{-1}) \qquad \text{for } j = 1, \ldots, J. \tag{5.10.12}$$

Also

$$\operatorname{cov}\{J^{(T)}(A_j), J^{(T)}(A_k)\}$$
$$= \frac{2\pi}{T} \int_0^{2\pi} A_j(2\pi - \alpha)\overline{A_k(\alpha)} f_{XX}(\alpha)^2 d\alpha + \frac{2\pi}{T} \int_0^{2\pi} A_j(\alpha)\overline{A_k(\alpha)} f_{XX}(\alpha)^2 d\alpha$$
$$+ \frac{2\pi}{T} \int_0^{2\pi} \int_0^{2\pi} A_j(\alpha)\overline{A_k(\beta)} f_{XXXX}(\alpha,\beta,-\alpha) d\alpha d\beta + O(T^{-2}). \tag{5.10.13}$$

Finally $J^{(T)}(A_j)$, $j = 1, \ldots, J$ are asymptotically jointly normal with the above first- and second-order moment structure.

From expression (5.10.12) we see that $J^{(T)}(A_j)$ is an asymptotically unbiased estimate of $J(A_j)$. From the fact that its variance tends to 0 as $T \to \infty$, we see that it is also a consistent estimate.

In the case of estimating the spectral measure $F_{XX}(\lambda)$, taking $A(\alpha)$ to be (5.10.6), expression (5.10.13) gives

$$\lim_{T \to \infty} T \operatorname{cov}\{F_{XX}{}^{(T)}(\lambda), F_{XX}{}^{(T)}(\mu)\}$$
$$= 2\pi \int_0^{\min(\lambda,\mu)} f_{XX}(\alpha) d\alpha + 2\pi \int_0^{\lambda} \int_0^{\mu} f_{XXXX}(\alpha,\beta,-\alpha) d\alpha d\beta$$
$$\text{for } 0 \leqslant \lambda, \mu \leqslant \pi. \tag{5.10.14}$$

In the case of estimating the autocovariance function $c_{XX}(u)$, where $A(\alpha)$ is given by (5.10.8), expression (5.10.13) gives

$$\lim_{T \to \infty} T \operatorname{cov}\{\hat{c}_{XX}{}^{(T)}(u), \hat{c}_{XX}{}^{(T)}(v)\}$$
$$= 2\pi \int_0^{2\pi} \exp\{-i(u + v)\alpha\} f_{XX}(\alpha)^2 d\alpha$$
$$+ 2\pi \int_0^{2\pi} \exp\{i(u - v)\alpha\} f_{XX}(\alpha)^2 d\alpha$$
$$+ 2\pi \int_0^{2\pi} \int_0^{2\pi} \exp\{i(u\alpha - v\beta)\} f_{XXXX}(\alpha,\beta,-\alpha) d\alpha d\beta$$
$$\text{for } u, v = 0, \pm 1, \ldots. \tag{5.10.15}$$

In the case of the broad-band spectral estimate of (5.10.10), expression (5.10.13) gives

$$\lim_{T\to\infty} T \operatorname{cov}\{f_{XX}{}^{(T)}(\lambda), f_{XX}{}^{(T)}(\mu)\}$$
$$= 2\pi \int_0^{2\pi} W(\lambda + \alpha)W(\mu - \alpha)f_{XX}(\alpha)^2 d\alpha$$
$$+ 2\pi \int_0^{2\pi} W(\lambda - \alpha)W(\mu - \alpha)f_{XX}(\alpha)^2 d\alpha$$
$$+ 2\pi \int_0^{2\pi}\int_0^{2\pi} W(\lambda - \alpha)W(\mu - \beta)f_{XXXX}(\alpha,\beta,-\alpha)d\alpha d\beta. \tag{5.10.16}$$

In this case of a constant weight function, the spectral estimates $f_{XX}{}^{(T)}(\lambda)$ and $f_{XX}{}^{(T)}(\mu)$ are not asymptotically independent as was the case for the weight functions considered earlier.

If estimates of $f_{XX}(\alpha)$ and $f_{XXXX}(\alpha,\beta,\gamma)$ are computed, then we may substitute them into expression (5.10.13) to obtain an estimate of var $J^{(T)}(A_j)$. This estimate may be used together with the asymptotic normality, to construct approximate confidence limits for the parameter.

In some situations the student may prefer to use the following estimate involving a continuous weighting,

$$\int_0^{2\pi} A(\alpha)I^{(T)}_{X-c_X{}^{(T)},X-c_X{}^{(T)}}(\alpha)d\alpha = (2\pi)^{-1}\sum_{u=-T+1}^{T-1} a(u)c_{XX}{}^{(T)}(u), \tag{5.10.17}$$

where $c_{XX}{}^{(T)}(u)$ is the sample autocovariance function of (5.9.4) and

$$a(u) = \int_0^{2\pi} \exp\{-iu\alpha\}A(\alpha)d\alpha. \tag{5.10.18}$$

For example, if $A(\alpha) = \exp\{iu\alpha\}$, this gives the sample autocovariance function $c_{XX}{}^{(T)}(u)$ itself, in contrast to the circular form obtained before.

The estimate (5.10.17) does not differ too much from the estimate (5.10.5). We have

Theorem 5.10.2 Let $A(\alpha)$, $0 \leqslant \alpha \leqslant 2\pi$, be bounded and of bounded variation. Let $X(t)$, $t = 0, \pm 1, \ldots$ satisfy Assumption 2.6.2(1). Then

$$E\left|\frac{2\pi}{T}\sum_{s=1}^{T-1} A\left(\frac{2\pi s}{T}\right)I_{XX}{}^{(T)}\left(\frac{2\pi s}{T}\right) - \int_0^{2\pi} A(\alpha)I^{(T)}_{X-c_X{}^{(T)},X-c_X{}^{(T)}}(\alpha)d\alpha\right| = O(T^{-1}). \tag{5.10.19}$$

We see that the two estimates will be close for large T and that their asymptotic distribution will be the same.

The spectral measure estimate, $F_{XX}^{(T)}(\lambda)$, given by (5.10.7) is sometimes useful for detecting periodic components in a series and for examining the plausibility of a proposed model especially that of pure noise. In Figure 5.10.1 we give $F_{XX}^{(T)}(\lambda)/F_{XX}^{(T)}(\pi)$, $0 < \lambda \leqslant \pi$, for the series of mean monthly sunspot numbers. The periodogram of this series was given in Section 5.2. The figure shows an exceedingly rapid increase at the lowest frequencies, followed by a steady increase to the value 1. We remark that if $f_{XX}(\lambda)$ were constant in a frequency band, then the increase of $F_{XX}(\lambda)$ would be linear in that frequency band. This does not appear to occur in Figure 5.10.1 except, possibly, at frequencies above $\pi/2$.

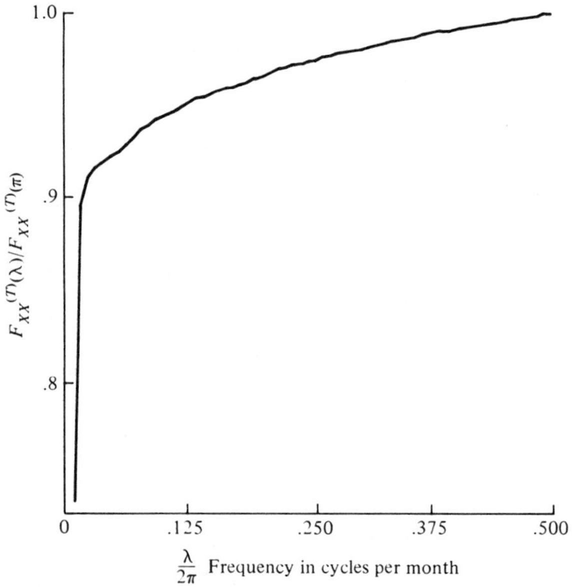

Figure 5.10.1 Plot of $F_{XX}^{(T)}(\lambda)/F_{XX}^{(T)}(\pi)$ for monthly mean sunspot numbers for the years 1750–1965.

The sample autocovariance function, $c_{XX}^{(T)}(u)$, $u = 0, \pm 1, \ldots$, of a series stretch also is often useful for examining the structure of a series. In Figures 5.10.2 and 5.10.3 we present portions of $c_{XX}^{(T)}(u)$ for the series of mean annual and mean monthly sunspot numbers, respectively. The most apparent character of these figures is the substantial correlation of values of the series that are multiples of approximately 10 years apart. The kink near lag 0 in Figure 5.10.3 suggests that measurement error is present in this data.

Asymptotic properties of estimates of the autocovariance function were considered by Slutsky (1934) in the case of a 0 mean Gaussian series. Bartlett

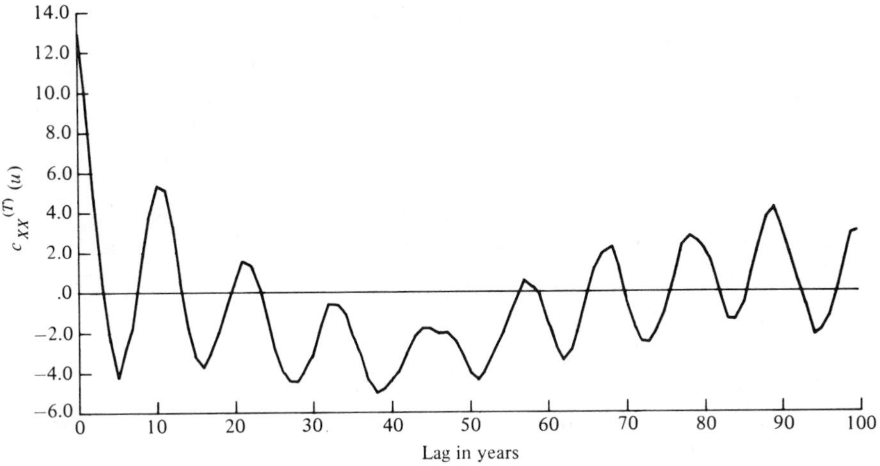

Figure 5.10.2 The autocovariance estimate, $c_{XX}^{(T)}(u)$, for annual mean sunspot numbers for the years 1750–1965.

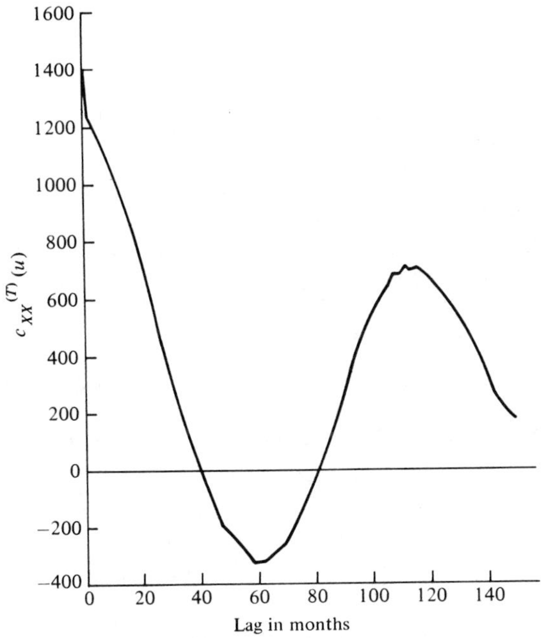

Figure 5.10.3 The autocovariance estimate, $c_{XX}^{(T)}(u)$, for monthly mean sunspot numbers for the years 1750–1965.

(1946) developed the asymptotic second-order moment structure in the case of a 0 mean linear process. Asymptotic normality was considered in Walker (1954), Lomnicki and Zaremba (1957b, 1959), Parzen (1957), Rosenblatt (1962), and Anderson and Walker (1964). Akaike (1962a) remarked that it might sometimes be reasonable to consider $c_{XX}^{(T)}(u)$, $u = 0, \pm 1, \ldots$ as a second-order stationary times series with power spectrum $2\pi T^{-1} f_{XX}(\lambda)^2$. This corresponds to retaining only the second term on the right in (5.10.15). Brillinger (1969c) indicated two forms of convergence with probability 1 and discussed the weak convergence of the estimate to a Gaussian process.

5.11 DEPARTURES FROM ASSUMPTIONS

In this section we discuss the effects of certain elementary departures from the assumptions adopted so far in this chapter. Among the important assumptions adopted are

$$EX(t) = c_X \qquad \text{for } t = 0, \pm 1, \ldots \tag{5.11.1}$$

and

$$\sum_{u=-\infty}^{\infty} |c_{XX}(u)| < \infty \tag{5.11.2}$$

where $c_{XX}(u) = \operatorname{cov}\{X(t+u), X(t)\}$ for $t, u = 0, \pm 1, \ldots$.

We first discuss a situation in which expression (5.11.2) is not satisfied. Suppose that the series under consideration is

$$X(t) = \sum_{j=1}^{J} R_j \cos(\omega_j t + \phi_j) + \varepsilon(t) \qquad \text{for } t = 0, \pm 1, \ldots \tag{5.11.3}$$

with R_j, ω_j constants, ϕ_j uniform on $(-\pi, \pi)$, $j = 1, \ldots, J$ and the series $\varepsilon(t)$ satisying Assumption 2.6.1. The autocovariance function of the series (5.11.3) is quickly seen to be

$$c_{XX}(u) = \tfrac{1}{2} \sum_{j=1}^{J} R_j^2 \cos \omega_j u + c_{\varepsilon\varepsilon}(u) \tag{5.11.4}$$

and so condition (5.11.2) is not satisfied. We note that the spectral measure, $F_{XX}(\lambda)$, whose existence was demonstrated in Theorem 2.5.2, is given by

$$F_{XX}(\lambda) = \tfrac{1}{2} \sum_{j=1}^{J} R_j^2 H(\lambda - \omega_j) + \int_0^{\lambda} f_{\varepsilon\varepsilon}(\alpha) d\alpha \qquad 0 \leqslant \lambda \leqslant \pi, \tag{5.11.5}$$

in this case where

$$\begin{aligned} H(\alpha) &= 1 \qquad \text{for } \alpha \geqslant 0 \\ &= 0 \qquad \alpha < 0 \end{aligned} \tag{5.11.6}$$

and $f_{\varepsilon\varepsilon}(\lambda)$ denotes the spectral density of the series $\varepsilon(t)$, $t = 0, \pm 1, \ldots$. The generalized derivative of expression (5.11.5) is

$$\tfrac{1}{2} \sum_{j=1}^{J} R_j^2 \delta(\lambda - \omega_j) + f_{\varepsilon\varepsilon}(\lambda) \tag{5.11.7}$$

$0 \leqslant \lambda \leqslant \pi$, $\delta(\lambda)$ being the Dirac delta function. The function (5.11.7) has infinite peaks at the frequencies ω_j, $j = 1, \ldots, J$, superimposed on a bounded continuous function, $f_{\varepsilon\varepsilon}(\lambda)$. A series of the character of expression (5.11.3) is said to be of **mixed spectrum.**

Turning to the analysis of such a series, we note from expression (5.11.3) that

$$\begin{aligned} d_X^{(T)}(\lambda) &= \sum_{t=0}^{T-1} X(t) \exp\{-i\lambda t\} \\ &= \tfrac{1}{2} \sum_{j=1}^{J} R_j[\exp\{i\phi_j\}\Delta^{(T)}(\lambda - \omega_j) + \exp\{-i\phi_j\}\Delta^{(T)}(\lambda + \omega_j)] \\ &\qquad + d_\varepsilon^{(T)}(\lambda) \end{aligned} \tag{5.11.8}$$

where $\Delta^{(T)}(\lambda)$ is given in expression (4.3.14). Now the function $\Delta^{(T)}(\lambda)$ has large amplitude only for $\lambda \doteqdot 0 \pmod{2\pi}$. This means that

$$\begin{aligned} d_X^{(T)}(\lambda) &\doteq \tfrac{1}{2} R_j T \exp\{i\phi_j\} && \text{for } \lambda \doteq \omega_j \\ &\doteq \tfrac{1}{2} R_j T \exp\{-i\phi_j\} && \text{for } \lambda \doteq -\omega_j \end{aligned} \tag{5.11.9}$$

while

$$d_X^{(T)}(\lambda) = d_\varepsilon^{(T)}(\lambda) + O(1) \tag{5.11.10}$$

for $|\lambda \pm \omega_j| > \delta/T$, $-\pi < \lambda \leqslant \pi$. The result (5.11.9) suggests that we may estimate the ω_j by examining the periodogram $I_{XX}^{(T)}(\lambda)$ for substantial peaks. At such a peak we might estimate R_j by

$$2|d_X^{(T)}(\omega_j)|/T = \sqrt{8\pi I_{XX}^{(T)}(\omega_j)/T}. \tag{5.11.11}$$

This is essentially the procedure suggestion of Schuster (1898) to use the periodogram as a tool for discovering hidden periodicities in a series. The result (5.11.10) suggests that we may estimate $f_{\varepsilon\varepsilon}(\lambda)$ by smoothing the periodogram $I_{XX}^{(T)}(\lambda)$, avoiding the ordinates at frequencies in the immediate neighborhoods of the ω_j. If ν periodogram ordinates $I_{XX}^{(T)}(2\pi s/T)$ (s an integer) are involved in a simple averaging to form an estimate, then it follows from Theorem 4.4.1 and expression (5.11.10) that this estimate will be asymptotically $f_{XX}(\lambda)\chi_{2\nu}^2/(2\nu)$ in the case $\lambda \not\equiv 0 \pmod{\pi}$ with similar results in the case $\lambda \equiv 0 \pmod{\pi}$.

Bartlett (1967) and Brillinger and Rosenblatt (1967b) discuss the above simple modification of periodogram smoothing that avoids peaks. It is clearly related to the technique of prefiltering discussed in Section 5.8.

Other references include: Hannan (1961b), Priestley (1962b, 1964) and Nicholls (1967). Albert (1964), Whittle (1952b), Hext (1966), and Walker (1971) consider the problem of constructing more precise estimates of the ω_j of (5.11.3).

We next turn to a situation in which the condition of constant mean (5.11.1) is violated. Suppose in the manner of the trend model of Section 2.12

$$X(t) = \sum_{j=1}^{J} \theta_j \phi_j(t) + \varepsilon(t) \tag{5.11.12}$$

for $t = 0, \pm 1, \ldots$ with $\phi_1(t), \ldots, \phi_J(t)$ known fixed functions, $\theta_1, \ldots, \theta_J$ being unknown constants, and $\varepsilon(t)$, $t = 0, \pm 1, \ldots$ being an unobservable 0 mean series satisfying Assumption 2.6.1. This sort of model was considered in Grenander (1954). One means of handling it is to determine the least squares estimates $\theta_1^{(T)}, \ldots, \theta_J^{(T)}$ of $\theta_1, \ldots, \theta_J$ by minimizing

$$\sum_{t=0}^{T-1} [X(t) - \theta_1\phi_1(t) - \cdots - \theta_J\phi_J(t)]^2 \tag{5.11.13}$$

and then to estimate $f_{\varepsilon\varepsilon}(\lambda)$ from the residual series

$$e(t) = X(t) - \theta_1^{(T)}\phi_1(t) - \cdots - \theta_J^{(T)}\phi_J(t) \tag{5.11.14}$$

$t = 0, \ldots, T - 1$. We proceed to an investigation of the asymptotic properties of such a procedure. We set down an assumption concerning the functions $\phi_1(t), \ldots, \phi_J(t)$.

Assumption 5.11.1 Given the real-valued functions $\phi_j(t)$, $t = 0, \pm 1, \ldots,$ $j = 1, \ldots, J$, there exists a sequence N_T, $T = 1, 2, \ldots$ with the properties $N_T \to \infty$, $N_{T+1}/N_T \to 1$ as $T \to \infty$ such that

$$\lim_{T\to\infty} N_T^{-1} \sum_{t=0}^{T-|u|} \phi_j(t + u)\phi_k(t) = m_{jk}(u) \tag{5.11.15}$$

for $u = 0, \pm 1, \ldots$ and $j, k = 1, \ldots, J$, and the sequence

$$N_T^{-1} \sum_{t=0}^{T-1} |\phi_{a_1}(t_1 + u_1)\cdots\phi_{a_{k-1}}(t + u_{k-1})\phi_{a_k}(t)|,$$

$T = 1, 2, \ldots$ is uniformly bounded for $a_1, \ldots, a_k = 1, \ldots, r$; $u_1, \ldots,$ $u_{k-1} = 0, \pm 1, \ldots$; $k = 1, 2, \ldots$.

As examples of functions satisfying this assumption we mention

$$\phi_j(t) = R_j \cos(\omega_j t + \phi_j) \tag{5.11.16}$$

for constant $R_j, \omega_j, \phi_j, j = 1, \ldots, J$. We see directly that

$$m_{jk}(u) = \tfrac{1}{2}R_j^2 \cos \omega_j u \qquad \text{if } j = k$$
$$= 0 \qquad \text{if } j \neq k \tag{5.11.17}$$

taking $N_T = T$. Other examples are given in Grenander (1954).

We suppose that $m_{jk}(u)$ is taken as the entry in row j and column k of the $J \times J$ matrix $\mathbf{m}_{\phi\phi}(u)$, $j, k = 1, \ldots, J$. It follows from Exercise 2.13.31 that there exists an $r \times r$ matrix-valued function $\mathbf{G}_{\phi\phi}(\lambda)$, $-\pi < \lambda \leqslant \pi$, whose entries are of bounded variation, such that

$$\mathbf{m}_{\phi\phi}(u) = \int_{-\pi}^{\pi} \exp\{iu\lambda\} d\mathbf{G}_{\phi\phi}(\lambda) \tag{5.11.18}$$

for $u = 0, \pm 1, \ldots$. We may now state

Theorem 5.11.1 Let $\varepsilon(t)$, $t = 0, \pm 1, \ldots$ be a real-valued series satisfying Assumption 2.6.2(1), having 0 mean and power spectrum $f_{\varepsilon\varepsilon}(\lambda)$, $-\infty < \lambda < \infty$. Let $\phi_j(t)$, $j = 1, \ldots, J$, $t = 0, \pm 1, \ldots$ satisfy Assumption 5.11.1 with $\mathbf{m}_{\phi\phi}(0)$ nonsingular. Let $X(t)$ be given by (5.11.12) for some constants $\theta_1, \ldots, \theta_J$. Let $\theta_1^{(T)}, \ldots, \theta_J^{(T)}$ be the least squares estimates of $\theta_1, \ldots, \theta_J$. Let $e(t)$ be given by (5.11.14), and

$$f_{ee}^{(T)}(\lambda) = \frac{2\pi}{T} \sum_{s=1}^{T-1} W^{(T)}\left(\lambda - \frac{2\pi s}{T}\right) I_{ee}^{(T)}\left(\frac{2\pi s}{T}\right) \tag{5.11.19}$$

where $W^{(T)}(\alpha) = \sum_j B_T^{-1} W(B_T^{-1}[\alpha + 2\pi j])$ and $W(\alpha)$ satisfies Assumption 5.6.1. Then the variate $\mathbf{\theta}^{(T)} = [\theta_1^{(T)} \cdots \theta_J^{(T)}]$ has mean $\mathbf{\theta} = [\theta_1 \cdots \theta_J]$. Its covariance matrix satisfies

$$\lim_{T\to\infty} N_T E\{(\mathbf{\theta}^{(T)} - \mathbf{\theta})^\tau(\mathbf{\theta}^{(T)} - \mathbf{\theta})\} = 2\pi \mathbf{m}_{\phi\phi}(0)^{-1} \int_{-\pi}^{\pi} f_{\varepsilon\varepsilon}(\alpha) d\mathbf{G}_{\phi\phi}(\alpha) \mathbf{m}_{\phi\phi}(0)^{-1} \tag{5.11.20}$$

and it is asymptotically normal. If $B_T T \to \infty$ as $T \to \infty$, then the variate $f_{ee}^{(T)}(\lambda)$ is asymptotically independent of $\mathbf{\theta}^{(T)}$ with mean

$$Ef_{ee}^{(T)}(\lambda) = \int_{-\pi}^{\pi} W^{(T)}(\lambda - \alpha) f_{\varepsilon\varepsilon}(\alpha) d\alpha + O(B_T^{-1}T^{-1}). \tag{5.11.21}$$

Its covariance function satisfies

$$\lim_{T\to\infty} B_T T \operatorname{cov}\{f_{ee}^{(T)}(\lambda), f_{ee}^{(T)}(\mu)\}$$
$$= 2\pi \int W(\alpha)^2 d\alpha\, f_{\varepsilon\varepsilon}(\lambda)^2 [\eta\{\lambda - \mu\} + \eta\{\lambda + \mu\}] \tag{5.11.22}$$

and finite collections of estimates $f_{ee}^{(T)}(\lambda_1), \ldots, f_{ee}^{(T)}(\lambda_K)$ are asymptotically jointly normal.

Under the limiting procedure adopted, the asymptotic behavior of $f_{ee}^{(T)}(\lambda)$ is seen to be the same as that of an estimate $f_{\varepsilon\varepsilon}^{(T)}(\lambda)$ based directly on the series $\varepsilon(t)$, $t = 0, \pm 1, \ldots$. We have already seen this in the case of a series of unknown mean, corresponding to $J = 1, \phi_1(t) = 1, \theta_1 = c_X, \theta_1^{(T)} = c_X^{(T)} = T^{-1}\sum_{t=0}^{T-1} X(t)$. The theorem has the following:

Corollary 5.11.1 Under the conditions of the theorem, $\boldsymbol{\theta}^{(T)}$ and $f_{ee}^{(T)}(\lambda)$ are consistent estimates of $\boldsymbol{\theta}$ and $f_{\varepsilon\varepsilon}(\lambda)$, respectively.

Other results of the character of this theorem will be presented in Chapter 6. Papers related to this problem include Grenander (1954), Rosenblatt (1956a), and Hannan (1968). Koopmans (1966) is also of interest. A common empirical procedure is to base a spectral estimate on the series of first differences, $e(t) = X(t) - X(t-1)$, $t = 1, \ldots, T-1$. This has the effect of removing a linear trend directly. (See Exercise 3.10.2.)

A departure of much more serious consequence, than those considered so far in this section, is one in which cov $\{X(t+u), X(t)\}$ depends on both t and u. Strictly speaking a power spectrum is no longer well defined, see Loynes (1968); however in the case that cov $\{X(t+u), X(t)\}$ depends only weakly on t we can sometimes garner important information from a stretch of series by spectral type calculations. We can proceed by forming spectral estimates of the types considered in this chapter, but based on segments of the data, rather than all the data, for which the assumption of stationarity does not appear to be too seriously violated. The particular spectral estimate which seems especially well suited to such calculations is the one constructed by averaging the squared output of a bank of band-pass filters (see Section 5.9). Papers discussing this approach include Priestley (1965) and Brillinger and Hatanaka (1969).

A departure from assumptions of an entirely different character is the following: suppose the series $X(t)$ is defined for all real numbers t, $-\infty < t < \infty$. (Until this point we have considered $X(t)$ defined for $t = 0, \pm 1, \ldots$.) Suppose

$$c_{XX}(u) = \text{cov}\,\{X(t+u), X(t)\} \tag{5.11.23}$$

is defined for $-\infty < t, u < \infty$ and satisfies

$$\sum_{u=-\infty}^{\infty} \sup_{u \le v \le u+1} |c_{XX}(v)| < \infty, \tag{5.11.24}$$

then both

$$f_{XX}(\lambda) = (2\pi)^{-1} \sum_{u=-\infty}^{\infty} c_{XX}(u) \exp\{-i\lambda u\} \tag{5.11.25}$$

and

$$g_{XX}(\lambda) = (2\pi)^{-1} \int_{-\infty}^{\infty} c_{XX}(u) \exp\{-i\lambda u\} du \qquad (5.11.26)$$

$-\infty < \lambda < \infty$, are defined. The function $f_{XX}(\lambda)$, $-\infty < \lambda < \infty$, is called the power spectrum of the discrete series $X(t)$, $t = 0, \pm 1, \ldots$, whereas $g_{XX}(\lambda)$, $-\infty < \lambda < \infty$, is called the power spectrum of the continuous series $X(t)$, $-\infty < t < \infty$. The spectrum $g_{XX}(\lambda)$ may be seen to have very much the same character, behavior, and interpretation as $f_{XX}(\lambda)$. The two spectra $f_{XX}(\lambda)$ and $g_{XX}(\lambda)$ are intimately related because from (5.11.26) we have

$$\begin{aligned} c_{XX}(u) &= \int_{-\infty}^{\infty} \exp\{iu\lambda\} g_{XX}(\lambda) d\lambda \\ &= \sum_{j=-\infty}^{\infty} \int_{2\pi j}^{2\pi(j+1)} \exp\{iu\lambda\} g_{XX}(\lambda) d\lambda \\ &= \int_0^{2\pi} \exp\{iu\lambda\} \sum_{j=-\infty}^{\infty} g_{XX}(\lambda + 2\pi j)\, d\lambda \end{aligned} \qquad (5.11.27)$$

for $u = 0, \pm 1, \ldots$. From (5.11.25)

$$c_{XX}(u) = \int_0^{2\pi} \exp\{iu\lambda\} f_{XX}(\lambda) d\lambda \qquad (5.11.28)$$

giving

$$f_{XX}(\lambda) = \sum_{j=-\infty}^{\infty} g_{XX}(\lambda + 2\pi j). \qquad (5.11.29)$$

We see from expression (5.11.29) that a frequency λ in the discrete series $X(t)$, $t = 0, \pm 1, \ldots$ relates to the frequencies λ, $\lambda \pm 2\pi, \ldots$ of the continuous series $X(t)$, $-\infty < t < \infty$. As $f_{XX}(\lambda) = f_{XX}(-\lambda)$, it also relates to the frequencies $-\lambda$, $-\lambda \pm 2\pi, \ldots$. For this reason the frequencies

$$\lambda + 2\pi j, \; -\lambda + 2\pi j \qquad j = 0, \pm 1, \ldots \qquad (5.11.30)$$

have been called **aliases** by Tukey. It will be impossible to distinguish their individual character by means of $f_{XX}(\lambda)$ alone. As an example of the meaning of this, consider the series

$$X(t) = R \cos(\omega t + \phi) \qquad (5.11.31)$$

$-\infty < t < \infty$, where ϕ is uniform on $(-\pi, \pi)$. Considering this continuous series we have

$$c_{XX}(u) = \operatorname{cov}\{X(t + u), X(t)\} = \tfrac{1}{2} R^2 \cos \omega u \qquad (5.11.32)$$

$-\infty < u < \infty$, and from the definition (5.11.26)

$$g_{XX}(\lambda) = \tfrac{1}{4} R^2 [\delta(\lambda - \omega) + \delta(\lambda + \omega)]. \qquad (5.11.33)$$

This function has infinite peaks at $\lambda = \pm\omega$. Now considering the function (5.11.32) for $t = 0, \pm 1, \ldots$, we have from (2.10.8) or (5.11.29)

$$f_{XX}(\lambda) = \tfrac{1}{4}R^2[\eta(\lambda - \omega) + \eta(\lambda + \omega)]. \tag{5.11.34}$$

This function has infinite peaks at the frequencies $\lambda = \pm\omega + 2\pi j$, $j = 0, \pm 1, \ldots$ and so ω cannot be determined directly, but only said to be one of these frequencies. An implication, for practice, of this discussion is that if a power spectral estimate $f_{XX}{}^{(T)}(\lambda)$ is computed for $0 \leqslant \lambda \leqslant \pi$, and is found to have a peak at the frequency ω, we cannot be sure which of the frequencies $\pm\omega + 2\pi j$, $j = 0, \pm 1, \ldots$ might be leading to the peak.

An example of such an occurrence with which the author was once concerned is the following: data were to be taken periodically on the number of electrons entering a conical horn on a spinning satellite of the Explorer series. The electron field being measured was highly directional in character and so the data could be expected to contain a substantial periodic component whose period was that of the satellite's rotation. It was planned that the satellite would rotate at such a rate and the data would be taken with such a time interval that the frequency of rotation of the satellite would fall in the interval $0 < \lambda < \pi$. Unfortunately the satellite ended up spinning substantially more rapidly than planned and so the frequency of rotation fell outside the interval $0 < \lambda < \pi$. The spectrum of the data was estimated and found to contain a substantial peak. It then had to be decided which of the aliased frequencies was the relevant one. This was possible on this occasion because optical information was available suggesting a crude value for the frequency.

Sometimes a prefiltering of the data can be carried out to reduce the difficulties of interpretation caused by aliasing. Suppose that the continuous time series $X(t)$, $-\infty < t < \infty$, is band-pass filtered to a band of the sort $[-\pi j - \pi, -\pi j]$, $[\pi j, \pi j + \pi]$ prior to recording values at the time $t = 0, \pm 1, \ldots$. In this case we see from (5.11.29) that

$$f_{XX}(\lambda) \doteq g_{XX}(\lambda + \pi j) \tag{5.11.35}$$

and the interpretation is consequently simplified.

We conclude this section by indicating some of the effects of sampling the series $X(t)$, $-\infty < t < \infty$, at a general time spacing $h > 0$. The values recorded for analysis in the time interval $[0,T)$ are now $X(uh)$, $u = 0, \ldots, U - 1$ where $U = T/h$. If the series is stationary with cov $\{X(uh), X(0)\} = c_{XX}(uh)$, $u = 0, \pm 1, \ldots$ and

$$\sum_{u=-\infty}^{\infty} |c_{XX}(uh)| < \infty, \tag{5.11.36}$$

we define the power spectrum $f_{XX}(\lambda)$, $-\infty < \lambda < \infty$, by

$$f_{XX}(\lambda) = \frac{h}{2\pi} \sum_{u=-\infty}^{\infty} c_{XX}(uh) \exp\{-i\lambda uh\} \tag{5.11.37}$$

and have the inverse relation

$$c_{XX}(uh) = \int_{-\pi/h}^{\pi/h} f_{XX}(\lambda) \exp\{i\lambda uh\} d\lambda. \tag{5.11.38}$$

The power spectrum $f_{XX}(\lambda)$ is seen to have period $2\pi/h$. As $f_{XX}(-\lambda) = f_{XX}(\lambda)$, its fundamental domain may be taken to be the interval $[0,\pi/h]$. The expression (5.11.29) is replaced by

$$f_{XX}(\lambda) = \sum_{j=-\infty}^{\infty} g_{XX}\left(\lambda + \frac{2\pi j}{h}\right). \tag{5.11.39}$$

The upper limit of the interval $[0,\pi/h]$, namely π/h, is called the **Nyquist frequency** or **folding frequency.** If the series $X(t)$, $-\infty < t < \infty$, possesses no components with frequency greater than the Nyquist frequency, then

$$f_{XX}(\lambda) = g_{XX}(\lambda) \tag{5.11.40}$$

for $|\lambda| \leqslant \pi/h$ and no aliasing complications arise.

When we come to estimate $f_{XX}(\lambda)$ from the stretch $X(uh)$, $u = 0, \ldots, U - 1$, $T = uh$, we define

$$d_X^{(T)}(\lambda) = \sum_{u=0}^{U-1} X(uh) \exp\{-i\lambda uh\} \tag{5.11.41}$$

$$I_{XX}^{(T)}(\lambda) = \frac{h}{2\pi U} |d_X^{(T)}(\lambda)|^2 \tag{5.11.42}$$

and proceed by smoothing $I_{XX}^{(T)}(\lambda)$, for example.

The problem of aliasing was alluded to in the discussion of Beveridge (1922). Discussions of it were given in Press and Tukey (1956) and Blackman and Tukey (1958).

5.12 THE USES OF POWER SPECTRUM ANALYSIS

In Chapter 1 of this work we documented some of the various fields of applied research wherein the frequency analysis of time series had proven useful. In this section we indicate some examples of particular uses of the power spectrum.

A Descriptive Statistic Given a stretch of data, $X(t)$, $t = 0, \ldots, T - 1$, the function $f_{XX}^{(T)}(\lambda)$ is often computed simply as a descriptive statistic. It condenses the data, but not too harshly. For stationary series its approximate sampling properties are elementary. Its form is often more elementary than that of the original record. It has been computed in the hope that an underlying mechanism generating the data will be suggested to the experimenter. Wiener (1957, 1958) discusses electroencephalograms in this manner. The spectrum $f_{XX}^{(T)}(\lambda)$ has been calculated as a direct measure of the power, in watts, of the various frequency components of an electric signal; see Bode (1945) for example. In the study of color (see Wright (1958)), the power spectrum is estimated as a key characteristic of the color of an object.

We have seen that the power spectrum behaves in an elementary manner when a series is filtered. This has led Nerlove (1964) and Godfrey and Karreman (1967) to use it to display the effect of various procedures that have been proposed for the seasonal adjustment of economic time series. Cartwright (1967) used it to display the effect of tidal filters.

Some further references taken from a variety of fields include: Condit and Grum (1964), Haubrich (1965), Yamanouchi (1961), Manwell and Simon (1966), Plageman et al (1969).

Informal Testing and Discrimination The use of power spectra for testing and discrimination has followed their use as descriptive statistics. In the study of color, workers have noted that the spectra of objects of different color do seem to vary in a systematic way; see Wright (1958). Carpenter (1965) and Bullard (1966) question whether earthquakes and explosions have substantially different power spectra in the hope that these two could be discriminated on the basis of spectra calculated from observed seismograms. Also, the spectra derived from the EEGs of healthy and neurologically ill patients have been compared in the hope of developing a diagnostic tool; see Bertrand and Lacape (1943), Wiener (1957), Yuzuriha (1960), Suhara and Suzuki (1964), Alberts et al (1965), and Barlow (1967).

We have seen that the power spectrum of a white noise series is constant. The power spectrum has, therefore, been used on occasion as an informal test statistic for pure noise; see Granger and Morgenstern (1963), Press and Tukey (1956), for example. It is especially useful if the alternate is some other form of stationary behavior. A common assumption of relationship between two series is that, up to an additive pure noise, one comes about from the other in some functional manner. After a functional form has been fit, its aptness can be measured by seeing how flat the estimated power spectrum of the residuals is; see also Macdonald and Ward (1963). The magnitude of this residual spectrum gives us a measure of the goodness of fit achieved. Frequency bands of poor fit are directly apparent.

A number of papers have examined economic theories through the spectrum; for example, Granger and Elliot (1968), Howrey (1968), and Sargent (1968).

Estimation Power spectra are of use in the estimation of parameters of interest. Sometimes we have an underlying model that leads to a functional form for the spectrum involving unknown parameters. The parameters may then be estimated from the experimental spectrum; see Whittle (1951, 1952a, 1961) and Ibragimov (1967). Many unknown parameters are involved if we want to fit a linear process to the data; see Ricker (1940) and Robinson (1967b). The spectrum is of use here. The shift of a peak in an observed spectrum from its standard position is used by astronomers to determine the direction of motion of a celestial body; see Bracewell (1965). The motion of a peak in a spectrum calculated on successive occasions was used by Munk and Snodgrass (1957) to determine the apparent presence of a storm in the Indian Ocean.

Search for Hidden Periodicities The original problem, leading to the definition of the second-order periodogram, was that of measuring the frequency of a (possibly) periodic phenomenon; see Schuster (1898). Peaks in $f_{XX}{}^{(T)}(\lambda)$ do spring into immediate view and their broadness gives a measure of the accuracy of determination of the underlying frequency. The determination of the dominant frequency of brain waves is an important step in the analysis of a patient with possible cerebral problems; see Gibbs and Grass (1947). Bryson and Dutton (1961) searched for the period of sunspots in tree ring records.

Smoothing and Prediction The accurate measurement of power spectra is an important stage on the way to determining Kolmogorov-Wiener smoothing and predicting formulas; see Kolmogorov (1941a), Wiener (1949), Whittle (1963a). The problems of signal enhancement, construction of optimum transmission forms for signals of harmonic nature (for example, human speech) fall into this area.

5.13 EXERCISES

5.13.1 If $Y(t) = \sum_u a(t-u)X(u)$, while $\hat{Y}(t) = \sum_u a(t-u)\hat{X}(u)$, where $X(t)$ is stationary with mean 0, prove that

$$EI_{\hat{Y}\hat{Y}}{}^{(T)}(\lambda) = (2\pi T)^{-1}\int_{-\pi}^{\pi}\left[\frac{\sin T(\lambda-\alpha)/2}{\sin(\lambda-\alpha)/2}\right]^2 |A(\alpha)|^2 f_{XX}(\alpha)d\alpha$$

while

$$EI_{\hat{Y}\hat{Y}}{}^{(T)}(\lambda) = |A(\lambda)|^2(2\pi T)^{-1}\int_{-\pi}^{\pi}\left[\frac{\sin T(\lambda-\alpha)/2}{\sin(\lambda-\alpha)/2}\right]^2 f_{XX}(\alpha)d\alpha.$$

5.13.2 Prove that

$$\left(\frac{2\pi}{T}\right) \sum_{j=0}^{T-1} I_{XX}^{(T)}\left(\frac{2\pi j}{T}\right) = T^{-1} \sum_{t=0}^{T-1} X(t)^2$$

and hence, if $I_{XX}^{(T)}(2\pi j/T)$ is smoothed across its whole domain, the value obtained is $m_{XX}^{(T)}(0)$. (This result may be employed as a check on the numerical accuracy of the computations.) Indicate a similar result concerning $c_{XX}^{(T)}(0)$.

5.13.3 Let $X(t)$, $t = 0, \pm 1, \pm 2, \ldots$ be a real-valued second-order stationary process with absolutely summable autocovariance function $c_{XX}(u)$ and power spectrum $f_{XX}(\lambda)$, $-\pi < \lambda \leqslant \pi$. If $f_{XX}(\lambda) \neq 0$, show that there exists a summable filter $b(u)$, such that the series $E(t) = \sum_u b(t - u)X(u)$ has constant power spectrum. *Hint:* Take the transfer function of $b(u)$ to be $[f_{XX}(\lambda)]^{-1/2}$ and use Theorem 3.8.3.

5.13.4 Prove that $\exp\{-I_{XX}^{(T)}(\lambda)/f_{XX}(\lambda)\}$ tends, in distribution, to a uniform variate on (0,1) as $T \to \infty$ under the conditions of Theorem 5.2.7.

5.13.5 Under the conditions of Theorem 5.2.7, prove that the statistics $(\pi T)^{-1}$ $[\mathrm{Re}\ d_X^{(T)}(\lambda)]^2$ and $(\pi T)^{-1}[\mathrm{Im}\ d_X^{(T)}(\lambda)]^2$ tend, in distribution, to independent $f_{XX}(\lambda)\ \chi_1^2$ variates.

5.13.6 Let $J_{XX}^{(T)}(\lambda)$ be the smaller of the two statistics of the previous exercise and $K_{XX}^{(T)}(\lambda)$ the larger. Under the previous conditions, prove that the interval $[J_{XX}^{(T)}(\lambda), K_{XX}^{(T)}(\lambda)]$ provides an approximate 42 percent confidence interval for $f_{XX}(\lambda)$. See Durbin's discussion of Hannan (1967b).

5.13.7 Prove that the result of Theorem 5.2.6 is exact, rather than asymptotic, if $X(0), \ldots, X(T-1)$ are mean 0, variance σ^2 independent normal variates.

5.13.8 Let $W(\alpha) = 0$ for $\alpha < A$, $\alpha > B$, $B > A$, A, B finite. Prove that the asymptotic variance given in (5.6.13) is minimized by setting $W(\alpha) = (B - A)^{-1}$ for $A \leqslant \alpha \leqslant B$.

5.13.9 Prove, under regularity conditions, that the periodogram is a consistent estimate of $f_{XX}(\lambda)$ if $f_{XX}(\lambda) = 0$.

5.13.10 Let $Y(t)$ be a series with power spectrum $f_{YY}(\lambda)$ and $Z(t)$ an independent series with power spectrum $f_{XX}(\lambda)$. Let $X(t) = Y(t)$ for $0 \leqslant t \leqslant T/2$ and $X(t) = Z(t)$ for $T/2 < t \leqslant T - 1$. Determine the approximate statistical properties of $I_{XX}^{(T)}(\lambda)$.

5.13.11 Prove that $I_{XX}^{(T)}(2\pi s/T) = (2\pi)^{-1} \sum_{u=0}^{T-1} \exp\left\{-i\frac{2\pi s u}{T}\right\} m_{XX}^{(T)}(u)$ for s an integer.

5.13.12 Prove that $m_{XX}^{(T)}(u) = \int_{-\pi}^{\pi} \exp\{i\alpha u\}\ I_{XX}^{(T)}(\alpha)d\alpha$.

5.13.13 Prove that $c_{XX}^{(T)}(u) = \int_{-\pi}^{\pi} \exp\{i\alpha u\}\ I^{(T)}_{X-c_X^{(T)}, X-c_X^{(T)}}(\alpha)d\alpha$.

5.13.14 Under the conditions of Theorem 5.6.2 prove that

$$\lim_{T \to \infty} T\left[\int_0^{2\pi} W^{(T)}(\alpha)^2 d\alpha\right]^{-1} \operatorname{var} f_{XX}^{(T)}(\lambda) = 2\pi[1 + \eta\{2\lambda\}][f_{XX}(\lambda)]^2.$$

5.13.15 (a) Prove that

$$I_{XX}^{(T)}(\lambda) = \int_0^{2\pi} D_{T-1}(\lambda - \alpha) I_{XX}^{(T)}(\alpha) d\alpha$$

where $D_{T-1}(\alpha)$ is given by (5.9.10). *Hint:* Use expression (3.2.5).

(b) Prove that

$$I_{XX}^{(T)}(\lambda) = \frac{2\pi}{S} \sum_{s=0}^{S-1} D_{T-1}\left(\lambda - \frac{2\pi s}{S}\right) I_{XX}^{(T)}\left(\frac{2\pi s}{S}\right)$$

for $S \geqslant 2T - 1$.

5.13.16 Prove that

$$2\pi \min_\lambda f_{XX}(\lambda) \leqslant c_{XX}(0) \leqslant 2\pi \max_\lambda f_{XX}(\lambda).$$

5.13.17 Under Assumption 2.6.1, prove that if

$$Y(t) = \sum_u a(t - u) X(u),$$

then

$$E|d_Y^{(T)}(\lambda) - A(\lambda) d_X^{(T)}(\lambda)|^2$$
$$= \int_{-\pi}^{\pi} \{[\sin T(\lambda - \alpha)/2]^2 / [\sin(\lambda - \alpha)/2]^2\} |A(\lambda) - A(\alpha)|^2 f_{XX}(\alpha) d\alpha$$
$$= o(T) \quad \text{uniformly in } \lambda.$$

If $A(\alpha)$ has a bounded first derivative, then it equals $O(1)$ uniformly in λ.

5.13.18 Suppose $X(t) = R \cos(\alpha t + \phi) + \varepsilon(t)$, $t = 0, \ldots, T - 1$ where the $\varepsilon(t)$ are independent $N(0,\sigma^2)$ variates. Show that the maximum likelihood estimate of α is approximately the value of λ that maximizes $I_{XX}^{(T)}(\lambda)$; see Walker (1969).

5.13.19 Let $X(t)$, $t = 0, \pm 1, \ldots$ be real-valued, satisfy Assumption 2.6.2(1), and have mean 0. Let $W(\alpha)$ satisfy Assumption 5.6.1. If $B_T T \to \infty$ as $T \to \infty$ show that

$$E\left[\frac{2\pi}{T} \sum_{s=0}^{T-1} W^{(T)}\left(\lambda - \frac{2\pi s}{T}\right) I_{XX}^{(T)}\left(\frac{2\pi s}{T}\right) - \int_0^{2\pi} W^{(T)}(\lambda - \alpha) I_{XX}^{(T)}(\alpha) d\alpha\right]^2$$
$$= O(B_T^{-2} T^{-2}).$$

5.13.20 Let $X(t)$, $t = 0, \pm 1, \ldots$ be a real-valued series satisfying Assumption 2.6.1. Let $c_X^{(T)} = T^{-1} \sum_{t=0}^{T-1} X(t)$. Show that $\sqrt{T}(c_X^{(T)} - c_X)$ is asymptotically independent of $\sqrt{T}(c_{XX}^{(T)}(u) - c_{XX}(u))$ which is asymptotically normal with mean 0 and variance

$$2\pi \int_0^{2\pi} (1 + \cos 2u\alpha) f_{XX}(\alpha)^2 d\alpha$$
$$+ 2\pi \int_0^{2\pi} \int_0^{2\pi} \exp\{iu(\alpha - \beta)\} f_{XXXX}(\alpha, \beta, -\alpha) d\alpha d\beta.$$

5.13.21 Under the conditions of Theorem 5.6.3, show that $\sqrt{T}(c_X^{(T)} - c_X)$ and $\sqrt{B_T T}[f_{XX}^{(T)}(\lambda) - Ef_{XX}^{(T)}(\lambda)]$ are asymptotically independent and normal.

5.13.22 Show that the expected value of the modified periodogram (5.3.13) is given by

$$\left(\int_{-\pi}^{\pi} |H^{(T)}(\alpha)|^2 d\alpha\right)^{-1}$$
$$\times \int_{-\pi}^{\pi} |H^{(T)}(\lambda - \alpha) - H^{(T)}(\lambda)H^{(T)}(-\alpha)/H^{(T)}(0)|^2 f_{XX}(\alpha)d\alpha$$

and tends to $f_{XX}(\lambda)$ for $\lambda \not\equiv 0 \pmod{2\pi}$.

5.13.23 Under the conditions of Theorem 5.2.6 prove that

$$\text{Prob}\left[\sup_{j=1}^{J} I_{XX}^{(T)}(\lambda_j(T))/f_{XX}(\lambda_j) < x\right] \to (1 - e^{-x})^J$$

as $T \to \infty$.

5.13.24 Let $f_{XX}^{(T)}(\lambda)$ be given by (5.6.1) with $W(\beta)$ bounded. Suppose

$$\sum |u||c_{XX}(u)| < \infty.$$

Show that

$$E \sup_{\lambda} | f_{XX}^{(T)}(\lambda) - Ef_{XX}^{(T)}(\lambda)| < \frac{K}{B_T T}$$

for some finite K. Conclude that $\sup_\lambda | f_{XX}^{(T)}(\lambda) - Ef_{XX}^{(T)}(\lambda)|$ tends to 0 in probability if $B_T T \to \infty$.

5.13.25 Under the conditions of Theorem 5.4.3 show that

$$\frac{c_X^{(T)} - c_X}{\sqrt{2\pi f_{XX}^{(T)}(0)/T}}$$

tends in distribution to a Student's t with $2m$ degrees of freedom. (This result may be used to set approximate confidence limits for c_X.)

5.13.26 Let $X(t)$, $t = 0, \pm 1, \ldots$ be a series with $EX(t) = 0$ and $\text{cov}\{X(t+u), X(t)\} = c_{XX}(u)$ for $t, u = 0, \pm 1, \ldots$. Suppose

$$\sum_u |c_{XX}(u)| < \infty.$$

Show that there exists a finite K such that

$$E\{|c_{XX}^{(T)}(u) - m_{XX}^{(T)}(u)|\} < \frac{K}{T}$$

for $u = 0, \pm 1, \ldots$ and $T = 1, 2, \ldots$.

5.13.27 Let the real-valued series $X(t)$, $t = 0, \pm 1, \ldots$ be generated by the autoregressive scheme

$$X(t) + a(1)X(t-1) + \cdots + a(m)X(t-m) = \varepsilon(t)$$

for $t = 0, \pm 1, \ldots$ where $\varepsilon(t)$ is a series of independent identically distributed random variables with mean 0 and finite fourth-order moment. Suppose all the roots of the equation

$$z^m + a(1)z^{m-1} + \cdots + a(m) = 0$$

satisfy $|z| > 1$. Let $a^{(T)}(1), \ldots, a^{(T)}(m)$ be the least squares estimates of $a(1), \ldots, a(m)$. They minimize

$$\sum_{t=m}^{T-1} [X(t) + a^{(T)}(1)X(t-1) + \cdots + a^{(T)}(m)X(t-m)]^2.$$

Show that $\sqrt{T}[a^{(T)}(1) - a(1), \ldots, a^{(T)}(m) - a(m)]$ tends in distribution to $N_m(\mathbf{0},[c_{XX}(j-k)]^{-1})$ as $T \to \infty$. This result is due to Mann and Wald (1943).

5.13.28 If a function $g(x)$ on [0,1] satisfies $|g(x) - g(y)| \leqslant G|x-y|^\alpha$ for some α, $0 < \alpha \leqslant 1$, show that

$$\left| \int_0^1 g(x)dx - n^{-1} \sum_{k=1}^{n} g\left(\frac{k}{n}\right) \right| \leqslant \frac{G}{n^\alpha}.$$

5.13.29 Show that $f_{XX}^{(T)}(\lambda)$, given by (5.6.1), satisfies

$$\int_0^{2\pi} f_{XX}^{(T)}(\alpha)d\alpha = c_{XX}^{(T)}(0).$$

5.13.30 Under the conditions of Theorem 5.2.1 and $c_X = 0$, show that

$$\lim_{T\to\infty} T[EI_{XX}^{(T)}(\lambda) - f_{XX}(\lambda)]$$
$$= (4\pi)^{-1} \int_{-\pi}^{\pi} [f_{XX}(\lambda+\alpha) - 2f_{XX}(\lambda) + f_{XX}(\lambda-\alpha)]/[\sin \alpha/2]^2 d\alpha.$$

5.13.31 Under the conditions of Theorem 5.10.1, show that

$$G_{XX}^{(T)}(\lambda) = \frac{2\pi}{T} \sum_{0<\frac{2\pi s}{T}\leq\lambda} I_{XX}^{(T)}\left(\frac{2\pi s}{T}\right) \Big/ f_{XX}\left(\frac{2\pi s}{T}\right)$$

is asymptotically normal with $EG_{XX}^{(T)}(\lambda) = \lambda + O(T^{-1})$ and

$$\operatorname{cov}\{G_{XX}^{(T)}(\lambda), G_{XX}^{(T)}(\mu)\}$$
$$\sim \frac{2\pi}{T}\min(\lambda,\mu) + \frac{2\pi}{T}\int_0^\lambda \int_0^\mu f_{XXXX}(\alpha,\beta,-\alpha) f_{XX}(\alpha)^{-1} f_{XX}(\beta)^{-1} d\alpha d\beta.$$

5.13.32 Use Exercise 2.13.31 to show that (5.11.20) may be written

$$E\{(\boldsymbol{\theta}^{(T)} - \boldsymbol{\theta})^\tau(\boldsymbol{\theta}^{(T)} - \boldsymbol{\theta})\} \backsim N_T^{-1} 2\pi \mathbf{m}_{\phi\phi}^{(T)}(0)^{-1} \int f_{\varepsilon\varepsilon}(\alpha)\mathbf{I}_{\phi\phi}^{(T)}(\alpha)d\alpha \mathbf{m}_{\phi\phi}^{(T)}(0)^{-1}.$$

5.13.33 With the notation of Section 5.11, show that $f_{XX}(\lambda) \geqslant g_{XX}(\lambda)$ for all λ.

6

ANALYSIS OF A LINEAR TIME INVARIANT RELATION BETWEEN A STOCHASTIC SERIES AND SEVERAL DETERMINISTIC SERIES

6.1 INTRODUCTION

Let $Y(t)$, $\varepsilon(t)$, $t = 0, \pm 1, \ldots$ be real-valued stochastic series and let $\mathbf{X}(t)$, $t = 0, \pm 1, \ldots$ be an r vector-valued fixed series. Then suppose μ is a constant and that $\{\mathbf{a}(u)\}$ is a $1 \times r$ filter. Hence, in this chapter we shall be concerned with the investigation of relations that have the form

$$Y(t) = \mu + \sum_{u=-\infty}^{\infty} \mathbf{a}(t - u)\mathbf{X}(u) + \varepsilon(t). \tag{6.1.1}$$

We will assume that the **error series,** $\varepsilon(t)$, is stationary with 0 mean and power spectrum $f_{\varepsilon\varepsilon}(\lambda)$. This power spectrum is called the **error spectrum,** it is seen to measure the extent to which the series $Y(t)$ is determinable from the series $\mathbf{X}(t)$ by linear filtering. We will assume throughout this text that values of the **dependent series,** $Y(t)$, and values of the **independent series,** $\mathbf{X}(t)$, are available for $t = 0, \ldots, T - 1$. Because $E\,\varepsilon(t) = 0$,

$$EY(t) = \mu + \sum_{u=-\infty}^{\infty} \mathbf{a}(t - u)\mathbf{X}(u). \tag{6.1.2}$$

That is, the expected value of $Y(t)$ is a filtered version of $\mathbf{X}(t)$. Note from relation (6.1.2) that the series $Y(t)$ is not generally stationary. However, for $k > 1$,

$$\text{cum}\,\{Y(t_1), \ldots, Y(t_k)\} = \text{cum}\,\{\varepsilon(t_1), \ldots, \varepsilon(t_k)\} \tag{6.1.3}$$

and so the cumulants of $Y(t)$ of order greater than 1 are stationary.

The transfer function of the filter $\mathbf{a}(u)$ is given by

$$\mathbf{A}(\lambda) = \sum_{u=-\infty}^{\infty} \mathbf{a}(u) \exp\{-iu\lambda\}. \tag{6.1.4}$$

Let us consider the behavior of this transfer function with respect to filterings of the series $Y(t)$, $\mathbf{X}(t)$. Let $\{\mathbf{b}(u)\}$ be an $r \times r$ filter with inverse $\{\mathbf{c}(u)\}$. Let $\{d(u)\}$ be a 1×1 filter. Set

$$\mathbf{X}_1(t) = \sum_{u=-\infty}^{\infty} \mathbf{c}(t-u)\mathbf{X}(u) \tag{6.1.5}$$

then

$$\mathbf{X}(t) = \sum_{u=-\infty}^{\infty} \mathbf{b}(t-u)\mathbf{X}_1(u). \tag{6.1.6}$$

Set

$$Y_1(t) = \sum_{u=-\infty}^{\infty} d(t-u)Y(u), \tag{6.1.7}$$

$$\mu_1 = \left[\sum_{u=-\infty}^{\infty} d(u)\right]\mu, \tag{6.1.8}$$

and

$$\varepsilon_1(t) = \sum_{u=-\infty}^{\infty} d(t-u)\varepsilon(u). \tag{6.1.9}$$

The relation (6.1.1) now yields

$$Y_1(t) = \mu_1 + \sum_{u=-\infty}^{\infty} \mathbf{a}_1(t-u)\mathbf{X}_1(u) + \varepsilon_1(t) \tag{6.1.10}$$

where

$$\mathbf{a}_1(t) = d*\mathbf{a}*\mathbf{b}(u). \tag{6.1.11}$$

That is, the relation between the filtered series $Y_1(t)$, $\mathbf{X}_1(t)$, $\varepsilon_1(t)$ has the same form as the relation (6.1.1). In terms of transfer functions, (6.1.11) may be written

$$\mathbf{A}_1(\lambda) = D(\lambda)\mathbf{A}(\lambda)\mathbf{B}(\lambda) \tag{6.1.12}$$

or

$$\mathbf{A}(\lambda) = \frac{\mathbf{A}_1(\lambda)\mathbf{C}(\lambda)}{D(\lambda)}. \tag{6.1.13}$$

We see that the transfer function relating $Y(t)$ to $\mathbf{X}(t)$ may be determined from the transfer function relating $Y_1(t)$ to $\mathbf{X}_1(t)$ provided the required inverses exist. We note in passing that similar relations exist even if $Y_1(t)$ of (6.1.7) involves **X** through a term

$$\sum_{u=-\infty}^{\infty} \mathbf{e}(t-u)\mathbf{X}(u) \tag{6.1.14}$$

for some $1 \times r$ filter $\{\mathbf{e}(u)\}$. These remarks will be especially important when we come to the problem of prefiltering the series prior to estimating $\mathbf{A}(\lambda)$.

Throughout this chapter we will consider the case of deterministic $\mathbf{X}(t)$ and real-valued stochastic $Y(t)$. Brillinger (1969a) considers the model

$$\mathbf{Y}(t) = \boldsymbol{\mu} + \sum_{u=-\infty}^{\infty} \mathbf{a}(t-u)\mathbf{X}(u) + \boldsymbol{\varepsilon}(t) \tag{6.1.15}$$

$t = 0, \pm 1, \ldots$ where $\mathbf{X}(t)$ is deterministic and $\mathbf{Y}(t)$, $\boldsymbol{\varepsilon}(t)$ are s vector-valued. In Chapter 8 the model (6.1.15) is considered with $\mathbf{X}(t)$ stochastic.

6.2 LEAST SQUARES AND REGRESSION THEORY

Two classical theorems form the basis of least squares and linear regression theory. The first is the Gauss-Markov Theorem or

Theorem 6.2.1 Let

$$\mathbf{Y} = \mathbf{aX} + \boldsymbol{\varepsilon} \tag{6.2.1}$$

where $\boldsymbol{\varepsilon}$ is a $1 \times n$ matrix of random variables with $E\boldsymbol{\varepsilon} = \mathbf{0}$, $E\boldsymbol{\varepsilon}^\tau\boldsymbol{\varepsilon} = \sigma^2\mathbf{I}$, $\mathbf{a}$ is a $1 \times k$ matrix of unknown parameters, $\mathbf{X}$ is a $k \times n$ matrix of known values. Then

$$(\mathbf{Y} - \mathbf{aX})(\mathbf{Y} - \mathbf{aX})^\tau \tag{6.2.2}$$

is minimized, for choice of $\mathbf{a}$, by $\hat{\mathbf{a}} = \mathbf{YX}^\tau(\mathbf{XX}^\tau)^{-1}$ if $\mathbf{XX}^\tau$ is nonsingular. The minimum achieved is $\mathbf{Y}(\mathbf{I} - \mathbf{X}^\tau(\mathbf{XX}^\tau)^{-1}\mathbf{X})\mathbf{Y}^\tau$. Also $E\hat{\mathbf{a}} = \mathbf{a}$, the covariance matrix of $\hat{\mathbf{a}}$, is given by $E(\hat{\mathbf{a}} - \mathbf{a})^\tau(\hat{\mathbf{a}} - \mathbf{a}) = \sigma^2(\mathbf{XX}^\tau)^{-1}$ and if $\hat{\sigma}^2 = (n-k)^{-1}$ $\mathbf{Y}(\mathbf{I} - \mathbf{X}^\tau(\mathbf{XX}^\tau)^{-1}\mathbf{X})\mathbf{Y}^\tau$ then $E\hat{\sigma}^2 = \sigma^2$. In addition, $\hat{\mathbf{a}}$ is the minimum variance linear unbiased estimate of $\mathbf{a}$.

These results may be found in Kendall and Stuart (1961) Chapter 19, for example. The **least squares estimate** of $\mathbf{a}$ is $\hat{\mathbf{a}}$. Turning to distributional aspects of the above $\hat{\mathbf{a}}$ and $\hat{\sigma}^2$ we have

Theorem 6.2.2 If, in addition to the conditions of Theorem 6.2.1, the n components of $\boldsymbol{\varepsilon}$ have independent normal distributions, then $\hat{\mathbf{a}}^\tau$ is $N_k(\mathbf{a}^\tau, \sigma^2(\mathbf{XX}^\tau)^{-1})$, and $\hat{\sigma}^2$ is $\sigma^2\chi^2_{n-k}/(n-k)$ independent of $\hat{\mathbf{a}}$.

It follows directly from Theorem 6.2.2 that

$$F = [(n - k)\hat{\mathbf{a}}\mathbf{X}\mathbf{X}^{\tau}\hat{\mathbf{a}}^{\tau}]/[k\mathbf{Y}(\mathbf{I} - \mathbf{X}^{\tau}(\mathbf{X}\mathbf{X}^{\tau})^{-1}\mathbf{X})\mathbf{Y}^{\tau}] \tag{6.2.3}$$

is noncentral F, degrees of freedom k over $n - k$ and noncentrality parameter $\mathbf{a}\mathbf{X}\mathbf{X}^{\tau}\mathbf{a}^{\tau}/\sigma^2$. We see that the hypothesis $\mathbf{a} = 0$ may be tested by noting that (6.2.3) has a central $F_{k,n-k}$ distribution when the hypothesis holds. A related statistic is

$$\hat{R}_{YX}{}^2 = \frac{\hat{\mathbf{a}}\mathbf{X}\mathbf{X}^{\tau}\hat{\mathbf{a}}^{\tau}}{\mathbf{Y}\mathbf{Y}^{\tau}} \tag{6.2.4}$$

the **squared sample multiple correlation coefficient.** It may be seen that $0 \leqslant \hat{R}_{YX}{}^2 \leqslant 1$. Also from (6.2.3) we see that

$$\hat{R}_{YX}{}^2 = [Fk/(n - k)]/[1 + Fk/(n - k)] \tag{6.2.5}$$

and so its distribution is determinable directly from the noncentral F.

Suppose $\hat{a}_j$ and a_j denote the jth entries of $\hat{\mathbf{a}}$ and $\mathbf{a}$ respectively and c_{jj} denotes the jth diagonal entry of $(\mathbf{X}\mathbf{X}^{\tau})^{-1}$. Then the confidence intervals for a_j may be derived through the pivotal quantity,

$$[\hat{a}_j - a_j]/[c_{jj}\mathbf{Y}(\mathbf{I} - \mathbf{X}^{\tau}(\mathbf{X}\mathbf{X}^{\tau})^{-1}\mathbf{X})\mathbf{Y}^{\tau}/(n - k)]^{1/2} \tag{6.2.6}$$

which has a t_{n-k} distribution.

These results apply to real-valued random variables and parameters. In fact, in the majority of cases of concern to us in time series analysis we will require extensions to the complex-valued case. We have

Theorem 6.2.3 Let

$$\mathbf{Y} = \mathbf{a}\mathbf{X} + \boldsymbol{\varepsilon} \tag{6.2.7}$$

where $\boldsymbol{\varepsilon}$ is a $1 \times n$ matrix of complex-valued random variables with $E\boldsymbol{\varepsilon} = \mathbf{0}$, $E\boldsymbol{\varepsilon}^{\tau}\boldsymbol{\varepsilon} = \mathbf{0}$, $E\bar{\boldsymbol{\varepsilon}}^{\tau}\boldsymbol{\varepsilon} = \sigma^2\mathbf{I}$. $\mathbf{a}$ is a $1 \times k$ matrix of unknown complex-valued parameters, $\mathbf{X}$ is a $k \times n$ matrix with known complex-valued entries and $\mathbf{Y}$ is a $1 \times n$ matrix of known complex-valued entries. Then

$$(\mathbf{Y} - \mathbf{a}\mathbf{X})\overline{(\mathbf{Y} - \mathbf{a}\mathbf{X})}^{\tau} \tag{6.2.8}$$

is minimized, for choice of $\mathbf{a}$, by $\hat{\mathbf{a}} = \mathbf{Y}\bar{\mathbf{X}}^{\tau}(\mathbf{X}\bar{\mathbf{X}}^{\tau})^{-1}$ if $\mathbf{X}\bar{\mathbf{X}}^{\tau}$ is nonsingular. The minimum achieved is $\mathbf{Y}(\mathbf{I} - \bar{\mathbf{X}}^{\tau}(\mathbf{X}\bar{\mathbf{X}}^{\tau})^{-1}\mathbf{X})\bar{\mathbf{Y}}^{\tau}$. Also $E\hat{\mathbf{a}} = \mathbf{a}$, $E(\hat{\mathbf{a}} - \mathbf{a})^{\tau} \times (\hat{\mathbf{a}} - \mathbf{a}) = \mathbf{0}$ and $E\overline{(\hat{\mathbf{a}} - \mathbf{a})}^{\tau}(\hat{\mathbf{a}} - \mathbf{a}) = (\mathbf{X}\bar{\mathbf{X}}^{\tau})^{-1}\sigma^2$. If $\hat{\sigma}^2 = (n - k)^{-1}$ $\mathbf{Y}(\mathbf{I} - \bar{\mathbf{X}}^{\tau}(\mathbf{X}\bar{\mathbf{X}}^{\tau})^{-1}\mathbf{X})\bar{\mathbf{Y}}^{\tau}$, then $E\hat{\sigma}^2 = \sigma^2$.

Turning to distributional aspects, we have

Theorem 6.2.4 If in addition to the conditions of Theorem 6.2.3, the components of $\boldsymbol{\varepsilon}$ have independent $N_1^C(0,\sigma^2)$ distributions, then $\hat{\mathbf{a}}^\tau$ is $N_k^C(\mathbf{a}^\tau,(\mathbf{X}\bar{\mathbf{X}}^\tau)^{-1}\sigma^2)$ and $\hat{\sigma}^2$ is $\sigma^2\chi^2_{2(n-k)}/[2(n-k)]$ independent of $\hat{\mathbf{a}}$.

We may conclude from this theorem that

$$G = [(n-k)\hat{\mathbf{a}}\mathbf{X}\bar{\mathbf{X}}^\tau\bar{\hat{\mathbf{a}}}^\tau]/[k\mathbf{Y}(\mathbf{I} - \bar{\mathbf{X}}^\tau(\mathbf{X}\bar{\mathbf{X}}^\tau)^{-1}\mathbf{X})\bar{\mathbf{Y}}^\tau] \tag{6.2.9}$$

is noncentral F, degrees of freedom $2k$ over $2(n-k)$ and noncentrality parameter $\mathbf{a}\mathbf{X}\bar{\mathbf{X}}^\tau\bar{\mathbf{a}}^\tau/\sigma^2$. This statistic could be used to test the hypothesis $\mathbf{a} = 0$. A related statistic is

$$|\hat{R}_{YX}|^2 = \frac{\hat{\mathbf{a}}\mathbf{X}\bar{\mathbf{X}}^\tau\bar{\hat{\mathbf{a}}}^\tau}{\mathbf{Y}\bar{\mathbf{Y}}^\tau} \tag{6.2.10}$$

the squared sample complex multiple correlation coefficient. It may be seen directly that $0 \leqslant |\hat{R}_{YX}|^2 \leqslant 1$. Also from (6.2.10) we see that

$$|\hat{R}_{YX}|^2 = [Gk/(n-k)]/[1 + Gk/(n-k)] \tag{6.2.11}$$

and so its distribution is determinable directly from the noncentral F. Theorems 6.2.3 and 6.2.4 above are indicated in Akaike (1965). Under the conditions of Theorem 6.2.4, Khatri (1965a) has shown that $\hat{\mathbf{a}}$ and $(n-k)\hat{\sigma}^2/n$ are the maximum likelihood estimates of $\mathbf{a}$ and σ^2.

An important use of the estimate $\hat{\mathbf{a}}$ is in predicting the expected value of y_0, the variate associated with given $\mathbf{x}_0$. In this connection we have

Theorem 6.2.5 Suppose the conditions of Theorem 6.2.4 are satisfied. Suppose also

$$y_0 = \mathbf{a}\mathbf{x}_0 + \varepsilon_0 \tag{6.2.12}$$

where ε_0 is independent of $\boldsymbol{\varepsilon}$ of (6.2.7). Let $\hat{y}_0 = \hat{\mathbf{a}}\mathbf{x}_0$, then $\hat{y}_0$ is distributed as $N_1^C(\mathbf{a}\mathbf{x}_0,\sigma^2\bar{\mathbf{x}}_0^\tau(\mathbf{X}\bar{\mathbf{X}}^\tau)^{-1}\mathbf{x}_0)$ and is independent of $\hat{\sigma}^2$.

On a variety of occasions we will wish to construct confidence regions for the entries of $\mathbf{a}$ of expression (6.2.7). In the real-valued case we saw that confidence intervals could be constructed through the quantity (6.2.6) which has a t distribution under the conditions of Theorem 6.2.2. In the present case complications arise because $\mathbf{a}$ has complex-valued entries.

Let $\hat{a}_j$ and a_j denote the jth entries of $\hat{\mathbf{a}}$ and $\mathbf{a}$ respectively. Let c_{jj} denote the jth diagonal entry of $(\mathbf{X}\bar{\mathbf{X}}^\tau)^{-1}$. Let w_j denote

$$\frac{c_{jj}\mathbf{Y}(\mathbf{I} - \bar{\mathbf{X}}^\tau(\mathbf{X}\bar{\mathbf{X}}^\tau)^{-1}\mathbf{X})\bar{\mathbf{Y}}^\tau}{n-k} = c_{jj}\hat{\sigma}^2. \tag{6.2.13}$$

The quantity

$$w_j^{-1/2}(\hat{a}_j - a_j) \tag{6.2.14}$$

has the form $v^{-1/2}z$ where z is $N_1^C(0,1)$ and v is independently $\chi^2_{2(n-k)}/\{2(n-k)\}$. Therefore

$$w_j^{-1}|\hat{a}_j - a_j|^2/2 \tag{6.2.15}$$

has an $F_{2;2(n-k)}$ distribution. A 100β percent confidence region for Re a_j, Im a_j may thus be determined from the inequality

$$\{\text{Re } \hat{a}_j - \text{Re } a_j\}^2 + \{\text{Im } \hat{a}_j - \text{Im } a_j\} \leqslant 2w_jF_{2;2(n-k)}(\beta) \tag{6.2.16}$$

where $F(\beta)$ denotes the upper 100β percent point of the F distribution. We note that this region has the form of a circle centered at Re $\hat{a}_j$, Im $\hat{a}_j$.

On other occasions it may be more relevant to set confidence intervals for $|a_j|$ and arg a_j. One means of doing this is to derive a region algebraically from expression (6.2.16). Let

$$v_j = \sqrt{2w_jF_{2;2(n-k)}(\beta)} \tag{6.2.17}$$

then (6.2.16) is, approximately, equivalent to the region

$$\begin{gathered} |\hat{a}_j| - v_j \leqslant |a_j| \leqslant |\hat{a}_j| + v_j \\ \arg \hat{a}_j - \arcsin \{v_j/|\hat{a}_j|\} \leqslant \arg a_j \leqslant \arg \hat{a}_j + \arcsin \{v_j/|\hat{a}_j|\}. \end{gathered} \tag{6.2.18}$$

This region was presented in Goodman (1957) and Akaike and Yamanouchi (1962).

The region (6.2.18) is only approximate. An exact 100γ percent interval for $|a_j|$ may be determined by noting that

$$w_j^{-1}|\hat{a}_j|^2/2 \tag{6.2.19}$$

is noncentral F with degrees of freedom 2 and $2(n-k)$ and noncentrality parameter $|a_j|^2/2$. Tables for the power of the F test (see Pearson and Hartley (1951) can now be used to construct an exact confidence interval for $|a_j|$. The charts in Fox (1956) may also be used.

Alternatively we could use expression (6.2.15) to construct an approximate 100γ percent confidence interval by determining its distribution by central F with degrees of freedom

$$(2 + |a_j|^2/2)^2/(2 + |a_j|^2) \tag{6.2.20}$$

and $2(n-k)$. This approximation to the noncentral F is given in Abramowitz and Stegun (1964); see also Laubscher (1960).

In the case of $\phi_j = \arg a_j$ we can determine an exact 100δ percent confidence interval by noting that

$$\{(\text{Im } \hat{a}_j) \cos \phi_j - (\text{Re } \hat{a}_j) \sin \phi_j\}w_j^{-1/2} \tag{6.2.21}$$

has a $t_{2(n-k)}$ distribution. It is interesting to note that this procedure is related to the Creasy-Fieller problem; see Fieller (1954) and Halperin (1967). The two exact confidence procedures suggested above are given in Groves and Hannan (1968).

If simultaneous intervals are required for several of the entries of **a**, then one can proceed through the complex generalization of the multivariate t. See Dunnett and Sobel (1954), Gupta (1963a), Kshirsagar (1961), and Dickey (1967) for a discussion of the multivariate t distribution. However, we content ourselves with defining the complex t distribution. Let z be distributed as $N_1{}^C(0,1)$ and independently let s^2 be distributed as $\chi_n{}^2/n$. Then z/s has a complex t distribution with n degrees of freedom. If $u = \text{Re } t$, $v = \text{Im } t$, then its density is given by

$$2^{1/2}\pi^{-1/2}[1 + u^2 + v^2]^{-1-n/2} \tag{6.2.22}$$

$-\infty < u, v < \infty$. A related reference is Hoyt (1947).

6.3 HEURISTIC CONSTRUCTION OF ESTIMATES

We can now construct estimates of the parameters of interest. Set

$$R(t) = \sum_{u=-\infty}^{\infty} \mathbf{a}(t-u)\mathbf{X}(u). \tag{6.3.1}$$

The model (6.1.1) then takes the form

$$Y(t) = \mu + R(t) + \varepsilon(t). \tag{6.3.2}$$

The values $\mathbf{X}(t)$, $t = 0, \ldots, T-1$ are available and therefore we can calculate the finite Fourier transform

$$\mathbf{d}_X{}^{(T)}(\lambda) = \sum_{t=0}^{T-1} \mathbf{X}(t) \exp\{-i\lambda t\}. \tag{6.3.3}$$

In the present situation, it is an r vector-valued statistic. Define

$$d_R{}^{(T)}(\lambda) = \sum_{t=0}^{T-1} R(t) \exp\{-i\lambda t\}. \tag{6.3.4}$$

The approximate relation between $d_R{}^{(T)}(\lambda)$ and $\mathbf{d}_X{}^{(T)}(\lambda)$ is given by

Lemma 6.3.1 Suppose that $|\mathbf{X}(t)| \leqslant M$, $t = 0, \pm 1, \ldots$ and that $\sum |u|\,|\mathbf{a}(u)| < \infty$. Then

$$|d_R{}^{(T)}(\alpha) - \mathbf{A}(\alpha)\mathbf{d}_X{}^{(T)}(\alpha)| \leqslant 4M \sum_u |u|\,|\mathbf{a}(u)| \tag{6.3.5}$$

and

$$|d_R^{(T)}(\alpha) - \mathbf{A}(\lambda)\mathbf{d}_X^{(T)}(\alpha)| \leqslant (4M + L) \sum_u |u|\ |\mathbf{a}(u)| \tag{6.3.6}$$

if $-\infty < \alpha < \infty$ with $|\alpha - \lambda| \leqslant LT^{-1}$.

Let $s(T)$ be an integer with $2\pi s(T)/T$ near λ. Suppose T is large. From expression (6.3.6)

$$d_Y^{(T)}\left(\frac{2\pi[s(T)+s]}{T}\right) \doteq \mathbf{A}(\lambda)\, \mathbf{d}_X^{(T)}\left(\frac{2\pi[s(T)+s]}{T}\right) + d_\varepsilon^{(T)}\left(\frac{2\pi[s(T)+s]}{T}\right) \tag{6.3.7}$$

for $s = 0, \pm 1, \ldots, \pm m$ say. If $\varepsilon(t)$ satisfies Assumption 2.6.1, then following Theorem 4.4.1 the quantities $d_\varepsilon^{(T)}(2\pi[s(T) + s]/T)$, $s = 0, \pm 1, \ldots, \pm m$ are approximately $N_1^C(0, 2\pi T f_{\varepsilon\varepsilon}(\lambda))$ variates. Relation (6.3.7) is seen to have the form of a multiple regression relation involving complex-valued variates. Noting Theorem 6.2.3, we define

$$\mathbf{I}_{YX}^{(T)}(\lambda) = (2\pi T)^{-1} d_Y^{(T)}(\lambda)\overline{\mathbf{d}_X^{(T)}(\lambda)}^\tau, \tag{6.3.8}$$

$$\mathbf{I}_{XX}^{(T)}(\lambda) = (2\pi T)^{-1} \mathbf{d}_X^{(T)}(\lambda)\overline{\mathbf{d}_X^{(T)}(\lambda)}^\tau, \tag{6.3.9}$$

$$\mathbf{f}_{YX}^{(T)}(\lambda) = (2m + 1)^{-1} \sum_{s=-m}^{m} \mathbf{I}_{YX}^{(T)}\left(\frac{2\pi[s(T)+s]}{T}\right), \tag{6.3.10}$$

$$\mathbf{f}_{XX}^{(T)}(\lambda) = (2m + 1)^{-1} \sum_{s=-m}^{m} \mathbf{I}_{XX}^{(T)}\left(\frac{2\pi[s(T)+s]}{T}\right). \tag{6.3.11}$$

Suppose that the $r \times r$ matrix $\mathbf{f}_{XX}^{(T)}(\lambda)$ is nonsingular. We now estimate $\mathbf{A}(\lambda)$ by

$$\mathbf{A}^{(T)}(\lambda) = \mathbf{f}_{YX}^{(T)}(\lambda)\, \mathbf{f}_{XX}^{(T)}(\lambda)^{-1} \tag{6.3.12}$$

and $f_{\varepsilon\varepsilon}(\lambda)$ by

$$g_{\varepsilon\varepsilon}^{(T)}(\lambda) = \frac{2m+1}{2m+1-r}\{f_{YY}^{(T)}(\lambda) - \mathbf{f}_{YX}^{(T)}(\lambda)\mathbf{f}_{XX}^{(T)}(\lambda)^{-1}\mathbf{f}_{XY}^{(T)}(\lambda)\}. \tag{6.3.13}$$

Theorem 6.2.4 suggests the approximating distributions $N_r^C(\mathbf{A}(\lambda)^\tau, (2m + 1)^{-1} f_{\varepsilon\varepsilon}(\lambda)\mathbf{f}_{XX}^{(T)}(\lambda)^{-1})$ for $\mathbf{A}^{(T)}(\lambda)^\tau$ and $[2(2m + 1 - r)]^{-1} f_{\varepsilon\varepsilon}(\lambda)\chi^2_{2(2m+1-r)}$ for $g_{\varepsilon\varepsilon}^{(T)}(\lambda)$.

In the next sections we generalize the estimates (6.3.12) and (6.3.13) and make precise the suggested approximate distributions.

As an estimate of μ we take

$$\mu^{(T)} = c_Y^{(T)} - \mathbf{A}^{(T)}(0)\mathbf{c}_X^{(T)} \tag{6.3.14}$$

where $c_Y^{(T)}$ and $\mathbf{c}_X^{(T)}$ are the sample means of the given Y and $\mathbf{X}$ values. We will find it convenient to use the statistic $\mu^{(T)} + \mathbf{A}^{(T)}(0)\mathbf{c}_X^{(T)} = c_Y^{(T)}$ in the statement of certain theorems below.

The heuristic approach given above is suggested in Akaike (1964, 1965), Duncan and Jones (1966) and Brillinger (1969a).

6.4 A FORM OF ASYMPTOTIC DISTRIBUTION

In this section we determine the asymptotic distribution of a class of elementary estimates, suggested by the heuristic arguments of Section 6.3, for the parameters $\mathbf{A}(\lambda)$ and $f_{\epsilon\epsilon}(\lambda)$. The form and statistical properties of these estimates will depend on whether or not $\lambda \equiv 0 \pmod{\pi}$. Consider three cases:

Case A λ satisfies $\lambda \not\equiv 0 \pmod{\pi}$
Case B λ satisfies $\lambda \equiv 0 \pmod{2\pi}$ or $= \pm\pi, \pm 3\pi, \ldots$ and T is even
Case C λ satisfies $\lambda = \pm\pi, \pm 3\pi, \ldots$ and T is odd.

Suppose $s(T)$ is an integer with $2\pi s(T)/T$ near λ. (We will later require $2\pi s(T)/T \to \lambda$ as $T \to \infty$.) Suppose m is a non-negative integer. Let $\mathbf{I}_{YX}^{(T)}(\lambda)$ be given by (6.3.8). Define

$$\mathbf{f}_{YX}^{(T)}(\lambda) = (2m+1)^{-1} \sum_{s=-m}^{m} \mathbf{I}_{YX}^{(T)}\left(\frac{2\pi[s(T)+s]}{T}\right) \quad \text{in Case A} \tag{6.4.1}$$

$$\mathbf{f}_{YX}^{(T)}(\lambda) = (2m)^{-1}\left\{\sum_{s=-m}^{-1} + \sum_{s=1}^{m}\right\} \mathbf{I}_{YX}^{(T)}\left(\lambda + \frac{2\pi s}{T}\right) \quad \text{in Case B} \tag{6.4.2}$$

and

$$\mathbf{f}_{YX}^{(T)}(\lambda) = (2m)^{-1} \sum_{s=1}^{m} \left\{\mathbf{I}_{YX}^{(T)}\left(\lambda - \frac{\pi}{T} + \frac{2\pi s}{T}\right) + \mathbf{I}_{YX}^{(T)}\left(\lambda + \frac{\pi}{T} - \frac{2\pi s}{T}\right)\right\} \quad \text{in Case C} \tag{6.4.3}$$

with similar definitions for $f_{YY}^{(T)}(\lambda)$ and $\mathbf{f}_{XX}^{(T)}(\lambda)$. These estimates are based on the discrete Fourier transforms of the data and so may be computed by a Fast Fourier Transform Algorithm.

As estimates of $\mathbf{A}(\lambda)$, $f_{\epsilon\epsilon}(\lambda)$, μ we take

$$\mathbf{A}^{(T)}(\lambda) = \mathbf{f}_{YX}^{(T)}(\lambda)\mathbf{f}_{XX}^{(T)}(\lambda)^{-1} \tag{6.4.4}$$

$$g_{\epsilon\epsilon}^{(T)}(\lambda) = C(m,r)[f_{YY}^{(T)}(\lambda) - \mathbf{f}_{YX}^{(T)}(\lambda)\mathbf{f}_{XX}^{(T)}(\lambda)^{-1}\mathbf{f}_{XY}^{(T)}(\lambda)] \tag{6.4.5}$$

with $C(m,r)$ a constant given by

$$C(m,r) = \frac{2m+1}{2m+1-r} \quad \text{in Case A}$$
$$= \frac{2m}{2m-r} \quad \text{in Case B and Case C,} \tag{6.4.6}$$

and finally take

$$\mu^{(T)} = c_Y{}^{(T)} - \mathbf{A}^{(T)}(0)\mathbf{c}_X{}^{(T)} \tag{6.4.7}$$

as an estimate of μ. A theorem indicating the behavior of the mean of $\mathbf{A}^{(T)}(\lambda)$ in the present situation is

Theorem 6.4.1 Let $\varepsilon(t)$, $t = 0, \pm 1, \ldots$ satisfy Assumption 2.6.1 and have mean 0. Let $\mathbf{X}(t)$, $t = 0, \pm 1, \ldots$ be uniformly bounded. Let $Y(t)$ be given by (6.1.1) where $\{\mathbf{a}(u)\}$ satisfies $\Sigma\,|u|\,|\mathbf{a}(u)| < \infty$. Let $\mathbf{A}^{(T)}(\lambda)$ be given by (6.4.4) where $\mathbf{f}_{YX}{}^{(T)}(\lambda)$ is given by (6.4.1). Then

$$E\mathbf{A}^{(T)}(\lambda) = \left\{\sum_{s=-m}^{m} \mathbf{A}\left(\frac{2\pi[s(T)+s]}{T}\right)\mathbf{I}_{XX}{}^{(T)}\left(\frac{2\pi[s(T)+s]}{T}\right)\right\}$$
$$\times \left\{\sum_{s=-m}^{m} \mathbf{I}_{XX}{}^{(T)}\left(\frac{2\pi[s(T)+s]}{T}\right)\right\}^{-1} + \mathbf{R}^{(T)} \tag{6.4.8}$$

in Case A where

$$\|\mathbf{R}^{(T)}\| \leq KT^{-1/2}\|\mathbf{f}_{XX}{}^{(T)}(\lambda)^{-1}\|^{1/2} \tag{6.4.9}$$

for finite K. There are similar expressions in Cases B and C.

We note from expression (6.4.8) that $E\mathbf{A}^{(T)}(\lambda)$ is principally a matrix weighted average of the transfer function $\mathbf{A}(\alpha)$. In addition, the expression suggests that the larger $\mathbf{f}_{XX}{}^{(T)}(\lambda)$ is, the smaller the departure from the weighted average will be. From Theorem 6.4.1 we may conclude

Corollary 6.4.1 Under the conditions of Theorem 6.4.1 and if $\|\mathbf{f}_{XX}{}^{(T)}(\lambda)^{-1}\|$ is bounded, as $T \to \infty$, $\mathbf{A}^{(T)}(\lambda)$ is asymptotically unbiased.

Turning to an investigation of asymptotic distributions we have

Theorem 6.4.2 Suppose the conditions of Theorem 6.4.1 are satisfied. Suppose also that $\mathbf{f}_{XX}{}^{(T)}(\lambda)$ is nonsingular for T sufficiently large and that $2\pi s(T)/T \to \lambda$ as $T \to \infty$. Then $\mathbf{A}^{(T)}(\lambda)^\tau$ is asymptotically $N_r{}^C(\mathbf{A}(\lambda)^\tau, (2m+1)^{-1}f_{\varepsilon\varepsilon}(\lambda)\mathbf{f}_{XX}{}^{(T)}(\lambda)^{-1})$ in Case A, is asymptotically $N_r(\mathbf{A}(\lambda)^\tau, (2m)^{-1}f_{\varepsilon\varepsilon}(\lambda)\mathbf{f}_{XX}{}^{(T)}(\lambda)^{-1})$ in Case B and Case C. Also $g_{\varepsilon\varepsilon}{}^{(T)}(\lambda)$ tends to

$f_{\epsilon\epsilon}(\lambda)\chi^2_{2(2m+1-r)}/[2(2m+1-r)]$ in Case A and to $f_{\epsilon\epsilon}(\lambda)\chi^2_{2m-r}/(2m-r)$ in Case B and Case C. The limiting normal and χ^2 distributions are independent. Finally $\mu^{(T)} + \mathbf{A}^{(T)}(0)\mathbf{c}_X{}^{(T)}$ is asymptotically $N_1(\mu + \mathbf{A}(0)\mathbf{c}_X{}^{(T)}, 2\pi T^{-1}f_{\epsilon\epsilon}(0))$ independently of $\mathbf{A}^{(T)}(\lambda)$, $g_{\epsilon\epsilon}{}^{(T)}(\lambda)$, $-\infty < \lambda < \infty$.

In the case $\lambda \not\equiv 0 \pmod{\pi}$, Theorem 6.4.2 suggests the approximation

$$\begin{aligned} \operatorname{var} g_{\epsilon\epsilon}{}^{(T)}(\lambda) &\doteq \{\operatorname{var} \chi^2_{2(2m+1-r)}/[2(2m+1-r)]\} f_{\epsilon\epsilon}(\lambda)^2 \\ &\doteq (2m+1-r)^{-1} f_{\epsilon\epsilon}(\lambda)^2. \end{aligned} \tag{6.4.10}$$

In the case $\lambda \equiv 0 \pmod{\pi}$, the theorem suggests the approximate variance $2f_{\epsilon\epsilon}(\lambda)^2/(2m-r)$.

The limiting distributions of the estimates of the gain and phase may be determined through

Corollary 6.4.2 Under the conditions of Theorem 6.4.2, functions of $\mathbf{A}^{(T)}(\lambda)$, $g_{\epsilon\epsilon}{}^{(T)}(\lambda)$, $\mu^{(T)} + \mathbf{A}^{(T)}(0)\mathbf{c}^{(T)}$ tend in distribution to the same functions based on the limiting variates of the theorem.

In Section 6.9 we will use Theorem 6.4.2 and its corollary to set up confidence regions for the parameters of interest. The statistic

$$|R_{YX}{}^{(T)}(\lambda)|^2 = \mathbf{f}_{YX}{}^{(T)}(\lambda)\mathbf{f}_{XX}{}^{(T)}(\lambda)^{-1}\mathbf{f}_{XY}{}^{(T)}(\lambda)/f_{YY}{}^{(T)}(\lambda) \tag{6.4.11}$$

is often of special interest as it provides a measure of the strength of a linear time invariant relation between the series $Y(t)$, $t = 0, \pm 1, \ldots$ and the series $\mathbf{X}(t)$, $t = 0, \pm 1, \ldots$. Its large sample distribution is indicated by

Theorem 6.4.3 Suppose the conditions of Theorem 6.4.1 are satisfied and suppose $|R_{YX}{}^{(T)}(\lambda)|^2$ is given by (6.4.11). Then, in Case A,

$$|R_{YX}{}^{(T)}(\lambda)|^2 = [Fr/(2m+1-r)]/[1 + Fr/(2m+1-r)] + o_{a.s.}(1) \tag{6.4.12}$$

as $T \to \infty$ where F is a noncentral F with degrees of freedom $2r$ over $2(2m+1-r)$ and noncentrality parameter $\mathbf{A}(\lambda)\mathbf{f}_{XX}{}^{(T)}(\lambda)\overline{\mathbf{A}(\lambda)}^\tau/f_{\epsilon\epsilon}(\lambda)$.

We will return to a discussion of this statistic in Chapter 8. The notation $o_{a.s.}(1)$ means that the term tends to 0 with probability 1.

6.5 EXPECTED VALUES OF ESTIMATES OF THE TRANSFER FUNCTION AND ERROR SPECTRUM

We now turn to an investigation of the expected values of estimates of slightly more general form than those in the previous section. Suppose that we are interested in estimating the parameters of the model (6.1.1) given the values $\mathbf{X}(t)$, $Y(t)$, $t = 0, \ldots, T-1$. Let $\mathbf{I}_{YX}^{(T)}(\lambda)$ be given by (6.3.8) with similar definitions for $I_{YY}^{(T)}(\lambda)$ and $\mathbf{I}_{XX}^{(T)}(\lambda)$. We will base our estimates on these statistics in the manner of (6.3.10); however, we will make our estimates more flexible by including a variable weighting of the terms in expression (6.3.10). Specifically, let $W(\alpha)$ be a weight function satisfying

Assumption 6.5.1 $W(\alpha)$, $-\infty < \alpha < \infty$, is bounded, even, non-negative, equal to 0 for $|\alpha| > \pi$ and such that

$$\int_{-\pi}^{\pi} W(\alpha) d\alpha = 1. \tag{6.5.1}$$

The principal restrictions introduced here on $W(\alpha)$, over those of Assumption 5.6.1, are the non-negativity and finite support.

In order to reflect the notion that the weight function should become more concentrated as the sample size T tends to ∞, we introduce a bandwidth parameter B_T that depends on T. Also in order that our estimate possess required symmetries we extend the weight function periodically. We therefore define

$$W^{(T)}(\alpha) = B_T^{-1} \sum_{j=-\infty}^{\infty} W(B_T^{-1}[\alpha + 2\pi j]). \tag{6.5.2}$$

We see that $W^{(T)}(\alpha)$ is non-negative,

$$W^{(T)}(\alpha + 2\pi) = W^{(T)}(\alpha), \tag{6.5.3}$$

and if $B_T \to 0$ as $T \to \infty$, then for T sufficiently large

$$\int_{0}^{2\pi} W^{(T)}(\alpha) d\alpha = 1. \tag{6.5.4}$$

The mass of $W^{(T)}(\alpha)$ is concentrated in intervals of width $2\pi B_T$ about $\alpha \equiv 0 \pmod{2\pi}$ as $T \to \infty$.

We now define

$$\mathbf{f}_{XX}^{(T)}(\lambda) = 2\pi T^{-1} \sum_{s=1}^{T-1} W^{(T)}\left(\lambda - \frac{2\pi s}{T}\right) \mathbf{I}_{XX}^{(T)}\left(\frac{2\pi s}{T}\right), \tag{6.5.5}$$

$$\mathbf{f}_{YX}^{(T)}(\lambda) = 2\pi T^{-1} \sum_{s=1}^{T-1} W^{(T)}\left(\lambda - \frac{2\pi s}{T}\right) \mathbf{I}_{YX}^{(T)}\left(\frac{2\pi s}{T}\right), \tag{6.5.6}$$

$$f_{YY}^{(T)}(\lambda) = 2\pi T^{-1} \sum_{s=1}^{T-1} W^{(T)}\left(\lambda - \frac{2\pi s}{T}\right) I_{YY}^{(T)}\left(\frac{2\pi s}{T}\right). \tag{6.5.7}$$

As estimates of $\mathbf{A}(\lambda)$, $f_{\epsilon\epsilon}(\lambda)$, μ we now take

$$\mathbf{A}^{(T)}(\lambda) = \mathbf{f}_{YX}^{(T)}(\lambda) \mathbf{f}_{XX}^{(T)}(\lambda)^{-1}, \tag{6.5.8}$$

$$g_{\epsilon\epsilon}^{(T)}(\lambda) = f_{YY}^{(T)}(\lambda) - \mathbf{f}_{YX}^{(T)}(\lambda) \mathbf{f}_{XX}^{(T)}(\lambda)^{-1} \mathbf{f}_{XY}^{(T)}(\lambda), \tag{6.5.9}$$

and

$$\mu^{(T)} = c_Y^{(T)} - \mathbf{A}^{(T)}(0)\mathbf{c}_X^{(T)} \tag{6.5.10}$$

respectively. If m is large, then definitions (6.5.9) and (6.4.5) are essentially equivalent.

Because of the conditions placed on $W(\alpha)$, we see that

$$\mathbf{A}^{(T)}(-\lambda) = \overline{\mathbf{A}^{(T)}(\lambda)}. \tag{6.5.11}$$

Also $\mathbf{A}^{(T)}(\lambda)$ and $g_{\varepsilon\varepsilon}^{(T)}(\lambda)$ have period 2π, while $g_{\varepsilon\varepsilon}^{(T)}(\lambda)$ is non-negative and symmetric about 0. Finally $\mu^{(T)}$ is real-valued as is its corresponding population parameter μ.

A statistic that will appear in our later investigations is $|R_{YX}^{(T)}(\lambda)|^2$ given by

$$|R_{YX}^{(T)}(\lambda)|^2 = 1 - \frac{g_{\varepsilon\varepsilon}^{(T)}(\lambda)}{f_{YY}^{(T)}(\lambda)}. \tag{6.5.12}$$

It will be seen to be a form of multiple correlation coefficient and may be seen to be bounded by 0 and 1, and it will appear in an essential manner in estimates of the variances of our statistics.

We make one important assumption concerning the sequence of fixed (as opposed to random) values $\mathbf{X}(t)$ and that is

Assumption 6.5.2 $\mathbf{X}(t)$, $t = 0, \pm 1, \ldots$ is uniformly bounded and if $\mathbf{f}_{XX}^{(T)}(\lambda)$ is given by (6.5.5), then there is a finite K such that

$$\|\mathbf{f}_{XX}^{(T)}(\lambda)\|, \|\mathbf{f}_{XX}^{(T)}(\lambda)^{-1}\| < K \tag{6.5.13}$$

for all λ and T sufficiently large.

Turning to the investigation of the large sample behavior of $\mathbf{A}^{(T)}(\lambda)$ we have

Theorem 6.5.1 Let $\varepsilon(t)$, $t = 0, \pm 1, \ldots$ satisfy Assumption 2.6.2(l), $\mathbf{X}(t)$, $t = 0, \pm 1, \ldots$ satisfy Assumption 6.5.2. Let $Y(t)$, $t = 0, \pm 1, \ldots$ be given by (6.1.1) where $\{\mathbf{a}(u)\}$ satisfies $\sum |u|\,|\mathbf{a}(u)| < \infty$. Let $W(\alpha)$ satisfy Assumption 6.5.1. Let $\mathbf{A}^{(T)}(\lambda)$ be given by (6.5.8), then

$$\begin{aligned} E\mathbf{A}^{(T)}(\lambda) &= \left\{\sum_{s=1}^{T-1} W^{(T)}\left(\lambda - \frac{2\pi s}{T}\right)\mathbf{A}\left(\frac{2\pi s}{T}\right)\mathbf{I}_{XX}^{(T)}\left(\frac{2\pi s}{T}\right)\right\} \\ &\quad \times \left\{\sum_{s=1}^{T-1} W^{(T)}\left(\lambda - \frac{2\pi s}{T}\right)\mathbf{I}_{XX}^{(T)}\left(\frac{2\pi s}{T}\right)\right\}^{-1} + O(T^{-1/2}) \\ &= \mathbf{A}(\lambda) + O(B_T) + O(T^{-1/2}) \end{aligned} \tag{6.5.14}$$

where the error terms are uniform in λ.

We see that the expected value of $\mathbf{A}^{(T)}(\lambda)$ is essentially a (matrix) weighted average of the population function $\mathbf{A}(\alpha)$ with weight concentrated in a neighborhood, of width $2\pi B_T$, of λ. Because it is a matrix weighted average, an entanglement of the various components of $\mathbf{A}(\alpha)$ has been introduced. If we wish to reduce the asymptotic bias, we should try to arrange for $\mathbf{A}(\alpha)$ to be near constant in the neighborhood of λ. The weights in (6.5.14) depend on the values $\mathbf{X}(t)$, $t = 0, \ldots, T-1$. It would be advantageous to make $\mathbf{I}_{XX}{}^{(T)}(\alpha)$ near constant as well and such that off-diagonal elements are near 0. The final expression of (6.5.14) suggests that the asymptotic bias of $\mathbf{A}^{(T)}(\lambda)$ is generally of the order of the band-width B_T. We have

Corollary 6.5.1 Under the conditions of Theorem 6.5.1 and if $B_T \to 0$ as $T \to \infty$, $\mathbf{A}^{(T)}(\lambda)$ is an asymptotically unbiased estimate of $\mathbf{A}(\lambda)$.

Let the entries of $\mathbf{A}(\lambda)$ and $\mathbf{A}^{(T)}(\lambda)$ be denoted by $A_j(\lambda)$ and $A_j{}^{(T)}(\lambda)$, $j = 1, \ldots, r$, respectively. On occasion we may be interested in the real-valued **gains**

$$G_j(\lambda) = |A_j(\lambda)| \tag{6.5.15}$$

and the real-valued **phases**

$$\phi_j(\lambda) = \arg A_j(\lambda). \tag{6.5.16}$$

These may be estimated by

$$G_j{}^{(T)}(\lambda) = |A_j{}^{(T)}(\lambda)| \tag{6.5.17}$$

and

$$\phi_j{}^{(T)}(\lambda) = \arg A_j{}^{(T)}(\lambda). \tag{6.5.18}$$

Theorem 6.5.2 Under the conditions of Theorem 6.5.1,

$$\begin{aligned} EG_j{}^{(T)}(\lambda) &= |EA_j{}^{(T)}(\lambda)| + O(B_T^{-1/2}T^{-1/2}) \\ &= G_j(\lambda) + O(B_T) + O(B_T^{-1/2}T^{-1/2}) \end{aligned} \tag{6.5.19}$$

and if $A_j(\lambda) \neq 0$, then

$$\begin{aligned} \vec{\text{ave}} \log G_j{}^{(T)}(\lambda) &= \log |EA_j{}^{(T)}(\lambda)| + O(B_T{}^{-1}T^{-1}) \\ &= \log G_j(\lambda) + O(B_T) + O(T^{-1/2}) + O(B_T{}^{-1}T^{-1}) \end{aligned} \tag{6.5.20}$$

$$\begin{aligned} \vec{\text{ave}}\, \phi_j{}^{(T)}(\lambda) &= \arg A_j{}^{(T)}(\lambda) + O(B_T{}^{-1}T^{-1}) \\ &= \phi_j(\lambda) + O(B_T) + O(T^{-1/2}) + O(B_T{}^{-1}T^{-1}). \end{aligned} \tag{6.5.21}$$

(In this theorem, $\vec{\text{ave}}$ denotes an expected value derived in a term by term manner from a Taylor expansion, see Brillinger and Tukey (1964).)

Corollary 6.5.2 Under the conditions of Theorem 6.5.2 and if $B_T \to 0$, $B_T T \to \infty$ as $T \to \infty$, $G_j^{(T)}(\lambda)$ is an asymptotically unbiased estimate of $G_j(\lambda)$.

Turning to the case of $g_{\varepsilon\varepsilon}^{(T)}(\lambda)$, our estimate of the error spectrum, we have

Theorem 6.5.3 Under the conditions of Theorem 6.5.1,

$$Eg_{\varepsilon\varepsilon}^{(T)}(\lambda) = f_{\varepsilon\varepsilon}(\lambda) + O(B_T) + O(B_T^{-1}T^{-1}) + O(T^{-1/2}). \qquad (6.5.22)$$

This result may be compared instructively with expression (5.8.22) in the case $P = 1$. In the limit we have

Corollary 6.5.3 Under the conditions of Theorem 6.5.3 and if $B_T \to 0$, $B_T T \to \infty$ as $T \to \infty$, $g_{\varepsilon\varepsilon}^{(T)}(\lambda)$ is an asymptotically unbiased estimate of $f_{\varepsilon\varepsilon}(\lambda)$.

In the case of $\mu^{(T)}$ we may prove

Theorem 6.5.4 Under the conditions of Theorem 6.5.1,

$$E\mu^{(T)} = \mu + O(B_T) + O(T^{-1/2}). \qquad (6.5.23)$$

From Theorem 6.5.4 follows

Corollary 6.5.4 Under the conditions of Theorem 6.5.4 and if $B_T \to 0$ as $T \to \infty$, $\mu^{(T)}$ is an asymptotically unbiased estimate of μ.

6.6 ASYMPTOTIC COVARIANCES OF THE PROPOSED ESTIMATES

In order to be able to assess the precision of our estimates we require the form of their second-order moments. A statistic that will appear in these moments is defined by

$$\mathbf{h}_{XX}^{(T)}(\lambda) = 2\pi T^{-1}\left\{\sum_{s=1}^{T-1} W^{(T)}\left(\lambda - \frac{2\pi s}{T}\right)^2 \mathbf{I}_{XX}^{(T)}\left(\frac{2\pi s}{T}\right)\right\}\left\{\int_0^{2\pi} W^{(T)}(\alpha)^2 d\alpha\right\}^{-1}. \qquad (6.6.1)$$

This statistic has the same form as $\mathbf{f}_{XX}^{(T)}(\lambda)$ given by expression (6.5.5) except that the weight function $W(\alpha)$ has been replaced by $W(\alpha)^2$. Typically the latter is more concentrated; however, in the case that $W(\alpha) = (2\pi)^{-1}$ for $|\alpha| \leqslant \pi$

$$\mathbf{h}_{XX}^{(T)}(\lambda) = \mathbf{f}_{XX}^{(T)}(\lambda). \qquad (6.6.2)$$

In a variety of cases it may prove reasonable to approximate $\mathbf{h}_{XX}^{(T)}(\lambda)$ by $\mathbf{f}_{XX}^{(T)}(\lambda)$. This has the advantage of reducing the number of computations required. Note that if $\mathbf{f}_{XX}^{(T)}(\lambda)$ is bounded, then the same is true for $\mathbf{h}_{XX}^{(T)}(\lambda)$. Thus we may now state

Theorem 6.6.1 Let $\varepsilon(t)$, $t = 0, \pm 1, \ldots$ satisfy Assumption 2.6.1 and have mean 0. Let $\mathbf{X}(t)$, $t = 0, \pm 1, \ldots$ satisfy Assumption 6.5.2. Let $Y(t)$, $t = 0, \pm 1, \ldots$ be given by (6.1.1) where $\{\mathbf{a}(u)\}$ satisfies $\Sigma\, |u|\, |\mathbf{a}(u)| < \infty$. Let $W(\alpha)$ satisfy Assumption 6.5.1. If $B_T \to 0$ as $T \to \infty$, then

$$\begin{aligned}
\operatorname{cov}&\{\mathbf{A}^{(T)}(\lambda)^\tau, \mathbf{A}^{(T)}(\mu)^\tau\} \\
&= \mathbf{f}_{XX}^{(T)}(\lambda)^{-1}\left\{(2\pi T^{-1})^2 \sum_s W^{(T)}\left(\lambda - \frac{2\pi s}{T}\right) W^{(T)}\left(\mu - \frac{2\pi s}{T}\right)\right. \\
&\quad \times \left.\mathbf{I}_{XX}^{(T)}\left(\frac{2\pi s}{T}\right) f_{\varepsilon\varepsilon}\left(\frac{2\pi s}{T}\right)\right\} \mathbf{f}_{XX}^{(T)}(\mu)^{-1} + O(T^{-1}) \\
&= \eta\{\lambda - \mu\} B_T^{-1} T^{-1} 2\pi \int W(\alpha)^2 d\alpha\, \mathbf{f}_{XX}^{(T)}(\lambda)^{-1} \mathbf{h}_{XX}^{(T)}(\lambda) \mathbf{f}_{XX}^{(T)}(\lambda)^{-1} f_{\varepsilon\varepsilon}(\lambda) \\
&\quad + O(T^{-1}).
\end{aligned} \tag{6.6.3}$$

In the case that (6.6.2) holds, the second expression of (6.6.3) has the form

$$\eta\{\lambda - \mu\} B_T^{-1} T^{-1} \mathbf{f}_{XX}^{(T)}(\lambda)^{-1} f_{\varepsilon\varepsilon}(\lambda) 2\pi \int W(\alpha)^2 d\alpha + O(T^{-1}), \tag{6.6.4}$$

an expression that may be estimated by

$$\eta\{\lambda - \mu\} B_T^{-1} T^{-1} \mathbf{f}_{XX}^{(T)}(\lambda)^{-1} g_{\varepsilon\varepsilon}^{(T)}(\lambda) 2\pi \int W(\alpha)^2 d\alpha. \tag{6.6.5}$$

We see from expression (6.6.3) that the asymptotic variance of $\mathbf{A}^{(T)}(\lambda)$ is of order $B_T^{-1}T^{-1}$ and so we have

Corollary 6.6.1 Under the conditions of the theorem and if $B_T T \to \infty$ as $T \to \infty$, $\mathbf{A}^{(T)}(\lambda)$ is a consistent estimate of $\mathbf{A}(\lambda)$.

We also note, from (6.6.3), that $\mathbf{A}^{(T)}(\lambda)$ and $\mathbf{A}^{(T)}(\mu)$ are asymptotically uncorrelated for $\lambda \not\equiv \mu \pmod{2\pi}$.

In practice we will record real-valued statistics. The asymptotic covariance structure of Re $\mathbf{A}^{(T)}(\lambda)$, Im $\mathbf{A}^{(T)}(\lambda)$ is given in Exercise 6.14.22. Alternatively we may record $G_j^{(T)}(\lambda)$, $\phi_j^{(T)}(\lambda)$ and so we now investigate their asymptotic covariances. We define $\Psi_{jk}^{(T)}(\lambda)$ to be the entry in the jth row and kth column of matrix

$$\mathbf{\Psi}^{(T)}(\lambda) = \mathbf{f}_{XX}^{(T)}(\lambda)^{-1} \mathbf{h}_{XX}^{(T)}(\lambda) \mathbf{f}_{XX}^{(T)}(\lambda)^{-1}. \tag{6.6.6}$$

Theorem 6.6.2 Under the conditions of Theorem 6.6.1 and if $A_j(\lambda)$, $A_k(\mu) \neq 0$

$$\vec{\text{cov}}\,\{\log_e G_j^{(T)}(\lambda), \log_e G_k^{(T)}(\mu)\}$$
$$= B_T^{-1}T^{-1}[\eta\{\lambda - \mu\} + \eta\{\lambda + \mu\}]\pi \int W(\alpha)^2 d\alpha\, f_{\varepsilon\varepsilon}(\lambda)$$
$$\times \operatorname{Re}\,\{A_j(\lambda)^{-1}\Psi_{jk}^{(T)}(\lambda)\overline{A_k(\lambda)^{-1}}\} + O(T^{-1}) \tag{6.6.7}$$

$$\vec{\text{cov}}\,\{\log_e G_j^{(T)}(\lambda), \phi_k^{(T)}(\mu)\} = O(T^{-1}) \tag{6.6.8}$$

$$\vec{\text{cov}}\,\{\phi_j^{(T)}(\lambda), \phi_k^{(T)}(\mu)\}$$
$$= B_T^{-1}T^{-1}[\eta\{\lambda - \mu\} - \eta\{\lambda + \mu\}]\pi \int W(\alpha)^2 d\alpha\, f_{\varepsilon\varepsilon}(\lambda)$$
$$\times \operatorname{Re}\,\{A_j(\lambda)^{-1}\Psi_{jk}^{(T)}(\lambda)\overline{A_k(\lambda)^{-1}}\} + O(T^{-1}) \tag{6.6.9}$$

for $j, k = 1, \ldots, r$.

Note that the asymptotic covariance structure of $\log G_j^{(T)}(\lambda)$ is the same as that of $\phi_j^{(T)}(\lambda)$ except in the cases $\lambda \equiv 0 \pmod \pi$. We can construct estimates of the covariances of Theorem 6.6.2 by substituting estimates for the unknowns $A_j(\lambda)$, $f_{\varepsilon\varepsilon}(\lambda)$; We note that $\log G_j^{(T)}(\lambda)$ and $\phi_k^{(T)}(\mu)$ are asymptotically uncorrelated for all j, k and λ, μ.

Turning to the investigation of $g_{\varepsilon\varepsilon}^{(T)}(\lambda)$, we have

Theorem 6.6.3 Under the conditions of Theorem 6.6.1,

$$\text{cov}\,\{g_{\varepsilon\varepsilon}^{(T)}(\lambda), g_{\varepsilon\varepsilon}^{(T)}(\mu)\}$$
$$= B_T^{-1}T^{-1}[\eta\{\lambda - \mu\} + \eta\{\lambda + \mu\}]2\pi \int W(\alpha)^2 d\alpha\, f_{\varepsilon\varepsilon}(\lambda)^2$$
$$+ O(T^{-1}) + O(B_T^{-1}T^{-2}). \tag{6.6.10}$$

In the limit we have,

Corollary 6.6.3 Under the conditions of Theorem 6.6.3 and if $B_T T \to \infty$ as $T \to \infty$

$$\lim_{T\to\infty} B_T T \,\text{cov}\,\{g_{\varepsilon\varepsilon}^{(T)}(\lambda), g_{\varepsilon\varepsilon}^{(T)}(\mu)\}$$
$$= 2\pi \int W(\alpha)^2 d\alpha\, [\eta\{\lambda - \mu\} + \eta\{\lambda + \mu\}] f_{\varepsilon\varepsilon}(\lambda)^2 \tag{6.6.11}$$

and

$$\lim_{T\to\infty} B_T T \,\vec{\text{var}}\, \log_e g_{\varepsilon\varepsilon}^{(T)}(\lambda) = 2\pi \int W(\alpha)^2 d\alpha\, [1 + \eta\{2\lambda\}]. \tag{6.6.12}$$

Expressions (6.6.11) and (6.6.12) should be compared with expressions (5.6.12) and (5.6.15). We see that under the indicated limiting processes the asymptotic behavior of the second-order moments of $g_{\varepsilon\varepsilon}^{(T)}(\lambda)$ is the same as if $g_{\varepsilon\varepsilon}^{(T)}(\lambda)$ were a power spectral estimate based directly on the values $\varepsilon(t)$, $t = 0, \ldots, T - 1$.

In the case of $\mu^{(T)} + \mathbf{A}^{(T)}(0)\mathbf{c}_X^{(T)} = c_Y^{(T)}$ we have

Theorem 6.6.4 Under the conditions of Theorem 6.6.1,

$$\operatorname{var}\{\mu^{(T)} + \mathbf{A}^{(T)}(0)\mathbf{c}_X^{(T)}\} = 2\pi T^{-1} f_{\varepsilon\varepsilon}(0) + o(T^{-1}). \tag{6.6.13}$$

We may use expression (6.6.15) below to obtain an expression for the large sample variance of $\mu^{(T)}$. (See Exercise 6.14.31.) This variance tends to 0 as $T \to \infty$ and so we have

Corollary 6.6.4 Under the conditions of the theorem and if $B_T T \to \infty$ as $T \to \infty$, $\mu^{(T)}$ is a consistent estimate of μ.

In the case of the joint behavior of $\mathbf{A}^{(T)}(\lambda)$, $g_{\varepsilon\varepsilon}^{(T)}(\lambda)$, $\mu^{(T)} + \mathbf{A}^{(T)}(0)\mathbf{c}_X^{(T)}$ we have

Theorem 6.6.5 Under the conditions of Theorem 6.6.1

$$\operatorname{cov}\{\mathbf{A}^{(T)}(\lambda)^\tau, g_{\varepsilon\varepsilon}^{(T)}(\mu)\} = O(T^{-1}), \tag{6.6.14}$$

$$\operatorname{cov}\{\mathbf{A}^{(T)}(\lambda)^\tau, \mu^{(T)} + \mathbf{A}^{(T)}(0)\mathbf{c}_X^{(T)}\} = O(T^{-1}), \tag{6.6.15}$$

and

$$\operatorname{cov}\{g_{\varepsilon\varepsilon}^{(T)}(\lambda), \mu^{(T)} + \mathbf{A}^{(T)}(0)\mathbf{c}_X^{(T)}\} = O(T^{-1}). \tag{6.6.16}$$

We see that $g_{\varepsilon\varepsilon}^{(T)}(\mu)$ is asymptotically uncorrelated with both $\mathbf{A}^{(T)}(\lambda)$ and $\mu^{(T)} + \mathbf{A}^{(T)}(0)\mathbf{c}_X^{(T)}$. Also $\mathbf{A}^{(T)}(\lambda)$ and $\mu^{(T)} + \mathbf{A}^{(T)}(0)\mathbf{c}_X^{(T)}$ are asymptotically uncorrelated.

In the case of the gains and phases we have

Theorem 6.6.6 Under the conditions of Theorem 6.6.1,

$$\vec{\operatorname{cov}}\{\log G_j^{(T)}(\lambda), g_{\varepsilon\varepsilon}^{(T)}(\mu)\} = O(T^{-1}) \tag{6.6.17}$$

$$\vec{\operatorname{cov}}\{\phi_j^{(T)}(\lambda), g_{\varepsilon\varepsilon}^{(T)}(\mu)\} = O(T^{-1}) \tag{6.6.18}$$

$$\vec{\operatorname{cov}}\{\log G_j^{(T)}(\lambda), \mu^{(T)} + \mathbf{A}^{(T)}(0)\mathbf{c}_X^{(T)}\} = O(T^{-1}) \tag{6.6.19}$$

and

$$\vec{\operatorname{cov}}\{\phi_j^{(T)}(\lambda), \mu^{(T)} + \mathbf{A}^{(T)}(0)\mathbf{c}_X^{(T)}\} = O(T^{-1}) \qquad \text{for } j = 1, \ldots, r. \tag{6.6.20}$$

6.7 ASYMPTOTIC NORMALITY OF THE ESTIMATES

We next turn to an investigation of the asymptotic distributions of the estimates $\mathbf{A}^{(T)}(\lambda)$, $g_{\varepsilon\varepsilon}^{(T)}(\lambda)$, $\mu^{(T)}$ under the limiting condition $B_T T \to \infty$ as $T \to \infty$.

Theorem 6.7.1 Let $\varepsilon(t)$, $t = 0, \pm 1, \ldots$ satisfy Assumption 2.6.1 and have mean 0. Let $\mathbf{X}(t)$, $t = 0, \pm 1, \ldots$ satisfy Assumption 6.5.2. Let $Y(t)$, $t = 0, \pm 1, \ldots$ be given by (6.1.1) where $\{\mathbf{a}(u)\}$ satifies $\sum |u|\,|\mathbf{a}(u)| < \infty$. Let

$W(\alpha)$ satisfy Assumption 6.5.1. If $B_T \to 0$, $B_T T \to \infty$ as $T \to \infty$, then $\mathbf{A}^{(T)}(\lambda_1)$, $g_{\varepsilon\varepsilon}^{(T)}(\lambda_1), \ldots, \mathbf{A}^{(T)}(\lambda_J)$, $g_{\varepsilon\varepsilon}^{(T)}(\lambda_J)$ are asymptotically jointly normal with covariance structure given by (6.6.3), (6.6.11), and (6.6.14). Finally $\mu^{(T)} + \mathbf{A}^{(T)}(0)\mathbf{c}_X^{(T)}$ is asymptotically independent of these variates with variance (6.6.13).

We see from expression (6.6.14) that $\mathbf{A}^{(T)}(\lambda)$ and $g_{\varepsilon\varepsilon}^{(T)}(\mu)$ are asymptotically independent for all λ, μ under the above conditions. From (6.6.3) we see that $\mathbf{A}^{(T)}(\lambda)$ and $\mathbf{A}^{(T)}(\mu)$ are asymptotically independent if $\lambda - \mu \not\equiv 0 \pmod{2\pi}$. From Exercise 6.14.22 we see Re $\mathbf{A}^{(T)}(\lambda)$ and Im $\mathbf{A}^{(T)}(\lambda)$ are asymptotically independent. All of these instances of asymptotic independence are in accord with the intuitive suggestions of Theorem 6.2.4.

The theorem indicates that $\mathbf{A}^{(T)}(\lambda)^\tau$ is asymptotically

$$N_r^C(E\mathbf{A}^{(T)}(\lambda)^\tau, B_T^{-1}T^{-1}2\pi \int W(\alpha)^2 d\alpha \; \mathbf{\Psi}^{(T)}(\lambda) f_{\varepsilon\varepsilon}(\lambda)) \tag{6.7.1}$$

if $\lambda \not\equiv 0 \pmod{\pi}$ where $\mathbf{\Psi}^{(T)}(\lambda)$ is given by (6.6.6). This result will later be used to set confidence regions for $\mathbf{A}(\lambda)$.

Following a theorem of Mann and Wald (1943a), we may conclude

Corollary 6.7.1 Under the conditions of Theorem 6.7.1 $\log_e G_j^{(T)}(\lambda)$, $\phi_j^{(T)}(\lambda)$, $g_{\varepsilon\varepsilon}^{(T)}(\lambda)$, $c_Y^{(T)} = \mu^{(T)} + \mathbf{A}^{(T)}(0)\mathbf{c}_X^{(T)}$ are asymptotically normal with covariance structure given by (6.6.7) to (6.6.10), (6.6.13), and (6.6.17) to (6.6.20) for $j = 1, \ldots, r$.

We note in particular that, under the indicated conditions, log $G_j^{(T)}(\lambda)$ and $\phi_j^{(T)}(\lambda)$ are asymptotically independent.

The asymptotic distribution of $\mathbf{A}^{(T)}(\lambda)$ given in this theorem is the same as that of Theorem 6.4.2 once one makes the identification

$$2m + 1 = \left\{\sum W^{(T)}\left(\lambda - \frac{2\pi s}{T}\right)^2\right\}^{-1} \left(\frac{T}{2\pi}\right)^2 \sim \frac{B_T T}{2\pi \int W(\alpha)^2 d\alpha}. \tag{6.7.2}$$

The asymptotic distribution of $g_{\varepsilon\varepsilon}^{(T)}(\lambda)$ is consistent with that of Theorem 6.4.2 in the case that (6.7.2) is large, for a χ^2 variate with a large number of degrees of freedom is near normal.

6.8 ESTIMATING THE IMPULSE RESPONSE

In previous sections we have considered the problem of estimating the transfer function $\mathbf{A}(\lambda)$. We now consider the problem of estimating the corresponding impulse response function $\{\mathbf{a}(u)\}$. In terms of $\mathbf{A}(\lambda)$ it is given by

$$\mathbf{a}(u) = (2\pi)^{-1} \int_0^{2\pi} \mathbf{A}(\lambda) \exp\{iu\lambda\} d\lambda \qquad u = 0, \pm 1, \ldots. \tag{6.8.1}$$

Let $\mathbf{A}^{(T)}(\lambda)$ be an estimate of $\mathbf{A}(\lambda)$ of the form considered previously. Let P_T be a sequence of positive integers tending to ∞ with T. As an estimate of $\mathbf{a}(u)$ we consider

$$\mathbf{a}^{(T)}(u) = P_T^{-1} \sum_{p=0}^{P_T-1} \mathbf{A}^{(T)}(2\pi p/P_T) \exp\{i2\pi pu/P_T\} \qquad u = 0, \pm 1, \ldots. \tag{6.8.2}$$

Note that because of the symmetry properties of $\mathbf{A}^{(T)}(\lambda)$, the range of summation in expression (6.8.2) may be reduced to $0 \leqslant p \leqslant (P_T - 1)/2$ in terms of Im $\mathbf{A}^{(T)}$, Re $\mathbf{A}^{(T)}$. Also, the estimate has period P_T and so, for example

$$\mathbf{a}^{(T)}(-u) = \mathbf{a}^{(T)}(P_T - u). \tag{6.8.3}$$

We may prove

Theorem 6.8.1 Let $\varepsilon(t)$, $t = 0, \pm 1, \ldots$ satisfy Assumption 2.6.1 and have mean 0. Let $\mathbf{X}(t)$, $t = 0, \pm 1, \ldots$ satisfy Assumption 6.5.2. Let $Y(t)$ be given by (6.1.1) where $\{\mathbf{a}(u)\}$ satisfies $\Sigma\, |u|\, |\mathbf{a}(u)| \leqslant \infty$. Let $W(\alpha)$ satisfy Assumption 6.5.1. Let $\mathbf{a}^{(T)}(u)$ be given by (6.8.2), then

$$\begin{aligned} E\mathbf{a}^{(T)}(u) &= P_T^{-1} \sum_{p=0}^{P_T-1} \mathbf{A}(2\pi p/P_T) \exp\{i2\pi pu/P_T\} + O(B_T) + O(T^{-1/2}) \\ &= \mathbf{a}(u) + \sum_{k\neq 0} \mathbf{a}(u + kP_T) + O(B_T) + O(T^{-1/2}). \end{aligned} \tag{6.8.4}$$

We see that for large P_T and small B_T the expected value of the suggested estimate is primarily the desired $\mathbf{a}(u)$. A consequence of the theorem is

Corollary 6.8.1 Under the conditions of Theorem 6.8.1 and if $B_T \to 0$, $P_T \to \infty$ as $T \to \infty$, $\mathbf{a}^{(T)}(u)$ is asymptotically unbiased.

Next turn to an investigation of the second-order moments of $\mathbf{a}^{(T)}(u)$. We have previously defined

$$\mathbf{\Psi}^{(T)}(\lambda) = \mathbf{f}_{XX}^{(T)}(\lambda)^{-1}\mathbf{h}_{XX}^{(T)}(\lambda)\mathbf{f}_{XX}^{(T)}(\lambda)^{-1} \doteq \mathbf{f}_{XX}^{(T)}(\lambda)^{-1} \tag{6.8.5}$$

and we now define

$$\mathbf{\Lambda}^{(T)}(u,v) = P_T^{-1} \sum_{p=0}^{P_T-1} \exp\{i2\pi p(u - v)/P_T\} f_{\varepsilon\varepsilon}(2\pi p/P_T)\mathbf{\Psi}^{(T)}(2\pi p/P_T). \tag{6.8.6}$$

This expression is bounded under the conditions we have set down. We now have

Theorem 6.8.2 Under the assumptions of Theorem 6.8.1 and if $B_T \leqslant P_T^{-1}$, $B_T \to 0$ as $T \to \infty$,

$$\text{cov}\,\{\mathbf{a}^{(T)}(u), \mathbf{a}^{(T)}(v)\} = P_T^{-1}B_T^{-1}T^{-1}2\pi \int W(\alpha)^2 d\alpha\, \mathbf{A}^{(T)}(u,v) + O(T^{-1})$$
$$\text{for } u, v = 0, \pm 1, \ldots . \quad (6.8.7)$$

Note from (6.8.6) that the asymptotic covariance matrix of $\mathbf{a}^{(T)}(u)$ does not depend on u. Also the asymptotic covariance matrix of $\mathbf{a}^{(T)}(u)$ with $\mathbf{a}^{(T)}(v)$ depends on the difference $u - v$ so in some sense the process $\mathbf{a}^{(T)}(u)$, $u = 0, \pm 1, \ldots$ may be considered to be a covariance stationary time series. In the limit we have

Corollary 6.8.2 Under the conditions of Theorem 6.8.2 and if $P_T B_T T \to \infty$ as $T \to \infty$, $\mathbf{a}^{(T)}(u)$ is a consistent estimate of $\mathbf{a}(u)$.

Turning to the joint behavior of $\mathbf{a}^{(T)}(u)$ and $g_{\varepsilon\varepsilon}^{(T)}(\lambda)$ we have

Theorem 6.8.3 Under the conditions of Theorem 6.8.1

$$\text{cov}\,\{\mathbf{a}^{(T)}(u)^\tau, g_{\varepsilon\varepsilon}^{(T)}(\lambda)\} = O(T^{-1}). \quad (6.8.8)$$

We see that $\mathbf{a}^{(T)}(u)$, $g_{\varepsilon\varepsilon}^{(T)}(\lambda)$ are asymptotically uncorrelated for all u, λ. In the case of the limiting distribution we may prove

Theorem 6.8.4 Under the conditions of Theorem 6.8.1 and if $P_T B_T \to 0$ as $T \to \infty$, $\mathbf{a}^{(T)}(u_1), \ldots, \mathbf{a}^{(T)}(u_J)$, $g_{\varepsilon\varepsilon}^{(T)}(\lambda_1), \ldots, g_{\varepsilon\varepsilon}^{(T)}(\lambda_K)$ are asymptotically normal with covariance structure given by (6.6.10), (6.8.7), and (6.8.8).

In Theorem 6.8.2 we required $B_T \leqslant P_T^{-1}$. From expression (6.8.7) we see that we should take $P_T B_T$ as large as possible. Setting $P_T = B_T^{-1}$ seems a sensible procedure, for the asymptotic variance of $\mathbf{a}^{(T)}(u)$ is then of order T^{-1}. However, in this case we are unable to identify its principal term from expression (6.8.7). In the case that $P_T B_T \to 0$, the first term in (6.8.7) is the dominant one. Finally we may contrast the asymptotic order of this variance with that of $\mathbf{A}^{(T)}(\lambda)$ which was $B_T^{-1}T^{-1}$.

6.9 CONFIDENCE REGIONS

The confidence regions that will be proposed in this section will be based on the asymptotic distributions obtained in Section 6.4. They will be constructed so as to be consistent with the asymptotic distributions of Section 6.7.

Suppose estimates $\mathbf{A}^{(T)}(\lambda)$, $\mu^{(T)}$, $g_{\varepsilon\varepsilon}^{(T)}(\lambda)$, $\mathbf{a}^{(T)}(u)$ have been constructed in the manner of Section 6.5 using a weight function $W(\alpha)$. A comparison of the asymptotic distributions obtained for $\mathbf{A}^{(T)}(\lambda)$ in Theorems 6.4.2 and 6.7.1 suggests that we set

$$(2m + 1)^{-1} \backsim B_T^{-1}T^{-1}2\pi \int W(\alpha)^2 d\alpha. \tag{6.9.1}$$

Following Theorem 6.7.1 we then approximate the distribution of $\mathbf{A}^{(T)}(\lambda)^\tau$ by

$$N_r^C(\mathbf{A}(\lambda)^\tau, (2m + 1)^{-1} f_{\varepsilon\varepsilon}(\lambda)\mathbf{f}_{XX}^{(T)}(\lambda)^{-1}) \quad \text{in Case A,} \tag{6.9.2}$$

and by

$$N_r(\mathbf{A}(\lambda)^\tau, (2m)^{-1} f_{\varepsilon\varepsilon}(\lambda)\mathbf{f}_{XX}^{(T)}(\lambda)^{-1}) \quad \text{in Case B, and Case C.} \tag{6.9.3}$$

At the same time the distribution of $g_{\varepsilon\varepsilon}^{(T)}(\lambda)$ is approximated by an independent

$$\frac{f_{\varepsilon\varepsilon}(\lambda)\chi^2_{2(2m+1-r)}}{2(2m + 1 - r)} \quad \text{in Case A} \tag{6.9.4}$$

and by

$$\frac{f_{\varepsilon\varepsilon}(\lambda)\chi^2_{2m-r}}{2m - r} \quad \text{in Case B, and Case C.} \tag{6.9.5}$$

A 100β percent confidence interval for $f_{\varepsilon\varepsilon}(\lambda)$ is therefore provided by

$$\frac{g_{\varepsilon\varepsilon}^{(T)}(\lambda)2(2m + 1 - r)}{\chi^2_{2(2m+1-r)}\left(\frac{1+\beta}{2}\right)} \leqslant f_{\varepsilon\varepsilon}(\lambda) \leqslant \frac{g_{\varepsilon\varepsilon}^{(T)}(\lambda)2(2m + 1 - r)}{\chi^2_{2(2m+1-r)}\left(\frac{1-\beta}{2}\right)} \tag{6.9.6}$$

in Case A with similar intervals in Cases B and C. A confidence interval for $\log f_{\varepsilon\varepsilon}(\lambda)$ is algebraically deducible from (6.9.6).

If we let c_{jj} denote the jth diagonal entry of

$$(2m + 1)^{-1}\mathbf{f}_{XX}^{(T)}(\lambda)^{-1} \tag{6.9.7}$$

and w_j denote $c_{jj}g_{\varepsilon\varepsilon}^{(T)}(\lambda)$, then following the discussion of Section 6.2 a 100β percent confidence region for Re $A_j^{(T)}(\lambda)$, Im $A_j^{(T)}(\lambda)$ may be determined from the inequality

$$\{\operatorname{Re} A_j(\lambda) - \operatorname{Re} A_j^{(T)}(\lambda)\}^2 + \{\operatorname{Im} A_j(\lambda) - \operatorname{Im} A_j^{(T)}(\lambda)\}^2 \leqslant 2w_j F_{2;2(2m+1-r)}(\beta). \tag{6.9.8}$$

This region is considered in Akaike (1965) and Groves and Hannan (1968). If a 100β percent simultaneous confidence region is desired for all the $A_j(\lambda)$, $j = 1, \ldots, r$ then following Exercise 6.14.17 we can consider the region

$$\begin{aligned} |\operatorname{Re} A_j(\lambda) - \operatorname{Re} A_j^{(T)}(\lambda)| &\leqslant (2rw_j F_{2r;2(2m+1-r)}(\beta))^{1/2} \\ |\operatorname{Im} A_j(\lambda) - \operatorname{Im} A_j^{(T)}(\lambda)| &\leqslant (2wr_j F_{2r;2(2m+1-r)}(\beta))^{1/2} \quad j = 1, \ldots, r. \end{aligned} \tag{6.9.9}$$

If we set

$$u_j = (2rw_j F_{2r;2(2m+1-r)}(\beta))^{1/2} \tag{6.9.10}$$

then the region (6.9.9) is approximately equivalent to the region

$$\begin{gathered} |A_j^{(T)}(\lambda)| - u_j \leq |A_j(\lambda)| \leq |A_j^{(T)}(\lambda)| + u_j \\ \arg A_j^{(T)}(\lambda) - \arcsin\{u_j/|A_j^{(T)}(\lambda)|\} \leq \arg A_j(\lambda) \leq \arg A_j^{(T)}(\lambda) \\ + \arcsin\{u_j/|A_j^{(T)}(\lambda)|\} \qquad j = 1, \ldots, r \end{gathered} \tag{6.9.11}$$

giving a simultaneous confidence region for the real-valued gains and phases. Regions of these forms are considered in Goodman (1965) and Bendat and Piersol (1966). The exact procedures based on (6.2.19) and (6.2.21) may also be of use in constructing separate intervals for $|A_j(\lambda)|$ or $\phi_j(\lambda)$. They involve approximating the distribution of

$$\frac{w_j^{-1}|A_j^{(T)}(\lambda)|^2}{2} \tag{6.9.12}$$

by a noncentral F with degrees of freedom 2, $2(2m + 1 - r)$ and noncentrality $|A_j(\lambda)|^2/2$ and on approximating the distribution of

$$w_j^{-1/2}\{\operatorname{Im} A_j^{(T)}(\lambda) \cos \phi_j(\lambda) - \operatorname{Re} A_j^{(T)}(\lambda) \sin \phi_j(\lambda)\} \tag{6.9.13}$$

by a $t_{2(2m+1-r)}$ distribution and then finding intervals by algebra.

On occasion we might be interested in examining the hypothesis $\mathbf{A}(\lambda) = \mathbf{0}$. This may be carried out by means of analogs of the statistics (6.2.9) and (6.2.10), namely

$$\frac{(2m+1)\mathbf{A}^{(T)}(\lambda)\mathbf{f}_{XX}^{(T)}(\lambda)\overline{\mathbf{A}^{(T)}(\lambda)}^{\tau}}{r g_{\varepsilon\varepsilon}^{(T)}(\lambda)} \tag{6.9.14}$$

and

$$|R_{YX}^{(T)}(\lambda)|^2 = \mathbf{f}_{YX}^{(T)}(\lambda)\mathbf{f}_{XX}^{(T)}(\lambda)^{-1}\mathbf{f}_{XY}^{(T)}(\lambda)/f_{YY}^{(T)}(\lambda). \tag{6.9.15}$$

In the case $\mathbf{A}(\lambda) = \mathbf{0}$, (6.9.14) is distributed asymptotically as $F_{2;2(2m+1-r)}$ and the latter statistic as

$$\left[\frac{m+1-r}{r} F_{2(2m+1-r);\,2r} + 1\right]^{-1} \tag{6.9.16}$$

respectively.

We now turn to the problem of setting confidence limits for the entries of $\mathbf{a}^{(T)}(u)$. The investigations of Section 6.8 suggest the evaluation of the statistic

$$\mathbf{A}^{(T)} = P_T^{-1} \sum_{p=0}^{P_T-1} g_{\varepsilon\varepsilon}^{(T)}\left(\frac{2\pi p}{P_T}\right)\mathbf{f}_{XX}^{(T)}\left(\frac{2\pi p}{P_T}\right)^{-1} \tag{6.9.17}$$

Let $\Lambda_{jj}^{(T)}$ signify the jth diagonal entry of $\mathbf{\Lambda}^{(T)}$. Theorem 6.8.4 now suggests

$$a_j^{(T)}(u) - [P_T^{-1}B_T^{-1}T^{-1}2\pi \int W(\alpha)^2 d\alpha\, \Lambda_{jj}^{(T)}]^{1/2} z\left(\frac{1+\beta}{2}\right) \leq a_j(u)$$
$$\leq a_j^{(T)}(u) + [P_T^{-1}B_T^{-1}T^{-1}2\pi \int W(\alpha)^2 d\alpha\, \Lambda_{jj}^{(T)}]^{1/2} z\left(\frac{1+\beta}{2}\right) \quad (6.9.18)$$

as an approximate 100β percent confidence interval for $a_j(u)$.

Simultaneous regions for $a_{j_1}(u_1), \ldots, a_{j_J}(u_J)$ may be constructed from (6.9.18) using Bonferroni's inequality; see Miller (1966).

6.10 A WORKED EXAMPLE

As a first example we investigate relations between the series, $B(t)$, of monthly mean temperatures in Berlin and the series, $V(t)$, of monthly mean temperatures in Vienna. Because these series have such definite annual variation we first adjust them seasonally. We do this by evaluating the mean value for each month along the course of each series and then subtracting that mean value from the corresponding month values. If $Y(t)$ denotes the adjusted series for Berlin, then it is given by

$$Y(j + 12k) = B(j + 12k) - K^{-1} \sum_{k=0}^{K-1} B(j + 12k) \quad (6.10.1)$$

$j = 0, \ldots, 11$; $k = 0, \ldots, K - 1$ and $K = T/12$. Let $X(t)$ likewise denote the series of adjusted values for Vienna. These series are given in Figures 6.10.1 and 2 for 1920–1930. The original series are given in Figure 1.1.1.

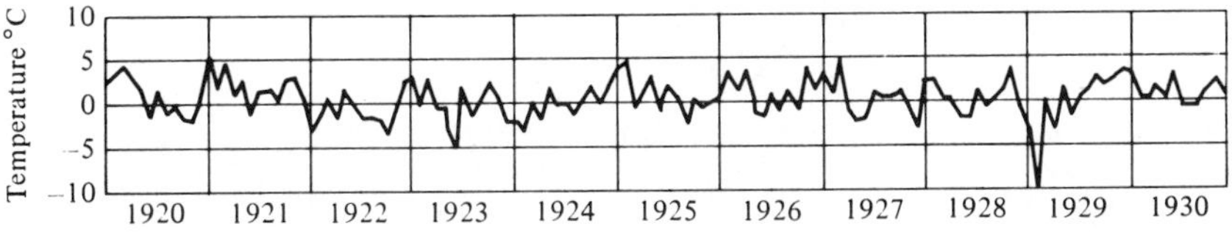

Figure 6.10.1 Seasonally adjusted series of monthly mean temperatures in °C at Berlin for the years 1920–1930.

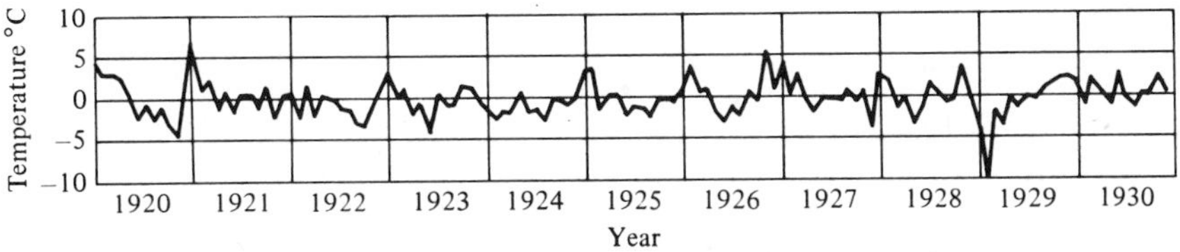

Figure 6.10.2 Seasonally adjusted series of monthly mean temperatures in °C at Vienna for the years 1920–1930.

The period for which we take these temperature series is 1780–1950. We determine the various statistics in the manner of Section 6.4. In fact we take $T = 2048$ and so are able to evaluate the required discrete Fourier transforms by means of the Fast Fourier Transform Algorithm. In forming the statistics $f_{YY}^{(T)}(\lambda)$, $f_{YX}^{(T)}(\lambda)$, $f_{XX}^{(T)}(\lambda)$ we take $m = 10$.

The results of the calculations are recorded in a series of figures. Figure 6.10.3 is a plot of $\log_{10} f_{YY}^{(T)}(\lambda)$ and $\log_{10} g_{\epsilon\epsilon}^{(T)}(\lambda)$, the first being the upper curve. If we use expressions (5.6.15) and (6.6.12) we find that the asymptotic standard errors of these values are both .095 for $\lambda \not\equiv 0 \pmod{\pi}$. Figure 6.10.4 is a plot of Re $A^{(T)}(\lambda)$ which fluctuates around the value .85; Figure 6.10.5 is a plot of Im $A^{(T)}(\lambda)$ which fluctuates around 0; Figure 6.10.6 is a plot of $G^{(T)}(\lambda)$ which fluctuates around .9; Figure 6.10.7 is a plot of $\phi^{(T)}(\lambda)$ which fluctuates around 0; Figure 6.10.8 is a plot of $|R_{YX}^{(T)}(\lambda)|^2$ which fluctuates around .7. Remember that this statistic is a measure of the degree to which Y is determinable from X in a linear manner. Figure 6.10.9 is a plot of $a^{(T)}(u)$ for $|u| \leq 50$. Following (6.8.7) the asymptotic standard error

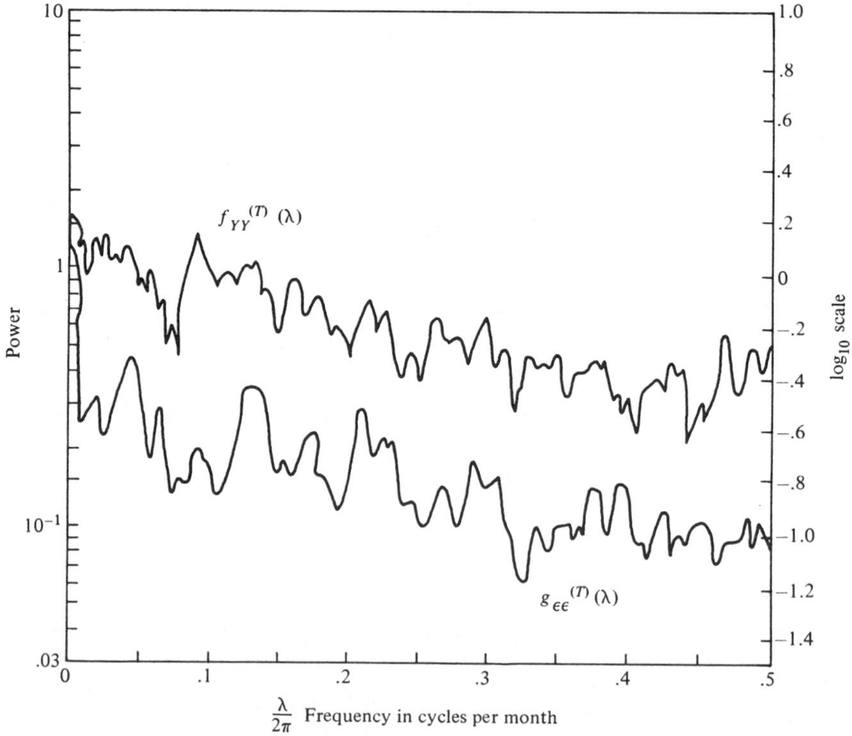

Figure 6.10.3 An estimate of the power spectrum of Berlin temperatures and an estimate of the error spectrum after fitting Vienna temperatures for the years 1780–1950.

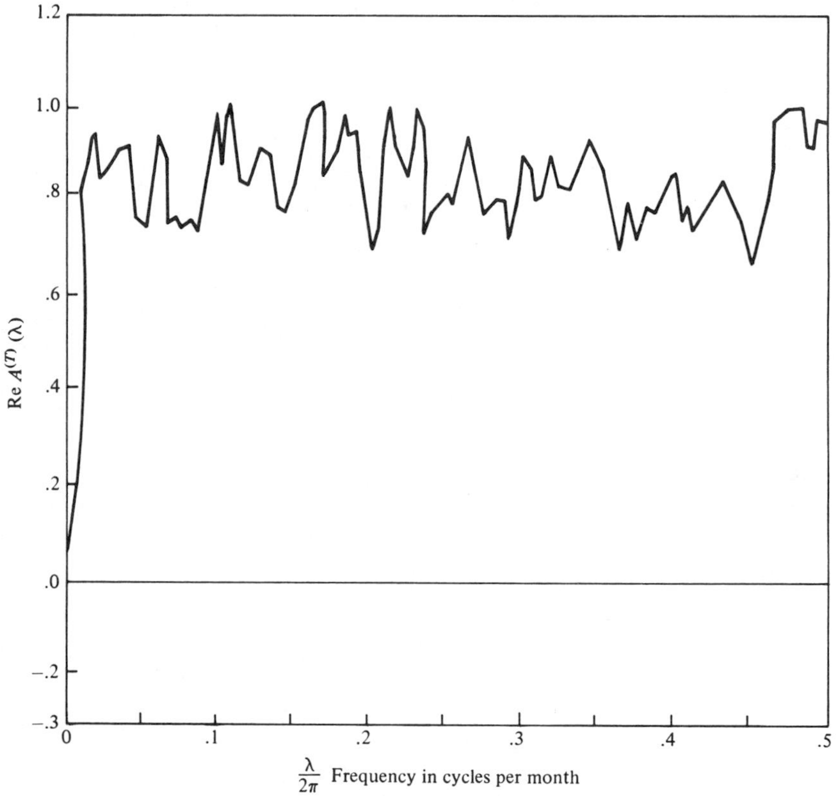

Figure 6.10.4 Re $A^{(T)}(\lambda)$, an estimate of the real part of the transfer function for fitting Berlin temperatures by Vienna temperatures.

of this statistic is .009. The value of $a^{(T)}(0)$ is .85. The other values are not significantly different from 0.

Our calculations appear to suggest the relation

$$Y(t) = .85X(t) + \varepsilon(t) \tag{6.10.2}$$

where the power spectrum of $\varepsilon(t)$ has the form of the lower curve in Figure 6.10.3. We fitted the instantaneous relation by least squares and found the simple regression coefficient of $Y(t)$ on $X(t)$ to be .81. If we assume the $\varepsilon(t)$ are independent and identically distributed, then the estimated standard error of this last is .015. The estimated error variance is 1.57.

As a second example of the techniques of this chapter we present the results of a frequency regression of the series of monthly mean temperatures recorded at Greenwhich on the monthly mean temperatures recorded at the thirteen other locations listed in Table 1.1.1. We prefilter these series by re-

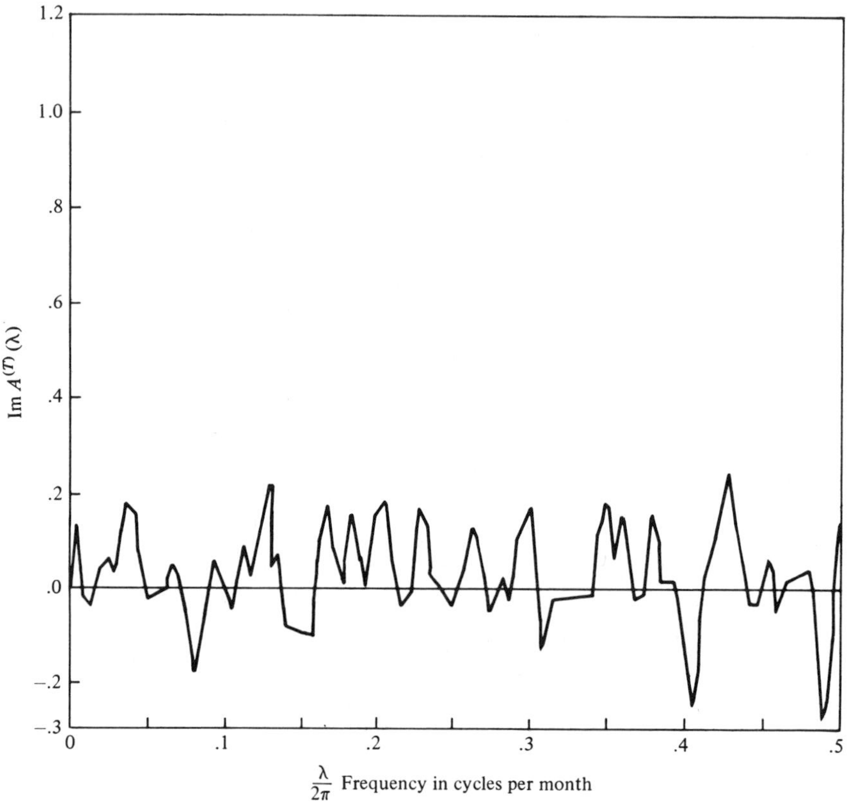

Figure 6.10.5 Im $A^{(T)}(\lambda)$, an estimate of the imaginary part of the transfer function for fitting Berlin temperatures by Vienna temperatures.

moving monthly means and a linear trend. Figure 1.1.1 presents the original data here.

We form estimates in the manner of (6.4.1) to (6.4.5) with $m = 57$. The Fourier transforms required for these calculations were computed using a Fast Fourier Transform Algorithm with $T = 2048$. Now: Figure 6.10.10 presents $G_j^{(T)}(\lambda)$, $\phi_j^{(T)}(\lambda)$ for $j = 1, \ldots, 13$; Figure 6.10.11 presents $\log_{10} g_{\varepsilon\varepsilon}^{(T)}(\lambda)$; Figure 6.10.12 presents $|R_{YX}^{(T)}(\lambda)|^2$ as defined by (6.4.11); The power spectrum of Greenwich is estimated in Figure 7.8.8.

Table 6.10.1 gives the results of an instantaneous multiple regression of the Greenwich series on the other thirteen series. The estimated error variance of this analysis is .269. The squared coefficient of multiple correlation of the analysis is .858.

The estimated gains, $G_j^{(T)}(\lambda)$, appear to fluctuate about horizontal levels as functions of λ. The highest levels correspond to Edinburgh, Basle, and De Bilt respectively. From Table 6.10.1 these are the stations having the

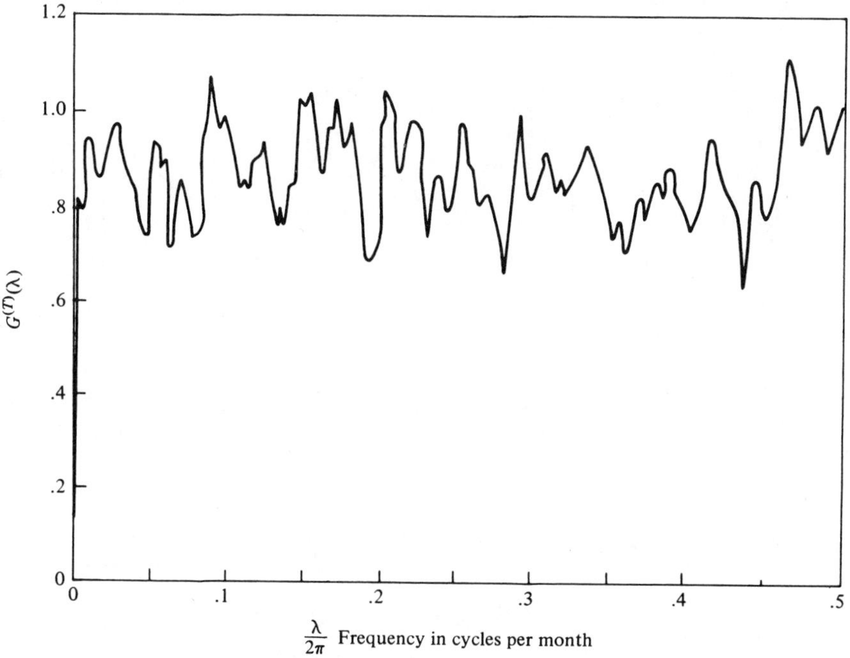

Figure 6.10.6 $G^{(T)}(\lambda)$, an estimate of the amplitude of the transfer function for fitting Berlin temperatures by Vienna temperatures.

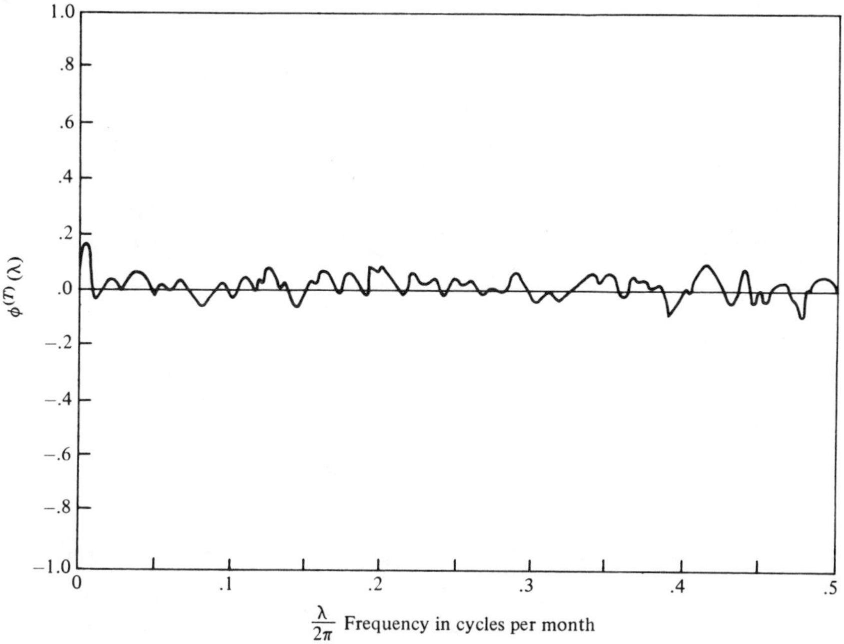

Figure 6.10.7 $\phi^{(T)}(\lambda)$, an estimate of the phase of the transfer function for fitting Berlin temperatures by Vienna temperatures.

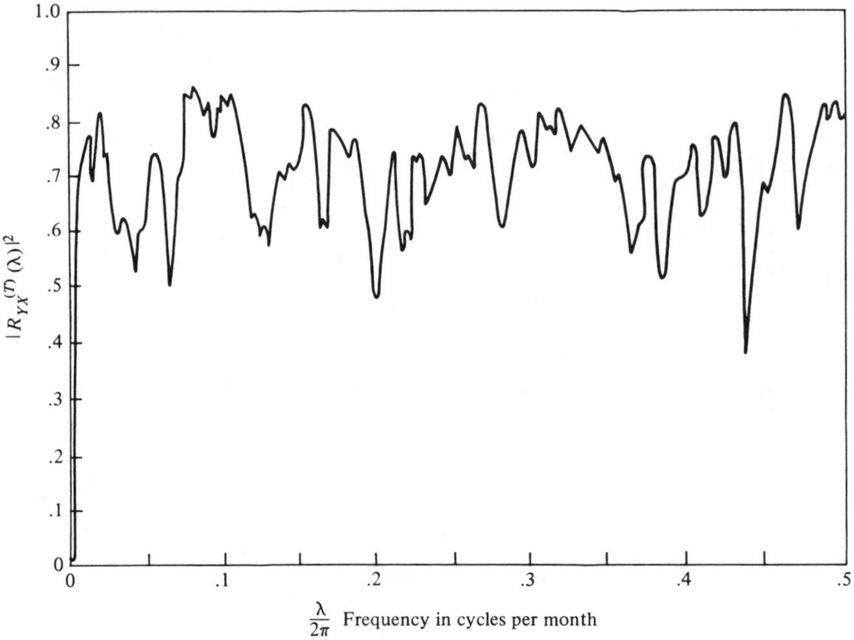

Figure 6.10.8 $|R_{YX}^{(T)}(\lambda)|^2$, an estimate of the coherence of Berlin and Vienna temperatures for the years 1780–1950.

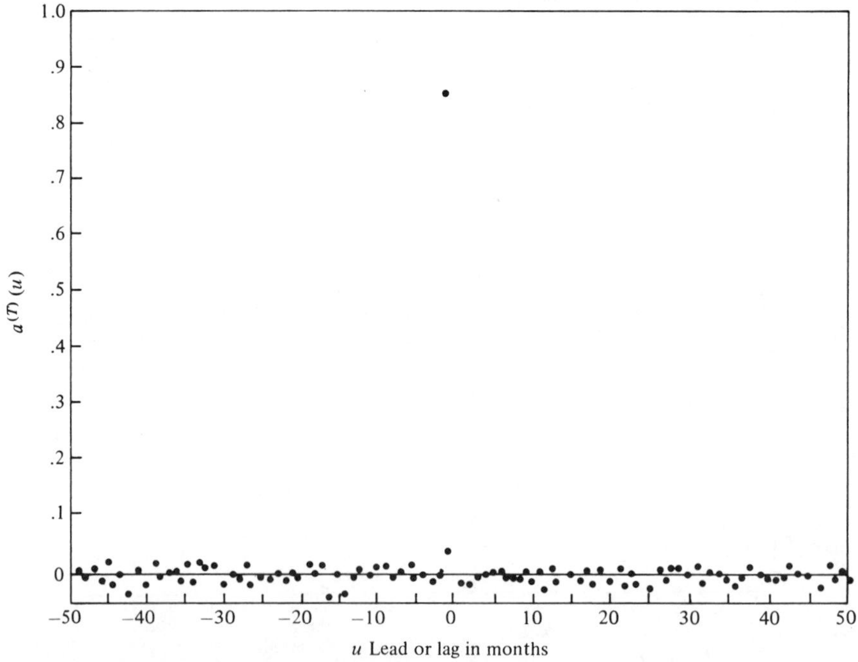

Figure 6.10.9 $a^{(T)}(u)$, an estimate of the filter coefficients for fitting Berlin temperatures by Vienna temperatures.

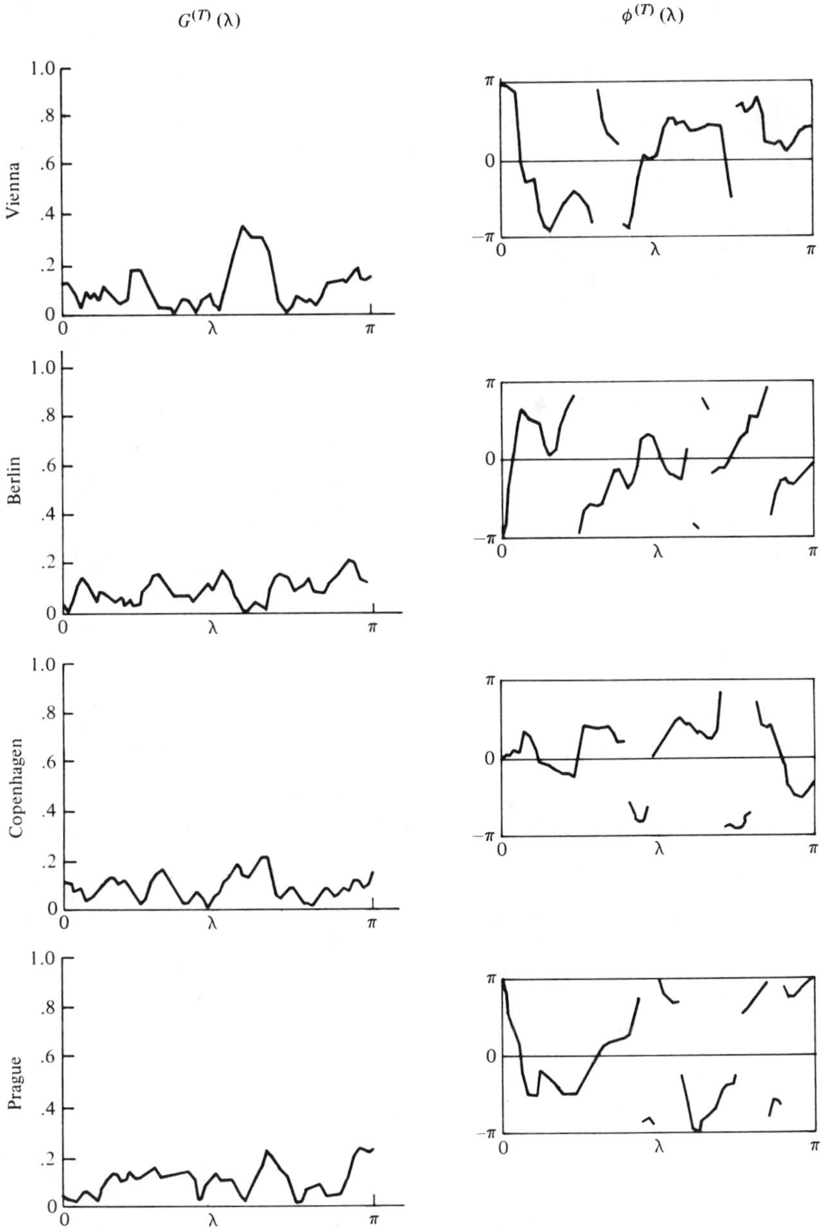

Figure 6.10.10 Estimated gains and phases for fitting seasonally adjusted Greenwich monthly mean temperatures by similar temperatures at thirteen other stations for the years 1780–1950.

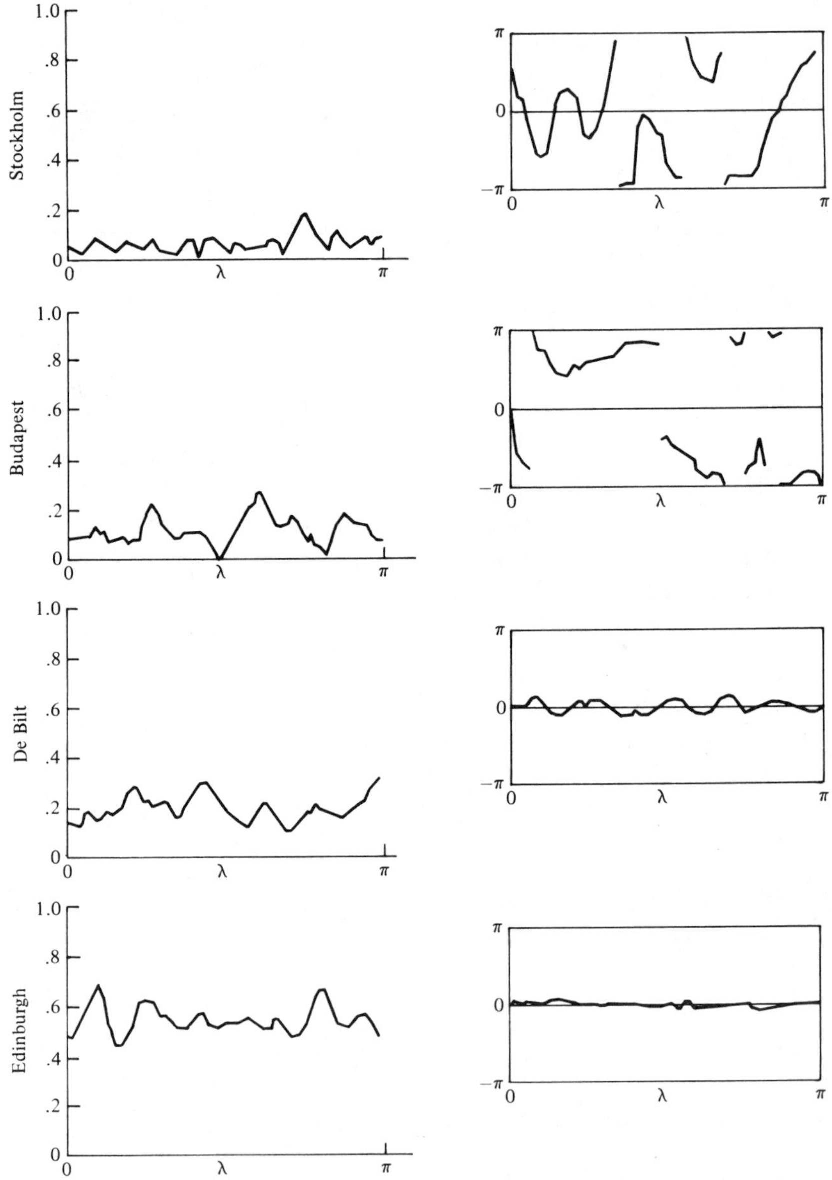
Stockholm
Budapest
De Bilt
Edinburgh
1.0
.8
.6
.4
.2
0
λ
π
−π

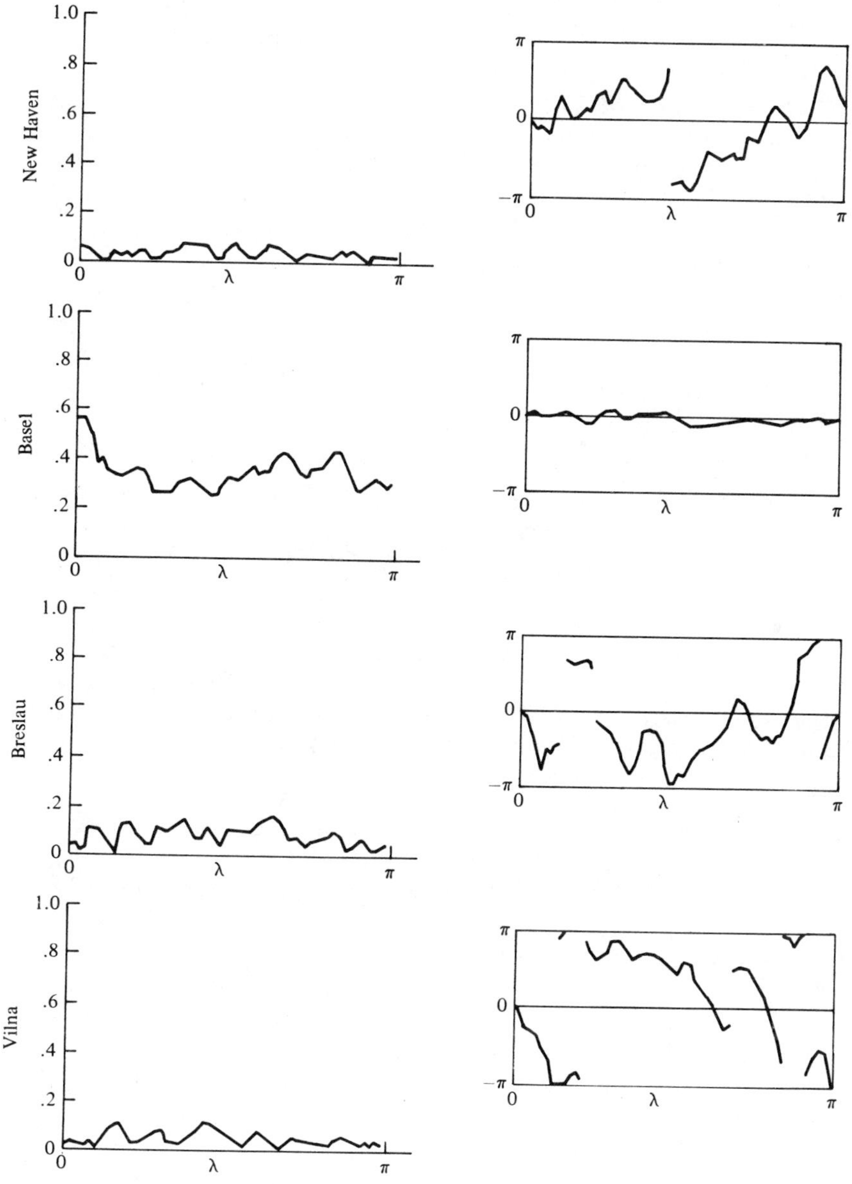
New Haven
Basel
Breslau
Vilna
1.0
.8
.6
.4
.2
0
λ
π
−π

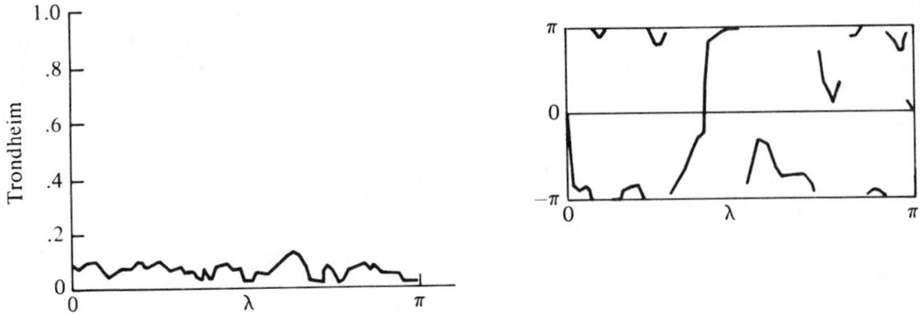

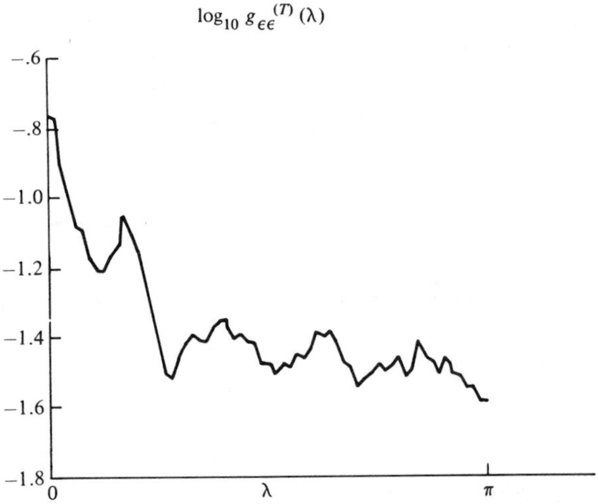

Figure 6.10.11 $\log_{10}\, g_{\epsilon\epsilon}^{(T)}(\lambda)$, the logarithm of the estimated error spectrum for fitting Greenwich temperatures by those at thirteen other stations.

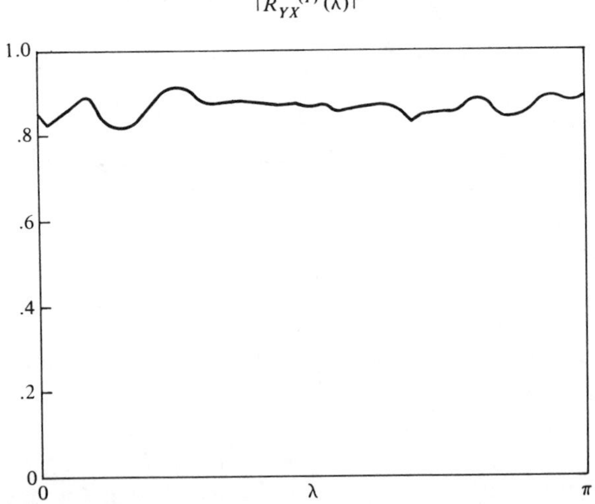

Figure 6.10.12 $|R_{YX}^{(T)}(\lambda)|^2$, an estimate of the multiple coherence of Greenwich temperatures with those at thirteen other stations.

Table 6.10.1. Regression Coefficients of Greenwich on Other Stations

Locations	*Sample Regression Coefficients*	*Estimated Standard Errors*
Vienna	−.071	.021
Berlin	−.125	.023
Copenhagen	.152	.022
Prague	−.040	.010
Stockholm	−.041	.016
Budapest	−.048	.019
De Bilt	.469	.022
Edinburgh	.305	.014
New Haven	.053	.009
Basle	.338	.016
Breslau	.030	.017
Vilna	−.024	.009
Trondheim	−.010	.013

largest sample regression coefficients but in the order De Bilt, Basle, and Edinburgh. The estimated phases $\phi_j^{(T)}(\lambda)$, corresponding to these stations, are each near constant at 0, suggesting there is no phase lead or lag and the relationship is instantaneous for these monthly values. As the estimated gains of the other stations decrease, the estimated phase function is seen to become more erratic. This was to be expected in view of expression (6.6.9) for the asymptotic variance of the phase estimate. Also, the estimated gain for New Haven, Conn., is least; this was to be expected in view of its great distance from Greenwich.

The estimated multiple coherence, $|R_{YX}^{(T)}(\lambda)|^2$, is seen to be near constant at the level .87. This is close to the value .858 obtained in the instantaneous multiple regression analysis. Finally, the estimated error spectrum, $g_{\epsilon\epsilon}^{(T)}(\lambda)$, is seen to fall off steadily as λ increases.

6.11 FURTHER CONSIDERATIONS

We turn to an investigation of the nature of the dependence of the various results we have obtained on the independent series $\mathbf{X}(t)$. We first consider the bias of $\mathbf{A}^{(T)}(\lambda)$. From expressions (6.4.8) and (6.5.14) we see that the expected value of $\mathbf{A}^{(T)}(\lambda)$ is primarily a matrix weighted average of $\mathbf{A}(\alpha)$ with weights depending on $\mathbf{I}_{XX}^{(T)}(\alpha)$. From the form of the expressions (6.4.8) and (6.5.14) it would be advantageous if we could arrange that the function $\mathbf{I}_{XX}^{(T)}(\alpha)$ be near constant in α and have off-diagonal terms near 0. Near 0 off-diagonal terms reduces the entanglement of the components of $\mathbf{A}(\alpha)$. Continuing, an examination of the error term in (6.4.8) suggests that the weighted average term will dominate in the case that $\|\mathbf{f}_{XX}^{(T)}(\lambda)^{-1}\|$ is small, that is, $\mathbf{f}_{XX}^{(T)}(\lambda)$ is far from being singular.

Next we consider the asymptotic second-order properties of $\mathbf{A}^{(T)}(\lambda)$. Expression (6.6.4) and the results of Theorem 6.4.2 indicate that in order to reduce the asymptotic variances of the entries of $\mathbf{A}^{(T)}(\lambda)$ if it is possible we should select an $\mathbf{X}(t)$, $t = 0, \pm 1, \ldots$ such that the diagonal entries of $\mathbf{f}_{XX}^{(T)}(\lambda)^{-1}$ are large. Suppose that $f_{jj}^{(T)}(\lambda)$, $j = 1, \ldots, r$, the diagonal entries of $\mathbf{f}_{XX}^{(T)}(\lambda)$ are given. Exercise 6.14.18 suggests that approximately

$$E|A_j^{(T)}(\lambda) - A_j(\lambda)|^2 \geqslant f_{jj}^{(T)}(\lambda)^{-1} \tag{6.11.1}$$

and that equality is achieved in the case that the off-diagonal elements are 0. We are again led to try to arrange that the off-diagonal elements of $\mathbf{f}_{XX}^{(T)}(\lambda)$ be near 0 and that its diagonal elements be large.

An additional advantage accrues from near 0 off-diagonal elements. From (6.6.4) we see that if they are near 0, then the statistics $A_j^{(T)}(\lambda)$, $A_k^{(T)}(\lambda)$, $1 \leqslant j < k \leqslant r$, will be nearly uncorrelated and asymptotically nearly independent. Their interpretation and approximate properties will be more elementary.

In order to obtain reasonable estimates of $\mathbf{A}^{(T)}(\lambda)$, $-\infty < \lambda < \infty$, we have been led to seek an $\mathbf{X}(t)$, $t = 0, \pm 1, \ldots$ such that $\mathbf{f}_{XX}^{(T)}(\alpha)$ is near constant in α, has off-diagonal terms near 0, and has large diagonal terms. We will see later that a choice of $\mathbf{X}(t)$ likely to lead to such an $\mathbf{f}_{XX}^{(T)}(\alpha)$ is a realization of a pure noise process having independent components with large variance.

On the bulk of occasions we will be presented with $\mathbf{X}(t)$, $t = 0, \pm 1, \ldots$ as a fait accompli; however as we have seen in Section 6.1 we can alter certain of the properties of $\mathbf{X}(t)$ by a filtering. We could evaluate

$$\mathbf{X}_1(t) = \sum_{u=-\infty}^{\infty} \mathbf{c}(t-u)\mathbf{X}(u) \tag{6.11.2}$$

$t = 0, \ldots, T-1$ for some $r \times r$ filter $\{\mathbf{c}(u)\}$ and then estimate the transfer function $\mathbf{A}_1(\lambda)$ relating $Y(t)$ to $\mathbf{X}_1(t)$, $t = 0, \pm 1, \ldots$. Let this estimate be $\mathbf{A}_1^{(T)}(\lambda)$. As an estimate of $\mathbf{A}(\lambda)$ we now consider

$$\mathbf{A}^{(T)}(\lambda) = \mathbf{A}_1^{(T)}(\lambda)\mathbf{C}(\lambda). \tag{6.11.3}$$

From (6.1.10) and (6.5.14)

$$E\mathbf{f}_{YX_1}^{(T)}(\lambda) \doteq \sum_{s=1}^{T-1} W^{(T)}\left(\lambda - \frac{2\pi s}{T}\right)\mathbf{A}\left(\frac{2\pi s}{T}\right)\mathbf{C}\left(\frac{2\pi s}{T}\right)^{-1}\mathbf{I}_{X_1X_1}^{(T)}\left(\frac{2\pi s}{T}\right) \tag{6.11.4}$$

suggesting that we should seek a filter $\mathbf{C}(\lambda)$ such that $\mathbf{A}(\lambda)\mathbf{C}(\lambda)^{-1}$ does not vary much with λ. Applying such an operation is called **prefiltering.** It can be absolutely essential even in simple situations.

Consider a common relationship in which $Y(t)$ is essentially a delayed version of $X(t)$; specifically suppose

$$Y(t) = \alpha X(t - v) + \varepsilon(t) \tag{6.11.5}$$

for some v. In this case

$$A(\lambda) = \alpha \exp\{-i\lambda v\} = \alpha \cos \lambda v - i\,\alpha \sin \lambda v \tag{6.11.6}$$

and so for example

$$\operatorname{Re}\left\{\sum_{s=1}^{T-1} W^{(T)}\left(\lambda - \frac{2\pi s}{T}\right) A\left(\frac{2\pi s}{T}\right) I_{XX}{}^{(T)}\left(\frac{2\pi s}{T}\right)\right\} = \alpha \sum_{s=1}^{T-1} W^{(T)}\left(\lambda - \frac{2\pi s}{T}\right) \cos \frac{2\pi s v}{T}\, I_{XX}{}^{(T)}\left(\frac{2\pi s}{T}\right). \tag{6.11.7}$$

In the case that v is large, $\cos 2\pi sv/T$ fluctuates rapidly in sign as s varies. Because of the smoothing involved, expression (6.11.7) will be near 0, rather than the desired

$$\alpha \cos \lambda v \sum_{s=1}^{T-1} W^{(T)}\left(\lambda - \frac{2\pi s}{T}\right) I_{XX}{}^{(T)}\left(\frac{2\pi s}{T}\right). \tag{6.11.8}$$

In terms of the previous discussion, we are led to prefilter the data using the transfer function $C(\lambda) = \exp\{-i\lambda v\}$, that is, to carry out the spectral calculations with the series $X_1(t) = X(t - v)$ instead of $X(t)$. We would then estimate $A(\lambda)$ here by $\exp\{-i\lambda v\} f_{YX_1}^{(T)}(\lambda)/f_{X_1X_1}^{(T)}(\lambda)$. This procedure was suggested by Darzell and Pearson (1960) and Yamanouchi (1961); see Akaike and Yamanouchi (1962). In practice the lag v must be guessed at before performing these calculations. One suggestion is to take the lag that maximizes the magnitude of the cross-covariance of the series $Y(t)$ and $X(t)$.

In Section 7.7 we will discuss a useful procedure for prefiltering in the case of vector-valued $\mathbf{X}(t)$. It is based on using least squares to fit a preliminary time domain model and then to carry out a spectral analysis of $\mathbf{X}(t)$ with the residuals of the fit.

We have so far considered means of improving the estimate $\mathbf{A}^{(T)}(\lambda)$. The other estimates $g_{\varepsilon\varepsilon}{}^{(T)}(\lambda)$, $\mu^{(T)}$, $\mathbf{a}^{(T)}(u)$ are based on this estimate in an intimate manner. We would therefore expect any improvement in $\mathbf{A}^{(T)}(\lambda)$ to result in an improvement of these additional statistics. In general terms we feel that the nearer the relation between $Y(t)$ and $\mathbf{X}(t)$, $t = 0, \pm 1, \ldots$ is to multiple regression of $Y(t)$ on $\mathbf{X}(t)$ with pure noise errors, the better the estimates will be. All prior knowledge should be used to shift the relation to one near this form.

A few comments on the computation of the statistics appear in order. The estimates have been based directly on the discrete Fourier transforms of the series involved. This was done to make their sampling properties more elementary. However, it will clearly make sense to save on computations by evaluating these Fourier transforms using the Fast Fourier Transform Algorithm. Another important simplification results from noting that the estimates of Section 6.4 can be determined directly from a standard multiple regression analysis involving real-valued variates. Consider the case $\lambda \not\equiv 0 \pmod{\pi}$. Following the discussion of Section 6.3, the model (6.1.1) leads to the approximate relation

$$d_Y^{(T)}\left(\frac{2\pi[s(T)+s]}{T}\right) \doteq \mathbf{A}(\lambda)\mathbf{d}_X^{(T)}\left(\frac{2\pi[s(T)+s]}{T}\right) + d_\varepsilon^{(T)}\left(\frac{2\pi[s(T)+s]}{T}\right) \tag{6.11.9}$$

$s = 0, \pm 1, \ldots, \pm m$ for $2\pi s(T)/T$ near λ. In terms of real-valued quantities this may be written

$$\begin{aligned}
\operatorname{Re} d_Y^{(T)}\left(\frac{2\pi[s(T)+s]}{T}\right) &\doteq \operatorname{Re} \mathbf{A}(\lambda) \operatorname{Re} \mathbf{d}_X^{(T)}\left(\frac{2\pi[s(T)+s]}{T}\right) \\
&\quad - \operatorname{Im} \mathbf{A}(\lambda) \operatorname{Im} \mathbf{d}_X^{(T)}\left(\frac{2\pi[s(T)+s]}{T}\right) \\
&\quad + \operatorname{Re} d_\varepsilon^{(T)}\left(\frac{2\pi[s(T)+s]}{T}\right) \\
\operatorname{Im} d_Y^{(T)}\left(\frac{2\pi[s(T)+s]}{T}\right) &\doteq \operatorname{Re} \mathbf{A}(\lambda) \operatorname{Im} \mathbf{d}_X^{(T)}\left(\frac{2\pi[s(T)+s]}{T}\right) \\
&\quad + \operatorname{Im} \mathbf{A}(\lambda) \operatorname{Re} \mathbf{d}_X^{(T)}\left(\frac{2\pi[s(T)+s]}{T}\right) \\
&\quad + \operatorname{Im} d_\varepsilon^{(T)}\left(\frac{2\pi[s(T)+s]}{T}\right)
\end{aligned} \tag{6.11.10}$$

$s = 0, \pm 1, \ldots, \pm m$. Because the values $\operatorname{Re} d_\varepsilon^{(T)}(2\pi[s(T)+s]/T)$, $\operatorname{Im} d_\varepsilon^{(T)}(2\pi[s(T)+s]/T)$, $s = 0, \pm 1, \ldots, \pm m$ are approximately uncorrelated $\pi T f_{\varepsilon\varepsilon}(\lambda)$ variates, (6.11.10) has the approximate form of a multiple regression analysis with regression coefficient matrix

$$[\operatorname{Re} \mathbf{A}(\lambda) \; \operatorname{Im} \mathbf{A}(\lambda)] \tag{6.11.11}$$

and error variance $\pi T f_{\varepsilon\varepsilon}(\lambda)$. Estimates of the parameters of interest will therefore drop out of a multiple regression analysis taking the **Y** matrix as

$$\left[\operatorname{Re} d_Y^{(T)}\left(\frac{2\pi[s(T)+s]}{T}\right) \; \operatorname{Im} d_Y^{(T)}\left(\frac{2\pi[s(T)+s]}{T}\right); \quad s = 0, \pm 1, \ldots, \pm m\right] \tag{6.11.12}$$

and the **X** matrix as

$$\begin{bmatrix} \operatorname{Re} \mathbf{d}_X{}^{(T)}\left(\frac{2\pi[s(T)+s]}{T}\right) & \operatorname{Im} \mathbf{d}_X{}^{(T)}\left(\frac{2\pi[s(T)+s]}{T}\right) \\ -\operatorname{Im} \mathbf{d}_X{}^{(T)}\left(\frac{2\pi[s(T)+s]}{T}\right) & \operatorname{Re} \mathbf{d}_X{}^{(T)}\left(\frac{2\pi[s(T)+s]}{T}\right) \end{bmatrix}; s = 0, \pm 1, \ldots, \pm m \tag{6.11.13}$$

Estimates in the case $\lambda \equiv 0 \pmod{\pi}$ follow in a similar manner.

We remark that the model (6.1.1) is of use even in the case that $\mathbf{X}(t)$, $t = 0, \pm 1, \ldots$ is not vector valued. For example if one wishes to investigate the possibility of a nonlinear relation between real-valued series $Y(t)$ and $X(t)$, $t = 0, \pm 1, \ldots$ one can consider setting

$$\begin{aligned} X_1(t) &= X(t) \\ X_2(t) &= [X(t)]^2 \\ X_3(t) &= X(t)X(t-1) \end{aligned} \tag{6.11.14}$$

and so forth in (6.1.1).

6.12 A COMPARISON OF THREE ESTIMATES OF THE IMPULSE RESPONSE

Suppose the model of this chapter takes the more elementary form

$$Y(t) = \mu + \sum_{u=-m}^{n} \mathbf{a}(u)\mathbf{X}(t-u) + \varepsilon(t) \qquad t = 0, \pm 1, \ldots \tag{6.12.1}$$

for some finite m, n. In this case the dependence of $Y(t)$ on the series $\mathbf{X}(t)$, $t = 0, \pm 1, \ldots$ is of finite duration only. We turn to a comparison of three plausible estimates of the coefficients $\mathbf{a}(-m), \ldots, \mathbf{a}(0), \ldots, \mathbf{a}(n)$ that now suggest themselves. These are the estimate of Section 6.8, a least squares estimate, and an asymptotically efficient linear estimate.

We begin by noting that it is enough to consider a model of the simple form

$$Y(t) = \mu + \mathbf{a}\mathbf{X}(t) + \varepsilon(t) \qquad t = 0, \pm 1, \ldots \tag{6.12.2}$$

for (6.12.1) may be rewritten

$$Y(t) = \mu + [\mathbf{a}(-m) \cdots \mathbf{a}(n)] \begin{bmatrix} \mathbf{X}(t+m) \\ \cdot \\ \cdot \\ \cdot \\ \mathbf{X}(t-n) \end{bmatrix} + \varepsilon(t) \tag{6.12.3}$$

which is of the form of (6.12.2) with expanded dimensions.

The estimate of **a** of (6.12.2) suggested by the material of Section 6.8 is

$$\mathbf{a}_1^{(T)} = P_T^{-1} \sum_{p=0}^{P-1} \mathbf{A}^{(T)}\left(\frac{2\pi p}{P}\right). \tag{6.12.4}$$

From (6.6.4)

$$\begin{aligned}\operatorname{cov}\left\{\mathbf{A}^{(T)}\left(\frac{2\pi p}{P}\right), \mathbf{A}^{(T)}\left(\frac{2\pi q}{P}\right)\right\} &\sim B_T^{-1}T^{-1}2\pi \int W(\alpha)^2 d\alpha\ f_{\varepsilon\varepsilon}\left(\frac{2\pi p}{P}\right)\mathbf{f}_{XX}^{(T)}\left(\frac{2\pi p}{P}\right) \\ &\qquad p = q \\ &\sim \mathbf{0} \qquad p \neq q \end{aligned} \tag{6.12.5}$$

and so the covariance matrix of $\mathbf{a}_1^{(T)}$ is approximately

$$B_T^{-1}T^{-1}2\pi \int W(\alpha)^2 d\alpha\ \ P_T^{-2} \sum_p f_{\varepsilon\varepsilon}\left(\frac{2\pi p}{P}\right)\mathbf{f}_{XX}^{(T)}\left(\frac{2\pi p}{P}\right)^{-1}. \tag{6.12.6}$$

The particular form of the model (6.12.2) suggests that we should also consider the least squares estimate found by minimizing

$$\sum_{t=0}^{T-1} [Y(t) - \mu - \mathbf{a}\mathbf{X}(t)]^2 \tag{6.12.7}$$

with respect to μ and **a**. This estimate is

$$\mathbf{a}_2^{(T)} = \mathbf{c}_{YX}^{(T)}(0)\mathbf{c}_{XX}^{(T)}(0)^{-1}. \tag{6.12.8}$$

We may approximate this estimate by

$$\begin{aligned}&\left\{\sum_{p=0}^{P-1} \mathbf{f}_{YX}^{(T)}\left(\frac{2\pi p}{P}\right)\right\}\left\{\sum_{p=0}^{P-1} \mathbf{f}_{XX}^{(T)}\left(\frac{2\pi p}{P}\right)\right\}^{-1} \\ &\quad = \left\{\sum_{p=0}^{P-1} \mathbf{A}^{(T)}\left(\frac{2\pi p}{P}\right)\mathbf{f}_{XX}^{(T)}\left(\frac{2\pi p}{P}\right)\right\}\left\{\sum_{p=0}^{P-1} \mathbf{f}_{XX}^{(T)}\left(\frac{2\pi p}{P}\right)\right\}^{-1}. \end{aligned} \tag{6.12.9}$$

Using (6.12.5), the covariance matrix of (6.12.9) will be approximately

$$\begin{aligned}B_T^{-1}T^{-1}2\pi \int W(\alpha)^2 d\alpha &\left\{\sum_{p=0}^{P-1} \mathbf{f}_{XX}^{(T)}\left(\frac{2\pi p}{P}\right)\right\}^{-1}\left\{\sum_{p=0}^{P-1} f_{\varepsilon\varepsilon}\left(\frac{2\pi p}{P}\right)\mathbf{f}_{XX}^{(T)}\left(\frac{2\pi p}{P}\right)\right\} \\ &\times \left\{\sum_{p=0}^{P-1} \mathbf{f}_{XX}^{(T)}\left(\frac{2\pi p}{P}\right)\right\}^{-1}. \end{aligned} \tag{6.12.10}$$

We note that both of the estimates (6.12.4) and (6.12.9) are weighted averages of the $\mathbf{A}^{(T)}(2\pi p/P)$ values. This suggests that we should consider, as a further estimate, the best linear combination of these values. Now Exercise 6.14.11 and expression (6.12.5) indicate that this is given approximately by

$$\begin{aligned}\mathbf{a}_3^{(T)} = &\left\{\sum_{p=0}^{P-1} \mathbf{A}^{(T)}\left(\frac{2\pi p}{P}\right)\mathbf{f}_{XX}^{(T)}\left(\frac{2\pi p}{P}\right)g_{\varepsilon\varepsilon}^{(T)}\left(\frac{2\pi p}{P}\right)^{-1}\right\} \\ &\times \left\{\sum_{p=0}^{P-1} \mathbf{f}_{XX}^{(T)}\left(\frac{2\pi p}{P}\right)g_{\varepsilon\varepsilon}^{(T)}\left(\frac{2\pi p}{P}\right)^{-1}\right\}^{-1} \end{aligned} \tag{6.12.11}$$

with approximate covariance matrix

$$B_T^{-1}T^{-1}2\pi \int W(\alpha)^2 d\alpha \left\{ \sum_{p=0}^{P-1} f_{\varepsilon\varepsilon}\left(\frac{2\pi p}{P}\right)^{-1} \mathbf{f}_{XX}^{(T)}\left(\frac{2\pi p}{P}\right) \right\}^{-1}. \quad (6.12.12)$$

In view of the source of $\mathbf{a}_3^{(T)}$, the matrix differences (6.12.6) − (6.12.12) and (6.12.10) − (6.12.12) will both be non-negative definite. In the case that $g_{\varepsilon\varepsilon}^{(T)}(\lambda)$ is near constant, as would be the case were the error series $\varepsilon(t)$ white noise, and T not too small, formulas (6.12.9) and (6.12.11) indicate that the least squares estimate $\mathbf{a}_2^{(T)}$ will be near the "efficient" estimate $\mathbf{a}_3^{(T)}$. In the case that $\mathbf{f}_{XX}^{(T)}(\lambda)g_{\varepsilon\varepsilon}^{(T)}(\lambda)$ is near constant, formulas (6.12.4) and (6.12.11) indicate that the estimate $\mathbf{a}_1^{(T)}$ will be near the estimate $\mathbf{a}_3^{(T)}$.

Hannan (1963b, 1967a, 1970) discusses the estimates $\mathbf{a}_1^{(T)}$, $\mathbf{a}_3^{(T)}$ in the case of stochastic $\mathbf{X}(t)$, $t = 0, \pm 1, \ldots$. Grenander and Rosenblatt (1957), Rosenblatt (1959), and Hannan (1970) discuss the estimates $\mathbf{a}_2^{(T)}$, $\mathbf{a}_3^{(T)}$ in the case of fixed $\mathbf{X}(t)$, $t = 0, \pm 1, \ldots$.

6.13 USES OF THE PROPOSED TECHNIQUE

The statistics of the present chapter have been calculated by many researchers in different situations. These workers found themselves considering a series $Y(t)$, $t = 0, \pm 1, \ldots$ which appeared to be coming about from a series $X(t)$, $t = 0, \pm 1, \ldots$ in a linear time invariant manner. The latter is the principal implication of the model (6.1.1). These researchers calculated various of the statistics $A^{(T)}(\lambda)$, $G^{(T)}(\lambda)$, $\phi^{(T)}(\lambda)$, $g_{\varepsilon\varepsilon}^{(T)}(\lambda)$, $|R_{YX}^{(T)}(\lambda)|^2$, $a^{(T)}(u)$.

An important area of application has been in the field of geophysics. Robinson (1967a) discusses the plausibility of a linear time invariant model relating a seismic disturbance $X(t)$, with $Y(t)$ its recorded form at some station. Tukey (1959c) relates a seismic record at one station with the seismic record at another station. Other references to applications in seismology include: Haubrich and MacKenzie (1965) and Pisarenko (1970). Turning to the field of oceanography, Hamon and Hannan (1963) and Groves and Hannan (1968) consider relations between sea level and pressure and wind stress at several stations. Groves and Zetler (1964) relate sea levels at San Francisco with those at Honolulu. Munk and Cartwright (1966) take $X(t)$ to be a theoretically specified mathematical function while $Y(t)$ is the series of tidal height. Kawashima (1964) considers the behavior of a boat on an ocean via cross-spectral analysis. Turning to the field of meteorology, Panofsky (1967) presents the results of spectral calculations for a variety of series including wind velocity and temperature. Madden (1964) considers certain electromagnetic data. Rodriguez-Iturbe and Yevjevich (1968) take $Y(t)$ to be rainfall recorded at a number of stations in the U.S.A. and $X(t)$ to be relative sunspot numbers. Brillinger (1969a) takes $Y(t)$ to be monthly rainfall in Sante Fe, New Mexico, and $X(t)$ to be monthly relative sunspot numbers.

Lee (1960) presents arguments to suggest that many electronic circuits behave in a linear time invariant manner. Akaike and Kaneshige (1964) take $Y(t)$ to be the output of a nonlinear circuit, $X(t)$ to be the input of the circuit, and evaluate certain of the statistics discussed in this chapter.

Goodman et al (1961) discuss industrial applications of the techniques of this chapter as do Jenkins (1963), Nakamura (1964), Nakamura and Murakami (1964). Takeda (1964) uses cross-spectral analysis in an investigation of aircraft behavior.

As examples of applications in economics we mention the books by Granger (1964) and Fishman (1969). Nerlove (1964) uses cross-spectral analysis to investigate the effectiveness of various seasonal adjustment procedures. Naylor et al (1967) examine the properties of a model of the textile industry.

Results discussed by Khatri (1965b) may be used to construct a test of the hypothesis Im $A(\lambda) = 0$, $-\infty < \lambda < \infty$, that is $a(u) = a(-u)$, $u = 0, \pm 1, \ldots$. The latter would occur if the relation between $Y(t)$ and $X(t)$ were time reversible.

A number of interesting physical problems lead to a consideration of integral equations of the form

$$g(t) = \beta f(t) + \int b(t-u) f(u) du \tag{6.13.1}$$

to be solved for $f(t)$, given $g(t)$, β, $b(t)$. A common means of solution is to set down a discrete approximation to the equation, such as

$$g(t) = \beta f(t) + \sum_{u=-N}^{N} b(u) f(t-u) \qquad t = 0, \pm 1, \ldots \tag{6.13.2}$$

which is then solved by matrix inversion. Suppose that we rewrite expression (6.13.2) in the form

$$Y(t) = \sum_u X(t-u) a(u) + \varepsilon(t) \tag{6.13.3}$$

with the series $\varepsilon(t)$ indicating the error resulting from having made a discrete approximation, with $X(0) = \beta + b(0)$, $X(u) = b(u)$, $u \neq 0$, and $Y(t) = g(t)$. The system (6.13.3) which we have been considering throughout this chapter, suggests that another way of solving the system (6.13.1) is to use cross-spectral analysis and to take $a^{(T)}(u)$, given by (6.8.2), as an approximation to the desired $f(t)$.

So far in this chapter we have placed principal emphasis on the estimation of $A(\lambda)$ and $a(u)$. However, we next mention a situation wherein the parameter of greatest interest is the error spectrum $f_{\varepsilon\varepsilon}(\lambda)$. The model that has been under consideration is

$$Y(t) = \mu + \sum_u a(t-u)X(u) + \varepsilon(t) \tag{6.13.4}$$

with $\Sigma\ |u|\ |a(u)| < \infty$. Suppose that we think of $a(t)$, $t = 0, \pm 1, \ldots$ as representing a transient signal of brief duration. Suppose we define

$$\begin{aligned} X(t) &= 1 \qquad t = 0 \\ &= 0 \qquad t \neq 0. \end{aligned} \tag{6.13.5}$$

Expression (6.13.4) now takes the simpler form

$$Y(t) = \mu + a(t) + \varepsilon(t). \tag{6.13.6}$$

The observed series, $Y(t)$, is the sum of a series of interest, $\mu + \varepsilon(t)$, and a possibly undesirable transient series $a(t)$. The procedures of this chapter provide a means of constructing an estimate of $f_{\varepsilon\varepsilon}(\lambda)$, the power spectrum of interest. We simply form $g_{\varepsilon\varepsilon}^{(T)}(\lambda)$, taking the observed values $Y(t)$, $t = 0, \ldots, T-1$ and $X(t)$ as given by (6.13.5). This estimate should be sensible even when brief undesirable transients get superimposed on the series of interest. In the case that the asymptotic procedure of Section 6.4 is adopted, the distribution of $g_{\varepsilon\varepsilon}^{(T)}(\lambda)$ is approximately a multiple of a chi-squared with $4m$ degrees of freedom. This is to be compared with the $4m + 2$ degrees of freedom the direct estimate $f_{\varepsilon\varepsilon}^{(T)}(\lambda)$ would have. Clearly not too much stability has been lost, in return for the gained robustness of the estimate.

6.14 EXERCISES

6.14.1 Let the conditions of Theorem 6.2.1 be satisfied, but with $E\boldsymbol{\varepsilon}^\tau\boldsymbol{\varepsilon} = \sigma^2\mathbf{I}$ replaced by $E\boldsymbol{\varepsilon}^\tau\boldsymbol{\varepsilon} = \boldsymbol{\Sigma}$. Show that $E\hat{\mathbf{a}} = \mathbf{a}$ as before, but now

$$E(\hat{\mathbf{a}} - \mathbf{a})^\tau(\hat{\mathbf{a}} - \mathbf{a}) = (\mathbf{X}\mathbf{X}^\tau)^{-1}\mathbf{X}\boldsymbol{\Sigma}\mathbf{X}^\tau(\mathbf{X}\mathbf{X}^\tau)^{-1}.$$

6.14.2 Let the conditions of Theorem 6.2.1 be satisfied, but with $E\boldsymbol{\varepsilon}^\tau\boldsymbol{\varepsilon} = \sigma^2\mathbf{I}$ replaced by $E\boldsymbol{\varepsilon}^\tau\boldsymbol{\varepsilon} = \sigma^2\mathbf{V}$. Prove that

$$(\mathbf{Y} - \mathbf{b}\mathbf{X})^\tau\mathbf{V}^{-1}(\mathbf{Y} - \mathbf{b}\mathbf{X})$$

is minimized by

$$\hat{\mathbf{b}} = \mathbf{Y}\mathbf{V}^{-1}\mathbf{X}^\tau(\mathbf{X}\mathbf{V}^{-1}\mathbf{X}^\tau)^{-1}.$$

Show that $\hat{\mathbf{b}}$ is unbiased with covariance matrix $\sigma^2(\mathbf{X}\mathbf{V}^{-1}\mathbf{X}^\tau)^{-1}$. Show that the least squares estimate $\hat{\mathbf{a}} = \mathbf{Y}\mathbf{X}^\tau(\mathbf{X}\mathbf{X}^\tau)^{-1}$ remains unbiased, but has covariance matrix $\sigma^2(\mathbf{X}\mathbf{X}^\tau)^{-1}\mathbf{X}\mathbf{V}\mathbf{X}^\tau(\mathbf{X}\mathbf{X}^\tau)^{-1}$.

6.14.3 In the notation of Theorem 6.2.3, prove that the unbiased, minimum variance linear estimate of $\boldsymbol{\alpha}^\tau\mathbf{a}$, for $\boldsymbol{\alpha}$ a k vector, is given by $\boldsymbol{\alpha}^\tau\hat{\mathbf{a}}$.

6.14.4 Let X be a complex-valued random variable. Prove that

$$\operatorname{var}|X| \leqslant \operatorname{cov}\{X, X\}.$$

6.14.5 In the notation of Theorem 6.2.3, let $|R|^2 = \hat{\mathbf{a}}\mathbf{X}\bar{\mathbf{X}}^\tau\bar{\hat{\mathbf{a}}}^\tau/\mathbf{Y}\bar{\mathbf{Y}}^\tau$. Prove that $0 \leqslant |R|^2 \leqslant 1$. Under the conditions of Theorem 6.2.4 prove that $(n-k)|R|^2/[k(1-|R|^2)]$ is distributed as noncentral F, degrees of freedom $2k$ and $2(n-k)$ and noncentrality parameter $\mathbf{a}\mathbf{X}\bar{\mathbf{X}}^\tau\bar{\mathbf{a}}^\tau/\sigma^2$.

6.14.6 Under the conditions of either Theorem 6.4.2 or Theorem 6.7.1, prove that $\phi^{(T)}(\lambda)$ is asymptotically uniform on $(0,2\pi]$ if $A(\lambda) = 0$.

6.14.7 Prove that the following is a consistent definition of asymptotic normality. A sequence of r vector-valued random variables $\mathbf{X}_n$ is asymptotically normal with mean $\boldsymbol{\theta}_n + \boldsymbol{\Psi}_n\boldsymbol{\mu}$ and covariance matrix $\boldsymbol{\Sigma}_n = \boldsymbol{\Psi}_n\boldsymbol{\Sigma}\boldsymbol{\Psi}_n$ if $\boldsymbol{\Psi}_n^{-1}(\mathbf{X}_n - \boldsymbol{\theta}_n)$ tends, in distribution, to $N_r(\boldsymbol{\mu},\boldsymbol{\Sigma})$ where $\boldsymbol{\theta}_n$ is a sequence of r vectors and $\boldsymbol{\Psi}_n$ a sequence of nonsingular $r \times r$ matrices.

6.14.8 With the notation of Exercise 6.14.2, show that $(\mathbf{X}\mathbf{X}^\tau)^{-1}\mathbf{X}\mathbf{V}\mathbf{X}^\tau(\mathbf{X}\mathbf{X}^\tau)^{-1} \geqslant (\mathbf{X}\mathbf{V}^{-1}\mathbf{X})^{-1}$.[$\mathbf{A} \geqslant \mathbf{B}$ here means $\mathbf{A} - \mathbf{B}$ is non-negative definite.]

6.14.9 Show that $\mathbf{f}_{YX}{}^{(T)}(\lambda)$ of (6.5.6) may be written in the form

$$\sum_{u=-T+1}^{T-1} w^{(T)}(u)\mathbf{c}_{YX}{}^{(T)}(u) \exp\{-iu\lambda\}$$

where $w^{(T)}(u)$ is given by

$$T^{-1}\sum_{s=1}^{T-1} \exp\left\{iu\left(\lambda - \frac{2\pi s}{T}\right)\right\} W^{(T)}\left(\lambda - \frac{2\pi s}{T}\right)$$

and $\mathbf{c}_{YX}{}^{(T)}(u)$ is given by

$$T^{-1}\sum_{0\le t,t+u\le T-1} [Y(t) - c_Y{}^{(T)}][\mathbf{X}(t+u) - \mathbf{c}_X{}^{(T)}]^\tau.$$

6.14.10 Let $\gamma(t)$, $t = 0, \pm 1, \ldots$ denote the series whose finite Fourier transform is $d_Y{}^{(T)}(\lambda) - \mathbf{A}^{(T)}(\lambda)\mathbf{d}_X{}^{(T)}(\lambda)$. Prove that $f_{\gamma\gamma}{}^{(T)}(\lambda) = g_{\varepsilon\varepsilon}{}^{(T)}(\lambda)$, that is, the estimate of the error spectrum may be considered to be a power spectral based on a series of residuals.

6.14.11 Let $\mathbf{Y}_j$, $j = 1, \ldots, J$ be $1 \times r$ matrix-valued random variables with $E\mathbf{Y}_j = \boldsymbol{\beta}$, $E\{(\mathbf{Y}_j - \boldsymbol{\beta})^\tau(\overline{\mathbf{Y}_k - \boldsymbol{\beta}})\} = \delta\{j-k\}\mathbf{V}_j$ for $1 \leqslant j \leqslant k \leqslant J$. Prove that the best linear unbiased estimate of $\boldsymbol{\beta}$ is given by

$$\hat{\boldsymbol{\beta}} = \sum_{j=1}^{J} \mathbf{Y}_j\mathbf{V}_j{}^{-1}\left[\sum_{j=1}^{J} \mathbf{V}_j{}^{-1}\right]^{-1}.$$

Hint: Use Exercises 6.14.2 and 6.14.8. Exercise 1.7.6 is the case $r = 1$. Show that $E\{(\hat{\boldsymbol{\beta}} - \boldsymbol{\beta})^\tau(\overline{\hat{\boldsymbol{\beta}} - \boldsymbol{\beta}})\} = [\sum_j \mathbf{V}_j{}^{-1}]^{-1}$.

6.14.12 Suppose the conditions of Theorem 6.2.4 hold. Prove that if two columns of the matrix $\mathbf{X}$ are orthogonal, then the corresponding entries of $\hat{\mathbf{a}}$ are statistically independent.

6.14.13 Demonstrate that $g_{\varepsilon\varepsilon}^{(T)}(\lambda)$, the estimate (6.4.5) of the error spectrum, is non-negative.

6.14.14 If $|R_{YX}^{(T)}(\lambda)|^2$ is defined by (6.4.11), show that it may be interpreted as the proportion of the sample power spectrum of the $Y(t)$ values explained by the $\mathbf{X}(t)$ values.

6.14.15 Show that the statistics $\mathbf{A}^{(T)}(\lambda)$, $g_{\varepsilon\varepsilon}^{(T)}(\lambda)$ do not depend on the values of the sample means $\mathbf{c}_X^{(T)}$, $c_Y^{(T)}$.

6.14.16 Prove that $f_{YY}^{(T)}(\lambda) \geqslant g_{\varepsilon\varepsilon}^{(T)}(\lambda)$ with the definitions of Sections 6.4.

6.14.17 Let $\boldsymbol{\alpha}$ be a k vector. Under the conditions of Theorem 6.2.4 show that

$$|\mathbf{a}\boldsymbol{\alpha} - \hat{\mathbf{a}}\boldsymbol{\alpha}| \leqslant (2kF_{2k;2(n-k)}(\beta))^{1/2}\hat{\sigma}(\boldsymbol{\alpha}(\mathbf{X}\overline{\mathbf{X}}^\tau)^{-1}\boldsymbol{\alpha}^\tau)^{1/2}$$

provides a 100β percent multiple confidence region for all linear combinations of the entries of $\mathbf{a}$. (This region is a complex analog of the Scheffé region; see Miller (1966) p. 49.)

6.14.18 We adopt the notation of Theorem 6.2.3. Let $\mathbf{X}_j$ denote the jth row of $\mathbf{X}$ and let $\mathbf{X}_j\overline{\mathbf{X}}_j^\tau = C_j$, $j = 1, \ldots, k$ with $C_1, \ldots, C_k$ given. Prove that

$$E|\hat{a}_j - a_j|^2 \geqslant \frac{1}{C_j}$$

and that the minimum is achieved when $\mathbf{X}_j\overline{\mathbf{X}}_k^\tau = 0$, $k \neq j$ that is, when the rows of $\mathbf{X}$ are orthogonal. (For the real case of this result see Rao (1965) p. 194.)

6.14.19 Let w be a $N_1^C(\mu,\sigma^2)$ variate and let $R = |w|$, $\rho = |\mu|$, $f = \arg w$, $\phi = \arg \mu$. Prove that the density function of R is

$$2R\sigma^{-2} \exp\{-(R^2 + \tau^2)/\sigma^2\} I_0(2R\rho/\sigma^2)$$

where $I_0(x)$ is the 0 order Bessel function of the first kind. Prove

$$ER^\nu = \sigma^\nu\Gamma(\nu/2 + 1)\,{}_1F_1(-\nu/2;1;\rho^2/\sigma^2)$$

for $\nu > 0$, where ${}_1F_1(a;b;x)$ is the confluent hypergeometric function. Evaluate ER if $\rho = 0$. Also prove the density function of f is

$$(2\pi)^{-1} \exp\{-\rho^2\sigma^{-2} \sin^2(f - \phi)\} [\sqrt{\pi}\rho^2\sigma^{-2} \cos(f - \phi) + {}_1F_1(-1/2;1/2;-\rho^2\sigma^{-2} \cos^2(f - \phi))].$$

See Middleton (1960) p. 417.

6.14.20 Let

$$\mathbf{y} = \mathbf{a}\mathbf{x} + \mathbf{e}$$

where $\mathbf{e}$ is an $s \times n$ matrix whose columns are independent $N_s^C(\mathbf{0},\boldsymbol{\Sigma})$ variates, $\mathbf{a}$ is an $s \times r$ matrix of unknown complex parameters, $\mathbf{x}$ is an $r \times n$ matrix of known complex entries and $\mathbf{y}$ is an $s \times n$ matrix of known complex variates. Let

$$\hat{\mathbf{a}} = \mathbf{y}\bar{\mathbf{x}}^\tau(\mathbf{x}\bar{\mathbf{x}}^\tau)^{-1}$$

and

$$\hat{\mathbf{\Sigma}} = (n - r)^{-1}\mathbf{y}(\mathbf{I} - \bar{\mathbf{x}}^{\tau}(\mathbf{x}\bar{\mathbf{x}}^{\tau})^{-1}\mathbf{x})\mathbf{y}^{\tau}.$$

Prove that vec $\hat{\mathbf{a}}$ is $N_{rs}^{C}(\text{vec } \mathbf{a}, \mathbf{\Sigma} \otimes (\mathbf{x}\bar{\mathbf{x}}^{\tau})^{-1})$ and $\hat{\mathbf{\Sigma}}$ is independent of $\hat{\mathbf{a}}$ and $(n - r)^{-1} W_s^C(n - r, \mathbf{\Sigma})$. The operations vec and $\otimes$ are defined in Section 8.2.

6.14.21 Let $\mathbf{x}$ and $\mathbf{y}$ be given $s \times n$ and $r \times n$ matrices respectively with complex entries. For given $c \times s$ $\mathbf{C}$, $r \times u$ $\mathbf{U}$, $c \times u$ $\mathbf{\Gamma}$, show that the $s \times r$ $\mathbf{a}$, constrained by $\mathbf{CaU} = \mathbf{\Gamma}$ that minimizes

$$\text{tr}\{[\mathbf{y} - \mathbf{ax}]\overline{[\mathbf{y} - \mathbf{ax}]^{\tau}}\}$$

is given by

$$\hat{\mathbf{a}}_* = \hat{\mathbf{a}} - \bar{\mathbf{C}}^{\tau}(\mathbf{C}\bar{\mathbf{C}}^{\tau})^{-1}[\mathbf{C}\hat{\mathbf{a}}\mathbf{U} - \mathbf{\Gamma}][\bar{\mathbf{U}}^{\tau}(\mathbf{x}\bar{\mathbf{x}}^{\tau})^{-1}\mathbf{U}]^{-1}\bar{\mathbf{U}}^{\tau}(\mathbf{x}\bar{\mathbf{x}}^{\tau})^{-1}$$

where $\hat{\mathbf{a}} = \mathbf{y}\bar{\mathbf{x}}^{\tau}(\mathbf{x}\bar{\mathbf{x}}^{\tau})^{-1}$, provided the indicated inverses exist.

6.14.22 Under the conditions of Theorem 6.6.1 prove that

$$\text{cov}\,\{\text{Re}\,\mathbf{A}^{(T)}(\lambda)^{\tau}, \text{Re}\,\mathbf{A}^{(T)}(\mu)^{\tau}\}$$
$$= B_T^{-1}T^{-1}\pi \int W(\alpha)^2 d\alpha\,[\eta\{\lambda - \mu\} + \eta\{\lambda + \mu\}]\,\text{Re}\,\mathbf{\Psi}^{(T)}(\lambda) + O(T^{-1})$$

$$\text{cov}\,\{\text{Re}\,\mathbf{A}^{(T)}(\lambda)^{\tau}, \text{Im}\,\mathbf{A}^{(T)}(\mu)^{\tau}\} = O(T^{-1})$$

$$\text{cov}\,\{\text{Im}\,\mathbf{A}^{(T)}(\lambda)^{\tau}, \text{Im}\,\mathbf{A}^{(T)}(\mu)^{\tau}\}$$
$$= B_T^{-1}T^{-1}\pi \int W(\alpha)^2 d\alpha\,[\eta\{\lambda - \mu\} - \eta\{\lambda + \mu\}]\,\text{Re}\,\mathbf{\Psi}^{(T)}(\lambda) + O(T^{-1}).$$

6.14.23 Under the conditions of Theorem 6.4.2 prove that

$$\{\mathbf{A}^{(T)}(\lambda) - \mathbf{A}(\lambda)\}\,\mathbf{f}_{XX}^{(T)}(\lambda)\overline{\{\mathbf{A}^{(T)}(\lambda) - \mathbf{A}(\lambda)\}}^{\tau}$$

tends to $(2m + 1)^{-1}f_{\varepsilon\varepsilon}(\lambda)\chi_{2r}^2$ independently of $g_{\varepsilon\varepsilon}^{(T)}(\lambda)$ for $\lambda \not\equiv 0 \pmod{\pi}$. Develop a corresponding result for the case of $\lambda \equiv 0 \pmod{\pi}$.

6.14.24 Suppose that in the formation of (6.4.2) one takes $m = T - 1$. Prove that the resulting $\mathbf{A}^{(T)}(\lambda)$ is

$$\left\{\sum_{t=0}^{T-1}[Y(t) - c_Y^{(T)}][\mathbf{X}(t) - \mathbf{c}_X^{(T)}]^{\tau}\right\}\left\{\sum_{t=0}^{T-1}[\mathbf{X}(t) - \mathbf{c}_X^{(T)}][\mathbf{X}(t) - \mathbf{c}_X^{(T)}]^{\tau}\right\}^{-1}$$

where $c_Y^{(T)}$ and $\mathbf{c}_X^{(T)}$ are the sample means of the Y and $\mathbf{X}$ values. Relate this result to the multiple regression coefficient of $Y(t)$ on $\mathbf{X}(t)$.

6.14.25 Under the conditions of Theorem 6.4.2 and if $\mathbf{A}(\lambda) = 0$, prove that for $\lambda \not\equiv 0 \pmod{\pi}$,

$$|R_{YX}^{(T)}(\lambda)|^2 = \mathbf{f}_{YX}^{(T)}(\lambda)\mathbf{f}_{XX}^{(T)}(\lambda)^{-1}\mathbf{f}_{XY}^{(T)}(\lambda)/f_{YY}^{(T)}(\lambda)$$

tends in distribution to

$$[(2m + 1 - r)r^{-1}F + 1]^{-1}$$

where F has an F distribution with degrees of freedom $2(2m + 1 - r)$ and $2r$.

6.14.26 Suppose that $\mathbf{Y}(t)$, $\boldsymbol{\varepsilon}(t)$, $t = 0, \pm 1, \ldots$ are s vector-valued stochastic series. Let $\boldsymbol{\mu}$ denote an s vector and $\mathbf{a}(t)$ denote an $s \times r$ matrix-valued function. Let $\mathbf{X}(t)$, $t = 0, \pm 1, \ldots$ denote an r vector-valued fixed series. Suppose

$$\mathbf{Y}(t) = \boldsymbol{\mu} + \sum_{u=-\infty}^{\infty} \mathbf{a}(t-u)\mathbf{X}(u) + \boldsymbol{\varepsilon}(t).$$

Develop estimates $\mathbf{A}^{(T)}(\lambda)$ of the transfer function of $\{\mathbf{a}(u)\}$ and $\mathbf{g}_{\varepsilon\varepsilon}{}^{(T)}(\lambda)$ of the spectral density matrix of $\boldsymbol{\varepsilon}(t)$; see Brillinger (1969a).

6.14.27 Suppose that $\mathbf{f}_{XX}{}^{(T)}(\lambda)$ tends to $\mathbf{f}_{XX}(\lambda)$ uniformly in λ as $T \to \infty$ and suppose that $\|\mathbf{f}_{XX}(\lambda)\|$, $\|\mathbf{f}_{XX}(\lambda)^{-1}\| < K$, $-\infty < \lambda < \infty$ for finite K. Prove that Assumption 6.5.2 is satisfied.

6.14.28 Prove that $\mathbf{f}_{XX}{}^{(T)}(\lambda)$ as defined by (6.5.5) is non-negative definite if $W(\alpha) \geqslant 0$. Also prove that $g_{\varepsilon\varepsilon}{}^{(T)}(\lambda)$ given by (6.5.9) is non-negative under this condition.

6.14.29 Let $\mathbf{X}_1(t) = \Sigma_u \mathbf{b}(t-u)\mathbf{X}(u)$ where $\{\mathbf{b}(u)\}$ is a summable $r \times r$ filter with transfer function $\mathbf{B}(\lambda)$. Suppose that $\mathbf{B}(\lambda)$ is nonsingular, $-\infty < \lambda < \infty$. Prove that $\mathbf{X}_1(t)$, $t = 0, \pm 1, \ldots$ satisfies Assumption 6.5.2 if $\mathbf{X}(t)$, $t = 0, \pm 1, \ldots$ satisfies Assumption 6.5.2.

6.14.30 Suppose $Y(t)$ and $\mathbf{X}(t)$ are related as in (6.1.1). Suppose that $X_j(t)$ is increased to $X_j(t) + \exp\{i\lambda t\}$ and the remaining components of $\mathbf{X}(t)$ are held fixed. Discuss how this procedure is useful in the interpretation of $A_j(\lambda)$.

6.14.31 Under the conditions of Theorem 6.6.1 show that

$$\operatorname{var} \mu^{(T)} = B_T{}^{-1}T^{-1}2\pi \int W(\alpha)^2 d\alpha \, f_{\varepsilon\varepsilon}(0) \times \{\mathbf{c}_X{}^{(T)\tau}\mathbf{f}_{XX}{}^{(T)}(0)^{-1}\mathbf{h}_{XX}{}^{(T)}(0)\mathbf{f}_{XX}{}^{(T)}(0)^{-1}\mathbf{c}_X{}^{(T)}\} + O(T^{-1}).$$

6.14.32 Suppose that we consider the full model (6.12.3), rather than the simpler form (6.12.2). Let $[\mathbf{a}_j{}^{(T)}(-m) \cdots \mathbf{a}_j{}^{(T)}(n)]$, $j = 1, 2, 3$, be the analogs here of the estimates $\mathbf{a}_1{}^{(T)}$, $\mathbf{a}_2{}^{(T)}$, $\mathbf{a}_3{}^{(T)}$ of Section 6.12. Show that the covariances $\operatorname{cov}\{\mathbf{a}_j{}^{(T)}(u), \mathbf{a}_j{}^{(T)}(v)\}$, $j = 1, 2, 3$ are approximately $B_T{}^{-1}T^{-1}2\pi \int W(\alpha)^2 d\alpha$ times

$$P^{-2}\sum_p \exp\left\{i\frac{2\pi p}{P}(u-v)\right\} f_{\varepsilon\varepsilon}\left(\frac{2\pi p}{P}\right)\mathbf{f}_{XX}{}^{(T)}\left(\frac{2\pi p}{P}\right)^{-1},$$

$$\left\{\sum_p \exp\left\{i\frac{2\pi p}{P}(u-v)\right\}\mathbf{f}_{XX}{}^{(T)}\left(\frac{2\pi p}{P}\right)\right\}^{-1} \times \left\{\sum_p \exp\left\{i\frac{2\pi p}{P}(u-v)\right\} f_{\varepsilon\varepsilon}\left(\frac{2\pi p}{P}\right)\mathbf{f}_{XX}{}^{(T)}\left(\frac{2\pi p}{P}\right)\right\} \times \left\{\sum_p \exp\left\{i\frac{2\pi p}{P}(u-v)\right\}\mathbf{f}_{XX}{}^{(T)}\left(\frac{2\pi p}{P}\right)\right\}^{-1},$$

$$\left\{\sum_p \exp\left\{i\frac{2\pi p}{P}(u-v)\right\} f_{\varepsilon\varepsilon}\left(\frac{2\pi p}{P}\right)^{-1}\mathbf{f}_{XX}{}^{(T)}\left(\frac{2\pi p}{P}\right)\right\}^{-1}.$$

7

ESTIMATING THE SECOND-ORDER SPECTRA OF VECTOR-VALUED SERIES

7.1 THE SPECTRAL DENSITY MATRIX AND ITS INTERPRETATION

In this chapter we extend the results of Chapter 5 to cover the case of the joint behavior of second-order statistics based on various components of a vector-valued stationary time series.

Let $\mathbf{X}(t)$, $t = 0, \pm 1, \ldots$ be an r vector-valued series with component series $X_a(t)$, $t = 0, \pm 1, \ldots$ for $a = 1, \ldots, r$. Suppose

$$E\mathbf{X}(t) = \mathbf{c}_X \tag{7.1.1}$$

$$\begin{aligned} E[\mathbf{X}(t+u) - \mathbf{c}_X][\mathbf{X}(t) - \mathbf{c}_X]^\tau &= \operatorname{cov}\{\mathbf{X}(t+u), \mathbf{X}(t)\} \\ &= \mathbf{c}_{XX}(u) \qquad \text{for } t, u = 0, \pm 1, \ldots. \end{aligned} \tag{7.1.2}$$

Indicate the individual entries of $\mathbf{c}_X$ by c_a, $a = 1, \ldots, r$ so $c_a = EX_a(t)$ is the mean of the series $X_a(t)$, $t = 0, \pm 1, \ldots$. Denote the entry in row a, column b of $\mathbf{c}_{XX}(u)$ by $c_{ab}(u)$, $a, b = 1, \ldots, r$, so $c_{ab}(u)$ is the cross-covariance function of the series $X_a(t)$ with the series $X_b(t)$. Note that

$$\begin{aligned} \mathbf{c}_{XX}(u)^\tau &= \operatorname{cov}\{\mathbf{X}(t), \mathbf{X}(t+u)\} \\ &= \mathbf{c}_{XX}(-u) \qquad \text{for } u = 0, \pm 1, \ldots. \end{aligned} \tag{7.1.3}$$

Supposing

$$\sum_{u=-\infty}^{\infty} |c_{ab}(u)| < \infty \qquad \text{for } a, b = 1, \ldots, r \tag{7.1.4}$$

we may define $\mathbf{f}_{XX}(\lambda)$, the **spectral density matrix at frequency** λ of the series $\mathbf{X}(t)$, $t = 0, \pm 1, \ldots$ by

$$\mathbf{f}_{XX}(\lambda) = (2\pi)^{-1} \sum_{u=-\infty}^{\infty} \exp\{-i\lambda u\}\, \mathbf{c}_{XX}(u). \tag{7.1.5}$$

$f_{ab}(\lambda)$, the entry in row a and column b of $\mathbf{f}_{XX}(\lambda)$, is seen to be the power spectrum of the series $X_a(t)$ if $a = b$ and to be the **cross-spectrum** of the series $X_a(t)$ with the series $X_b(t)$ if $a \neq b$. $\mathbf{f}_{XX}(\lambda)$ has period 2π with respect to λ. Also, because the entries of $\mathbf{c}_{XX}(u)$ are real-valued

$$\overline{\mathbf{f}_{XX}(\lambda)} = \mathbf{f}_{XX}(-\lambda) = \mathbf{f}_{XX}(\lambda)^{\tau} \tag{7.1.6}$$

from (7.1.3). The matrix $\mathbf{f}_{XX}(\lambda)$ is Hermitian from the last expression. These properties mean that the basic domain of definition of $\mathbf{f}_{XX}(\lambda)$ may be the interval $[0,\pi]$. We have already seen in Theorem 2.5.1 that $\mathbf{f}_{XX}(\lambda)$ is non-negative definite, $\mathbf{f}_{XX}(\lambda) \geqslant \mathbf{0}$, for $-\infty < \lambda < \infty$, extending the result that the power spectrum of a real-valued series is non-negative.

Example 2.8.2 shows the effect of filtering on the spectral density matrix. Suppose

$$\mathbf{Y}(t) = \sum_{u=-\infty}^{\infty} \mathbf{a}(t-u)\mathbf{X}(u) \qquad t = 0, \pm 1, \ldots \tag{7.1.7}$$

for an $s \times r$ matrix-valued filter with transfer function

$$\mathbf{A}(\lambda) = \sum_{u=-\infty}^{\infty} \mathbf{a}(u) \exp\{-i\lambda u\} \qquad -\infty < \lambda < \infty \tag{7.1.8}$$

then the spectral density matrix of the series $\mathbf{Y}(t)$ is given by

$$\mathbf{f}_{YY}(\lambda) = \mathbf{A}(\lambda)\mathbf{f}_{XX}(\lambda)\overline{\mathbf{A}(\lambda)}^{\tau} \qquad \text{for } -\infty < \lambda < \infty. \tag{7.1.9}$$

The definition of the spectral density matrix may be inverted to obtain

$$\mathbf{c}_{YY}(u) = \int_{-\pi}^{\pi} \exp\{i\alpha u\}\mathbf{f}_{YY}(\alpha)d\alpha \qquad \text{for } u = 0, \pm 1, \ldots. \tag{7.1.10}$$

Expressions (7.1.9) and (7.1.10) imply that the covariance matrix of the s vector-valued variate $\mathbf{Y}(t)$ is given by

$$\begin{aligned} \mathbf{c}_{YY}(0) &= \operatorname{cov}\{\mathbf{Y}(t), \mathbf{Y}(t)\} \\ &= \int_{-\pi}^{\pi} \mathbf{A}(\alpha)\mathbf{f}_{XX}(\alpha)\overline{\mathbf{A}(\alpha)}^{\tau} d\alpha. \end{aligned} \tag{7.1.11}$$

With a goal of obtaining an interpretation of $\mathbf{f}_{XX}(\lambda)$ we consider the implication of this result for the $2r$ vector-valued filter with transfer function

$$\mathbf{A}(\alpha) = \begin{bmatrix} 1 \\ \cdot \\ \cdot \\ \cdot \\ 1 \\ -i \operatorname{sgn} \lambda \\ \cdot \\ \cdot \\ \cdot \\ -i \operatorname{sgn} \lambda \end{bmatrix} \qquad \text{for } |\alpha \pm \lambda| < \Delta \tag{7.1.12}$$

and $= \mathbf{0}$ for all other essentially different frequencies. (For (7.1.12) we are using the definition of Theorem 2.7.1 of the filter.) If Δ is small, the output of this filter is the $2r$ vector-valued series

$$\begin{bmatrix} \mathbf{X}(t,\lambda) \\ \mathbf{X}^H(t,\lambda) \end{bmatrix} \qquad t = 0, \pm 1, \ldots \tag{7.1.13}$$

involving the component of frequency λ discussed in Section 4.6. By inspection, expression (7.1.11) takes the approximate form

$$4\Delta \begin{bmatrix} \operatorname{Re} \mathbf{f}_{XX}(\lambda) & \operatorname{Im} \mathbf{f}_{XX}(\lambda) \\ -\operatorname{Im} \mathbf{f}_{XX}(\lambda) & \operatorname{Re} \mathbf{f}_{XX}(\lambda) \end{bmatrix} = 4\Delta \mathbf{f}_{XX}(\lambda)^R \qquad \text{if } \lambda \not\equiv 0 \pmod{\pi} \tag{7.1.14}$$

and the approximate form

$$2\Delta \begin{bmatrix} \mathbf{f}_{XX}(\lambda) & \mathbf{0} \\ \mathbf{0} & \mathbf{f}_{XX}(\lambda) \end{bmatrix} = 2\Delta f_{XX}(\lambda)^R \qquad \text{if } \lambda \equiv 0 \pmod{\pi}. \tag{7.1.15}$$

Both approximations lead to the useful interpretation of $\operatorname{Re} \mathbf{f}_{XX}(\lambda)$ as proportional to the covariance matrix of $\mathbf{X}(t,\lambda)$ (the component of frequency λ in $\mathbf{X}(t)$), and the interpretation of $\operatorname{Im} \mathbf{f}_{XX}(\lambda)$ as proportional to the cross-covariance matrix of $\mathbf{X}(t,\lambda)$ with its Hilbert transform $\mathbf{X}^H(t,\lambda)$. $\operatorname{Re} f_{ab}(\lambda)$, the co-spectrum of $X_a(t)$ with $X_b(t)$, is proportional to the covariance of the component of frequency λ in the series $X_a(t)$ with the corresponding component in the series $X_b(t)$. $\operatorname{Im} f_{ab}(\lambda)$, the quadrature spectrum, is proportional to the covariance of the Hilbert transform of the component of frequency λ in the series $X_a(t)$ with the component of frequency λ in the series $X_b(t)$. Being covariances, both of these are measures of degree of linear relationship.

When interpreting the spectral density matrix, $\mathbf{f}_{XX}(\lambda)$, it is also useful to recall the second-order properties of the Cramér representation. In Theorem 4.6.2 we saw that $\mathbf{X}(t)$ could be represented as

$$\mathbf{X}(t) = \int_0^{2\pi} \exp\{i\lambda t\} d\mathbf{Z}_X(\lambda) \qquad \text{for } t = 0, \pm 1, \ldots \tag{7.1.16}$$

where the function $\mathbf{Z}_X(\lambda)$ is stochastic with the property

$$\operatorname{cov}\{d\mathbf{Z}_X(\lambda), d\mathbf{Z}_X(\mu)\} = \eta(\lambda - \mu)\mathbf{f}_{XX}(\lambda)d\lambda d\mu \qquad (7.1.17)$$

where $\eta(\cdot)$ is the 2π periodic extension of the Dirac delta function. From (7.1.17) it is apparent that $\mathbf{f}_{XX}(\lambda)$ may be interpreted as being proportional to the covariance matrix of the complex-valued differential $d\mathbf{Z}_X(\lambda)$. Both interpretations will later suggest plausible estimates for $\mathbf{f}_{XX}(\lambda)$.

7.2 SECOND-ORDER PERIODOGRAMS

Suppose that the stretch $\mathbf{X}(t)$, $t = 0, \ldots, T-1$ of T consecutive values of an r vector-valued series is available for analysis and the series is stationary with mean function $\mathbf{c}_X$ and spectral density matrix $\mathbf{f}_{XX}(\lambda)$, $-\infty < \lambda < \infty$. Suppose also we are interested in estimating $\mathbf{f}_{XX}(\lambda)$. Consider basing an estimate on the finite Fourier transform

$$\begin{aligned}\mathbf{d}_X^{(T)}(\lambda) &= [d_a^{(T)}(\lambda)] \\ &= \left[\sum_t h_a\left(\frac{t}{T}\right)X_a(t)\exp\{-i\lambda t\}\right] \qquad -\infty < \lambda < \infty \qquad (7.2.1)\end{aligned}$$

where $h_a(t)$ is a tapering function vanishing for $|t|$ sufficiently large, $a = 1, \ldots, r$. Following Theorem 4.4.2, this variate is asymptotically

$$\begin{aligned}&N_r^C(\mathbf{0}, 2\pi T[H_{ab}(0)f_{ab}(\lambda)]) && \text{if } \lambda \not\equiv 0 \pmod{\pi} \\ &N_r(T[H_a(0)c_a], 2\pi T[H_{ab}(0)f_{ab}(\lambda)]) && \text{if } \lambda = 0, \pm 2\pi, \ldots \\ &N_r(\mathbf{0}, 2\pi T[H_{ab}(0)f_{ab}(\lambda)]) && \text{if } \lambda = \pm\pi, \pm 3\pi, \ldots \qquad (7.2.2)\end{aligned}$$

where

$$TH_a(0) = T\int h_a(t)dt \backsim \sum_t h_a\left(\frac{t}{T}\right) = H_a^{(T)}(0) \qquad (7.2.3)$$

and

$$TH_{ab}(0) = T\int h_a(t)h_b(t)dt \backsim \sum_t h_a\left(\frac{t}{T}\right)h_b\left(\frac{t}{T}\right) = H_{ab}^{(T)}(0)$$
$$\text{for } a, b = 1, \ldots, r. \qquad (7.2.4)$$

These distributions suggests a consideration of the statistic

$$\begin{aligned}\mathbf{I}_{XX}^{(T)}(\lambda) &= [I_{ab}^{(T)}(\lambda)] \\ &= [\{2\pi H_{ab}^{(T)}(0)\}^{-1}d_a^{(T)}(\lambda)\overline{d_b^{(T)}(\lambda)}] \qquad (7.2.5)\end{aligned}$$

as an estimate of $\mathbf{f}_{XX}(\lambda)$ in the case $\lambda \neq 0, \pm 2\pi, \ldots$. The entries of $\mathbf{I}_{XX}^{(T)}(\lambda)$ are the **second order periodograms** of the tapered values $h_a(t/T)X_a(t)$, $t = 0, \pm 1, \ldots$. This statistic is seen to have the same symmetry and periodicity properties as $\mathbf{f}_{XX}(\lambda)$. In connection with it we have

Theorem 7.2.1 Let $\mathbf{X}(t)$, $t = 0, \pm 1, \ldots$ be an r vector-valued series with mean function $E\mathbf{X}(t) = \mathbf{c}_X$ and cross-covariance function cov $\{\mathbf{X}(t+u), \mathbf{X}(t)\} = \mathbf{c}_{XX}(u)$ for $t, u = 0, \pm 1, \ldots$. Suppose

$$\sum_u |\mathbf{c}_{XX}(u)| < \infty. \tag{7.2.6}$$

Let $h_a(u)$, $-\infty < u < \infty$, satisfy Assumption 4.3.1 for $a = 1, \ldots, r$. Let $\mathbf{I}_{XX}^{(T)}(\lambda)$ be given by (7.2.5). Then

$$EI_{ab}^{(T)}(\lambda) = \left\{\int_{-\pi}^{\pi} H_a^{(T)}(\alpha)H_b^{(T)}(-\alpha)d\alpha\right\}^{-1}$$
$$\times \left\{\int_{-\pi}^{\pi} H_a^{(T)}(\alpha)H_b^{(T)}(-\alpha)f_{ab}(\lambda-\alpha)d\alpha + H_a^{(T)}(\lambda)H_b^{(T)}(-\lambda)c_ac_b\right\}$$
$$\text{for } -\infty < \lambda < \infty;\ a, b = 1, \ldots, r. \tag{7.2.7}$$

The character of the tapering function $h_a(t/T)$ is such that its Fourier transform $H_a^{(T)}(\lambda)$ is concentrated in the neighborhood of the frequencies $\lambda = 0, \pm 2\pi, \ldots$ for large T. It follows that in the case $\lambda \not\equiv 0 \pmod{2\pi}$, the final term in (7.2.7) will be of reduced magnitude for T large. The first term on the right side of (7.2.7) is seen to be a weighted average of the cross-spectrum f_{ab} of interest with weight concentrated in the neighborhood of λ and with relative weight determined by the tapers. In the limit we have

Corollary 7.2.1 Under the conditions of Theorem 7.2.1 and if $\int h_a(u)h_b(u)du \neq 0$ for $a, b = 1, \ldots, r$

$$\lim_{T\to\infty} E\mathbf{I}_{XX}^{(T)}(\lambda) = \mathbf{f}_{XX}(\lambda) \tag{7.2.8}$$

if $\lambda \not\equiv 0 \pmod{2\pi}$ or if $\mathbf{c}_X = \mathbf{0}$.

The estimate is asymptotically unbiased if $\lambda \not\equiv 0 \pmod{2\pi}$ or if $\mathbf{c}_X = \mathbf{0}$. If c_a, c_b are far from 0, then substantial bias may be present in the estimate $\mathbf{I}_{XX}^{(T)}(\lambda)$ as shown by the term in c_a, c_b of (7.2.7). This effect may be reduced by subtracting an estimate of the mean of $\mathbf{X}(t)$ before forming the finite Fourier transform. We could consider the statistics

$$\sum_t h_a\left(\frac{t}{T}\right)\{X_a(t) - c_a^{(T)}\}\exp\{-i\lambda t\} = d_a^{(T)}(\lambda) - d_a^{(T)}(0)H_a^{(T)}(\lambda)/H_a^{(T)}(0) \tag{7.2.9}$$

with

$$c_a^{(T)} = \sum_t h_a\left(\frac{t}{T}\right)X_a(t) \Big/ \sum_t h_a\left(\frac{t}{T}\right) \qquad \text{for } a = 1, \ldots, r \tag{7.2.10}$$

and then the estimate

$$\mathbf{I}^{(T)}_{X-c_X^{(T)},X-c_X^{(T)}}(\lambda) = [\{2\pi H_{ab}^{(T)}(0)\}^{-1}\{d_a^{(T)}(\lambda) - d_a^{(T)}(0)H_a^{(T)}(\lambda)/H_a^{(T)}(0)\} \times \overline{\{d_b^{(T)}(\lambda) - d_b^{(T)}(0)H_b^{(T)}(\lambda)/H_b^{(T)}(0)\}}] \quad (7.2.11)$$

of $\mathbf{f}_{XX}(\lambda)$.

The asymptotic form of the covariance of two entries of $\mathbf{I}_{XX}^{(T)}(\lambda)$ in the case that the series has mean $\mathbf{0}$ is indicated by

Theorem 7.2.2 Let $\mathbf{X}(t)$, $t = 0, \pm 1, \ldots$ be an r vector-valued series satisfying Assumption 2.6.2(1). Let $h_a(u)$, $a = 1, \ldots, r$ satisfy Assumption 4.3.1. Let $\mathbf{I}_{XX}^{(T)}(\lambda)$ be given by (7.2.5), then

$$\begin{aligned}\operatorname{cov}\,&\{I^{(T)}_{a_1b_1}(\lambda), I^{(T)}_{a_2b_2}(\mu)\}\\ &= H^{(T)}_{a_1b_1}(0)^{-1}H^{(T)}_{a_2b_2}(0)^{-1}\,\{H^{(T)}_{a_1a_2}(\lambda-\mu)\overline{H^{(T)}_{b_1b_2}(\lambda-\mu)}f_{a_1a_2}(\lambda)f_{b_1b_2}(-\lambda)\\ &\quad + H^{(T)}_{a_1b_2}(\lambda+\mu)\,\overline{H^{(T)}_{b_1a_2}(\lambda+\mu)}f_{a_1b_2}(\lambda)f_{b_1a_2}(-\lambda)\} + T^{-1}R_T(\lambda,\mu) \qquad (7.2.12)\end{aligned}$$

where $|R_T(\lambda,\mu)| \leq K_1|H_a^{(T)}(\lambda)|\,|H_a^{(T)}(\mu)| + K_2|H_a^{(T)}(\lambda)| + K_3|H_a^{(T)}(\mu)| + K_4$ for constants $K_1, \ldots, K_4$ and $a = a_1, a_2, b_1, b_2$, $-\infty < \lambda, \mu < \infty$.

The statistical dependence of $I^{(T)}_{a_1b_1}$ and $I^{(T)}_{a_2b_2}$ is seen to fall off as the functions $H_{ab}^{(T)}$ fall off. In the limit the theorem becomes

Corollary 7.2.2 Under the conditions of Theorem 7.2.2

$$\begin{aligned}\lim_{T\to\infty} \operatorname{cov}\,\{I^{(T)}_{a_1b_1}(\lambda), I^{(T)}_{a_2b_2}(\mu)\} &= \eta\{\lambda-\mu\}f_{a_1a_2}(\lambda)f_{b_1b_2}(-\lambda)\\ &\quad + \eta\{\lambda+\mu\}f_{a_1b_2}(\lambda)f_{a_1b_2}(-\lambda) \qquad \text{for } \lambda, \mu \not\equiv 0 \pmod{2\pi}. \quad (7.2.13)\end{aligned}$$

In the case of untapered values, $h_a(u) = 1$ for $0 \leq u < 1$, and $= 0$ otherwise, Exercise 7.10.14 shows that we have

$$\begin{aligned}\operatorname{cov}\{I^{(T)}_{a_1b_1}(\lambda), I^{(T)}_{a_2b_2}(\mu)\} &= \eta\{\lambda-\mu\}f_{a_1a_2}(\lambda)f_{b_1b_2}(-\lambda)\\ &\quad + \eta\{\lambda+\mu\}f_{a_1b_2}(\lambda)f_{b_1a_2}(-\lambda) + \frac{2\pi}{T}f_{a_1b_1a_2b_2}(\lambda,-\lambda,-\mu)\\ &\quad + \eta\{\lambda-\mu\}O(T^{-1}) + \eta\{\lambda+\mu\}O(T^{-1}) + O(T^{-2}) \qquad (7.2.14)\end{aligned}$$

for frequencies λ, μ of the form $2\pi r/T$, $2\pi s/T$ where r, s are integers with $r, s \not\equiv 0 \pmod{T}$.

We complete the present discussion of the asymptotic properties of the matrix of second-order periodograms by indicating its asymptotic distribution.

Theorem 7.2.3 Let $\mathbf{X}(t)$, $t = 0, \pm 1, \ldots$ be an r vector-valued series satisfying Assumption 2.6.1. Let $h_a(t)$, $a = 1, \ldots, r$ satisfy Assumption 4.3.1. Let

$\mathbf{I}_{XX}^{(T)}(\lambda)$ be given by (7.2.5). Suppose $2\lambda_j$, $\lambda_j \pm \lambda_k \not\equiv 0 \pmod{2\pi}$ for $1 \leqslant j < k \leqslant J$. Then $\mathbf{I}_{XX}^{(T)}(\lambda_j)$, $j = 1, \ldots, J$ are asymptotically independent $W_r^C(1,\mathbf{f}_{XX}(\lambda_j))$ variates. Also if $\lambda = \pm\pi, \pm 3\pi, \ldots$, then $\mathbf{I}_{XX}^{(T)}(\lambda)$ is asymptotically $W_r(1,\mathbf{f}_{XX}(\lambda))$ independently of the previous variates.

The Wishart distribution was given in Section 4.2 with its density function and various properties. The limiting distribution of this theorem is seen to involve $\mathbf{f}_{XX}(\lambda)$ in a direct manner. However, being a Wishart with just 1 degree of freedom, the distribution is well spread out about $\mathbf{f}_{XX}(\lambda)$. Therefore $\mathbf{I}_{XX}^{(T)}(\lambda)$ cannot be considered a reasonable estimate.

It is interesting to note that the limiting distributions of Theorem 7.2.3 do not involve the particular tapering functions employed. In the limit the taper used does not matter; however, as expression (7.2.7) shows, the taper does affect the large sample bias before we actually get to the limit. Consequently, if there may be peaks close together in $\mathbf{f}_{XX}(\lambda)$, we should taper the data to improve the resolution.

The frequencies considered in Theorem 7.2.3 did not depend on T. The following theorem considers the asymptotic distribution in the case of a number of frequencies tending to λ as $T \to \infty$. We revert to the untapered case in

Theorem 7.2.4 Let $\mathbf{X}(t)$, $t = 0, \pm 1, \ldots$ be an r vector-valued series satisfying Assumption 2.6.1. Let

$$\mathbf{I}_{XX}^{(T)}(\lambda) = (2\pi T)^{-1}\left(\sum_{t=0}^{T-1} \mathbf{X}(t) \exp\{-i\lambda t\}\right)\overline{\left(\sum_{t=0}^{T-1} \mathbf{X}(t) \exp\{-i\lambda t\}\right)}^{\tau} \tag{7.2.15}$$

for $-\infty < \lambda < \infty$. Let $s_j(T)$ be an integer with $\lambda_j(T) = 2\pi s_j(T)/T$ tending to λ_j as $T \to \infty$ for $j = 1, \ldots, J$. Suppose $2\lambda_j(T)$, $\lambda_j(T) \pm \lambda_k(T) \not\equiv 0 \pmod{2\pi}$ for $1 \leqslant j < k \leqslant J$ and T sufficiently large. Then $\mathbf{I}_{XX}^{(T)}(\lambda_j(T))$, $j = 1, \ldots, J$ are asymptotically independent $W_r^C(1,\mathbf{f}_{XX}(\lambda_j))$, $j = 1, \ldots, J$. Also if $\lambda = \pm\pi, \pm 3\pi, \ldots$, $\mathbf{I}_{XX}^{(T)}(\lambda)$ is asymptotically $W_r(1,\mathbf{f}_{XX}(\lambda))$ independently of the previous variates.

The most important case of this theorem occurs when $\lambda_j = \lambda$ for $j = 1, \ldots, J$. The theorem then indicates a source of J asymptotically independent estimates of $\mathbf{f}_{XX}(\lambda)$. The conclusions of this theorem were very much to be expected in light of Theorem 4.4.1 which indicated that the $\sum_t \mathbf{X}(t) \exp\{-it\lambda_j(T)\}$, $j = 1, \ldots, J$, are asymptotically independent $N_r^C(\mathbf{0},2\pi T\mathbf{f}_{XX}(\lambda_j))$ variates.

In order to avoid technical details we have made Theorem 7.2.4 refer to the untapered case. Exercise 4.8.20 and Brillinger (1970b) present results applying to frequencies depending on T as well as in the tapered case. The essential requirement for asymptotic independence indicated by them is that $\lambda_j(T) - \lambda_k(T)$, $1 \leqslant j < k \leqslant J$ do not tend to 0 too quickly.

In particular, Theorems 7.2.3 and 7.2.4 give the marginal distributions previously determined in Chapter 5 for a periodogram $I_{aa}^{(T)}(\lambda)$.

The following theorem shows how we may construct L asymptotically independent estimates of $\mathbf{f}_{XX}(\lambda)$ in the case that the data have been tapered. We split the data into L disjoint segments of V observations, taper and form a periodogram for each stretch.

Theorem 7.2.5 Let $\mathbf{X}(t)$, $t = 0, \pm 1, \ldots$ be an r vector-valued series satisfying Assumption 2.6.1. Let $h_a(u)$, $-\infty < u < \infty$ satisfy Assumption 4.3.1 and vanish for $u < 0$, $u \geqslant 1$. Let

$$\mathbf{I}_{XX}^{(V)}(\lambda,l) = [\{2\pi H_{ab}^{(V)}(0)\}^{-1} d_a^{(V)}(\lambda,l)\overline{d_b^{(V)}(\lambda,l)}] \tag{7.2.16}$$

$l = 0, \ldots, L-1$ where

$$d_a^{(V)}(\lambda,l) = \sum_{v=0}^{V-1} h_a\left(\frac{v}{V}\right) X_a(v + lV) \exp\{-i\lambda(v + lV)\}. \tag{7.2.17}$$

Then the $\mathbf{I}_{XX}^{(V)}(\lambda,l)$, $l = 0, \ldots, L-1$ are asymptotically independent

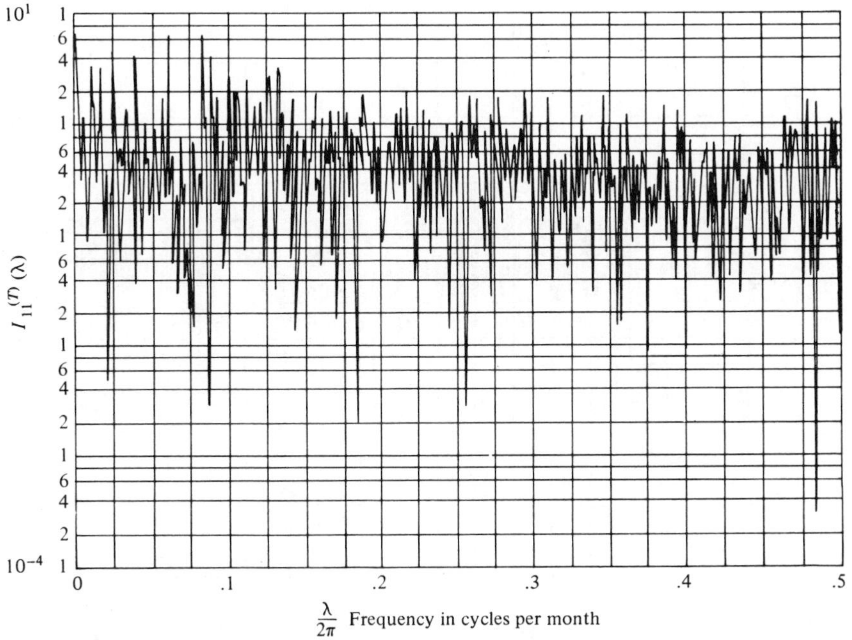

Figure 7.2.1 Periodogram of seasonally adjusted monthly mean temperatures at Berlin for the years 1780–1950. (Logarithmic plot.)

$W_r^C(1,\mathbf{f}_{XX}(\lambda))$ variates if $\lambda \not\equiv 0 \pmod{\pi}$ and asymptotically independent $W_r(1,\mathbf{f}_{XX}(\lambda))$ variates if $\lambda = \pm\pi, \pm 3\pi, \ldots$, as $V \to \infty$.

Once again the limiting distribution is seen not to involve the tapers employed; however, the tapers certainly appeared in the standardization of $d_a^{(V)}(\lambda,l)$ to form $\mathbf{I}_{XX}^{(V)}(\lambda,l)$.

Goodman (1963) introduced the complex Wishart distribution as an approximation for the distribution of spectral estimates in the case of vector-valued series. Brillinger (1969c) developed $W_r^C(1,\mathbf{f}_{XX}(\lambda))$ as the limiting distribution of the matrix of second-order periodograms.

In Figures 7.2.1 to 7.2.5 we give the periodograms and cross-periodogram for a bivariate series of interest. The series $X_1(t)$ is the seasonally adjusted series of mean monthly temperatures for Berlin (1780–1950). The series $X_2(t)$ is the seasonally adjusted series of mean monthly temperatures for Vienna (1780–1950). Figures 7.2.1 and 7.2.2 give $I_{11}^{(T)}(\lambda)$, $I_{22}^{(T)}(\lambda)$, the periodograms of the series. The cross-periodogram is illustrated in the remaining figures which give Re $I_{12}^{(T)}(\lambda)$, Im $I_{12}^{(T)}(\lambda)$, arg $I_{12}^{(T)}(\lambda)$ in turn. All of the figures are erratic, a characteristic consistent with Theorem 7.2.3, which suggested that second-order periodograms were not generally reasonable estimates of second-order spectra.

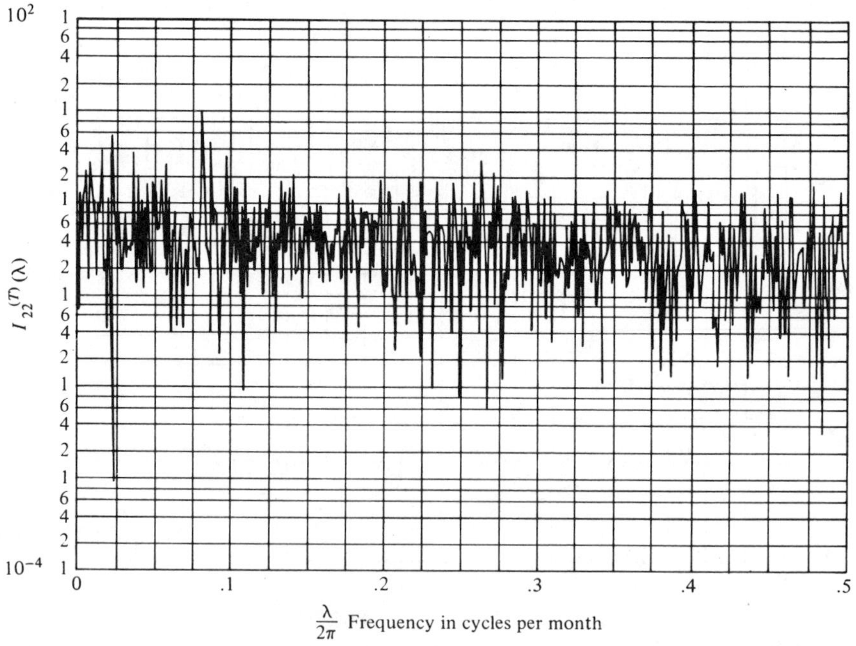

Figure 7.2.2 Periodogram of seasonally adjusted monthly mean temperatures at Vienna for the years 1780–1950. (Logarithmic plot.)

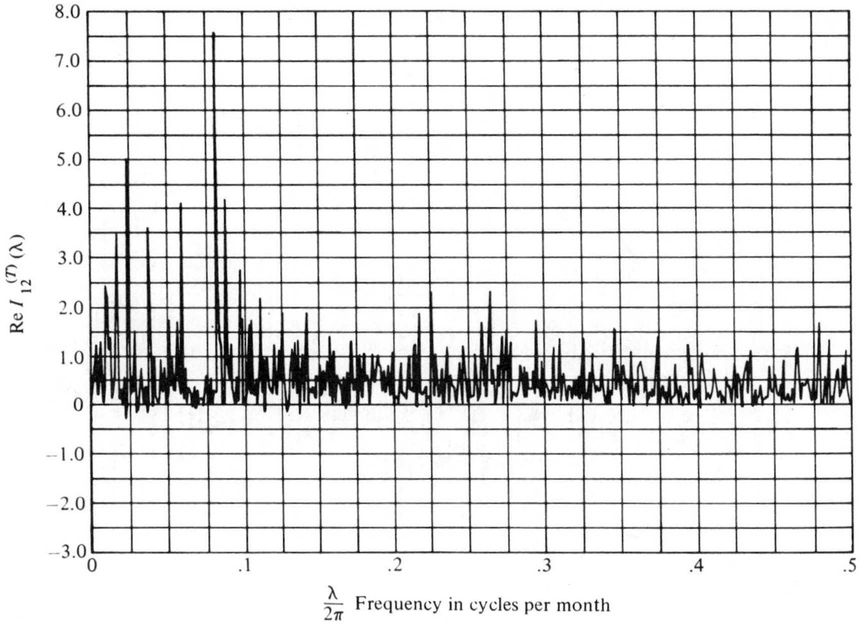

Figure 7.2.3 Real part of the cross-periodogram of temperatures at Berlin with those at Vienna.

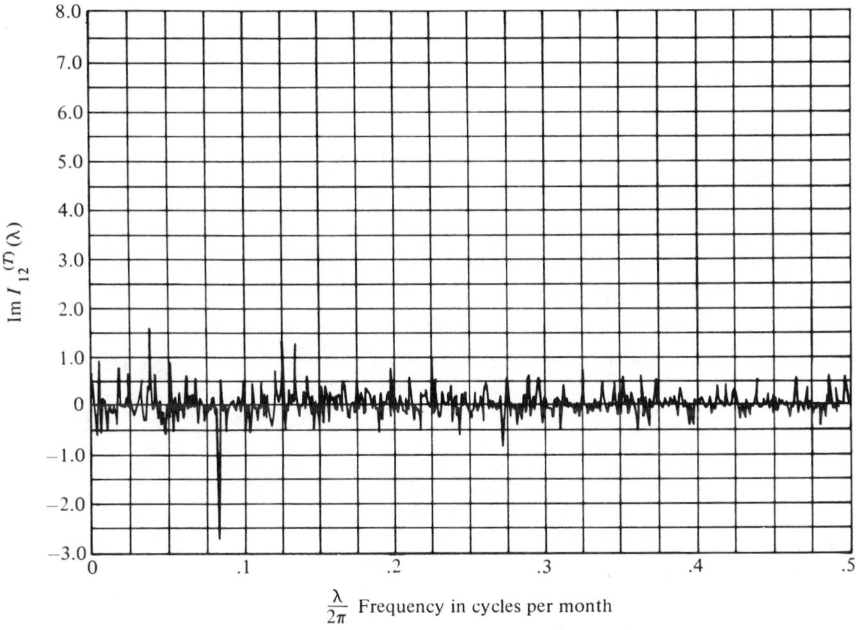

Figure 7.2.4 Imaginary part of the cross-periodogram of temperatures at Berlin with those at Vienna.

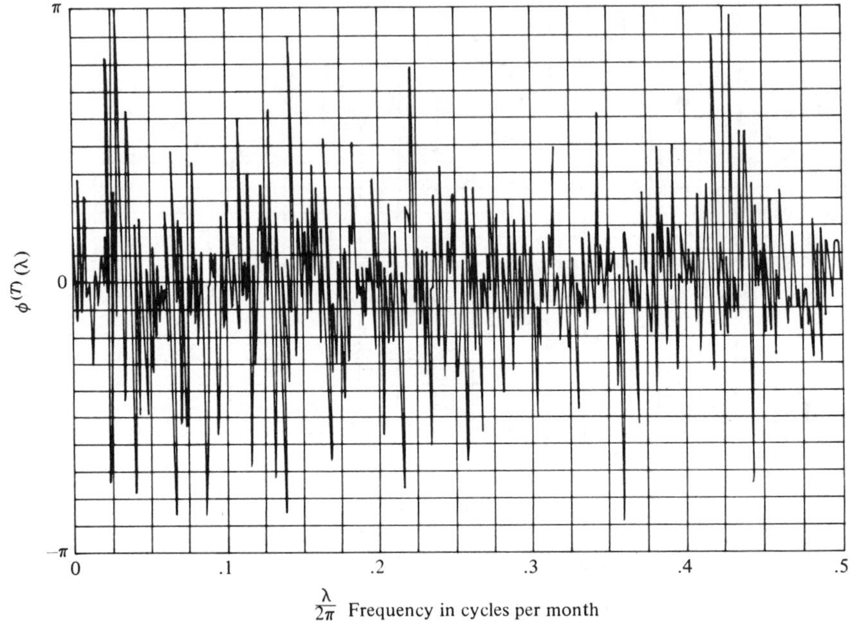

Figure 7.2.5 Phase of the cross-periodogram of temperatures at Berlin with those at Vienna.

7.3 ESTIMATING THE SPECTRAL DENSITY MATRIX BY SMOOTHING

Theorem 7.2.4 suggests a means of constructing an estimate of $\mathbf{f}_{XX}(\lambda)$ with a degree of flexibility. If

$$\mathbf{I}_{XX}^{(T)}(\lambda) = (2\pi T)^{-1}\left(\sum_{t=0}^{T-1} \mathbf{X}(t) \exp\{-i\lambda t\}\right)\overline{\left(\sum_{t=0}^{T-1} \mathbf{X}(t) \exp\{-i\lambda t\}\right)}^{\tau} \tag{7.3.1}$$

then, from that theorem, for $s(T)$ an integer with $2\pi s(T)/T$ near $\lambda \not\equiv 0 \pmod{\pi}$, the distribution of the variates $\mathbf{I}_{XX}^{(T)}(2\pi[s(T) + s]/T)$, $s = 0, \pm 1, \ldots, \pm m$ may be approximated by $2m + 1$ independent $W_r^C(1,\mathbf{f}_{XX}(\lambda))$ distributions. The preceding suggests the consideration of the estimate

$$\mathbf{f}_{XX}^{(T)}(\lambda) = (2m + 1)^{-1} \sum_{s=-m}^{m} \mathbf{I}_{XX}^{(T)}\left(\frac{2\pi[s(T) + s]}{T}\right) \quad \text{if } \lambda \not\equiv 0 \pmod{\pi}. \tag{7.3.2}$$

A further examination of the results of the theorem suggests the form

$$\mathbf{f}_{XX}^{(T)}(\lambda) = (2m)^{-1} \sum_{s=1}^{m} \left\{ \mathbf{I}_{XX}^{(T)}\left(\lambda + \frac{2\pi s}{T}\right) + \mathbf{I}_{XX}^{(T)}\left(\lambda - \frac{2\pi s}{T}\right) \right\}$$

$$= m^{-1} \sum_{s=1}^{m} \operatorname{Re} \mathbf{I}_{XX}^{(T)}\left(\lambda + \frac{2\pi s}{T}\right) \qquad \text{if } \lambda = 0, \pm 2\pi, \ldots \text{ or if } \lambda = \pm\pi, \pm 3\pi, \ldots \text{ and } T \text{ is even} \tag{7.3.3}$$

and the form

$$\mathbf{f}_{XX}^{(T)}(\lambda) = (2m)^{-1} \sum_{s=1}^{m} \left\{ \mathbf{I}_{XX}^{(T)}\left(\lambda - \frac{\pi}{T} + \frac{2\pi s}{T}\right) + \mathbf{I}_{XX}^{(T)}\left(\lambda + \frac{\pi}{T} - \frac{2\pi s}{T}\right) \right\}$$

$$= m^{-1} \sum_{s=1}^{m} \operatorname{Re} \mathbf{I}_{XX}^{(T)}\left(\lambda - \frac{\pi}{T} + \frac{2\pi s}{T}\right) \qquad \text{if } \lambda = \pm\pi, \pm 3\pi, \ldots \text{ and } T \text{ is odd.} \tag{7.3.4}$$

The estimate given by (7.3.2) to (7.3.4) is seen to have the same symmetry and periodicity properties as $\mathbf{f}_{XX}(\lambda)$ and to be based on the values, $\mathbf{d}_X^{(T)}(2\pi s/T)$, $s \not\equiv 0 \pmod{T}$ of the discrete Fourier transform. In connection with it we have

Theorem 7.3.1 Let $\mathbf{X}(t)$, $t = 0, \pm 1, \ldots$ be an r vector-valued series with mean function $\mathbf{c}_X$ and cross-covariance function $\mathbf{c}_{XX}(u) = \operatorname{cov}\{\mathbf{X}(t+u), \mathbf{X}(t)\}$ for $t, u = 0, \pm 1, \ldots$. Suppose

$$\sum_{u=-\infty}^{\infty} |\mathbf{c}_{XX}(u)| < \infty. \tag{7.3.5}$$

Let $\mathbf{f}_{XX}^{(T)}(\lambda)$ be given by (7.3.2) to (7.3.4). Then

$$E\mathbf{f}_{XX}^{(T)}(\lambda) = \int_{-\pi}^{\pi} A_T{}^m(\alpha) \mathbf{f}_{XX}\left(\frac{2\pi s(T)}{T} - \alpha\right) d\alpha \qquad \text{if } \lambda \not\equiv 0 \pmod{\pi}, \tag{7.3.6}$$

$$= \int_{-\pi}^{\pi} B_T{}^m(\alpha) \mathbf{f}_{XX}(\lambda - \alpha) d\alpha \qquad \text{if } \lambda = 0, \pm 2\pi, \ldots \text{ or if } \lambda = \pm\pi, \pm 3\pi, \ldots \text{ and } T \text{ is even,} \tag{7.3.7}$$

and

$$= \int_{-\pi}^{\pi} C_T{}^m(\alpha) \mathbf{f}_{XX}(\lambda - \alpha) d\alpha \qquad \text{if } \lambda = \pm\pi, \pm 3\pi, \ldots \text{ and } T \text{ is odd} \tag{7.3.8}$$

where

$$A_T^m(\alpha) = (2m+1)^{-1}(2\pi T)^{-1} \sum_{s=-m}^{m} \left|\Delta^{(T)}\left(\alpha + \frac{2\pi s}{T}\right)\right|^2, \qquad (7.3.9)$$

$$B_T^m(\alpha) = (2m)^{-1}(2\pi T)^{-1} \sum_{s=1}^{m} \left|\Delta^{(T)}\left(\alpha + \frac{2\pi s}{T}\right)\right|^2 + \left|\Delta^{(T)}\left(\alpha - \frac{2\pi s}{T}\right)\right|^2 \qquad (7.3.10)$$

and

$$C_T^m(\alpha) = (2m)^{-1}(2\pi T)^{-1} \sum_{s=1}^{m} \left|\Delta^{(T)}\left(\alpha - \frac{\pi}{T} + \frac{2\pi s}{T}\right)\right|^2 + \left|\Delta^{(T)}\left(\alpha + \frac{\pi}{T} - \frac{2\pi s}{T}\right)\right|^2 \quad \text{for} -\infty < \alpha < \infty. \qquad (7.3.11)$$

The functions $A_T^m(\alpha)$, $B_T^m(\alpha)$, $C_T^m(\alpha)$ are non-negative weight functions. The first has peaks at $\alpha = 0, \pm 2\pi, \pm 4\pi, \ldots$ and has width there of approximately $4\pi m/T$. The second and third are also concentrated in intervals of approximate width $4\pi m/T$ about the frequencies $\alpha = 0, \pm 2\pi, \ldots$; however, they dip at these particular frequencies. They are graphed in Figure 5.4.1 for T = 11. In any case, $E\mathbf{f}_{XX}^{(T)}(\lambda)$ should be near the desired $\mathbf{f}_{XX}(\lambda)$ in the case that $\mathbf{f}_{XX}(\alpha)$ is near constant in a band of width $4\pi m/T$ about λ. In the limit we have

Corollary 7.3.1 Under the conditions of Theorem 7.3.1 and if $2\pi s(T)/T \to \lambda$ as $T \to \infty$

$$\lim_{T\to\infty} E\mathbf{f}_{XX}^{(T)}(\lambda) = \mathbf{f}_{XX}(\lambda) \quad \text{for } -\infty < \lambda < \infty. \qquad (7.3.12)$$

The estimate is asymptotically unbiased as is clearly desirable. We next turn to a consideration of second-order properties.

Theorem 7.3.2 Let $\mathbf{X}(t)$, $t = 0, \pm 1, \ldots$ be an r vector-valued series satisfying Assumption 2.6.2(1). Let $\mathbf{f}_{XX}^{(T)}(\lambda)$ be given by (7.3.2) to (7.3.4) with $\lambda - 2\pi s(T)/T = O(T^{-1})$. Then

$$\operatorname{cov}\{f_{a_1b_1}^{(T)}(\lambda), f_{a_2b_2}^{(T)}(\mu)\}$$
$$= \frac{\eta\{\lambda-\mu\} f_{a_1a_2}(\lambda) f_{b_1b_2}(-\lambda) + \eta\{\lambda+\mu\} f_{a_1b_2}(\lambda) f_{b_1a_2}(-\lambda)}{2m+1} + O(T^{-1}) \quad \text{if } \lambda \not\equiv 0 \pmod{\pi}$$
$$= \frac{\eta\{\lambda-\mu\} f_{a_1a_2}(\lambda) f_{b_1b_2}(-\lambda) + \eta\{\lambda+\mu\} f_{a_1b_2}(\lambda) f_{b_1a_2}(-\lambda)}{2m} + O(T^{-1})$$
$$\text{if } \lambda \equiv 0 \pmod{\pi} \quad \text{for } -\infty < \lambda, \mu < \infty. \qquad (7.3.13)$$

The second-order moments are seen to fall off in magnitude as m increases. By choice of m, the statistician has a means of reducing the asymptotic variability of the estimates to a desired level. The statistics are seen to be asymptotically uncorrelated in the case that $\lambda \pm \mu \not\equiv 0 \pmod{2\pi}$. In addition, expression (7.3.13) has a singularity at the frequencies λ, $\mu = 0$, $\pm\pi, \pm 2\pi, \ldots$. This results from two things: not knowing the mean $\mathbf{c}_X$ and the fact that $\mathbf{f}_{XX}(\lambda)$ is real at these particular frequencies. We remark that the estimate $\mathbf{f}_{XX}^{(T)}(\lambda)$ is not consistent under the conditions of this theorem. However, in the next section we will develop a consistent estimate.

Turning to the development of a large sample approximation to the distribution of $\mathbf{f}_{XX}^{(T)}(\lambda)$, we may consider

Theorem 7.3.3 Let $\mathbf{X}(t)$, $t = 0, \pm 1, \ldots$ be an r vector-valued series satisfying Assumption 2.6.1. Let $\mathbf{f}_{XX}^{(T)}(\lambda)$ be given by (7.3.2) to (7.3.4) with $2\pi s(T)/T \to \lambda$ as $T \to \infty$. Then $\mathbf{f}_{XX}^{(T)}(\lambda)$ is asymptotically distributed as $(2m+1)^{-1} W_r^C(2m+1, \mathbf{f}_{XX}(\lambda))$ if $\lambda \not\equiv 0 \pmod{\pi}$ and as $(2m)^{-1} W_r(2m, \mathbf{f}_{XX}(\lambda))$ if $\lambda \equiv 0 \pmod{\pi}$. Also $\mathbf{f}_{XX}^{(T)}(\lambda_j)$, $j = 1, \ldots, J$ are asymptotically independent if $\lambda_j \pm \lambda_k \not\equiv 0 \pmod{2\pi}$ for $1 \leqslant j < k \leqslant J$.

Asymptotically, the marginal distributions of the diagonal entries of $\mathbf{f}_{XX}(\lambda)$ are seen to be those obtained previously. The diagonal elements $f_{aa}^{(T)}(\lambda)$ are asymptotically the scaled chi-squared variates of Theorem 5.4.3. The standardized off-diagonal elements asymptotically have the densities of Exercise 7.10.15.

The approximation of the distribution of $\mathbf{f}_{XX}^{(T)}(\lambda)$ by a complex Wishart distribution was suggested by Goodman (1963). Wahba (1968) considers the approximation in the case of a Gaussian series and $m \to \infty$. Brillinger (1969c) considers the present case with mean $\mathbf{0}$.

Theorems of the same character as Theorems 7.3.1 to 7.3.3 may be developed in the case of tapered values if we proceed by splitting the data into L disjoint segments of V observations. Specifically we set

$$d_a^{(V)}(\lambda, l) = \sum_{v=0}^{V-1} h_a\left(\frac{v}{V}\right) X_a(v + lV) \exp\{-i\lambda(v + lV)\}$$
$$\text{for } -\infty < \lambda < \infty;\ l = 0, \ldots, L-1, \quad (7.3.14)$$

where $h_a(u)$, $-\infty < u < \infty$ vanishes for $u < 0$, $u \geqslant 1$. Next we set

$$\mathbf{I}_{XX}^{(V)}(\lambda, l) = [\{2\pi H_{ab}^{(V)}(0)\}^{-1} d_a^{(V)}(\lambda, l) \overline{d_b^{(V)}(\lambda, l)}]$$
$$\text{for } l = 0, \ldots, L-1. \quad (7.3.15)$$

Following Theorem 7.2.5, the estimates $\mathbf{I}_{XX}^{(V)}(\lambda, l)$, $l = 0, \ldots, L-1$ are asymptotically independent $W_r^C(1, \mathbf{f}_{XX}(\lambda))$ variates if $\lambda \not\equiv 0 \pmod{\pi}$ and $W_r(1, \mathbf{f}_{XX}(\lambda))$ variates if $\lambda = \pm\pi, \pm 3\pi, \ldots$. This suggests a consideration of the estimate

$$\mathbf{f}_{XX}^{(LV)}(\lambda) = L^{-1} \sum_{l=0}^{L-1} I_{XX}^{(V)}(\lambda,l). \tag{7.3.16}$$

In connection with the above we have

Theorem 7.3.4 Suppose the conditions of Theorem 7.3.1 are satisfied. Suppose also the functions $h_a(u)$, $a = 1, \ldots, r$ satisfy Assumption 4.3.1, vanish for $u < 0$, $u \geqslant 1$, and satisfy $\int h_a(u)h_b(u)\,du \neq 0$. Let $\mathbf{f}_{XX}^{(LV)}$ be given by (7.3.16). Then

$$\begin{aligned} Ef_{ab}^{(LV)}(\lambda) = &\left\{\int_{-\pi}^{\pi} H_a^{(V)}(\alpha)H_b^{(V)}(-\alpha)d\alpha\right\}^{-1} \left\{\int_{-\pi}^{\pi} H_a^{(V)}(\alpha)H_b^{(V)}(-\alpha)\right. \\ &\left. \times f_{ab}(\lambda - \alpha)d\alpha + H_a^{(V)}(\lambda)H_b^{(V)}(-\lambda)c_a c_b\right\} \\ &\rightarrow f_{ab}(\lambda) \end{aligned} \tag{7.3.17}$$

as $V \rightarrow \infty$ for $\lambda \not\equiv 0 \pmod{2\pi}$ and $a, b = 1, \ldots, r$.

This theorem is an immediate consequence of Theorem 7.2.1 and its corollary. It is interesting to note that the weighted average of f_{ab} appearing in expression (7.3.17) is concentrated in an interval of width proportional to V^{-1}.

Theorem 7.3.5 Let $\mathbf{X}(t)$, $t = 0, \pm 1, \ldots$ be an r vector-valued series satisfying Assumption 2.6.1. Let $h_a(u)$, $a = 1, \ldots, r$ satisfy Assumption 4.3.1, vanish for $u < 0$, $u \geqslant 1$ and be such that $\int h_a(u)h_b(u)\,du \neq 0$. Let $\mathbf{f}_{XX}^{(LV)}(\lambda)$ be given by (7.3.16). Then

$$\begin{aligned} &\lim_{V\rightarrow\infty} \operatorname{cov}\{f_{a_1b_1}^{(LV)}(\lambda), f_{a_2b_2}^{(LV)}(\mu)\} \\ &\qquad = \frac{\eta\{\lambda - \mu\}f_{a_1a_2}(\lambda)f_{b_1b_2}(-\lambda) + \eta\{\lambda + \mu\}f_{a_1b_2}(\lambda)f_{b_1a_2}(-\lambda)}{L} \end{aligned} \tag{7.3.18}$$

for $\lambda, \mu \not\equiv 0 \pmod{2\pi}$ and $a_1, a_2, b_1, b_2 = 1, \ldots, r$.

The second-order moments are here reduced from those of (7.2.13) by the factor $1/L$. The statistician may choose L appropriately large enough for his purposes in many cases. Finally we have

Theorem 7.3.6 Under the conditions of Theorem 7.2.5 and if $\mathbf{f}_{XX}^{(LV)}(\lambda)$ is given by (7.3.16), $\mathbf{f}_{XX}^{(LV)}(\lambda)$ is asymptotically $L^{-1}W_r^C(L,\mathbf{f}_{XX}(\lambda))$ if $\lambda \not\equiv 0 \pmod{\pi}$ and asymptotically $L^{-1}W_r(L,\mathbf{f}_{XX}(\lambda))$ if $\lambda = \pm\pi, \pm 3\pi, \ldots$ as $V \rightarrow \infty$.

Again the Wishart distribution is suggested as an approximation for the distribution of an estimate of the spectral density matrix. One difficulty with the above estimation procedure is that it does not provide an estimate in the case of $\lambda \equiv 0 \pmod{2\pi}$. An estimate for this case may possibly be obtained by extrapolating estimates at nearby frequencies. Note also the estimate of Exercise 7.10.23.

Exercise 7.10.24 indicates the asymptotic distribution of the estimate

$$\mathbf{f}_{XX}^{(T)}(\lambda) = \sum_{s=-m}^{m} W_s \mathbf{I}_{XX}^{(T)}\left(\frac{2\pi[s(T) + s]}{T}\right) \tag{7.3.19}$$

involving an unequal weighting of periodogram values.

7.4 CONSISTENT ESTIMATES OF THE SPECTRAL DENSITY MATRIX

The estimates of the previous section were not generally consistent, that is, $\mathbf{f}_{XX}^{(T)}(\lambda)$ did not tend in probability to $\mathbf{f}_{XX}(\lambda)$ as $T \to \infty$, typically. However, the estimates did involve a parameter (m or L) that affected their asymptotic variability. A consideration of the specific results obtained suggests that if we were to allow this parameter to tend to ∞ as $T \to \infty$, then we might obtain a consistent estimate. In this section we shall see that this is in fact the case. The results to be obtained will not be important so much for the specific computations to be carried out, as for their suggestion of alternate plausible large sample approximations for the moments and distribution of the estimate.

Suppose the stretch $\mathbf{X}(t)$, $t = 0, \ldots, T-1$ of an r vector-valued series is available for analysis. Suppose the discrete Fourier transform

$$\mathbf{d}_X^{(T)}\left(\frac{2\pi s}{T}\right) = \sum_{t=0}^{T-1} \mathbf{X}(t) \exp\left\{-i\frac{2\pi st}{T}\right\} \qquad s = 0, \pm 1, \ldots \tag{7.4.1}$$

is computed. The corresponding second-order periodograms are then given by

$$I_{ab}^{(T)}\left(\frac{2\pi s}{T}\right) = (2\pi T)^{-1} d_a^{(T)}\left(\frac{2\pi s}{T}\right)\overline{d_b^{(T)}\left(\frac{2\pi s}{T}\right)} \qquad s = 0, \pm 1, \ldots. \tag{7.4.2}$$

We shall form an estimate of $f_{ab}(\lambda)$ by taking a weighted average of this statistic concentrating weight in a neighborhood of λ having width $O(B_T)$ where B_T is a band-width parameter tending to 0 as $T \to \infty$.

Let $W_{ab}(\alpha)$, $-\infty < \alpha < \infty$, be a weight function satisfying

$$\int_{-\infty}^{\infty} W_{ab}(\alpha) d\alpha = 1. \tag{7.4.3}$$

Let B_T, $T = 1, 2, \ldots$ be a bounded sequence of non-negative scale parameters. As an estimate of $f_{ab}(\lambda)$ consider

$$f_{ab}^{(T)}(\lambda) = \frac{2\pi}{T} \sum_{s \not\equiv 0 (\text{mod } T)} W_{ab}\left(B_T^{-1}\left[\lambda - \frac{2\pi s}{T}\right]\right) I_{ab}^{(T)}\left(\frac{2\pi s}{T}\right)$$
$$\text{for } -\infty < \lambda < \infty;\ a, b = 1, \ldots, r. \quad (7.4.4)$$

In view of the 2π period of $I_{ab}^{(T)}(\alpha)$, the estimate may be written

$$f_{ab}^{(T)}(\lambda) = \frac{2\pi}{T} \sum_{s=1}^{T-1} W_{ab}^{(T)}\left(\lambda - \frac{2\pi s}{T}\right) I_{ab}^{(T)}\left(\frac{2\pi s}{T}\right) \quad (7.4.5)$$

where

$$W_{ab}^{(T)}(\alpha) = B_T^{-1} \sum_{j=-\infty}^{\infty} W_{ab}(B_T^{-1}[\alpha + 2\pi j]). \quad (7.4.6)$$

The estimate (7.4.4) is seen to weight periodogram values heavily at frequencies within $O(B_T)$ of λ. This suggests that we will later require $B_T \to 0$ as $T \to \infty$.

As an estimate of $\mathbf{f}_{XX}(\lambda)$ we now take

$$\mathbf{f}_{XX}^{(T)}(\lambda) = [f_{ab}^{(T)}(\lambda)]. \quad (7.4.7)$$

This estimate has the same symmetry and periodicity properties as does $\mathbf{f}_{XX}(\lambda)$ in the case that the functions $W_{ab}(\alpha)$ are even, $W_{ab}(-\alpha) = W_{ab}(\alpha)$. In addition, if the matrix $[W_{ab}(\alpha)]$ is non-negative definite for all α, then $\mathbf{f}_{XX}^{(T)}(\lambda)$ will be non-negative definite as is $\mathbf{f}_{XX}(\lambda)$; see Exercise 7.10.26. We now set down

Theorem 7.4.1 Let $\mathbf{X}(t)$, $t = 0, \pm 1, \ldots$ be an r vector-valued series with mean function $E\mathbf{X}(t) = \mathbf{c}_X$ and covariance function cov $\{\mathbf{X}(t + u), \mathbf{X}(t)\} = \mathbf{c}_{XX}(u)$, for $t, u = 0, \pm 1, \ldots$. Suppose

$$\sum_{u=-\infty}^{\infty} |\mathbf{c}_{XX}(u)| < \infty. \quad (7.4.8)$$

Let $f_{ab}^{(T)}(\lambda)$ be given by (7.4.5) where $W_{ab}(\alpha)$ satisfies Assumption 5.6.1, $a, b = 1, \ldots, r$. Then

$$Ef_{ab}^{(T)}(\lambda) = \frac{2\pi}{T} \int_{-\pi}^{\pi} \sum_{s=1}^{T-1} W_{ab}^{(T)}\left(\lambda - \frac{2\pi s}{T}\right) \times (2\pi T)^{-1} \left[\frac{\sin T\left(\frac{2\pi s}{T} - \alpha\right)/2}{\sin\left(\frac{2\pi s}{T} - \alpha\right)/2}\right]^2 f_{ab}(\alpha) d\alpha$$
$$\to f_{ab}(\lambda) \quad \text{if } B_T \to 0 \text{ as } T \to \infty \quad \text{for } a, b = 1, \ldots, r. \quad (7.4.9)$$

If in addition

$$\sum_{u=-\infty}^{\infty} |u|\, |c_{XX}(u)| < \infty, \tag{7.4.10}$$

then

$$\begin{aligned} Ef_{ab}^{(T)}(\lambda) &= \frac{2\pi}{T} \sum_{s=1}^{T-1} W_{ab}^{(T)}\left(\lambda - \frac{2\pi s}{T}\right) f_{ab}\left(\frac{2\pi s}{T}\right) + O(T^{-1}) \\ &= \int_{-\infty}^{\infty} W_{ab}(\alpha) f_{ab}(\lambda - B_T\alpha)\, d\alpha + O(B_T^{-1}T^{-1}) \end{aligned}$$

$$\text{for } -\infty < \lambda < \infty;\ a, b = 1, \ldots, r. \tag{7.4.11}$$

The error term is uniform in λ.

Expressions (7.4.9) and (7.4.11) show that the expected value of the proposed estimate is a weighted average of $f_{ab}(\alpha)$, $-\infty < \alpha < \infty$, with weight concentrated in a band of width $O(B_T)$ about λ. In the case that $B_T \to 0$ as $T \to \infty$, the estimate is asymptotically unbiased. We may proceed as in Theorem 3.3.1 to develop the asymptotic bias of the estimate (7.4.5) as a function of B_T. Specifically we have

Theorem 7.4.2 Let $f_{ab}(\lambda)$ have bounded derivatives of order $\leqslant P$. Suppose

$$\int_{-\infty}^{\infty} |\alpha|^P |W_{ab}(\alpha)|\, d\alpha < \infty. \tag{7.4.12}$$

Then, if $B_T \to 0$ as $T \to \infty$,

$$\begin{aligned} &\int_{-\infty}^{\infty} W_{ab}(\alpha) f_{ab}(\lambda - B_T\alpha)\, d\alpha \\ &\quad = f_{ab}(\lambda) + \sum_{p=1}^{P-1} \frac{1}{p!} B_T{}^p \int \alpha^p W(\alpha)\, d\alpha\, \frac{d^p f_{ab}(\lambda)}{d\lambda^p} + O(B_T{}^P) \end{aligned} \tag{7.4.13}$$

for $-\infty < \lambda < \infty$; $a, b = 1, \ldots, r$.

If $P = 3$, the above theorems and the fact that $W(-\alpha) = W(\alpha)$ give

$$Ef_{ab}^{(T)}(\lambda) = f_{ab}(\lambda) + \tfrac{1}{2} B_T{}^2 \int \alpha^2 W(\alpha)\, d\alpha\, \frac{d^2 f_{ab}(\lambda)}{d\lambda^2} + O(B_T{}^3) + O(B_T{}^{-1}T^{-1}). \tag{7.4.14}$$

From this, and expression (7.4.13), we see that in connection with the bias of the estimate $f_{ab}^{(T)}(\lambda)$ it is desirable that $f_{ab}(\alpha)$ be near constant in the neighborhood of λ, that B_T be small and that $\int \alpha^p W(\alpha)\, d\alpha$, $p = 2, 4, \ldots$ be small. The next theorem will show that we cannot take B_T too small if we wish the estimate to be consistent.

Theorem 7.4.3 Let $X(t)$, $t = 0, \pm 1, \ldots$ be an r vector-valued series satisfying Assumption 2.6.2(1). Let $W_{ab}(\alpha)$, $-\infty < \alpha < \infty$, satisfy Assumption 5.6.1, $a, b = 1, \ldots, r$. Let $f_{ab}^{(T)}(\lambda)$ be given by (7.4.5). Let $B_T T \to \infty$. Then

$$\text{cov}\,\{f_{a_1 b_1}^{(T)}(\lambda), f_{a_2 b_2}^{(T)}(\mu)\}$$

$$= \left(\frac{2\pi}{T}\right)^2 \left(\sum_{s=1}^{T-1} W_{a_1 b_1}^{(T)}\left(\lambda - \frac{2\pi s}{T}\right) W_{a_2 b_2}^{(T)}\left(\mu - \frac{2\pi s}{T}\right) f_{a_1 a_2}\left(\frac{2\pi s}{T}\right) f_{b_1 b_2}\left(-\frac{2\pi s}{T}\right)\right.$$

$$+ \sum_{s=1}^{T-1} W_{a_1 b_1}^{(T)}\left(\lambda - \frac{2\pi s}{T}\right) W_{a_2 b_2}^{(T)}\left(\mu + \frac{2\pi s}{T}\right) f_{a_1 b_2}\left(\frac{2\pi s}{T}\right) f_{b_1 a_2}\left(-\frac{2\pi s}{T}\right)\bigg)$$

$$+ O(T^{-1})$$

$$= \frac{2\pi}{T}\left\{\int_{-\pi}^{\pi} W_{a_1 b_1}^{(T)}(\lambda - \alpha) W_{a_2 b_2}^{(T)}(\mu - \alpha) f_{a_1 b_1}(\alpha) f_{a_2 b_2}(-\alpha)\,d\alpha\right.$$

$$+ \int_{-\pi}^{\pi} W_{a_1 b_1}^{(T)}(\lambda - \alpha) W_{a_2 b_2}^{(T)}(\mu + \alpha) f_{a_1 b_2}(\alpha) f_{b_1 a_2}(-\alpha)\,d\alpha\bigg\}$$

$$+ O(B_T^{-2} T^{-2}) + O(T^{-1}) \tag{7.4.15}$$

for $a_1, a_2, b_1, b_2 = 1, \ldots, r$. The error term is uniform in λ, μ.

Given the character of the $W^{(T)}$ functions, this covariance is seen to have greatest magnitude for $\lambda \pm \mu \equiv 0 \pmod{2\pi}$. The averages in (7.4.15) are approximately concentrated in a band of width $O(B_T)$ about λ, μ, and so the covariance approximately equals

$$(\eta\{\lambda - \mu\} f_{a_1 a_2}(\lambda) f_{b_1 b_2}(-\lambda) + \eta\{\lambda + \mu\} f_{a_1 b_2}(\lambda) f_{b_1 a_2}(-\lambda))$$

$$\times \left(\frac{2\pi}{T}\right)^2 \left(\sum_{s=1}^{T-1} W_{a_1 b_1}^{(T)}\left(\lambda - \frac{2\pi s}{T}\right) W_{a_2 b_2}^{(T)}\left(\lambda - \frac{2\pi s}{T}\right)\right). \tag{7.4.16}$$

In the limit we have

Corollary 7.4.3 Under the conditions of Theorem 7.4.3 and if $B_T \to 0$, $B_T T \to \infty$ as $T \to \infty$

$$\lim_{T \to \infty} B_T T \,\text{cov}\,\{f_{a_1 b_1}^{(T)}(\lambda), f_{a_2 b_2}^{(T)}(\mu)\}$$

$$= 2\pi \left(\int W_{a_1 b_1}(\alpha) W_{a_2 b_2}(\alpha)\,d\alpha\right)$$

$$\times (\eta\{\lambda - \mu\} f_{a_1 a_2}(\lambda) f_{b_1 b_2}(-\lambda) + \eta\{\lambda + \mu\} f_{a_1 b_2}(\lambda) f_{a_2 b_1}(-\lambda))$$

$$\text{for } -\infty < \lambda, \mu < \infty \text{ and } a_1, a_2, b_1, b_2 = 1, \ldots, r. \tag{7.4.17}$$

We see that the second-order moments are $O(B_T^{-1} T^{-1})$ and so tend to 0 as $T \to \infty$. We have already seen that the estimate is asymptotically unbiased. It therefore follows that it is consistent. We see that estimates evaluated at frequencies λ, μ with $\lambda \pm \mu \not\equiv 0 \pmod{2\pi}$ are asymptotically uncorrelated.

The first statement of expression (7.4.15) may be used to give an expression for the large sample covariance in the case where $B_T = 2\pi/T$. Suppose $W_{ab}(\alpha)$ vanishes for $|\alpha|$ sufficiently large and $\lambda = 2\pi s(T)/T$ with $s(T)$ an integer. For large T, the estimate (7.4.4) then takes the form

$$\frac{2\pi}{T}\sum_{s\neq 0} W_{ab}(s(T) - s)I_{ab}^{(T)}\left(\frac{2\pi s}{T}\right) = \frac{2\pi}{T}\sum_{s\neq s(T)} W_{ab}(s)I_{ab}^{(T)}\left(\frac{2\pi[s(T) - s]}{T}\right). \tag{7.4.18}$$

The estimate (7.3.2) had this form with $W_{ab}(s) = T/2\pi(2m + 1)$ for $|s| \leqslant m$. Expression (7.4.16) may be seen to give the following approximate form for the covariance here

$$\left(\frac{2\pi}{T}\right)^2(\eta\{\lambda - \mu\} f_{a_1b_1}(\lambda)f_{a_2b_2}(-\lambda) + \eta\{\lambda + \mu\} f_{a_1b_2}(\lambda)f_{b_1a_2}(-\lambda)$$
$$\times \sum W_{a_1b_1}(s)W_{a_2b_2}(s). \tag{7.4.19}$$

The results of Theorem 5.5.2 are particular cases of (7.4.19).

Expression (7.4.17) may be combined with expression (7.4.14) to obtain a form for the large sample mean squared error of $f_{ab}^{(T)}(\lambda)$. Specifically, if $\lambda \not\equiv 0 \pmod{\pi}$ it is

$$E|f_{ab}^{(T)}(\lambda) - f_{ab}(\lambda)|^2$$
$$\backsim B_T^{-1}T^{-1}2\pi \int W_{ab}(\alpha)^2 d\alpha\, f_{aa}(\lambda)f_{bb}(\lambda) + \tfrac{1}{4}B_T^4\left\{\int \alpha^2 W(\alpha)d\alpha\right\}^2\left(\frac{d^2 f_{ab}(\lambda)}{d\lambda^2}\right)^2. \tag{7.4.20}$$

Exercise 7.10.30 indicates that B_T should be taken to fall off as $T^{-1/5}$ if we wish to minimize this asymptotic mean-squared error.

Turning to the asymptotic distribution itself, we have

Theorem 7.4.4 Suppose Theorem 7.4.1 and Assumption 2.6.1 are satisfied. Then $\mathbf{f}_{XX}^{(T)}(\lambda_1), \ldots, \mathbf{f}_{XX}^{(T)}(\lambda_J)$ are asymptotically jointly normal with covariance structure given by (7.4.17) as $T \to \infty$ with $B_T T \to \infty$, $B_T \to 0$.

An examination of expression (7.4.17) shows that the estimates $\mathbf{f}_{XX}^{(T)}(\lambda)$, $\mathbf{f}_{XX}^{(T)}(\mu)$ are asymptotically independent if $\lambda \pm \mu \not\equiv 0 \pmod{2\pi}$. In the case that $\lambda \equiv 0 \pmod{\pi}$, the estimate $\mathbf{f}_{XX}^{(T)}(\lambda)$ is real-valued and its limiting distribution is seen to be real normal.

In Section 7.3, taking an estimate to be the average of $2m + 1$ periodogram ordinates, we obtained a Wishart with $2m + 1$ degrees of freedom as the limiting distribution. That result is consistent with the result just obtained in Theorem 7.4.4. The estimate (7.4.4) is essentially a weighted average of periodogram ordinates at frequencies within $O(B_T)$ of λ. There are $O(B_T T)$ such ordinates, in contrast with the previous $2m + 1$. Now the

Wishart is approximately normal for large degrees of freedom. As we have assumed $B_T T \to \infty$, the two approximations are essentially the same. We may set up a formal equivalence between the approximations. Suppose the same weight function is used in all the estimates, $W_{ab}(\alpha) = W(\alpha)$ for $a, b = 1, \ldots, r$. Comparing expression (7.4.16) with expression (7.3.13) suggests the identification

$$2m + 1 = \left\{\sum_{s=1}^{T-1} W^{(T)}\left(\lambda - \frac{2\pi s}{T}\right)\right\}^2 \Big/ \left\{\sum_{s=1}^{T-1} W^{(T)}\left(\lambda - \frac{2\pi s}{T}\right)^2\right\}$$
$$\backsim T \Big/ \left\{2\pi \int_0^{2\pi} W^{(T)}(\alpha)^2 d\alpha\right\}$$
$$\backsim B_T T / \{2\pi \textstyle\int W(\alpha)^2 d\alpha\}. \tag{7.4.21}$$

Having formed an estimate in the manner of (7.4.4) or (7.4.18) we may consider approximating the distribution of that estimate by $(2m + 1)^{-1} W_r^C(2m + 1, \mathbf{f}_{XX}(\lambda))$ if $\lambda \not\equiv 0 \pmod{\pi}$ and by $(2m)^{-1} W_r(2m, \mathbf{f}_{XX}(\lambda))$ if $\lambda \equiv 0 \pmod{\pi}$ taking $2m + 1$ to be given by (7.4.21).

Rosenblatt (1959) discussed the asymptotic first- and second-order moment structure and the joint asymptotic distribution of consistent estimates of second-order spectra. Parzen (1967c) was also concerned with the asymptotic theory and certain empirical aspects. We end this section by remarking that we will develop the asymptotic distribution of spectral estimates based on tapered data in Section 7.7.

7.5 CONSTRUCTION OF CONFIDENCE LIMITS

Having determined certain limiting distributions for estimates, $f_{ab}^{(T)}(\lambda)$, of second-order spectra we turn to a discussion of the use of these distributions in setting confidence limits for the parameter $f_{ab}(\lambda)$. We begin with the estimate of Section 7.3. In the case of $\lambda \not\equiv 0 \pmod{\pi}$, the estimate is given by

$$f_{ab}^{(T)}(\lambda) = (2m + 1)^{-1} \sum_{s=-m}^{m} I_{ab}^{(T)}\left(\frac{2\pi[s(T) + s]}{T}\right) \tag{7.5.1}$$

for $s(T)$ an integer with $2\pi s(T)/T$ near λ. Its consideration resulted from Theorem 7.2.4 which suggested that the variates

$$I_{ab}^{(T)}\left(\frac{2\pi[s(T) + s]}{T}\right) \qquad s = 0, \pm 1, \ldots, \pm m \tag{7.5.2}$$

might be considered to be $2m + 1$ independent estimates of $f_{ab}(\lambda)$. Having a number of approximately independent estimates of a parameter of interest, a means of setting approximate confidence limits is clear. Consider for example the case of $\theta = \operatorname{Re} f_{ab}(\lambda)$. Set

$$\hat{\theta}_s = \operatorname{Re} I_{ab}^{(T)}\left(\frac{2\pi[s(T) + s]}{T}\right) \qquad \text{for } s = 0, \pm 1, \ldots, \pm m. \tag{7.5.3}$$

Our estimate of θ is now

$$\hat{\theta} = \operatorname{Re} f_{ab}^{(T)}(\lambda) = (2m + 1)^{-1} \sum_{s=-m}^{m} \hat{\theta}_s. \tag{7.5.4}$$

Set

$$\hat{\sigma}^2 = (2m)^{-1} \sum_s (\hat{\theta}_s - \hat{\theta})^2. \tag{7.5.5}$$

Even when the basic variates $\hat{\theta}_s$ are not normal, it has often proved reasonable statistical practice (see Chap. 31 in Kendall and Stuart (1961)) to approximate the distribution of a variate such as

$$\frac{\hat{\theta} - \theta}{\hat{\sigma}/\sqrt{2m + 1}} \tag{7.5.6}$$

by a Student's t distribution with $2m$ degrees of freedom. This leads to the following approximate 100β percent confidence interval for $\theta = \operatorname{Re} f_{ab}(\lambda)$

$$\hat{\theta} - \frac{\hat{\sigma}}{\sqrt{2m + 1}} t_{2m}\left(\frac{1 + \beta}{2}\right) < \theta < \hat{\theta} + \frac{\hat{\sigma}}{\sqrt{2m + 1}} t_{2m}\left(\frac{1 + \beta}{2}\right) \tag{7.5.7}$$

where $t_\nu(\gamma)$ denotes the 100γ percentile of Student's t distribution with ν degrees of freedom. In the case of $\lambda \equiv 0 \pmod{\pi}$ we again proceed from Theorem 7.2.4.

By setting

$$\hat{\theta}_s = \operatorname{Im} I_{ab}^{(T)}\left(\frac{2\pi[s(T) + s]}{T}\right) \tag{7.5.8}$$

for $s = 0, \pm 1, \ldots, \pm m$ we may likewise obtain an approximate confidence interval for the quad-spectrum, $\operatorname{Im} f_{ab}(\lambda)$.

A closely related means of setting approximate confidence limits follows from Theorem 7.2.5. Here the statistics $I_{ab}^{(V)}(\lambda,l)$, $l = 1, \ldots, L$ for $\lambda \not\equiv 0 \pmod{2\pi}$ provide L approximately independent estimates of $f_{ab}(\lambda)$. Proceeding as above, we set $\theta = \operatorname{Re} f_{ab}(\lambda)$,

$$\hat{\theta}_l = I_{ab}^{(V)}(\lambda,l) \qquad \text{for } l = 1, \ldots, L \tag{7.5.9}$$

and set

$$\hat{\theta} = \operatorname{Re} f_{ab}^{(T)}(\lambda) = L^{-1} \sum_{l=1}^{L} \hat{\theta}_l \tag{7.5.10}$$

$$\hat{\sigma}^2 = (L - 1)^{-1} \sum_l (\hat{\theta}_l - \hat{\theta})^2. \tag{7.5.11}$$

We then approximate the distribution of

$$\frac{\hat{\theta} - \theta}{\hat{\sigma}/\sqrt{L}} \tag{7.5.12}$$

by a Student's t distribution with $L - 1$ degrees of freedom and thence obtain the desired limits. Similar steps lead to approximate limits in the case of the quad-spectrum, $\text{Im}\, f_{ab}(\lambda)$.

The results of Theorem 7.4.4 and Exercise 7.10.8 suggest a different means of proceeding. Suppose $\lambda \not\equiv 0 \pmod{\pi}$ and that the estimate $f_{ab}^{(T)}(\lambda)$ is given by (7.4.4). Then the exercise suggests that the distribution of $\text{Re}\, f_{ab}^{(T)}(\lambda)$ is approximately normal with mean $\text{Re}\, f_{ab}(\lambda)$ and variance

$$\sigma^2 = (B_T T)^{-1}\pi \int W(\alpha)^2 d\alpha\, [f_{aa}(\lambda) f_{bb}(\lambda) + \{\text{Re}\, f_{ab}(\lambda)\}^2 - \{\text{Im}\, f_{ab}(\lambda)\}^2]. \tag{7.5.13}$$

Expression (7.5.13) can be estimated by

$$\hat{\sigma}^2 = (B_T T)^{-1}\pi \int W(\alpha)^2 d\alpha\, [f_{aa}^{(T)}(\lambda) f_{bb}^{(T)}(\lambda) + \{\text{Re}\, f_{ab}^{(T)}(\lambda)\}^2 - \{\text{Im}\, f_{ab}^{(T)}(\lambda)\}^2] \tag{7.5.14}$$

and the following approximate 100β percent confidence interval can be set down

$$\text{Re}\, f_{ab}^{(T)}(\lambda) - \hat{\sigma} z\left(\frac{1+\beta}{2}\right) < \text{Re}\, f_{ab}(\lambda) < \text{Re}\, f_{ab}^{(T)}(\lambda) + \hat{\sigma} z\left(\frac{1+\beta}{2}\right) \tag{7.5.15}$$

where $z(\gamma)$ denotes the 100γ percent point of the distribution of a standard normal variate. We may obtain an approximate interval for the quad-spectrum $\text{Im}\, f_{ab}(\lambda)$ in a similar manner.

Finally, we note that the approximations suggested in Freiberger (1963) may prove useful in constructing confidence intervals for $\text{Re}\, f_{ab}(\lambda)$, $\text{Im}\, f_{ab}(\lambda)$. Rosenblatt (1960) and Gyires (1961) relate to these approximations.

7.6 THE ESTIMATION OF RELATED PARAMETERS

Let $\mathbf{X}(t)$, $t = 0, \pm 1, \ldots$ denote an r vector-valued stationary series with covariance function $\mathbf{c}_{XX}(u)$, $u = 0, \pm 1, \ldots$ and spectral density matrix $\mathbf{f}_{XX}(\lambda)$, $-\infty < \lambda < \infty$. Sometimes we are interested in estimating parameters of the process having the form

$$J_{ab}(A) = \int_0^{2\pi} A(\alpha) f_{ab}(\alpha) d\alpha \tag{7.6.1}$$

for some function $A(\alpha)$ and $a, b = 1, \ldots, r$. Examples of such a parameter include the covariance functions

$$c_{ab}(u) = \int_0^{2\pi} \exp\{iu\alpha\} f_{ab}(\alpha)d\alpha \qquad \text{for } u = 0, \pm 1, \ldots \tag{7.6.2}$$

and the spectral measures

$$F_{ab}(\lambda) = \int_0^{\lambda} f_{ab}(\alpha)d\alpha \qquad \text{for } 0 \leqslant \lambda \leqslant 2\pi, \tag{7.6.3}$$

$a, b = 1, \ldots, r$. If $I_{ab}^{(T)}(\lambda)$ indicates a periodogram of a stretch of data,

$$I_{ab}^{(T)}(\lambda) = (2\pi T)^{-1}\left(\sum_{t=0}^{T-1} X_a(t)\exp\{-i\lambda t\}\right)\overline{\left(\sum_{t=0}^{T-1} X_b(t)\exp\{-i\lambda t\}\right)}, \tag{7.6.4}$$

then an obvious estimate of $J_{ab}(A)$ is provided by

$$J_{ab}^{(T)}(A) = \frac{2\pi}{T}\sum_{s=1}^{T-1} A\left(\frac{2\pi s}{T}\right) I_{ab}^{(T)}\left(\frac{2\pi s}{T}\right) \qquad \text{for } a, b = 1, \ldots, r. \tag{7.6.5}$$

In connection with this estimate we have

Theorem 7.6.1 Let $\mathbf{X}(t)$, $t = 0, \pm 1, \ldots$ be an r vector-valued series satisfying Assumption 2.6.1. Let $A_j(\alpha)$, $0 \leqslant \alpha \leqslant 2\pi$, be of bounded variation for $j = 1, \ldots, J$. Then

$$\begin{aligned} EJ_{ab}^{(T)}(A_j) &= \frac{2\pi}{T}\sum_{s=1}^{T-1} A_j\left(\frac{2\pi s}{T}\right) f_{ab}\left(\frac{2\pi s}{T}\right) + o(1) \\ &= \int_0^{2\pi} A_j(\alpha) f_{ab}(\alpha)d\alpha + o(1) \end{aligned} \tag{7.6.6}$$

for $j = 1, \ldots, J$. Also

$$\begin{aligned} &\operatorname{cov}\{J_{a_1b_1}^{(T)}(A_j), J_{a_2b_2}^{(T)}(A_k)\} \\ &= \frac{2\pi}{T}\int_0^{2\pi} A_j(\alpha)\overline{A_k(\alpha)} f_{a_1a_2}(\alpha) f_{b_1b_2}(-\alpha)d\alpha \\ &\quad + \frac{2\pi}{T}\int_0^{2\pi} A_j(\alpha)\overline{A_k(2\pi - \alpha)} f_{a_1b_2}(\alpha) f_{b_1a_2}(-\alpha)d\alpha \\ &\quad + \frac{2\pi}{T}\int_0^{2\pi}\int_0^{2\pi} A_j(\alpha)\overline{A_k(\beta)} f_{a_1b_1a_2b_2}(\alpha, -\alpha, -\beta)d\alpha d\beta + o(T^{-1}). \end{aligned} \tag{7.6.7}$$

Finally, $J_{ab}^{(T)}(A_j)$, $j = 1, \ldots, J$; $a, b = 1, \ldots, r$ are asymptotically jointly normal with the above first- and second-order moment structure.

From Theorem 7.6.1, we see that $J_{ab}^{(T)}(A_j)$ is an asymptotically unbiased and consistent estimate of $J_{ab}(A_j)$. It is based on the discrete Fourier transform and so can possibly be computed taking advantage of the Fast Fourier

Transform Algorithm. Were Assumption 2.6.2(1) adopted the error terms would be $O(T^{-1})$, $O(T^{-2})$ in the manner of Theorem 5.10.1.

In the case of the estimate

$$F_{ab}^{(T)}(\lambda) = \frac{2\pi}{T} \sum_{0 < \frac{2\pi s}{T} \leq \lambda} I_{ab}^{(T)}\left(\frac{2\pi s}{T}\right) \tag{7.6.8}$$

of the spectral measure, $F_{ab}(\lambda)$, corresponding to $A(\alpha) = 1$ for $0 \leqslant \alpha \leqslant \lambda$ and $= 0$ otherwise, expression (7.6.7) gives

$$\lim_{T\to\infty} T \operatorname{cov} \{F_{a_1b_1}^{(T)}(\lambda), F_{a_2b_2}^{(T)}(\mu)\} = 2\pi \int_0^{\min(\lambda,\mu)} f_{a_1a_2}(\alpha) f_{b_1b_2}(-\alpha) d\alpha + 2\pi \int_0^{\lambda}\int_0^{\mu} f_{a_1b_1a_2b_2}(\alpha,-\alpha,-\beta) d\alpha d\beta \tag{7.6.9}$$

for $0 \leqslant \lambda, \mu \leqslant \pi$; $a_1, b_1, a_2, b_2 = 1, \ldots, r$. We will return to the discussion of the convergence of $F_{ab}^{(T)}(\lambda)$ later in this section. In the case of the estimate

$$\hat{c}_{ab}^{(T)}(u) = \frac{2\pi}{T} \sum_{s=1}^{T-1} \exp\left\{\frac{i2\pi su}{T}\right\} I_{ab}^{(T)}\left(\frac{2\pi s}{T}\right) = T^{-1} \sum_{t=0}^{T-1} [\hat{X}_b(t+u) - c_a^{(T)}][\hat{X}_b(t) - c_b^{(T)}] \tag{7.6.10}$$

of $c_{ab}(u)$, corresponding to $A(\alpha) = \exp\{iu\alpha\}$ and with $\hat{X}_a(t)$ denoting the T periodic extension of the sequence $X(0), \ldots, X(T-1)$, expression (7.6.7) gives

$$\lim_{T\to\infty} T \operatorname{cov} \{\hat{c}_{a_1b_1}^{(T)}(u), \hat{c}_{a_2b_2}^{(T)}(v)\}$$

$$= 2\pi \int_0^{2\pi} \exp\{i(u-v)\alpha\} f_{a_1a_2}(\alpha) f_{b_1b_2}(-\alpha) d\alpha + 2\pi \int_0^{2\pi} \exp\{-i(u+v)\alpha\} f_{a_1b_2}(\alpha) f_{b_1a_2}(-\alpha) d\alpha + 2\pi \int_0^{2\pi}\int_0^{2\pi} \exp\{i(u\alpha - v\beta)\} f_{a_1b_1a_2b_2}(\alpha,-\alpha,-\beta) d\alpha d\beta$$

$$\text{for } u, v = 0, \pm 1, \ldots. \tag{7.6.11}$$

Exercise 7.10.36 shows that the autocovariance estimate

$$\mathbf{c}_{XX}^{(T)}(u) = T^{-1} \sum_{0 \leq t, t+u \leq T-1} [\mathbf{X}(t+u) - \mathbf{c}_X^{(T)}][\mathbf{X}(t) - \mathbf{c}_X^{(T)}]^\tau \tag{7.6.12}$$

is also asymptotically normal with the covariance structure (7.6.11).

It will sometimes be useful to consider the parameters

$$R_{ab}(\lambda) = \frac{f_{ab}(\lambda)}{[f_{aa}(\lambda) f_{bb}(\lambda)]^{1/2}} \tag{7.6.13}$$

$-\infty < \lambda < \infty$; $1 \leqslant a < b \leqslant r$. $R_{ab}(\lambda)$ is called the **coherency of the series $X_a(t)$ with the series $X_b(t)$ at frequency** λ. Its modulus squared, $|R_{ab}(\lambda)|^2$, is called the **coherence of the series $X_a(t)$ with the series $X_b(t)$ at frequency** λ. The interpretation of the parameter $R_{ab}(\lambda)$ will be considered in Chapter 8. It is a complex-valued analog of the coefficient of correlation. We may estimate it by

$$R_{ab}^{(T)}(\lambda) = \frac{f_{ab}^{(T)}(\lambda)}{[f_{aa}^{(T)}(\lambda) f_{bb}^{(T)}(\lambda)]^{1/2}}. \tag{7.6.14}$$

In the case that the spectral estimates are of the form considered in Section 7.4 we have

Theorem 7.6.2 Under the conditions of Theorem 7.4.3 and if $R_{ab}^{(T)}(\lambda)$ is given by (7.6.14)

$$\vec{\text{ave}}\, R_{ab}^{(T)}(\lambda) = R_{ab}(\lambda) + O(B_T) + O(B_T^{-1}T^{-1}) \tag{7.6.15}$$

$$\begin{aligned}\vec{\text{cov}}\, \{R_{ab}^{(T)}(\lambda), R_{cd}^{(T)}(\mu)\} \\ = [\eta\{\lambda - \mu\}(R_{ac}R_{db} - \tfrac{1}{2}R_{dc}R_{ac}R_{cb} - \tfrac{1}{2}R_{dc}R_{ad}R_{db} \\ - \tfrac{1}{2}R_{ab}R_{ac}R_{da} - \tfrac{1}{2}R_{ab}R_{bc}R_{db} + \tfrac{1}{4}R_{ab}R_{dc}R_{ac}R_{ca} + \tfrac{1}{4}R_{ab}R_{dc}R_{ad}R_{da} \\ + \tfrac{1}{4}R_{ab}R_{dc}R_{bc}R_{cb} + \tfrac{1}{4}R_{ab}R_{dc}R_{bd}R_{db}) + \eta\{\lambda + \mu\}(R_{ad}R_{cb} \\ - \tfrac{1}{2}R_{cd}R_{ac}R_{cb} - \tfrac{1}{2}R_{cd}R_{ad}R_{db} - \tfrac{1}{2}R_{ab}R_{ad}R_{ca} - \tfrac{1}{2}R_{ab}R_{bc}R_{db} \\ + \tfrac{1}{4}R_{ab}R_{cd}R_{ac}R_{ca} + \tfrac{1}{4}R_{ab}R_{cd}R_{ad}R_{da} + \tfrac{1}{4}R_{ab}R_{cd}R_{bc}R_{cb} \\ + \tfrac{1}{4}R_{ab}R_{cd}R_{bd}R_{db})]2\pi \int W(\alpha)^2 d\alpha\, B_T^{-1}T^{-1} + O(B_T^{-2}T^{-2})\end{aligned} \tag{7.6.16}$$

for $a, b, c, d = 1, \ldots, r$. Also the variates $R_{ab}^{(T)}(\lambda)$, $a, b = 1, \ldots, r$ are asymptotically jointly normal with covariance structure indicated by expression (7.6.16) where we have written R_{ab} for $R_{ab}(\lambda)$, $a, b = 1, \ldots, r$.

The asymptotic covariance structure of estimated correlation coefficients is presented in Pearson and Filon (1898), Hall (1927), and Hsu (1949) for the case of vector-valued variates with real components. We could clearly develop an alternate form of limiting distribution taking the estimate and limiting Wishart distributions of Theorem 7.3.3. This distribution is given by Fisher (1962) for the case of vector-valued variates with real components. The theorem has this useful corollary:

Corollary 7.6.2 Under the conditions of Theorem 7.6.2,

$$\vec{\text{ave}}\, |R_{ab}^{(T)}(\lambda)|^2 = |R_{ab}(\lambda)|^2 + O(B_T) + O(B_T^{-1}T^{-1}) \tag{7.6.17}$$

$$\begin{aligned}\vec{\text{cov}}\, \{|R_{ab}^{(T)}(\lambda)|^2, |R_{ab}^{(T)}(\mu)|^2\} = [\eta\{\lambda - \mu\} + \eta\{\lambda + \mu\}]|R_{ab}(\lambda)|^2 \\ \times [1 - |R_{ab}(\lambda)|^2]^2 4\pi \int W(\alpha)^2 d\alpha\, B_T^{-1}T^{-1} + O(B_T^{-2}T^{-2})\end{aligned} \tag{7.6.18}$$

and, for given J, the variates $R_{ab}^{(T)}(\lambda_1), \ldots, R_{ab}^{(T)}(\lambda_J)$ are asymptotically jointly normal with covariance structure given by (7.6.18) for $1 \leqslant a < b \leqslant r$.

In Section 8.5 we will discuss further aspects of the asymptotic distribution of $|R_{ab}^{(T)}(\lambda)|^2$, and in Section 8.9, we will discuss the construction of approximate confidence intervals for $|R_{ab}(\lambda)|$.

Let $D[0,\pi]$ signify the space of right continuous functions having left-hand limits. This space can be endowed with a metric which makes it complete and separable; see Billingsley (1968), Chap. 3. Let $D_C^{r\times r}[0,\pi]$ denote the space of $r \times r$ matrix-valued functions whose entries are complex-valued functions that are right continuous and have left-hand limits. This space is isomorphic with $D^{2r^2}[0,\pi]$ and may be endowed with a metric making it complete and separable. Continuing, if P_T, $T = 1, 2, \ldots$ denotes a sequence of probability measures on $D_C^{r\times r}[0,\pi]$, we shall say that the sequence **converges weakly** to a probability measure P on $D_C^{r\times r}[0,\pi]$ if

$$\int h dP_T \rightarrow \int h dP \tag{7.6.19}$$

as $T \rightarrow \infty$ for all real-valued bounded continuous functions, h, on $D_C^{r\times r}[0,\pi]$. In this circumstance, if P_T is determined by the random element $\mathbf{X}_T$ and P is determined by the random element $\mathbf{X}$, we shall also say that the sequence $\mathbf{X}_T$, $T = 1, 2, \ldots$ **converges in distribution** to $\mathbf{X}$.

The random function $\mathbf{F}_{XX}^{(T)}(\lambda)$, $0 \leqslant \lambda \leqslant \pi$, clearly lies in $D_C^{r\times r}[0,\pi]$ as does the function $\sqrt{T}[\mathbf{F}_{XX}^{(T)}(\lambda) - \mathbf{F}_{XX}(\lambda)]$. We may now state

Theorem 7.6.3 Let $\mathbf{X}(t)$, $t = 0, \pm 1, \ldots$ be an r vector-valued series satisfying Assumption 2.6.2(1). Let $\mathbf{F}_{XX}^{(T)}(\lambda)$ be given by (7.6.8). Then the sequence of processes $\{\sqrt{T}[\mathbf{F}_{XX}^{(T)}(\lambda) - \mathbf{F}_{XX}(\lambda)];\ 0 \leqslant \lambda \leqslant \pi\}$ converges in distribution to an $r \times r$ matrix-valued Gaussian process $\{\mathbf{Y}(\lambda);\ 0 \leqslant \lambda \leqslant \pi\}$ with mean $\mathbf{0}$ and

$$\begin{aligned}\operatorname{cov}\{Y_{a_1b_1}(\lambda), Y_{a_2b_2}(\mu)\} &= 2\pi \int_0^{\min(\lambda,\mu)} f_{a_1a_2}(\alpha) f_{b_1b_2}(-\alpha) d\alpha \\ &\quad + 2\pi \int_0^{\lambda} \int_0^{\mu} f_{a_1b_1a_2b_2}(\alpha, -\alpha, -\beta) d\alpha d\beta \end{aligned} \tag{7.6.20}$$

for $0 \leqslant \lambda, \mu \leqslant \pi$ and $a_1, a_2, b_1, b_2 = 1, \ldots, r$.

We may use the results of Chapter 4 in Cramér and Leadbetter (1967) to see that the sample paths of the limit process $\{\mathbf{Y}(\lambda);\ 0 \leqslant \lambda \leqslant \pi\}$ are continuous with probability 1. In the case that the series $\mathbf{X}(t)$, $t = 0, \pm 1, \ldots$ is Gaussian, the fourth-order spectra are identically 0 and the covariance function (7.6.20) is simplified. In this case, by setting $A_1(\alpha) = 1$ for $\mu_1 \leqslant \alpha \leqslant \lambda_1$ and $A_2(\alpha) = 1$ for $\mu_2 \leqslant \alpha \leqslant \lambda_2$ and both $= 0$ otherwise, we see from (7.6.7) that

$$\operatorname{cov}\{Y_{a_1b_1}(\lambda_1) - Y_{a_1b_1}(\mu_1), Y_{a_2b_2}(\lambda_2) - Y_{a_2b_2}(\mu_2)\} = 0 \quad \text{for } \mu_1 \leqslant \lambda_1 \leqslant \mu_2 \leqslant \lambda_2. \tag{7.6.21}$$

That is, the limiting Gaussian process has independent increments.

A key implication of Theorem 7.6.3 is that if h is a function on $D_C^{r\times r}[0,\pi]$ whose set of discontinuities has probability 0 with respect to the process $\{\mathbf{Y}(\lambda); 0 \leqslant \lambda \leqslant \pi\}$, then $h(\sqrt{T}[\mathbf{F}_{XX}^{(T)}(\cdot) - \mathbf{F}_{XX}(\cdot)])$ converges in distribution to $h(\mathbf{Y}(\cdot))$; see in Billingsley (1968) p. 31. The metric for $D_C^{r\times r}[0,\pi]$ used above is often not convenient. Luckily, as the limit process of the theorem is continuous, a result of M. L. Straf applies to indicate that if h is continuous in the norm

$$\sum_{a,b} \sup_{\lambda} |Y_{ab}(\lambda)| \tag{7.6.22}$$

and the $h(\sqrt{T}[\mathbf{F}_{XX}^{(T)}(\cdot) - \mathbf{F}_{XX}(\cdot)])$ are (measurable) random variables, then h converges in distribution to $h(\mathbf{Y}(\cdot))$. For example this implies that

$$\sqrt{T} \sup_{0 \leq \lambda \leq \pi} |F_{aa}^{(T)}(\lambda) - F_{aa}(\lambda)| \tag{7.6.23}$$

tends in distribution to

$$\sup_{0 \leq \lambda \leq \pi} |Y_{aa}(\lambda)| \tag{7.6.24}$$

where $Y_{aa}(\lambda)$ is a Gaussian process with 0 mean and

$$\begin{aligned}\operatorname{cov}\{Y_{aa}(\lambda), Y_{aa}(\mu)\} &= 2\pi \int_0^{\min(\lambda,\mu)} f_{aa}(\alpha)^2 d\alpha \\ &+ 2\pi \int_0^{\lambda} \int_0^{\mu} f_{aaaa}(\alpha,-\alpha,-\beta) d\alpha d\beta \\ &\text{for } 0 \leqslant \lambda, \mu \leqslant \pi,\ a = 1, \ldots, r.\end{aligned} \tag{7.6.25}$$

The estimate considered in the theorem has the disadvantage of being discontinuous even though the corresponding population parameter is continuous and indeed differentiable. A continuous estimate is provided by

$$\int_0^{\lambda} \mathbf{I}_{X-c_X^{(T)}X-c_X^{(T)}}^{(T)}(\alpha) d\alpha. \tag{7.6.26}$$

It may be shown that the process

$$\sqrt{T}\left\{\int_0^{\lambda} \mathbf{I}_{X-c_X^{(T)}X-c_X^{(T)}}^{(T)}(\alpha) d\alpha - \mathbf{F}_{XX}(\alpha)\right\} \tag{7.6.27}$$

converges in distribution to a 0 mean Gaussian process with covariance function (7.6.20).

If $r = 1$ and the series $X(t)$, $t = 0, \pm 1, \ldots$ is a 0 mean linear process, then Grenander and Rosenblatt (1957) demonstrated the weak convergence of the process

$$\sqrt{T}\left\{\int_0^{\lambda} I_{XX}^{(T)}(\alpha) d\alpha - F_{XX}(\lambda)\right\}. \tag{7.6.28}$$

They also considered the weak convergence of the process

$$\sqrt{T}\left\{\int_0^\lambda f_{XX}^{(T)}(\alpha)d\alpha - F_{XX}(\lambda)\right\} \tag{7.6.29}$$

where $f_{XX}^{(T)}(\lambda)$ is an estimate of the spectral density involving a weight function. The case of a 0 mean Gaussian process with square integrable spectral density was considered by Ibragimov (1963) and Malevich (1964, 1965). MacNeil (1971) considered the case of a 0 mean bivariate Gaussian process. Brillinger (1969c) considers the case of a 0 mean r vector-valued process satisfying Assumption 2.6.2(1) and shows convergence in a finer topology. Clevenson (1970) considered the weak convegence of the discontinuous process of the theorem in the case of a 0 mean Gaussian series.

7.7 FURTHER CONSIDERATIONS IN THE ESTIMATION OF SECOND-ORDER SPECTRA

We begin this section by developing the asymptotic distribution of a consistent estimate of the spectral density matrix based on tapered data. Suppose that we wish to estimate the spectral density matrix, $\mathbf{f}_{XX}(\lambda)$, of an r vector-valued series $\mathbf{X}(t)$, $t = 0, \pm 1, \ldots$ with mean function $\mathbf{c}_X$. Let $h_a(u)$, $-\infty < u < \infty$, denote a tapering function satisfying Assumption 4.3.1 for $a = 1, \ldots, r$. Suppose that the tapered values $h_a(t/T)X_a(t)$, $t = 0, \pm 1, \ldots$ are available for analysis. Suppose the mean function is estimated by

$$c_a^{(T)} = \frac{\sum_t h_a\left(\frac{t}{T}\right)X_a(t)}{\sum_t h_a\left(\frac{t}{T}\right)} \qquad \text{for } a = 1, \ldots, r. \tag{7.7.1}$$

Let

$$d_a^{(T)}(\lambda) = \sum_t h_a\left(\frac{t}{T}\right)X_a(t) \exp\{-i\lambda t\} \qquad \text{for } -\infty < \lambda < \infty,\ a = 1, \ldots, r. \tag{7.7.2}$$

We will base our estimate of $\mathbf{f}_{XX}(\lambda)$ on the Fourier transforms of mean-adjusted tapered values, specifically on

$$\begin{aligned} d_{X_a - c_a^{(T)}}^{(T)}(\lambda) &= \sum_t h_a\left(\frac{t}{T}\right)[X_a(t) - c_a^{(T)}] \exp\{-i\lambda t\} \\ &= d_a^{(T)}(\lambda) - \frac{d_a^{(T)}(0)H_a^{(T)}(\lambda)}{H_a^{(T)}(0)} \end{aligned} \tag{7.7.3}$$

where

$$H_a^{(T)}(\lambda) = \sum_t h_a\left(\frac{t}{T}\right) \exp\{-i\lambda t\} \qquad \text{for } a = 1, \ldots, r. \tag{7.7.4}$$

Following the discussion of Section 7.2, we next form the second-order periodograms

$$I_{X_a-c_a^{(T)}, X_b-c_b^{(T)}}^{(T)}(\lambda) = \{2\pi H_{ab}^{(T)}(0)\}^{-1} d_{X_a-c_a^{(T)}}^{(T)}(\lambda)\, \overline{d_{X_b-c_b^{(T)}}(\lambda)} \quad (7.7.5)$$

where

$$H_{ab}^{(T)}(\lambda) = \sum_t h_a\left(\frac{t}{T}\right)h_b\left(\frac{t}{T}\right) \exp\{-i\lambda t\} \qquad \text{for } a, b = 1, \ldots, r. \quad (7.7.6)$$

From (7.7.3) we see that expression (7.7.5) may be written

$$(2\pi)^{-1} \sum_u \exp\{-i\lambda u\} c_{ab}^{(T)}(u) \quad (7.7.7)$$

where

$$c_{ab}^{(T)}(u) = H_{ab}^{(T)}(0)^{-1} \sum_t h_a\left(\frac{t+u}{T}\right) h_b\left(\frac{t}{T}\right) [X_a(t+u) - c_a^{(T)}][X_b(t) - c_b^{(T)}] \quad (7.7.8)$$

is an estimate of the cross-covariance function $c_{ab}(u)$.

Suppose $W_{ab}(\alpha)$, $-\infty < \alpha < \infty$, $a, b = 1, \ldots, r$ are weight functions satisfying $\int W_{ab}(\alpha)d\alpha = 1$. In this present case involving arbitrary tapering functions, no particular advantage accrues from a smoothing of the periodogram values at the particular frequencies $2\pi s/T$, $s = 0, \pm 1, \ldots$. For this reason we consider the following estimate involving a continuous weighting

$$f_{ab}^{(T)}(\lambda) = \int_{-\infty}^{\infty} B_T^{-1} W_{ab}(B_T^{-1}[\lambda - \alpha]) I_{X_a-c_a^{(T)}, X_b-c_b^{(T)}}^{(T)}(\alpha)d\alpha \qquad \text{for } a, b = 1, \ldots, r \quad (7.7.9)$$

where the values B_T, $T = 1, 2, \ldots$ are positive and bounded. Using expression (7.7.7) we see that (7.7.9) may be written

$$f_{ab}^{(T)}(\lambda) = (2\pi)^{-1} \sum_u w_{ab}(B_T u) c_{ab}^{(T)}(u) \exp\{-i\lambda u\} \quad (7.7.10)$$

where

$$w_{ab}(u) = \int_{-\infty}^{\infty} W_{ab}(\alpha) \exp\{iu\alpha\} d\alpha \qquad \text{for } -\infty < u < \infty. \quad (7.7.11)$$

We will require this function to satisfy the following:

Assumption 7.7.1 The function $w(u)$, $-\infty < u < \infty$, is real-valued, bounded, symmetric, $w(0) = 1$, and such that

$$\int |w(u)|du, \int |u|\, |w(u)|du < \infty. \quad (7.7.12)$$

Exercise 3.10.7 shows the estimate (7.7.10) may be computed using a Fast Fourier Transform. We have

Theorem 7.7.1 Let $X(t)$, $t = 0, \pm 1, \ldots$ be an r vector-valued series satisfying Assumption 2.6.2(1). Let $h_a(u)$, $-\infty < u < \infty$, satisfy Assumption 4.3.1 for $a = 1, \ldots, r$ and be such that $\int h_a(u)h_b(u)du \neq 0$. Let $w_{ab}(u)$, $-\infty < u < \infty$, satisfy Assumption 7.7.1 for $a, b = 1, \ldots, r$. Let $B_T T \to \infty$ as $T \to \infty$. Then

$$\begin{aligned} Ef_{ab}^{(T)}(\lambda) &= \int W_{ab}(\alpha) f_{ab}(\lambda - B_T\alpha)d\alpha + O(B_T^{-1}T^{-1}) \\ &= (2\pi)^{-1} \sum_u w_{ab}(B_T u)c_{ab}(u) \exp\{-i\lambda u\} + O(B_T^{-1}T^{-1}). \end{aligned} \tag{7.7.13}$$

Also

$$\begin{aligned} \lim_{T\to\infty} B_T T \operatorname{cov}\{f_{a_1b_1}^{(T)}(\lambda), f_{a_2b_2}^{(T)}(\mu)\} \\ = 2\pi\{\textstyle\int h_{a_1}(t)h_{b_1}(t)dt\}^{-1}\{\int h_{a_2}(t)h_{b_2}(t)dt\}^{-1} \\ \times \textstyle\int h_{a_1}(t)h_{a_2}(t)h_{b_1}(t)h_{b_2}(t)dt[\eta\{\lambda - \mu\} f_{a_1a_2}(\lambda)f_{b_1b_2}(-\lambda) \\ + \eta\{\lambda + \mu\} f_{a_1b_2}(\lambda)f_{b_1a_2}(-\lambda)] \textstyle\int W_{a_1b_1}(\alpha)W_{a_2b_2}(\alpha)d\alpha. \end{aligned} \tag{7.7.14}$$

Finally, the variates $f_{a_1b_1}^{(T)}(\lambda_1), \ldots, f_{a_Kb_K}^{(T)}(\lambda_K)$ are asymptotically normal with the above covariance structure.

A comparison of expressions (7.7.14) and (7.4.17) shows that, asymptotically, the effect of tapering is to multiply the limiting variance by a factor

$$\frac{\int h_{a_1}(t)h_{a_2}(t)h_{b_1}(t)h_{b_2}(t)dt}{(\int h_{a_1}(t)h_{b_1}(t)dt)(\int h_{a_2}(t)h_{b_2}(t)dt)}. \tag{7.7.15}$$

This factor equals 1 in the case of no tapering, that is, $h_a(t) = 1$ for $0 \leqslant t < 1$ and $= 0$ for other t. In the case that the same tapering function is used for all series, that is, $h_a(t) = h(t)$ for $a = 1, \ldots, r$, the factor becomes

$$\frac{\int h(t)^4 dt}{\{\int h(t)^2 dt\}^2}. \tag{7.7.16}$$

Following Schwarz's inequality this is $\geqslant 1$ and so the limiting variance is increased by tapering. However, the hope is that there has been a sufficient reduction in bias to compensate for any increase in variance. We also have

Corollary 7.7.1 Under the conditions of Theorem 7.7.1, and if $B_T \to 0$ as $T \to \infty$, the estimate is asymptotically unbiased.

Historically, the first cross-spectral estimate widely considered had the form (7.7.10) (see Goodman (1957) and Rosenblatt (1959)), although tapering was not generally employed. Its asymptotic properties are seen to be essentially the same as those of the estimate of Section 7.4. It is investigated in Akaike and Yamanouchi (1962), Jenkins (1963a), Murthy (1963), and Granger (1964). Freiberger (1963) considers approximations to its distribution in the case that the series is bivariate Gaussian.

The discussion of Section 7.1 suggests an alternate class of estimates of the second-order spectrum $f_{ab}(\lambda)$. Let $Y_a(t)$ denote the series resulting from band-pass filtering the series $X_a(t)$ with a filter having transfer function $A(\alpha) = 1$ for $|\alpha \pm \lambda| < \Delta$, and $= 0$ otherwise, $-\pi < \alpha, \lambda \leqslant \pi$. Consider estimating $\operatorname{Re} f_{ab}(\lambda)$ by

$$(4\Delta T)^{-1} \sum_{t=0}^{T-1} Y_a(t) Y_b(t) \quad \text{or} \quad (4\Delta T)^{-1} \sum_{t=0}^{T-1} Y_a(t)^H Y_b(t)^H \tag{7.7.17}$$

or the average of the two. Consider estimating $\operatorname{Im} f_{ab}(\lambda)$ by

$$(4\Delta T)^{-1} \sum_{t=0}^{T-1} Y_a(t)^H Y_b(t) \quad \text{or} \quad -(4\Delta T)^{-1} \sum_{t=0}^{T-1} Y_a(t) Y_b(t)^H \tag{7.7.18}$$

or the average of the two. These estimation procedures have the advantage of allowing us to investigate whether or not the structure of the series is slowly evolving in time; see Brillinger and Hatanaka (1969). This type of estimate was suggested in Blanc-Lapierre and Fortet (1953). One useful means of forming the required series is through the technique of complex demodulation; see Section 2.7 and Brillinger (1964b).

Brillinger (1968) was concerned with estimating the cross-spectrum of a 0 mean bivariate Gaussian series from the values $\operatorname{sgn} X_1(t)$, $\operatorname{sgn} X_2(t)$, $t = 0, \ldots, T-1$. The asymptotic distribution of the estimate (7.7.10), without tapering, was derived.

On some occasions we may wish a measure of the extent to which $f_{ab}^{(T)}(\lambda)$ may deviate from its expected value simultaneously as a function of λ and T. We begin by examining the behavior of the second-order periodograms. Theorem 4.5.1 indicated that for a 0 mean series and under regularity conditions

$$\overline{\lim_{T\to\infty}} \sup_{\lambda} |d_a^{(T)}(\lambda)| / (T \log T)^{1/2} \leqslant 2\left(2\pi \int h_a(t)^2 dt \sup_{\lambda} f_{aa}(\lambda)\right)^{1/2} \tag{7.7.19}$$

with probability 1. This gives us directly

Theorem 7.7.2 Let $\mathbf{X}(t)$, $t = 0, \pm 1, \ldots$ be an r vector-valued series satisfying Assumption 2.6.3 and having mean $\mathbf{0}$. Let $h_a(u)$, $-\infty < u < \infty$, satisfy Assumption 4.3.1, $a = 1, \ldots, r$. Let $\mathbf{I}_{XX}^{(T)}(\lambda)$ be given by (7.2.5). Then

$$\overline{\lim_{T\to\infty}} \sup_{\lambda} |I_{ab}^{(T)}(\lambda)| / \log T \leqslant 4\left\{\left|\int h_a(t) h_b(t) dt\right|\right\}^{-1} \left\{\int h_a(t)^2 dt \int h_b(t)^2 dt\right\}^{1/2} \times \left\{\sup_{\lambda} f_{aa}(\lambda) \sup_{\lambda} f_{bb}(\lambda)\right\}^{1/2} \tag{7.7.20}$$

with probability 1 for $a, b = 1, \ldots, r$.

Whittle (1959) determined a bound for the second-order periodogram that held in probability; see also Walker (1965). Parthasarathy (1960) found a probability 1 bound for the case of a single periodogram ordinate; he found that a single ordinate could grow at the rate log log T, rather than the log T of (7.7.20). Before turning to an investigation of the behavior of $f_{ab}^{(T)}(\lambda) - Ef_{ab}^{(T)}(\lambda)$ we set down a further assumption of the character of Assumption 2.6.3 concerning the series $\mathbf{X}(t)$, $t = 0, \pm 1, \ldots$.

Assumption 7.7.2 $\mathbf{X}(t)$, $t = 0, \pm 1, \ldots$ is an r vector-valued series satisfying Assumption 2.6.1. Also, with C_n given by (2.6.7),

$$\sum_{L=1}^{\infty} \Big(\sum_{\boldsymbol{\nu}} C_{n_1} \cdots C_{n_P}\Big) z^L / L! < \infty \tag{7.7.21}$$

for z in a neighborhood of 0. In (7.7.21) the inner summation is over all indecomposable partitions $\boldsymbol{\nu} = (\nu_1, \ldots, \nu_P)$ of the table

$$\begin{matrix} 1 & 2 \\ 3 & 4 \\ \cdot & \cdot \\ \cdot & \cdot \\ \cdot & \cdot \\ 2L-1 & 2L \end{matrix} \tag{7.7.22}$$

with ν_p having $n_p > 1$ elements, $p = 1, \ldots, P$.

In the case of a Gaussian series, $C_n = 0$ for $n > 2$ and the series of (7.7.21) becomes

$$\sum_{L=1}^{\infty} 2^{L-1}(L-1)!\, C_2^L z^L / L! \tag{7.7.23}$$

This is finite for $2C_2|z| < 1$ and so Assumption 7.7.1 is satisfied in this case so long as Assumption 2.6.1 is satisfied. We may now set down

Theorem 7.7.3 $\mathbf{X}(t)$, $t = 0, \pm 1, \ldots$ is an r vector-valued series satisfying Assumption 7.7.2 $h_a(u)$, $-\infty < u < \infty$, satisfies Assumption 4.3.1 for $a = 1, \ldots, r$. The $w_{ab}(u)$, $-\infty < u < \infty$, satisfy Assumption 7.7.1 and vanish for $|u|$ sufficiently large, $a, b = 1, \ldots, r$. $f_{ab}^{(T)}(\lambda)$ is given by (7.7.10). Let $\eta > 0$ be given and such that $\sum_T B_T^\eta < \infty$. Then

$$\overline{\lim_{T \to \infty}} \sup_{\lambda} |f_{ab}^{(T)}(\lambda) - Ef_{ab}^{(T)}(\lambda)| (B_T T / \log 1/B_T)^{1/2}$$
$$\leq \Big[(1+\eta) 8\pi \int W_{ab}(\alpha)^2 d\alpha \Big\{\int h_a(t) h_b(t) dt\Big\}^{-2} \times \int h_a(t)^2 h_b(t)^2 dt \sup_{\lambda} f_{aa}(\lambda) \sup_{\lambda} f_{bb}(\lambda)\Big]^{1/2} \tag{7.7.24}$$

with probability 1 for $a, b = 1, \ldots, r$.

If $\Sigma\ |u|\ |c_{ab}(u)| < \infty$, and $\int |\alpha||W_{ab}(\alpha)|d\alpha < \infty$, then Theorem 3.3.1 and expression (7.7.13) show that

$$Ef_{ab}{}^{(T)}(\lambda) = f_{ab}(\lambda) + O(B_T) + O(B_T{}^{-1}T^{-1}) \tag{7.7.25}$$

and so we can say

$$f_{ab}{}^{(T)}(\lambda) = f_{ab}(\lambda) + O(B_T) + O([B_TT/\log 1/B_T]^{-1/2}) \tag{7.7.26}$$

with probability 1, the error terms being uniform in λ.

We see that in the case of Theorem 7.7.3, $f_{ab}{}^{(T)}(\lambda)$ is a strongly consistent estimate of $f_{ab}(\lambda)$. Woodroofe and Van Ness (1967) showed, under regularity conditions including $X(t)$ being a linear process, that

$$\overline{\lim_{T\to\infty}}\ (B_TT/\log 1/B_T)^{1/2} \sup_{\lambda} |f_{aa}{}^{(T)}(\lambda) - f_{aa}(\lambda)|/f_{aa}(\lambda) = [4\pi \int W(\alpha)^2 d\alpha]^{1/2} \tag{7.7.27}$$

in probability. The data is not tapered here. They also investigated the limiting distribution of the maximum deviation.

The following cruder result may be developed under the weaker Assumption 2.6.1:

Theorem 7.7.4 Let $\mathbf{X}(t)$, $t = 0, \pm 1, \ldots$ be an r vector-valued series satisfying Assumption 2.6.1. Let $h_a(u)$, $-\infty < u < \infty$, satisfy Assumption 4.3.1. Let $w_{ab}(u)$, $-\infty < u < \infty$ satisfy Assumption 7.7.1 and vanish for $|u|$ sufficiently large, $a, b = 1, \ldots, r$. Let $f_{ab}{}^{(T)}(\lambda)$ be given by (7.7.10). Let $B_TT \to \infty$, $B_T \to 0$ as $T \to \infty$. Then for any $\varepsilon > 0$,

$$(B_TT)^{1/2}B_T^{\epsilon} \sup_{\lambda} |f_{ab}{}^{(T)}(\lambda) - Ef_{ab}{}^{(T)}(\lambda)| \to 0 \tag{7.7.28}$$

in probability as $T \to \infty$. If, in addition, $\sum_T B_T^m < \infty$ for some $m > 0$, then the event (7.7.28) occurs with probability 1 as $T \to \infty$.

In Theorem 7.7.4 the multiplier $(B_TT/\log 1/B_T)^{1/2}$ of (7.7.27) has become replaced by the smaller $(B_TT)^{1/2}B_T^{\epsilon}$.

If we wish to use the estimate (7.4.5) and are content with a result concerning the maximum over a discrete set of points, we have

Theorem 7.7.5 Let $\mathbf{X}(t)$, $t = 0, \pm 1, \ldots$ be an r vector-valued series satisfying Assumption 2.6.1. Let $W_{ab}(\alpha)$, $-\infty < \alpha < \infty$, satisfy Assumption 5.6.1. Let $f_{ab}{}^{(T)}(\lambda)$ be given by (7.4.5). Let $B_T \to 0$, P_T, $B_TT \to \infty$ as $T \to \infty$. Then for any $\varepsilon > 0$

$$(B_TT)^{1/2}P_T^{-\epsilon} \sup_{p=0}^{P_T-1} \left| f_{ab}{}^{(T)}\left(\frac{2\pi p}{P_T}\right) - Ef_{ab}{}^{(T)}\left(\frac{2\pi p}{P_T}\right)\right| \to 0 \tag{7.7.29}$$

in probability as $T \to \infty$. If, in addition, $\sum_T P_T^{-m} < \infty$, for some $m > 0$, then the event (7.7.29) occurs with probability 1 as $T \to \infty$.

In Section 5.8 we discussed the importance of prefiltering a stretch of data prior to forming a spectral estimate. Expression (7.7.13) again makes this clear. The expected value of $f_{ab}^{(T)}(\lambda)$ is not generally $f_{ab}(\lambda)$, rather it is a weighted average of $f_{ab}(\alpha)$, $-\infty < \alpha < \infty$, with weight concentrated in the neighborhood of λ. If $f_{ab}(\alpha)$ has any substantial peaks or valleys, the weighted average could be far from $f_{ab}(\lambda)$. In practice it appears to be the case that cross-spectra vary more substantially than power spectra. Consider a commonly occurring situation in which a series $X_2(t)$ is essentially a delayed version of a series $X_1(t)$, for example

$$X_2(t) = \alpha X_1(t - v) + \varepsilon(t) \tag{7.7.30}$$

$t = 0, \pm 1, \ldots$ for constants α, v and $\varepsilon(t)$ an error series orthogonal to the series $X_1(t)$. Then the cross-spectrum is given by

$$\begin{aligned} f_{21}(\lambda) &= \exp\{-i\lambda v\}\, \alpha f_{11}(\lambda) \\ &= \cos\{\lambda v\}\, \alpha f_{11}(\lambda) - i \sin\{\lambda v\}\, \alpha f_{11}(\lambda). \end{aligned} \tag{7.7.31}$$

Here, if v has any appreciable magnitude at all, the function $f_{21}(\lambda)$ will be rapidly altering in sign as λ varies. Any weighted average of it, such as (7.7.13) will be near 0. We could well be led to conclude that there was no relation between the series, when in fact there was a strong linear relation. Akaike (1962) has suggested that a situation of this character be handled by delaying the series $X_1(t)$ by approximately v time units. That is, we analyze the series $[X_1(t - v^*), X_2(t)]$, $t = 0, \pm 1, \ldots$ with v^* near v, instead of the original stretch of series. This is a form of prefiltering. Akaike suggests that in practice one might determine v^* as the lag where $|c_{21}^{(T)}(u)|$ is greatest. If the estimated delay is anywhere near v at all, the cross-spectrum being estimated now should be a much less rapidly varying function.

In Section 5.8 it was suggested that a prewhitening filter be determined by fitting an autoregressive model to a time series of interest. Nettheim (1966) has suggested an analagous procedure in the estimation of a cross-spectrum. We fit a model such as

$$X_2(t) \doteq a(m)X_1(t - m) + \cdots + a(0)X_1(t) + \cdots + a(-n)X_1(t + n) \tag{7.7.32}$$

by least squares and estimate the cross-spectrum of the residuals with $X_1(t)$. In the full r vector-valued situation we could determine r vectors $\mathbf{a}^{(T)}(1), \ldots, \mathbf{a}^{(T)}(m)$ to minimize

$$\begin{aligned} \sum_{t=m}^{T-1} &[\mathbf{X}(t) - \mathbf{a}^{(T)}(1)\mathbf{X}(t-1) - \cdots - \mathbf{a}^{(T)}(m)\mathbf{X}(t-m)]^{\tau} \\ &\times [\mathbf{X}(t) - \mathbf{a}^{(T)}(1)\mathbf{X}(t-1) - \cdots - \mathbf{a}^{(T)}(m)\mathbf{X}(t-m)]. \end{aligned} \tag{7.7.33}$$

We then form $\mathbf{f}_{\varepsilon\varepsilon}^{(T)}(\lambda)$, a spectral estimate based on the residuals

$$\hat{\varepsilon}(t) = \mathbf{X}(t) - \mathbf{a}^{(T)}(1)\mathbf{X}(t-1) - \cdots - \mathbf{a}^{(T)}(m)\mathbf{X}(t-m) \qquad t = m, \ldots, T-1 \quad (7.7.34)$$

and then estimate $\mathbf{f}_{XX}(\lambda)$ by

$$\mathbf{A}^{(T)}(\lambda)^{-1}\mathbf{f}_{\hat{\varepsilon}\hat{\varepsilon}}^{(T)}(\lambda)(\overline{\mathbf{A}^{(T)}(\lambda)^{-1}})^{\tau} \quad (7.7.35)$$

where

$$\mathbf{A}^{(T)}(\lambda) = \mathbf{I} - \mathbf{a}^{(T)}(1)\exp\{-i\lambda\} - \cdots - \mathbf{a}^{(T)}(m)\exp\{-i\lambda m\} \qquad \text{for } -\infty < \lambda < \infty. \quad (7.7.36)$$

Generally it is wise to use prior knowledge to suggest a statistical model for a series of interest, to fit the model, and then to compute a spectral estimate based on the residuals.

Nothing much remains to be said about the complication of **aliasing** after the discussion of Section 5.11. We simply note that the population parameter $\mathbf{f}_{XX}(\lambda)$ and its estimates both possess the periodicity and symmetry properties

$$\mathbf{f}_{XX}(\lambda + 2\pi) = \mathbf{f}_{XX}(\lambda); \qquad \mathbf{f}_{XX}(-\lambda) = \mathbf{f}_{XX}(\lambda)^{\tau}$$
$$\mathbf{f}_{XX}^{(T)}(\lambda + 2\pi) = \mathbf{f}_{XX}^{(T)}(\lambda); \qquad \mathbf{f}_{XX}^{(T)}(-\lambda) = \mathbf{f}_{XX}^{(T)}(\lambda)^{\tau}. \quad (7.7.37)$$

It follows that the population parameter and corresponding estimate will be essentially the same for all the frequencies

$$\pm\lambda, \pm\lambda \pm 2\pi, \pm\lambda \pm 4\pi, \ldots . \quad (7.7.38)$$

If possible we should band-pass filter the series prior to digitization in order to essentially eliminate any frequency components that might cause confusion in the interpretation of the spectral estimate.

7.8 A WORKED EXAMPLE

For an example of the estimate developed in Section 7.3 we return to the series considered in Section 7.2. There $X_1(t)$ was the seasonally adjusted series of mean monthly temperatures for Berlin (1780 to 1950) and $X_2(t)$ was the seasonally adjusted series of mean monthly temperatures for Vienna (1780 to 1950). The periodograms and cross-periodogram for this data were given in Figures 7.2.1 to 7.2.4.

Figures 7.8.1 to 7.8.4 of the present section give $f_{11}^{(T)}(\lambda)$, $f_{22}^{(T)}(\lambda)$, $\mathrm{Re} f_{12}^{(T)}(\lambda)$, $\mathrm{Im} f_{12}^{(T)}(\lambda)$ using estimates of the form (5.4.1) and (7.3.2) with $m = 10$. If we consider $\log_{10}$ power spectral estimates, expression (5.6.15) suggests that the standard errors are both approximately .095. It is interesting to contrast the forms of $\mathrm{Re}\, f_{12}^{(T)}(\lambda)$ and $\mathrm{Im}\, f_{12}^{(T)}(\lambda)$; $\mathrm{Re}\, f_{12}^{(T)}(\lambda)$ is

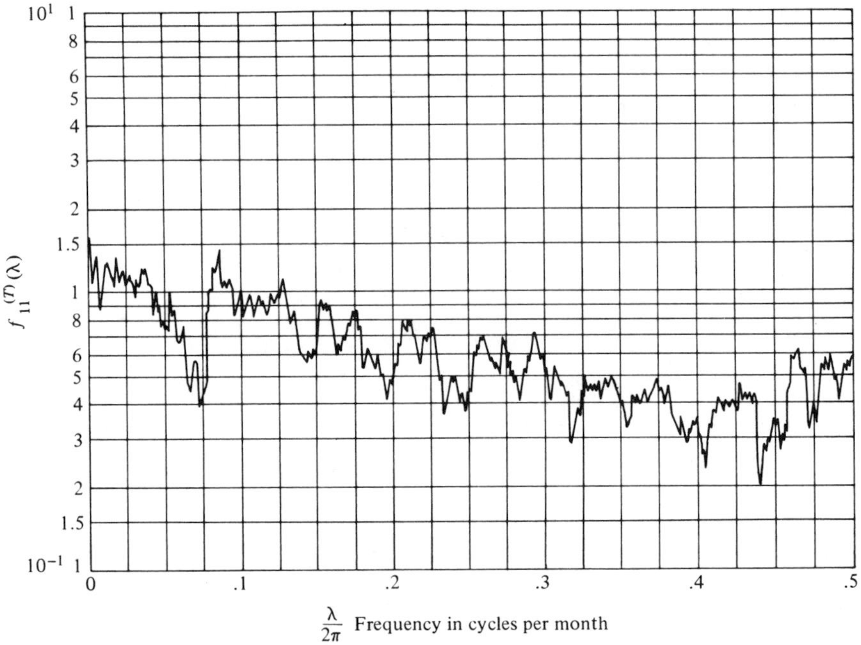

Figure 7.8.1 $f_{11}{}^{(T)}(\lambda)$ for seasonally adjusted monthly mean temperatures at Berlin for the years 1780–1950 with 21 periodogram ordinates averaged. (Logarithmic plot.)

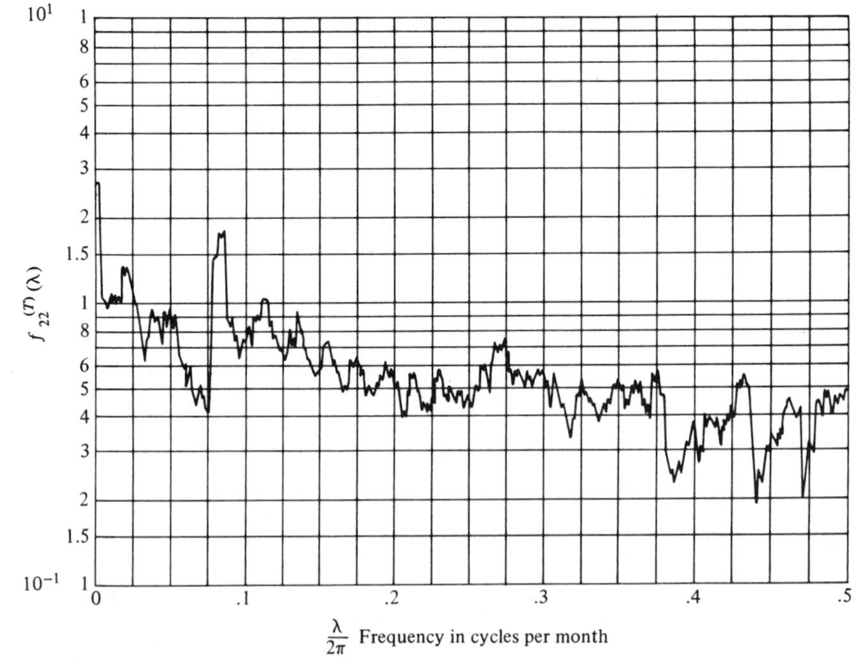

Figure 7.8.2 $f_{22}{}^{(T)}(\lambda)$ for seasonally adjusted monthly mean temperatures at Vienna for the years 1780–1950 with 21 periodogram ordinates averaged. (Logarithmic plot.)

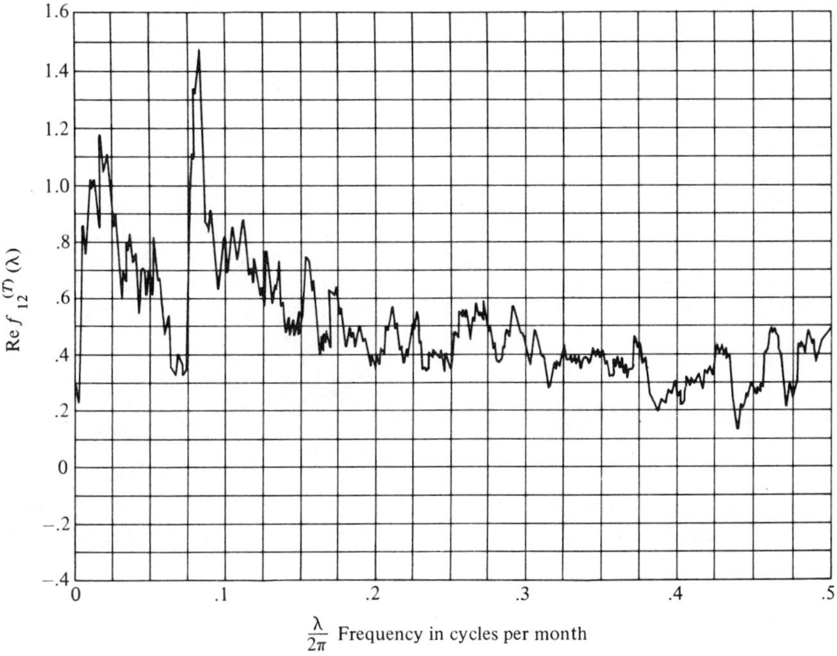

Figure 7.8.3 Re $f_{12}^{(T)}(\lambda)$, estimate of the cospectrum of Berlin and Vienna temperatures for the years 1780–1950 with 21 periodogram ordinates averaged.

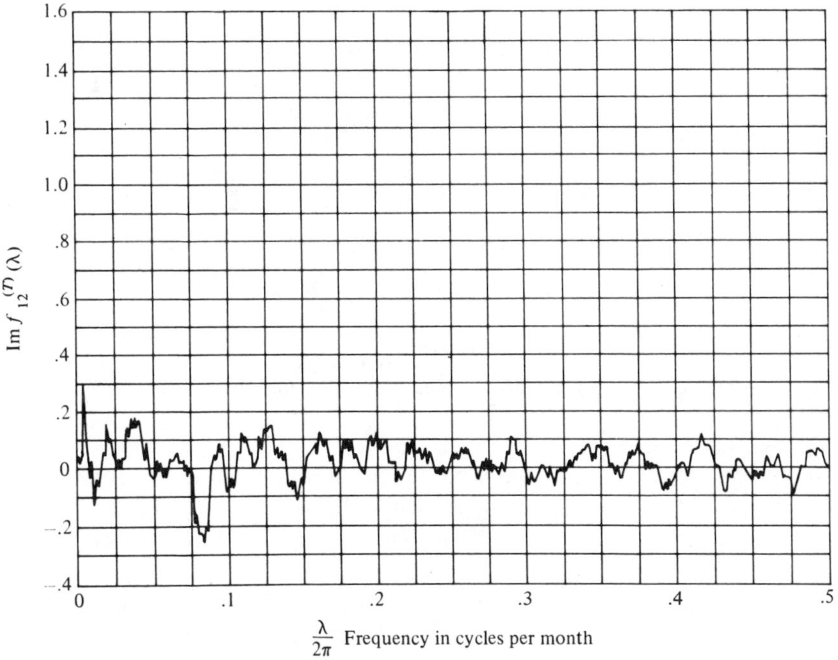

Figure 7.8.4 Im $f_{12}^{(T)}(\lambda)$, estimate of the quadspectrum of Berlin and Vienna temperatures for the years 1780–1950 with 21 periodogram ordinates averaged.

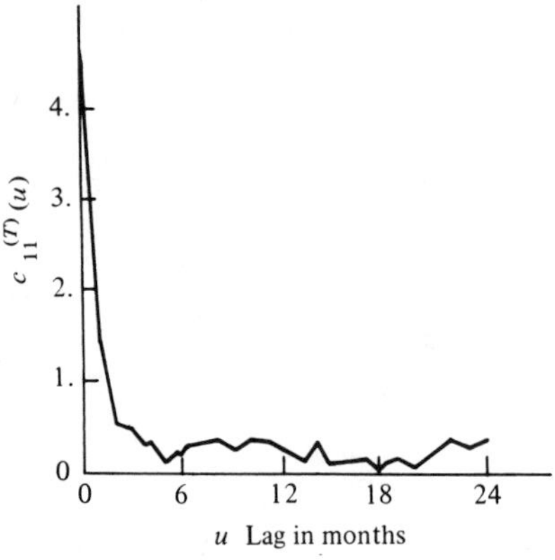

Figure 7.8.5 $c_{11}^{(T)}(u)$, estimate of the autocovariance function of Berlin temperatures.

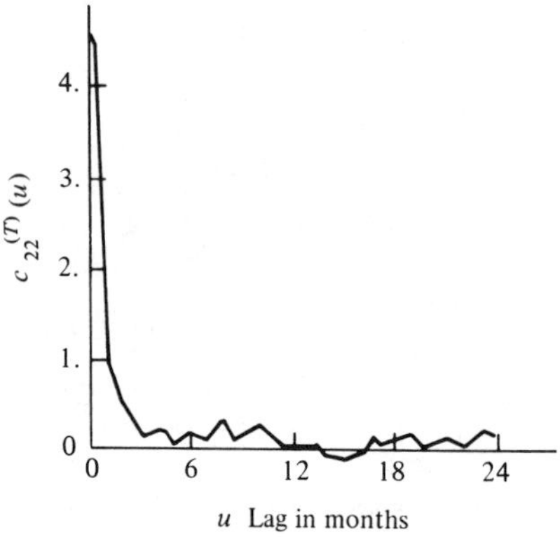

Figure 7.8.6 $c_{22}^{(T)}(u)$, estimate of the autocovariance function of Vienna temperatures.

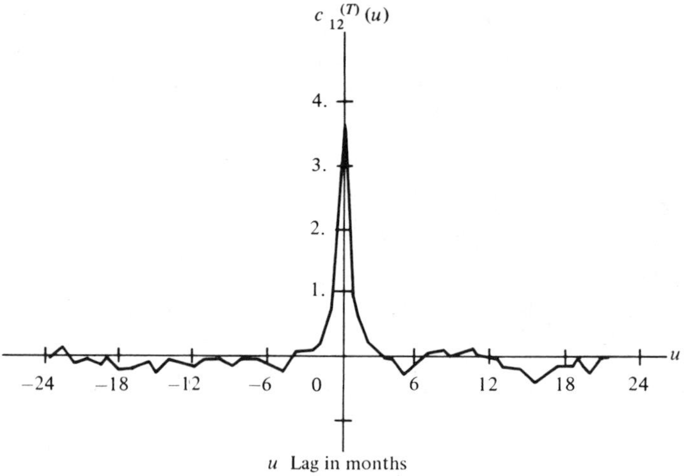

Figure 7.8.7 $c_{12}^{(T)}(u)$, estimate of the crosscovariance function of Berlin and Vienna temperatures.

everywhere positive, of appreciable magnitude at several frequencies and approximately constant otherwise, while $\text{Im } f_{12}^{(T)}(\lambda)$ simply fluctuates a little about the value 0 suggesting that $\text{Im } f_{12}(\lambda) = 0$. Other statistics for this example were given in Section 6.10.

For completeness we also give estimates of the auto- and cross-covariance functions of these two series. Figure 7.8.5 is an estimate of the autocovariance function of the series of Berlin mean monthly temperatures, with seasonal effects removed. Likewise Figure 7.8.6 is an estimate of the autocovariance function of the Vienna series. Figure 7.8.7 is the function $c_{12}^{(T)}(u)$ for $u = 0, \pm 1, \ldots$.

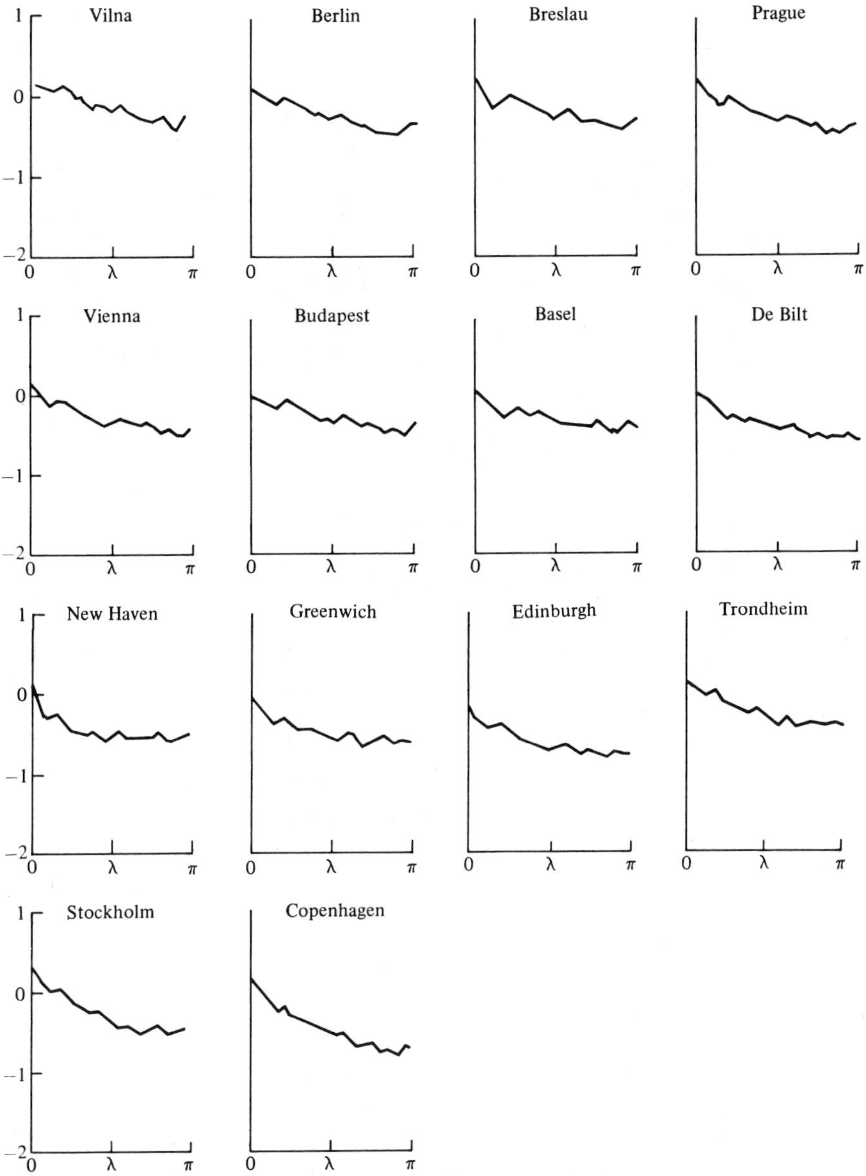

Figure 7.8.8 Logarithm of estimated power spectrum of seasonally adjusted monthly mean temperatures at various stations, with 115 periodogram ordinates averaged.

Table 7.8.1 Covariance Matrix of the Temperature Series

	Vienna	*Berlin*	*Copenhagen*	*Prague*	*Stockholm*	*Budapest*	*De Bilt*	*Edinburgh*	*New Haven*	*Basel*	*Breslau*	*Vilna*	*Trondheim*	*Greenwich*
Vienna	4.272													
Berlin	3.438	4.333												
Copenhagen	2.312	2.962	2.939											
Prague	3.986	3.756	2.635	6.030										
Stockholm	2.056	2.950	3.052	2.325	4.386									
Budapest	3.808	3.132	2.047	3.558	1.843	4.040								
De Bilt	2.665	3.209	2.315	2.960	2.170	2.261	3.073							
Edinburgh	.941	1.482	1.349	1.182	1.418	.627	1.509	2.050						
New Haven	.045	.288	.520	.076	.672	.009	.206	.404	2.939					
Basel	3.099	3.051	1.946	3.212	1.576	2.776	2.747	1.179	.178	3.694				
Breslau	3.868	4.227	2.868	4.100	2.805	3.646	3.053	1.139	.165	3.123	5.095			
Vilna	3.126	3.623	2.795	3.152	3.349	2.993	2.392	.712	.057	1.962	3.911	6.502		
Trondheim	1.230	2.165	2.358	1.496	3.312	.984	1.656	1.429	.594	.884	1.801	2.185	3.949	
Greenwich	1.805	2.255	1.658	2.005	1.570	1.450	2.300	1.564	.440	2.310	2.059	1.261	1.255	2.355

Table 7.8.2 Sample Correlation Matrix of the Seasonally Adjusted Series

	1	2	3	4	5	6	7	8	9	10	11	12	13
1 *Vienna*													
2 *Berlin*	.80												
3 *Copenhagen*	.65	.83											
4 *Prague*	.79	.73	.63										
5 *Stockholm*	.48	.68	.85	.45									
6 *Budapest*	.92	.75	.59	.72	.44								
7 *De Bilt*	.74	.88	.77	.69	.59	.64							
8 *Edinburgh*	.32	.50	.56	.34	.48	.22	.61						
9 *New Haven*	.01	.08	.18	.02	.19	.00	.07	.16					
10 *Basel*	.78	.76	.59	.68	.39	.72	.82	.43	.05				
11 *Breslau*	.83	.90	.74	.74	.59	.80	.77	.36	.04	.72			
12 *Vilna*	.59	.68	.64	.50	.63	.58	.54	.20	.01	.40	.68		
13 *Trondheim*	.30	.52	.69	.31	.80	.25	.48	.50	.17	.23	.40	.43	
14 *Greenwich*	.57	.71	.67	.53	.49	.47	.86	.72	.17	.78	.59	.32	.41
	1	2	3	4	5	6	7	8	9	10	11	12	13

All of these figures are consistent with a hypothesis of an instantaneous relation between the two series. (Instantaneous here means small time lead or lag relative to an interval of one month, because the data is monthly.)

As a full vector-valued example we consider the series of mean monthly temperatures recorded at the stations listed in Table 1.1.1. The series were initially seasonally adjusted by removing monthly means. Table 7.8.1 gives $\mathbf{c}_{XX}^{(T)}(0)$, the estimated 0 lag autocovariance matrix. Table 7.8.2 gives the 0 lag correlations of the series. Except for the New Haven series the series are seen to be quite intercorrelated.

The spectral density matrix was estimated through a statistic of the form (7.3.2) with $m = 57$. Because there are so many second-order spectra we do not present all the estimates. Figure 7.8.8 gives the $\log_{10}$ of the estimated power spectra. These are all seen to have essentially the same shape. Figure 7.8.9 gives the sample coherences, $|R_{1j}^{(T)}(\lambda)|^2$, taking $X_1(t)$ to be the Greenwich series and letting j run across the remaining series. The horizontal line in each of the diagrams corresponds to the 0 lag correlation squared. The plots are seen to be vaguely constant, fluctuating about the horizontal line in each case. This last is suggestive of instantaneous dependence of the series for if $c_{ab}(u) = 0$ for $u \neq 0$, then $|R_{ab}(\lambda)|^2 = |c_{ab}(0)|^2/|c_{aa}(0)c_{bb}(0)|$ for $-\infty < \lambda < \infty$. The correlation is seen to be greatest for the De Bilt series followed by Basel. The correlation is lowest for New Haven, Conn., on the opposite side of the Atlantic.

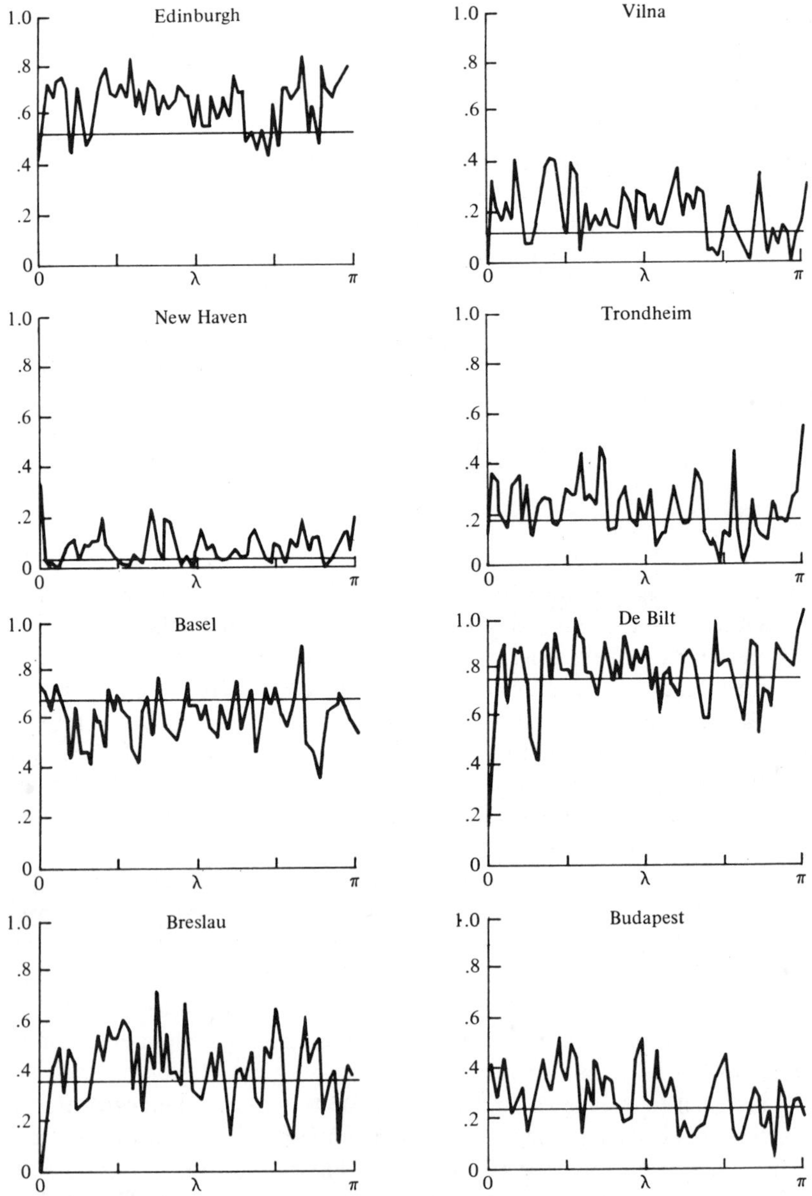

Figure 7.8.9 Estimated coherences of seasonally adjusted Greenwich monthly mean temperatures with similar temperatures at 13 other stations for the years 1780–1950.

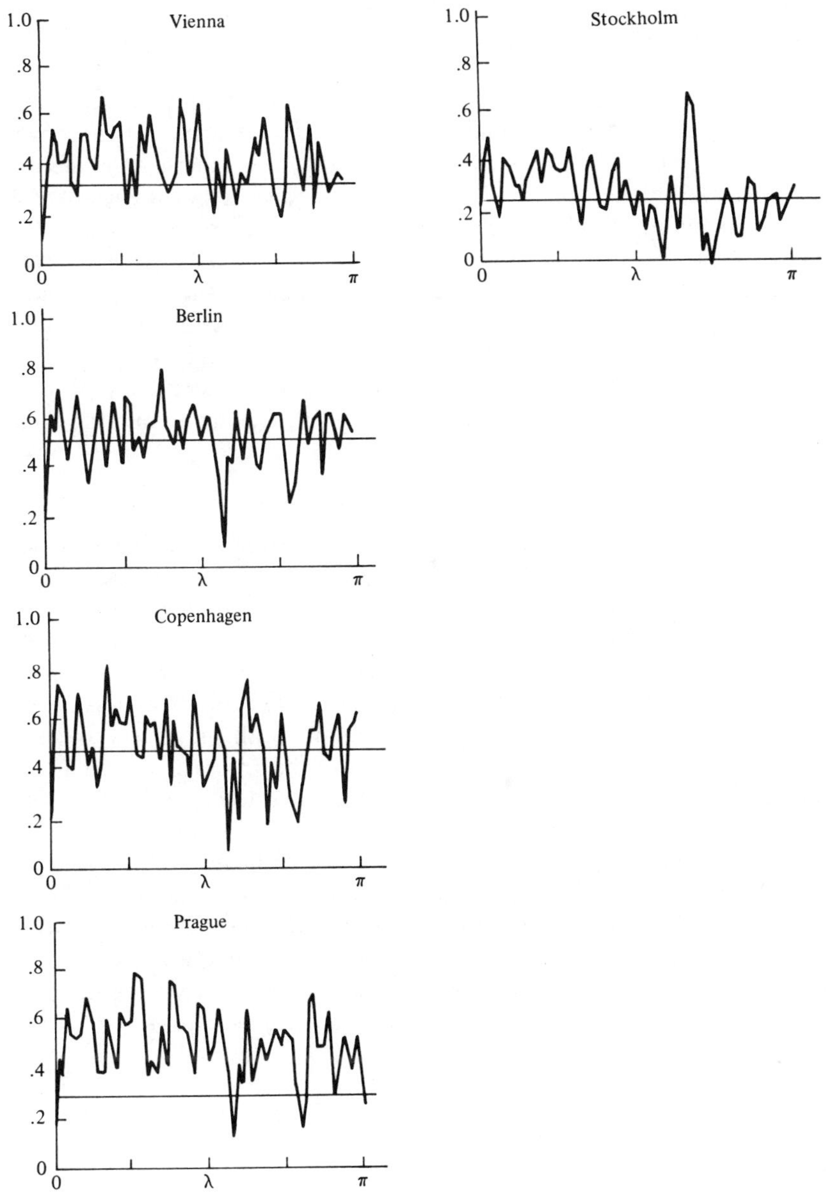

Figure 9.6.1 gives $\log_{10}$ of the estimated power spectra for an estimate of the form (7.3.2) with $m = 25$. These curves are more variable as to be expected from the sampling theory developed in this chapter.

7.9 THE ANALYSIS OF SERIES COLLECTED IN AN EXPERIMENTAL DESIGN

On occasion the subscripts $a = 1, \ldots, r$ of an r vector-valued series $\mathbf{X}(t) = [X_a(t)]$, $t = 0, \pm 1, \ldots$ may have an inherent structure of their own as in the case where the series have been collected in an experimental design. Consider for example the case of a **balanced one-way classification,** where K of the series fall into each of J classes. Here we would probably denote the series by $X_{jk}(t)$, $t = 0, \pm 1, \ldots$; $k = 1, \ldots, K$; $j = 1, \ldots, J$ with $r = JK$. Such series would arise if we were making up J batches of sheeting and drawing K pieces of sheeting from each batch. If we were interested in the uniformity of the sheeting we could let t refer to position from an origin along a cross-section of the sheeting and let $X_{jk}(t)$ denote the thickness at position t on sheet k selected from batch j. A model that might come to mind for this situation is

$$X_{jk}(t) = \mu + \alpha(t) + \beta_j(t) + \varepsilon_{jk}(t) \tag{7.9.1}$$

where μ is a constant; where $\alpha(t)$, $t = 0, \pm 1, \ldots$ is a 0 mean stationary series with power spectrum $f_{\alpha\alpha}(\lambda)$; where the series $\beta_j(t)$, $t = 0, \pm 1, \ldots$, $j = 1, \ldots, J$ are 0 mean stationary series each with power spectrum $f_{\beta\beta}(\lambda)$, $-\infty < \lambda < \infty$; and where the series $\varepsilon_{jk}(t)$, $t = 0, \pm 1, \ldots$; $k = 1, \ldots, K$; $j = 1, \ldots, J$ are 0 mean stationary series each with power spectrum $f_{\varepsilon\varepsilon}(\lambda)$, $-\infty < \lambda < \infty$. The parameter μ relates to the mean thickness of the sheeting. Series $\alpha(t)$, $t = 0, \pm 1, \ldots$ is common to all the sheets and the series $\beta_j(t)$, $t = 0, \pm 1, \ldots$ relates to the effect of the jth batch, if such an individual effect exists. It is common to all sheets selected from batch j. The series $\varepsilon_{jk}(t)$, $t = 0, \pm 1, \ldots$ is an error series. Taking note of the language of the random effects model of experimental design (Scheffé (1959)) we might call $f_{\alpha\alpha}(\lambda)$, $f_{\beta\beta}(\lambda)$, $f_{\varepsilon\varepsilon}(\lambda)$ **components of the power spectrum at frequency** λ. Spectrum $f_{\beta\beta}(\lambda)$ might be called the **between batch power spectrum at frequency** λ and $f_{\varepsilon\varepsilon}(\lambda)$ the **within batch power spectrum at frequency** λ.

Under the above assumptions, we note that $EX_{jk}(t) = \mu$, $t = 0, \pm 1, \ldots$ and the series have power spectra and cross-spectra as follows:

$$f_{X_{jk}, X_{jk}}(\lambda) = f_{\alpha\alpha}(\lambda) + f_{\beta\beta}(\lambda) + f_{\varepsilon\varepsilon}(\lambda) \tag{7.9.2}$$

$$f_{X_{jk}, X_{jk'}}(\lambda) = f_{\alpha\alpha}(\lambda) + f_{\beta\beta}(\lambda) \qquad \text{if } k \neq k' \tag{7.9.3}$$

and

$$f_{X_{jk}, X_{j'k'}}(\lambda) = f_{\alpha\alpha}(\lambda) \qquad \text{if } j \neq j',\ k \neq k'. \tag{7.9.4}$$

The coherency between series corresponding to sheets selected from the same batch is seen to be

$$R_{X_{jk}, X_{jk'}}(\lambda) = \frac{f_{\alpha\alpha}(\lambda) + f_{\beta\beta}(\lambda)}{f_{\alpha\alpha}(\lambda) + f_{\beta\beta}(\lambda) + f_{\varepsilon\varepsilon}(\lambda)}. \tag{7.9.5}$$

This might be called the **intraclass coherency at frequency** λ. The coherency between the series corresponding to sheets selected from different batches is seen to be $f_{\alpha\alpha}(\lambda)/[f_{\alpha\alpha}(\lambda) + f_{\beta\beta}(\lambda) + f_{\varepsilon\varepsilon}(\lambda)]$.

We might be interested in a measure of the extent to which sheets from the same batch are related at frequency λ. One such measure is the coherency (7.9.5). In the extreme case of $\alpha(t)$, $\beta_j(t)$ identically 0, this measure is identically 0. In another extreme case where $\varepsilon_{jk}(t)$ is identically 0, this measure is 1. We turn to the problem of estimating $f_{\alpha\alpha}(\lambda)$, $f_{\beta\beta}(\lambda)$, and $f_{\varepsilon\varepsilon}(\lambda)$.

From the model (7.9.1) we see that

$$d_{X_{jk}}^{(T)}(\lambda) = \mu\Delta^{(T)}(\lambda) + d_\alpha^{(T)}(\lambda) + d_{\beta_j}^{(T)}(\lambda) + d_{\epsilon_{jk}}^{(T)}(\lambda) \tag{7.9.6}$$

where

$$d_{X_{jk}}^{(T)}(\lambda) = \sum_{t=0}^{T-1} X_{jk}(t) \exp\{-i\lambda t\} \qquad \text{for } -\infty < \lambda < \infty, \tag{7.9.7}$$

with similar definitions for $d_\alpha^{(T)}$, $d_\beta^{(T)}$, $d_\varepsilon^{(T)}$. From Theorem 4.4.2 the variate $d_\alpha^{(T)}(\lambda)$ is approximately $N_1^C(0,2\pi Tf_{\alpha\alpha}(\lambda))$, the variates $d_{\beta_j}^{(T)}(\lambda), j = 1, \ldots, J$ are approximately independent $N_1^C(0,2\pi Tf_{\beta\beta}(\lambda))$ variates for $\lambda \not\equiv 0 \pmod{\pi}$, while the variates $d_{\epsilon_{jk}}^{(T)}(\lambda)$, $k = 1, \ldots, K$, $j = 1, \ldots, J$ are approximately independent $N_1^C(0, 2\pi Tf_{\epsilon\epsilon}(\lambda))$ variates for $\lambda \not\equiv 0 \pmod{\pi}$. The model (7.9.6) therefore has the approximate form of the random effects model of analysis of variance in a balanced one-way classification; see Scheffé (1959). This suggests that we evaluate the statistic

$$d_{X_{j.}}^{(T)}(\lambda) = K^{-1} \sum_{k=1}^{K} d_{X_{jk}}^{(T)}(\lambda) = \sum_{t=0}^{T-1} \left(K^{-1} \sum_{k=1}^{K} X_{jk}(t)\right) \exp\{-i\lambda t\} \tag{7.9.8}$$

and then estimate $f_{\varepsilon\varepsilon}(\lambda)$ by

$$I_{\varepsilon\varepsilon}^{(T)}(\lambda) = J^{-1}(K-1)^{-1} \sum_{j=1}^{J}\sum_{k=1}^{K} (2\pi T)^{-1}|d_{X_{jk}}^{(T)}(\lambda) - d_{X_{j.}}^{(T)}(\lambda)|^2. \tag{7.9.9}$$

We estimate $Kf_{\beta\beta}(\lambda) + f_{\varepsilon\varepsilon}(\lambda)$ by

$$KI_{\beta\beta}^{(T)}(\lambda) + I_{\epsilon\epsilon}^{(T)}(\lambda) = J^{-1}K \sum_{j=1}^{J} (2\pi T)^{-1}|d_{X_{j.}}^{(T)}(\lambda) - d_{X_{..}}^{(T)}(\lambda)|^2; \tag{7.9.10}$$

and finally estimate $JKf_{\alpha\alpha}(\lambda) + Kf_{\beta\beta}(\lambda) + f_{\varepsilon\varepsilon}(\lambda)$ by

$$JKI_{\alpha\alpha}^{(T)}(\lambda) + KI_{\beta\beta}^{(T)}(\lambda) + I_{\varepsilon\varepsilon}^{(T)}(\lambda) = (2\pi T)^{-1}|d_{X_{..}}^{(T)}(\lambda)|^2 \tag{7.9.11}$$

in the case that $\lambda \not\equiv 0 \pmod{2\pi}$.

Theorem 7.9.1 Let JK series $X_{jk}(t)$, $t = 0, \pm 1, \ldots$; $k = 1, \ldots, K$; $j = 1, \ldots, J$ be given of the form (7.9.1) where μ is a constant, where $\alpha(t)$,

$\beta_j(t)$, $\varepsilon_{jk}(t)$, $t = 0, \pm 1, \ldots; k = 1, \ldots, K; j = 1, \ldots, J$ are independent 0 mean series satisfying Assumption 2.6.1 and having power spectra $f_{\alpha\alpha}(\lambda)$, $f_{\beta\beta}(\lambda)$, $f_{\varepsilon\varepsilon}(\lambda)$ respectively. Let $I_{\varepsilon\varepsilon}^{(T)}(\lambda)$, $KI_{\beta\beta}^{(T)}(\lambda) + I_{\varepsilon\varepsilon}^{(T)}(\lambda)$, $JKI_{\alpha\alpha}^{(T)}(\lambda) + KI_{\beta\beta}^{(T)}(\lambda) + I_{\varepsilon\varepsilon}^{(T)}(\lambda)$ be (7.9.9) to (7.9.11). Then if $\lambda \not\equiv 0 \pmod{\pi}$, these statistics are asymptotically independent $f_{\varepsilon\varepsilon}(\lambda)\chi^2_{2J(K-1)}/[2J(K-1)]$, $[Kf_{\beta\beta}(\lambda) + f_{\varepsilon\varepsilon}(\lambda)]\chi_{2J}^2/(2J)$, $[JKf_{\alpha\alpha}(\lambda) + Kf_{\beta\beta}(\lambda) + f_{\varepsilon\varepsilon}(\lambda)]\chi_2^2/2$. Also for $s_l(T)$ an integer with $\lambda_l(T) = 2\pi s_l(T)/T \to \lambda_l$ as $T \to \infty$ with $2\lambda_l(T)$, $\lambda_l(T) \pm \lambda_m(T) \not\equiv 0 \pmod{2\pi}$ for $1 \leqslant l < m \leqslant L$ for T sufficiently large, the statistics $I_{\varepsilon\varepsilon}^{(T)}(\lambda_l(T))$, $KI_{\beta\beta}^{(T)}(\lambda_l(T)) + I_{\varepsilon\varepsilon}^{(T)}(\lambda_l(T))$, $JKI_{\alpha\alpha}^{(T)}(\lambda_l(T)) + KI_{\beta\beta}^{(T)}(\lambda_l(T)) + I_{\varepsilon\varepsilon}^{(T)}(\lambda_l(T))$, $l = 1, \ldots, L$ are asymptotically independent.

It follows from Theorem 7.9.1 that the estimate

$$K^{-1}[\{KI_{\beta\beta}^{(T)}(\lambda) + I_{\varepsilon\varepsilon}^{(T)}(\lambda)\} - I_{\varepsilon\varepsilon}^{(T)}(\lambda)] \tag{7.9.12}$$

of $f_{\beta\beta}(\lambda)$ will be distributed asymptotically as the difference of two independent chi-squared variates. It also follows that the ratio

$$\frac{KI_{\beta\beta}^{(T)}(\lambda) + I_{\varepsilon\varepsilon}^{(T)}(\lambda)}{I_{\varepsilon\varepsilon}^{(T)}(\lambda)} \tag{7.9.13}$$

will be distributed asymptotically as

$$\left(K\frac{f_{\beta\beta}(\lambda)}{f_{\varepsilon\varepsilon}(\lambda)} + 1\right)F_{2J;2J(K-1)} \tag{7.9.14}$$

as $T \to \infty$. This last result may be used to set approximate confidence intervals for the ratio of power spectra $f_{\beta\beta}(\lambda)/f_{\varepsilon\varepsilon}(\lambda)$.

We have seen previously that advantages accrue from the smoothing of periodogram type statistics. The same is true in the present context. For $s(T)$ an integer with $2\pi s(T)/T$ near $\lambda \not\equiv 0 \pmod{2\pi}$ consider the statistics

$$f_{\varepsilon\varepsilon}^{(T)}(\lambda) = (2m+1)^{-1} \sum_{s=-m}^{m} I_{\varepsilon\varepsilon}^{(T)}\left(\frac{2\pi[s(T)+s]}{T}\right) \tag{7.9.15}$$

and

$$Kf_{\beta\beta}^{(T)}(\lambda) + f_{\varepsilon\varepsilon}^{(T)}(\lambda) = (2m+1)^{-1} \sum_{s=-m}^{m} \left\{KI_{\beta\beta}^{(T)}\left(\frac{2\pi[s(T)+s]}{T}\right) + I_{\varepsilon\varepsilon}^{(T)}\left(\frac{2\pi[s(T)+s]}{T}\right)\right\}. \tag{7.9.16}$$

It follows from Theorem 7.9.1 that these will be asymptotically distributed as independent $f_{\varepsilon\varepsilon}(\lambda)\chi^2_{2J(K-1)(2m+1)}/[2J(K-1)(2m+1)]$ and $[Kf_{\beta\beta}(\lambda) + f_{\varepsilon\varepsilon}(\lambda)]\chi^2_{2J(2m+1)}/[2J(2m+1)]$ respectively.

The discussion of this section may clearly be extended to apply to time series collected in more complicated experimental designs. The calculations and asymptotic distributions will parallel those going along with a normal random effects model for the design concerned. Shumway (1971) considered the model

$$Y_j(t) = s(t) + n_j(t) \tag{7.9.17}$$

$j = 1, \ldots, N$ where $s(t)$ is a fixed unknown signal and $n_j(t)$ a random noise series. He suggests the consideration of F ratios computed in the frequency domain. Brillinger (1973) considers the model (7.9.1) also in the case that the series $\alpha(t)$, $\beta_j(t)$ are fixed and in the case that a transient series is present.

7.10 EXERCISES

7.10.1 Given the series $[X_1(t),X_2(t)]$, $t = 0, \pm 1, \ldots$ with absolutely summable cross-covariance function $c_{12}(u) = \operatorname{cov}\{X_1(t+u),X_2(t)\}$ $t, u = 0, \pm 1, \ldots$ show that $f_{21}(\lambda) = f_{12}(-\lambda)$.

7.10.2 Under the conditions of the previous exercise, show that the co-spectrum of the series $X_1(t)^H$ with the series $X_2(t)$ is the quad-spectrum of the series $X_1(t)$ with the series $X_2(t)$.

7.10.3 Under the conditions of the first exercise, show that $f_{12}(\lambda)$, $-\infty < \lambda < \infty$, is real-valued in the case that $c_{12}(u) = c_{21}(u)$.

7.10.4 Suppose the auto- and cross-covariance functions of the stationary series $[X_1(t),X_2(t)]$, $t = 0, \pm 1, \ldots$ are absolutely summable. Use the identity

$$|I_{12}^{(T)}(\lambda)|^2 = I_{11}^{(T)}(\lambda)I_{22}^{(T)}(\lambda)$$

to prove that

$$|f_{12}(\lambda)|^2 \leqslant f_{11}(\lambda) f_{22}(\lambda).$$

7.10.5 If $d_j^{(T)}(\lambda) = \sum_{t=0}^{T-1} e^{-i\lambda t}X_j(t)$, $I_{12}^{(T)}(\lambda) = (2\pi T)^{-1}d_1^{(T)}(\lambda)\overline{d_2^{(T)}(\lambda)}$ and $\sum_u |u||c_{12}(u)| < \infty$, show that

$$EI_{12}^{(T)}(\lambda) = f_{12}(\lambda) + O(T^{-1})$$

for $\lambda \not\equiv 0 \pmod{2\pi}$.

7.10.6 Prove

(a) $$\frac{2\pi}{T}\sum_{s=0}^{T-1} I_{12}^{(T)}\left(\frac{2\pi s}{T}\right) = \frac{1}{T}\sum_{t=0}^{T-1} X_1(t)X_2(t) = m_{12}^{(T)}(0)$$

(b) $$\frac{2\pi}{T}\sum_{s=1}^{T-1} I_{12}^{(T)}\left(\frac{2\pi s}{T}\right) = \frac{1}{T}\sum_{t=0}^{T-1} [X_1(t) - c_1^{(T)}][X_2(t) - c_2^{(T)}] = c_{12}^{(T)}(0).$$

7.10.7 Let $[X_1(t),X_2(t)]$, $t = 0, \pm 1, \ldots$ be a stationary series.

(a) If $Y_1(t) = X_1(t) + X_2(t)$, $Y_2(t) = X_1(t) - X_2(t)$, show how the co-spectrum, $\operatorname{Re} f_{12}(\lambda)$, may be estimated from the power spectra of $Y_1(t)$ and $Y_2(t)$.

(b) If $Y_1(t) = X_1(t+1) - X_1(t-1)$, and $Y_2(t) = X_2(t)$, show how the quad-spectrum, $\operatorname{Im} f_{12}(\lambda)$, may be estimated from the co-spectrum of $Y_1(t)$ and $Y_2(t)$.

7.10.8 Under the conditions of Theorem 7.4.4 show that

$$(B_T T)^{1/2}\left(2\pi \int W(\alpha)^2 d\alpha\right)^{-1/2}[\text{Re } f_{12}^{(T)}(\lambda), \text{Im } f_{12}^{(T)}(\lambda)]$$

is asymptotically bivariate normal with variances

$$[1 + \eta\{2\lambda\}][f_{11}(\lambda) f_{22}(\lambda) + \{\text{Re } f_{12}(\lambda)\}^2 - \{\text{Im } f_{12}(\lambda)\}^2]/2,$$
$$[1 - \eta\{2\lambda\}][f_{11}(\lambda) f_{22}(\lambda) - \{\text{Re } f_{12}(\lambda)\}^2 + \{\text{Im } f_{12}(\lambda)\}^2]/2,$$

and covariance

$$[1 - \eta\{2\lambda\}][-\{\text{Re } f_{12}(\lambda)\}\{\text{Im } f_{12}(\lambda)\}].$$

7.10.9 Under the conditions of Theorem 7.4.4, show that $|f_{12}^{(T)}(\lambda)|$, $|f_{12}^{(T)}(\mu)|$ are asymptotically bivariate normal with covariance structure

$$\lim_{T\to\infty} B_T T \operatorname{cov}\{|f_{12}^{(T)}(\lambda)|, |f_{12}^{(T)}(\mu)|\}$$
$$= \pi[\eta\{\lambda - \mu\} + \eta\{\lambda + \mu\}][f_{11}(\lambda) f_{22}(\lambda) + |f_{12}(\lambda)|^2] \int W(\alpha)^2 d\alpha.$$

7.10.10 Under the conditions of Theorem 7.4.4, show that $\phi_{12}^{(T)}(\lambda) = \arg f_{12}^{(T)}(\lambda)$, $\phi_{12}^{(T)}(\mu) = \arg f_{12}^{(T)}(\mu)$ are asymptotically bivariate normal with covariance structure given by

$$\lim_{T\to\infty} B_T T \operatorname{cov}\{\phi_{12}^{(T)}(\lambda), \phi_{12}^{(T)}(\mu)\}$$
$$= \pi[\eta\{\lambda - \mu\} - \eta\{\lambda + \mu\}][f_{11}(\lambda) f_{22}(\lambda) - |f_{12}(\lambda)|^2]\, |f_{12}(\lambda)|^{-2} \int W(\alpha)^2 d\alpha.$$

7.10.11 Under the condition $\Sigma\, |c_{12}(u)| < \infty$, show that the expected value of $I^{(T)}_{X_1 - c_1^{(T)}, X_2 - c_2^{(T)}}(\lambda)$ is

$$(2\pi T)^{-1} \int_0^{2\pi} \{|\Delta^{(T)}(\lambda - \alpha)|^2 - T^{-1}\Delta^{(T)}(\lambda)\Delta^{(T)}(-\alpha)\Delta^{(T)}(-\lambda + \alpha)$$
$$- T^{-1}\Delta^{(T)}(-\lambda)\Delta^{(T)}(\alpha)\Delta^{(T)}(\lambda - \alpha) + T^{-2}|\Delta^{(T)}(\lambda)|^2|\Delta^{(T)}(\alpha)|^2\} f_{12}(\alpha) d\alpha.$$

7.10.12 Let $[X_1(t), X_2(t)]$, $t = 0, \pm 1, \ldots$ be a bivariate series satisfying Assumption 2.6.1. Let

$$\mathbf{c}_X^{(V)}(l) = \begin{bmatrix} c_1^{(V)}(l) \\ c_2^{(V)}(l) \end{bmatrix} = V^{-1} \sum_{v=0}^{V-1} \begin{bmatrix} X_1(v + lV) \\ X_2(v + lV) \end{bmatrix}$$

for $l = 1, \ldots, L$. Show that $\mathbf{c}_X^{(V)}(l)$, $l = 1, \ldots, L$ are asymptotically independent

$$N_2\left(\begin{bmatrix} c_1 \\ c_2 \end{bmatrix}, 2\pi V^{-1} \begin{bmatrix} f_{11}(0) & f_{12}(0) \\ f_{21}(0) & f_{22}(0) \end{bmatrix}\right)$$

variates. Conclude that with $T = LV$

$$\frac{V}{2\pi} \sum_l [c_1^{(V)}(l) - c_1^{(T)}][c_2^{(V)}(l) - c_2^{(T)}]/(L - 1)$$

is a plausible estimate of $f_{12}(0)$.

7.10.13 Show that the results of Theorems 7.2.3, 7.2.4, 7.2.5, and 7.3.3 are exact rather than asymptotic when $[X_1(t), X_2(t)]$, $t = 0, \pm 1, \ldots$ is a sequence of independent identically distributed bivariate normals.

7.10.14 Let the series $[X_1(t),X_2(t)]$, $t = 0, \pm 1, \ldots$ satisfy Assumption 2.6.2(1) and have mean $\mathbf{0}$. Then

$$\begin{aligned}\operatorname{cov}\{I_{12}^{(T)}(\lambda), I_{12}^{(T)}(\mu)\} &= T^{-2}|\Delta^{(T)}(\lambda - \mu)|^2 f_{11}(\lambda) f_{22}(\lambda)\\ &+ T^{-2}|\Delta^{(T)}(\lambda + \mu)|^2 f_{12}(\lambda) f_{21}(-\lambda)\\ &+ 2\pi T^{-1} f_{1212}(\lambda,-\lambda,-\mu) + T^{-2}R^{(T)}(\lambda,\mu)\end{aligned}$$

where there are finite K, L such that

$$|R^{(T)}(\lambda,\mu)| \leqslant K\{|\Delta^{(T)}(\lambda + \mu)| + |\Delta^{(T)}(\lambda - \mu)|\} + L.$$

7.10.15 Suppose the conditions of Theorem 7.3.3 are satisfied. Let $\rho = f_{ab}(\lambda)/\sqrt{f_{aa}(\lambda) f_{bb}(\lambda)}$, $a \neq b$. Then $x = f_{ab}^{(T)}(\lambda)/\sqrt{f_{aa}^{(T)}(\lambda) f_{bb}^{(T)}(\lambda)}$ is asymptotically distributed with density function

$$\frac{|x|^{2m}}{\pi\Gamma(2m+1)2^{2m+1}\sqrt{1 - |\rho|^2}} \exp\{\operatorname{Re}(x\bar{\rho})/(1 - |\rho|^2\} K_{2m}(|x|/(1 - |\rho|^2))$$

if $\lambda \not\equiv 0 \pmod{\pi}$,

and density function

$$\frac{|x|^{(2m-1)/2}}{\pi\Gamma(m)2^{(2m-1)/2}\sqrt{1 - \rho^2}} \exp\{x\rho/(1 - \rho^2)\} K_{(2m-1)/2}(|x|/(1 - \rho^2))$$

if $\lambda \equiv 0 \pmod{\pi}$.

Hint: Use Exercise 4.8.33.

7.10.16 Let $\mathbf{c}_{XX}(u)$, $u = 0, \pm 1, \ldots$ denote the autocovariance matrix of the stationary r vector-valued series $\mathbf{X}(t)$, $t = 0, \pm 1, \ldots$. Show that the matrix $\mathbf{c}_{XX}(0) - \mathbf{c}_{XX}(u)^\tau \mathbf{c}_{XX}(0)^{-1}\mathbf{c}_{XX}(u)$ is non-negative definite for $u = 0, \pm 1, \ldots$.

7.10.17 Let $\mathbf{f}_{XX}(\lambda)$, $-\infty < \lambda > \infty$, denote the spectral density matrix of the stationary r vector-valued series $\mathbf{X}(t)$, $t = 0, \pm 1, \ldots$. Show that Im $\mathbf{f}_{XX}(\lambda) = \mathbf{0}$ for $\lambda \equiv 0 \pmod{\pi}$.

7.10.18 Let the autocovariance function of Exercise 7.10.16 satisfy

$$\sum_u |u|^l |\mathbf{c}_{XX}(u)| < \infty.$$

Suppose Det $\mathbf{f}_{XX}(\lambda) \neq 0$, $-\infty < \lambda < \infty$. Show that there exists an l summable $r \times r$ filter $\{\mathbf{a}(u)\}$, such that the series

$$\mathbf{Y}(t) = \sum_u \mathbf{a}(t - u)\mathbf{X}(u)$$

satisfies $\mathbf{f}_{YY}(\lambda) = (2\pi)^{-1}\mathbf{I}$, $-\infty < \lambda < \infty$, and $\mathbf{c}_{YY}(0) = \mathbf{I}$, $\mathbf{c}_{YY}(u) = \mathbf{0}$ for $u \neq 0$.

7.10.19 Let $\mathbf{X}(t)$, $t = 0, \pm 1, \ldots$ be the vector-valued series of Example 2.9.7. Show that

$$\mathbf{f}_{XX}(\lambda) = (2\pi)^{-1}\mathbf{A}(\lambda)^{-1}\mathbf{B}(\lambda)\mathbf{c}_{\epsilon\epsilon}(0)\overline{\mathbf{B}(\lambda)}^\tau\overline{\mathbf{A}(\lambda)^{-1}}^\tau.$$

7.10.20 Under the conditions of Theorem 4.5.2, show that there exists a finite L such that with probability 1

$$\overline{\lim_{T\to\infty}} \sup_\lambda |\mathbf{I}_{YY}^{(T)}(\lambda) - \mathbf{A}(\lambda)\mathbf{I}_{XX}^{(T)}(\lambda)\overline{\mathbf{A}(\lambda)}^\tau| < L.$$

7.10.21 Show that Theorem 7.2.1 takes the form

$$E\mathbf{I}_{XX}^{(T)}(\lambda) = (2\pi T)^{-1}\int_0^{2\pi}\left[\frac{\sin T\alpha/2}{\sin \alpha/2}\right]^2 \mathbf{f}_{XX}(\lambda - \alpha)d\alpha + (2\pi T)^{-1}\left[\frac{\sin T\lambda/2}{\sin \lambda/2}\right]^2 \mathbf{c}_X\mathbf{c}_X{}^\tau$$

in the case of untapered data.

7.10.22 Let $\mathbf{X}(t)$, $t = 0, \pm 1, \ldots$ be an r vector-valued series satisfying Assumption 2.6.2(1). Let $\mathbf{I}_{XX}^{(T)}(\lambda)$ be given by (7.3.1). Let $\mu = 2\pi r/T$, $\lambda = 2\pi s/T$ for r, s integers. If $\lambda, \mu \not\equiv 0 \pmod{2\pi}$ show that

(a) $\operatorname{cov}\{I_{a_1b_1}^{(T)}(\lambda), I_{a_2b_2}^{(T)}(\mu)\}$
$= \eta\{\lambda - \mu\} f_{a_1a_2}(\lambda) f_{b_1b_2}(-\lambda) + \eta\{\lambda + \mu\} f_{a_1b_2}(\lambda) f_{b_1a_2}(-\lambda) + O(T^{-1})$

(b) $\operatorname{cov}\{\operatorname{Re} I_{a_1b_1}^{(T)}(\lambda), \operatorname{Re} I_{a_2b_2}^{(T)}(\mu)\}$
$= \frac{1}{2}[\eta\{\lambda - \mu\} + \eta\{\lambda + \mu\}]$
$\times \operatorname{Re}\{ f_{a_1a_2}(\lambda) f_{b_1b_2}(-\lambda) + f_{a_1b_2}(\lambda) f_{b_1a_2}(-\lambda)\} + O(T^{-1})$

(c) $\operatorname{cov}\{\operatorname{Re} I_{a_1b_1}^{(T)}(\lambda), \operatorname{Im} I_{a_2b_2}^{(T)}(\mu)\}$
$= -\frac{1}{2}[\eta\{\lambda - \mu\} - \eta\{\lambda + \mu\}]$
$\times \operatorname{Im}\{ f_{a_1a_2}(\lambda) f_{b_1b_2}(-\lambda) - f_{a_1b_2}(\lambda) f_{b_1a_2}(-\lambda)\} + O(T^{-1})$

(d) $\operatorname{cov}\{\operatorname{Im} I_{a_1b_1}^{(T)}(\lambda), \operatorname{Im} I_{a_2b_2}^{(T)}(\mu)\}$
$= \frac{1}{2}[\eta\{\lambda - \mu\} - \eta\{\lambda + \mu\}]$
$\times \operatorname{Re}\{ f_{a_1a_2}(\lambda) f_{b_1b_2}(-\lambda) - f_{a_1b_2}(\lambda) f_{b_1a_2}(-\lambda)\} + O(T^{-1})$.

7.10.23 Suppose the conditions of Theorem 7.2.5 are staisfied. Set

$$d_a^{(V)}(\lambda,\cdot) = L^{-1}\sum_{l=0}^{L-1} d_a^{(V)}(\lambda,l).$$

Then

$$\mathbf{f}_{XX}^{(LV)}(\lambda) = (L-1)^{-1}\sum_{l=0}^{L-1}[\{2\pi H_{ab}^{(V)}(0)\}^{-1}\{d_a^{(V)}(\lambda,l) - d_a^{(V)}(\lambda,\cdot)\} \times \overline{\{d_b^{(V)}(\lambda,l) - d_b^{(V)}(\lambda,\cdot)\}}]$$

is asymptotically $(L-1)^{-1}W_r{}^C(L-1,\mathbf{f}_{XX}(\lambda))$ if $\lambda \not\equiv 0 \pmod{\pi}$ and asymptotically $(L-1)^{-1}W_r(L-1,\mathbf{f}_{XX}(\lambda))$ if $\lambda \equiv 0 \pmod{\pi}$.

7.10.24 Consider the estimate

$$\mathbf{f}_{XX}^{(T)}(\lambda) = \sum_{s=-m}^{m} W_s\mathbf{I}_{XX}^{(T)}\left(\frac{2\pi[s(T)+s]}{T}\right)$$

where $2\pi s(T)/T \to \lambda \not\equiv 0 \pmod{\pi}$ and $\sum_s W_s = 1$. Under the conditions of Theorem 7.3.3, show that $\mathbf{f}_{XX}^{(T)}(\lambda)$ is distributed asymptotically as

$$\sum_{s=-m}^{m} W_s\mathbf{W}_s$$

where the $\mathbf{W}_s$, $s = 0, \pm 1, \ldots, \pm m$ are independent $W_r{}^C(1,\mathbf{f}_{XX}(\lambda))$ variates. Indicate the mean and covariance matrix of the limiting distribution.

7.10.25 Suppose the estimate

$$(2m+1)^{-1} \sum_{s=-m}^{m} \mathbf{I}_{XX}^{(T)}\left(\lambda + \frac{2\pi s}{T}\right)$$

is used in the case of T even and $\lambda = \pi$. Under the conditions of Theorem 7.3.3, show that it is asymptotically $(2m+1)^{-1}W_r(2m+1,\mathbf{f}_{XX}(\pi))$.

7.10.26 Show that the estimate (7.4.5) is non-negative definite if the matrix $[W_{ab}(\alpha)]$ is non-negative definite for $-\infty < \alpha < \infty$. *Hint:* Use Schur's result that $[A_{ab}B_{ab}]$ is non-negative if $[A_{ab}]$, $[B_{ab}]$ are; see Bellman (1960) p. 94.

7.10.27 Show that the matrix $[f_{ab}^{(T)}(\lambda)]$ of estimates (7.7.9) is non-negative definite if the matrix $[W_{ab}(\alpha)]$ is non-negative definite, $-\infty < \alpha < \infty$, and if $h_a(u) = h(u)$ for $a = 1, \ldots, r$.

7.10.28 Under the conditions of Theorem 7.3.2, show that $f_{ab}^{(T)}(\lambda)$ is consistent if $f_{aa}(\lambda)$ or $f_{bb}(\lambda) = 0$.

7.10.29 Under the conditions of Theorem 7.3.3, show that $\sqrt{T}(\mathbf{c}_X^{(T)} - \mathbf{c}_X)$ and $\mathbf{f}_{XX}^{(T)}(0)$ are asymptotically independent $N_r(\mathbf{0},2\pi\mathbf{f}_{XX}(0))$ and $(2m)^{-1} W_r(2m,\mathbf{f}_{XX}(0))$ respectively. If

$$\mathfrak{J}^2 = (2\pi)^{-1}T(\mathbf{c}_X^{(T)} - \mathbf{c}_X)^\tau \mathbf{f}_{XX}^{(T)}(0)^{-1}(\mathbf{c}_X^{(T)} - \mathbf{c}_X)$$

conclude that $\mathfrak{J}^2/r$ is asymptotically $F_{r,2m}$. This result may be used to construct approximate confidence regions for $\mathbf{c}_X$.

7.10.30 Under the conditions of Theorem 7.4.3, show that in order to minimize the mean-squared error $E|f_{ab}^{(T)}(\lambda) - f_{ab}(\lambda)|^2$ asymptotically, one should have $B_T = O(T^{-1/5})$; see Bartlett (1966) p. 316.

7.10.31 Under the conditions of Theorem 7.6.2, prove that $R_{ab}^{(T)}(\lambda)$ and $R_{cd}^{(T)}(\lambda)$ are asymptotically independent if R_{ab}, R_{ac}, R_{ad}, R_{bc}, R_{bd}, R_{cd} are 0.

7.10.32 Show that in the case that the series $\mathbf{X}(t)$, $t = 0, \pm 1, \ldots$ is not necessarily Gaussian, the covariance (7.6.21) equals

$$2\pi \int_{\mu_1}^{\lambda_1} \int_{\mu_2}^{\lambda_2} f_{a_1b_1a_2b_2}(\alpha,-\alpha,-\beta)d\alpha d\beta.$$

7.10.33 If the series $X(t)$, $t = 0, \pm 1, \ldots$ is stationary, real-valued and Gaussian, prove that the covariance structure of the limit process of Theorem 7.6.3 is the same as that of

$$\sqrt{2\pi} \int_0^\lambda f_{XX}(\alpha)dB(\alpha)$$

where $B(\alpha)$ is Brownian motion on $[0,\pi]$.

7.10.34 If the series $X(t)$, $t = 0, \pm 1, \ldots$ is a real-valued white noise process with variance σ^2 and fourth cumulant κ_4, show that the limit process of Theorem 7.6.3 has covariance function

$$(2\pi)^{-1}\sigma^4 \min(\lambda,\mu) + (2\pi)^{-2}\kappa_4\lambda\mu.$$

7.10.35 If $X(t)$, $t = 0, \pm 1, \ldots$ is a real-valued linear process, under the conditions of Theorem 7.6.3 show that

$$T^{1/2}[F_{XX}^{(T)}(\lambda)F_{XX}(\pi) - F_{XX}(\lambda)F_{XX}^{(T)}(\pi)]$$

converges weakly to a Gaussian process whose covariance function does not involve the fourth-order spectrum of the series $X(t)$.

7.10.36 Let $\mathbf{X}(t)$, $t = 0, \pm 1, \ldots$ be an r vector-valued series satisfying Assumption 2.6.1. Show that $c_{ab}^{(T)}(u)$ given by (7.6.10) and $c_{ab}^{(T)}(u)$ given by (7.6.12) have the same limiting normal distributions. See also Exercise 4.8.37.

7.10.37 Let

$$J_{ab}^{(V)}(A,l) = \frac{2\pi}{V}\sum_{v=1}^{V-1} A\left(\frac{2\pi v}{V}\right) I_{ab}^{(V)}\left(\frac{2\pi v}{V}, l\right)$$

where

$$I_{ab}^{(V)}(\lambda,l) = (2\pi V)^{-1}\left(\sum_{v=0}^{V-1} X_a(v + lV)\exp\{-i\lambda v\}\right)\left(\sum_{v=0}^{V-1} X_b(v + lV)\exp\{i\lambda v\}\right)$$

$l = 0, \ldots, L-1$; $a, b = 1, \ldots, r$. Under the conditions of Theorem 7.6.1 show that $J_{ab}^{(V)}(A,l)$, $l = 0, \ldots, L-1$ are asymptotically independent normal with mean $\int_0^{2\pi} A(\alpha) f_{ab}(\alpha) d\alpha$ as $V \to \infty$. This result may be used to set approximate confidence limits for $J_{ab}(A)$.

7.10.38 With the notation of Section 7.9, show that the following identity holds

$$\sum_{j=1}^{J}\sum_{k=1}^{K} |d_{X_{jk}}^{(T)}(\lambda)|^2 = \sum_{j=1}^{J}\sum_{k=1}^{K} |d_{X_{jk}}^{(T)}(\lambda) - d_{X_{j.}}^{(T)}(\lambda)|^2 + K\sum_{j=1}^{J} |d_{X_{j.}}^{(T)}(\lambda) - d_{X_{..}}^{(T)}(\lambda)|^2 + JK|d_{X_{..}}^{(T)}(\lambda)|^2.$$

7.10.39 Suppose

$$d_a^{(T_a)}(\lambda) = \sum_{t=0}^{T_a-1} X_a(t)\exp\{-i\lambda t\}$$

and

$$I_{ab}^{(T)}(\lambda) = (2\pi T_a T_b)^{-1} d_a^{(T_a)}(\lambda)\overline{d_b^{(T_b)}(\lambda)}$$

for $-\infty < \lambda < \infty$; $a, b = 1, \ldots, r$. Show that the matrix $\mathbf{I}_{XX}^{(T)}(\lambda) = [I_{ab}^{(T)}(\lambda)]$ is non-negative definite.

7.10.40 Let the series $\mathbf{X}(t)$, $t = 0, \pm 1, \ldots$ satisfy Assumption 2.6.2(1). Show that expression (7.2.14) holds with the $O(T^{-1})$, $O(T^{-2})$ terms uniform in r, $s \not\equiv 0 \pmod{T}$.

7.10.41 Use the results of the previous exercise to show that, under the conditions of Theorem 7.4.3,

$$\operatorname{cov}\{f_{a_1b_1}^{(T)}(\lambda), f_{a_2b_2}^{(T)}(\mu)\}$$

$$= \left(\frac{2\pi}{T}\right)^2 \left(\sum_{s=1}^{T-1} W_{a_1b_1}^{(T)}\left(\lambda - \frac{2\pi s}{T}\right) W_{a_2b_2}^{(T)}\left(\mu - \frac{2\pi s}{T}\right) f_{a_1a_2}\left(\frac{2\pi s}{T}\right) f_{b_1b_2}\left(-\frac{2\pi s}{T}\right)\right.$$

$$\left. + \sum_{s=1}^{T-1} W_{a_1b_1}^{(T)}\left(\lambda - \frac{2\pi s}{T}\right) W_{a_2b_2}^{(T)}\left(\mu + \frac{2\pi s}{T}\right) f_{a_1b_2}\left(\frac{2\pi s}{T}\right) f_{b_1a_2}\left(-\frac{2\pi s}{T}\right)\right)$$

$$+ \left(\frac{2\pi}{T}\right)^3 \sum_{r=1}^{T-1} \sum_{s=1}^{T-1} W_{a_1b_1}^{(T)}\left(\lambda - \frac{2\pi r}{T}\right) W_{a_2b_2}^{(T)}\left(\mu - \frac{2\pi s}{T}\right)$$

$$\times f_{a_1b_1a_2b_2}\left(\frac{2\pi r}{T}, -\frac{2\pi r}{T}, -\frac{2\pi s}{T}\right) + O(B_T^{-1}T^{-2}).$$

7.10.42 Let $\mathbf{X}(t)$, $t = 0, \pm 1, \ldots$ satisfy Assumption 2.6.2(1). Let $A(\alpha)$ be of bounded variation. Let $W(\alpha)$ satisfy Assumption 6.4.1. Suppose $P_T = P \to \infty$, with $P_T B_T \leqslant 1$, $P_T B_T T \to \infty$ as $T \to \infty$. Let

$$J_{ab}^{(P)}(A) = \frac{2\pi}{P} \sum_{p=0}^{P-1} A\left(\frac{2\pi p}{P}\right) f_{ab}^{(T)}\left(\frac{2\pi p}{P}\right).$$

Show that $J_{ab}^{(P)}(A)$ is asymptotically normal with

$$EJ_{ab}^{(P)}(A) = \int_0^{2\pi} A(\alpha) f_{ab}(\alpha) d\alpha + O(B_T) + O(P_T^{-1}) + O(B_T^{-1}T^{-1})$$

and

$$\operatorname{cov}\{J_{a_1b_1}^{(P)}(A_j), J_{a_2b_2}^{(P)}(A_k)\}$$

$$\backsim \frac{2\pi}{P_T} \frac{2\pi}{B_T T} [\textstyle\int W(\alpha)^2 d\alpha] \left[\int_0^{2\pi} A_j(\alpha)\overline{A_k(\alpha)} f_{a_1a_2}(\alpha) f_{b_1b_2}(-\alpha) d\alpha\right.$$

$$\left. + \int_0^{2\pi} A_j(\alpha)\overline{A_k(2\pi - \alpha)} f_{a_1b_2}(\alpha) f_{b_1a_2}(-\alpha) d\alpha\right]$$

$$+ \frac{2\pi}{T} \int_0^{2\pi} \int_0^{2\pi} A_j(\alpha)\overline{A_k(\beta)} f_{a_1b_1a_2b_2}(\alpha, -\alpha, -\beta) d\alpha d\beta.$$

Hint: Use the previous exercise.

8

ANALYSIS OF A LINEAR TIME INVARIANT RELATION BETWEEN TWO VECTOR-VALUED STOCHASTIC SERIES

8.1 INTRODUCTION

Consider an $(r + s)$ vector-valued stationary series

$$\begin{bmatrix} \mathbf{X}(t) \\ \mathbf{Y}(t) \end{bmatrix} \tag{8.1.1}$$

$t = 0, \pm 1, \ldots$ with $\mathbf{X}(t)$ r vector-valued and $\mathbf{Y}(t)$ s vector-valued. We assume the series (8.1.1) satisfies Assumption 2.6.1 and we define the means

$$\begin{aligned} E\mathbf{X}(t) &= \mathbf{c}_X \\ E\mathbf{Y}(t) &= \mathbf{c}_Y, \end{aligned} \tag{8.1.2}$$

the covariances

$$\begin{aligned} E\{[\mathbf{X}(t+u) - \mathbf{c}_X][\mathbf{X}(t) - \mathbf{c}_X]^\tau\} &= \mathbf{c}_{XX}(u) \\ E\{[\mathbf{X}(t+u) - \mathbf{c}_X][\mathbf{Y}(t) - \mathbf{c}_Y]^\tau\} &= \mathbf{c}_{XY}(u) \\ E\{[\mathbf{Y}(t+u) - \mathbf{c}_Y][\mathbf{Y}(t) - \mathbf{c}_Y]^\tau\} &= \mathbf{c}_{YY}(u) \qquad u = 0, \pm 1, \ldots, \end{aligned} \tag{8.1.3}$$

and the second-order spectral densities

$$\begin{aligned} \mathbf{f}_{XX}(\lambda) &= (2\pi)^{-1} \sum_{u=-\infty}^{\infty} \mathbf{c}_{XX}(u) \exp\{-i\lambda u\} \\ \mathbf{f}_{XY}(\lambda) &= (2\pi)^{-1} \sum_{u=-\infty}^{\infty} \mathbf{c}_{XY}(u) \exp\{-i\lambda u\} \\ \mathbf{f}_{YY}(\lambda) &= (2\pi)^{-1} \sum_{u=-\infty}^{\infty} \mathbf{c}_{YY}(u) \exp\{-i\lambda u\} \qquad \text{for } -\infty < \lambda < \infty. \end{aligned} \tag{8.1.4}$$

The problem we investigate in this chapter is the selection of an s vector $\mathbf{\mu}$ and an $s \times r$ filter $\{\mathbf{a}(u)\}$ such that the value

$$\mathbf{\mu} + \sum_{u=-\infty}^{\infty} \mathbf{a}(t-u)\mathbf{X}(u) \tag{8.1.5}$$

is near the value $\mathbf{Y}(t)$ in some sense. We develop statistical properties of estimates of the desired $\mathbf{\mu}$, $\mathbf{a}(u)$ based on a sample of values $\mathbf{X}(t)$, $\mathbf{Y}(t)$, $t = 0, \ldots, T-1$. The problems considered in this chapter differ from those of Chapter 6 in that the independent series, $\mathbf{X}(t)$, $t = 0, \pm 1, \ldots$, is taken to be stochastic rather than fixed.

In the next section we review a variety of results concerning analagous multivariate problems.

8.2 ANALOGOUS MULTIVARIATE RESULTS

We remind the reader of the ordering for Hermitian matrices given by

$$\mathbf{A} \geqslant \mathbf{B} \tag{8.2.1}$$

if the matrix $\mathbf{A} - \mathbf{B}$ is non-negative definite. This ordering is discussed in Bellman (1960), Gelfand (1961), and Siotani (1967), for example. The inequality (8.2.1) implies, among other things, that

$$\text{Det } \mathbf{A} \geqslant \text{Det } \mathbf{B}, \tag{8.2.2}$$

$$\text{tr } \mathbf{A} \geqslant \text{tr } \mathbf{B}, \tag{8.2.3}$$

$$A_{jj} \geqslant B_{jj}, \tag{8.2.4}$$

and

$$\mu_j(\mathbf{A}) \geqslant \mu_j(\mathbf{B}), \tag{8.2.5}$$

where $\mu_j(\mathbf{A})$, $\mu_j(\mathbf{B})$ denote the jth largest latent values of $\mathbf{A}$, $\mathbf{B}$, respectively.

In the theorem below, when we talk of minimizing a Hermitian matrix-valued function $\mathbf{A}(\theta)$ with respect to θ, we mean finding the value θ_0 such that

$$\mathbf{A}(\theta) \geqslant \mathbf{A}(\theta_0) \tag{8.2.6}$$

for all θ. $\mathbf{A}(\theta_0)$ is called the minimum value of $\mathbf{A}(\theta)$. We note that if θ_0 minimizes $\mathbf{A}(\theta)$, then from (8.2.2) to (8.2.5) it also minimizes simultaneously the functionals Det $\mathbf{A}(\theta)$, tr $\mathbf{A}(\theta)$, $A_{jj}(\theta)$, and $\mu_j(\mathbf{A}(\theta))$.

We next introduce some additional notation. Let $\mathbf{Z}$ be an arbitrary matrix with columns $\mathbf{Z}_1, \ldots, \mathbf{Z}_I$. We use the notation

$$\text{vec } \mathbf{Z} = \begin{bmatrix} \mathbf{Z}_1 \\ \cdot \\ \cdot \\ \cdot \\ \mathbf{Z}_I \end{bmatrix} \tag{8.2.7}$$

for the column vector obtained from $\mathbf{Z}$ by placing its columns under one another successively. Given arbitrary matrices $\mathbf{U}, \mathbf{V}$ we define their **Kronecker product,** $\mathbf{U} \otimes \mathbf{V}$, to be the block matrix

$$\mathbf{U} \otimes \mathbf{V} = \begin{bmatrix} \mathbf{U}V_{11} & \cdots & \mathbf{U}V_{1K} \\ \cdot & \cdots & \cdot \\ \cdot & \cdots & \cdot \\ \cdot & \cdots & \cdot \\ \mathbf{U}V_{J1} & \cdots & \mathbf{U}V_{JK} \end{bmatrix} \tag{8.2.8}$$

if $\mathbf{V}$ is $J \times K$. An important relation connecting the two notations of this paragraph is

$$(\mathbf{U} \otimes \mathbf{V}) \operatorname{vec} \mathbf{Z} = \operatorname{vec} (\mathbf{U}\mathbf{Z}\mathbf{V}^\tau), \tag{8.2.9}$$

if the dimensions of the matrices that appear are appropriate; see Exercise 8.16.26. Neudecker (1968) and Nissen (1968) discuss statistical applications of these definitions.

We now turn to the consideration of $(r + s)$ vector-valued random variables of the form

$$\begin{bmatrix} \mathbf{X} \\ \mathbf{Y} \end{bmatrix} \tag{8.2.10}$$

with $\mathbf{X}$, r vector-valued and $\mathbf{Y}$, s vector-valued. Suppose the variate (8.2.10) has mean

$$\begin{bmatrix} \boldsymbol{\mu}_X \\ \boldsymbol{\mu}_Y \end{bmatrix} \tag{8.2.11}$$

and covariance matrix

$$\begin{bmatrix} \boldsymbol{\Sigma}_{XX} & \boldsymbol{\Sigma}_{XY} \\ \boldsymbol{\Sigma}_{YX} & \boldsymbol{\Sigma}_{YY} \end{bmatrix}. \tag{8.2.12}$$

Consider the problem of choosing the s vector $\boldsymbol{\mu}$ and $s \times r$ matrix $\mathbf{a}$ to minimize the $s \times s$ Hermitian matrix

$$E\{[\mathbf{Y} - \boldsymbol{\mu} - \mathbf{a}\mathbf{X}][\mathbf{Y} - \boldsymbol{\mu} - \mathbf{a}\mathbf{X}]^\tau\}. \tag{8.2.13}$$

We have

Theorem 8.2.1 Let an $(r + s)$ vector-valued variate of the form (8.2.10), with mean (8.2.11) and covariance matrix (8.2.12), be given. Suppose $\boldsymbol{\Sigma}_{XX}$ is nonsingular. Then $\boldsymbol{\mu}$ and $\mathbf{a}$ minimizing (8.2.13) are given by

$$\boldsymbol{\mu} = \boldsymbol{\mu}_Y - \boldsymbol{\Sigma}_{YX}\boldsymbol{\Sigma}_{XX}^{-1}\boldsymbol{\mu}_X \tag{8.2.14}$$

and

$$\mathbf{a} = \boldsymbol{\Sigma}_{YX}\boldsymbol{\Sigma}_{XX}^{-1}. \tag{8.2.15}$$

The minimum achieved is

$$\mathbf{\Sigma}_{YY} - \mathbf{\Sigma}_{YX}\mathbf{\Sigma}_{XX}^{-1}\mathbf{\Sigma}_{XY}. \tag{8.2.16}$$

We call **a**, given by (8.2.15), the **regression coefficient of Y on X.** The variate

$$\boldsymbol{\mu}_Y + \mathbf{\Sigma}_{YX}\mathbf{\Sigma}_{XX}^{-1}(\mathbf{X} - \boldsymbol{\mu}_X) \tag{8.2.17}$$

is called the **best linear predictor** of **Y** based on **X.** From Theorem 8.2.1, we see that the $\boldsymbol{\mu}$ and **a** values given also minimize the determinant, trace, diagonal entries, and latent values of the matrix (8.2.13). References to this theorem include: Whittle (1963a) Chap. 4, Goldberger (1964) p. 280, Rao (1965), and Khatri (1967). In the case $s = 1$, the square of the correlation coefficient of Y with the best linear predictor of Y is called the **squared coefficient of multiple correlation.** It is given by

$$R_{YX}^2 = \frac{\mathbf{\Sigma}_{YX}\mathbf{\Sigma}_{XX}^{-1}\mathbf{\Sigma}_{XY}}{\Sigma_{YY}}. \tag{8.2.18}$$

In the case of vector-valued **Y,** $\mathbf{\Sigma}_{YY}^{-1/2}\mathbf{\Sigma}_{YX}\mathbf{\Sigma}_{XX}^{-1}\mathbf{\Sigma}_{XY}\mathbf{\Sigma}_{YY}^{-1/2}$ has been proposed. It will appear in our discussion of canonical correlations given in Chapter 10. Real-valued functions of it, such as trace and determinant, will sometimes be of use. The matrix appears in Khatri (1964). Tate (1966) makes remarks concerning multivariate analogs of the correlation coefficient; see also Williams (1967) and Hotelling (1936).

We may define an **error variate** by

$$\boldsymbol{\varepsilon} = \mathbf{Y} - \boldsymbol{\mu}_Y - \mathbf{\Sigma}_{YX}\mathbf{\Sigma}_{XX}^{-1}(\mathbf{X} - \boldsymbol{\mu}_X). \tag{8.2.19}$$

This variate represents the residual after approximating **Y** by the best linear function of **X.** The covariance matrix of $\boldsymbol{\varepsilon}$ is given by

$$\mathbf{\Sigma}_{\varepsilon\varepsilon} = \mathbf{\Sigma}_{YY} - \mathbf{\Sigma}_{YX}\mathbf{\Sigma}_{XX}^{-1}\mathbf{\Sigma}_{XY} \tag{8.2.20}$$

that is, the matrix (8.2.16). The covariance of ε_j with ε_k is called the **partial covariance of Y_j with Y_k.** It measures the linear relation of Y_j with Y_k after the linear effects of **X** have been removed. Similarly the correlation coefficient of ε_j with ε_k is called the **partial correlation of Y_j with Y_k.** These parameters are discussed in Kendall and Stuart (1961) Chap. 27, and Morrison (1967) Chap. 3.

In the case that the variate (8.2.10) has a multivariate normal distribution, the predictor suggested by Theorem 8.2.1 is best within a larger class of predictors.

Theorem 8.2.2 Suppose the variate (8.2.10) is multivariate normal with mean (8.2.11) and covariance matrix (8.2.12). Suppose $\mathbf{\Sigma}_{XX}$ is nonsingular.

The s vector-valued function $\boldsymbol{\phi}(\mathbf{X})$, with $E\{\boldsymbol{\phi}(\mathbf{X})^\tau\boldsymbol{\phi}(\mathbf{X})\} < \infty$, that minimizes

$$E\{[\mathbf{Y} - \boldsymbol{\phi}(\mathbf{X})][\mathbf{Y} - \boldsymbol{\phi}(\mathbf{X})]^\tau\} \tag{8.2.21}$$

is given by

$$\boldsymbol{\phi}(\mathbf{X}) = \boldsymbol{\mu}_Y + \boldsymbol{\Sigma}_{YX}\boldsymbol{\Sigma}_{XX}^{-1}(\mathbf{X} - \boldsymbol{\mu}_X). \tag{8.2.22}$$

The minimum achieved is (8.2.16).

In the case that the variate has a normal distribution, the conditional distribution of $\mathbf{Y}$ given $\mathbf{X}$ is

$$N_s(\boldsymbol{\mu}_Y + \boldsymbol{\Sigma}_{YX}\boldsymbol{\Sigma}_{XX}^{-1}(\mathbf{X} - \boldsymbol{\mu}_X), \boldsymbol{\Sigma}_{YY} - \boldsymbol{\Sigma}_{YX}\boldsymbol{\Sigma}_{XX}^{-1}\boldsymbol{\Sigma}_{XY}) \tag{8.2.23}$$

and so we see that the partial correlation of Y_j with Y_k is the conditional correlation of Y_j with Y_k given $\mathbf{X}$.

We turn to some details of the estimation of the parameters of the above theorems. Suppose that a sample of values

$$\begin{bmatrix} \mathbf{X}_j \\ \mathbf{Y}_j \end{bmatrix} \tag{8.2.24}$$

$j = 1, \ldots, n$ of the variate of Theorem 8.2.1 are available. For convenience assume $\boldsymbol{\mu}_X = \mathbf{0}$ and $\boldsymbol{\mu}_Y = \mathbf{0}$. Define the $r \times n$ matrix $\mathbf{x}$ and the $s \times n$ matrix $\mathbf{y}$ by

$$\mathbf{x} = [\mathbf{X}_1 \cdots \mathbf{X}_n] \tag{8.2.25}$$

$$\mathbf{y} = [\mathbf{Y}_1 \cdots \mathbf{Y}_n]. \tag{8.2.26}$$

We may estimate the covariance matrix (8.2.12) by

$$\hat{\boldsymbol{\Sigma}}_{XX} = \frac{\mathbf{x}\mathbf{x}^\tau}{n},$$

$$\hat{\boldsymbol{\Sigma}}_{XY} = \frac{\mathbf{x}\mathbf{y}^\tau}{n},$$

and

$$\hat{\boldsymbol{\Sigma}}_{YY} = \frac{\mathbf{y}\mathbf{y}^\tau}{n}. \tag{8.2.27}$$

The regression coefficient of $\mathbf{Y}$ on $\mathbf{X}$ may be estimated by

$$\hat{\mathbf{a}} = \hat{\boldsymbol{\Sigma}}_{YX}\hat{\boldsymbol{\Sigma}}_{XX}^{-1} \tag{8.2.28}$$

and the error matrix (8.2.20) may be estimated by

$$\begin{aligned} \hat{\boldsymbol{\Sigma}}_{\varepsilon\varepsilon} &= (n - r)^{-1}\mathbf{y}[\mathbf{I} - \mathbf{x}^\tau(\mathbf{x}\mathbf{x}^\tau)^{-1}\mathbf{x}]\mathbf{y}^\tau \\ &= (n - r)^{-1}n(\hat{\boldsymbol{\Sigma}}_{YY} - \hat{\boldsymbol{\Sigma}}_{YX}\hat{\boldsymbol{\Sigma}}_{XX}^{-1}\hat{\boldsymbol{\Sigma}}_{XY}). \end{aligned} \tag{8.2.29}$$

The reason for the divisor $(n - r)$ rather then n will become apparent in the course of the statement of the next theorem. We have

Theorem 8.2.3 Suppose the values (8.2.24), $j = 1, \ldots, n$, are a sample from a multivariate normal distribution with mean **0** and covariance matrix (8.2.12). Let $\hat{\mathbf{a}}$ be given by (8.2.28) and $\hat{\mathbf{\Sigma}}_{\varepsilon\varepsilon}$ by (8.2.29). Then for any (rs) vector $\boldsymbol{\alpha}$

$$\frac{\boldsymbol{\alpha}^T(\text{vec }[\hat{\mathbf{a}} - \mathbf{a}])}{[\boldsymbol{\alpha}^T\{\hat{\mathbf{\Sigma}}_{\varepsilon\varepsilon} \otimes (\mathbf{x}\mathbf{x}^T)^{-1}\}\boldsymbol{\alpha}]^{1/2}} \tag{8.2.30}$$

is distributed as t_{n-r}. Also $E\hat{\mathbf{a}} = \mathbf{a}$,

$$\text{cov }\{\text{vec }\hat{\mathbf{a}}, \text{vec }\hat{\mathbf{a}}\} = (n - r - 1)^{-1}\, \mathbf{\Sigma}_{\varepsilon\varepsilon} \otimes \mathbf{\Sigma}_{XX}^{-1} \tag{8.2.31}$$

and if $n \rightarrow \infty$, $\hat{\mathbf{a}}$ is asymptotically normal with these moments. Also $\hat{\mathbf{\Sigma}}_{\varepsilon\varepsilon}$ is independent of $\hat{\mathbf{a}}$ and distributed as $(n - r)^{-1}W_s(n - r, \mathbf{\Sigma}_{\varepsilon\varepsilon})$. In the case $s = 1$, $\hat{R}_{YX}{}^2 = \hat{\mathbf{\Sigma}}_{YX}\hat{\mathbf{\Sigma}}_{XX}{}^{-1}\hat{\mathbf{\Sigma}}_{XY}/\hat{\Sigma}_{YY}$ has density function

$$(1 - R_{YX}{}^2)^{n/2}{}_2F_1\left(\frac{n}{2},\frac{n}{2},\frac{r}{2}; R_{XY}{}^2\hat{R}_{YX}{}^2\right)\frac{\Gamma(n/2)}{\Gamma([n - r]/2)\Gamma(r/2)} \times (\hat{R}_{YX})^{r-2}(1 - \hat{R}_{YX}{}^2)^{(n-r-2)/2}. \tag{8.2.32}$$

The function appearing in (8.2.32) is a generalized hypergeometric function; see Abramowitz and Stegun (1964). Percentage points and moments of $\hat{R}_{YX}{}^2$ are given in Amos and Koopmans (1962), Ezekiel and Fox (1959) and Kramer (1963). Olkin and Pratt (1958) construct an unbiased estimate of $R_{YX}{}^2$. The distributions of further statistics may be determined from the fact that the matrix

$$\begin{bmatrix} \hat{\mathbf{\Sigma}}_{XX} & \hat{\mathbf{\Sigma}}_{XY} \\ \hat{\mathbf{\Sigma}}_{YX} & \hat{\mathbf{\Sigma}}_{YY} \end{bmatrix}$$

is distributed as

$$n^{-1}W_{r+s}\left(n, \begin{bmatrix} \mathbf{\Sigma}_{XX} & \mathbf{\Sigma}_{XY} \\ \mathbf{\Sigma}_{YX} & \mathbf{\Sigma}_{YY} \end{bmatrix}\right). \tag{8.2.33}$$

The distribution of $\hat{\mathbf{a}}$ is given in Kshirsagar (1961). Its density function is proportional to

$$\{\text{Det }[\mathbf{\Sigma}_{XX}{}^{-1} + (\hat{\mathbf{a}} - \mathbf{a})^T\mathbf{\Sigma}_{\varepsilon\varepsilon}{}^{-1}(\hat{\mathbf{a}} - \mathbf{a})]\}^{-(n+s)/2}. \tag{8.2.34}$$

This is a form of multivariate t distribution; see Dickey (1967).

Estimates of the partial correlations may be based on the entries of $\hat{\mathbf{\Sigma}}_{\varepsilon\varepsilon}$ in a manner paralleling their definition. For example an estimate of the partial correlation of Y_j and Y_k with **X** held linearly constant is

$$\hat{R}_{Y_j,Y_k\cdot X} = \frac{[\hat{\mathbf{\Sigma}}_{\varepsilon\varepsilon}]_{jk}}{\{[\hat{\mathbf{\Sigma}}_{\varepsilon\varepsilon}]_{jj}[\hat{\mathbf{\Sigma}}_{\varepsilon\varepsilon}]_{kk}\}^{1/2}}. \tag{8.2.35}$$

with $[\hat{\mathbf{\Sigma}}_{\varepsilon\varepsilon}]_{jk}$ denoting the entry in row j, column k of $\hat{\mathbf{\Sigma}}_{\varepsilon\varepsilon}$.

From the distribution of $\hat{\mathbf{\Sigma}}_{\varepsilon\varepsilon}$ given in Theorem 8.2.3, we see that this expression is distributed as the sample correlation coefficient of ε_j with ε_k based on $n - r$ observations. The density function of its square will be given by expression (8.2.32) with $R_{YX}{}^2$, $\hat{R}_{YX}{}^2$, n, r replaced by $R^2_{Y_j,Y_k\cdot X}$, $\hat{R}^2_{Y_j,Y_k\cdot X}$, $n - r$, 1, respectively. The large sample variance of this $\hat{R}^2$ is approximately $4R^2[1 - R^2]/n$. The distribution of correlation coefficients developed in Fisher (1962) may be modified to obtain the joint distribution of all the partial correlations. The asymptotic joint covariance structure may be deduced from the results of Pearson and Filon (1898), Hall (1927), and Hsu (1949). Further results and approximations to the distributions of estimates of squared correlation coefficients are given in Kendall and Stuart (1961) p. 341, Gajjar (1967), Hodgson (1968), Alexander and Vok (1963), Giri (1965), and Gurland (1966).

There are complex variate analogs of the preceding theorems. For example:

Theorem 8.2.4 Let the $(r + s)$ vector-valued variate

$$\begin{bmatrix} \mathbf{X} \\ \mathbf{Y} \end{bmatrix} \tag{8.2.36}$$

have complex entries, mean $\mathbf{0}$ and be such that

$$E\left\{\begin{bmatrix} \mathbf{X}\overline{\mathbf{X}}^\tau & \mathbf{X}\overline{\mathbf{Y}}^\tau \\ \mathbf{Y}\overline{\mathbf{X}}^\tau & \mathbf{Y}\overline{\mathbf{Y}}^\tau \end{bmatrix}\right\} = \begin{bmatrix} \mathbf{\Sigma}_{XX} & \mathbf{\Sigma}_{XY} \\ \mathbf{\Sigma}_{YX} & \mathbf{\Sigma}_{YY} \end{bmatrix} \tag{8.2.37}$$

and

$$E\left\{\begin{bmatrix} \mathbf{X}\mathbf{X}^\tau & \mathbf{X}\mathbf{Y}^\tau \\ \mathbf{Y}\mathbf{X}^\tau & \mathbf{Y}\mathbf{Y}^\tau \end{bmatrix}\right\} = \mathbf{0}. \tag{8.2.38}$$

Suppose $\mathbf{\Sigma}_{XX}$ is nonsingular. Then the $\boldsymbol{\mu}$ and $\mathbf{a}$ minimizing

$$E\{[\mathbf{Y} - \boldsymbol{\mu} - \mathbf{a}\mathbf{X}]\overline{[\mathbf{Y} - \boldsymbol{\mu} - \mathbf{a}\mathbf{X}]^\tau}\} \tag{8.2.39}$$

are given by

$$\boldsymbol{\mu} = 0$$

$$\mathbf{a} = \mathbf{\Sigma}_{YX}\mathbf{\Sigma}_{XX}{}^{-1}. \tag{8.2.40}$$

The minimum achieved is

$$\mathbf{\Sigma}_{YY} - \mathbf{\Sigma}_{YX}\mathbf{\Sigma}_{XX}{}^{-1}\mathbf{\Sigma}_{XY}. \tag{8.2.41}$$

We call $\mathbf{a}$, given by (8.2.40), the **complex regression coefficient of Y on X.** It is a consequence that the indicated $\boldsymbol{\mu}$, $\mathbf{a}$ also minimize the determinant, trace, and diagonal entries of (8.2.39). In the case $s = 1$ the minimum

(8.2.41) may be written

$$[1 - |R_{YX}|^2]\Sigma_{YY}, \tag{8.2.42}$$

where we define

$$|R_{YX}|^2 = \frac{\mathbf{\Sigma}_{YX}\mathbf{\Sigma}_{XX}^{-1}\mathbf{\Sigma}_{XY}}{\Sigma_{YY}}. \tag{8.2.43}$$

This parameter is clearly an extension to the complex-valued case of the squared coefficient of multiple correlation. Because the minimum (8.2.41) must lie between $\mathbf{\Sigma}_{YY}$ and $\mathbf{0}$, it follows that $0 \leqslant |R_{YX}|^2 \leqslant 1$, the value 1 occurring when the minimum is 0. On occasion we may wish to partition $|R_{YX}|^2$ into

$$\frac{[\text{Re } \mathbf{\Sigma}_{YX}]\mathbf{\Sigma}_{XX}^{-1}[\text{Re } \mathbf{\Sigma}_{XY}]}{\Sigma_{YY}}, \tag{8.2.44}$$

and

$$\frac{[\text{Im } \mathbf{\Sigma}_{YX}]\mathbf{\Sigma}_{XX}^{-1}[\text{Im } \mathbf{\Sigma}_{XY}]}{\Sigma_{YY}}, \tag{8.2.45}$$

where we have $\mathbf{\Sigma}_{YX} = \text{Re } \mathbf{\Sigma}_{YX} + i \text{ Im } \mathbf{\Sigma}_{YX}$. These expressions are measures of the degree of linear relation of Y with Re **X** and Im **X** respectively.

Returning now to the case of vector-valued **Y**, a direct measure of the degree of approximation of **Y** by a linear function of **X** is provided by the error variate

$$\boldsymbol{\varepsilon} = \mathbf{Y} - \boldsymbol{\mu}_Y - \mathbf{\Sigma}_{YX}\mathbf{\Sigma}_{XX}^{-1}(\mathbf{X} - \boldsymbol{\mu}_X), \tag{8.2.46}$$

which has mean **0** and is such that

$$\begin{aligned} E\boldsymbol{\varepsilon}\bar{\boldsymbol{\varepsilon}}^\tau &= \mathbf{\Sigma}_{\varepsilon\varepsilon} \\ &= \mathbf{\Sigma}_{YY} - \mathbf{\Sigma}_{YX}\mathbf{\Sigma}_{XX}^{-1}\mathbf{\Sigma}_{XY}, \end{aligned} \tag{8.2.47}$$

and

$$E\boldsymbol{\varepsilon}\boldsymbol{\varepsilon}^\tau = \mathbf{0}. \tag{8.2.48}$$

Analogs of the partial covariance and partial correlation may be based on the matrix (8.2.47) in an immediate manner.

Suppose now that a sample of values

$$\begin{bmatrix} \mathbf{X}_j \\ \mathbf{Y}_j \end{bmatrix} \qquad j = 1, \ldots, n \tag{8.2.49}$$

of the variate of Theorem 8.2.4 are available. Define matrices **x** and **y** as in (8.2.25) and (8.2.26). We are led to construct the statistics

$$\hat{\mathbf{\Sigma}}_{XX} = \frac{\mathbf{x}\bar{\mathbf{x}}^\tau}{n} = \frac{\sum_j \mathbf{X}_j\bar{\mathbf{X}}_j^\tau}{n}$$

$$\hat{\mathbf{\Sigma}}_{XY} = \frac{\mathbf{x}\bar{\mathbf{y}}^\tau}{n} = \frac{\sum_j \mathbf{X}_j\bar{\mathbf{Y}}_j^\tau}{n}$$

$$\hat{\mathbf{\Sigma}}_{YY} = \frac{\mathbf{y}\bar{\mathbf{y}}^\tau}{n} = \frac{\sum_j \mathbf{Y}_j\bar{\mathbf{Y}}_j^\tau}{n}, \tag{8.2.50}$$

and

$$\hat{\mathbf{a}} = \hat{\mathbf{\Sigma}}_{YX}\hat{\mathbf{\Sigma}}_{XX}^{-1}, \tag{8.2.51}$$

$$\hat{\mathbf{\Sigma}}_{\varepsilon\varepsilon} = (n - r)^{-1}\mathbf{y}[\mathbf{I} - \bar{\mathbf{x}}^\tau(\mathbf{x}\bar{\mathbf{x}}^\tau)^{-1}\mathbf{x}]\bar{\mathbf{y}}^\tau$$

$$= (n - r)^{-1}n(\hat{\mathbf{\Sigma}}_{YY} - \hat{\mathbf{\Sigma}}_{YX}\hat{\mathbf{\Sigma}}_{XX}^{-1}\hat{\mathbf{\Sigma}}_{XY}), \tag{8.2.52}$$

which leads us to

Theorem 8.2.5 Suppose values of the form (8.2.49), $j = 1, \ldots, n$, are a sample from a complex multivariate normal distribution with mean $\mathbf{0}$ and covariance matrix (8.2.37). Let $\hat{\mathbf{a}}$ be given by (8.2.51) and $\hat{\mathbf{\Sigma}}_{\varepsilon\varepsilon}$ by (8.2.52). Then for any (rs) vector $\boldsymbol{\alpha}$

$$\frac{\boldsymbol{\alpha}^\tau(\text{vec}\,[\hat{\mathbf{a}} - \mathbf{a}])}{[\boldsymbol{\alpha}^\tau(\hat{\mathbf{\Sigma}}_{\varepsilon\varepsilon} \otimes (\mathbf{x}\bar{\mathbf{x}}^\tau)^{-1})\bar{\boldsymbol{\alpha}}]^{1/2}} \tag{8.2.53}$$

is distributed as $t_{2\,(n-r)}^C$. Also $E\hat{\mathbf{a}} = \mathbf{a}$,

$$\text{cov}\,\{\text{vec}\,\hat{\mathbf{a}}, \text{vec}\,\hat{\mathbf{a}}\} = (n - r)^{-1}\mathbf{\Sigma}_{\varepsilon\varepsilon} \otimes \mathbf{\Sigma}_{XX}^{-1} \tag{8.2.54}$$

and if $n \to \infty$, vec $\hat{\mathbf{a}}$ is asymptotically $N_{rs}^C(\text{vec}\,\mathbf{a}, n^{-1}\mathbf{\Sigma}_{\varepsilon\varepsilon} \otimes \mathbf{\Sigma}_{XX}^{-1})$. Continuing $\hat{\mathbf{\Sigma}}_{\varepsilon\varepsilon}$ is independent of $\hat{\mathbf{a}}$ and distributed as $(n - r)^{-1}W_s^C(n - r, \mathbf{\Sigma}_{\varepsilon\varepsilon})$. Finally in the case $s = 1$ the density function of $|\hat{R}_{YX}|^2 = \hat{\mathbf{\Sigma}}_{YX}\hat{\mathbf{\Sigma}}_{XX}^{-1}\hat{\mathbf{\Sigma}}_{XY}/\hat{\Sigma}_{YY}$ is

$$(1 - |R_{YX}|^2)^n {}_2F_1(n,n;r;|R_{YX}|^2\hat{R}|_{YX}|^2)\frac{\Gamma(n)}{\Gamma(n - r)\Gamma(r)}|\hat{R}_{YX}|^{2r-2}(1 - |\hat{R}_{YX}|^2)^{n-r-1}. \tag{8.2.55}$$

We note that the distribution of $|\hat{R}_{YX}|^2$ in the complex case is identical with the real case distribution having twice the sample size and twice the $\mathbf{X}$ dimension. The heuristic approach described in Section 8.4 will suggest the reason for this occurrence. A useful consequence is that we may use tables and results derived for the real case. The density function (8.2.55) is given in Goodman (1963); see also James (1964) expression (112), and Khatri (1965a). In the case $|R_{YX}|^2 = 0$, expression (8.2.55) becomes

$$\frac{\Gamma(n)}{\Gamma(n - r)\Gamma(r)}|\hat{R}_{YX}|^{2r-2}(1 - |\hat{R}_{YX}|^2)^{n-r-1}. \tag{8.2.56}$$

This is the same as the null distribution of (6.2.10) derived under the assumption of fixed **X**. Percentage points in this case may therefore be derived from F percentage points as they were in Chapter 6. Amos and Koopmans (1962) and Groves and Hannan (1968) provide a variety of non-null percentage points for $|R_{YX}|^2$.

Confidence regions for the entries of **a** may be constructed from expression (8.2.53) in the manner of Section 6.2.

By analogy with (8.2.34) the density function of **a** will be proportional to

$$\{\text{Det}\,[\mathbf{\Sigma}_{XX}{}^{-1} + \overline{(\hat{\mathbf{a}} - \mathbf{a})}^{\tau}\mathbf{\Sigma}_{\varepsilon\varepsilon}{}^{-1}(\hat{\mathbf{a}} - \mathbf{a})]\}^{-(n+s)}. \tag{8.2.57}$$

Wahba (1966) determined this density in the case $s = 1$.

Sometimes it is of interest to consider the following complex analogs of the partial correlations

$$R_{Y_j,Y_k\cdot X} = \frac{[\mathbf{\Sigma}_{\varepsilon\varepsilon}]_{jk}}{\{[\mathbf{\Sigma}_{\varepsilon\varepsilon}]_{jj}[\mathbf{\Sigma}_{\varepsilon\varepsilon}]_{kk}\}^{1/2}} \tag{8.2.58}$$

for $1 \leqslant j \neq k \leqslant s$. A natural estimate of (8.2.58) is provided by

$$\hat{R}_{Y_j,Y_k\cdot X} = \frac{[\hat{\mathbf{\Sigma}}_{\varepsilon\varepsilon}]_{jk}}{\{[\hat{\mathbf{\Sigma}}_{\varepsilon\varepsilon}]_{jj}[\hat{\mathbf{\Sigma}}_{\varepsilon\varepsilon}]_{kk}\}^{1/2}}. \tag{8.2.59}$$

We see from the distribution of $\hat{\mathbf{\Sigma}}_{\varepsilon\varepsilon}$ given in Theorem 8.2.5 that this last is distributed as the sample complex correlation coefficient of ε_j with ε_k based on $n - r$ observations. Its modulus-square will have density function (8.2.55) with the replacement of R_{YX}, $\hat{R}_{YX}$, n, r by $R_{Y_j,Y_k\cdot X}$, $\hat{R}_{Y_j,Y_k\cdot X}$, $n - r$, 1, respectively. The asymptotic covariances of pairs of these estimates may be deduced from expression (7.6.16).

8.3 DETERMINATION OF AN OPTIMUM LINEAR FILTER

We return to the notation of Section 8.1 and the problem of determining an s vector, $\boldsymbol{\mu}$, and an $s \times r$ filter, $\{\mathbf{a}(u)\}$, so that

$$\boldsymbol{\mu} + \sum_{u=-\infty}^{\infty} \mathbf{a}(t - u)\mathbf{X}(u) \tag{8.3.1}$$

is close to $\mathbf{Y}(t)$. Suppose we measure closeness by the $s \times s$ Hermitian matrix

$$E\{[\mathbf{Y}(t) - \boldsymbol{\mu} - \sum_u \mathbf{a}(t - u)\mathbf{X}(u)][\mathbf{Y}(t) - \boldsymbol{\mu} - \sum_u \mathbf{a}(t - u)\mathbf{X}(u)]^{\tau}\}. \tag{8.3.2}$$

We then have

Theorem 8.3.1 Consider an $(r + s)$ vector-valued second-order stationary time series of the form (8.1.1) with mean (8.1.2) and autocovariance function (8.1.3). Suppose $\mathbf{c}_{XX}(u)$, $\mathbf{c}_{YY}(u)$ are absolutely summable and suppose $\mathbf{f}_{XX}(\lambda)$, given by (8.1.4), is nonsingular, $-\infty < \lambda < \infty$. Then the $\boldsymbol{\mu}$ and $\mathbf{a}(u)$ that minimize (8.3.2) are given by

$$\boldsymbol{\mu} = \mathbf{c}_Y - \left(\sum_u \mathbf{a}(u)\right)\mathbf{c}_X = \mathbf{c}_Y - \mathbf{A}(0)\mathbf{c}_X, \tag{8.3.3}$$

and

$$\mathbf{a}(u) = (2\pi)^{-1}\int_0^{2\pi} \mathbf{A}(\alpha)\exp\{iu\alpha\}\,d\alpha, \tag{8.3.4}$$

where

$$\mathbf{A}(\lambda) = \mathbf{f}_{YX}(\lambda)\mathbf{f}_{XX}(\lambda)^{-1}. \tag{8.3.5}$$

The filter $\{\mathbf{a}(u)\}$ is absolutely summable. The minimum achieved is

$$\int_0^{2\pi} [\mathbf{f}_{YY}(\alpha) - \mathbf{f}_{YX}(\alpha)\mathbf{f}_{XX}(\alpha)^{-1}\mathbf{f}_{XY}(\alpha)]\,d\alpha. \tag{8.3.6}$$

$\mathbf{A}(\lambda)$, given by expression (8.3.5), is the transfer function of the $s \times r$ filter achieving the indicated minimum. We call $\mathbf{A}(\lambda)$ the **complex regression coefficient** of $\mathbf{Y}(t)$ on $\mathbf{X}(t)$ at frequency λ.

The s vector-valued series

$$\boldsymbol{\varepsilon}(t) = \mathbf{Y}(t) - \boldsymbol{\mu} - \sum_u \mathbf{a}(t-u)\mathbf{X}(u) \qquad t = 0, \pm 1, \ldots, \tag{8.3.7}$$

where $\boldsymbol{\mu}$ and $\mathbf{a}(u)$ are given in Theorem, 8.3.1 is called the **error series.** It is seen to have 0 mean and spectral density matrix

$$\mathbf{f}_{\varepsilon\varepsilon}(\lambda) = \mathbf{f}_{YY}(\lambda) - \mathbf{f}_{YX}(\lambda)\mathbf{f}_{XX}(\lambda)^{-1}\mathbf{f}_{XY}(\lambda). \tag{8.3.8}$$

$\mathbf{f}_{\varepsilon\varepsilon}(\lambda)$ is called the **error spectrum.** We may write (8.3.8) in the form

$$\mathbf{f}_{\varepsilon\varepsilon}(\lambda) = \mathbf{f}_{YY}(\lambda)^{1/2}[\mathbf{I} - \mathbf{f}_{YY}(\lambda)^{-1/2}\mathbf{f}_{YX}(\lambda)\mathbf{f}_{XX}(\lambda)^{-1}\mathbf{f}_{XY}(\lambda)\mathbf{f}_{YY}(\lambda)^{-1/2}]\mathbf{f}_{YY}(\lambda)^{1/2}, \tag{8.3.9}$$

and thus we are led to measure the linear association of $\mathbf{Y}(t)$ with $\mathbf{X}(t)$ by the $s \times s$ matrix

$$\mathbf{f}_{YY}(\lambda)^{-1/2}\mathbf{f}_{YX}(\lambda)\mathbf{f}_{XX}(\lambda)^{-1}\mathbf{f}_{XY}(\lambda)\mathbf{f}_{YY}(\lambda)^{-1/2}. \tag{8.3.10}$$

In the case that $s = 1$, (8.3.10) is called the **multiple coherence** of $Y(t)$ with $\mathbf{X}(t)$ at frequency λ. We denote it by $|R_{YX}(\lambda)|^2$ and write

$$|R_{YX}(\lambda)|^2 = \frac{\mathbf{f}_{YX}(\lambda)\mathbf{f}_{XX}(\lambda)^{-1}\mathbf{f}_{XY}(\lambda)}{f_{YY}(\lambda)}. \tag{8.3.11}$$

(In the case, $r, s = 1$, we define the **coherency** $R_{YX}(\lambda) = f_{YX}(\lambda)/[f_{XX}(\lambda)f_{YY}(\lambda)]^{1/2}$.) The multiple coherence satisfies the inequalities

$$0 \leqslant |R_{YX}(\lambda)|^2 \leqslant 1 \tag{8.3.12}$$

(see Exercise 8.16.35) and measures the extent to which the real-valued $Y(t)$ is determinable from the r vector-valued $\mathbf{X}(t)$ by linear time invariant operations. We write

$$f_{\varepsilon\varepsilon}(\lambda) = [1 - |R_{YX}(\lambda)|^2] f_{YY}(\lambda) \tag{8.3.13}$$

and see that $|R_{YX}(\lambda)|^2 = 0$ corresponds to the incoherent case in which $X(t)$ does not reduce the error variance. The value $|R_{YX}(\lambda)|^2 = 1$ corresponds to the perfectly coherent case in which the error series is reduced to 0. The coefficient of multiple coherence was defined in Goodman (1963); see also Koopmans (1964a,b).

Returning to the case of general s we call the cross-spectrum between the ath and bth components of the error series, $\varepsilon_a(t)$ and $\varepsilon_b(t)$, the **partial cross-spectrum of $Y_a(t)$ with $Y_b(t)$ after removing the linear effects of $\mathbf{X}(t)$**. It is given by

$$f_{Y_aY_b\cdot X}(\lambda) = f_{Y_aY_b}(\lambda) - \mathbf{f}_{Y_aX}(\lambda)\mathbf{f}_{XX}(\lambda)^{-1}\mathbf{f}_{XY_b}(\lambda) = f_{\varepsilon_a\varepsilon_b}(\lambda) \tag{8.3.14}$$

$-\infty < \lambda < \infty$. We call the coherency of these components, the **partial coherency of $Y_a(t)$ with $Y_b(t)$ after removing the linear effects of $\mathbf{X}(t)$**. It is given by

$$R_{Y_aY_b\cdot X}(\lambda) = \frac{f_{Y_aY_b\cdot X}(\lambda)}{[f_{Y_aY_a\cdot X}(\lambda)\, f_{Y_bY_b\cdot X}(\lambda)]^{1/2}}. \tag{8.3.15}$$

These last parameters are of use in determining the extent to which an apparent time invariant linear relation between the series $Y_a(t)$ and $Y_b(t)$ is due to the linear relation of each to a series $\mathbf{X}(t)$; see Gersch (1972). We can likewise define the **partial complex regression coefficient of $Y_a(t)$ on $Y_b(t)$ after removing the linear effects of $\mathbf{X}(t)$** to be

$$\frac{f_{Y_aY_b\cdot X}(\lambda)}{f_{Y_bY_b\cdot X}(\lambda)}. \tag{8.3.16}$$

As would have been expected from the situation in the real variate case it turns out that expression (8.3.16) is the entry corresponding to $Y_b(t)$ in the matrix-valued complex regression coefficient of $Y_a(t)$ on the $r + 1$ vector-valued series

$$\begin{bmatrix} Y_b(t) \\ \mathbf{X}(t) \end{bmatrix}.$$

This gives us an interpretation for the individual entries of a matrix-valued complex regression coefficient.

The above parameters of the partial cross-spectral analysis of time series were introduced by Tick (1963) and Wonnacott, see Granger (1964) p. xiii. They are studied further in Koopmans (1964b), Goodman (1965), Akaike (1965), Parzen (1967c), and Jenkins and Watt (1968).

As an example of the values of these various parameters consider the model

$$\mathbf{Y}(t) = \boldsymbol{\mu} + \sum_u \mathbf{a}(t-u)\mathbf{X}(u) + \boldsymbol{\varepsilon}(t), \tag{8.3.17}$$

where $\mathbf{X}(t)$ is r vector-valued, stationary with spectral density matrix $\mathbf{f}_{XX}(\lambda)$; $\boldsymbol{\varepsilon}(t)$ is s vector-valued, stationary, mean $\mathbf{0}$, with spectral density matrix, $\mathbf{f}_{\varepsilon\varepsilon}(\lambda)$, and independent of $\mathbf{X}(t)$ at all lags; $\boldsymbol{\mu}$ is an s vector; and $\{\mathbf{a}(u)\}$ is an absolutely summable $s \times r$ matrix-valued filter. We quickly see that the complex regression coefficient of $\mathbf{Y}(t)$ on $\mathbf{X}(t)$ is given by

$$\mathbf{A}(\lambda) = \sum_u \mathbf{a}(u) \exp\{-i\lambda u\}. \tag{8.3.18}$$

Also

$$f_{Y_aY_b\cdot X}(\lambda) = f_{\varepsilon_a\varepsilon_b}(\lambda), \tag{8.3.19}$$

and so

$$R_{Y_aY_b\cdot X}(\lambda) = R_{\varepsilon_a\varepsilon_b}(\lambda). \tag{8.3.20}$$

In the case that the series (8.1.1) is Gaussian a direct interpretation may be placed on $\boldsymbol{\mu}$ and $\mathbf{a}(u)$ of Theorem 8.3.1. We have

Theorem 8.3.2 Under the conditions of Theorem 8.3.1 and if the series (8.1.1) is Gaussian, $\boldsymbol{\mu}$ and $\mathbf{a}(u)$ of (8.3.3) and (8.3.4) are given by

$$E\{\mathbf{Y}(t) \mid \mathbf{X}(v), v = 0, \pm 1, \ldots\} = \boldsymbol{\mu} + \sum_{u=-\infty}^{\infty} \mathbf{a}(t-u)\mathbf{X}(u). \tag{8.3.21}$$

Also

$$\operatorname{cov}\{\mathbf{Y}(t+u), \mathbf{Y}(t) \mid \mathbf{X}(v), v = 0, \pm 1, \ldots\}$$
$$= \int_0^{2\pi} [\mathbf{f}_{YY}(\lambda) - \mathbf{f}_{YX}(\lambda)\mathbf{f}_{XX}(\lambda)^{-1}\mathbf{f}_{XY}(\lambda)] \exp\{i\lambda u\} d\lambda. \tag{8.3.22}$$

General references to the previous development, in the case $r, s = 1$, include: Wiener (1949), Solodovnikov (1950), Koopmans (1964a), and Blackman (1965). There are a variety of connections between the approach of this section and that of Chapter 6. The principal difference in assumption is that the series $\mathbf{X}(t)$ is now stochastic rather than fixed. The model of Chapter 6 was

$$\mathbf{Y}(t) = \boldsymbol{\mu} + \sum_u \mathbf{a}(t-u)\mathbf{X}(u) + \boldsymbol{\varepsilon}(t) \tag{8.3.23}$$

with $\boldsymbol{\mu}$ constant, $\mathbf{a}(u)$ a summable filter, and $\boldsymbol{\varepsilon}(t)$ a $\mathbf{0}$ mean error series. Exercise 8.16.33 is to show that such a model holds under the conditions of Theorem 8.3.1.

We end this section with an example of the application of Theorem 8.3.1. Suppose that $\boldsymbol{\eta}(t)$ and $\mathbf{Y}(t)$ are independent s vector-valued, $\mathbf{0}$ mean stationary series. Suppose that the series $\mathbf{X}(t)$ is given by

$$\mathbf{X}(t) = \mathbf{Y}(t) + \boldsymbol{\eta}(t). \tag{8.3.24}$$

The series $\mathbf{Y}(t)$ may be thought of as a signal immersed in a noise series $\boldsymbol{\eta}(t)$. Suppose that we wish to approximate $\mathbf{Y}(t)$ by a filtered version of $\mathbf{X}(t)$. The spectral density matrix of $\mathbf{X}(t)$ and $\mathbf{Y}(t)$ is given by

$$\begin{bmatrix} \mathbf{f}_{YY}(\lambda) + \mathbf{f}_{\eta\eta}(\lambda) & \mathbf{f}_{YY}(\lambda) \\ \mathbf{f}_{YY}(\lambda) & \mathbf{f}_{YY}(\lambda) \end{bmatrix}. \tag{8.3.25}$$

Following expression (8.3.5) the transfer function of the best linear filter for determining $\mathbf{Y}(t)$ from $\mathbf{X}(t)$ is given by

$$\mathbf{A}(\lambda) = \mathbf{f}_{YY}(\lambda)(\mathbf{f}_{YY}(\lambda) + \mathbf{f}_{\eta\eta}(\lambda))^{-1}. \tag{8.3.26}$$

This $\mathbf{A}(\lambda)$ is called the **matched filter** for the signal $\mathbf{Y}(t)$ in the noise $\boldsymbol{\eta}(t)$. We see its general character is one of not passing the frequency components of $\mathbf{X}(t)$ in frequency intervals where $\mathbf{f}_{\eta\eta}(\lambda)$ is very large relative to $\mathbf{f}_{YY}(\lambda)$, while the components are passed virtually unaltered in intervals where $\mathbf{f}_{\eta\eta}(\lambda)$ is small relative to $\mathbf{f}_{YY}(\lambda)$. In the case $s = 1$, the parameter $f_{YY}(\lambda)/f_{\eta\eta}(\lambda)$ is called the **signal to noise ratio** at frequency λ.

8.4 HEURISTIC INTERPRETATION OF PARAMETERS AND CONSTRUCTION OF ESTIMATES

Let the $(r + s)$ vector-valued series

$$\begin{bmatrix} \mathbf{X}(t) \\ \mathbf{Y}(t) \end{bmatrix} \tag{8.4.1}$$

$t = 0, \pm 1, \ldots$ satisfy Assumption 2.6.1 and suppose its values are available for $t = 0, \ldots, T - 1$. We evaluate the finite Fourier transform of these values

$$\begin{bmatrix} \mathbf{d}_X^{(T)}(\lambda) \\ \mathbf{d}_Y^{(T)}(\lambda) \end{bmatrix} = \sum_{t=0}^{T-1} \exp\{-i\lambda t\} \begin{bmatrix} \mathbf{X}(t) \\ \mathbf{Y}(t) \end{bmatrix} \tag{8.4.2}$$

$-\infty < \lambda < \infty$. Following Theorem 4.4.2, for large T, this variate will be distributed approximately as

$$N_{r+s}^C\left(\mathbf{0}, 2\pi T \begin{bmatrix} \mathbf{f}_{XX}(\lambda) & \mathbf{f}_{XY}(\lambda) \\ \mathbf{f}_{YX}(\lambda) & \mathbf{f}_{YY}(\lambda) \end{bmatrix}\right) \tag{8.4.3}$$

if $\lambda \not\equiv 0 \pmod{\pi}$.

Referring to the discussion of Theorem 8.2.4, we now see that $\mathbf{A}(\lambda)$, the complex regression coefficient of $\mathbf{Y}(t)$ on $\mathbf{X}(t)$ at frequency λ may be interpreted, approximately, as the complex regression coefficient of $\mathbf{d}_Y^{(T)}(\lambda)$ on $\mathbf{d}_X^{(T)}(\lambda)$. It is therefore of use in the prediction of the value of $\mathbf{d}_Y^{(T)}(\lambda)$ from that of $\mathbf{d}_X^{(T)}(\lambda)$ in a linear manner. The error spectrum, $\mathbf{f}_{\varepsilon\varepsilon}(\lambda)$, is approximately proportional to the covariance matrix of the error variate of this prediction problem. Likewise the partial complex regression coefficient of $Y_a(t)$ on $Y_b(t)$ after removing the linear effects of $\mathbf{X}(t)$ is nearly the complex regression coefficient of $d_{Y_a}{}^{(T)}(\lambda)$ on $d_{Y_b}{}^{(T)}(\lambda)$ after removing the linear effects of $\mathbf{d}_X^{(T)}(\lambda)$. Continuing, suppose $s = 1$. We see that $|R_{YX}(\lambda)|^2$, the multiple coherence of $Y(t)$ with $\mathbf{X}(t)$ at frequency λ, may, following the discussion of Theorem 8.2.4, be interpreted as the complex analog of the squared coefficient of multiple correlation of $d_Y^{(T)}(\lambda)$ with $\mathbf{d}_X^{(T)}(\lambda)$. Finally the partial coherency of $Y_a(t)$ with $Y_b(t)$ after removing the linear effects of $\mathbf{X}(t)$ may be interpreted as the complex analog of the partial correlation of $d_{Y_a}{}^{(T)}(\lambda)$ with $d_{Y_b}{}^{(T)}(\lambda)$ after removing the linear effects of $\mathbf{d}_X^{(T)}(\lambda)$. In the case that the series (8.4.1) is Gaussian these partial parameters will be approximately conditional parameters given the value $\mathbf{d}_X^{(T)}(\lambda)$.

Similar interpretations may be given in the case $\lambda \equiv 0 \pmod{\pi}$. Real-valued statistics and distributions will be involved in this case.

Let us next turn to the construction of estimates of the various parameters. Suppose $s(T)$ is an integer with $2\pi s(T)/T$ near λ, where we take $\lambda \not\equiv 0 \pmod{\pi}$. Following Theorem 4.4.1, the values

$$\begin{bmatrix} \mathbf{d}_X^{(T)}\left(\dfrac{2\pi[s(T)+s]}{T}\right) \\ \mathbf{d}_Y^{(T)}\left(\dfrac{2\pi[s(T)+s]}{T}\right) \end{bmatrix} \tag{8.4.4}$$

$s = 0, \pm 1, \ldots, \pm m$ will be approximately independent realizations of the variate (8.4.3). Following the discussion of Theorem 8.2.5, specifically expression (8.2.50), we can consider forming the statistics

$$\begin{bmatrix} \mathbf{f}_{XX}^{(T)}(\lambda) & \mathbf{f}_{XY}^{(T)}(\lambda) \\ \mathbf{f}_{YX}^{(T)}(\lambda) & \mathbf{f}_{YY}^{(T)}(\lambda) \end{bmatrix} = (2m+1)^{-1}(2\pi T)^{-1} \sum_{s=-m}^{m} \begin{bmatrix} \mathbf{d}_X^{(T)}(2\pi[s(T)+s]/T) \\ \mathbf{d}_Y^{(T)}(2\pi[s(T)+s]/T) \end{bmatrix} \times \overline{\begin{bmatrix} \mathbf{d}_X^{(T)}(2\pi[s(T)+s]/T) \\ \mathbf{d}_Y^{(T)}(2\pi[s(T)+s]/T) \end{bmatrix}}^{\tau}, \tag{8.4.5}$$

$$\mathbf{A}^{(T)}(\lambda) = \mathbf{f}_{YX}^{(T)}(\lambda)\mathbf{f}_{XX}^{(T)}(\lambda)^{-1}, \tag{8.4.6}$$

and

$$\mathbf{g}_{\varepsilon\varepsilon}^{(T)}(\lambda) = (2m+1-r)^{-1}(2m+1) \times [\mathbf{f}_{YY}^{(T)}(\lambda) - \mathbf{f}_{YX}^{(T)}(\lambda)\mathbf{f}_{XX}^{(T)}(\lambda)^{-1}\mathbf{f}_{XY}^{(T)}(\lambda)], \tag{8.4.7}$$

in turn, the latter two being estimates of $\mathbf{A}(\lambda)$, $\mathbf{f}_{\varepsilon\varepsilon}(\lambda)$, respectively. Theorem 8.2.5 suggests approximations to the distributions of these statistics. In

Section 8.6 we will make the definition (8.4.5) more flexible by including weights in the summation.

Heuristic approaches to the linear analysis of multivatiate series are given in Tick (1963), Akaike (1965), and Groves and Hannan (1968). A discussion of the parameters and estimates is given in Fishman (1969).

We may also provide an interpretation of the parameters of Section 8.3 by means of the frequency components $\mathbf{X}(t,\lambda)$, $\mathbf{Y}(t,\lambda)$, $t = 0, \pm 1, \ldots$ and their Hilbert transforms $\mathbf{X}^H(t,\lambda)$, $\mathbf{Y}^H(t,\lambda)$, $t = 0, \pm 1, \ldots$. From the discussion of Section 7.1 we see that the covariance matrix of the variate

$$\begin{bmatrix} \mathbf{X}(t,\lambda) \\ \mathbf{Y}(t,\lambda) \\ \mathbf{X}^H(t,\lambda) \\ \mathbf{Y}^H(t,\lambda) \end{bmatrix} \tag{8.4.8}$$

is, approximately, proportional to

$$\begin{bmatrix} \operatorname{Re} \mathbf{f}_{XX}(\lambda) & \operatorname{Re} \mathbf{f}_{XY}(\lambda) & \operatorname{Im} \mathbf{f}_{XX}(\lambda) & \operatorname{Im} \mathbf{f}_{XY}(\lambda) \\ \operatorname{Re} \mathbf{f}_{YX}(\lambda) & \operatorname{Re} \mathbf{f}_{YY}(\lambda) & \operatorname{Im} \mathbf{f}_{YX}(\lambda) & \operatorname{Im} \mathbf{f}_{YY}(\lambda) \\ -\operatorname{Im} \mathbf{f}_{XX}(\lambda) & -\operatorname{Im} \mathbf{f}_{XY}(\lambda) & \operatorname{Re} \mathbf{f}_{XX}(\lambda) & \operatorname{Re} \mathbf{f}_{XY}(\lambda) \\ -\operatorname{Im} \mathbf{f}_{YX}(\lambda) & -\operatorname{Im} \mathbf{f}_{YY}(\lambda) & \operatorname{Re} \mathbf{f}_{YX}(\lambda) & \operatorname{Re} \mathbf{f}_{YY}(\lambda) \end{bmatrix}$$
$$= \begin{bmatrix} \mathbf{f}_{XX}(\lambda) & \mathbf{f}_{XY}(\lambda) \\ \mathbf{f}_{YX}(\lambda) & \mathbf{f}_{YY}(\lambda) \end{bmatrix}^R. \tag{8.4.9}$$

Now

$$\mathbf{A}(\lambda) = \mathbf{f}_{YX}(\lambda)\mathbf{f}_{XX}(\lambda)^{-1} \tag{8.4.10}$$

and so

$$[\operatorname{Re} \mathbf{A}(\lambda) \quad \operatorname{Im} \mathbf{A}(\lambda)] = [\operatorname{Re} \mathbf{f}_{YX}(\lambda) \quad \operatorname{Im} \mathbf{f}_{YX}(\lambda)] \begin{bmatrix} \operatorname{Re} \mathbf{f}_{XX}(\lambda) & \operatorname{Im} \mathbf{f}_{XX}(\lambda) \\ -\operatorname{Im} \mathbf{f}_{XX}(\lambda) & \operatorname{Re} \mathbf{f}_{XX}(\lambda) \end{bmatrix}^{-1}. \tag{8.4.11}$$

We now see that $\operatorname{Re} \mathbf{A}(\lambda)$ may be interpreted as the coefficient of $\mathbf{X}(t,\lambda)$ in the regression of $\mathbf{Y}(t,\lambda)$ on

$$\begin{bmatrix} \mathbf{X}(t,\lambda) \\ \mathbf{X}^H(t,\lambda) \end{bmatrix}. \tag{8.4.12}$$

Likewise we see that $\operatorname{Im} \mathbf{A}(\lambda)$ may be interpreted as the coefficient of $\mathbf{X}^H(t,\lambda)$ in the same regression.

The covariance matrix of the error variate of this regression analysis is

$$\begin{aligned} \operatorname{Re} \Bigg\{ & \mathbf{f}_{YY}(\lambda) - [\operatorname{Re} \mathbf{f}_{YX}(\lambda) \quad \operatorname{Im} \mathbf{f}_{YX}(\lambda)] \\ & \times \begin{bmatrix} \operatorname{Re} \mathbf{f}_{XX}(\lambda) & \operatorname{Im} \mathbf{f}_{XX}(\lambda) \\ -\operatorname{Im} \mathbf{f}_{XX}(\lambda) & \operatorname{Re} \mathbf{f}_{XX}(\lambda) \end{bmatrix}^{-1} \begin{bmatrix} \operatorname{Re} \mathbf{f}_{XY}(\lambda) \\ -\operatorname{Im} \mathbf{f}_{XY}(\lambda) \end{bmatrix} \Bigg\} \\ &= \operatorname{Re} \{\mathbf{f}_{YY}(\lambda) - \mathbf{f}_{YX}(\lambda)\mathbf{f}_{XX}(\lambda)^{-1}\mathbf{f}_{XY}(\lambda)\} \\ &= \operatorname{Re} \mathbf{f}_{\varepsilon\varepsilon}(\lambda). \end{aligned} \tag{8.4.13}$$

We see, therefore, that the real parts of the partial coherencies may be interpreted as partial correlations involved in the regression of $\mathbf{Y}(t,\lambda)$ on the variate (8.4.12). Similar considerations indicate that the imaginary parts may be interpreted as partial correlations of the regression of $\mathbf{Y}^H(t,\lambda)$ on (8.4.12).

If $s = 1$, then the squared coefficient of multiple correlation of the regression of $Y(t)$ on the variate (8.4.12) is

$$\begin{aligned}
[\text{Re } \mathbf{f}_{YX}(\lambda) \quad \text{Im } \mathbf{f}_{YX}(\lambda)]\begin{bmatrix} \text{Re } \mathbf{f}_{XX}(\lambda) & \text{Im } \mathbf{f}_{XX}(\lambda) \\ -\text{Im } \mathbf{f}_{XX}(\lambda) & \text{Re } \mathbf{f}_{XX}(\lambda)\end{bmatrix}^{-1}\begin{bmatrix} \text{Re } \mathbf{f}_{XY}(\lambda) \\ -\text{Im } \mathbf{f}_{XY}(\lambda)\end{bmatrix}\Big/ f_{YY}(\lambda) \\
= \frac{\mathbf{f}_{YX}(\lambda)\mathbf{f}_{XX}(\lambda)^{-1}\mathbf{f}_{XY}(\lambda)}{f_{YY}(\lambda)} \\
= |R_{YX}(\lambda)|^2. \qquad (8.4.14)
\end{aligned}$$

We see that the coefficient of multiple coherence may be interpreted as the squared coefficient of multiple correlation of $Y(t)$ with expression (8.4.12).

We end this section with a discussion of some useful parameters. The entries are generally complex-valued. In practice we may wish to deal with the real-valued Re $A_{ab}(\lambda)$, Im $A_{ab}(\lambda)$, or the real-valued modulus $G_{ab}(\lambda) = |A_{ab}(\lambda)|$, and argument $\phi_{ab}(\lambda) = \arg A_{ab}(\lambda)$. Consider the case $r, s = 1$. $G(\lambda) = |A(\lambda)|$ is called the **gain of $Y(t)$ over $X(t)$ at frequency** λ. The function $G(\lambda)$ is non-negative and we see that

$$G(-\lambda) = G(\lambda) \qquad (8.4.15)$$

and

$$G(\lambda + 2\pi) = G(\lambda) \qquad -\infty < \lambda < \infty. \qquad (8.4.16)$$

If

$$Y(t) = \sum_u a(t - u)X(u) \qquad (8.4.17)$$

then

$$\begin{aligned} f_{YY}(\lambda) &= |A(\lambda)|^2 f_{XX}(\lambda) \\ &= G(\lambda)^2 f_{XX}(\lambda). \qquad (8.4.18)\end{aligned}$$

Expression (8.4.18) suggests the source of the term gain. We see that the amplitude of the component of frequency λ in $X(t)$ is multiplied by $G(\lambda)$ in the case of $Y(t)$.

In the example $Y(t) = \alpha X(t - u)$, we see that

$$G(\lambda) = |\alpha|. \qquad (8.4.19)$$

The gain here has the nature of the absolute value of a regression coefficient and is constant with respect to λ.

The function $\phi(\lambda) = \arg A(\lambda)$ is called the **phase between $Y(t)$ and $X(t)$ at**

frequency λ. The fundamental range of values of $\phi(\lambda)$ is the interval $(-\pi,\pi]$. Because $f_{XX}(\lambda) \geqslant 0$, $\phi(\lambda)$ is given by

$$\phi(\lambda) = \arg f_{YX}(\lambda). \tag{8.4.20}$$

We see

$$\phi(-\lambda) = -\phi(\lambda) \tag{8.4.21}$$

and so $\phi(0) = 0$. Also

$$\phi(\lambda + 2\pi) = \phi(\lambda). \tag{8.4.22}$$

Suppose

$$Y(t) = \sum_u a(t - u)X(u). \tag{8.4.23}$$

In terms of the Cramér representations

$$\begin{aligned} \int e^{i\lambda t} dZ_Y(\lambda) &= \int e^{i\lambda t} A(\lambda) dZ_X(\lambda) \\ &= \int e^{i\lambda t} G(\lambda) e^{i\phi(\lambda)} dZ_X(\lambda) \end{aligned} \tag{8.4.24}$$

and so $\phi(\lambda)$ may be interpreted as the angle between the component of frequency λ in $X(t)$ and the corresponding component in $Y(t)$.

If, for example, $Y(t) = \alpha X(t - u)$, we see

$$\phi(\lambda) \equiv -\lambda u \ (\text{mod } 2\pi) \qquad \text{if } \alpha > 0 \tag{8.4.25}$$

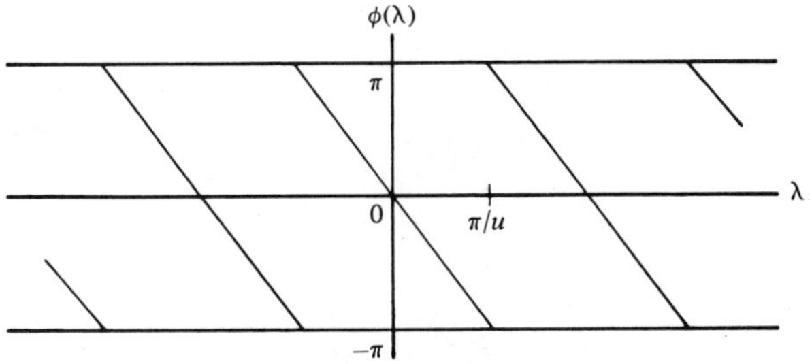

Figure 8.4.1 $\phi(\lambda)$, phase angle corresponding to delay of u time units when $\alpha > 0$.

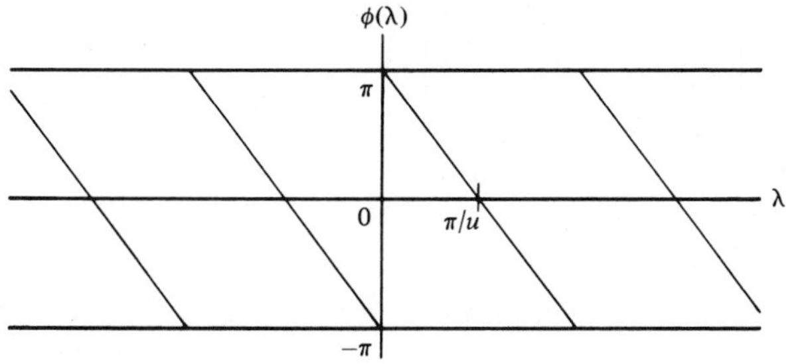

Figure 8.4.2 $\phi(\lambda)$, phase angle corresponding to delay of u time units when $\alpha < 0$.

and

$$\phi(\lambda) \equiv \pi - \lambda u \pmod{2\pi} \qquad \text{if } \alpha < 0. \tag{8.4.26}$$

These two functions are plotted in Figures 8.4.1 and 8.4.2 respectively taking $(-\pi,\pi]$ as the fundamental range of values for $\phi(\lambda)$.

On occasion the function

$$-\frac{d\phi(\lambda)}{d\lambda} = -\frac{d}{d\lambda}\log f_{YX}(\lambda) \tag{8.4.27}$$

is more easily interpreted. It is called the **group delay of $Y(t)$ over $X(t)$ at frequency** λ.

In the case of the example, we see that the group delay is u for all values of α. That is, it is the amount that $Y(t)$ is delayed with respect to $X(t)$.

We note that the group delay is defined uniquely, whereas $\phi(\lambda)$ is defined only up to an arbitrary multiple of 2π.

8.5 A LIMITING DISTRIBUTION FOR ESTIMATES

In this section we determine the limiting distribution of the estimates constructed in the previous section under the conditions $T \to \infty$, but m is fixed. Let

$$\mathbf{Z}(t) = \begin{bmatrix} \mathbf{X}(t) \\ \mathbf{Y}(t) \end{bmatrix} \tag{8.5.1}$$

and so

$$\mathbf{f}_{ZZ}(\lambda) = \begin{bmatrix} \mathbf{f}_{XX}(\lambda) & \mathbf{f}_{XY}(\lambda) \\ \mathbf{f}_{YX}(\lambda) & \mathbf{f}_{YY}(\lambda) \end{bmatrix}. \tag{8.5.2}$$

Set

$$\mathbf{I}_{ZZ}^{(T)}(\lambda) = \begin{bmatrix} \mathbf{I}_{XX}^{(T)}(\lambda) & \mathbf{I}_{XY}^{(T)}(\lambda) \\ \mathbf{I}_{YX}^{(T)}(\lambda) & \mathbf{I}_{YY}^{(T)}(\lambda) \end{bmatrix}. \tag{8.5.3}$$

Let m be a non-negative integer and $s(T)$, $T = 1, 2, \ldots$ a sequence of integers with $2\pi s(T)/T \to \lambda$ as $T \to \infty$. In the manner of Section 7.3, set

$$\begin{aligned}
\mathbf{f}_{ZZ}^{(T)}(\lambda) &= \begin{bmatrix} \mathbf{f}_{XX}^{(T)}(\lambda) & \mathbf{f}_{XY}^{(T)}(\lambda) \\ \mathbf{f}_{YX}^{(T)}(\lambda) & \mathbf{f}_{YY}^{(T)}(\lambda) \end{bmatrix} \\
&= (2m+1)^{-1} \sum_{s=-m}^{m} \mathbf{I}_{ZZ}^{(T)}\left(\frac{2\pi[s(T)+s]}{T}\right) \qquad \text{if } \lambda \not\equiv 0 \pmod{\pi} \\
&= m^{-1} \sum_{s=1}^{m} \operatorname{Re} \mathbf{I}_{ZZ}^{(T)}\left(\lambda + \frac{2\pi s}{T}\right) \qquad \begin{array}{l} \text{if } \lambda = 0, \pm 2\pi, \ldots \\ \text{or if } \lambda = \pm\pi, \pm 3\pi, \ldots \\ \text{and } T \text{ is even} \end{array} \\
&= m^{-1} \sum_{s=1}^{m} \operatorname{Re} \mathbf{I}_{ZZ}^{(T)}\left(\lambda - \frac{\pi}{T} + \frac{2\pi s}{T}\right) \qquad \begin{array}{l} \text{if } \lambda = \pm\pi, \pm 3\pi, \ldots \\ \text{and } T \text{ is odd.} \end{array}
\end{aligned} \tag{8.5.4}$$

We now construct the estimates

$$\mathbf{A}^{(T)}(\lambda) = \mathbf{f}_{YX}{}^{(T)}(\lambda)\mathbf{f}_{XX}{}^{(T)}(\lambda)^{-1}, \tag{8.5.5}$$

$$\mathbf{g}_{\varepsilon\varepsilon}{}^{(T)}(\lambda) = C(m,r)[\mathbf{f}_{YY}{}^{(T)}(\lambda) - \mathbf{f}_{YX}{}^{(T)}(\lambda)\mathbf{f}_{XX}{}^{(T)}(\lambda)^{-1}\mathbf{f}_{XY}{}^{(T)}(\lambda)] \tag{8.5.6}$$

where

$$\begin{aligned} C(m,r) &= \frac{2m+1}{2m+1-r} \qquad \text{if } \lambda \not\equiv 0 \pmod{2\pi} \\ &= \frac{2m}{2m-r} \qquad \text{if } \lambda \equiv 0 \pmod{\pi}. \end{aligned} \tag{8.5.7}$$

We notice that if m is large, then $C(m,r) \doteq 1$ and definition (8.5.6) is simplified. We also form

$$R^{(T)}_{Y_aY_b \cdot X}(\lambda) = \frac{g^{(T)}_{\varepsilon_a\varepsilon_b}(\lambda)}{[g^{(T)}_{\varepsilon_a\varepsilon_a}(\lambda)g^{(T)}_{\varepsilon_b\varepsilon_b}(\lambda)]^{1/2}} \tag{8.5.8}$$

for $a, b = 1, \ldots, s$ and if $s = 1$ we form

$$|R_{YX}{}^{(T)}(\lambda)|^2 = \frac{\mathbf{f}_{YX}{}^{(T)}(\lambda)\mathbf{f}_{XX}{}^{(T)}(\lambda)^{-1}\mathbf{f}_{XY}{}^{(T)}(\lambda)}{f_{YY}{}^{(T)}(\lambda)}. \tag{8.5.9}$$

We now state,

Theorem 8.5.1 Let the $(r + s)$ vector-valued series (8.5.1) satisfy Assumption 2.6.1 and have spectral density matrix (8.5.2). Let (8.5.2) be estimated by (8.5.4) where m, $s(T)$ are integers with $2\pi s(T)/T \to \lambda$ as $T \to \infty$. Let

$$\begin{bmatrix} \mathbf{W}_{XX} & \mathbf{W}_{XY} \\ \mathbf{W}_{YX} & \mathbf{W}_{YY} \end{bmatrix} \tag{8.5.10}$$

be distributed as $(2m+1)^{-1}W^C_{r+s}(2m+1,\mathbf{f}_{ZZ}(\lambda))$ if $\lambda \not\equiv 0 \pmod{\pi}$, as $(2m)^{-1}W_{r+s}(2m, \mathbf{f}_{ZZ}(\lambda))$ if $\lambda \equiv 0 \pmod{\pi}$. Then $\mathbf{A}^{(T)}(\lambda) - \mathbf{A}(\lambda)$, $\mathbf{g}_{\varepsilon\varepsilon}{}^{(T)}(\lambda)$ tend in distribution to $\mathbf{W}_{YX}\mathbf{W}_{XX}{}^{-1}$, $\mathbf{W}_{\varepsilon\varepsilon} = C(m,r)(\mathbf{W}_{YY} - \mathbf{W}_{YX}\mathbf{W}_{XX}{}^{-1}\mathbf{W}_{XY})$ respectively. Also $R^{(T)}_{Y_aY_b \cdot X}(\lambda)$ tends to $W_{\varepsilon_a\varepsilon_b}/[W_{\varepsilon_a\varepsilon_a}W_{\varepsilon_b\varepsilon_b}]^{1/2}$, $j, k = 1, \ldots, s$ and if $s = 1$, $|R_{YX}{}^{(T)}(\lambda)|^2$ tends to $\mathbf{W}_{YX}\mathbf{W}_{XX}{}^{-1}\mathbf{W}_{XY}/W_{YY}$.

The density function of the limiting distribution of $\mathbf{A}^{(T)}(\lambda)$ is deducible from (8.2.57) and (8.2.34). This was given in Wahba (1966) for the case $s = 1$, $\lambda \not\equiv 0 \pmod{\pi}$. A more useful result comes from noting that for any (rs) vector $\boldsymbol{\alpha}$

$$\frac{\boldsymbol{\alpha}^\tau(\text{vec}\,[\mathbf{A}^{(T)}(\lambda) - \mathbf{A}(\lambda)])}{[(2m+1)^{-1}\boldsymbol{\alpha}^\tau(\mathbf{g}_{\varepsilon\varepsilon}{}^{(T)}(\lambda) \otimes \mathbf{f}_{XX}{}^{(T)}(\lambda)^{-1})\bar{\boldsymbol{\alpha}}]^{1/2}} \tag{8.5.11}$$

has the limiting distribution $t^C_{2(2m+1-r)}$ in the case $\lambda \not\equiv 0 \pmod{\pi}$. Similar results hold in the case $\lambda \equiv 0 \pmod{\pi}$.

We conclude from Exercise 4.8.8 that under the conditions of Theorem 8.5.1, $\mathbf{g}_{\varepsilon\varepsilon}{}^{(T)}(\lambda)$ is asymptotically $(2m+1-r)^{-1}W_s{}^C(2m+1-r,\mathbf{f}_{\varepsilon\varepsilon}(\lambda))$ if

$\lambda \not\equiv 0 \pmod{\pi}$, asymptotically $(2m - r)^{-1} W_s(2m - r, \mathbf{f}_{\varepsilon\varepsilon}(\lambda))$ if $\lambda \equiv 0 \pmod{\pi}$. It is also asymptotically independent of $\mathbf{A}^{(T)}(\lambda)$. We note, from Theorem 7.3.3, that the asymptotic distribution of $\mathbf{g}_{\varepsilon\varepsilon}^{(T)}(\lambda)$ has the nature of the asymptotic distribution of a spectral estimate based directly on the values $\boldsymbol{\varepsilon}(t)$, $t = 0, \ldots, T - 1$ with the parameter $2m$ in that case replaced by $2m - r$ in the present case.

The partial coherencies $R_{Y_aY_b \cdot X}^{(T)}(\lambda)$, $a, b = 1, \ldots, s$ are based directly on the matrix $\mathbf{g}_{\varepsilon\varepsilon}^{(T)}(\lambda)$. We conclude from the above remarks that under the conditions of Theorem 8.5.1, their asymptotic distribution will be that of unconditional coherencies with the parameter $2m$ replaced by $2m - r$. In the case of vector-valued normal variates this result was noted by Fisher (1924). The distribution for a single $R_{Y_aY_b \cdot X}^{(T)}(\lambda)$ is given by (8.2.32) and (8.2.55) with $r = 1$.

Turning to the asymptotic distribution of the coefficient of multiple coherence in the case $s = 1$, set $|R_{YX}|^2 = |R_{YX}(\lambda)|^2$, $|\hat{R}_{YX}|^2 = |R_{YX}^{(T)}(\lambda)|^2$. Then the limiting distribution of $|R_{YX}^{(T)}(\lambda)|^2$ will be given by (8.2.55) with $n = 2m + 1$, if $\lambda \not\equiv 0 \pmod{\pi}$, by (8.2.32) with $n = 2m$, if $\lambda \equiv 0 \pmod{\pi}$.

Goodman (1963) suggested the above limiting distribution for the coherence. See also Goodman (1965), Khatri (1965), and Groves and Hannan (1968). Enochson and Goodman (1965) investigate the accuracy of approximating the distribution of $\tanh^{-1} |R_{YX}^{(T)}(\lambda)|$ by a normal distribution with mean

$$\tanh^{-1} |R_{YX}(\lambda)| + \frac{r}{2(2m + 1 - r)} \tag{8.5.12}$$

and variance $1/[2(2m - r)]$. The approximation seems reasonable.

8.6 A CLASS OF CONSISTENT ESTIMATES

In this section we develop a general class of estimates of the parameters that have been defined in Section 8.3. Suppose the values

$$\begin{bmatrix} \mathbf{X}(t) \\ \mathbf{Y}(t) \end{bmatrix} \tag{8.6.1}$$

$t = 0, \ldots, T - 1$ are available. Define $\mathbf{d}_X^{(T)}(\lambda)$, $\mathbf{d}_Y^{(T)}(\lambda)$, $-\infty < \lambda < \infty$, in the manner of (8.4.2). Define the matrix of cross-periodograms

$$\mathbf{I}_{XY}^{(T)}(\lambda) = (2\pi T)^{-1} \mathbf{d}_X^{(T)}(\lambda) \overline{\mathbf{d}_Y^{(T)}(\lambda)}^\tau \tag{8.6.2}$$

$-\infty < \lambda < \infty$ with similar definitions for $\mathbf{I}_{XX}^{(T)}(\lambda)$, $\mathbf{I}_{YY}^{(T)}(\lambda)$. Let $W(\alpha)$ be a weight function satisfying Assumption 5.4.1.

We now estimate

$$\begin{bmatrix} \mathbf{f}_{XX}(\lambda) & \mathbf{f}_{XY}(\lambda) \\ \mathbf{f}_{YX}(\lambda) & \mathbf{f}_{YY}(\lambda) \end{bmatrix} \tag{8.6.3}$$

the matrix of second-order spectra by

$$\begin{bmatrix} \mathbf{f}_{XX}^{(T)}(\lambda) & \mathbf{f}_{XY}^{(T)}(\lambda) \\ \mathbf{f}_{YX}^{(T)}(\lambda) & \mathbf{f}_{YY}^{(T)}(\lambda) \end{bmatrix}$$

$$= 2\pi T^{-1} \sum_{s=1}^{T-1} W^{(T)}\left(\lambda - \frac{2\pi s}{T}\right) \begin{bmatrix} \mathbf{I}_{XX}^{(T)}\left(\frac{2\pi s}{T}\right) & \mathbf{I}_{XY}^{(T)}\left(\frac{2\pi s}{T}\right) \\ \mathbf{I}_{YX}^{(T)}\left(\frac{2\pi s}{T}\right) & \mathbf{I}_{YY}^{(T)}\left(\frac{2\pi s}{T}\right) \end{bmatrix} \quad (8.6.4)$$

having taken note of the heuristic estimate (8.4.5). We estimate $\mathbf{A}(\lambda)$ by

$$\mathbf{A}^{(T)}(\lambda) = \mathbf{f}_{YX}^{(T)}(\lambda)\mathbf{f}_{XX}^{(T)}(\lambda)^{-1}. \quad (8.6.5)$$

The typical entry, $A_{ab}(\lambda)$, of $\mathbf{A}(\lambda)$ is generally complex-valued. On occasion we may wish to consider its amplitude $G_{ab}(\lambda)$ and its argument $\phi_{ab}(\lambda)$. Based on this estimate we take

$$\phi_{ab}^{(T)}(\lambda) = \arg A_{ab}^{(T)}(\lambda) \quad (8.6.6)$$

and

$$G_{ab}^{(T)}(\lambda) = |A_{ab}^{(T)}(\lambda)| \quad (8.6.7)$$

for $a = 1, \ldots, s$ and $b = 1, \ldots, r$. We estimate the error spectral density matrix $\mathbf{f}_{\varepsilon\varepsilon}(\lambda)$ by

$$\mathbf{g}_{\varepsilon\varepsilon}^{(T)}(\lambda) = \mathbf{f}_{YY}^{(T)}(\lambda) - \mathbf{f}_{YX}^{(T)}(\lambda)\mathbf{f}_{XX}^{(T)}(\lambda)^{-1}\mathbf{f}_{XY}^{(T)}(\lambda). \quad (8.6.8)$$

We estimate the partial coherency $R_{Y_aY_b\cdot X}(\lambda)$ by

$$R_{Y_aY_b\cdot X}^{(T)}(\lambda) = \frac{g_{\varepsilon_a\varepsilon_b}^{(T)}(\lambda)}{[g_{\varepsilon_a\varepsilon_a}^{(T)}(\lambda)g_{\varepsilon_b\varepsilon_b}^{(T)}(\lambda)]^{1/2}}. \quad (8.6.9)$$

In the case $s = 1$ we estimate $|R_{YX}(\lambda)|^2$, the multiple coherence of $Y(t)$ with $\mathbf{X}(t)$ by

$$|R_{YX}^{(T)}(\lambda)|^2 = \frac{\mathbf{f}_{YX}^{(T)}(\lambda)\mathbf{f}_{XX}^{(T)}(\lambda)^{-1}\mathbf{f}_{XY}^{(T)}(\lambda)}{f_{YY}^{(T)}(\lambda)} \quad (8.6.10)$$

$-\infty < \lambda < \infty$. The various estimates are seen to be sample analogs of corresponding population definitions.

Turning to the asymptotic first-order moments of the various statistics we have

Theorem 8.6.1 Let the $(r + s)$ vector-valued series (8.6.1) satisfy Assumption 2.6.2(1) and have spectral density matrix (8.6.3). Suppose $\mathbf{f}_{XX}(\lambda)$ is nonsingular. Let $W(\alpha)$ satisfy Assumption 5.6.1. Suppose the statistics $\mathbf{A}^{(T)}(\lambda)$, $\phi_{ab}^{(T)}(\lambda)$, $G_{ab}^{(T)}(\lambda)$, $\mathbf{g}_{\varepsilon\varepsilon}^{(T)}(\lambda)$, $R_{Y_aY_b\cdot X}^{(T)}(\lambda)$ are given by (8.6.5) to (8.6.9). Then if $B_T \to 0$, $B_TT \to \infty$ as $T \to \infty$

$$\vec{\text{ave}}\,\mathbf{A}^{(T)}(\lambda) = \left\{\int_0^{2\pi} W^{(T)}(\lambda-\alpha)\mathbf{A}(\alpha)\mathbf{f}_{XX}(\alpha)d\alpha\right\}\left\{\int_0^{2\pi} W^{(T)}(\lambda-\alpha)\mathbf{f}_{XX}(\alpha)d\alpha\right\}^{-1} + O(B_T^{-1}T^{-1}) \tag{8.6.11}$$

$$\vec{\text{ave}}\,\phi_{ab}^{(T)}(\lambda) = \arg\{\vec{\text{ave}}\,A_{ab}^{(T)}(\lambda)\} + O(B_T^{-1}T^{-1}) \tag{8.6.12}$$

$$\vec{\text{ave}}\,G_{ab}^{(T)}(\lambda) = |\vec{\text{ave}}\,A_{ab}^{(T)}(\lambda)| + O(B_T^{-1}T^{-1}) \tag{8.6.13}$$

and

$$\vec{\text{ave}}\,\mathbf{g}_{\varepsilon\varepsilon}^{(T)}(\lambda) = \int W^{(T)}(\lambda-\alpha)\mathbf{f}_{YY}(\alpha)d\alpha - \{\int W^{(T)}(\lambda-\alpha)\mathbf{f}_{YX}(\alpha)d\alpha\} \times \{\int W^{(T)}(\lambda-\alpha)\mathbf{f}_{XX}(\alpha)d\alpha\}^{-1}\{\int W^{(T)}(\lambda-\alpha)\mathbf{f}_{XY}(\alpha)d\alpha\} + O(B_T^{-1}T^{-1}). \tag{8.6.14}$$

If it is the case that $s = 1$ and $f_{YY}(\lambda) \neq 0$, then

$$\text{ave}\,|R_{YX}^{(T)}(\lambda)|^2 = \{\int W^{(T)}(\lambda-\alpha)\mathbf{f}_{YX}(\alpha)d\alpha\}\{\int W^{(T)}(\lambda-\alpha)\mathbf{f}_{XX}(\alpha)d\alpha\}^{-1} \times \{\int W^{(T)}(\lambda-\alpha)\,\mathbf{f}_{XY}(\alpha)d\alpha\}/\int W^{(T)}(\lambda-\alpha)\,f_{YY}(\alpha)d\alpha + O(B_T^{-1}T^{-1}). \tag{8.6.15}$$

We see that, in each case, the asymptotic means of the various statistics are nonlinear matrix weighted averages of the population values of interest. The asymptotic bias will therefore depend on how near constant these averages are in the neighborhood of λ. In the limit we have

Corollary 8.6.1 Under the conditions of Theorem 8.6.1

$$\lim_{T\to\infty} \vec{\text{ave}}\,\mathbf{A}^{(T)}(\lambda) = \mathbf{A}(\lambda), \tag{8.6.16}$$

$$\lim_{T\to\infty} \vec{\text{ave}}\,\phi_{ab}^{(T)}(\lambda) = \phi_{ab}(\lambda), \tag{8.6.17}$$

$$\lim_{T\to\infty} \vec{\text{ave}}\,G_{ab}^{(T)}(\lambda) = G_{ab}(\lambda), \tag{8.6.18}$$

$$\lim_{T\to\infty} \vec{\text{ave}}\,\mathbf{g}_{\varepsilon\varepsilon}^{(T)}(\lambda) = \mathbf{g}_{\varepsilon\varepsilon}(\lambda), \tag{8.6.19}$$

$$\lim_{T\to\infty} \vec{\text{ave}}\,R_{Y_aY_b\cdot X}^{(T)}(\lambda) = R_{Y_aY_b\cdot X}(\lambda), \tag{8.6.20}$$

and in the case $s = 1$

$$\lim_{T\to\infty} \vec{\text{ave}}\,|R_{YX}^{(T)}(\lambda)|^2 = |R_{YX}(\lambda)|^2. \tag{8.6.21}$$

The various estimates are asymptotically unbiased in an extended sense. We can develop expansions in powers of B_T of the asymptotic means; see Exercise 8.16.25. The important thing that we note from such expressions is that the nearer the derivatives of the population second-order spectra are to 0, the less the asymptotic bias. Nettheim (1966) expanded in powers of $B_T^{-1}T^{-1}$ in the Gaussian case.

Estimates of the parameters under consideration were investigated in Goodman (1965), Akaike (1965), Wahba (1966), Parzen (1967), and Jenkins and Watt (1968). The case $r,s = 1$ was considered in Goodman (1957), Tukey (1959a,b), Akaike and Yamanouchi (1962), Jenkins (1963a,b), Akaike (1964), Granger (1964), and Parzen (1964).

8.7 SECOND-ORDER ASYMPTOTIC MOMENTS OF THE ESTIMATES

We now turn to the development of certain second-order properties of the statistics of the previous section.

Theorem 8.7.1 Under the conditions of Theorem 8.6.1 and if $\mathbf{f}_{XX}(\alpha)$ is not singular in a neighborhood of λ or μ, then

$$\overrightarrow{\text{cov}}\,\{\text{vec}\,\mathbf{A}^{(T)}(\lambda),\ \text{vec}\,\mathbf{A}^{(T)}(\mu)\} \backsim \eta\{\lambda - \mu\}(\mathbf{f}_{\varepsilon\varepsilon}(\lambda) \otimes \mathbf{f}_{XX}(\lambda)^{-1})B_T^{-1}T^{-1}2\pi \int W(\alpha)^2 d\alpha \tag{8.7.1}$$

$$\overrightarrow{\text{cov}}\,\{g_{\varepsilon_a\varepsilon_b}^{(T)}(\lambda),\ g_{\varepsilon_c\varepsilon_d}^{(T)}(\mu)\} \backsim [\eta\{\lambda - \mu\} f_{\varepsilon_a\varepsilon_c}(\lambda) f_{\varepsilon_b\varepsilon_d}(-\lambda) + \eta\{\lambda + \mu\} f_{\varepsilon_a\varepsilon_d}(\lambda) f_{\varepsilon_b\varepsilon_c}(-\lambda)] \times B_T^{-1}T^{-1}2\pi \int W(\alpha)^2 d\alpha \tag{8.7.2}$$

$$\overrightarrow{\text{cov}}\,\{R_{Y_aY_b\cdot X}^{(T)}(\lambda),\ R_{Y_cY_d\cdot X}^{(T)}(\mu)\} \backsim (7.6.16) \tag{8.7.3}$$

for $a, b, c, d = 1, \ldots, s$ with $R_{no} = R_{Y_nY_o\cdot X}(\lambda)$ for $n, o = 1, \ldots, s$. If $s = 1$, then

$$\overrightarrow{\text{cov}}\,\{|R_{YX}^{(T)}(\lambda)|^2,\ |R_{YX}^{(T)}(\mu)|^2\} \backsim [\eta\{\lambda - \mu\} + \eta\{\lambda + \mu\}]\,|R_{YX}(\lambda)|^2 \times [1 - |R_{YX}(\lambda)|^2]^2 B_T^{-1}T^{-1}4\pi \int W(\alpha)^2 d\alpha. \tag{8.7.4}$$

To consider various aspects of these results, let $\mathbf{\Psi}(\lambda)$ denote the matrix $\mathbf{f}_{XX}(\lambda)^{-1}$, then from (8.7.1) and the perturbation expansions given in Exercise 8.16.24, we conclude

$$\overrightarrow{\text{cov}}\,\{A_{ab}^{(T)}(\lambda),\ A_{cd}^{(T)}(\mu)\} \backsim \eta\{\lambda - \mu\} f_{\varepsilon_a\varepsilon_b}(\lambda)\Psi_{cd}(\lambda)B_T^{-1}T^{-1}2\pi \int W(\alpha)^2 d\alpha, \tag{8.7.5}$$

$$\overrightarrow{\text{cov}}\,\{\log G_{ab}^{(T)}(\lambda),\ \log G_{cd}^{(T)}(\mu)\} \backsim [\eta\{\lambda - \mu\} + \eta\{\lambda + \mu\}] \times \text{Re}\,\{A_{ab}(\lambda)^{-1} f_{\varepsilon_a\varepsilon_c}(\lambda)\Psi_{bd}(\lambda)\overline{A_{cd}(\lambda)^{-1}}\}\, B_T^{-1}T^{-1}\pi \int W(\alpha)^2 d\alpha, \tag{8.7.6}$$

and

$$\overrightarrow{\text{cov}}\,\{\phi_{ab}^{(T)}(\lambda),\ \phi_{cd}^{(T)}(\mu)\} \backsim [\eta\{\lambda - \mu\} - \eta\{\lambda + \mu\}] \times \text{Re}\,\{A_{ab}(\lambda)^{-1} f_{\varepsilon_a\varepsilon_c}(\lambda)\Psi_{bd}(\lambda)\overline{A_{cd}(\lambda)^{-1}}\}\, B_T^{-1}T^{-1}\pi \int W(\alpha)^2 d\alpha, \tag{8.7.7}$$

for $a, c = 1, \ldots, s$; $b, d = 1, \ldots, r$.

Let us use the notation X'_b to denote the set of X_d, $d = 1, \ldots, r$ excluding X_b. Then we have from Exercise 8.16.37

$$\Psi_{bb}(\lambda) = \frac{1}{f_{X_bX_b\cdot X'_b}(\lambda)} = \frac{1}{[1 - |R_{X_bX'_b}(\lambda)|^2]f_{X_bX_b}(\lambda)}. \tag{8.7.8}$$

We also have

$$f_{\varepsilon_a\varepsilon_a}(\lambda) = f_{Y_aY_a\cdot X}(\lambda) = [1 - |R_{Y_aX}(\lambda)|^2]f_{Y_aY_a}(\lambda), \tag{8.7.9}$$

and so

$$\text{var}\, A_{ab}^{(T)}(\lambda) \sim B_T^{-1}T^{-1}2\pi \int W(\alpha)^2 d\alpha \frac{[1 - |R_{Y_aX}(\lambda)|^2]f_{Y_aY_a}(\lambda)}{[1 - |R_{X_bX'_b}(\lambda)|^2]f_{X_bX_b}(\lambda)}. \tag{8.7.10}$$

From the standpoint of variability we see, from (8.7.10), that the estimate $A_{ab}^{(T)}(\lambda)$ will be best if the multiple coherence of $Y_a(t)$ with $\mathbf{X}(t)$ is near 1 and if the multiple coherence of $X_b(t)$ with $X_1(t), \ldots, X_{b-1}(t), X_{b+1}(t), \ldots, X_r(t)$ is near 0.

Turning to a consideration of the estimated gain and phase we first note the relations

$$A_{ab}(\lambda) = \frac{f_{Y_aX_b\cdot X'_b}(\lambda)}{f_{X_bX_b\cdot X'_b}(\lambda)} \tag{8.7.11}$$

$$|R_{Y_aX_b\cdot X'_b}(\lambda)|^2 = \frac{|f_{Y_aX_b\cdot X'_b}(\lambda)|^2}{f_{Y_aY_a\cdot X'_b}(\lambda)f_{X_bX_b\cdot X'_b}(\lambda)}, \tag{8.7.12}$$

and

$$f_{Y_aY_a\cdot X'_b}(\lambda) = [1 - |R_{Y_aX_b\cdot X'_b}(\lambda)]^2|f_{Y_aY_a\cdot X}(\lambda). \tag{8.7.13}$$

We have from expressions to (8.7.6) to (8.7.8), (8.7.11), and (8.7.13) the following:

$$\text{var}\log G_{ab}^{(T)}(\lambda) \sim B_T^{-1}T^{-1}\pi \int W(\alpha)^2 d\alpha\, [\eta\{\lambda - \mu\} + \eta\{\lambda + \mu\}] \times [|R_{Y_aX_b\cdot X'_b}(\lambda)|^{-2} - 1] \tag{8.7.14}$$

and

$$\text{var}\, \phi_{ab}^{(T)}(\lambda) \sim B_T^{-1}T^{-1}\pi \int W(\alpha)^2 d\alpha\, [\eta\{\lambda - \mu\} - \eta\{\lambda + \mu\}] \times [|R_{Y_aX_b\cdot X'_b}(\lambda)|^{-2} - 1]. \tag{8.7.15}$$

We see that the variability of $\log G_{ab}^{(T)}(\lambda)$, $\phi_{ab}^{(T)}(\lambda)$ will be small if the partial coherence of $Y_a(t)$ with $X_b(t)$ after removing the linear effects of $X_1(t), \ldots, X_{b-1}(t), X_{b+1}(t), \ldots, X_r(t)$ is near 1. In the case that $r, s = 1$, the partial coherence in expressions (8.7.14) and (8.7.15) is replaced by the bivariate coherence $|R_{YX}(\lambda)|^2$.

We note that if $\lambda \pm \mu \not\equiv 0 \pmod{2\pi}$, then the asymptotic covariance structure of the log gain and phase is identical.

Turning to the estimated error spectral density matrix we note, from (8.7.2) and (7.4.17), that the second-order asymptotic behavior of $g_{\varepsilon\varepsilon}^{(T)}(\lambda)$ is exactly the same as if it were a direct spectral estimate $f_{\varepsilon\varepsilon}^{(T)}(\lambda)$ based on the values $\varepsilon(t)$, $t = 0, \ldots, T-1$.

We note from (8.7.3) that the asymptotic behavior of estimated partial coherencies is the same as that of the estimated coherencies of an s vector-valued series whose population coherencies are the partial coherencies $R_{Y_aY_b\cdot X}(\lambda)$, $a, b = 1, \ldots, s$. Taking $a = c$, $b = d$ we may deduce from (8.7.3) in the manner of Corollary 7.6.2 that

$$\vec{\operatorname{cov}}\,\{|R_{Y_aY_b\cdot X}^{(T)}(\lambda)|^2, |R_{Y_aY_b\cdot X}^{(T)}(\mu)|^2\} \backsim [\eta\{\lambda-\mu\} + \eta\{\lambda+\mu\}]\,|R_{Y_aY_b\cdot X}(\lambda)|^2 \times [1 - |R_{Y_aY_b\cdot X}(\lambda)|^2]^2 B_T^{-1}T^{-1}4\pi \int W(\alpha)^2 d\alpha. \quad (8.7.16)$$

The asymptotic covariance structure of $|R_{Y_aY_b\cdot X}(\lambda)|^2$ is seen to be the same for all values of s, r. An examination of expression (8.7.16) suggests the consideration of the variance stabilizing transformation

$$\tanh^{-1} |R_{Y_aY_b\cdot X}^{(T)}(\lambda)| \quad (8.7.17)$$

whose behavior is indicated in Table 8.7.1 and Figure 8.7.1. We see that values of $|R|$ near 0 are not changed much, while values near 1 are greatly increased. Now

$$\vec{\operatorname{ave}}\,\{\tanh^{-1} |R_{Y_aY_b\cdot X}^{(T)}(\lambda)|\} = \tanh^{-1} |R_{Y_aY_b\cdot X}(\lambda)| + O(B_T) + O(B_T^{-1}T^{-1}) \quad (8.7.18)$$

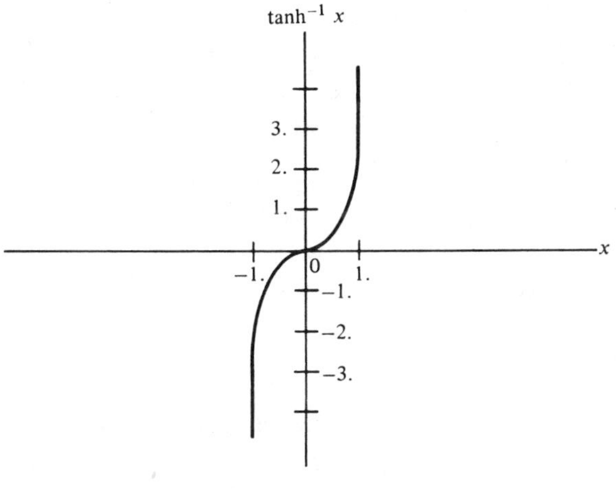

Figure 8.7.1 Graph of the transformation $y = \tanh^{-1} x$.

$$\overrightarrow{\text{cov}}\ \{\tanh^{-1}|R^{(T)}_{Y_aY_b\cdot X}(\lambda)|,\ \tanh^{-1}|R^{(T)}_{Y_aY_b\cdot X}(\mu)|\}$$
$$= [\eta\{\lambda - \mu\} + \eta\{\lambda + \mu\}]B_T^{-1}T^{-1}\pi \int W(\alpha)^2 d\alpha + O(B_T^{-2}T^{-2}). \quad (8.7.19)$$

Table 8.7.1 Values of the Hyperbolic Tangent

x	$\tanh^{-1} x$
.00	.0000
.05	.0500
.10	.1003
.15	.1511
.20	.2027
.25	.2554
.30	.3095
.35	.3654
.40	.4236
.45	.4847
.50	.5493
.55	.6184
.60	.6931
.65	.7753
.70	.8673
.75	.9730
.80	1.0986
.85	1.2562
.90	1.4722
.95	1.8318
1.00	∞

In the case $s = 1$, the partial coherence is the multiple coherence, $|R_{YX}(\lambda)|^2$, its estimate becomes the estimate $|R_{YX}^{(T)}(\lambda)|^2$. It follows that expressions (8.7.18) and (8.7.19) are valid for $|R_{YX}^{(T)}(\lambda)|^2$ as well. Enochson and Goodman (1965) have investigated the effect of this transformation and have suggested the approximations

$$E \tanh^{-1}|R_{YX}^{(T)}(\lambda)| \doteq \tanh^{-1}|R_{YX}(\lambda)| + \frac{r}{2(n-r)} \quad (8.7.20)$$

$$\text{var} \tanh^{-1}|R_{YX}(\lambda)| \doteq \frac{1}{2(n-r-1)}, \quad (8.7.21)$$

if $r > 1$, $\lambda \not\equiv 0 \pmod{\pi}$, where

$$n = \frac{T}{2\pi \int W^{(T)}(\alpha)^2 d\alpha} = \frac{B_T T}{2\pi \int W(\alpha)^2 d\alpha}. \quad (8.7.22)$$

Were the estimate of Section 8.5 employed, then $n = 2m + 1$.

Parzen (1967) derived the asymptotic mean and variance of $A_{ab}^{(T)}(\lambda)$, $\log G_{ab}^{(T)}(\lambda)$, and $\phi_{ab}^{(T)}(\lambda)$ in the case $s = 1$. Jenkins and Watt (1968), pp. 484, 492, indicated the asymptotic covariance structure of $\mathbf{A}^{(T)}(\lambda)$ and $|R_{YX}^{(T)}(\lambda)|^2$. In the case $r, s = 1$ Jenkins (1963a) derived the asymptotic variances of the phase, gain, and coherence.

8.8 ASYMPTOTIC DISTRIBUTION OF THE ESTIMATES

We now indicate limiting distributions for the statistics of interest. We begin with

Theorem 8.8.1 Under the conditions of Theorem 8.6.1 and if $\mathbf{f}_{XX}(\lambda^{(l)})$ is not singular for $l = 1, \ldots, L$ the estimates $\mathbf{A}^{(T)}(\lambda^{(l)})$, $\mathbf{g}_{\varepsilon\varepsilon}^{(T)}(\lambda^{(l)})$, $R_{Y_aY_b}^{(T)}(\lambda^{(l)})$, $a, b = 1, \ldots, s$ are asymptotically normally distributed with covariance structure given by (8.7.1) to (8.7.3). $\mathbf{A}^{(T)}(\lambda)$ and $\mathbf{g}_{\varepsilon\varepsilon}^{(T)}(\lambda)$ are asymptotically independent.

This theorem will be of use in constructing confidence regions of interest. We conclude from Theorem 8.8.1 and expression (8.7.1) that if $\lambda \not\equiv 0 \pmod{\pi}$, then vec $\mathbf{A}^{(T)}(\lambda)$ is asymptotically

$$N_{rs}^C(\text{vec } E\mathbf{A}^{(T)}(\lambda), \mathbf{\Sigma}_T), \tag{8.8.1}$$

where

$$\mathbf{\Sigma}_T = B_T^{-1}T^{-1}2\pi \int W(\alpha)^2 d\alpha \, (\mathbf{f}_{\varepsilon\varepsilon}(\lambda) \otimes \mathbf{f}_{XX}(\lambda)^{-1}). \tag{8.8.2}$$

It follows from Exercise 4.8.2 that the individual entries of $\mathbf{A}^{(T)}(\lambda)$ will be asymptotically complex normal as conjectured in Parzen (1967). Theorem 8.8.1 has the following:

Corollary 8.8.1 Under the conditions of Theorem 8.8.1, functions of $\mathbf{A}^{(T)}(\lambda)$, $\mathbf{g}_{\varepsilon\varepsilon}^{(T)}(\lambda)$, $R_{Y_aY_b}^{(T)}(\lambda)$ with nonsingular first derivative will be asymptotically normal.

In particular, we may conclude that $\log G_{ab}^{(T)}(\lambda)$ will be asymptotically normal with variance

$$[1 + \eta\{2\lambda\}][|R_{Y_aX_b\cdot X'_b}(\lambda)|^{-2} - 1]B_T^{-1}T^{-1}\pi \int W(\alpha)^2 d\alpha. \tag{8.8.3}$$

$\phi_{ab}^{(T)}(\lambda)$ will be asymptotically normal with variance

$$[1 - \eta\{2\lambda\}][|R_{Y_aX_b\cdot X'_b}(\lambda)|^{-2} - 1]B_T^{-1}T^{-1}\pi \int W(\alpha)^2 d\alpha \tag{8.8.4}$$

and $\log G_{ab}^{(T)}(\lambda)$, $\phi_{ab}^{(T)}(\lambda)$ will be asymptotically independent $a = 1, \ldots, s$

and $b = 1, \ldots, r$. Also $\tanh^{-1} |R^{(T)}_{Y_a Y_b \cdot X}(\lambda)|$ will be asymptotically normal with variance

$$[1 + \eta\{2\lambda\}]B_T^{-1}T^{-1}\pi \int W(\alpha)^2 d\alpha \tag{8.8.5}$$

$a, b = 1, \ldots, s$; and if $s = 1$, $\tanh^{-1} |R_{YX}{}^{(T)}(\lambda)|$ will be asymptotically normal with variance (8.8.5) as well. Experience with variance stabilizing transformations (see Kendall and Stuart (1966) p. 93) suggests that the transformed variate may be more nearly normal than the untransformed one. We will use the transformed variate to set confidence intervals for the population coherence in the next section.

We note that the limiting distribution of $\mathbf{A}^{(T)}(\lambda)$ given in Theorem 8.8.1 is consistent with that of Theorem 8.5.1 for large m, if we make the identification

$$2m + 1 = \frac{T}{2\pi \int W^{(T)}(\alpha)^2 d\alpha} = \frac{B_T T}{2\pi \int W(\alpha)^2 d\alpha}. \tag{8.8.6}$$

The distributions of the other variates are also consistent, since the Wishart distribution is near the normal when the degrees of freedom are large.

8.9 CONFIDENCE REGIONS FOR THE PROPOSED ESTIMATES

The asymptotic distributions derived in the previous section may be used to construct confidence regions for the parameters of interest. Throughout this section we make the identification (8.8.6).

We begin by constructing an approximate confidence region for $A_{ab}(\lambda)$. Suppose $\lambda \not\equiv 0 \pmod{\pi}$. Expression (8.5.11) lead us to approximate the distribution of

$$\frac{A_{ab}{}^{(T)}(\lambda) - A_{ab}(\lambda)}{[(2m + 1)^{-1} g^{(T)}_{\varepsilon_a \varepsilon_a}(\lambda) \Psi_{bb}{}^{(T)}(\lambda)]^{1/2}} \tag{8.9.1}$$

by $t^C_{2(2m+1-r)}$ where $\mathbf{\Psi}^{(T)}(\lambda) = \mathbf{f}_{XX}{}^{(T)}(\lambda)^{-1}$. This approximation may be manipulated in the manner of Section 6.9 to obtain a confidence region for either $\{\text{Re } A_{ab}(\lambda), \text{Im } A_{ab}(\lambda)\}$ or $\{\log G_{ab}(\lambda), \phi_{ab}(\lambda)\}$. In the case $\lambda \equiv 0 \pmod{\pi}$ we approximate the distribution of (8.9.1) by $t_{2(2m-r)}$.

If we let $\mathbf{A}_a{}^{(T)}(\lambda)$, $\mathbf{A}_a(\lambda)$ denote the ath row of $\mathbf{A}^{(T)}(\lambda)$, $\mathbf{A}(\lambda)$ respectively, then a confidence region for $\mathbf{A}_a(\lambda)$ may be obtained by approximating the distribution of

$$\{(2m + 1)\overline{[\mathbf{A}_a{}^{(T)}(\lambda) - \mathbf{A}_a(\lambda)]}\mathbf{f}_{XX}{}^{(T)}(\lambda)[\mathbf{A}_a{}^{(T)}(\lambda) - \mathbf{A}_a(\lambda)]^\tau/(2r)\}/f^{(T)}_{\varepsilon_a \varepsilon_a}(\lambda) \tag{8.9.2}$$

by $F_{2r;2(2m+1-r)}$ in the case $\lambda \not\equiv 0 \pmod{\pi}$. Exercise 6.14.17 indicates a means to construct approximate multiple confidence regions for all linear combinations of the entries of $\mathbf{A}_a(\lambda)$. This leads us to a consideration of the 100β percent region of the form

$$|A_{ab}^{(T)}(\lambda) - A_{ab}(\lambda)|^2 \leqslant 2F_{2r;2(2m+1-r)}(\beta)(2m+1)\Psi_{bb}(\lambda)g_{\varepsilon_a\varepsilon_a}^{(T)}(\lambda) \tag{8.9.3}$$

$b = 1, \ldots, r$ in the case $\lambda \not\equiv 0 \pmod{\pi}$. This last may be converted directly into a simultaneous region for $\phi_{ab}(\lambda)$, $\log G_{ab}(\lambda)$, $b = 1, \ldots, r$ in the manner of expression (6.9.11).

Turning to a consideration of $\mathbf{f}_{\varepsilon\varepsilon}(\lambda)$ we note that the parameters $f_{\varepsilon_a\varepsilon_b}(\lambda)$, $1 \leqslant a \leqslant b \leqslant s$ are algebraically equivalent to the parameters $f_{\varepsilon_a\varepsilon_a}(\lambda)$, $a = 1, \ldots, s$; $R_{Y_aY_b\cdot X}(\lambda)$, $1 \leqslant a < b \leqslant s$. We will indicate confidence intervals for these.

Theorem 8.5.1 leads us to approximate the distribution of $g_{\varepsilon_a\varepsilon_a}^{(T)}(\lambda)/f_{\varepsilon_a\varepsilon_a}(\lambda)$

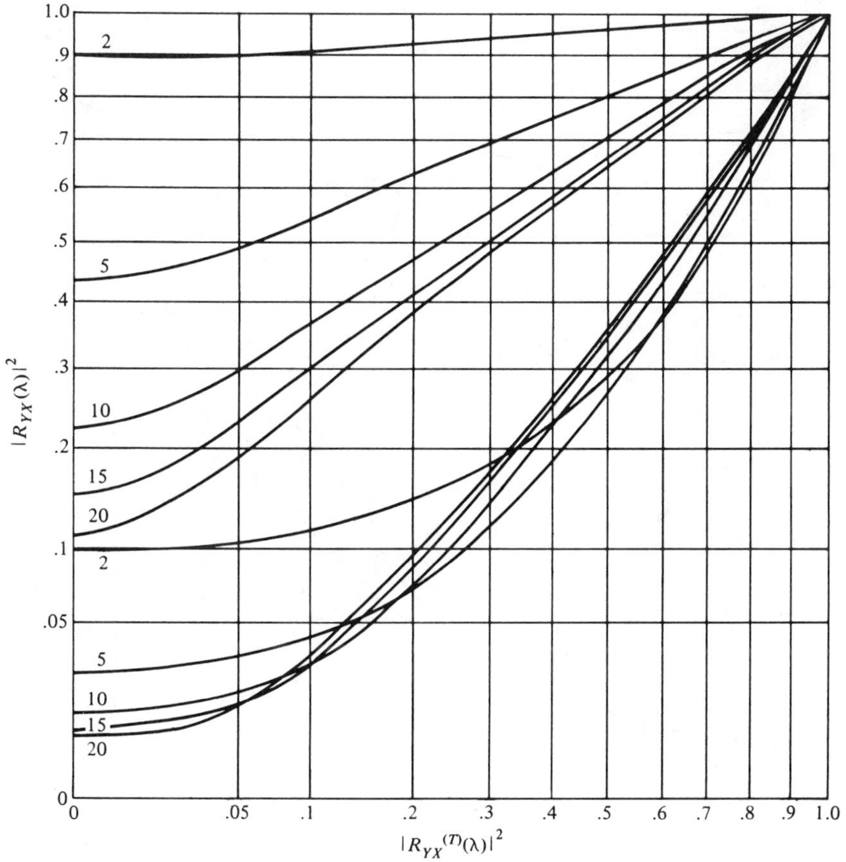

Figure 8.9.1 Confidence intervals of size 80 percent for the coherence, indexed by the number of periodograms averaged.

by $\chi^2_{2(2m+1-r)}/\{2(2m+1-r)\}$ if $\lambda \not\equiv 0 \pmod{\pi}$, by $\chi^2_{2m-r}/\{2m-r\}$ if $\lambda \equiv 0 \pmod{\pi}$. Confidence intervals for $f_{\varepsilon_a\varepsilon_a}(\lambda)$ may be obtained from these approximations in the manner of expression (5.7.5).

In the case of a single $R_{Y_aY_b\cdot X}(\lambda)$, Theorem 8.8.1 leads us to consider the $100(1-\alpha)$ percent confidence interval

$$\tanh^{-1} |R^{(T)}_{Y_aY_b\cdot X}(\lambda)| + ([1+\eta\{2\lambda\}]B_T^{-1}T^{-1}2\pi \int W(\alpha)^2 d\alpha)^{1/2} z\left(\frac{\alpha}{2}\right)$$
$$\leqslant \tanh^{-1} |R_{Y_aY_b\cdot X}(\lambda)|$$
$$\leqslant \tanh^{-1} |R^{(T)}_{Y_aY_b\cdot X}(\lambda)| - ([1+\eta\{2\lambda\}]B_T^{-1}T^{-1}2\pi \int W(\alpha)^2 d\alpha)^{1/2} z\left(\frac{\alpha}{2}\right). \tag{8.9.4}$$

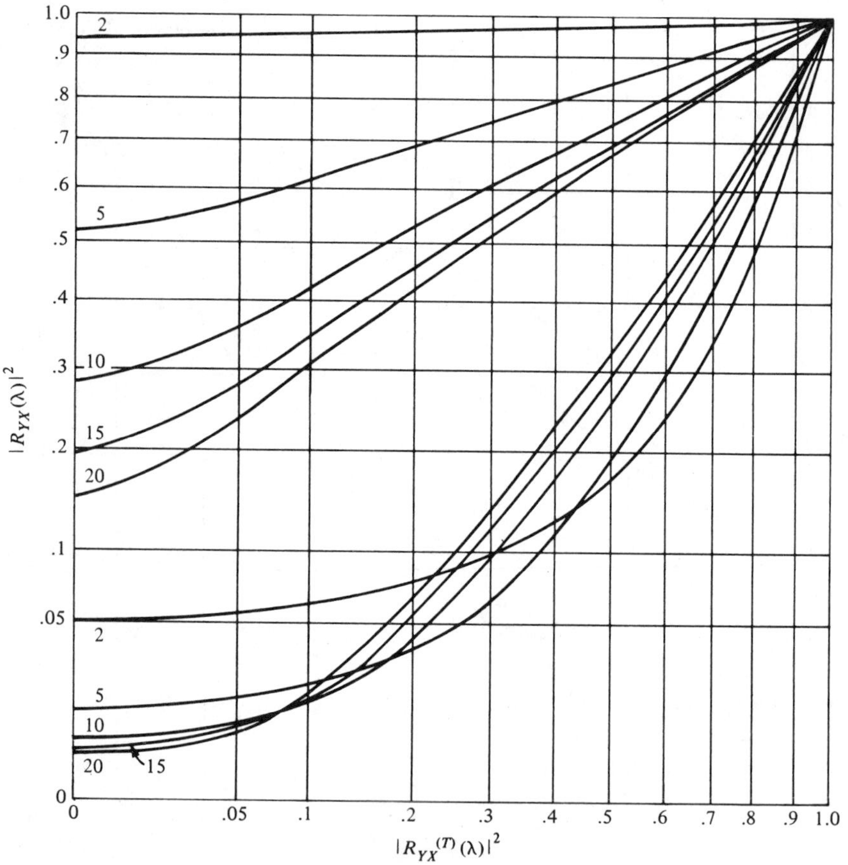

Figure 8.9.2 Confidence intervals of size 90 percent for the coherence, indexed by the number of periodograms averaged.

Alternately we could consult the tables of Amos and Koopmans (1962) for the distribution of the complex analog of the coefficient of correlation with the sample size reduced by r or use Figures 8.9.1 and 8.9.2 prepared from that reference.

In the case of a multiple coherence, we can consider the approximate $100(1 - \alpha)$ percent confidence interval

$$\tanh^{-1} |R_{YX}{}^{(T)}(\lambda)| + ([1 + \eta\{2\lambda\}]B_T{}^{-1}T^{-1}2\pi \int W(\alpha)^2 d\alpha)^{1/2} z\left(\frac{\alpha}{2}\right)$$
$$\leqslant \tanh^{-1} |R_{YX}(\lambda)|$$
$$\leqslant \tanh^{-1} |R_{YX}{}^{(T)}(\lambda)| - ([1 + \eta\{2\lambda\}]B_T{}^{-1}T^{-1}2\pi \int W(\alpha)^2 d\alpha)^{1/2} z\left(\frac{\alpha}{2}\right). \tag{8.9.5}$$

Alternately we could consult the tables of Alexander and Vok (1963).

The setting of confidence regions of the sort considered in this section is carried out in Goodman (1965), Enochson and Goodman (1965), Akaike (1965), and Groves and Hannan (1968). In the case $|R_{YX}(\lambda)|^2 = 0$, $\lambda \not\equiv 0 \pmod{\pi}$, the approximate 100α percent point of $|R_{YX}{}^{(T)}(\lambda)|^2$ is given by the elementary expression $1 - (1 - \alpha)^{1/2m}$; see Exercise 8.16.22.

8.10 ESTIMATION OF THE FILTER COEFFICIENTS

Suppose that the $(r + s)$ vector-valued series (8.1.1) satisfies

$$\mathbf{Y}(t) = \boldsymbol{\mu} + \sum_u \mathbf{a}(t - u)\mathbf{X}(u) + \boldsymbol{\varepsilon}(t) \tag{8.10.1}$$

$t = 0, \pm 1, \ldots$ where $\boldsymbol{\varepsilon}(t)$, $t = 0, \pm 1, \ldots$ is a stationary series independent of the series $\mathbf{X}(t)$. Theorem 8.3.1 leads us to consider the time domain coefficients

$$\mathbf{a}(u) = (2\pi)^{-1} \int \mathbf{A}(\alpha) \exp\{iu\alpha\} d\alpha \tag{8.10.2}$$

where $\mathbf{A}(\lambda) = \mathbf{f}_{YX}(\lambda)\mathbf{f}_{XX}(\lambda)^{-1}$.

Suppose now that $\mathbf{A}^{(T)}(\lambda)$ is an estimate of $\mathbf{A}(\lambda)$ of the form considered previously in this chapter. We can consider estimating $\mathbf{a}(u)$ by the statistic

$$\mathbf{a}^{(T)}(u) = P_T{}^{-1} \sum_{p=0}^{P_T - 1} \mathbf{A}^{(T)}(2\pi p/P_T) \exp\{i2\pi pu/P_T\} \tag{8.10.3}$$

where P_T is a sequence of integers tending to ∞ as $T \to \infty$.

We would expect the distribution of $\mathbf{a}^{(T)}(u)$ to be centered near

$$P_T{}^{-1} \sum_{p=0}^{P_T - 1} \{E\mathbf{f}_{YX}{}^{(T)}(2\pi p/P_T)\}^{-1}\{E\mathbf{f}_{XX}{}^{(T)}(2\pi p/P_T)\} \exp\{i2\pi p/P_T\}. \tag{8.10.4}$$

After the discussion following Theorem 7.4.2, the latter will be near

$$P_T^{-1} \sum_{p=0}^{P_T-1} \mathbf{A}(2\pi p/P_T) \exp\{i2\pi pu/P_T\} \tag{8.10.5}$$

in the case that the population parameters $\mathbf{f}_{YX}(\alpha)$, $\mathbf{f}_{XX}(\alpha)$ do not vary much in intervals of length $O(B_T)$. Expression (8.10.5) may be written

$$\mathbf{a}(u) + \sum_{k\neq 0} \mathbf{a}(u + kP_T) \tag{8.10.6}$$

which is near the desired $\mathbf{a}(u)$ in the case that the filter coefficients fall off to $\mathbf{0}$ sufficiently rapidly. These remarks suggest that, if anything, the procedure of prefiltering will be especially necessary in the present context.

Turning next to second-order moment considerations, expression (8.7.1) suggests that

$$\begin{aligned}
&\overrightarrow{\operatorname{cov}} \{\operatorname{vec} \mathbf{a}^{(T)}(u), \operatorname{vec} \mathbf{a}^{(T)}(v)\} \\
&\quad \backsim P_T^{-2} \sum_{p=0}^{P_T-1} \mathbf{f}_{\varepsilon\varepsilon}(2\pi p/P_T) \otimes \mathbf{f}_{XX}(2\pi p/P_T)^{-1} \\
&\qquad \times \exp\{i2\pi p(u-v)/P_T\} B_T^{-1}T^{-1}2\pi \textstyle\int W(\alpha)^2 d\alpha \\
&\quad \backsim P_T^{-1}B_T^{-1}T^{-1}[\textstyle\int W(\alpha)^2 d\alpha] \\
&\qquad \times \left[\int_0^{2\pi} \mathbf{f}_{\varepsilon\varepsilon}(\alpha) \otimes \mathbf{f}_{XX}(\alpha)^{-1} \exp\{i\alpha(u-v)\} d\alpha\right]
\end{aligned} \tag{8.10.7}$$

provided P_T is not too large. In fact we have

Theorem 8.10.1 Let the $(r + s)$ vector-valued series (8.1.1) satisfy (8.10.1) where the series $\mathbf{X}(t)$, $\boldsymbol{\varepsilon}(t)$ satisfy Assumption 2.6.2(1) and are independent. Suppose $\mathbf{f}_{XX}(\lambda)$ is nonsingular and has a bounded second derivative. Let $W(\alpha)$ satisfy Assumption 6.4.1. Let $\mathbf{A}^{(T)}(\lambda)$ be given by (8.6.5) and $\mathbf{a}^{(T)}(u)$ by (8.10.3) for $u = 0, \pm 1, \ldots$. Suppose $P_T \to \infty$ with $P_T B_T \geqslant 1$, $P_T^{1+\varepsilon} B_T^{-1} T^{-1} \to 0$ for an $\varepsilon > 0$. Then $\mathbf{a}^{(T)}(u_1), \ldots, \mathbf{a}^{(T)}(u_J)$ are asymptotically jointly normal with means given by (8.10.4) and covariances given by (8.10.7).

We note that, to first order, the asymptotic covariance matrix of vec $\mathbf{a}^{(T)}(u)$ does not depend on u. We may consider estimating it by

$$P_T^{-2} \sum_{p=0}^{P_T-1} \mathbf{g}_{\varepsilon\varepsilon}^{(T)}\left(\frac{2\pi p}{P_T}\right) \otimes \mathbf{f}_{XX}^{(T)}\left(\frac{2\pi p}{P_T}\right)^{-1} B_T^{-1}T^{-1}2\pi \textstyle\int W(\alpha)^2 d\alpha \tag{8.10.8}$$

where $\mathbf{g}_{\varepsilon\varepsilon}^{(T)}(\lambda)$ is given by (8.6.8). If we let $\mathbf{\Psi}^{(T)}(\lambda)$ denote $\mathbf{f}_{XX}^{(T)}(\lambda)^{-1}$ and set

$$\Gamma_{jk}^{(T)} = P_T^{-1} \sum_{p=0}^{P_T-1} g_{jj}^{(T)}\left(\frac{2\pi p}{P_T}\right) \Psi_{kk}^{(T)}\left(\frac{2\pi p}{P_T}\right) \tag{8.10.9}$$

then we can set down the following approximate $100(1-\alpha)$ percent confidence interval for $a_{jk}(u)$:

$$a_{jk}^{(T)}(u) + \{P_T^{-1}B_T^{-1}T^{-1}\Gamma_{jk}^{(T)}2\pi \int W(\alpha)^2 d\alpha\}^{1/2} z\left(\frac{\alpha}{2}\right) \leqslant a_{jk}(u) \leqslant a_{jk}^{(T)}(u) - \{P_T^{-1}B_T^{-1}T^{-1}\Gamma_{jk}^{(T)}2\pi \int W(\alpha)^2 d\alpha\}^{1/2} z\left(\frac{\alpha}{2}\right). \quad (8.10.10)$$

If one sets $P_T = B_T^{-1}$, then the asymptotic variance is of order T^{-1}.

Hannan (1967a) considered the estimation of the $\mathbf{a}(u)$ in the case that $\mathbf{a}(v) = 0$ for v sufficiently large and for a linear process error series $\boldsymbol{\varepsilon}(t)$, $t = 0, \pm 1, \ldots$. Wahba (1966, 1969) considers the Gaussian case with fixed $\boldsymbol{P}$.

It is of interest to consider also least squares estimates of the $\mathbf{a}(u)$, $u = 0, \pm 1, \ldots$ obtained by minimizing the sum of squares

$$\sum_{t=q}^{T-p-1} \operatorname{tr}\left(\left\{\mathbf{Y}(t) - \boldsymbol{\mu} - \sum_{u=-p}^{q} \mathbf{a}(u)\mathbf{X}(t-u)\right\} \times \left\{\mathbf{Y}(t) - \boldsymbol{\mu} - \sum_{u=-p}^{q} \mathbf{a}(u)\mathbf{X}(t-u)\right\}^{\tau}\right) \quad (8.10.11)$$

for some $p, q \geqslant 0$. We approach the investigation of these estimates through a consideration of the model

$$\mathbf{Y}(t) = \boldsymbol{\mu} + \mathbf{a}\mathbf{X}(t) + \boldsymbol{\varepsilon}(t) \quad (8.10.12)$$

for $t = 0, \pm 1, \ldots$. Here we assume that $\boldsymbol{\mu}$ is an unknown s vector; $\mathbf{a}$ is an unknown $s \times r$ matrix; $\mathbf{X}(t)$, $t = 0, \pm 1, \ldots$ is an observable stationary r vector-valued series; and $\boldsymbol{\varepsilon}(t)$, $t = 0, \pm 1, \ldots$ an unobservable $\mathbf{0}$ mean stationary s vector-valued error series having spectral density matrix $\mathbf{f}_{\varepsilon\varepsilon}(\lambda)$, $-\infty < \lambda < \infty$. The series $\mathbf{Y}(t)$, $t = 0, \pm 1, \ldots$ is assumed observable. Given a stretch of values

$$\begin{bmatrix} \mathbf{X}(t) \\ \mathbf{Y}(t) \end{bmatrix} \qquad t = 0, \ldots, T-1 \quad (8.10.13)$$

we consider the problem of estimating $\boldsymbol{\mu}$, $\mathbf{a}$, and $\mathbf{f}_{\varepsilon\varepsilon}(\lambda)$, $-\infty < \lambda < \infty$. The model (8.10.12) is broader than might be thought on initial reflection. For example consider a model

$$\mathbf{Y}(t) = \boldsymbol{\mu} + \sum_{u=-p}^{q} \mathbf{a}(u)\boldsymbol{\mathfrak{X}}(t-u) + \boldsymbol{\varepsilon}(t) \quad (8.10.14)$$

for $t = 0, \pm 1, \ldots$ where $\boldsymbol{\mathfrak{X}}(t)$, $t = 0, \pm 1, \ldots$ is a stationary r' vector-valued series and the series $\boldsymbol{\varepsilon}(t)$, $t = 0, \pm 1, \ldots$ is an independent stationary series. This model may be rewritten in the form (8.10.12) with the definitions

$$\mathbf{a} = [\mathbf{a}(-p) \cdots \mathbf{a}(q)] \quad (8.10.15)$$

and

$$\mathbf{X}(t) = \begin{bmatrix} \mathfrak{X}(t+p) \\ \cdot \\ \cdot \\ \cdot \\ \mathfrak{X}(t-q) \end{bmatrix} \tag{8.10.16}$$

for $t = 0, \pm 1, \ldots$. These last matrices have the dimensions $s \times r'(p+q-1)$ and $r'(p+q-1) \times 1$ respectively. A particular case of the model (8.10.14) is the autoregressive scheme

$$\mathfrak{X}(t) = \mathbf{a}(1)\mathfrak{X}(t-1) + \cdots + \mathbf{a}(q)\mathfrak{X}(t-q) + \boldsymbol{\varepsilon}(t) \tag{8.10.17}$$

in which $\boldsymbol{\varepsilon}(t)$, $t = 0, \pm 1, \ldots$ is a $\mathbf{0}$ mean white noise process. The results below may therefore be used to obtain estimates and the asymptotic properties of those estimates for the models (8.10.14) and (8.10.17).

Given the stretch of values (8.10.13), the least squares estimates $\boldsymbol{\mu}^{(T)}$, $\mathbf{a}^{(T)}$ of $\boldsymbol{\mu}$ and $\mathbf{a}$ are given by

$$\mathbf{c}_Y^{(T)} = \boldsymbol{\mu}^{(T)} + \mathbf{a}^{(T)}\mathbf{c}_X^{(T)} \tag{8.10.18}$$

and

$$\mathbf{c}_{YX}^{(T)}(0) = \mathbf{a}^{(T)}\mathbf{c}_{XX}^{(T)}(0). \tag{8.10.19}$$

As an estimate of $\mathbf{f}_{\varepsilon\varepsilon}(\lambda)$, we could consider

$$\mathbf{g}_{\varepsilon\varepsilon}^{(T)}(\lambda) = \frac{2\pi}{T}\sum_{s=1}^{T-1} W^{(T)}\left(\lambda - \frac{2\pi s}{T}\right)\mathbf{I}_{ee}^{(T)}\left(\frac{2\pi s}{T}\right) \tag{8.10.20}$$

where $\mathbf{e}(t)$ is the residual series given by

$$\mathbf{e}(t) = \mathbf{Y}(t) - \boldsymbol{\mu}^{(T)} - \mathbf{a}^{(T)}\mathbf{X}(t) \quad \text{for } t = 0, \pm 1, \ldots. \tag{8.10.21}$$

In connection with these estimates we have

Theorem 8.10.2 Let the s vector-valued series $\mathbf{Y}(t)$, $t = 0, \pm 1, \ldots$ satisfy (8.10.12) where $\mathbf{X}(t)$, $t = 0, \pm 1, \ldots$ is a 0 mean r vector-valued series satisfying Assumption 2.6.1 having autocovariance function $\mathbf{c}_{XX}(u)$, $u = 0, \pm 1, \ldots$ and spectral density matrix $\mathbf{f}_{XX}(\lambda), -\infty < \lambda < \infty$; $\boldsymbol{\varepsilon}(t)$, $t = 0, \pm 1, \ldots$ is an independent s vector-valued series satisfying Assumption 2.6.1, having spectral density matrix $\mathbf{f}_{\varepsilon\varepsilon}(\lambda) = [f_{ab}(\lambda)]$; and $\boldsymbol{\mu}, \mathbf{a}$ are $s \times 1$ and $s \times r$ matrices. Let $\boldsymbol{\mu}^{(T)}$, $\mathbf{a}^{(T)}$ be given by (8.10.18) and (8.10.19). Let $\mathbf{f}_{ee}^{(T)}(\lambda) = [f_{ab}^{(T)}(\lambda)]$ be given by (8.10.20) where $W(\alpha)$, $-\infty < \alpha < \infty$, satisfies Assumption 5.6.1 and $B_T T \to \infty$ as $T \to \infty$. Then $\boldsymbol{\mu}^{(T)}$ is asymptotically $N_s(\boldsymbol{\mu}, T^{-1}2\pi\mathbf{f}_{\varepsilon\varepsilon}(0))$; vec $\mathbf{a}^{(T)}$ is asymptotically independent $N_{rs}(\text{vec } \mathbf{a}, 2\pi T^{-1} \int \mathbf{f}_{\varepsilon\varepsilon}(\alpha) \otimes [\mathbf{c}_{XX}(0)^{-1}\mathbf{f}_{XX}(\alpha)\mathbf{c}_{XX}(0)^{-1}]d\alpha)$. Also $\mathbf{g}_{\varepsilon\varepsilon}^{(T)}(\lambda)$ is asymptotically independent normal with

$$E\mathbf{g}_{\varepsilon\varepsilon}^{(T)}(\lambda) = \int W(\alpha)\mathbf{f}_{\varepsilon\varepsilon}(\lambda - B_T\alpha)d\alpha + O(B_T) + O(B_T^{-1}T^{-1}) \tag{8.10.22}$$

and

$$\lim_{T\to\infty} B_T T \operatorname{cov}^{\rightarrow} \{g_{a_1b_1}^{(T)}(\lambda_1), g_{a_2b_2}^{(T)}(\lambda_2)\}$$
$$= [2\pi \textstyle\int W(\alpha)^2 d\alpha][\eta\{\lambda_1 - \lambda_2\} f_{a_1a_2}(\lambda_1) f_{b_1b_2}(-\lambda_1)$$
$$+ \eta\{\lambda_1 + \lambda_2\} f_{a_1b_2}(\lambda_1) f_{b_1a_2}(-\lambda_1)]$$
$$\text{for } -\infty < \lambda_1, \lambda_2 < \infty. \quad (8.10.23)$$

The asymptotic distribution of $\mathbf{g}_{\varepsilon\varepsilon}^{(T)}(\lambda)$ is seen to be the same as that of $\mathbf{f}_{\varepsilon\varepsilon}^{(T)}(\lambda)$, the variate based directly on the error series $\boldsymbol{\varepsilon}(t)$, $t = 0, \pm 1, \ldots$. In the case of the model (8.10.14) the limiting distributions are seen to involve the parameters

$$\mathbf{c}_{XX}(0) = \begin{bmatrix} \mathbf{c}_{\mathfrak{X}\mathfrak{X}}(0) & \cdots & \mathbf{c}_{\mathfrak{X}\mathfrak{X}}(p+q) \\ \cdot & & \cdot \\ \cdot & & \cdot \\ \cdot & & \cdot \\ \mathbf{c}_{\mathfrak{X}\mathfrak{X}}(-p-q) & \cdots & \mathbf{c}_{\mathfrak{X}\mathfrak{X}}(0) \end{bmatrix} \quad (8.10.24)$$

and

$$\mathbf{f}_{XX}(\lambda) = \begin{bmatrix} \mathbf{f}_{\mathfrak{X}\mathfrak{X}}(\lambda) & \exp\{-i\lambda\}\mathbf{f}_{\mathfrak{X}\mathfrak{X}}(\lambda) \cdots \\ \exp\{i\lambda\}\mathbf{f}_{\mathfrak{X}\mathfrak{X}}(\lambda) & \\ \cdot & \\ \cdot & \\ \cdot & \\ \exp\{i\lambda(p+q)\}\mathbf{f}_{\mathfrak{X}\mathfrak{X}}(\lambda) & \end{bmatrix}. \quad (8.10.25)$$

In the case that $\boldsymbol{\varepsilon}(t)$, $t = 0, \pm 1, \ldots$ is a white noise series with spectral density matrix $\mathbf{f}_{\varepsilon\varepsilon}(\lambda) = (2\pi)^{-1}\boldsymbol{\Sigma}$, $-\infty < \lambda < \infty$. Theorem 8.10.2 indicates that vec $[\mathbf{a}^{(T)}(-p), \ldots, \mathbf{a}^{(T)}(q)]$ is asymptotically normal with mean vec $[\mathbf{a}(-p), \ldots, \mathbf{a}(q)]$ and covariance matrix $T^{-1}\boldsymbol{\Sigma} \otimes \mathbf{c}_{XX}(0)^{-1}$. This gives the asymptotic distribution of the least squares estimates of the parameters of an autoregressive scheme. We considered corresponding results in the case of fixed $\mathbf{X}(t)$ in Section 6.12. We could also have here considered an analog of the "best" linear estimate (6.12.11).

8.11 PROBABILITY 1 BOUNDS

In Section 7.7 we derived a probability 1 bound for the deviations of a spectral estimate from its expected value as $T \to \infty$. That result may be used to develop a bound for the deviation of $\mathbf{A}^{(T)}(\lambda)$ from $\{E\mathbf{f}_{YX}^{(T)}(\lambda)\}\{E\mathbf{f}_{XX}^{(T)}(\lambda)\}^{-1}$. We may also bound the deviation of $\mathbf{A}^{(T)}(\lambda)$ from $\mathbf{A}(\lambda)$ and the other statistics considered from their corresponding population parameters. Specifically, we have

Theorem 8.11.1 Let the $(r + s)$ vector-valued series (8.1.1) satisfy Assumption 2.6.1. Let the conditions of Theorem 8.6.1 be satisfied. Let $D_T = (B_T T)^{1/2} B_T^{\varepsilon}$ for some $\varepsilon > 0$. Suppose $\Sigma_T B_T^m < \infty$ for some $m > 0$. Then

$$\mathbf{A}^{(T)}(\lambda) = \{E\mathbf{f}_{YX}{}^{(T)}(\lambda)\}\{E\mathbf{f}_{XX}{}^{(T)}(\lambda)\}^{-1} + O(D_T{}^{-1}) \tag{8.11.1}$$

almost surely as $T \rightarrow \infty$. In addition

$$\mathbf{A}^{(T)}(\lambda) = \mathbf{A}(\lambda) + O(B_T) + O(D_T{}^{-1}) \tag{8.11.2}$$

$$\phi_{jk}{}^{(T)}(\lambda) = \phi_{jk}(\lambda) + O(B_T) + O(D_T{}^{-1}) \tag{8.11.3}$$

$$G_{jk}{}^{(T)}(\lambda) = G_{jk}(\lambda) + O(B_T) + O(D_T{}^{-1}) \tag{8.11.4}$$

$$\mathbf{g}_{\varepsilon\varepsilon}{}^{(T)}(\lambda) = \mathbf{g}_{\varepsilon\varepsilon}(\lambda) + O(B_T) + O(D_T{}^{-1}) \tag{8.11.5}$$

$$R_{Y_jY_k\cdot X}(\lambda) = R_{Y_jY_k\cdot X}(\lambda) + O(B_T) + O(D_T{}^{-1}) \tag{8.11.6}$$

$$|R_{YX}{}^{(T)}(\lambda)|^2 = |R_{YX}(\lambda)|^2 + O(B_T) + O(D_T{}^{-1}) \tag{8.11.7}$$

almost surely as $T \rightarrow \infty$ for $-\infty < \lambda < \infty$, $j = 1, \ldots, s$; $k = 1, \ldots, r$. The error terms are uniform in λ.

We conclude from this theorem that if B_T, $D_T{}^{-1} \rightarrow 0$ as $T \rightarrow \infty$, then the various statistics are strongly consistent estimates of their corresponding population parameters.

8.12 FURTHER CONSIDERATIONS

The statistics discussed in this chapter are generally complex-valued. Thus, if we have computer programs that handle complex-valued quantities there will be no difficulty. However, since this is often not the case, it is worth noting that the statistics may all be evaluated using programs based on real-valued quantities. For example, consider the estimate of the complex regression coefficient:

$$\mathbf{A}^{(T)}(\lambda) = \mathbf{f}_{YX}{}^{(T)}(\lambda)\mathbf{f}_{XX}{}^{(T)}(\lambda)^{-1}. \tag{8.12.1}$$

This gives

$$\mathbf{A}^{(T)}(\lambda)^R = \mathbf{f}_{YX}{}^{(T)}(\lambda)^R\{\mathbf{f}_{XX}{}^{(T)}(\lambda)^R\}^{-1} \tag{8.12.2}$$

if one uses the operation of Section 3.7. Taking the first s rows of (8.12.2) gives

$$[\mathrm{Re}\,\mathbf{A}^{(T)}(\lambda) \quad \mathrm{Im}\,\mathbf{A}^{(T)}(\lambda)] = [\mathrm{Re}\,\mathbf{f}_{YX}{}^{(T)}(\lambda) \quad \mathrm{Im}\,\mathbf{f}_{YX}{}^{(T)}(\lambda)] \times \begin{bmatrix} \mathrm{Re}\,\mathbf{f}_{XX}{}^{(T)}(\lambda) & \mathrm{Im}\,\mathbf{f}_{XX}{}^{(T)}(\lambda) \\ -\mathrm{Im}\,\mathbf{f}_{XX}{}^{(T)}(\lambda) & \mathrm{Re}\,\mathbf{f}_{XX}{}^{(T)}(\lambda) \end{bmatrix}^{-1} \tag{8.12.3}$$

a set of equations that involves only real-valued quantities. The principal complication introduced by this reduction is a doubling of the dimension of the **X** variate. Exercise 3.10.11 indicates an identity that we could use in an alternate approach to equation (8.12.1).

Likewise we may set down sample parallels of expressions (8.4.13) and (8.4.14) to determine the error spectral density, partial coherency, and multiple coherence statistics.

We next mention that there are interesting frequency domain analogs of the important problems of **errors in variables** and of **systems of simultaneous equations.**

Suppose that a series $\mathbf{Y}(t)$, $t = 0, \pm 1, \ldots$ is given by

$$\mathbf{Y}(t) = \boldsymbol{\mu} + \sum_u \mathbf{a}(t-u)\boldsymbol{\mathfrak{X}}(u) + \boldsymbol{\varepsilon}(t) \qquad t = 0, \pm 1, \ldots \tag{8.12.4}$$

where the r vector-valued series $\boldsymbol{\mathfrak{X}}(t)$, $t = 0, \pm 1, \ldots$ is not observed directly and where $\boldsymbol{\varepsilon}(t)$, $t = 0, \pm 1, \ldots$ is an error series independent of the series $\boldsymbol{\mathfrak{X}}(t)$. Suppose, however, that the series

$$\mathbf{X}(t) = \boldsymbol{\mathfrak{X}}(t) + \boldsymbol{\eta}(t) \tag{8.12.5}$$

is observed where $\boldsymbol{\eta}(t)$, $t = 0, \pm 1, \ldots$ is an error series independent of $\boldsymbol{\mathfrak{X}}(t)$. The problem of estimating $\boldsymbol{\mu}$, $\{\mathbf{a}(u)\}$ in a situation such as this is a problem of errors in variables. Considerable literature exists concerning this problem for series not serially correlated; see Durbin (1954), Kendall and Stuart (1961), for example. If the series involved are stationary then we may write

$$\mathbf{d}_Y{}^{(T)}\left(\frac{2\pi s}{T}\right) \doteq \mathbf{A}(\lambda)\mathbf{d}_{\mathfrak{X}}{}^{(T)}\left(\frac{2\pi s}{T}\right) + \mathbf{d}_\varepsilon{}^{(T)}\left(\frac{2\pi s}{T}\right) \tag{8.12.6}$$

$$\mathbf{d}_X{}^{(T)}\left(\frac{2\pi s}{T}\right) = \mathbf{d}_{\mathfrak{X}}{}^{(T)}\left(\frac{2\pi s}{T}\right) + \mathbf{d}_\eta{}^{(T)}\left(\frac{2\pi s}{T}\right) \tag{8.12.7}$$

with the variates approximately uncorrelated for distinct s. Because of this weak correlation we can now consider applying the various classical procedures for approaching the problem of errors in variables. The solution of the problem (8.12.4-5) will involve separate errors in variables solution for each of a number of frequencies λ lying in $[0,\pi]$.

Perhaps the nicest results occur when an r vector-valued **instrumental series** $\mathbf{Z}(t)$, $t = 0, \pm 1, \ldots$ is available for analysis as well as the series $\mathbf{Y}(t)$, $\mathbf{X}(t)$. This is a series that is correlated with the series $\boldsymbol{\mathfrak{X}}(t)$, $t = 0, \pm 1, \ldots$, but uncorrelated with the series $\boldsymbol{\varepsilon}(t)$ and $\boldsymbol{\eta}(t)$. In the stationary case we have, from expressions (8.12.5) and (8.12.4),

$$\mathbf{f}_{YZ}(\lambda) = \mathbf{A}(\lambda)\mathbf{f}_{XZ}(\lambda). \tag{8.12.8}$$

The statistic

$$\mathbf{A}^{(T)}(\lambda) = \mathbf{f}_{YZ}{}^{(T)}(\lambda)\mathbf{f}_{XZ}{}^{(T)}(\lambda)^{-1} \tag{8.12.9}$$

now suggests itself as an estimate of $\mathbf{A}(\lambda)$. Hannan (1963a) and Parzen (1967b) are references related to this procedure. Akaike (1966) suggests a procedure useful when the series $\boldsymbol{\eta}(t)$ is Gaussian, but the series $\boldsymbol{\mathfrak{X}}(t)$ is not.

A variety of models in econometrics lead to systems of simultaneous equations taking the form

$$\sum_u \mathbf{a}(t-u)\mathbf{Y}(u) = \sum_u \mathbf{b}(t-u)\mathbf{Z}(u) + \boldsymbol{\varepsilon}(t), \tag{8.12.10}$$

where $\mathbf{Y}(t)$, $\boldsymbol{\varepsilon}(t)$ are s vector-valued series and $\mathbf{Z}(t)$ is an r vector-valued series independent of the series $\boldsymbol{\varepsilon}(t)$; see Malinvaud (1964). A model of the form (8.12.10) is called a **structural equation system.** It is exceedingly general becoming, for example, an autoregressive scheme in one case and a linear system

$$\mathbf{Y}(t) = \sum_u \mathbf{a}(t-u)\mathbf{X}(u) + \boldsymbol{\varepsilon}(t) \tag{8.12.11}$$

with the series $\mathbf{X}(t)$, $\boldsymbol{\varepsilon}(t)$ correlated in another. This correlation may be due to the presence of feed-back loops in the system. The econometrician is often interested in the estimation of the coefficients of a single equation of the system (8.12.10) and a variety of procedures for doing this have now been proposed (Malinvaud (1964)) in the case that the series are not serially correlated.

In the stationary case we can consider setting down the expression

$$\mathbf{A}(\lambda)\mathbf{d}_Y{}^{(T)}\left(\frac{2\pi s}{T}\right) \doteq \mathbf{B}(\lambda)\mathbf{d}_Z{}^{(T)}\left(\frac{2\pi s}{T}\right) + \mathbf{d}_\varepsilon{}^{(T)}\left(\frac{2\pi s}{T}\right) \tag{8.12.12}$$

for $2\pi s/T$ near λ with the variates approximately uncorrelated for distinct s. It is now apparent that complex analogs of the various econometric estimation procedures may be applied to the system (8.12.12) in order to estimate coefficients of interest. The character of this procedure involves analyzing a system of simultaneous equations separately in a number of narrow frequency bands. Brillinger and Hatanaka (1969) set down the system (8.12.10) and recommend a frequency analysis of it. Akaike (1969) and Priestly (1969) consider the problem of estimation in a system when feed-back is present.

In fact, as Durbin (1954) remarks, the errors in variables model (8.12.4) and (8.12.5) with instrumental series $\mathbf{Z}(t)$ may be considered within the simultaneous equation framework. We simply write the model in the form

$$\mathbf{Y}(t) - \sum_u \mathbf{a}(t-u)\mathbf{X}(u) = \boldsymbol{\mu} + \boldsymbol{\zeta}(t) \tag{8.12.13}$$

$$\mathbf{X}(t) = \sum_u \mathbf{d}(t-u)\mathbf{Z}(u) + \boldsymbol{\gamma}(t) \tag{8.12.14}$$

and look on the pair $\mathbf{Y}(t)$, $\mathbf{X}(t)$ as being $\mathbf{Y}(t)$ of (8.12.10).

8.13 ALTERNATE FORMS OF ESTIMATES

The estimates that we have constructed of the gain, phase, and coherence have in each case been the sample analog of the population definition. For example, we defined

$$G(\lambda) = \frac{|f_{YX}(\lambda)|}{f_{XX}(\lambda)} \tag{8.13.1}$$

and then constructed the estimate

$$G^{(T)}(\lambda) = \frac{|f_{YX}{}^{(T)}(\lambda)|}{f_{XX}{}^{(T)}(\lambda)}. \tag{8.13.2}$$

On some occasions it may prove advantageous not to proceed in such a direct manner.

For example expressions (8.6.11) and (8.6.13) indicate that asymptotic bias occurs for $G^{(T)}(\lambda)$ if the spectra $f_{YX}(\alpha)$ and $f_{XX}(\alpha)$ are not flat for α near λ. This suggests that if possible we should prefilter $X(t)$ and $Y(t)$ to obtain series for which the second-order spectra are near constant. The gain relating these filtered series should be estimated and an estimate of $G(\lambda)$ be constructed.

In another vein expression (8.7.14) indicated that

$$\vec{\text{var}} \log G^{(T)}(\lambda) = [1 + \eta\{2\lambda\}][|R_{YX}(\lambda)|^{-2} - 1]B_T{}^{-1}T^{-1}\pi \int W(\alpha)^2 d\alpha + O(B_T{}^{-2}T^{-2}). \tag{8.13.3}$$

This suggests that in situations where $|R_{YX}(\lambda)|^2$ is near constant, with respect to λ, we could consider carrying out a further smoothing and estimate $\log G(\lambda)$ by

$$(2N + 1)^{-1} \sum_{n=-N}^{N} \log G^{(T)}(\lambda + n\Delta_T) \tag{8.13.4}$$

for some N, Δ_T where it is supposed that $G^{(T)}(\alpha)$ has been constructed in the manner of Section 8.6.

We note in passing the possibility suggested by (8.4.18) of estimating $G(\lambda)^2$ by

$$\frac{f_{YY}{}^{(T)}(\lambda)}{f_{XX}{}^{(T)}(\lambda)}. \tag{8.13.5}$$

Exercise 8.16.12 indicates that this is not generally a reasonable procedure.

We have proposed

$$\phi^{(T)}(\lambda) = \arg f_{YX}{}^{(T)}(\lambda) \tag{8.13.6}$$

as an estimate of the phase, $\phi(\lambda)$. Expression (8.6.12) indicates that $\overrightarrow{\text{ave}}\,\phi^{(T)}(\lambda)$ is principally a nonlinear average of the phase with unequal weights. This occurrence leads us, when possible, to prefilter the series prior to estimating the phase in order to obtain a flatter cross-spectrum.

Alternately we could consider nonlinear estimates that are not as affected by variation in weights. For example, we could consider an estimate of the form

$$\arg\left((2N+1)^{-1}\sum_{n=-N}^{N}\exp\{i\arg f_{YX}^{(T)}(\lambda+n\Delta_T)\}\right) \tag{8.13.7}$$

or of the form

$$(2N+1)^{-1}\sum_{n=-N}^{N}\arg f_{YX}^{(T)}(\lambda+n\Delta_T). \tag{8.13.8}$$

The fact that the phase angle is only defined up to an arbitrary multiple of 2π means we must be careful in the determination of the value of $\arg f_{YX}^{(T)}(\lambda + n\Delta_T)$ when forming (8.13.8).

This indetermination also leads to complications in the pictorial display of $\phi^{(T)}(\lambda)$. If either $\phi(\lambda)$ is changing rapidly or var $\phi^{(T)}(\lambda)$ is large, then an extremely erratic picture can result. For example, Figure 7.2.5 is a plot of the estimated phase angle between the series of seasonally adjusted mean month-

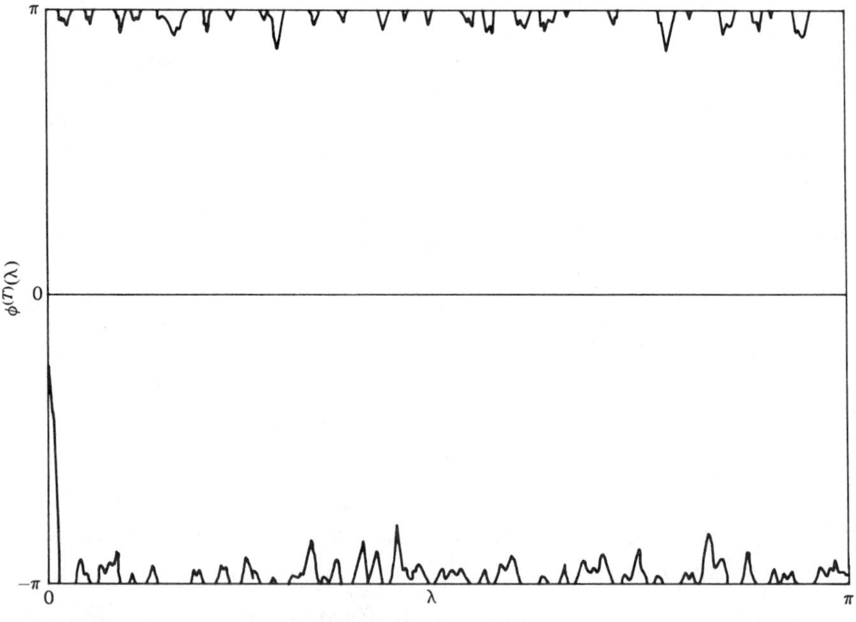

Figure 8.13.1 $\phi^{(T)}(\lambda)$, the estimated phase angle between seasonally adjusted mean monthly Berlin temperatures and the negative of seasonally adjusted mean monthly Vienna temperatures. (15 periodograms averaged in estimation.)

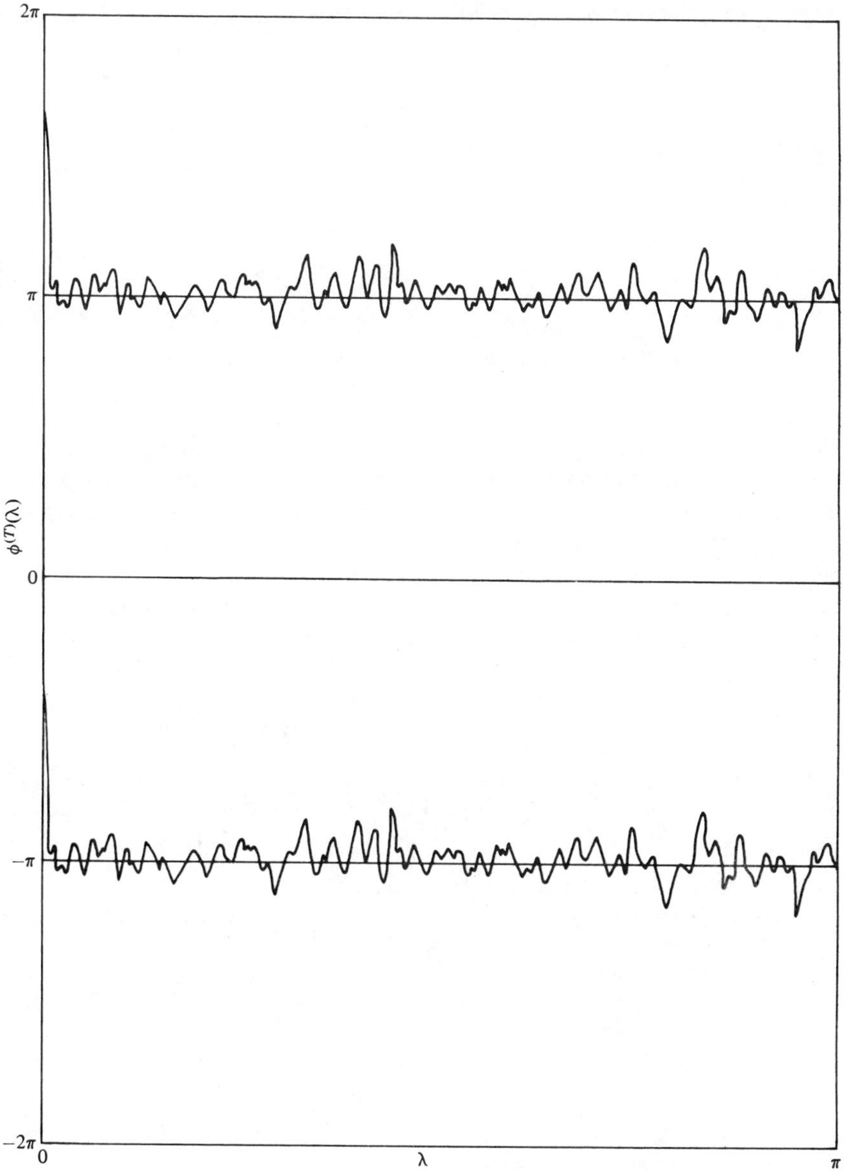

Figure 8.13.2 Another manner of plotting the data of Figure 8.13.1. (The range of $\phi^{(T)}(\lambda)$ is taken to be $[-2\pi, 2\pi]$.)

ly temperatures at Berlin and Vienna determined from the cross-periodogram. It is difficult to interpret this graph because when the phase takes a small jump of the form $\pi - \varepsilon$ to $\pi + \varepsilon$, $\phi^{(T)}(\lambda)$ when plotted in the range $(-\pi,\pi]$ moves from $\pi - \varepsilon$ to $-\pi - \varepsilon$. One means of reducing the impact of this

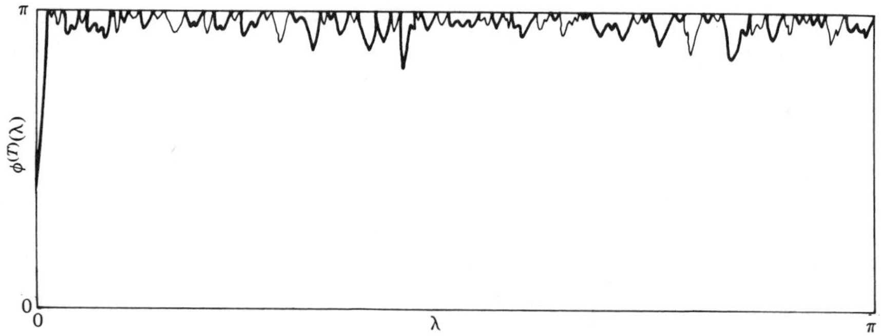

Figure 8.13.3 Another manner of plotting the data of Figure 8.13.1. (The heavy line corresponds to $\phi^{(T)}(\lambda)$ in $[\pi, 2\pi]$.)

effect is to plot each phase twice, taking its two values in the interval $(-2\pi,2\pi]$. If the true phase is near π, then an especially improved picture is obtained. For example, Figure 8.13.1 is the estimated phase when 15 periodograms are averaged between seasonally adjusted monthly Berlin temperatures and the negative of seasonally adjusted monthly Vienna temperatures taking the range of $\phi^{(T)}(\lambda)$ to be $(-\pi,\pi]$. If this range is increased to

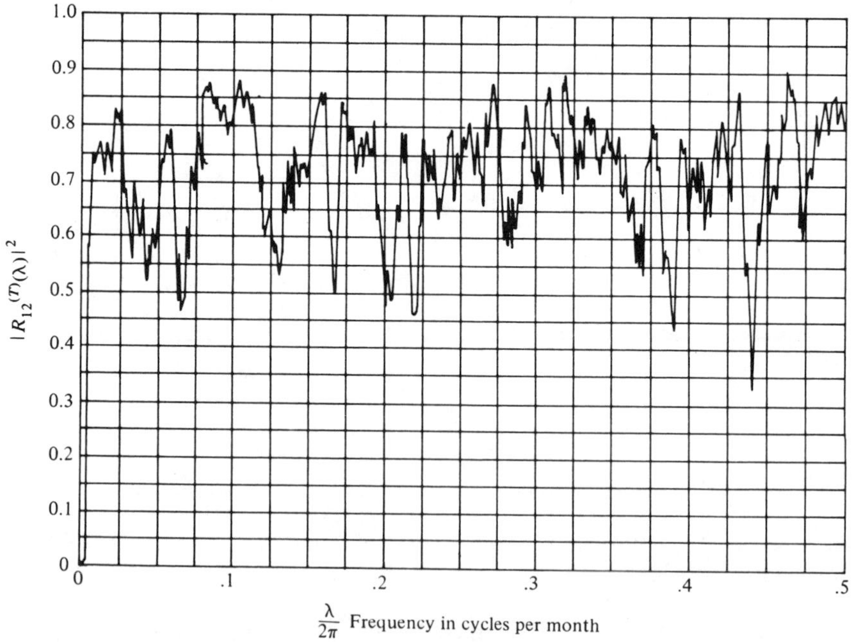

Figure 8.13.4 $|R_{YX}{}^{(T)}(\lambda)|^2$, estimated coherence of seasonally adjusted mean monthly Berlin temperatures and seasonally adjusted mean monthly Vienna temperatures. (15 periodograms averaged in estimation.)

$(-2\pi,2\pi]$, as suggested, Figure 8.13.2 results. J. W. Tukey has proposed making a plot on the range $[0,\pi]$ using different symbols or lines for phases whose principal values are in $[0,\pi]$ from those whose values are in$(\pi,2\pi]$. If this is done for the Berlin-Vienna data, then Figure 8.13.3 is obtained.

Another procedure is to plot an estimate of the group delay expression (8.4.27), then the difficulty over arbitrary multiples of 2π does not arise. Generally speaking it appears to be the case that the best form of plot depends on the $\phi(\lambda)$ at hand.

We next turn to alternate estimates of the coherence. The bias of $|R_{YX}^{(T)}(\lambda)|^2$ may be reduced if we carry out a prefiltering of the filtered series, and then algebraically deduce an estimate of the desired coherence.

Alternatively we can take note of the variance stabilizing properties of the $\tanh^{-1}$ transformation and by analogy with expression (8.13.4) consider as an estimate of $\tanh^{-1} |R_{YX}(\lambda)|$:

$$(2N+1)^{-1} \sum_{n=-N}^{N} \tanh^{-1} |R_{YX}^{(T)}(\lambda + n\Delta_T)|. \tag{8.13.9}$$

We note that the effect of the $\tanh^{-1}$ transformation is to increase values of $|R_{YX}^{(T)}(\alpha)|$ that are near 1 while retaining the values near 0. High coherences are therefore weighted more heavily if we form (8.13.9). Figure

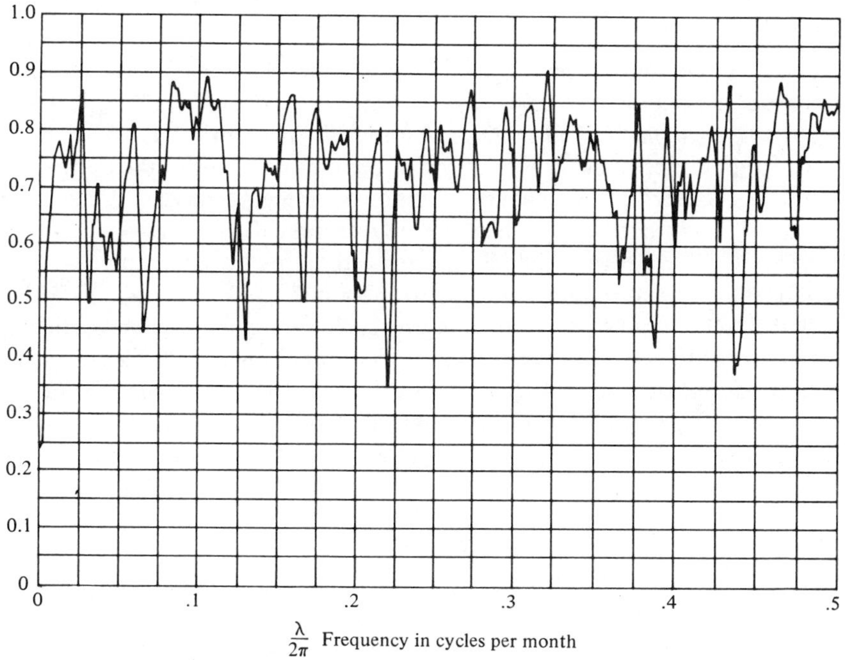

Figure 8.13.5 Coherence estimate based on the form (8.13.9) with $m = 5$, $N = 2$ for Berlin and Vienna temperature series.

8.13.4 is a plot of $|R_{YX}^{(T)}(\lambda)|^2$, for the previously mentioned Berlin and Vienna series, based on second-order spectra of the form (8.5.4) with $m = 7$. Figure 8.13.6 results from expression (8.13.9) basing $|R_{YX}^{(T)}(\alpha)|$ on second-order spectra of the form (8.5.4) with $m = 5$ and then taking $N = 2$. The estimates in the two pictures therefore have comparable bandwidth and stability. It is apparent that the peaks of Figure 8.13.5 are less jagged than those of Figure 8.13.4. The nonlinear combination of correlation coefficients is considered in Fisher and Mackenzie (1922). See also Rao (1965) p. 365.

Tick (1967) argues that it may well be the case that $|f_{YX}(\alpha)|^2$ is near constant, whereas $f_{YX}(\alpha)$ is not. (This would be the case if $Y(t) = X(t - u)$ for large u.) He is then led to propose estimates of the form

$$\frac{(2N+1)\sum_{-N}^{N} |f_{YX}^{(T)}(\lambda + n\Delta_T)|^2}{\sum_{-N}^{N} f_{XX}^{(T)}(\lambda + n\Delta_T)|^2 \sum_{-N}^{N} f_{YY}^{(T)}(\lambda + n\Delta_T)|^2}; \tag{8.13.10}$$

he also proposes estimates of the form

$$(2N+1)^{-1} \sum_{n=-N}^{N} |R_{YX}^{(T)}(\lambda + n\Delta_T)|^2 \tag{8.13.11}$$

in the case that $|R_{YX}(\alpha)|^2$ is near constant, but the second-order spectra are not.

Jones (1969) considered the maximum likelihood estimation of $|R_{YX}(\lambda)|^2$ from the marginal distribution of $f_{XX}^{(T)}(\lambda)$, $f_{YY}^{(T)}(\lambda)$, and $|f_{YX}^{(T)}(\lambda)|^2$, deriving the latter from the limiting distribution of Theorem 8.5.1.

The importance of using some form of prefiltering, prior to the estimation of the parameters of this chapter, cannot be overemphasized. We saw, in Section 7.7, the need to do this when we estimated the cross-spectrum of two series. *A fortiori* we should do it when estimating the complex regression coefficient, coherency, and error spectrum. Akaike and Yamanouchi (1962) and Tick (1967) put forth compelling reasons for prefiltering. In particular there appear to be a variety of physical examples in which straight-forward data processing leads to a coherency estimate that is near 0 when for physical reasons the population value is not. Techniques of prefiltering are discussed in Section 7.7, the simplest being to lag one series relative to the other.

8.14 A WORKED EXAMPLE

As a worked example of the suggested calculations in the case $r, s = 1$ we refer the reader to the Berlin and Vienna monthly temperature series previously considered in Chapters 6 and 7. The spectra and cross-spectrum of this series are presented as Figures 7.8.1 to 7.8.4. The estimates are equal to those in expression (8.5.4) with $m = 10$. Figure 6.10.3 gives $g_{\varepsilon\varepsilon}^{(T)}(\lambda)$, Figure 6.10.4 gives Re $A^{(T)}(\lambda)$, Figure 6.10.5 gives Im $A^{(T)}(\lambda)$, Figure 6.10.6 gives $G^{(T)}(\lambda)$, Figure 6.10.7 gives $\phi^{(T)}(\lambda)$ and Figure 6.10.8 gives $|R_{YX}^{(T)}(\lambda)|^2$. Finally Figure 6.10.9 gives $a^{(T)}(u)$. The estimated standard errors of these various statistics are given in Section 6.10.

As a worked example in the case $r = 13$ and $s = 1$ we refer the reader back to Section 6.10 where the results of a frequency analysis of the sort under study are presented: the series $Y(t)$ refers to the seasonally adjusted monthly mean temperatures at Greenwich, England and $\mathbf{X}(t)$ refers to seasonally adjusted mean monthly temperatures at 13 other stations. Figure 6.10.10 gives the gains, $G_a^{(T)}(\lambda)$ and phases, $\phi_a^{(T)}(\lambda)$. Figure 6.10.11 gives the error spectrum, $\log_{10} g_{\varepsilon\varepsilon}^{(T)}(\lambda)$. Figure 6.10.12 gives the multiple coherence, $|R_{YX}^{(T)}(\lambda)|^2$.

8.15 USES OF THE ANALYSIS OF THIS CHAPTER

The uses that the techniques of this chapter have been put to are intimately entwined with the uses of the analysis of Chapter 6. We have already noted that many of the statistics of the present chapter are the same as statistics of Chapter 6, however, the principal difference in assumption between the chapters is that in the present chapter the series $\mathbf{X}(t)$, $t = 0, \pm 1, \ldots$ is taken as stochastic, whereas in Chapter 6 it was taken as fixed. In consequence, the statistical properties developed in this chapter refer to averages across the space of all realizations of $\mathbf{X}(t)$ whereas those of Chapter 6 refer to the particular realization at hand.

One area in which researchers have tended to assume $\mathbf{X}(t)$ stochastic, is the statistical theory of filtering and prediction. See Wiener (1949), Solodovnikov (1960), Lee (1960), Whittle (1963a), and Robinson (1967b) for example. The optimum predictors developed work best across the space of all realizations of $\mathbf{X}(t)$ that may come to hand and statistical properties of empirical predictors refer to this broad population.

The reader may look back to Section 6.10 for a listing of situations in which the various statistics of this chapter have been calculated. In fact the authors of the papers listed typically introduced the statistics in terms of stochastic $\mathbf{X}(t)$. Brillinger and Hatanaka (1970), Gersch (1972) estimate partial coherences and spectra.

The choice of whether to make $\mathbf{X}(t)$ fixed or stochastic is clearly tied up with the choice of population to which we wish to extend inferences based on a given sample. Luckily, as we have seen, the practical details of the two situations are not too different if the sample is large.

8.16 EXERCISES

8.16.1 Under the conditions of Theorem 8.2.2 and if $s = 1$, prove that $\phi(\mathbf{X})$ is the function with finite second moment having maximum correlation with Y; see Rao (1965) p. 221, and Brillinger (1966a).

8.16.2 Under the conditions of Theorem 8.2.2 prove that the conditional distribution of $\mathbf{Y}$ given $\mathbf{X}$ is multivariate normal with mean (8.2.14) and covariance matrix (8.2.16).

8.16.3 Let $A_{YX}(\lambda)$ denote the complex regression coefficient of $Y(t)$ on the series $X(t)$ and $A_{XY}(\lambda)$ denote the complex regression coefficient of $X(t)$ on the series $Y(t)$ in the case $s, r = 1$. Show that

$$A_{YX}(\lambda)A_{XY}(\lambda) = |R_{YX}(\lambda)|^2.$$

Hence, note that $A_{XY}(\lambda) = A_{YX}(\lambda)^{-1}$ only if the coherence between $X(t)$ and $Y(t)$ is 1.

8.16.4 If $\mathbf{A}(\lambda)$, the complex regression coefficient of $\mathbf{Y}(t)$ on the series $\mathbf{X}(t)$, is constant for all λ, show that it is equal to the ordinary regression coefficient of $\mathbf{Y}(t)$ on $\mathbf{X}(t)$.

8.16.5 Let $\rho(t)$, $t = 0, \pm 1, \ldots$ be a white noise process, that is, a second-order stationary process with constant power spectrum. Suppose $X(t) = \sum_u b(t-u)\rho(u)$, $Y(t) = \sum_u c(t-u)\rho(u)$. Determine $A(\lambda)$, $\phi(\lambda)$, $G(\lambda)$, $R_{YX}(\lambda)$, and $|R_{YX}(\lambda)|^2$.

8.16.6 Under the conditions of Theorem 8.3.1 and if $s = 1$, prove that $|R_{YX}(\lambda)|^2 = 1$, $-\infty < \lambda < \infty$, if and only if $Y(t)$ is a linear filtered version of $\mathbf{X}(t)$.

8.16.7 Under the conditions of Theorem 8.3.1 and if $s, r = 1$ determine the coherency between $Y(t)$ and its best linear predictor based on $X(t)$. Also determine the coherency between the error series $\varepsilon(t)$ and $X(t)$.

8.16.8 Under the conditions of Theorem 8.3.1 and if $s, r = 1$, prove that $R_{YX}(\lambda)$, $|R_{YX}(\lambda)|^2$ are the Fourier transforms of absolutely summable functions if $f_{XX}(\lambda), f_{YY}(\lambda) \neq 0$, $-\infty < \lambda < \infty$.

8.16.9 If $Y(t) = X^H(t)$, prove that $\phi(\lambda) = \pi/2$. Find $\phi(\lambda)$ if $Y(t) = X^H(t-u)$ for some integer u.

8.16.10 Prove that

$$[\operatorname{cor}\{X(t), Y(t)\}]^2 = c_{YX}(0)^2/[c_{XX}(0)c_{YY}(0)] \leqslant \int |R_{YX}(\alpha)|^2 f_{XX}(\alpha)\, d\alpha \Big/ \int f_{XX}(\alpha)\, d\alpha \leqslant \sup_\alpha |R_{YX}(\alpha)|^2$$

in the case $r, s = 1$.

8.16.11 Prove $|I_{YX}^{(T)}(\lambda)|^2 = I_{XX}^{(T)}(\lambda)I_{YY}^{(T)}(\lambda)$ and so $|I_{YX}^{(T)}(\lambda)|^2/[I_{XX}^{(T)}(\lambda) I_{YY}^{(T)}(\lambda)]$ is not a reasonable estimate of $|R_{YX}(\lambda)|^2$ in the case $r, s = 1$.

8.16.12 Prove that

$$\frac{f_{YY}(\lambda)}{f_{XX}(\lambda)} = |A(\lambda)|^2 + \frac{f_{\varepsilon\varepsilon}(\lambda)}{f_{XX}(\lambda)}$$

and so $[f_{YY}^{(T)}(\lambda)/f_{XX}^{(T)}(\lambda)]^{1/2}$ is not generally a reasonable estimate of $G(\lambda)$ in the case $r, s = 1$.

8.16.13 Discuss the reason why $X(t)$ and $Y(t)$ may have coherence 1 and yet it is not the case that $|R_{YX}^{(T)}(\lambda)|^2 = 1$.

8.16.14 Suppose that we estimate the spectral density matrix, $\mathbf{f}_{ZZ}(\lambda)$, by the second expression of (8.5.4) with $m = T - 1$. Show that

$$\mathbf{A}^{(T)}(\lambda) = \mathbf{c}_{YX}^{(T)}(0)\mathbf{c}_{XX}^{(T)}(0)^{-1}$$

and if $r, s = 1$

$$|R_{YX}^{(T)}(\lambda)|^2 = \frac{c_{YX}^{(T)}(0)^2}{c_{XX}^{(T)}(0)c_{YY}^{(T)}(0)}.$$

Discuss the effect on these expressions if $Y(t)$ had previously been lagged u time units with respect to $X(t)$.

8.16.15 Under the conditions of Section 8.6, if $W(\alpha) \geqslant 0$, and if $r, s = 1$ prove that $|R_{YX}^{(T)}(\lambda)|^2 \leqslant 1$.

8.16.16 Under the conditions of Theorem 8.7.1 and $r, s = 1$, prove that

$$\begin{aligned}\lim_{T\to\infty} B_T T \,\overrightarrow{\text{var}}\, \text{Re}\, A^{(T)}(\lambda) &= \lim_{T\to\infty} B_T T \,\overrightarrow{\text{var}}\, \text{Im}\, A^{(T)}(\lambda) \\ &= [1 + \eta\{2\lambda\}] f_{YY}(\lambda) f_{XX}(\lambda)^{-1}[1 - |R_{YX}(\lambda)|^2]\pi \int W(\alpha)^2 d\alpha\end{aligned}$$

while

$$\lim_{T\to\infty} B_T T \,\overrightarrow{\text{cov}}\, \{\text{Re}\, A^{(T)}(\lambda), \text{Im}\, A^{(T)}(\lambda)\} = 0.$$

8.16.17 Under the conditions of Theorem 8.7.1, and $r, s = 1$ except that $f_{YX}(\lambda) = 0$, show that

$$\begin{aligned}\overrightarrow{\text{cov}}\, \{G^{(T)}(\lambda), G^{(T)}(\mu)\} &= [\eta\{\lambda - \mu\} + \eta\{\lambda + \mu\}] \\ &\quad \times f_{YY}(\lambda) f_{XX}(\lambda)^{-1} B_T^{-1} T^{-1} \pi \int W(\alpha)^2 d\alpha + O(B_T^{-2} T^{-2}).\end{aligned}$$

8.16.18 Under the conditions of Theorem 8.8.1, and $r, s = 1$ except that $f_{YX}(\lambda) = 0$, show that $\phi^{(T)}(\lambda)$ is asymptotically uniformly distributed on $(-\pi,\pi]$.

8.16.19 Develop a sample analog of the error series $\boldsymbol{\varepsilon}(t)$ and of expression (8.3.8).

8.16.20 Under the conditions of Theorem 8.7.1 and $r, s = 1$, show that

$$\begin{aligned}\overrightarrow{\text{cov}}\, \{\text{Re}\, R_{YX}^{(T)}(\lambda), \text{Re}\, R_{YX}^{(T)}(\mu)\} &\doteq [\eta\{\lambda - \mu\} + \eta\{\lambda + \mu\}] \\ &\quad \times [1 - (\text{Re}\, R_{YX}(\lambda))^2][1 - |R_{YX}(\lambda)|^2] B_T^{-1} T^{-1} \pi \int W(\alpha)^2 d\alpha\end{aligned}$$

$$\begin{aligned}\overrightarrow{\text{cov}}\, \{\text{Re}\, R_{YX}^{(T)}(\lambda), \text{Im}\, R_{YX}^{(T)}(\mu)\} &\doteq -[\eta\{\lambda - \mu\} - \eta\{\lambda + \mu\}] \\ &\quad \times [\text{Re}\, R_{YX}(\lambda)][\text{Im}\, R_{YX}(\lambda)][1 - |R_{YX}(\lambda)|^2] B_T^{-1} T^{-1} \pi \int W(\alpha)^2 d\alpha\end{aligned}$$

$$\begin{aligned}\overrightarrow{\text{cov}}\, \{\text{Im}\, R_{YX}^{(T)}(\lambda), \text{Im}\, R_{YX}^{(T)}(\mu)\} &\doteq [\eta\{\lambda - \mu\} - \eta\{\lambda + \mu\}] \\ &\quad \times [1 - (\text{Im}\, R_{YX}(\lambda))^2][1 - |R_{YX}(\lambda)|^2] B_T^{-1} T^{-1} \pi \int W(\alpha)^2 d\alpha.\end{aligned}$$

8.16.21 Under the conditions of Theorem 8.2.3, show that the conditional variance of the sample squared coefficient of multiple correlation given the **X** values in approximately

$$\frac{2R_{YX}^2(2 - R_{YX}^2)(1 - R_{YX}^2)}{n}.$$

Contrast this with the unconditional value $4R_{YX}^2(1 - R_{YX}^2)/n$; see Hooper (1958).

8.16.22 For the random variable whose density is given by expression (8.2.56) show that $E|\hat{R}_{YX}|^2 = r/n$ and

$$\text{Prob}\,[|\hat{R}_{YX}|^2 < x] = 1 - (1 - x)^{n-1} \sum_{j=0}^{r-1} \binom{n-1}{j} x^j(1 - x)^{-j}$$

for $0 < x < 1$; see Abramowitz and Stegun (1964) p. 944. In the case $r = 1$, this leads to the simple expression $x = 1 - (1 - \alpha)^{1/(n-1)}$ for the 100α percent point of $|\hat{R}_{YX}|^2$.

8.16.23 For a real-valued series $Y(t)$ and a vector-valued series $\mathbf{X}(t)$, show that the multiple coherence is unaltered by nonsingular linear filtering of the series separately.

8.16.24 Show that the following perturbation expansions are valid for small α, β, γ, ε:

(a) $$\frac{(b + \beta)}{(a + \alpha)} = \frac{b}{a} + \frac{b}{a}\left\{\frac{\beta}{b} - \frac{\alpha}{a}\right\} + \cdots$$

(b) $$\arg\{e + \varepsilon\} = \arg e + \frac{1}{2i}\left\{\frac{\varepsilon}{e} - \frac{\bar{\varepsilon}}{\bar{e}}\right\} + \cdots$$

(c) $$\log|e + \varepsilon| = \log|e| + \frac{1}{2}\left\{\frac{\varepsilon}{e} + \frac{\bar{\varepsilon}}{\bar{e}}\right\} + \cdots$$

(d) $$\frac{(b + \beta)}{[(a + \alpha)(c + \gamma)]^{1/2}} = \frac{b}{[ac]^{1/2}} + \frac{b}{[ac]^{1/2}}\left\{\frac{\beta}{b} - \frac{\alpha}{2a} - \frac{\gamma}{2c}\right\} + \cdots$$

(e) $$\frac{|b + \beta|^2}{(a + \alpha)(c + \gamma)} = \frac{|b|^2}{ac} + \frac{|b|^2}{ac}\left\{\frac{\beta}{b} + \frac{\bar{\beta}}{\bar{b}} - \frac{\alpha}{a} - \frac{\gamma}{c}\right\} + \cdots$$

(f) $$(c + \gamma) - \frac{|b + \beta|^2}{(a + \alpha)} = c - \frac{|b|^2}{a} + \left\{\gamma + \frac{|b|^2}{a}\left(\frac{\beta}{b} + \frac{\bar{\beta}}{\bar{b}} - \frac{\alpha}{a}\right)\right\} + \cdots.$$

8.16.25 Let the bivariate time series $[X(t),Y(t)]$ satisfy Assumption 2.6.2(3). Let $W(\alpha)$ satisfy Assumption 5.6.1 and (5.8.21) with $P = 3$. Suppose the remaining conditions of Theorem 8.6.1 are satisfied, then

$$\vec{\text{ave}}\, A^{(T)}(\lambda) = A + W_2 f_{XX}^{-1}[f''_{YX} - Af''_{XX}]B_T^2/2 + O(B_T^3) + O(B_T^{-1}T^{-1})$$

$$\vec{\text{ave}}\, \phi^{(T)}(\lambda) = \phi + W_2 \operatorname{Im}\{f''_{YX} f_{YX}^{-1}\}B_T^2/2 + O(B_T^3) + O(B_T^{-1}T^{-1})$$

$$\vec{\text{ave}}\, G^{(T)}(\lambda) = G + W_2 \operatorname{Re}\{f_{YX}^{-1}[f''_{YX} - Af''_{XX}]\}B_T^2/2 + O(B_T^3) + O(B_T^{-1}T^{-1})$$

$$\vec{\text{ave}}\, R_{YX}^{(T)}(\lambda) = R_{YX} + W_2 f_{XX}^{-1/2} f_{YY}^{-1/2}[f''_{YX} - \tfrac{1}{2} f_{XX}^{-1} f_{YX} f''_{XX} - \tfrac{1}{2} f_{YY}^{-1} f_{YX} f''_{YY}]B_T^2/2 + O(B_T^3) + O(B_T^{-1}T^{-1})$$

where f'' denotes the second derivative.

8.16.26 Prove that

(a) $(\mathbf{A} \otimes \mathbf{B})^\tau = \mathbf{A}^\tau \otimes \mathbf{B}^\tau$
(b) $(\mathbf{A} \otimes \mathbf{B})^{-1} = \mathbf{A}^{-1} \otimes \mathbf{B}^{-1}$
(c) $(\mathbf{A}_1 \otimes \mathbf{B}_1)(\mathbf{A}_2 \otimes \mathbf{B}_2) = (\mathbf{A}_1\mathbf{A}_2) \otimes (\mathbf{B}_1\mathbf{B}_2)$
(d) $(\mathbf{A} \otimes \mathbf{B}) \text{ vec } \mathbf{X} = \text{vec } (\mathbf{AXB}^\tau)$

if the dimensions of the matrices are appropriate.

8.16.27 In connection with the matrix just after (8.2.18) prove that

$$\mathbf{0} \leqslant \mathbf{\Sigma}_{YY}^{-1/2} \mathbf{\Sigma}_{YX} \mathbf{\Sigma}_{XX}{}^{-1} \mathbf{\Sigma}_{XY} \mathbf{\Sigma}_{YY}^{-1/2} \leqslant \mathbf{I}.$$

8.16.28 Given the error variate (8.2.19), under the conditions of Theorem 8.2.1 prove:

(a) $E\boldsymbol{\varepsilon} = \mathbf{0}$
(b) $E\boldsymbol{\varepsilon}\boldsymbol{\varepsilon}^\tau = \mathbf{\Sigma}_{YY} - \mathbf{\Sigma}_{YX} \mathbf{\Sigma}_{XX}{}^{-1} \mathbf{\Sigma}_{XY}$
(c) $E\boldsymbol{\varepsilon}\mathbf{X}^\tau = \mathbf{0}$.

8.16.29 Prove that the partial correlation of Y_1 with Y_2 after removing the linear effects of $\mathbf{X}$ does not involve any covariances based on Y_j, $j > 2$.

8.16.30 Prove that $\mathbf{a}$ given by (8.2.15) maximizes the squared vector correlation coefficient

$$\frac{[\text{Det cov } \{\mathbf{Y}, \mathbf{aX}\}]^2}{[\text{Det cov } \{\mathbf{Y}, \mathbf{Y}\}][\text{Det cov } \{\mathbf{aX}, \mathbf{aX}\}]}.$$

8.16.31 Under the conditions of Theorem 8.2.1 and if the $s \times s$ $\mathbf{\Gamma} \geqslant 0$, determine $\boldsymbol{\mu}$ and $\boldsymbol{\alpha}$ that minimize

$$E\{[\mathbf{Y} - \boldsymbol{\mu} - \boldsymbol{\alpha}\mathbf{X}]\mathbf{\Gamma}[\mathbf{Y} - \boldsymbol{\mu} - \boldsymbol{\alpha}\mathbf{X}]^\tau\}.$$

8.16.32 Let $\mathbf{X}(t)$, $t = 0, \pm 1, \ldots$ be an r vector-valued autoregressive process of order m. Prove that the partial covariance function

$$\mathbf{\Sigma}_{X(0),X(u)\cdot X(1)\ldots X(u-1)}$$

vanishes for $u > m$.

8.16.33 Under the conditions of Theorem 8.3.1, prove that there exist $\boldsymbol{\mu}$, absolutely summable $\{\mathbf{a}(u)\}$ and a second-order stationary series $\boldsymbol{\varepsilon}(t)$ that is orthogonal to $\mathbf{X}(t)$ and has absolutely summable autocovariance function, such that $\mathbf{Y}(t) = \boldsymbol{\mu} + \sum_u \mathbf{a}(t - u)\mathbf{X}(u) + \boldsymbol{\varepsilon}(t)$.

8.16.34 Let the series of Theorem 8.3.1 be an m dependent process, that is, such that values of the process more than m time units apart are statistically independent. Show that $\mathbf{a}(u) = 0$ for $|u| > m$.

8.16.35 Under the conditions of Theorem 8.3.1, prove that $|R_{Y_aY_b\cdot X}(\lambda)|^2 \leqslant 1$. If $s = 1$, prove that $|R_{YX}(\lambda)|^2 \leqslant 1$.

8.16.36 Prove that in the case $s = 1$

$$|R_{YX}(\lambda)|^2 = 1 - \frac{f_{YY\cdot X}(\lambda)}{f_{YY}(\lambda)}.$$

8.16.37 Show that the inverse of the matrix (8.2.47) of partial covariances is the $s \times s$ lower diagonal matrix of the inverse of the covariance matrix (8.2.37).

8.16.38 If $s = 1$, determine the coherency between $Y(t)$ and the best linear predictor based on the series $\mathbf{X}(t)$, $t = 0, \pm 1, \ldots$.

8.16.39 Prove that

$$1 - |R_{YX}(\lambda)|^2 = [1 - |R_{YX_1}(\lambda)|^2][1 - |R_{YX_2 \cdot X_1}(\lambda)|^2] \cdots [1 - |R_{YX_r \cdot X_1 \ldots X_{r-1}}(\lambda)|^2].$$

8.16.40 Let $\rho_{YX}(0)^2$ denote the instantaneous squared multiple correlation of $Y(t)$ with $\mathbf{X}(t)$. Show that

$$\rho_{YX}(0)^2 \leqslant \frac{\int |R_{YX}(\lambda)|^2 f_{YY}(\lambda) d\lambda}{\int f_{YY}(\lambda) d\lambda} \leqslant \max_{\lambda} |R_{YX}(\lambda)|^2.$$

8.16.41 Under the conditions of Theorem 8.3.2, prove that the conditional spectral density matrix of $\mathbf{Y}(t)$ given the series $\mathbf{X}(t)$, $t = 0, \pm 1, \ldots$ is

$$\mathbf{f}_{YY}(\lambda) - \mathbf{f}_{YX}(\lambda)\mathbf{f}_{XX}(\lambda)^{-1}\mathbf{f}_{XY}(\lambda).$$

8.16.42 Suppose the weight function $W(\alpha)$ used in forming the estimate (8.6.4) is non-negative. Show that $|R^{(T)}_{Y_aY_b \cdot X}(\lambda)|^2$, $|R_{YX}{}^{(T)}(\lambda)|^2 \leqslant 1$.

8.16.43 Suppose the conditions of Theorem 8.5.1 are satisfied. Suppose $f_{Y_aX_b \cdot X'_b}(\lambda) = 0$. Show that the asymptotic distribution of $\phi_{ab}{}^{(T)}(\lambda)$ is the uniform distribution on $(-\pi, \pi)$.

8.16.44 Let the conditions of Theorem 8.3.1 be satisfied. Show that the complex regression coefficient of the real-valued series $Y_a(t)$ on the series $\mathbf{X}(t)$ is the same as the ath row of the complex regression coefficient of the s vector-valued series $\mathbf{Y}(t)$ on the series $\mathbf{X}(t)$ for $a = 1, \ldots, s$. Discuss the implications of this result.

8.16.45 Under the conditions of Theorem 8.2.1, show that $\mathbf{a} = \mathbf{\Sigma}_{YX}\mathbf{\Sigma}_{XX}{}^{-1}$ maximizes $\mathbf{\Sigma}_{Y,aX}(\mathbf{\Sigma}_{aX,aX})^{-1}\mathbf{\Sigma}_{aX,Y}$.

8.16.46 Let $\mathbf{W}$ be distributed as $W_r{}^C(n, \mathbf{\Sigma})$. Show that vec $\mathbf{W}$ has covariance matrix $n\mathbf{\Sigma} \otimes \mathbf{\Sigma}^{\tau}$.

8.16.47 (a) If $\mathbf{W}$ is distributed as $W_r(n, \mathbf{\Sigma})$ show that

$$E\mathbf{W}^{-1} = (n - r - 1)^{-1}\mathbf{\Sigma}^{-1}.$$

(b) If $\mathbf{W}$ is distributed as $W_r{}^C(n, \mathbf{\Sigma})$ show that

$$E\mathbf{W}^{-1} = (n - r)^{-1}\mathbf{\Sigma}^{-1}.$$

See Wahba (1966).

9

PRINCIPAL COMPONENTS IN THE FREQUENCY DOMAIN

9.1 INTRODUCTION

In the previous chapter we considered the problem of approximating a stationary series by a linear filtered version of another stationary series. In this chapter we investigate the problem of approximating a series by a filtered version of itself, but restraining the filter to have reduced rank.

Specifically, consider the r vector-valued series $\mathbf{X}(t)$, $t = 0, \pm 1, \ldots$ with mean

$$E\mathbf{X}(t) = \mathbf{c}_X \tag{9.1.1}$$

absolutely summable autocovariance function

$$E\{[\mathbf{X}(t+u) - \mathbf{c}_X][\mathbf{X}(t) - \mathbf{c}_X]^\tau\} = \mathbf{c}_{XX}(u) \qquad u = 0, \pm 1, \ldots \tag{9.1.2}$$

and spectral density matrix

$$\mathbf{f}_{XX}(\lambda) = (2\pi)^{-1} \sum_{u=-\infty}^{\infty} \mathbf{c}_{XX}(u) \exp\{-i\lambda u\} \qquad -\infty < \lambda < \infty. \tag{9.1.3}$$

Suppose we are interested in transmitting the values of the $\mathbf{X}(t)$ series from one location to another; however, only $q \leq r$ channels are available for the transmission. Imagine forming the series

$$\boldsymbol{\zeta}(t) = \sum_u \mathbf{b}(t-u)\mathbf{X}(u) \qquad t = 0, \pm 1, \ldots \tag{9.1.4}$$

with $\{\mathbf{b}(u)\}$ a $q \times r$ matrix-valued filter, transmitting the series $\boldsymbol{\zeta}(t)$ over the q available series and then, on receipt of this series, forming

$$\mathbf{X}^*(t) = \boldsymbol{\mu} + \sum_u \mathbf{c}(t-u)\boldsymbol{\zeta}(u) \tag{9.1.5}$$

as an estimate of $\mathbf{X}(t)$ for some r vector-valued $\boldsymbol{\mu}$ and $r \times q$ filter $\{\mathbf{c}(u)\}$. In this chapter we will be concerned with the choice of $\boldsymbol{\mu}$ and the filters $\{\mathbf{b}(u)\}$, $\{\mathbf{c}(u)\}$ so that $\mathbf{X}^*(t)$ is near $\mathbf{X}(t)$.

The relation between $\mathbf{X}^*(t) - \boldsymbol{\mu}$ and $\mathbf{X}(t)$ is of linear time invariant form with transfer function

$$\mathbf{A}(\lambda) = \mathbf{C}(\lambda)\mathbf{B}(\lambda) \tag{9.1.6}$$

where $\mathbf{B}(\lambda)$, $\mathbf{C}(\lambda)$ indicate the transfer functions of $\{\mathbf{b}(u)\}$, $\{\mathbf{c}(u)\}$ respectively. We now see that the problem posed is that of determining an $r \times r$ matrix $\mathbf{A}(\lambda)$ of reduced rank so that the difference

$$\mathbf{X}(t) - \mathbf{X}^*(t) + \boldsymbol{\mu} = \int \exp\{i\lambda t\}d\mathbf{Z}_X(\lambda) - \int \mathbf{A}(\lambda)\exp\{i\lambda t\}d\mathbf{Z}_X(\lambda) \tag{9.1.7}$$

is small.

We might view the problem as that of determining a q vector-valued series $\boldsymbol{\zeta}(t)$ that contains much of the information in $\mathbf{X}(t)$. Here, we note that Bowley (1920) once remarked "Index numbers are used to measure the change in some quantity which we cannot observe directly, which we know to have a definite influence on many other quantities which we can so observe, tending to increase all, or diminish all, while this influence is concealed by the action of many causes affecting the separate quantities in various ways." Perhaps, $\boldsymbol{\zeta}(t)$ above plays the role of an index number series following some hidden series influencing $\mathbf{X}(t)$. As we have described in its derivation, the above series $\boldsymbol{\zeta}(t)$ is the q vector-valued series that is best for getting back $\mathbf{X}(t)$ through linear time invariant operations.

Alternatively suppose we define the error series $\boldsymbol{\varepsilon}(t)$ by

$$\boldsymbol{\varepsilon}(t) = \mathbf{X}(t) - \mathbf{X}^*(t) \tag{9.1.8}$$

and then write

$$\mathbf{X}(t) = \boldsymbol{\mu} + \sum_u \mathbf{c}(t-u)\boldsymbol{\zeta}(u) + \boldsymbol{\varepsilon}(t) \qquad t = 0, \pm 1, \ldots. \tag{9.1.9}$$

Then $\mathbf{X}(t)$ is represented as a filtered version of a series $\boldsymbol{\zeta}(t)$ of reduced dimension plus an error series. A situation in which we might wish to set down such a model is the following: let $\boldsymbol{\zeta}(t)$ represent the impulse series of q earthquakes occurring simultaneously at various locations; let $\mathbf{X}(t)$ represent the signals received by r seismometers; and let $\mathbf{c}(u)$ represent the transmission effects of the earth on the earthquakes. Seismologists are interested in investigating the series $\boldsymbol{\zeta}(t)$, $t = 0, \pm 1, \ldots$; see for example Ricker (1940) and Robinson (1967b).

An underlying thread of these problems is the approximation of a series of interest by a related series of lower dimension. In Section 9.2 we review some aspects of the classical principal component analysis of vector-valued variates.

9.2 PRINCIPAL COMPONENT ANALYSIS OF VECTOR-VALUED VARIATES

Let **X** be an r vector-valued random variable with mean $\boldsymbol{\mu}_X$ and covariance matrix $\boldsymbol{\Sigma}_{XX}$. Consider the problem of determining the r vector $\boldsymbol{\mu}$, the $q \times r$ matrix **B** and the $r \times q$ matrix **C** to minimize simultaneously all the latent roots of the symmetric matrix

$$E\{(\mathbf{X} - \boldsymbol{\mu} - \mathbf{CBX})(\mathbf{X} - \boldsymbol{\mu} - \mathbf{CBX})^\tau\}. \tag{9.2.1}$$

When we determine these values it will follow, as we mentioned in Section 8.2, that they also minimize monotonic functions of the latent roots of (9.2.1) such as trace, determinant, and diagonal entries.

Because any $r \times r$ matrix **A** of rank $q \leqslant r$ may be written in the form **CB** with **B**, $q \times r$, and **C**, $r \times q$ (Exercise 3.10.36), we are also determining **A** of rank $\leqslant q$ to minimize the latent values of

$$E\{(\mathbf{X} - \boldsymbol{\mu} - \mathbf{AX})(\mathbf{X} - \boldsymbol{\mu} - \mathbf{AX})^\tau\}. \tag{9.2.2}$$

We now state

Theorem 9.2.1 Let **X** be an r vector-valued variate with $E\mathbf{X} = \boldsymbol{\mu}_X$, $E\{(\mathbf{X} - \boldsymbol{\mu}_X)(\mathbf{X} - \boldsymbol{\mu}_X)^\tau\} = \boldsymbol{\Sigma}_{XX}$. The $r \times 1$ $\boldsymbol{\mu}$, $q \times r$ **B** and $r \times q$ **C** that minimize simultaneously all latent values of (9.2.1) are given by

$$\mathbf{B} = \begin{bmatrix} \mathbf{V}_1{}^\tau \\ \cdot \\ \cdot \\ \cdot \\ \mathbf{V}_q{}^\tau \end{bmatrix}, \tag{9.2.3}$$

$$\mathbf{C} = [\mathbf{V}_1 \cdots \mathbf{V}_q] = \mathbf{B}^\tau, \tag{9.2.4}$$

and

$$\boldsymbol{\mu} = \boldsymbol{\mu}_X - \mathbf{CB}\boldsymbol{\mu}_X, \tag{9.2.5}$$

where $\mathbf{V}_j$ is the jth latent vector of $\boldsymbol{\Sigma}_{XX}$, $j = 1, \ldots, r$. If μ_j indicates the corresponding latent root, then the matrix (9.2.1) corresponding to these values is

$$\sum_{j>q} \mu_j \mathbf{V}_j \mathbf{V}_j{}^\tau. \tag{9.2.6}$$

Theorem 9.2.1 is a particular case of one proved by Okamoto and Kanazawa (1968); see also Okamoto (1969). The fact that the above **B, C,** $\boldsymbol{\mu}$ minimize the trace of (9.2.1) was proved by Kramer and Mathews (1956), Rao (1964, 1965), and Darroch (1965).

The variate

$$\zeta_j = \mathbf{V}_j{}^\tau \mathbf{X} \tag{9.2.7}$$

is called the **jth principal component of X**, $j = 1, \ldots, r$. In connection with the principal components we have

Corollary 9.2.1 Under the conditions of Theorem 9.2.1

$$\begin{aligned} \operatorname{cov}\{\mathbf{V}_j{}^\tau\mathbf{X}, \mathbf{V}_k{}^\tau\mathbf{X}\} &= 0 \qquad j \neq k \\ &= \mu_j \qquad j = k. \end{aligned} \tag{9.2.8}$$

The principal components of **X** are seen to provide linear combinations of the entries of **X** that are uncorrelated. We could have characterized the *j*th principal component as the linear combination $\zeta_j = \boldsymbol{\alpha}^\tau\mathbf{X}$, with $\boldsymbol{\alpha}^\tau\boldsymbol{\alpha} = 1$, which has maximum variance and is uncorrelated with ζ_k, $k < j$ (see Hotelling (1933), Anderson (1957) Chap. 11, Rao (1964, 1965), and Morrison (1967) Chap. 7); however, the above approach fits in better with our later work.

We next review details of the estimation of the above parameters. For convenience assume $\boldsymbol{\mu}_X = \mathbf{0}$, then $\boldsymbol{\mu}$ of expression (9.2.5) is **0**. Suppose that a sample of values $\mathbf{X}_j$, $j = 1, \ldots, n$ of the variate of Theorem 9.2.1 is available. Define the $r \times n$ matrix **x** by

$$\mathbf{x} = [\mathbf{X}_1 \cdots \mathbf{X}_n]. \tag{9.2.9}$$

Estimate the covariance matrix $\boldsymbol{\Sigma}_{XX}$ by

$$\hat{\boldsymbol{\Sigma}}_{XX} = \frac{\mathbf{x}\mathbf{x}^\tau}{n}. \tag{9.2.10}$$

We may now estimate μ_j by $\hat{\mu}_j$ the *j*th largest latent root of $\hat{\boldsymbol{\Sigma}}_{XX}$ and estimate $\mathbf{V}_j$ by $\hat{\mathbf{V}}_j$ the corresponding latent vector of $\hat{\boldsymbol{\Sigma}}_{XX}$. We have

Theorem 9.2.2 Suppose the values $\mathbf{X}_j$, $j = 1, \ldots, n$ are a sample from $N_r(\mathbf{0}, \boldsymbol{\Sigma}_{XX})$. Suppose the latent roots μ_j, $j = 1, \ldots, r$ of $\boldsymbol{\Sigma}_{XX}$ are distinct. Then the variate $\{\hat{\boldsymbol{\mu}}_j, \hat{\mathbf{V}}_j;\, j = 1, \ldots, r\}$ is asymptotically normal with $\{\hat{\mu}_j;\, j = 1, \ldots, r\}$ asymptotically independent of $\{\hat{\mathbf{V}}_j; j = 1, \ldots, r\}$. The asymptotic moments are given by

$$\vec{\text{ave}}\, \hat{\mu}_j = \mu_j + O(n^{-1}), \tag{9.2.11}$$

$$\vec{\text{ave}}\, \hat{\mathbf{V}}_j = \mathbf{V}_j + O(n^{-1}), \tag{9.2.12}$$

$$\vec{\text{cov}}\ \{\hat{\mu}_j, \hat{\mu}_k\} = \delta\{j-k\}2\mu_j^2/n + O(n^{-2}), \tag{9.2.13}$$

and

$$\begin{aligned}\vec{\text{cov}}\ \{\hat{\mathbf{V}}_j, \hat{\mathbf{V}}_k\} = \delta\{j-k\}\mu_j \left\{\sum_{l \neq j} \mu_l (\mu_j - \mu_l)^{-2}\mathbf{V}_l\mathbf{V}_l^{\tau}\right\}/n \\ - [1 - \delta\{j-k\}]\mu_k\mu_j(\mu_j - \mu_k)^{-2}\mathbf{V}_k\mathbf{V}_j^{\tau}/n + O(n^{-2}),\end{aligned} \tag{9.2.14}$$

for $j, k = 1, \ldots, r$.

This theorem was derived by Girshick (1939). Anderson (1963) developed the limiting distribution in the case that the latent roots of $\mathbf{\Sigma}_{XX}$ are not all distinct. Expression (9.2.13) implies the useful result

$$\vec{\text{var}}\ \log_e \mu_j = \frac{2}{n} + O(n^{-2}); \tag{9.2.15}$$

$\log_e$ here denotes the natural logarithm. James (1964) has derived the exact distribution of $\mu_1, \ldots, \mu_r$ under the conditions of the theorem. This distribution turns out to depend only on $\mu_1, \ldots, \mu_r$. James has also obtained asymptotic expressions for the likelihood function of $\mu_1, \ldots, \mu_r$ more detailed than that indicated by the theorem; see James (1964), Anderson (1965), and James (1966). Dempster (1969), p. 303, indicates the exact distribution of vectors dual to $\mathbf{V}_1, \ldots, \mathbf{V}_r$. Tumura (1965) derives a distribution equivalent to that of $\mathbf{V}_1, \ldots, \mathbf{V}_r$. Chambers (1967) indicates further cumulants of the asymptotic distribution for distributions having finite moments. These cumulants may be used to construct Cornish-Fisher approximations to the distributions. Because the μ_j have the approximate form of sample variances it may prove reasonable to approximate their distributions by scaled χ^2 distributions, for example, to take μ_j to be $\mu_j\chi_n^2/n$. Madansky and Olkin (1969) indicate approximate confidence bounds for the collection $\mu_1, \ldots, \mu_r$; see also Mallows (1961). We could clearly use Tukey's jack-knife procedure (Brillinger (1964c, 1966b)) to obtain approximate confidence regions for the latent roots and vectors.

Sugiyama (1966) determines the distribution of the largest root and corresponding vector. Krishnaiah and Waikar (1970) give the joint distribution of several roots. Golub (1969) discusses the computations involved in the present situation. Izenman (1972) finds the asymptotic distribution of

$$\hat{\mathbf{C}}\hat{\mathbf{B}} = \sum_{j=1}^{q} \hat{\mathbf{V}}_j\hat{\mathbf{V}}_j^{\tau} \tag{9.2.16}$$

in the normal case.

In our work with time series we will require complex variate analogs of the above results. We begin with

Theorem 9.2.3 Let X be an r vector-valued variate with $E\mathbf{X} = \boldsymbol{\mu}_X$, $E\{(\mathbf{X} - \boldsymbol{\mu}_X)\overline{(\mathbf{X} - \boldsymbol{\mu}_X)}^\tau\} = \boldsymbol{\Sigma}_{XX}$, $E\{(\mathbf{X} - \boldsymbol{\mu}_X)(\mathbf{X} - \boldsymbol{\mu}_X)^\tau\} = \mathbf{0}$. The $r \times 1$ $\boldsymbol{\mu}$, $q \times r$ $\mathbf{B}$, and $r \times q$ $\mathbf{C}$ that simultaneously minimize all the latent values of

$$E\{(\mathbf{X} - \boldsymbol{\mu} - \mathbf{CBX})\overline{(\mathbf{X} - \boldsymbol{\mu} - \mathbf{CBX})}^\tau\} \tag{9.2.17}$$

are given by

$$\mathbf{B} = \begin{bmatrix} \overline{\mathbf{V}}_1{}^\tau \\ \cdot \\ \cdot \\ \cdot \\ \overline{\mathbf{V}}_q{}^\tau \end{bmatrix}, \tag{9.2.18}$$

$$\mathbf{C} = [\mathbf{V}_1 \cdots \mathbf{V}_q] = \overline{\mathbf{B}}^\tau, \tag{9.2.19}$$

and

$$\boldsymbol{\mu} = \boldsymbol{\mu}_X - \mathbf{CB}\boldsymbol{\mu}_X, \tag{9.2.20}$$

where $\mathbf{V}_j$ is the jth latent vector of $\boldsymbol{\Sigma}_{XX}$, $j = 1, \ldots, r$. If μ_j denotes the corresponding latent root, then the extreme value of (9.2.17) is

$$\sum_{j>q} \mu_j \mathbf{V}_j \overline{\mathbf{V}}_j{}^\tau. \tag{9.2.21}$$

We note that as the matrix $\boldsymbol{\Sigma}_{XX}$ is Hermitian non-negative definite, the μ_j will be non-negative. The degree of approximation achieved depends directly on how near the μ_j, $j > q$ are to 0. Note that we have been led to approximate $\mathbf{X}$ by

$$\boldsymbol{\mu}_X + \mathbf{A}(\mathbf{X} - \boldsymbol{\mu}_X), \tag{9.2.22}$$

where

$$\mathbf{A} = \mathbf{V}_1\overline{\mathbf{V}}_1{}^\tau + \cdots + \mathbf{V}_q\overline{\mathbf{V}}_q{}^\tau. \tag{9.2.23}$$

We have previously seen a related result in Theorem 4.7.1.

Theorem 9.2.3 leads us to consider the variates $\zeta_j = \overline{\mathbf{V}}_j{}^\tau\mathbf{X}$, $j = 1, \ldots, r$. These are called the **principal components** of $\mathbf{X}$. In the case that $\mathbf{X}$ is $N_r{}^C(\mathbf{0},\boldsymbol{\Sigma}_{XX})$, we see that $\zeta_1, \ldots, \zeta_r$ are independent $N_1{}^C(0,\mu_j)$, $j = 1, \ldots, r$, variates.

Now we will estimate these parameters. Let $\mathbf{X}_j$, $j = 1, \ldots, n$ be a sample from $N_r{}^C(\mathbf{0},\boldsymbol{\Sigma}_{XX})$ and define $\mathbf{x}$ by expression (9.2.9). Then we estimate $\boldsymbol{\Sigma}_{XX}$ by

$$\hat{\boldsymbol{\Sigma}}_{XX} = \frac{\mathbf{x}\overline{\mathbf{x}}^\tau}{n}. \tag{9.2.24}$$

This matrix has a complex Wishart distribution. We signify its latent roots and vectors by $\hat{\mu}_j$, $\hat{\mathbf{V}}_j$ respectively $j = 1, \ldots, r$. The matrix $\hat{\mathbf{\Sigma}}_{XX}$ is Hermitian non-negative definite, therefore the $\hat{\mu}_j$ will be non-negative. We have

Theorem 9.2.4 Suppose the values $\mathbf{X}_1, \ldots, \mathbf{X}_n$ are a sample from $N_r^C(\mathbf{0},\mathbf{\Sigma}_{XX})$. Suppose the latent roots of $\mathbf{\Sigma}_{XX}$ are distinct. Then the variate $\{\hat{\mu}_j, \hat{\mathbf{V}}_j; j = 1, \ldots, r\}$ is asymptotically normal with $\{\hat{\mu}_j; j = 1, \ldots, r\}$ asymptotically independent of $\{\hat{\mathbf{V}}_j; j = 1, \ldots, r\}$. The asymptotic moments are given by

$$\overrightarrow{\text{ave}}\, \hat{\mu}_j = \mu_j + O(n^{-1}), \tag{9.2.25}$$

$$\overrightarrow{\text{ave}}\, \hat{\mathbf{V}}_j = \mathbf{V}_j + O(n^{-1}), \tag{9.2.26}$$

$$\overrightarrow{\text{cov}}\, \{\hat{\mu}_j, \hat{\mu}_k\} = \delta\{j - k\}\mu_j^2/n + O(n^{-2}), \tag{9.2.27}$$

$$\begin{aligned} \overrightarrow{\text{cov}}\, \{\hat{\mathbf{V}}_j, \hat{\mathbf{V}}_k\} &= \mu_j \sum_{l \neq j} \mu_l(\mu_j - \mu_l)^{-2}\mathbf{V}_l\overline{\mathbf{V}}_l^{\tau}/n + O(n^{-2}) && \text{if } j = k \\ &= O(n^{-2}) && \text{if } j \neq k \end{aligned} \tag{9.2.28}$$

and

$$\begin{aligned} \overrightarrow{\text{cov}}\, \{\hat{\mathbf{V}}_j, \overline{\hat{\mathbf{V}}}_k\} &= -\mu_k\mu_j(\mu_j - \mu_k)^{-2}\mathbf{V}_k\mathbf{V}_j^{\tau}/n + O(n^{-2}) && \text{if } j \neq k \\ &= O(n^{-2}) && \text{if } j = k, \end{aligned} \tag{9.2.29}$$

for $j, k = 1, \ldots, r$.

Theorem 9.2.4 results from two facts: the indicated latent roots and vectors are differentiable functions of the entries of $\hat{\mathbf{\Sigma}}_{XX}$ and $\hat{\mathbf{\Sigma}}_{XX}$ is asymptotically normal as $n \to \infty$; see Gupta (1965).

We see from expression (9.2.27) that

$$\overrightarrow{\text{var}}\, \log_e \hat{\mu}_j = \frac{1}{n} + O(n^{-2}). \tag{9.2.30}$$

Also by analogy with the real-valued case we might consider approximating the distribution of $\hat{\mu}_j$ by

$$\frac{\mu_j \chi_{2n}^2}{2n}. \tag{9.2.31}$$

The approximation in expression (9.2.31) would be especially good if the off-diagonal elements of $\mathbf{\Sigma}_{XX}$ were small, and if the diagonal elements were quite different. James (1964) has given the exact distribution of $\hat{\mu}_1, \ldots, \hat{\mu}_r$ in the complex normal case. Expression (9.2.29) with $j = k$ indicates that the asymptotic distribution of the $\hat{\mathbf{V}}_j$ is complex normal. Also from (9.2.28) we see that the sampling variability of the $\hat{\mathbf{V}}_j$ will be high if some of the μ_j are nearly equal.

9.3 THE PRINCIPAL COMPONENT SERIES

We return to the problem of determining the r vector $\mathbf{\mu}$, the $q \times r$ filter $\{\mathbf{b}(u)\}$ and the $r \times q$ filter $\{\mathbf{c}(u)\}$ so that if

$$\mathbf{\zeta}(t) = \sum_u \mathbf{b}(t-u)\mathbf{X}(u) \tag{9.3.1}$$

then the r vector-valued series

$$\mathbf{X}(t) - \mathbf{\mu} - \sum_u \mathbf{c}(t-u)\mathbf{\zeta}(u) \tag{9.3.2}$$

is small. If we measure the size of this series by

$$E\left\{\left[\overline{\mathbf{X}(t) - \mathbf{\mu} - \sum_u \mathbf{c}(t-u)\mathbf{\zeta}(u)}\right]^T\left[\mathbf{X}(t) - \mathbf{\mu} - \sum_u \mathbf{c}(t-u)\mathbf{\zeta}(u)\right]\right\}, \tag{9.3.3}$$

we have

Theorem 9.3.1 Let $\mathbf{X}(t)$, $t = 0, \pm 1, \ldots$ be an r vector-valued second-order stationary series with mean $\mathbf{c}_X$, absolutely summable autocovariance function $\mathbf{c}_{XX}(u)$ and spectral density matrix $\mathbf{f}_{XX}(\lambda)$, $-\infty < \lambda < \infty$. Then the $\mathbf{\mu}$, $\{\mathbf{b}(u)\}$, $\{\mathbf{c}(u)\}$ that minimize (9.3.3) are given by

$$\mathbf{\mu} = \mathbf{c}_X - \left(\sum_u \mathbf{c}(u)\right)\left(\sum_u \mathbf{b}(u)\right)\mathbf{c}_X, \tag{9.3.4}$$

$$\mathbf{b}(u) = (2\pi)^{-1}\int_0^{2\pi} \mathbf{B}(\alpha) \exp\{iu\alpha\}\,d\alpha, \tag{9.3.5}$$

and

$$\mathbf{c}(u) = (2\pi)^{-1}\int_0^{2\pi} \mathbf{C}(\alpha) \exp\{iu\alpha\}\,d\alpha, \tag{9.3.6}$$

where

$$\mathbf{B}(\lambda) = \begin{bmatrix} \overline{\mathbf{V}_1(\lambda)}^T \\ \cdot \\ \cdot \\ \cdot \\ \overline{\mathbf{V}_q(\lambda)}^T \end{bmatrix} \tag{9.3.7}$$

and

$$\mathbf{C}(\lambda) = [\mathbf{V}_1(\lambda)\cdots\mathbf{V}_q(\lambda)] = \overline{\mathbf{B}(\lambda)}^T. \tag{9.3.8}$$

Here $\mathbf{V}_j(\lambda)$ denotes the jth latent vector of $\mathbf{f}_{XX}(\lambda)$, $j = 1, \ldots, r$. If $\mu_j(\lambda)$ denotes the corresponding latent root, $j = 1, \ldots, r$, then the minimum obtained is

$$\int_0^{2\pi} \left\{ \sum_{j>q} \mu_j(\alpha) \right\} d\alpha. \tag{9.3.9}$$

Theorem 9.3.1 is given in Brillinger (1969d).

Let $\mathbf{A}(\lambda)$ denote the transfer function of the filter resulting from applying the filter $\{\mathbf{b}(u)\}$ followed by the filter $\{\mathbf{c}(u)\}$. Note that

$$\begin{aligned} \mathbf{A}(\lambda) &= \mathbf{C}(\lambda)\mathbf{B}(\lambda) \\ &= \mathbf{V}_1(\lambda)\overline{\mathbf{V}_1(\lambda)}^\tau + \cdots + \mathbf{V}_q(\lambda)\overline{\mathbf{V}_q(\lambda)}^\tau \end{aligned} \tag{9.3.10}$$

which has rank $\leq q$. Now let the series $\mathbf{X}(t)$, $t = 0, \pm 1, \ldots$ have Cramér representation

$$\mathbf{X}(t) = \int \exp\{i\lambda t\} d\mathbf{Z}_X(\lambda), \tag{9.3.11}$$

then the series $\boldsymbol{\zeta}(t)$ corresponding to the extremal choice has the form

$$\boldsymbol{\zeta}(t) = \int \mathbf{B}(\lambda) \exp\{i\lambda t\} d\mathbf{Z}_X(\lambda) \tag{9.3.12}$$

with $\mathbf{B}(\lambda)$ given by (9.3.7). The jth component, $\zeta_j(t)$, is given by

$$\zeta_j(t) = \int \overline{\mathbf{V}_j(\lambda)}^\tau \exp\{i\lambda t\} d\mathbf{Z}_X(\lambda). \tag{9.3.13}$$

This series is called the ***j*th principal component series of** $\mathbf{X}(t)$. In connection with the principal component series we have

Theorem 9.3.2 Under the conditions of Theorem 9.3.1, the jth principal component series, $\zeta_j(t)$, has power spectrum $\mu_j(\lambda)$, $-\infty < \lambda < \infty$. Also $\zeta_j(t)$ and $\zeta_k(t)$, $j \neq k$, have 0 coherency for all frequencies.

The series $\boldsymbol{\zeta}(t)$ has spectral density matrix

$$\begin{bmatrix} \mu_1(\lambda) & & & \\ & \cdot & & 0 \\ & & \cdot & \\ 0 & & & \cdot \\ & & & \mu_q(\lambda) \end{bmatrix}. \tag{9.3.14}$$

Let $\mathbf{X}^*(t)$, $t = 0, \pm 1, \ldots$ denote the best approximant series as given in Theorem 9.3.1. We define the **error series** by

$$\boldsymbol{\varepsilon}(t) = \mathbf{X}(t) - \mathbf{X}^*(t). \tag{9.3.15}$$

In terms of the Cramér representation this series has the form

$$\int [\mathbf{I} - \mathbf{A}(\lambda)] \exp\{i\lambda t\} d\mathbf{Z}_X(\lambda) = \int \left\{ \sum_{j>q} \mathbf{V}_j(\lambda)\overline{\mathbf{V}_j(\lambda)}^\tau \right\} \exp\{i\lambda t\} d\mathbf{Z}_X(\lambda). \tag{9.3.16}$$

We see that $\boldsymbol{\varepsilon}(t)$ has mean $\mathbf{0}$ and spectral density matrix

$$\mathbf{f}_{\varepsilon\varepsilon}(\lambda) = \sum_{j>q} \mu_j(\lambda)\mathbf{V}_j(\lambda)\overline{\mathbf{V}_j(\lambda)}^{\tau}. \tag{9.3.17}$$

The degree of approximation of $\mathbf{X}(t)$ by $\mathbf{X}^*(t)$ is therefore directly related to how near the $\mu_j(\lambda), j > q$, are to $0, -\infty < \lambda < \infty$. We also see that both the cross-spectral matrix between $\boldsymbol{\varepsilon}(t)$ and $\boldsymbol{\zeta}(t)$ and the cross-spectral matrix between $\boldsymbol{\varepsilon}(t)$ and $\mathbf{X}^*(t)$ are identically $\mathbf{0}$.

We next mention a few algebraic properties of the principal component series. Because

$$\mathbf{f}_{XX}(-\lambda) = \overline{\mathbf{f}_{XX}(\lambda)} = \mathbf{f}_{XX}(\lambda)^{\tau}, \tag{9.3.18}$$

we have

$$\mu_j(-\lambda) = \mu_j(\lambda) \tag{9.3.19}$$

while

$$\mathbf{V}_j(-\lambda) = \overline{\mathbf{V}_j(\lambda)} \qquad \text{for } j = 1, \ldots, r. \tag{9.3.20}$$

Also because

$$\mathbf{f}_{XX}(\lambda + 2\pi) = \mathbf{f}_{XX}(\lambda), \tag{9.3.21}$$

we see

$$\mu_j(\lambda + 2\pi) = \mu_j(\lambda), \tag{9.3.22}$$

and

$$\mathbf{V}_j(\lambda + 2\pi) = \mathbf{V}_j(\lambda) \qquad \text{for } j = 1, \ldots, r. \tag{9.3.23}$$

Unfortunately the principal component series do not generally transform in an elementary manner when the series $\mathbf{X}(t)$ is filtered. Specifically, suppose

$$\mathbf{Y}(t) = \sum_u \mathbf{d}(t - u)\mathbf{X}(u) \tag{9.3.24}$$

for some $r \times r$ filter $\{\mathbf{d}(u)\}$ with transfer function $\mathbf{D}(\lambda)$. The spectral density matrix of the series $\mathbf{Y}(t)$ is

$$\mathbf{f}_{YY}(\lambda) = \mathbf{D}(\lambda)\mathbf{f}_{XX}(\lambda)\overline{\mathbf{D}(\lambda)}^{\tau}. \tag{9.3.25}$$

The latent roots and vectors of this matrix are not generally related in any elementary manner to those of $\mathbf{f}_{XX}(\lambda)$. However, one case in which there is a convenient relation is when the matrix $\mathbf{D}(\lambda)$ is unitary. In this case

$$\mu_j(\mathbf{f}_{YY}(\lambda)) = \mu_j(\mathbf{f}_{XX}(\lambda)) \tag{9.3.26}$$

while

$$\mathbf{V}_j(\mathbf{f}_{YY}(\lambda)) = \overline{\mathbf{D}(\lambda)}^{\tau}\mathbf{V}_j(\mathbf{f}_{XX}(\lambda)). \tag{9.3.27}$$

We may derive certain regularity properties of the filters $\{\mathbf{b}(u)\}$, $\{\mathbf{c}(u)\}$ of Theorem 9.3.1 under additional conditions. We have

Theorem 9.3.3 Suppose the conditions of Theorem 9.3.1 are satisfied. Also, suppose

$$\sum_u [1 + |u|^P]|\mathbf{c}_{XX}(u)| < \infty \tag{9.3.28}$$

for some $P \geqslant 0$ and suppose that the latent roots of $\mathbf{f}_{XX}(\lambda)$ are distinct. Then $\{\mathbf{b}(u)\}$ and $\{\mathbf{c}(u)\}$ given in Theorem 9.3.1 satisfy

$$\sum_u [1 + |u|^P]|\mathbf{b}(u)| < \infty \tag{9.3.29}$$

and

$$\sum_u [1 + |u|^P]|\mathbf{c}(u)| < \infty. \tag{9.3.30}$$

In qualitative terms, the weaker the time dependence of the series $\mathbf{X}(t)$, the more rapidly the filter coefficients fall off to $\mathbf{0}$ as $|u| \to \infty$. With reference to the covariance functions of the principal component series and the error series we have

Corollary 9.3.3 Under the conditions of Theorem 9.3.3

$$\sum_u [1 + |u|^P]|\mathbf{c}_{\zeta\zeta}(u)| < \infty \tag{9.3.31}$$

and

$$\sum_u [1 + |u|^P]|\mathbf{c}_{\varepsilon\varepsilon}(u)| < \infty. \tag{9.3.32}$$

The principal component series might have been introduced in an alternate manner to that of Theorem 9.3.1. We have

Theorem 9.3.4 Suppose the conditions of Theorem 9.3.1 are satisfied. $\zeta_j(t)$, $t = 0, \pm 1, \ldots$ given by (9.3.13) is the real-valued series of the form

$$\int_0^{2\pi} \mathbf{B}_j(\lambda) \exp\{i\lambda t\}\, d\mathbf{Z}_X(\lambda) \tag{9.3.33}$$

(with the $1 \times r$ $\mathbf{B}_j(\lambda)$ satisfying $\mathbf{B}_j(\lambda)\overline{\mathbf{B}_j(\lambda)}^\tau = 1$), that has maximum variance and coherency 0 with $\zeta_k(t)$, $k < j$, $j = 1, \ldots, r$. The maximum variance achieved by $\zeta_j(t)$ is

$$\int_0^{2\pi} \mu_j(\alpha) d\alpha. \tag{9.3.34}$$

This approach was adopted in Brillinger (1964a) and Goodman (1967); it provides a recursive, rather than direct, definition of the principal component series.

The principal component series satisfy stronger optimality properties of the nature of those of Theorem 9.2.3. For convenience, assume $E\mathbf{X}(t) = \mathbf{0}$ in the theorem below.

Theorem 9.3.5 Let $\mathbf{X}(t)$, $t = 0, \pm 1, \ldots$, be an r vector-valued series with mean $\mathbf{0}$, absolutely summable autocovariance function, and spectral density matrix $\mathbf{f}_{XX}(\lambda)$, $-\infty < \lambda < \infty$. Then the $q \times r$ $\{\mathbf{b}(u)\}$, and $r \times q$ $\{\mathbf{c}(u)\}$ that minimize the jth latent root of the spectral density matrix of the series

$$\mathbf{X}(t) - \sum_u \mathbf{c}(t-u)\boldsymbol{\zeta}(u) \tag{9.3.35}$$

where

$$\boldsymbol{\zeta}(t) = \sum_u \mathbf{b}(t-u)\mathbf{X}(u) \tag{9.3.36}$$

are given by (9.3.5), (9.3.6). The jth extremal latent root is $\mu_{j+q}(\lambda)$.

The latent roots and vectors of spectral density matrices appear in the work of Wiener (1930), Whittle (1953), Pinsker (1964), Koopmans (1964b), and Rozanov (1967). Another related result is Lemma 11, Dunford and Schwartz (1963) p. 1341.

9.4 THE CONSTRUCTION OF ESTIMATES AND ASYMPTOTIC PROPERTIES

Suppose that we have a stretch, $\mathbf{X}(t)$, $t = 0, \ldots, T-1$, of an r vector-valued series $\mathbf{X}(t)$ with spectral density matrix $\mathbf{f}_{XX}(\lambda)$, and we wish to construct estimates of the latent roots and vectors $\mu_j(\lambda)$, $\mathbf{V}_j(\lambda)$, $j = 1, \ldots, r$ of this matrix. An obvious way of proceeding is to construct an estimate $\mathbf{f}_{XX}^{(T)}(\lambda)$ of the spectral density matrix and to estimate $\mu_j(\lambda)$, $\mathbf{V}_j(\lambda)$ by the corresponding latent root and vector of $\mathbf{f}_{XX}^{(T)}(\lambda)$, $j = 1, \ldots, r$. We turn to an investigation of certain of the statistical properties of estimates constructed in this way.

In Chapter 7 we discussed procedures for forming estimates of a spectral density matrix and the asymptotic properties of these estimates. One estimate discussed had the form

$$\mathbf{f}_{XX}^{(T)}(\lambda) = 2\pi T^{-1} \sum_{s=1}^{T-1} W^{(T)}\left(\lambda - \frac{2\pi s}{T}\right) \mathbf{I}_{XX}^{(T)}\left(\frac{2\pi s}{T}\right) \tag{9.4.1}$$

where $\mathbf{I}_{XX}^{(T)}(\alpha)$ was the matrix of second-order periodograms

$$\mathbf{I}_{XX}^{(T)}(\alpha) = (2\pi T)^{-1}\left[\sum_{t=0}^{T-1} \mathbf{X}(t) \exp\{-i\alpha t\}\right]\overline{\left[\sum_{t=0}^{T-1} \mathbf{X}(t) \exp\{-i\alpha t\}\right]}^{\tau} \tag{9.4.2}$$

and $W^{(T)}(\alpha)$ was a weight function of the form

$$W^{(T)}(\alpha) = \sum_{j=-\infty}^{\infty} W(B_T^{-1}[\alpha + 2\pi j]) \tag{9.4.3}$$

$W(\alpha)$ being concentrated in the neighborhood of $\alpha = 0$ and B_T, $T = 1, 2, \ldots$ a sequence of non-negative bandwidth parameters. We may now state

Theorem 9.4.1 Let $\mathbf{X}(t)$, $t = 0, \pm 1, \ldots$ be an r vector-valued series satisfying Assumption 2.6.2(1). Let $\nu_j^{(T)}(\lambda)$, $\mathbf{U}_j^{(T)}(\lambda)$, $j = 1, \ldots, r$ be the latent roots and vectors of the matrix

$$\int_0^{2\pi} W^{(T)}(\lambda - \alpha)\mathbf{f}_{XX}(\alpha)d\alpha. \tag{9.4.4}$$

Let $\mathbf{f}_{XX}^{(T)}(\lambda)$ be given by (9.4.1) where $W(\alpha)$ satisfies Assumption 5.6.1. Let $\mu_j^{(T)}(\lambda)$, $\mathbf{V}_j^{(T)}(\lambda)$, $j = 1, \ldots, r$, be the latent roots and vectors of $\mathbf{f}_{XX}^{(T)}(\lambda)$. If $B_T T \to \infty$ as $T \to \infty$, then

$$E\mu_j^{(T)}(\lambda) = \nu_j^{(T)}(\lambda) + O(B_T^{-1/2}T^{-1/2}). \tag{9.4.5}$$

If, in addition, the latent roots of $\mathbf{f}_{XX}(\lambda)$ are distinct, then

$$\vec{\text{ave}}\ \mu_j^{(T)}(\lambda) = \nu_j^{(T)}(\lambda) + O(B_T^{-1}T^{-1}) \tag{9.4.6}$$

and

$$\vec{\text{ave}}\ \mathbf{V}_j^{(T)}(\lambda) = \mathbf{U}_j^{(T)}(\lambda) + O(B_T^{-1}T^{-1}) \tag{9.4.7}$$

for $j = 1, \ldots, r$.

Theorem 9.4.1 suggests that for large values of $B_T T$, the distributions of the latent roots and vectors $\mu_j^{(T)}(\lambda)$, $\mathbf{V}_j^{(T)}(\lambda)$ will be centered at the corresponding latent roots and vectors of the matrix average (9.4.4). If in addition $B_T \to 0$ as $T \to \infty$, then clearly

$$E\mu_j^{(T)}(\lambda),\ \vec{\text{ave}}\ \mu_j^{(T)}(\lambda) \to \mu_j(\lambda) \tag{9.4.8}$$

and

$$\vec{\text{ave}}\ \mathbf{V}_j^{(T)}(\lambda) \to \mathbf{V}_j(\lambda) \qquad \text{for } j = 1, \ldots, r \text{ as } T \to \infty. \tag{9.4.9}$$

The latent roots and vectors of (9.4.4) will be near the desired $\mu_j(\lambda)$, $\mathbf{V}_j(\lambda)$ in the case that $\mathbf{f}_{XX}(\alpha)$, $-\infty < \alpha < \infty$, is near constant. This suggests once again the importance of prefiltering the data in order to obtain near constant spectra prior to estimating parameters of interest. Some aspects of the relation between $\nu_j^{(T)}(\lambda)$, $\mathbf{U}_j^{(T)}(\lambda)$ and $\mu_j(\lambda)$, $\mathbf{V}_j(\lambda)$ are indicated in the following:

Theorem 9.4.2 Let the $r \times r$ spectral density matrix $\mathbf{f}_{XX}(\lambda)$ be given by

$$\mathbf{f}_{XX}(\lambda) = (2\pi)^{-1} \sum_{u=-\infty}^{\infty} \mathbf{c}_{XX}(u) \exp\{-i\lambda u\} \tag{9.4.10}$$

where

$$\sum_{u=-\infty}^{\infty} |u|^3 |\mathbf{c}_{XX}(u)| < \infty. \tag{9.4.11}$$

Let $W^{(T)}(\alpha)$ be given by (9.4.3) where $W(\alpha) = W(-\alpha)$ and

$$\int_{-\infty}^{\infty} |\alpha|^3 |W(\alpha)| d\alpha < \infty. \tag{9.4.12}$$

Suppose the latent roots $\mu_j(\lambda)$, $j = 1, \ldots, r$, of $\mathbf{f}_{XX}(\lambda)$ are distinct. Let $B_T \to 0$ as $T \to \infty$, then

$$\nu_j^{(T)}(\lambda) = \mu_j(\lambda) + \tfrac{1}{2} B_T^2 \overline{\mathbf{V}_j(\lambda)}^\tau \frac{d^2\mathbf{f}_{XX}(\lambda)}{d\lambda^2} \mathbf{V}_j(\lambda) \int \alpha^2 W(\alpha) d\alpha + O(B_T^3) \tag{9.4.13}$$

and

$$\mathbf{U}_j^{(T)}(\lambda) = \mathbf{V}_j(\lambda) + \tfrac{1}{2} B_T^2 \sum_{k \neq j} \left\{ \overline{\mathbf{V}_k(\lambda)}^\tau \frac{d^2\mathbf{f}_{XX}(\lambda)}{d\lambda^2} \mathbf{V}_j(\lambda) \right\} \mathbf{V}_k(\lambda) [\mu_j(\lambda) - \mu_k(\lambda)]^{-1}$$
$$\times \int \alpha^2 W(\alpha) d\alpha + O(B_T^3) \tag{9.4.14}$$

for $j = 1, \ldots, r$.

Theorems 9.4.1 and 9.4.2 indicate that the asymptotic biases of the estimates $\mu_j^{(T)}(\lambda)$, $\mathbf{V}_j^{(T)}(\lambda)$ depend in an intimate manner on the bandwidth B_T appearing in the weight function $W^{(T)}(\alpha)$ and on the smoothness of the population spectral density $\mathbf{f}_{XX}(\alpha)$ for α in the neighborhood of λ.

Turning to an investigation of the asymptotic distribution of the $\mu_j^{(T)}(\lambda)$, $\mathbf{V}_j^{(T)}(\lambda)$ we have

Theorem 9.4.3 Under the conditions of Theorem 9.4.1 and if the latent roots of $\mathbf{f}_{XX}(\lambda_m)$ are distinct, $m = 1, \ldots, M$, the variates $\mu_j^{(T)}(\lambda_m)$, $\mathbf{V}_j^{(T)}(\lambda_m)$, $j = 1, \ldots, r$, $m = 1, \ldots, M$ are asymptotically jointly normal with asymptotic covariance structure

$$\lim_{T\to\infty} B_T T \operatorname{cov}\{\mu_j^{(T)}(\lambda_m), \mu_k^{(T)}(\lambda_n)\}$$
$$= 2\pi \int W(\alpha)^2 d\alpha\, [\eta\{\lambda_m - \lambda_n\} + \eta\{\lambda_m + \lambda_n\}] \mu_j(\lambda_m)^2 \qquad \text{if } j = k$$
$$= 0 \qquad \text{if } j \neq k \tag{9.4.15}$$

$$\lim_{T\to\infty} B_T T \operatorname{cov}\{\mu_j^{(T)}(\lambda_m), \mathbf{V}_k^{(T)}(\lambda_n)\} = \mathbf{0} \tag{9.4.16}$$

and

$$\lim_{T\to\infty} B_T T \,\vec{\operatorname{cov}}\, \{\mathbf{V}_j^{(T)}(\lambda_m), \mathbf{V}_k^{(T)}(\lambda_n)\}$$
$$= 2\pi \int W(\alpha)^2 d\alpha\, \eta\{\lambda_m - \lambda_n\} \mu_j(\lambda_m) \sum_{l \neq j} \mu_l(\lambda_m)[\mu_j(\lambda_m) - \mu_l(\lambda_m)]^{-2} \mathbf{V}_l(\lambda_m)\overline{\mathbf{V}_l(\lambda_m)}^\tau \quad \text{if } j = k$$
$$= -2\pi \int W(\alpha)^2 d\alpha\, \eta\{\lambda_m + \lambda_n\} \mu_k(\lambda_m)\mu_j(\lambda_m)[\mu_j(\lambda_m) - \mu_k(\lambda_m)]^{-2} \mathbf{V}_k(\lambda_m)\mathbf{V}_j(\lambda_m)^\tau \quad \text{if } j \neq k \qquad (9.4.17)$$

for $j, k = 1, \ldots, r$; $m, n = 1, \ldots, M$.

The limiting expressions appearing in Theorem 9.4.3 parallel those of Theorems 9.2.2 and 9.2.4. The asymptotic independence indicated for variates at frequencies λ_m, λ_n with $\lambda_m \pm \lambda_n \not\equiv 0 \pmod{2\pi}$ was expected due to the corresponding asymptotic independence of $\mathbf{f}_{XX}^{(T)}(\lambda_m)$, $\mathbf{f}_{XX}^{(T)}(\lambda_n)$. The asymptotic independence of the different latent roots and vectors was perhaps unexpected.

Expression (9.4.15) implies that

$$\vec{\operatorname{var}} \log_{10} \mu_j^{(T)}(\lambda) \sim B_T^{-1}T^{-1}(\log_{10} e)^2 2\pi \int W(\alpha)^2 d\alpha \quad \text{if } \lambda \not\equiv 0 \pmod{\pi}$$
$$\sim B_T^{-1}T^{-1}(\log_{10} e)^2 4\pi \int W(\alpha)^2 d\alpha \quad \text{if } \lambda \equiv 0 \pmod{\pi}. \qquad (9.4.18)$$

This last is of identical character with the corresponding result, (5.6.15), for the variance of the logarithm of a power spectrum estimate. It was anticipated due to the interpretation, given in Theorem 9.3.2, of $\mu_j(\lambda)$ as the power spectrum of the jth principal component series. Expression (9.4.18) suggests that we should take $\log \mu_j^{(T)}(\lambda)$ as the basic statistic rather than $\mu_j^{(T)}(\lambda)$.

An alternate form of limiting distribution results if we consider the spectral estimate of Section 7.3

$$\mathbf{f}_{XX}^{(T)}(\lambda) = (2m+1)^{-1} \sum_{s=-m}^{m} \mathbf{I}_{XX}^{(T)}\left(\frac{2\pi[s(T)+s]}{T}\right) \quad \text{for } \frac{2\pi s(T)}{T} \doteq \lambda \not\equiv 0 \pmod{\pi}$$
$$= (2m)^{-1} \sum_{s=1}^{m} \left\{\mathbf{I}_{XX}^{(T)}\left(\lambda + \frac{2\pi s}{T}\right) + \mathbf{I}_{XX}^{(T)}\left(\lambda - \frac{2\pi s}{T}\right)\right\} \quad \text{for } \lambda \equiv 0 \pmod{2\pi} \text{ or for } \lambda = \pm\pi, \pm 3\pi, \ldots \text{ and } T \text{ even}$$
$$= (2m)^{-1} \sum_{s=1}^{m} \left\{\mathbf{I}_{XX}^{(T)}\left(\lambda - \frac{\pi}{T} + \frac{2\pi s}{T}\right) + \mathbf{I}_{XX}^{(T)}\left(\lambda + \frac{\pi}{T} - \frac{2\pi s}{T}\right)\right\} \quad \text{for } \lambda = \pm\pi, \pm 3\pi, \ldots \text{ and } T \text{ odd}. \qquad (9.4.19)$$

In Theorem 7.3.3, we saw that this estimate was distributed asymptotically as $(2m+1)^{-1}W_r{}^C(2m+1,\mathbf{f}_{XX}(\lambda))$, $(2m)^{-1}W_r(2m,\mathbf{f}_{XX}(\lambda))$, $(2m)^{-1}W_r(2m,\mathbf{f}_{XX}(\lambda))$ as $T\to\infty$ in the three cases. This result leads us directly to

Theorem 9.4.4 Let $\mathbf{X}(t)$, $t = 0, \pm 1, \ldots$ be an r vector-valued series satisfying Assumption 2.6.1. Let m be fixed and $[2\pi s(T)/T]\to\lambda$ as $T\to\infty$. Let $\mu_j{}^{(T)}(\lambda)$, $\mathbf{V}_j{}^{(T)}(\lambda)$, $j = 1, \ldots, r$ be the latent roots and vectors of the matrix (9.4.19). Then they tend, in distribution, to the latent roots and vectors of a $(2m+1)^{-1}W_r{}^C(2m+1,\mathbf{f}_{XX}(\lambda))$ variate if $\lambda\not\equiv 0 \pmod{\pi}$ and of a $(2m)^{-1}$ $W_r(2m,\mathbf{f}_{XX}(\lambda))$ variate if $\lambda\equiv 0 \pmod{\pi}$. Estimates at frequencies λ_n, $n = 1, \ldots, N$ with $\lambda_n \pm \lambda_{n'} \not\equiv 0 \pmod{2\pi}$ are asymptotically independent.

The distribution of the latent roots of matrices with real or complex Wishart distributions has been given in James (1964).

The distributions obtained in Theorems 9.4.3 and 9.4.4 are not inconsistent. If, as in Sections 5.7 and 7.4, we make the identification

$$2m+1 \sim \frac{1}{\sum_s \left[\frac{2\pi}{T} W^{(T)}\left(\lambda - \frac{2\pi s}{T}\right)\right]^2} \backsim \frac{B_T T}{2\pi \int W(\alpha)^2 d\alpha}, \tag{9.4.20}$$

and m is large, then as Theorem 9.2.2 and 9.2.4 imply, the latent roots and vectors are approximately normal with the appropriate first- and second-order moment structure.

The results developed in this section may be used to set approximate confidence limits for the $\mu_j(\lambda)$, $V_{pj}(\lambda)$, $j, p = 1, \ldots, r$. For example, the result of Theorem 9.4.3 and the discussion of Section 5.7 suggest the following approximate 100γ percent confidence interval for $\log_{10}\mu_j(\lambda)$:

$$\log_{10}\mu_j{}^{(T)}(\lambda) - z\left(\frac{1+\gamma}{2}\right)(.4343)\sqrt{\frac{2\pi\int W(\alpha)^2 d\alpha}{B_T T}} < \log_{10}\mu_j(\lambda)$$
$$< \log_{10}\mu_j{}^{(T)}(\lambda) + z\left(\frac{1+\gamma}{2}\right)(.4343)\sqrt{\frac{2\pi\int W(\alpha)^2 d\alpha}{B_T T}}. \tag{9.4.21}$$

At the same time the result of Exercise 9.7.5 suggests that it might prove reasonable to approximate the distribution of

$$\frac{|V_{pj}{}^{(T)}(\lambda) - V_{pj}(\lambda)|^2}{\hat\sigma_T{}^2} \tag{9.4.22}$$

by $\chi_2{}^2/2$ where

$$\hat\sigma_T{}^2 = B_T{}^{-1}T^{-1}2\pi \int W(\alpha)^2 d\alpha\, \mu_j{}^{(T)}(\lambda) \times \sum_{l\neq j} \mu_l{}^{(T)}(\lambda)[\mu_j{}^{(T)}(\lambda) - \mu_l{}^{(T)}(\lambda)]^{-2}\, |V_{pl}{}^{(T)}(\lambda)|^2. \tag{9.4.23}$$

This approximation might then be used to determine confidence regions for

$$\{\text{Re } V_{pj}(\lambda), \text{Im } V_{pj}(\lambda)\} \quad \text{or} \quad \{|V_{pj}(\lambda)|, \arg V_{pj}(\lambda)\}$$

in the manner of Section 6.2. Much of the material of this section was presented in Brillinger (1969d).

9.5 FURTHER ASPECTS OF PRINCIPAL COMPONENTS

The principal component series introduced in Section 9.3 may be interpreted in terms of the usual principal components of multivariate analysis. Given the r vector-valued stationary series $\mathbf{X}(t)$, $t = 0, \pm 1, \ldots$ with spectral density matrix $\mathbf{f}_{XX}(\lambda)$, let $\mathbf{X}(t,\lambda)$ denote the component of frequency λ of $\mathbf{X}(t)$ (see Section 4.6). Then, see Sections 4.6, 7.1, the $2r$ vector-valued variate, with real-valued entries

$$\begin{bmatrix} \mathbf{X}(t,\lambda) \\ \mathbf{X}(t,\lambda)^H \end{bmatrix}, \tag{9.5.1}$$

has covariance matrix proportional to

$$\begin{bmatrix} \text{Re } \mathbf{f}_{XX}(\lambda) & \text{Im } \mathbf{f}_{XX}(\lambda) \\ -\text{Im } \mathbf{f}_{XX}(\lambda) & \text{Re } \mathbf{f}_{XX}(\lambda) \end{bmatrix} = \mathbf{f}_{XX}(\lambda)^R. \tag{9.5.2}$$

A standard principal component analysis of the variate (9.5.1) would lead us to consider the latent roots and vectors of (9.5.2). From Lemma 3.7.1 these are given by

$$\mu_j(\lambda), \begin{bmatrix} \text{Re } \mathbf{V}_j(\lambda) \\ \text{Im } \mathbf{V}_j(\lambda) \end{bmatrix} \qquad \mu_j(\lambda), \begin{bmatrix} -\text{Im } \mathbf{V}_j(\lambda) \\ \text{Re } \mathbf{V}_j(\lambda) \end{bmatrix} \tag{9.5.3}$$

$j = 1, \ldots, r$ where $\mu_j(\lambda)$, $\mathbf{V}_j(\lambda)$, $j = 1, \ldots, r$ are the latent roots and vectors of $\mathbf{f}_{XX}(\lambda)$ and appear in Theorem 9.3.1. We see therefore that a frequency domain principal component analysis of a stationary series $\mathbf{X}(t)$ is a standard principal component analysis carried out on the individual frequency components of $\mathbf{X}(t)$ and their Hilbert transforms.

A variety of uses suggest themselves for the sort of procedures discussed in Section 9.3. To begin, as in the introduction of this chapter, we may be interested in transmitting an r vector-valued series over a reduced number, $q < r$, of communication channels. Theorem 9.3.1 indicates one solution to this problem. Alternately we may be interested in examining a succession of real-valued series providing the information in a series of interest in a useful manner. This is often the case when the value of r is large. Theorem 9.3.4 suggests the consideration of the series corresponding to the largest latent roots, followed by the consideration of the series corresponding to the second largest latent roots and so on in such a situation.

At the other extreme, we may consider the series corresponding to the smallest latent roots. Suppose we feel the series $\mathbf{X}(t)$, $t = 0, \pm 1, \ldots$ may satisfy some linear time invariant identity of the form

$$\sum_u \mathbf{b}(t-u)\mathbf{X}(u) = K \tag{9.5.4}$$

where $\mathbf{b}(u)$ is $1 \times r$ and unknown and K is constant. Thus

$$\text{var}\,\{\sum_u \mathbf{b}(t-u)\mathbf{X}(u)\} = 0 \tag{9.5.5}$$

and it is reasonable to take $\mathbf{b}(u)$ to correspond to the rth principal component series derived from the smallest latent roots. This is an extension of a suggestion of Bartlett (1948a) concerning the multivariate case.

On another occasion, we may be concerned with some form of factor analytic model such as

$$\mathbf{X}(t) = \boldsymbol{\mu} + \sum_u \mathbf{c}(t-u)\boldsymbol{\zeta}(u) + \boldsymbol{\varepsilon}(t) \tag{9.5.6}$$

$t = 0, \pm 1, \ldots$ where the q vector-valued series $\boldsymbol{\zeta}(t)$, $t = 0, \pm 1, \ldots$ represents q "hidden" factor series and the $r \times q$ filter $\{\mathbf{c}(u)\}$ represents the loadings of the factors. We may wish to determine the $\boldsymbol{\zeta}(t)$, $t = 0, \pm 1, \ldots$ as being the essence of $\mathbf{X}(t)$ in some sense. The procedures of Section 9.3 suggest one means of doing this. In the case that the series are not autocorrelated, the procedure reduces to factor analysis, used so often by psychometricians; see Horst (1966). They generally interpret the individual principal components and try to make the interpretation easier by rotating (or transforming linearly) the most important components. In the present time series situation, the problem of interpretation is greatly complicated by the fact that if $\mathbf{V}_j(\lambda)$ is a standardized latent vector corresponding to a latent root $\mu_j(\lambda)$, then so is $\alpha_j(\lambda)\mathbf{V}_j(\lambda)$ for $\alpha_j(\lambda)$ with modulus 1.

Another complication that arises relates to the fact that the latent roots and vectors of a spectral density matrix are not invariant under linear filtering of the series. Hence, the series with greater variability end up weighted more heavily in the principal components. If the series are not recorded in comparable scales, difficulties arise. One means of reducing these complications is to carry out the computations on the estimated matrix of coherencies, $[R_{jk}{}^{(T)}(\lambda)]$, rather than on the matrix of spectral densities.

We conclude this section by reminding the reader that we saw, in Section 4.7, that the Cramér representation resulted from a form of principal component analysis carried out in the time domain. Other time domain principal component analyses appear in the work of Craddock (1965), Hannan (1961a), Stone (1947), Yaglom (1965), and Craddock and Flood (1969).

9.6 A WORKED EXAMPLE

We consider the estimation of the coefficients of the principal component series for the 14 vector-valued series of monthly mean temperatures at the one American and 13 European stations indicated in Chapter 1. In the discussion of Theorem 9.4.2 we saw that the estimates $\mu_j^{(T)}(\lambda)$, $\mathbf{V}_j^{(T)}(\lambda)$ could be substantially biased if the spectral density matrix was far from constant with respect to λ. For this reason the series were prefiltered initially by removing the seasonal effects. Figure 9.6.1 presents estimates of the power spectra of the seasonally adjusted series, taking an estimate of the form (9.4.19) with $m = 25$.

Figure 9.6.2 gives $\log_{10} \mu_j^{(T)}(\lambda)$, $j = 1, \ldots, 14$. The $\mu_j^{(T)}(\lambda)$ are the latent roots of the estimated spectral density matrix $\mathbf{f}_{XX}^{(T)}(\lambda)$. In fact, because of the unavailability of a computer program evaluating the latent roots and vectors of a complex Hermitian matrix, the $\mu_j^{(T)}(\lambda)$ and the $\mathbf{V}_j^{(T)}(\lambda)$ were derived from the following matrix with real-valued entries

$$\begin{bmatrix} \operatorname{Re} \mathbf{f}_{XX}^{(T)}(\lambda) & \operatorname{Im} \mathbf{f}_{XX}^{(T)}(\lambda) \\ -\operatorname{Im} \mathbf{f}_{XX}^{(T)}(\lambda) & \operatorname{Re} \mathbf{f}_{XX}^{(T)}(\lambda) \end{bmatrix} \tag{9.6.1}$$

making use of Lemma 3.7.1. The curves of Figure 9.6.2 are seen to fall off as λ increases in much the same manner as the power spectra appearing in Figure 9.6.1. Following expressions (9.4.18) and (9.4.20), the standard error of these estimates is approximately

$$\frac{\log_{10} e}{\sqrt{2m+1}} \doteq .062. \tag{9.6.2}$$

Figures 9.6.3 and 9.6.4 give the estimated gain and phase, $|V_{pj}^{(T)}(\lambda)|$ and $\arg V_{pj}^{(T)}(\lambda)$, for the first two principal components. For the first component, the gains are surprisingly constant with respect to λ. They are not near 0 except in the case of New Haven. The phases take on values near 0 or $\pi/2$, simultaneously for most series. In interpreting the latter we must remember the fact that the latent vectors are determined only up to an arbitrary multiplier of modulus 1. This is why 4 dots stand out in most of the plots. It appears that the first principal component series is essentially proportional to the average of the 13 European series, with no time lags involved. The gains and phases of the second component series are seen to be much more erratic and not at all easy to interpret. The gain for New Haven is noticeably large for λ near 0. The discussion at the end of Section 9.4 and Exercise 9.7.7 suggest two possible means of constructing approximate standard errors for the estimates.

Table 9.6.1 gives $\log_{10}$ of the latent values of the matrix $\mathbf{c}_{XX}^{(T)}(0)$ of Table

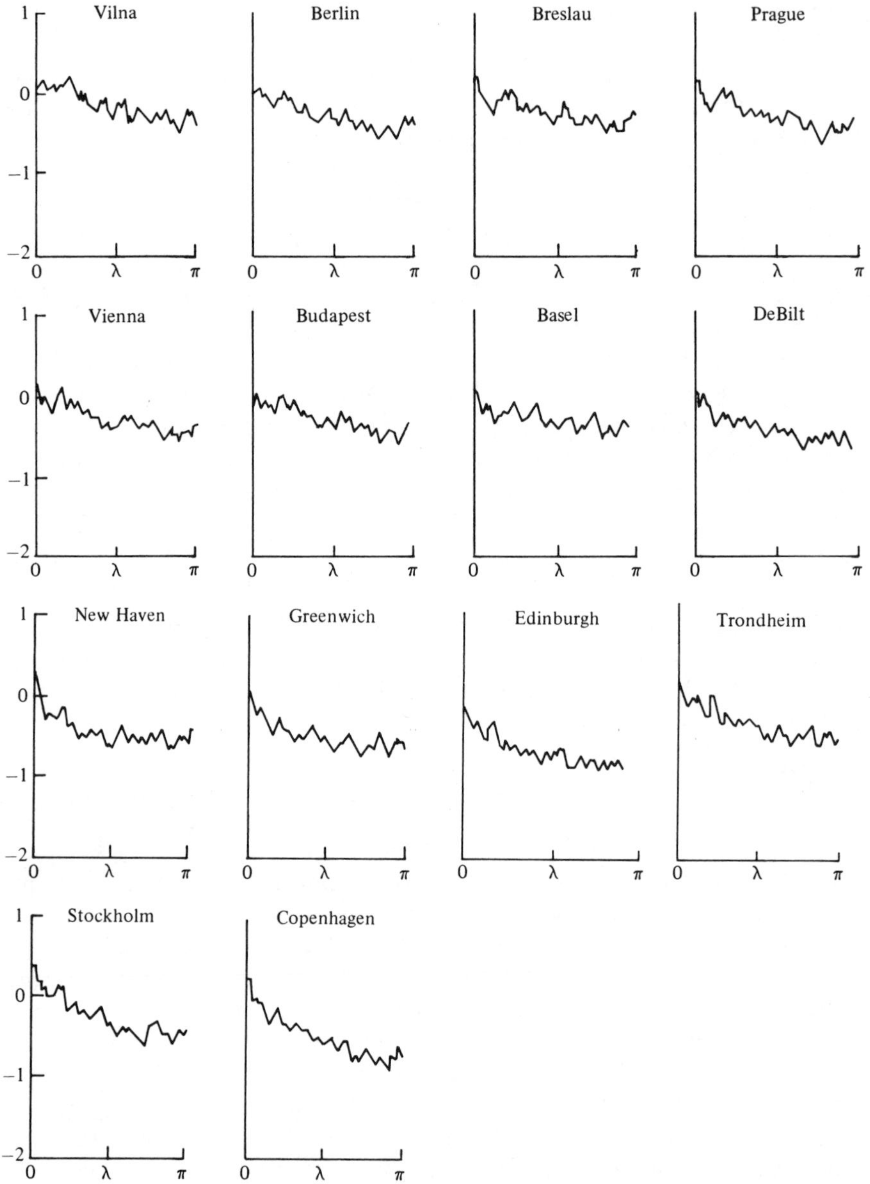

Figure 9.6.1 Logarithm of estimated power spectrum of seasonally adjusted monthly mean temperatures at various stations with 51 periodogram ordinates averaged.

7.8.1. Table 9.6.2 gives the corresponding latent vectors. In view of the apparent character of the first principal component series, suggested above, it

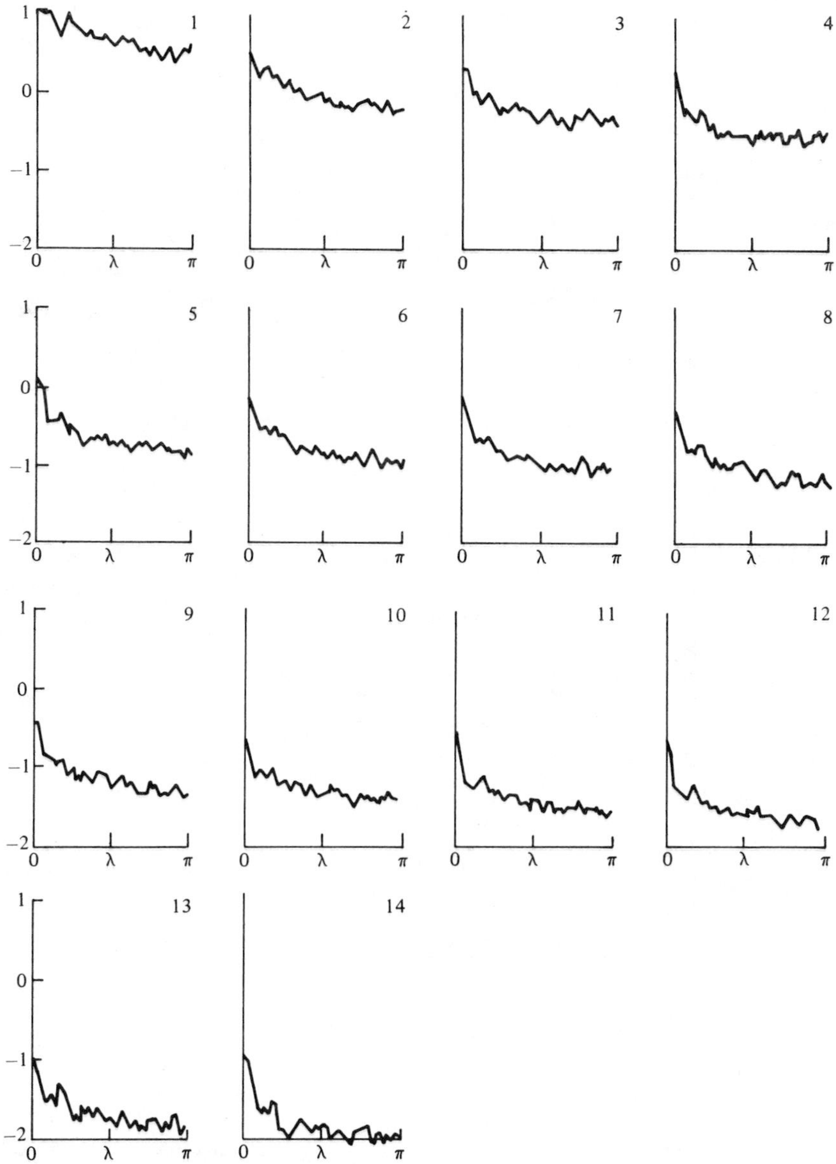

Figure 9.6.2 Logarithm of estimate of the power spectrum for the principal component series.

makes sense to consider these quantities. An examination of Table 9.6.2 suggests that the first vector corresponds to a simple average of the 13

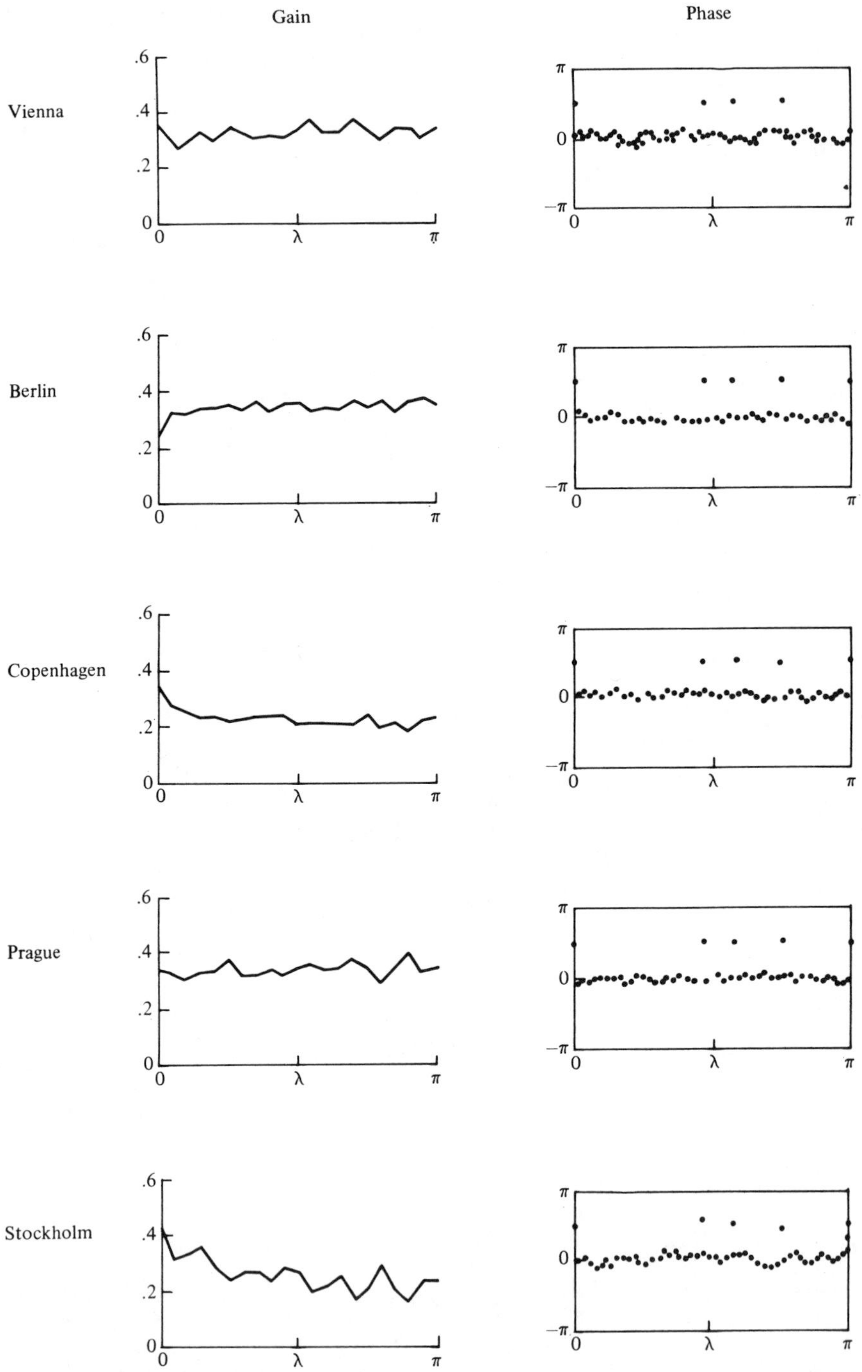

Figure 9.6.3 Estimated gains and phases for the first principal component series.

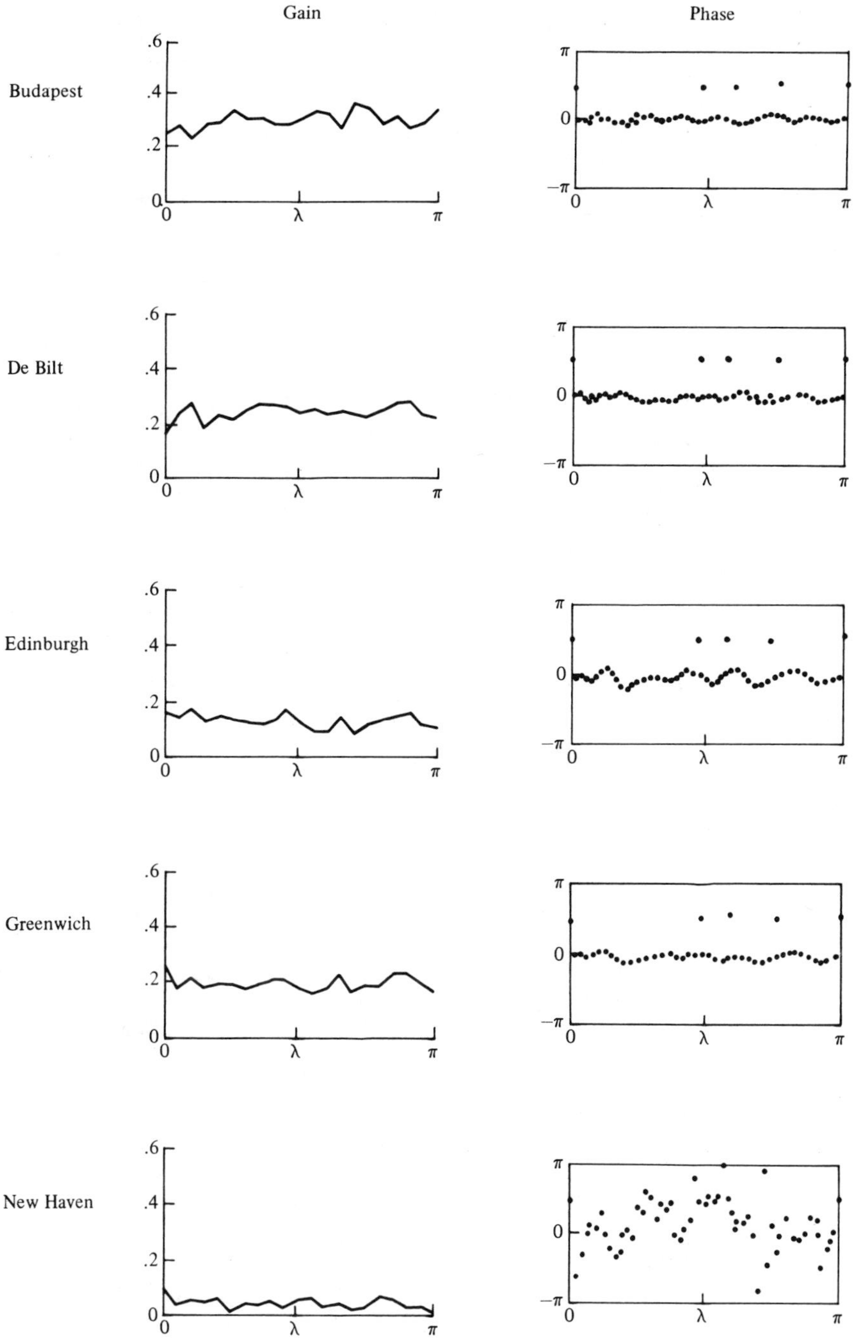
Gain
Phase
Budapest
De Bilt
Edinburgh
Greenwich
New Haven
.6
.4
.2
0
λ
π
−π

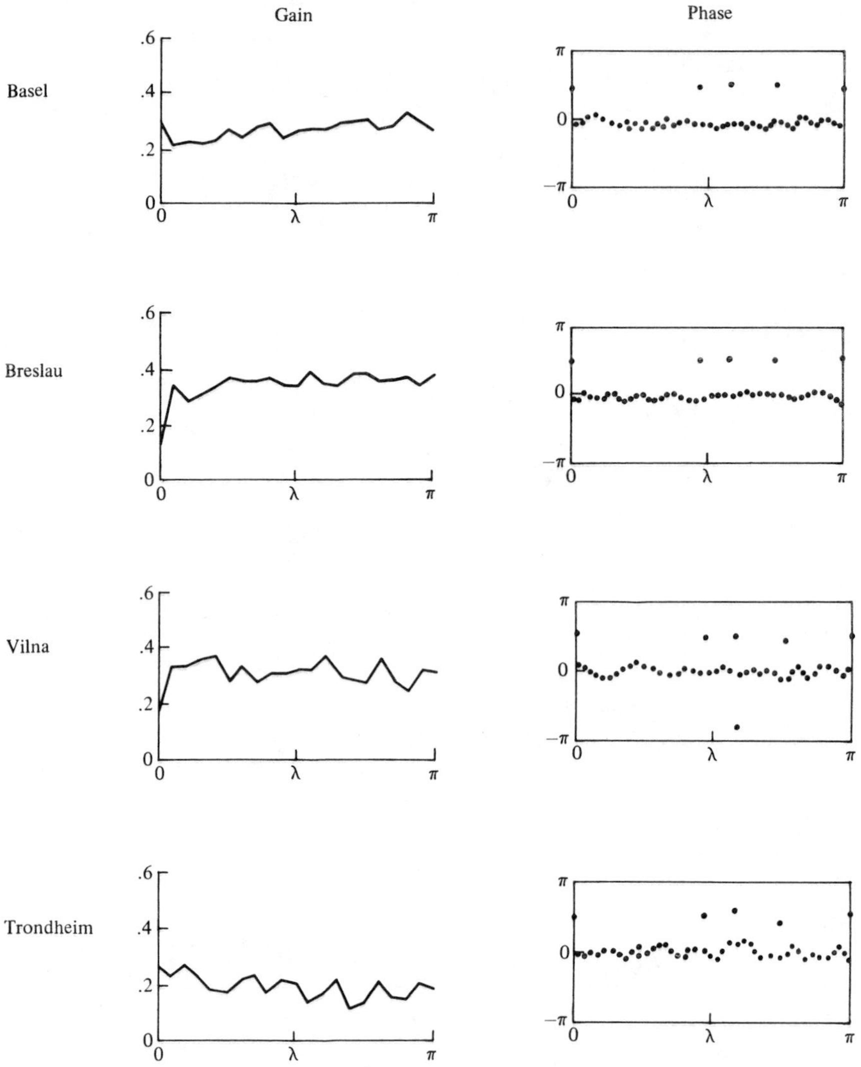
Gain
Phase
Basel
Breslau
Vilna
Trondheim
.6
.4
.2
0
λ
π
−π

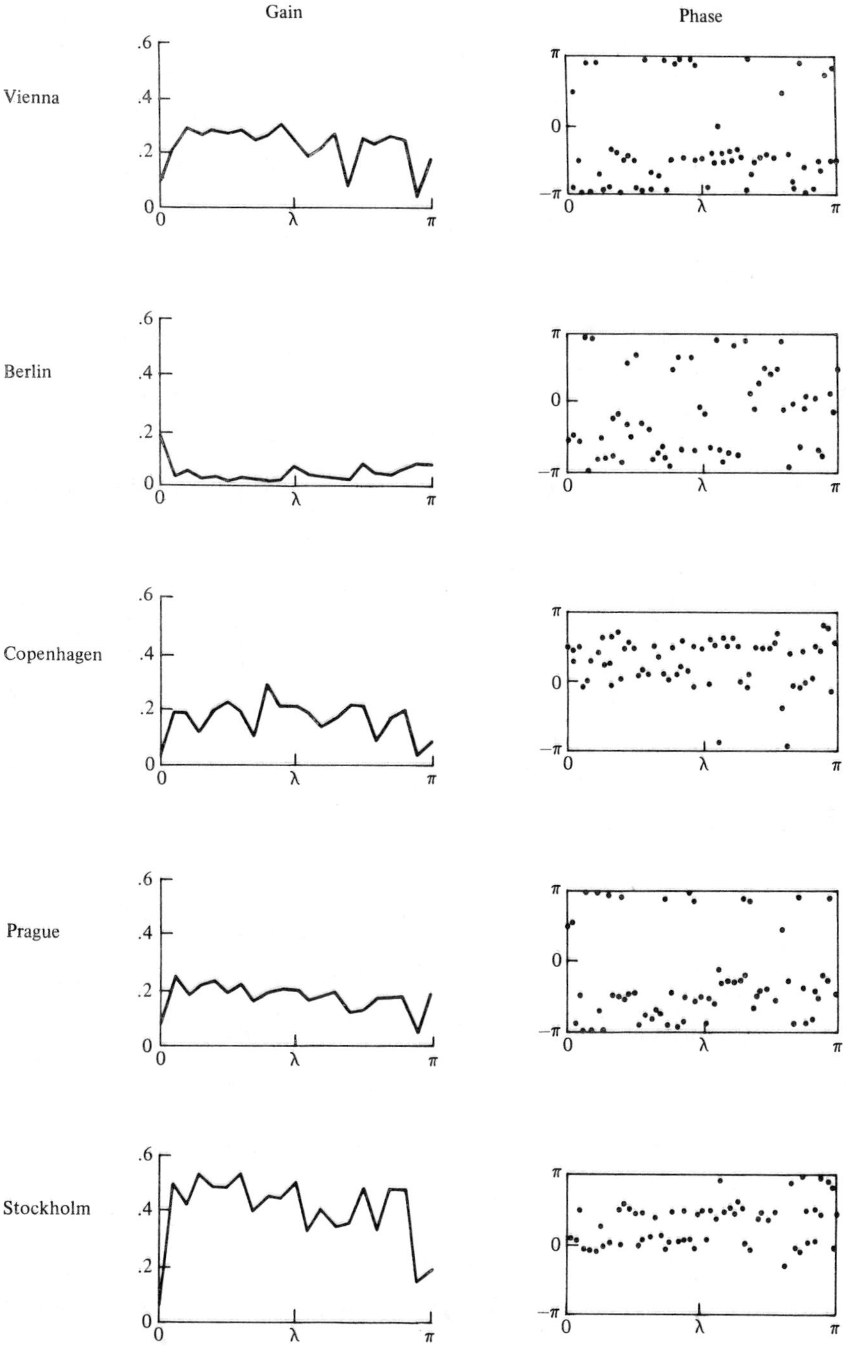

Figure 9.6.4 Estimated gains and phases for the second principal component series.

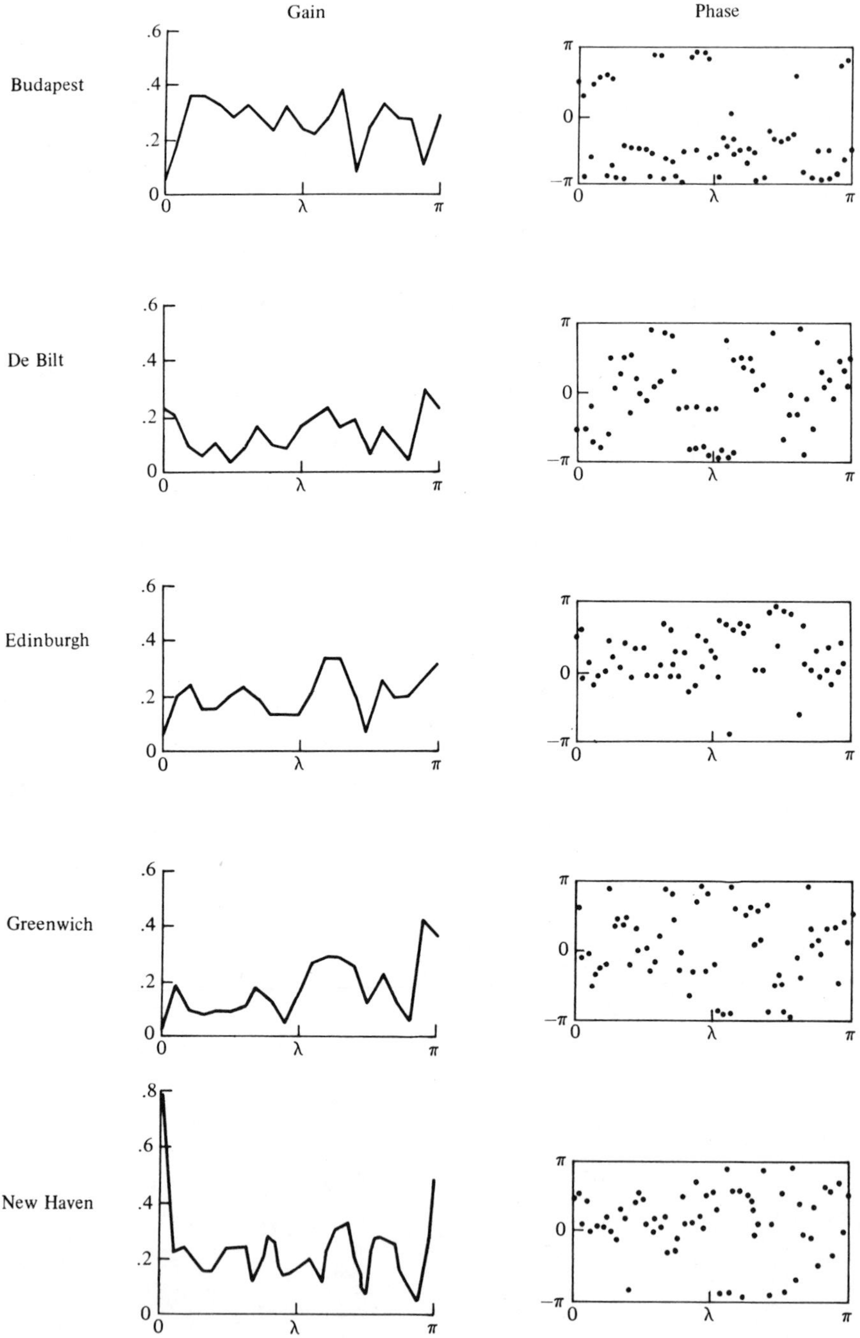
Gain
Phase
Budapest
De Bilt
Edinburgh
Greenwich
New Haven
.8
.6
.4
.2
0
π
0
−π
0
λ
π

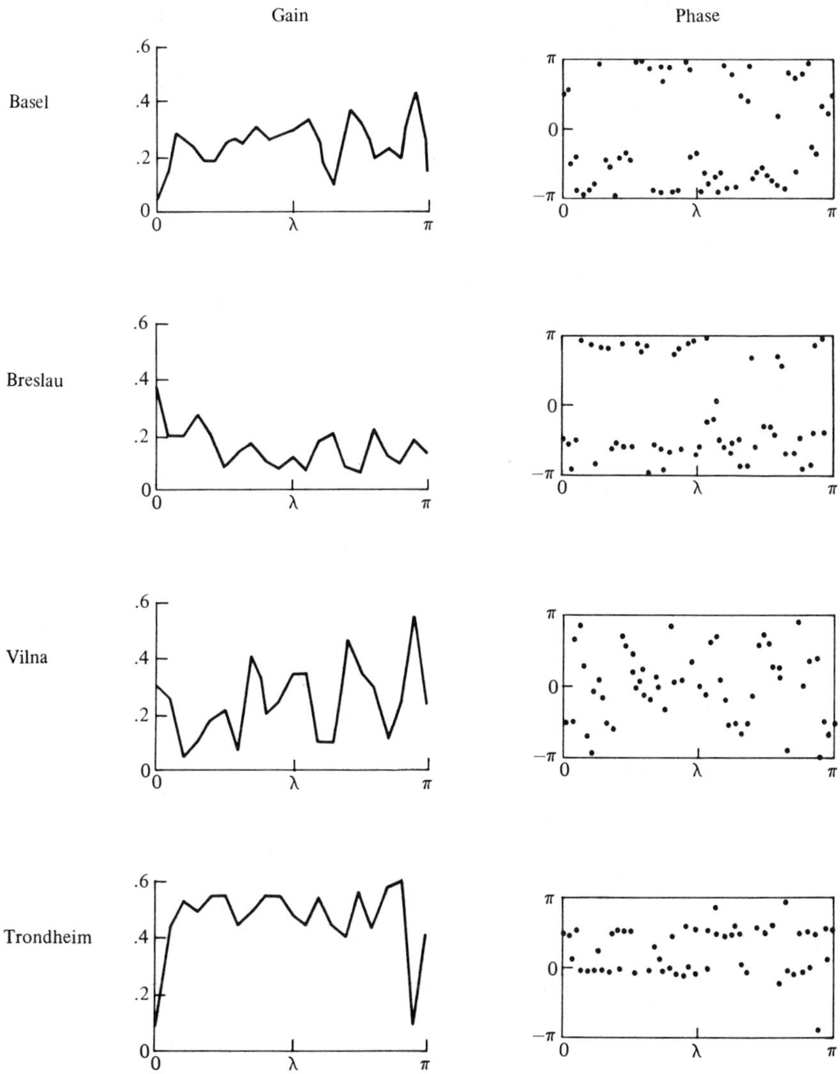

series obtained from the 14 by excluding New Haven. The second vector appears to correspond to New Haven primarily.

Table 9.6.1 $\log_{10}$ Latent Values of the Temperature Series

1.591
1.025
.852
.781
.369
.267
.164
.009
−.121
−.276
−.345
−.511
−.520
−.670

9.7 EXERCISES

9.7.1 Let $\mu_j(\lambda)$, $j = 1, \ldots, r$ denote the latent roots of the $r \times r$ non-negative definite matrix $\mathbf{f}_{XX}(\lambda)$. Let $\nu_j^{(T)}(\lambda)$, $j = 1, \ldots, r$ denote the latent roots of the matrix

$$\int_0^{2\pi} W^{(T)}(\lambda - \alpha)\, \mathbf{f}_{XX}(\alpha) d\alpha$$

where $W^{(T)}(\alpha) \geqslant 0$. Show that

$$0 \leqslant \nu_j^{(T)}(\lambda) \leqslant \int_0^{2\pi} W^{(T)}(\lambda - \alpha) \mu_j(\alpha) d\alpha.$$

9.7.2 Suppose the conditions of Theorem 9.3.1 are satisfied. Suppose $\mathbf{c}_{XX}(u) = \mathbf{0}$ for $u \neq 0$. Show that $\{\mathbf{b}(u)\}$, $\{\mathbf{c}(u)\}$ given in the theorem satisfy $\mathbf{b}(u), \mathbf{c}(u) = \mathbf{0}$ for $u \neq 0$.

9.7.3 Under the conditions of Theorem 9.3.1, show that the coherency of the series $X_j(t)$ and $\zeta_k(t)$ is

$$\mu_k(\lambda) |V_{jk}(\lambda)|^2 / f_{jj}(\lambda).$$

9.7.4 For the variates of Theorems 9.2.2, 9.2.4 show that $E\hat{\mu}_j = \mu_j + O(n^{-1/2})$, $j = 1, \ldots, r$.

9.7.5 Under the conditions of Theorem 9.4.3, show that $V_{pj}^{(T)}(\lambda)$ is asymptotically $N_1^C(V_{pj}(\lambda), \sigma_T^2)$ where

$$\sigma_T^2 = B_T^{-1} T^{-1} 2\pi \int W(\alpha)^2 d\alpha\, \mu_j(\lambda) \sum_{l \neq j} \mu_l(\lambda) [\mu_j(\lambda) - \mu_l(\lambda)]^{-2} |V_{pl}(\lambda)|^2$$

for $j, p = 1, \ldots, r$.

9.7.6 Suppose that the data is tapered, with tapering function $h(t/T)$, prior to calculating the estimates of Theorem 9.4.3. Under the conditions of that

Table 9.6.2 Latent Vectors of the Temperature Series

Vienna	*Berlin*	*Copenhagen*	*Prague*	*Stockholm*	*Budapest*	*De Bilt*	*Edinburgh*	*Greenwich*	*New Haven*	*Basel*	*Breslau*	*Vilna*	*Trondheim*
.281	.315	.242	.312	.248	.252	.256	.250	.357	.071	.251	.321	.283	.186
−.173	−.055	.060	−.169	.118	−.194	−.001	−.173	.185	.822	−.061	−.142	−.167	.173
−.129	−.041	−.028	−.098	−.098	−.194	.133	−.129	.441	−.481	.104	−.157	−.393	.018
−.211	.012	.186	−.234	.450	−.205	−.061	−.211	−.200	−.288	−.307	−.063	.362	.497
−.112	.063	−.081	−.594	−.160	−.088	.117	−.112	.336	.020	.125	.046	.570	−.343
−.018	.082	.138	−.343	.252	.022	.122	−.018	.257	−.050	.255	.065	−.411	.286
.365	.013	−.047	−.567	−.007	.522	−.136	.365	−.278	.001	−.011	.152	−.193	.061
−.278	.424	.069	−.064	−.091	−.275	.092	−.278	−.195	.011	−.213	.686	−.233	−.169
−.034	−.126	.366	−.002	.610	.046	−.127	−.034	−.011	−.034	.001	−.067	−.117	−.660
−.095	.230	−.136	−.033	.119	−.227	.112	−.095	−.491	.019	.740	−.177	.029	−.009
.174	.399	.388	−.073	−.204	.023	.507	.174	−.164	.015	−.284	−.482	−.057	−.098
−.367	.052	.633	.013	−.392	.217	−.437	−.367	.003	−.018	.239	−.036	.048	.095
.599	−.302	.368	−.083	−.168	−.571	−.091	.599	−.059	−.001	.102	.174	.020	.034
−.271	−.619	.177	−.004	−.076	.188	.610	−.272	−.197	.015	.083	.220	.023	.035

theorem, show that the asymptotic covariances (9.4.15), and (9.4.17) are multiplied by $\int h(t)^4dt/[\int h(t)^2dt]^2$.

9.7.7 Under the conditions of Theorem 9.4.3, show that log $|V_{pj}^{(T)}(\lambda)|$ and arg $\{V_{pj}^{(T)}(\lambda)\}$ are asymptotically distributed as independent $N(\log |V_{pj}(\lambda)|, (\frac{1}{2})\sigma_T^2|V_{pj}(\lambda)|^{-2})$ and $N(\arg \{V_{pj}(\lambda)\}, (\frac{1}{2})\sigma_T^2|V_{pj}(\lambda)|^{-2})$ variates respectively where σ_T^2 is given in Exercise 9.7.5.

9.7.8 (a) Show that if in the estimate (9.4.19) we smooth across the whole frequency domain, the proposed analysis reduces to a standard principal component analysis of the sample covariance matrix $\mathbf{c}_{XX}^{(T)}(0)$.

(b) Let the series $\mathbf{X}(t)$, $t = 0, \pm 1, \ldots$ be Gaussian and satisfy Assumption 2.6.2(1). Let μ_j, $\mathbf{V}_j$, $j = 1, \ldots, r$ denote the latent roots and vectors of $\mathbf{c}_{XX}(0)$. Suppose the roots are distinct. Let $\hat{\mu}_j$, $\hat{\mathbf{V}}_j$, $j = 1, \ldots, r$ indicate the latent roots and vectors of $\mathbf{c}_{XX}^{(T)}(0)$. Use (7.6.11) and the expansions of the proof of Theorem 9.2.4 to show that the $\hat{\mu}_j$, $\hat{\mathbf{V}}_j$, $j = 1, \ldots, r$ are asymptotically jointly normal with

$$\overrightarrow{\text{ave}}\, \hat{\mu}_j = \mu_j + O(T^{-1})$$

$$\overrightarrow{\text{ave}}\, \hat{\mathbf{V}}_j = \mathbf{V}_j + O(T^{-1})$$

$$\text{cov}\,\{\hat{\mu}_j, \hat{\mu}_k\} \sim \delta\{j-k\}\frac{4\pi}{T}\int_0^{2\pi} |\mathbf{V}_j^\tau \mathbf{f}_{XX}(\alpha)\mathbf{V}_j|^2 d\alpha$$

$$\text{cov}\,\{\hat{\mu}_j, \hat{\mathbf{V}}_k\} \sim \frac{2\pi}{T}\int_0^{2\pi} \Big[\sum_{m\neq k} \{(\mathbf{V}_j^\tau \mathbf{f}_{XX}(\alpha)\mathbf{V}_m)(\mathbf{V}_j^\tau \mathbf{f}_{XX}(-\alpha)\mathbf{V}_k) + (\mathbf{V}_j^\tau \mathbf{f}_{XX}(\alpha)\mathbf{V}_k)(\mathbf{V}_j^\tau \mathbf{f}_{XX}(-\alpha)\mathbf{V}_m)\}(\mu_k - \mu_m)^{-1}\mathbf{V}_m\Big] d\alpha$$

$$\text{cov}\,\{\hat{\mathbf{V}}_j, \hat{\mathbf{V}}_k\} \sim \frac{2\pi}{T}\int_0^{2\pi} \Big[\sum_{l\neq j}\sum_{m\neq k} \{(\mathbf{V}_l^\tau \mathbf{f}_{XX}(\alpha)\mathbf{V}_m)(\mathbf{V}_j^\tau \mathbf{f}_{XX}(-\alpha)\mathbf{V}_k) + (\mathbf{V}_l^\tau \mathbf{f}_{XX}(\alpha)\mathbf{V}_k)(\mathbf{V}_j^\tau \mathbf{f}_{XX}(-\alpha)\mathbf{V}_m)\}(\mu_j - \mu_l)^{-1}(\mu_k - \mu_m)^{-1}\mathbf{V}_l^\tau \mathbf{V}_m\Big] d\alpha.$$

10

THE CANONICAL ANALYSIS OF TIME SERIES

10.1 INTRODUCTION

In this chapter we consider the problem of approximating one stationary time series by a filtered version of a second series where the filter employed has reduced rank. Specifically consider the $(r + s)$ vector-valued stationary series

$$\begin{bmatrix} \mathbf{X}(t) \\ \mathbf{Y}(t) \end{bmatrix} \tag{10.1.1}$$

$t = 0, \pm 1, \ldots$ with $\mathbf{X}(t)$ r vector-valued and $\mathbf{Y}(t)$ s vector-valued.

Suppose we are interested in reducing the series $\mathbf{X}(t)$ to be q vector-valued forming, for example, the series

$$\boldsymbol{\zeta}(t) = \sum_u \mathbf{b}(t - u)\mathbf{X}(u) \tag{10.1.2}$$

$t = 0, \pm 1, \ldots$ with $\{\mathbf{b}(u)\}$ a $q \times r$ matrix-valued filter, and suppose we wish to do this so that the s vector-valued series

$$\mathbf{Y}^*(t) = \boldsymbol{\mu} + \sum_u \mathbf{c}(t - u)\boldsymbol{\zeta}(u) \tag{10.1.3}$$

is near $\mathbf{Y}(t)$ for some s vector $\boldsymbol{\mu}$ and $s \times q$ filter $\{\mathbf{c}(u)\}$. If the series $\mathbf{Y}(t)$ were identical with $\mathbf{X}(t)$, then we would have the problem discussed in the previous chapter whose solution led to a principal component analysis of the spectral density matrix. If $q = \min(r,s)$, then we are not requiring any real reduction in dimension and we have the multiple regression problem discussed in Chapter 8.

The relation connecting $\mathbf{Y}^*(t) - \boldsymbol{\mu}$ to $\mathbf{X}(t)$ is linear and time invariant with transfer function

$$\mathbf{A}(\lambda) = \mathbf{C}(\lambda)\mathbf{B}(\lambda) \tag{10.1.4}$$

where $\mathbf{B}(\lambda)$ and $\mathbf{C}(\lambda)$ are the transfer functions of $\{\mathbf{b}(u)\}$ and $\{\mathbf{c}(u)\}$, respectively. Note that under the indicated requirements the matrix $\mathbf{A}(\lambda)$ has rank $\leqslant q$. Conversely if it were known that $\mathbf{A}(\lambda)$ had rank $\leqslant q$, then we could find a $q \times r$ $\mathbf{B}(\lambda)$ and a $s \times q$ $\mathbf{C}(\lambda)$ so that relation (10.1.4) holds. The problem indicated is approximating $\mathbf{Y}(t)$ by a filtered version of $\mathbf{X}(t)$ where the filter employed has rank $\leqslant q$.

In the next section we discuss an analog of this problem for vector-valued variates. A general reference to the work of this chapter is Brillinger (1969d).

10.2 THE CANONICAL ANALYSIS OF VECTOR-VALUED VARIATES

Let

$$\begin{bmatrix} \mathbf{X} \\ \mathbf{Y} \end{bmatrix} \tag{10.2.1}$$

be an $(r + s)$ vector-valued variate with $\mathbf{X}$ r vector-valued and $\mathbf{Y}$ s vector-valued. Suppose the mean of (10.2.1) is

$$\begin{bmatrix} \boldsymbol{\mu}_X \\ \boldsymbol{\mu}_Y \end{bmatrix} \tag{10.2.2}$$

and its covariance matrix is

$$\begin{bmatrix} \mathbf{\Sigma}_{XX} & \mathbf{\Sigma}_{XY} \\ \mathbf{\Sigma}_{YX} & \mathbf{\Sigma}_{YY} \end{bmatrix}. \tag{10.2.3}$$

Consider the problem of determining the s vector $\boldsymbol{\mu}$, the $q \times r$ matrix $\mathbf{B}$ and the $s \times q$ matrix $\mathbf{C}$ so that the variate

$$\mathbf{Y} - \boldsymbol{\mu} - \mathbf{CBX} \tag{10.2.4}$$

is small. Let us measure the size of this variate by the real number

$$E\{[\mathbf{Y} - \boldsymbol{\mu} - \mathbf{CBX}]^\tau \mathbf{\Gamma}^{-1}[\mathbf{Y} - \boldsymbol{\mu} - \mathbf{CBX}]\} \tag{10.2.5}$$

for some symmetric positive definite $\mathbf{\Gamma}$. We have

Theorem 10.2.1 Let an $(r + s)$ vector-valued variate of the form (10.2.1) with mean (10.2.2) and covariance matrix (10.2.3) be given. Suppose $\mathbf{\Sigma}_{XX}$, $\mathbf{\Gamma}$

are nonsingular. Then the $s \times 1$ $\boldsymbol{\mu}$, $q \times r$ $\mathbf{B}$ and $s \times q$ $\mathbf{C}$, $q \leqslant r,s$, that minimize (10.2.5) are given by

$$\mathbf{B} = \begin{bmatrix} \mathbf{V}_1^{\tau} \\ \cdot \\ \cdot \\ \cdot \\ \mathbf{V}_q^{\tau} \end{bmatrix} \boldsymbol{\Gamma}^{-1/2}\boldsymbol{\Sigma}_{YX}\boldsymbol{\Sigma}_{XX}^{-1}, \tag{10.2.6}$$

$$\mathbf{C} = \boldsymbol{\Gamma}^{1/2}[\mathbf{V}_1 \cdots \mathbf{V}_q], \tag{10.2.7}$$

and

$$\boldsymbol{\mu} = \boldsymbol{\mu}_Y - \mathbf{C}\mathbf{B}\boldsymbol{\mu}_X \tag{10.2.8}$$

where $\mathbf{V}_j$ is the jth latent vector of $\boldsymbol{\Gamma}^{-1/2}\boldsymbol{\Sigma}_{YX}\boldsymbol{\Sigma}_{XX}^{-1}\boldsymbol{\Sigma}_{XY}\boldsymbol{\Gamma}^{-1/2}$, $j = 1, \ldots, s$. If μ_j denotes the corresponding latent root, $j = 1, \ldots, s$, then the minimum obtained is

$$\text{tr}\,\{(\boldsymbol{\Sigma}_{YY} - \boldsymbol{\Sigma}_{YX}\boldsymbol{\Sigma}_{XX}^{-1}\boldsymbol{\Sigma}_{XY})\boldsymbol{\Gamma}^{-1}\} + \sum_{j>q} \mu_j = \text{tr}\,\{\boldsymbol{\Sigma}_{YY}\boldsymbol{\Gamma}^{-1}\} - \sum_{j\leq q} \mu_j. \tag{10.2.9}$$

The case $\boldsymbol{\Gamma} = \mathbf{I}$ is of particular importance. Then we are led to evaluate the latent roots and vectors of the matrix $\boldsymbol{\Sigma}_{YX}\boldsymbol{\Sigma}_{XX}^{-1}\boldsymbol{\Sigma}_{XY}$. If μ_j and $\mathbf{V}_j$ denote these, then the covariance matrix of the error series

$$\mathbf{Y} - \boldsymbol{\mu} - \mathbf{C}\mathbf{B}\mathbf{X} \tag{10.2.10}$$

with

$$\mathbf{B} = \begin{bmatrix} \mathbf{V}_1^{\tau} \\ \cdot \\ \cdot \\ \cdot \\ \mathbf{V}_q^{\tau} \end{bmatrix} \boldsymbol{\Sigma}_{YX}\boldsymbol{\Sigma}_{XX}^{-1}, \tag{10.2.11}$$

$$\mathbf{C} = [\mathbf{V}_1 \cdots \mathbf{V}_q], \tag{10.2.12}$$

and

$$\boldsymbol{\mu} = \boldsymbol{\mu}_Y - \mathbf{C}\mathbf{B}\boldsymbol{\mu}_X, \tag{10.2.13}$$

is

$$\boldsymbol{\Sigma}_{YY} - \boldsymbol{\Sigma}_{YX}\boldsymbol{\Sigma}_{XX}^{-1}\boldsymbol{\Sigma}_{XY} + \sum_{j>q} \mu_j\mathbf{V}_j\mathbf{V}_j^{\tau}. \tag{10.2.14}$$

If we take $q = r$, then we are led to the multiple regression results of Theorem 8.2.1. If $s = r$ and $\mathbf{Y} = \mathbf{X}$, then we are led to the principal component results of Theorem 9.2.1. A result related to Theorem 10.2.1 is given in Rao (1965) p. 505.

A closely related problem to this theorem is that of determining the $q \times 1$ $\boldsymbol{\mu}$, $q \times r$ $\mathbf{D}$, and $q \times s$ $\mathbf{E}$ so that the q vector-valued variate

$$\mathbf{EY} - \boldsymbol{\mu} - \mathbf{DX} \tag{10.2.15}$$

is small. This problem leads us to

Theorem 10.2.2 Let an $(r + s)$ vector-valued variate of the form (10.2.1) with mean (10.2.2) and covariance matrix (10.2.3) be given. Suppose $\mathbf{\Sigma}_{XX}$ and $\mathbf{\Sigma}_{YY}$ are nonsingular. The $q \times 1$ $\boldsymbol{\mu}$, $q \times r$ $\mathbf{D}$, and $q \times s$ $\mathbf{E}$ with $\mathbf{E\Sigma}_{YY}\mathbf{E}^\tau = \mathbf{I}$, $\mathbf{D\Sigma}_{XX}\mathbf{D}^\tau = \mathbf{I}$ that minimize

$$E\{[\mathbf{EY} - \boldsymbol{\mu} - \mathbf{DX}]^\tau[\mathbf{EY} - \boldsymbol{\mu} - \mathbf{DX}]\} \tag{10.2.16}$$

are given by

$$\mathbf{D} = \begin{bmatrix} \mathbf{U}_1{}^\tau \\ \cdot \\ \cdot \\ \cdot \\ \mathbf{U}_q{}^\tau \end{bmatrix} \mathbf{\Sigma}_{XX}^{-1/2}, \tag{10.2.17}$$

$$\mathbf{E} = \begin{bmatrix} \mathbf{V}_1{}^\tau \\ \cdot \\ \cdot \\ \cdot \\ \mathbf{V}_q{}^\tau \end{bmatrix} \mathbf{\Sigma}_{YY}^{-1/2}, \tag{10.2.18}$$

and

$$\boldsymbol{\mu} = \mathbf{E}\boldsymbol{\mu}_Y - \mathbf{D}\boldsymbol{\mu}_X, \tag{10.2.19}$$

where $\mathbf{V}_j$ denotes the jth latent vector of $\mathbf{\Sigma}_{YY}^{-1/2}\mathbf{\Sigma}_{YX}\mathbf{\Sigma}_{XX}{}^{-1}\mathbf{\Sigma}_{XY}\mathbf{\Sigma}_{YY}^{-1/2}$, where $\mathbf{U}_j$ denotes the jth latent vector of $\mathbf{\Sigma}_{XX}^{-1/2}\mathbf{\Sigma}_{XY}\mathbf{\Sigma}_{YY}{}^{-1}\mathbf{\Sigma}_{YX}\mathbf{\Sigma}_{XX}^{-1/2}$. If μ_j denotes the jth latent root of either matrix, then the minimum achieved is

$$2q - 2 \sum_{j \leq q} \mu_j^{1/2}.$$

We see that the covariance matrix of the variate

$$\begin{bmatrix} \mathbf{DX} \\ \mathbf{EY} \end{bmatrix} \tag{10.2.20}$$

is given by

$$\begin{bmatrix}
1 & & & \sqrt{\mu_1} & & \\
 & \ddots & O & & \ddots & O \\
O & & 1 & O & & \sqrt{\mu_q} \\
\sqrt{\mu_1} & & & 1 & & \\
 & \ddots & O & & \ddots & O \\
O & & \sqrt{\mu_q} & O & & 1
\end{bmatrix}. \tag{10.2.21}$$

This result leads us to define the **canonical variates**

$$\begin{aligned} \zeta_j &= \boldsymbol{\alpha}_j^{\tau}\mathbf{X} \\ \omega_j &= \boldsymbol{\beta}_j^{\tau}\mathbf{Y} \end{aligned} \tag{10.2.22}$$

with $\boldsymbol{\alpha}_j$ and $\boldsymbol{\beta}_j$ proportional to $\boldsymbol{\Sigma}_{XX}^{-1/2}\mathbf{U}_j$ and $\boldsymbol{\Sigma}_{YY}^{-1/2}\mathbf{V}_j$ respectively. The coefficients of the canonical variates satisfy

$$\boldsymbol{\Sigma}_{XX}{}^{-1}\boldsymbol{\Sigma}_{XY}\boldsymbol{\Sigma}_{YY}{}^{-1}\boldsymbol{\Sigma}_{YX}\boldsymbol{\alpha}_j = \mu_j\boldsymbol{\alpha}_j \qquad \text{for } j = 1, \ldots, r \tag{10.2.23}$$

and

$$\boldsymbol{\Sigma}_{YY}{}^{-1}\boldsymbol{\Sigma}_{YX}\boldsymbol{\Sigma}_{XX}{}^{-1}\boldsymbol{\Sigma}_{XY}\boldsymbol{\beta}_j = \mu_j\boldsymbol{\beta}_j, \qquad \text{for } j = 1, \ldots, s. \tag{10.2.24}$$

We standardize them so that

$$\boldsymbol{\alpha}_j{}^{\tau}\boldsymbol{\alpha}_j,\ \boldsymbol{\beta}_k{}^{\tau}\boldsymbol{\beta}_k = 1 \qquad j = 1, \ldots, r;\ k = 1. \ldots, s. \tag{10.2.25}$$

We note that the standardization $\boldsymbol{\alpha}_j{}^{\tau}\boldsymbol{\Sigma}_{XX}\boldsymbol{\alpha}_j,\ \boldsymbol{\beta}_k{}^{\tau}\boldsymbol{\Sigma}_{YY}\boldsymbol{\beta}_k = 1$ is sometimes adopted. However, sampling properties of the empirical variates are simplified by adopting (10.2.25). We define

$$\begin{aligned} \rho_j &= \mu_j{}^{1/2}, && \text{for } j = 1, \ldots, \min(r,s) \\ &= 0 && \text{otherwise.} \end{aligned} \tag{10.2.26}$$

Corollary 10.2.2 Under the conditions of Theorem 10.2.2

$$\operatorname{cor}\{\zeta_j, \zeta_k\} = \delta\{j - k\} \qquad \text{for } j, k = 1, \ldots, r, \tag{10.2.27}$$

$$\operatorname{cor}\{\zeta_j, \omega_k\} = \delta\{j - k\}\rho_j \qquad \text{for } j = 1, \ldots, r;\ k = 1, \ldots, s \tag{10.2.28}$$

and

$$\operatorname{cor}\{\omega_j, \omega_k\} = \delta\{j - k\} \qquad \text{for } j, k = 1, \ldots, s. \tag{10.2.29}$$

The value $\rho_j = \mu_j^{1/2}$ is called the jth **canonical correlation** in view of (10.2.28). We note that the variates introduced in this theorem could alternately have been deduced by setting $\mathbf{\Gamma} = \mathbf{\Sigma}_{YY}$ in Theorem 10.2.1.

Canonical variates were introduced by Hotelling (1936) as linear combinations of the entries of **X** and **Y** that have extremal correlation. Related references include: Obukhov (1938, 1940), Anderson (1957), Morrison (1967), Rao (1965), and Kendall and Stuart (1966). In the case that the variate (10.2.1) is Gaussian, the first canonical variate is extremal within a broader class of variates; see Lancaster (1966). Canonical variates are useful: in studying relations between two vector-valued variates (Hotelling (1936)), in discriminating between several populations (Glahn (1968), Dempster (1969) p. 186, and Kshirsager (1971)), in searching for common factors (Rao (1965) p. 496), in predicting variables from other variables (Dempster (1969) p. 176, Glahn (1968)), and in the analysis of systems of linear equations (Hooper (1959) and Hannan (1967c)).

Let us consider certain aspects of the estimation of the above parameters. Assume, for convenience, that $\boldsymbol{\mu}_X$ and $\boldsymbol{\mu}_Y = \mathbf{0}$. Suppose that a sample of values

$$\begin{bmatrix} \mathbf{X}_j \\ \mathbf{Y}_j \end{bmatrix} \tag{10.2.30}$$

$j = 1, \ldots, n$ of the variate of Theorem 10.2.2 is available. As an estimate of (10.2.3) we take

$$\hat{\mathbf{\Sigma}}_{XX} = \frac{\sum_j \mathbf{X}_j\mathbf{X}_j^\tau}{n}$$
$$\hat{\mathbf{\Sigma}}_{XY} = \frac{\sum_j \mathbf{X}_j\mathbf{Y}_j^\tau}{n}$$
$$\hat{\mathbf{\Sigma}}_{YY} = \frac{\sum_j \mathbf{Y}_j\mathbf{Y}_j^\tau}{n}. \tag{10.2.31}$$

We then determine estimates of μ_j, $\boldsymbol{\alpha}_j$, $\boldsymbol{\beta}_j$ from the equations

$$\hat{\mathbf{\Sigma}}_{XX}^{-1}\hat{\mathbf{\Sigma}}_{XY}\hat{\mathbf{\Sigma}}_{YY}^{-1}\hat{\mathbf{\Sigma}}_{YX}\hat{\boldsymbol{\alpha}}_j = \hat{\mu}_j\hat{\boldsymbol{\alpha}}_j \tag{10.2.32}$$

and

$$\hat{\mathbf{\Sigma}}_{YY}^{-1}\hat{\mathbf{\Sigma}}_{YX}\hat{\mathbf{\Sigma}}_{XX}^{-1}\hat{\mathbf{\Sigma}}_{XY}\hat{\boldsymbol{\beta}}_j = \hat{\mu}_j\hat{\boldsymbol{\beta}}_j \tag{10.2.33}$$

with the standardizations

$$\hat{\boldsymbol{\alpha}}_j^\tau\hat{\boldsymbol{\alpha}}_j \quad \text{and} \quad \hat{\boldsymbol{\beta}}_j^\tau\hat{\boldsymbol{\beta}}_j = 1. \tag{10.2.34}$$

Below we set

$$\mathbf{a}_j = \frac{\boldsymbol{\alpha}_j}{(\boldsymbol{\alpha}_j^\tau\mathbf{\Sigma}_{XX}\boldsymbol{\alpha}_j)^{1/2}} \tag{10.2.35}$$

$$\mathbf{b}_j = \frac{\boldsymbol{\beta}_j}{(\boldsymbol{\beta}_j^\tau \boldsymbol{\Sigma}_{YY} \boldsymbol{\beta}_j)^{1/2}} \tag{10.2.36}$$

in order to obtain

Theorem 10.2.3 Suppose the values (10.2.30) are a sample of size n from

$$N_{r+s}\left(\mathbf{0}, \begin{bmatrix} \boldsymbol{\Sigma}_{XX} & \boldsymbol{\Sigma}_{XY} \\ \boldsymbol{\Sigma}_{YX} & \boldsymbol{\Sigma}_{YY} \end{bmatrix}\right). \tag{10.2.37}$$

Suppose $r \geqslant s$ and suppose the latent roots μ_j, $j = 1, \ldots, s$ are distinct. Then the variate $\{\hat{\mu}_j, \hat{\boldsymbol{\alpha}}_j, \hat{\boldsymbol{\beta}}_j; j = 1, \ldots, s\}$ is asymptotically normal with $\{\hat{\mu}_j; j = 1, \ldots, s\}$ asymptotically independent of $\{\hat{\boldsymbol{\alpha}}_j, \hat{\boldsymbol{\beta}}_j; j = 1, \ldots, s\}$. The asymptotic moments are given by

$$\overrightarrow{\text{ave}}\, \hat{\mu}_j = \mu_j + O(n^{-1}), \tag{10.2.38}$$

$$\overrightarrow{\text{ave}}\, \hat{\boldsymbol{\alpha}}_j = \boldsymbol{\alpha}_j + O(n^{-1}), \tag{10.2.39}$$

$$\overrightarrow{\text{ave}}\, \hat{\boldsymbol{\beta}}_j = \boldsymbol{\beta}_j + O(n^{-1}), \tag{10.2.40}$$

$$\overrightarrow{\text{cov}}\, \{\hat{\mu}_j, \hat{\mu}_k\} = \delta\{j - k\} 4\mu_j(1 - \mu_j)^2/n + O(n^{-2}), \tag{10.2.41}$$

$$\begin{aligned} \overrightarrow{\text{cov}}\, \{\hat{\boldsymbol{\alpha}}_j, \hat{\boldsymbol{\alpha}}_k\} = {} & \delta\{j - k\}(\boldsymbol{\alpha}_j^\tau \boldsymbol{\Sigma}_{XX} \boldsymbol{\alpha}_j)(1 - \mu_j) \sum_{l \neq j} (\rho_j^2 - \rho_l^2)^{-2} \\ & \times (\rho_j^2 + \rho_l^2 - 2\rho_j^2\rho_l^2)\mathbf{a}_l \mathbf{a}_l^\tau/n + (1 - \delta\{j - k\})(\rho_j^2 - \rho_k^2)^{-2} \\ & \times (-\rho_j^4\rho_k^2 - \rho_j^2\rho_k^4 + 4\rho_j^2\rho_k^2 + \rho_j^4 + \rho_k^4 - \rho_j^2 - \rho_k^2) \\ & \times \boldsymbol{\alpha}_k \boldsymbol{\alpha}_j^\tau/n + O(n^{-2}), \end{aligned} \tag{10.2.42}$$

for $j, k = 1, \ldots, s$; $l = 1, \ldots, r$;

$$\begin{aligned} \overrightarrow{\text{cov}}\, \{\hat{\boldsymbol{\alpha}}_j, \hat{\boldsymbol{\beta}}_k\} = {} & \delta\{j - k\}(\boldsymbol{\alpha}_j^\tau \boldsymbol{\Sigma}_{XX} \boldsymbol{\alpha}_j)^{1/2}(\boldsymbol{\beta}_j^\tau \boldsymbol{\Sigma}_{YY} \boldsymbol{\beta}_j)^{1/2}(1 - \mu_j) \\ & \times \sum_{l \neq j} (\rho_j^2 - \rho_l^2)^{-2}(2 - \rho_j^2 - \rho_l^2)\mathbf{a}_l \mathbf{b}_l^\tau/n + (1 - \delta\{j - k\}) \\ & \times (\boldsymbol{\alpha}_j^\tau \boldsymbol{\Sigma}_{XX} \boldsymbol{\alpha}_j)^{1/2} (\boldsymbol{\beta}_k^\tau \boldsymbol{\Sigma}_{YY} \boldsymbol{\beta}_k)^{1/2}(\rho_j^2 - \rho_k^2)^{-2} \\ & \times (-\rho_j^5\rho_k - \rho_j^3\rho_k^3 + \rho_j^4 + \rho_j^2\rho_k^2 + 2\rho_j^3\rho_k + 2\rho_j\rho_k^3 \\ & - \rho_j^2 - \rho_k^2)\mathbf{a}_k \mathbf{b}_j^\tau/n + O(n^{-2}), \end{aligned} \tag{10.2.43}$$

for $j, k, l = 1, \ldots, s$; and

$$\begin{aligned} \overrightarrow{\text{cov}}\, \{\hat{\boldsymbol{\beta}}_j, \hat{\boldsymbol{\beta}}_k\} = {} & \delta\{j - k\}(\boldsymbol{\beta}_j^\tau \boldsymbol{\Sigma}_{YY} \boldsymbol{\beta}_j)(1 - \mu_j) \sum_{l \neq j} (\rho_j^2 - \rho_l^2)^{-2} \\ & \times (\rho_j^2 + \rho_l^2 - 2\rho_j^2\rho_l^2)\mathbf{b}_l \mathbf{b}_l^\tau/n + (1 - \delta\{j - k\}) \\ & \times (\rho_j^2 - \rho_k^2)^{-2} \\ & \times (-\rho_j^4\rho_k^2 - \rho_j^2\rho_k^4 + 4\rho_j^2\rho_k^2 + \rho_j^4 + \rho_k^4 - \rho_j^2 - \rho_k^2) \\ & \times \boldsymbol{\beta}_k \boldsymbol{\beta}_j^\tau/n + O(n^{-2}), \end{aligned} \tag{10.2.44}$$

for $j, k, l = 1, \ldots, s$.

The asymptotic variances of the statistics may now be estimated by substituting estimates for the parameters appearing in expressions (10.2.41) to (10.2.44). In the case of $\hat{\mu}_j$ we note that

$$\overrightarrow{\text{var}} \tanh^{-1}\hat{\mu}_j^{1/2} = \frac{1}{n} + O(n^{-2}) \tag{10.2.45}$$

and so it is simpler to consider the transformed variate $\tanh^{-1} \hat{\mu}_j^{1/2}$. In practice it is probably most sensible to estimate the asymptotic second-order moments by means of the jack-knife procedure; see Brillinger (1964c, 1966b).

If $s = 1$ we note that the canonical correlation squared, $\rho_1^2 = \mu_1$, is the squared coefficient of multiple correlation discussed in Section 8.2.

The asymptotic covariance of $\hat{\mu}_j$ with $\hat{\mu}_k$ was derived in Hotelling (1936); Hsu (1941) found the asymptotic distribution; Lawley (1959) derived higher cumulants; Chambers (1966) derived further terms in the expansion of the asymptotic means; Dempster (1966) considered the problem of bias reduction; Hooper (1958) derived the asymptotic covariance structure under an assumption of fixed $\mathbf{X}_j$, $j = 1, \ldots, n$; the exact distribution of the sample canonical correlations depends only upon the population canonical correlations and has been given in Constantine (1963) and James (1964). The distribution of the vectors was found in Tumura (1965). Golub (1969) discusses the computations involved. Izenman (1972) finds the asymptotic distribution of an estimate for **CB** of (10.2.4) in the normal case.

We will require complex analogs of the previous results. Consider the $(r + s)$ vector-valued variate

$$\begin{bmatrix} \mathbf{X} \\ \mathbf{Y} \end{bmatrix} \tag{10.2.46}$$

with complex entries. Suppose it has mean

$$\begin{bmatrix} E\mathbf{X} \\ E\mathbf{Y} \end{bmatrix} = \begin{bmatrix} \boldsymbol{\mu}_X \\ \boldsymbol{\mu}_Y \end{bmatrix}, \tag{10.2.47}$$

covariance matrix

$$E\left\{\begin{bmatrix} \mathbf{X} - \boldsymbol{\mu}_X \\ \mathbf{Y} - \boldsymbol{\mu}_Y \end{bmatrix}\overline{\begin{bmatrix} \mathbf{X} - \boldsymbol{\mu}_X \\ \mathbf{Y} - \boldsymbol{\mu}_Y \end{bmatrix}}^{\tau}\right\} = \begin{bmatrix} \boldsymbol{\Sigma}_{XX} & \boldsymbol{\Sigma}_{XY} \\ \boldsymbol{\Sigma}_{YX} & \boldsymbol{\Sigma}_{YY} \end{bmatrix}, \tag{10.2.48}$$

and

$$E\left\{\begin{bmatrix} \mathbf{X} - \boldsymbol{\mu}_X \\ \mathbf{Y} - \boldsymbol{\mu}_Y \end{bmatrix}\begin{bmatrix} \mathbf{X} - \boldsymbol{\mu}_X \\ \mathbf{Y} - \boldsymbol{\mu}_Y \end{bmatrix}^{\tau}\right\} = \mathbf{0}. \tag{10.2.49}$$

We then have

Theorem 10.2.4 Let an $(r + s)$ vector-valued variate of the form (10.2.46) with mean (10.2.47) and covariance matrix (10.2.48) be given. Suppose

$\mathbf{\Sigma}_{XX}$, $\mathbf{\Gamma}$ are nonsingular with $\mathbf{\Gamma} > 0$. Then the $s \times 1$ $\boldsymbol{\mu}$, $q \times r$ $\mathbf{B}$ and $s \times q$ $\mathbf{C}$, $q \leqslant r, s$ that minimize

$$E\{\overline{[\mathbf{Y} - \boldsymbol{\mu} - \mathbf{CBX}]}^\tau \mathbf{\Gamma}^{-1}[\mathbf{Y} - \boldsymbol{\mu} - \mathbf{CBX}]\} \tag{10.2.50}$$

are given by

$$\mathbf{B} = \begin{bmatrix} \overline{\mathbf{V}}_1^\tau \\ \cdot \\ \cdot \\ \cdot \\ \overline{\mathbf{V}}_q^\tau \end{bmatrix} \mathbf{\Gamma}^{-1/2}\mathbf{\Sigma}_{YX}\mathbf{\Sigma}_{XX}{}^{-1}, \tag{10.2.51}$$

$$\mathbf{C} = \mathbf{\Gamma}^{1/2}[\mathbf{V}_1 \cdots \mathbf{V}_q], \tag{10.2.52}$$

and

$$\boldsymbol{\mu} = \boldsymbol{\mu}_Y - \mathbf{CB}\boldsymbol{\mu}_X, \tag{10.2.53}$$

where $\mathbf{V}_j$ is the jth latent vector of $\mathbf{\Gamma}^{-1/2}\mathbf{\Sigma}_{YX}\mathbf{\Sigma}_{XX}{}^{-1}\mathbf{\Sigma}_{XY}\mathbf{\Gamma}^{-1/2}$, $j = 1, \ldots, r$. If μ_j indicates the corresponding latent root, the minimum obtained is

$$\operatorname{tr}\{(\mathbf{\Sigma}_{YY} - \mathbf{\Sigma}_{YX}\mathbf{\Sigma}_{XX}{}^{-1}\mathbf{\Sigma}_{XY})\mathbf{\Gamma}^{-1}\} + \sum_{j>q} \mu_j. \tag{10.2.54}$$

We note that the $\mathbf{V}_j$ are arbitrary to the extent of a complex multiplier of modulus 1. Next we have

Theorem 10.2.5 Let an $(r + s)$ vector-valued variate of the form (10.2.46) with mean (10.2.47) and covariance matrix (10.2.48) be given. Suppose $\mathbf{\Sigma}_{XX}$, $\mathbf{\Sigma}_{YY}$ are nonsingular. Then the $q \times 1$ $\boldsymbol{\mu}$, $q \times r$ $\mathbf{D}$, $q \times s$ $\mathbf{E}$ with $\mathbf{E}\mathbf{\Sigma}_{YY}\overline{\mathbf{E}}^\tau = \mathbf{I}$, $\mathbf{D}\mathbf{\Sigma}_{XX}\overline{\mathbf{D}}^\tau = \mathbf{I}$, that minimize

$$E\{\overline{[\mathbf{EY} - \boldsymbol{\mu} - \mathbf{DX}]}^\tau[\mathbf{EY} - \boldsymbol{\mu} - \mathbf{DX}]\} \tag{10.2.55}$$

are given by

$$\mathbf{D} = \begin{bmatrix} \overline{\mathbf{U}}_1^\tau \\ \cdot \\ \cdot \\ \cdot \\ \overline{\mathbf{U}}_q^\tau \end{bmatrix} \mathbf{\Sigma}_{XX}^{-1/2}, \tag{10.2.56}$$

$$\mathbf{E} = \begin{bmatrix} \overline{\mathbf{V}}_1^\tau \\ \cdot \\ \cdot \\ \cdot \\ \overline{\mathbf{V}}_q^\tau \end{bmatrix} \mathbf{\Sigma}_{YY}^{-1/2}, \tag{10.2.57}$$

and

$$\boldsymbol{\mu} = \mathbf{E}\boldsymbol{\mu}_Y - \mathbf{D}\boldsymbol{\mu}_X, \tag{10.2.58}$$

where $\mathbf{V}_j$ signifies the jth latent vector of $\mathbf{\Sigma}_{YY}^{-1/2}\mathbf{\Sigma}_{YX}\mathbf{\Sigma}_{XX}{}^{-1}\mathbf{\Sigma}_{XY}\mathbf{\Sigma}_{YY}^{-1/2}$, and where $\mathbf{U}_j$ signifies the jth latent vector of $\mathbf{\Sigma}_{XX}^{-1/2}\mathbf{\Sigma}_{XY}\mathbf{\Sigma}_{YY}{}^{-1}\mathbf{\Sigma}_{YX}\mathbf{\Sigma}_{XX}^{-1/2}$.

As in the real case, we are led to consider the variates

$$\zeta_j = \bar{\boldsymbol{\alpha}}_j{}^{\tau}\mathbf{X}$$
$$\omega_j = \bar{\boldsymbol{\beta}}_j{}^{\tau}\mathbf{Y} \tag{10.2.59}$$

where $\boldsymbol{\alpha}_j$ and $\boldsymbol{\beta}_j$ are proportional to $\mathbf{\Sigma}_{XX}^{-1/2}\mathbf{U}_j$ and $\mathbf{\Sigma}_{YY}^{-1/2}\mathbf{V}_j$, respectively. We standardize them so that

$$\bar{\boldsymbol{\alpha}}_j{}^{\tau}\boldsymbol{\alpha}_j,\ \bar{\boldsymbol{\beta}}_j{}^{\tau}\boldsymbol{\beta}_j = 1. \tag{10.2.60}$$

Thus we have

Corollary 10.2.5 Under the conditions of Theorem 10.2.5

$$\operatorname{cov}\{\zeta_j, \zeta_k\} = \delta\{j-k\}(\bar{\boldsymbol{\alpha}}_j{}^{\tau}\mathbf{\Sigma}_{XX}\boldsymbol{\alpha}_j), \tag{10.2.61}$$

$$\operatorname{cov}\{\zeta_j, \bar{\zeta}_k\} = 0, \tag{10.2.62}$$

for $j, k = 1, \ldots, r$;

$$\operatorname{cov}\{\zeta_j, \omega_k\} = \delta\{j-k\}(\bar{\boldsymbol{\alpha}}_j{}^{\tau}\mathbf{\Sigma}_{XY}\boldsymbol{\beta}_j), \tag{10.2.63}$$

$$\operatorname{cov}\{\zeta_j, \bar{\omega}_k\} = 0, \tag{10.2.64}$$

for $j = 1, \ldots, r$; $k = 1, \ldots, s$;

$$\operatorname{cov}\{\omega_j, \omega_k\} = \delta\{j-k\}(\bar{\boldsymbol{\beta}}_j{}^{\tau}\mathbf{\Sigma}_{YY}\boldsymbol{\beta}_j) \tag{10.2.65}$$

$$\operatorname{cov}\{\omega_j, \bar{\omega}_k\} = 0, \tag{10.2.66}$$

for $j, k = 1, \ldots, s$.

If we let μ_j denote the jth latent root of $\mathbf{\Sigma}_{YY}^{-1/2}\mathbf{\Sigma}_{YX}\mathbf{\Sigma}_{XX}{}^{-1}\mathbf{\Sigma}_{XY}\mathbf{\Sigma}_{YY}^{-1/2}$ then it appears as

$$\mu_j = \frac{|\bar{\boldsymbol{\alpha}}_j{}^{\tau}\mathbf{\Sigma}_{XY}\boldsymbol{\beta}_j|^2}{(\bar{\boldsymbol{\alpha}}_j{}^{\tau}\mathbf{\Sigma}_{XX}\boldsymbol{\alpha}_j)(\bar{\boldsymbol{\beta}}_j{}^{\tau}\mathbf{\Sigma}_{YY}\boldsymbol{\beta}_j)} \tag{10.2.67}$$

for $j = 1, \ldots, \min(r,s)$. We call the variates $\zeta_j, \omega_j, j = 1, \ldots, \min(r,s)$ the ***j*th pair of canonical variates.** The coefficient $\rho_j = \mu_j{}^{1/2} \geqslant 0$ is called the ***j*th canonical correlation coefficient.** We set $\rho_j = 0$ for $j > \min(r,s)$ and we take a determination of $\boldsymbol{\alpha}_j$ and $\boldsymbol{\beta}_j$ so that

$$\rho_j = \frac{\bar{\boldsymbol{\alpha}}_j{}^{\tau}\mathbf{\Sigma}_{XY}\boldsymbol{\beta}_j}{[(\bar{\boldsymbol{\alpha}}_j{}^{\tau}\mathbf{\Sigma}_{XX}\boldsymbol{\alpha}_j)(\bar{\boldsymbol{\beta}}_j{}^{\tau}\mathbf{\Sigma}_{YY}\boldsymbol{\beta}_j)]^{1/2}} \tag{10.2.68}$$

for $j = 1, \ldots, \min(r,s)$.

Canonical variates of complex-valued random variables appear in Pinsker (1964) p. 134.

Suppose now $\boldsymbol{\mu}_X$ and $\boldsymbol{\mu}_Y = \mathbf{0}$ and a sample of values

$$\begin{bmatrix} \mathbf{X}_j \\ \mathbf{Y}_j \end{bmatrix} \tag{10.2.69}$$

$j = 1, \ldots, n$ of the variate of Theorem 10.2.5 is available. As an estimate of expression (10.2.48) we take

$$\hat{\mathbf{\Sigma}}_{XX} = \frac{\sum_j \mathbf{X}_j \bar{\mathbf{X}}_j^\tau}{n}$$

$$\hat{\mathbf{\Sigma}}_{XY} = \frac{\sum_j \mathbf{X}_j \bar{\mathbf{Y}}_j^\tau}{n}$$

$$\hat{\mathbf{\Sigma}}_{YY} = \frac{\sum_j \mathbf{Y}_j \bar{\mathbf{Y}}_j^\tau}{n}. \tag{10.2.70}$$

We then determine estimates of μ_j, $\boldsymbol{\alpha}_j$, $\boldsymbol{\beta}_j$ from the equations

$$\hat{\mathbf{\Sigma}}_{XX}^{-1}\hat{\mathbf{\Sigma}}_{XY}\hat{\mathbf{\Sigma}}_{YY}^{-1}\hat{\mathbf{\Sigma}}_{YX}\hat{\boldsymbol{\alpha}}_j = \hat{\mu}_j\hat{\boldsymbol{\alpha}}_j \tag{19.2.71}$$

and

$$\hat{\mathbf{\Sigma}}_{YY}^{-1}\hat{\mathbf{\Sigma}}_{YX}\hat{\mathbf{\Sigma}}_{XX}^{-1}\hat{\mathbf{\Sigma}}_{XY}\hat{\boldsymbol{\beta}}_j = \hat{\mu}_j\hat{\boldsymbol{\beta}}_j \tag{10.2.72}$$

with the normalizations

$$\bar{\hat{\boldsymbol{\alpha}}}_j^\tau\hat{\boldsymbol{\alpha}}_j,\ \bar{\hat{\boldsymbol{\beta}}}_j^\tau\hat{\boldsymbol{\beta}}_j = 1 \tag{10.2.73}$$

and

$$\hat{\mu}_j^{1/2} = \frac{\bar{\hat{\boldsymbol{\alpha}}}_j^\tau\hat{\mathbf{\Sigma}}_{XY}\hat{\boldsymbol{\beta}}_j}{[(\bar{\hat{\boldsymbol{\alpha}}}_j^\tau\hat{\mathbf{\Sigma}}_{XX}\hat{\boldsymbol{\alpha}}_j)(\bar{\hat{\boldsymbol{\beta}}}_j^\tau\hat{\mathbf{\Sigma}}_{YY}\hat{\boldsymbol{\beta}}_j)]^{1/2}} \geqslant 0. \tag{10.2.74}$$

In Theorem 10.2.6 we set

$$\mathbf{a}_j = \frac{\boldsymbol{\alpha}_j}{(\bar{\boldsymbol{\alpha}}_j^\tau\mathbf{\Sigma}_{XX}\boldsymbol{\alpha}_j)^{1/2}} \tag{10.2.75}$$

and

$$\mathbf{b}_j = \frac{\boldsymbol{\beta}_j}{(\bar{\boldsymbol{\beta}}_j^\tau\mathbf{\Sigma}_{YY}\boldsymbol{\beta}_j)^{1/2}}. \tag{10.2.76}$$

Theorem 10.2.6 Suppose the values (10.2.69) are a sample of size n from

$$N_{r+s}^C\left(\mathbf{0}, \begin{bmatrix} \mathbf{\Sigma}_{XX} & \mathbf{\Sigma}_{XY} \\ \mathbf{\Sigma}_{YX} & \mathbf{\Sigma}_{YY} \end{bmatrix}\right). \tag{10.2.77}$$

Suppose $r \geqslant s$ and suppose the latent roots μ_j, $j = 1, \ldots, s$ are distinct. Then the variate $\{\hat{\mu}_j, \hat{\boldsymbol{\alpha}}_j, \hat{\boldsymbol{\beta}}_j; j = 1, \ldots, s\}$ is asymptotically normal with $\{\hat{\mu}_j; j = 1, \ldots, s\}$ asymptotically independent of $\{\hat{\boldsymbol{\alpha}}_j, \hat{\boldsymbol{\beta}}_j; j = 1, \ldots, s\}$. The asymptotic moments are given by

$$\overrightarrow{\text{ave}}\ \hat{\mu}_j = \mu_j + O(n^{-1}), \tag{10.2.78}$$

$$\overrightarrow{\text{ave}}\ \hat{\boldsymbol{\alpha}}_j = \boldsymbol{\alpha}_j + O(n^{-1}), \tag{10.2.79}$$

$$\overrightarrow{\text{ave}}\ \hat{\boldsymbol{\beta}}_j = \boldsymbol{\beta}_j + O(n^{-1}), \tag{10.2.80}$$

$$\overrightarrow{\text{cov}}\ \{\hat{\mu}_j, \hat{\mu}_k\} = \delta\{j - k\}2\mu_j(1 - \mu_j)^2/n + O(n^{-2}), \tag{10.2.81}$$

$$\begin{aligned}\overrightarrow{\text{cov}}\ \{\hat{\alpha}_j, \hat{\alpha}_k\} = \delta\{j - k\}(\bar{\boldsymbol{\alpha}}_j^\tau \boldsymbol{\Sigma}_{XX}\boldsymbol{\alpha}_j)(1 - \mu_j)\sum_{l \neq j} (\rho_j^2 - \rho_l^2)^{-2} \\ \times (\rho_j^2 + \rho_l^2 - 2\rho_j^2\rho_l^2)\mathbf{a}_l\bar{\mathbf{a}}_l^\tau/n + O(n^{-2}),\end{aligned} \tag{10.2.82}$$

and

$$\begin{aligned}\overrightarrow{\text{cov}}\ \{\hat{\alpha}_j, \bar{\hat{\alpha}}_k\} = (1 - \delta\{j - k\})(\rho_j^2 - \rho_k^2)^{-2}(-\rho_j^4\rho_k^2 - \rho_j^2\rho_k^4 + 4\rho_j^2\rho_k^2 \\ + \rho_j^4 + \rho_k^4 - \rho_j^2 - \rho_k^2)\boldsymbol{\alpha}_k\boldsymbol{\alpha}_j^\tau/n + O(n^{-2}),\end{aligned} \tag{10.2.83}$$

for $j, k = 1, \ldots, s$; $l = 1, \ldots, r$;

$$\begin{aligned}\overrightarrow{\text{cov}}\ \{\hat{\alpha}_j, \hat{\beta}_k\} = \delta\{j - k\}(\bar{\boldsymbol{\alpha}}_j^\tau \boldsymbol{\Sigma}_{XX}\boldsymbol{\alpha}_j)^{1/2}(\bar{\boldsymbol{\beta}}_k^\tau \boldsymbol{\Sigma}_{YY}\boldsymbol{\beta}_k)^{1/2}(1 - \mu_j) \\ \times \sum_{l \neq j} (\rho_j^2 - \rho_l^2)^{-2}(2 - \rho_j^2 - \rho_l^2)\mathbf{a}_l\bar{\mathbf{b}}_l^\tau/n + O(n^{-2}),\end{aligned} \tag{10.2.84}$$

$$\begin{aligned}\overrightarrow{\text{cov}}\ \{\hat{\alpha}_j, \bar{\hat{\boldsymbol{\beta}}}_k\} = (1 - \delta\{j - k\})(\bar{\boldsymbol{\alpha}}_j^\tau \boldsymbol{\Sigma}_{XX}\boldsymbol{\alpha}_j)^{1/2}(\bar{\boldsymbol{\beta}}_k^\tau \boldsymbol{\Sigma}_{YY}\boldsymbol{\beta}_k)^{1/2} \\ \times (\rho_j^2 - \rho_k^2)^{-2}\,(-\rho_j^5\rho_k - \rho_j^3\rho_k^3 + \rho_j^4 + \rho_j^2\rho_k^2 + 2\rho_j^3\rho_k \\ + 2\rho_j\rho_k^3 - \rho_j^2 - \rho_k^2)\mathbf{a}_k\mathbf{b}_j^\tau/n + O(n^{-2}),\end{aligned} \tag{10.2.85}$$

for $j, k, l = 1, \ldots, s$; and

$$\begin{aligned}\overrightarrow{\text{cov}}\ \{\hat{\boldsymbol{\beta}}_j, \hat{\boldsymbol{\beta}}_k\} = \delta\{j - k\}(\bar{\boldsymbol{\beta}}_j^\tau \boldsymbol{\Sigma}_{YY}\boldsymbol{\beta}_j)(1 - \mu_j)\sum_{l \neq j} (\rho_j^2 - \rho_l^2)^{-2} \\ \times (\rho_j^2 + \rho_l^2 - 2\rho_j^2\rho_l^2)\mathbf{b}_l\bar{\mathbf{b}}_l^\tau/n + O(n^{-2}),\end{aligned} \tag{10.2.86}$$

$$\begin{aligned}\overrightarrow{\text{cov}}\ \{\hat{\boldsymbol{\beta}}_j, \bar{\hat{\boldsymbol{\beta}}}_k\} = (1 - \delta\{j - k\})(\rho_j^2 - \rho_k^2)^{-2} \\ \times (-\rho_j^4\rho_k^2 - \rho_j^2\rho_k^4 + 4\rho_j^2\rho_k^2 + \rho_j^4 + \rho_k^4 - \rho_j^2 - \rho_k^2) \\ \times \boldsymbol{\beta}_k\boldsymbol{\beta}_j^\tau/n + O(n^{-2}),\end{aligned} \tag{10.2.87}$$

for $j, k, l = 1, \ldots, s$.

We note that the asymptotic distribution of the variate

$$\begin{bmatrix} \hat{\boldsymbol{\alpha}}_j \\ \hat{\boldsymbol{\beta}}_j \end{bmatrix} \tag{10.2.88}$$

is complex normal, $j = 1, \ldots, s$. We also note that

$$\overrightarrow{\text{var}} \tanh^{-1} \hat{\mu}_j^{1/2} = \frac{1}{2n} + O(n^{-2}). \tag{10.2.89}$$

James (1964) gives the exact distribution of the $\hat{\mu}_j$, $j = 1, \ldots, s$ in this complex case.

10.3 THE CANONICAL VARIATE SERIES

Consider the problem, referred to in the introduction of this chapter, of determining the s vector $\mathbf{\mu}$, the $q \times r$ filter $\{\mathbf{b}(u)\}$ and the $s \times q$ filter $\{\mathbf{c}(u)\}$ so that if

$$\zeta(t) = \sum_u \mathbf{b}(t-u)\mathbf{X}(u), \tag{10.3.1}$$

then

$$\mathbf{Y}^*(t) = \mathbf{\mu} + \sum_u \mathbf{c}(t-u)\zeta(u) \tag{10.3.2}$$

is near $\mathbf{Y}(t)$, $t = 0, \pm 1, \ldots$. Suppose we measure the degree of nearness by

$$E\{[\mathbf{Y}(t) - \mathbf{Y}^*(t)]^\tau[\mathbf{Y}(t) - \mathbf{Y}^*(t)]\} \tag{10.3.3}$$

which may be written

$$\int_0^{2\pi} \text{tr} \{\mathbf{f}_{Y-Y^*,Y-Y^*}(\alpha)\} d\alpha \tag{10.3.4}$$

in the case that $E\mathbf{Y}(t) = E\mathbf{Y}^*(t)$. We have

Theorem 10.3.1 Let

$$\begin{bmatrix} \mathbf{X}(t) \\ \mathbf{Y}(t) \end{bmatrix} \tag{10.3.5}$$

$t = 0, \pm 1, \ldots$ be an $(r + s)$ vector-valued, second-order stationary series with mean

$$\begin{bmatrix} \mathbf{c}_X \\ \mathbf{c}_Y \end{bmatrix} \tag{10.3.6}$$

absolutely summable autocovariance function and spectral density matrix

$$\begin{bmatrix} \mathbf{f}_{XX}(\lambda) & \mathbf{f}_{XY}(\lambda) \\ \mathbf{f}_{YX}(\lambda) & \mathbf{f}_{YY}(\lambda) \end{bmatrix} \tag{10.3.7}$$

$-\infty < \lambda < \infty$. Suppose $\mathbf{f}_{XX}(\lambda)$ is nonsingular. Then, for given $q \leqslant r, s$, the

$s \times 1$ $\mathbf{\mu}$, $q \times r$ $\{\mathbf{b}(u)\}$ and $s \times q$ $\{\mathbf{c}(u)\}$ that minimize (10.3.3) are given by

$$\mathbf{\mu} = \mathbf{c}_Y - \left(\sum_u \mathbf{c}(u)\right)\left(\sum_u \mathbf{b}(u)\right)\mathbf{c}_X, \tag{10.3.8}$$

$$\mathbf{b}(u) = (2\pi)^{-1} \int_0^{2\pi} \mathbf{B}(\alpha) \exp\{iu\alpha\} d\alpha, \tag{10.3.9}$$

and

$$\mathbf{c}(u) = (2\pi)^{-1} \int_0^{2\pi} \mathbf{C}(\alpha) \exp\{iu\alpha\} d\alpha, \tag{10.3.10}$$

where

$$\mathbf{B}(\lambda) = \begin{bmatrix} \overline{\mathbf{V}_1(\lambda)}^\tau \\ \cdot \\ \cdot \\ \cdot \\ \overline{\mathbf{V}_q(\lambda)}^\tau \end{bmatrix} \mathbf{f}_{YX}(\lambda)\mathbf{f}_{XX}(\lambda)^{-1}, \tag{10.3.11}$$

and

$$\mathbf{C}(\lambda) = [\mathbf{V}_1(\lambda) \cdots \mathbf{V}_q(\lambda)]. \tag{10.3.12}$$

Here $\mathbf{V}_j(\lambda)$ denotes the jth latent vector of the matrix $\mathbf{f}_{YX}(\lambda)\mathbf{f}_{XX}(\lambda)^{-1}\mathbf{f}_{XY}(\lambda)$, $j = 1, \ldots, s$. If $\mu_j(\lambda)$ denotes the corresponding latent root, $j = 1, \ldots, s$, then the minimum achieved is

$$\int_0^{2\pi} \operatorname{tr} \{\mathbf{f}_{YY}(\alpha) - \mathbf{f}_{YX}(\alpha)\mathbf{f}_{XX}(\alpha)^{-1}\mathbf{f}_{XY}(\alpha)\} d\alpha + \int_0^{2\pi} \left\{\sum_{j>q} \mu_j(\alpha)\right\} d\alpha. \tag{10.3.13}$$

The previous theorem has led us to consider the latent roots and vectors of certain matrices based on the spectral density matrix of a given series. Theorem 10.3.1 is seen to provide a generalization of Theorems 8.3.1 and 9.3.1 which correspond to taking $q = s$ and $\mathbf{Y}(t) = \mathbf{X}(t)$ with probability 1, respectively.

We see that the error series

$$\mathbf{\varepsilon}(t) = \mathbf{Y}(t) - \mathbf{Y}^*(t) \tag{10.3.14}$$

has mean $\mathbf{0}$ and spectral density matrix

$$\begin{aligned}\mathbf{f}_{\varepsilon\varepsilon}(\lambda) &= \mathbf{f}_{YY}(\lambda) - \mathbf{C}(\lambda)\mathbf{B}(\lambda)\mathbf{f}_{XY}(\lambda) - \mathbf{f}_{YX}(\lambda)\overline{\mathbf{B}(\lambda)\mathbf{C}(\lambda)}^\tau \\ &\quad + \mathbf{C}(\lambda)\mathbf{B}(\lambda)\mathbf{f}_{XX}(\lambda)\overline{\mathbf{B}(\lambda)\mathbf{C}(\lambda)}^\tau \\ &= \mathbf{f}_{YY}(\lambda) - \sum_{j=1}^{q} \mu_j(\lambda)\mathbf{V}_j(\lambda)\overline{\mathbf{V}_j(\lambda)}^\tau \\ &= \mathbf{f}_{YY}(\lambda) - \mathbf{f}_{YX}(\lambda)\mathbf{f}_{XX}(\lambda)^{-1}\mathbf{f}_{XY}(\lambda) + \sum_{j>q} \mu_j(\lambda)\mathbf{V}_j(\lambda)\overline{\mathbf{V}_j(\lambda)}^\tau\end{aligned} \tag{10.3.15}$$

for $-\infty < \lambda < \infty$; this spectral density matrix is the sum of two parts of different character. The first part

$$\mathbf{f}_{YY}(\lambda) - \mathbf{f}_{YX}(\lambda)\mathbf{f}_{XX}(\lambda)^{-1}\mathbf{f}_{XY}(\lambda) \tag{10.3.16}$$

appeared in Section 8.3 as the error spectral density matrix resulting from regressing $\mathbf{Y}(t)$ on the series $\mathbf{X}(t)$, $t = 0, \pm 1, \ldots$. It represents a lower bound beyond which no improvement in degree of approximation is possible by choice of q, and also measures the inherent degree of linear approximation of $\mathbf{Y}(t)$ by the series $\mathbf{X}(t)$, $t = 0, \pm 1, \ldots$. The second part

$$\sum_{j>q} \mu_j(\lambda)\mathbf{V}_j(\lambda)\overline{\mathbf{V}_j(\lambda)}^\tau \tag{10.3.17}$$

will be small, for given q, in the case that the latent roots $\mu_j(\lambda)$, $j > q$, are small. As a function of q it decreases with increasing q and becomes 0 when $q \geqslant r$ or s.

The criterion (10.3.3) has the property of weighting the various components of $\mathbf{Y}(t)$ equally. This may not be desirable in the case that the different components have substantially unequal variances or a complicated correlation structure. For some purposes it may be more reasonable to minimize a criterion such as

$$\int_0^{2\pi} \operatorname{tr} \{\mathbf{f}_{YY}(\alpha)^{-1/2}\mathbf{f}_{Y-Y^*,Y-Y^*}(\alpha)\mathbf{f}_{YY}(\alpha)^{-1/2}\}\, d\alpha. \tag{10.3.18}$$

In this case we have

Corollary 10.3.1 Under the conditions of Theorem 10.3.1, expression (10.3.18) is minimized by the $\{\mathbf{b}(u)\}$ and $\{\mathbf{c}(u)\}$ of the theorem now based on $\mathbf{V}_j(\lambda)$, $j = 1, \ldots, s$, the latent vectors of $\mathbf{f}_{YY}(\lambda)^{-1/2}\mathbf{f}_{YX}(\lambda)\mathbf{f}_{XX}(\lambda)^{-1}\mathbf{f}_{XY}(\lambda)$ $\mathbf{f}_{YY}(\lambda)^{-1/2}$.

The procedure suggested by this corollary has the advantage of being invariant under nonsingular filtering of the series involved; see Exercise 10.6.5. The latent vectors of the matrix of this corollary essentially appear in the following:

Theorem 10.3.2 Suppose the conditions of Theorem 10.3.1 are satisfied. The real-valued series $\zeta_j(t)$, $\eta_j(t)$, $t = 0, \pm 1, \ldots$ of the form

$$\zeta_j(t) = \int_0^{2\pi} \overline{\mathbf{A}_j(\alpha)}^\tau \exp\{i\alpha t\}\, d\mathbf{Z}_X(\alpha) \tag{10.3.19}$$

and

$$\eta_j(t) = \int_0^{2\pi} \overline{\mathbf{B}_j(\alpha)}^\tau \exp\{i\alpha t\}\, d\mathbf{Z}_Y(\alpha) \tag{13.3.20}$$

with the standardizations $\overline{\mathbf{A}_j(\alpha)}^\tau \mathbf{A}_j(\alpha) = 1$, $\overline{\mathbf{B}_j(\alpha)}^\tau \mathbf{B}_j(\alpha) = 1$, having maximum coherence, $|R_{\zeta_j\eta_j}(\lambda)|^2$, and coherence 0 with the series $\zeta_k(t)$, $\eta_k(t)$, $t = 0, \pm 1, \ldots, k < j$, $j = 1, \ldots, \min(r,s)$, are given by the solutions of the equations:

$$\mathbf{f}_{XX}(\lambda)^{-1}\mathbf{f}_{XY}(\lambda)\mathbf{f}_{YY}(\lambda)^{-1}\mathbf{f}_{YX}(\lambda)\mathbf{A}_j(\lambda) = \mu_j(\lambda)\mathbf{A}_j(\lambda) \tag{10.3.21}$$

and

$$\mathbf{f}_{YY}(\lambda)^{-1}\mathbf{f}_{YX}(\lambda)\mathbf{f}_{XX}(\lambda)^{-1}\mathbf{f}_{XY}(\lambda)\mathbf{B}_j(\lambda) = \mu_j(\lambda)\mathbf{B}_j(\lambda) \tag{10.3.22}$$

$j = 1, \ldots, \min(r,s)$, where $\mu_1(\lambda) \geqslant \mu_2(\lambda) \geqslant \cdots$. The maximum coherence achieved is $\mu_j(\lambda)$, $j = 1, \ldots, \min(r,s)$.

The solutions of the equations (10.3.21) and (10.3.22) are intimately connected to the latent roots and vectors of the matrix of Corollary 10.3.1, which satisfy

$$\mathbf{f}_{YY}(\lambda)^{-1/2}\mathbf{f}_{YX}(\lambda)\mathbf{f}_{XX}(\lambda)^{-1}\mathbf{f}_{XY}(\lambda)\mathbf{f}_{YY}(\lambda)^{-1/2}\mathbf{V}_j(\lambda) = \nu_j(\lambda)\mathbf{V}_j(\lambda). \tag{10.3.23}$$

This last gives

$$\mathbf{f}_{XX}(\lambda)^{-1}\mathbf{f}_{XY}(\lambda)\mathbf{f}_{YY}(\lambda)^{-1}\mathbf{f}_{YX}(\lambda)[\mathbf{f}_{XX}(\lambda)^{-1}\mathbf{f}_{XY}(\lambda)\mathbf{f}_{YY}(\lambda)^{-1/2}\mathbf{V}_j(\lambda)] = \nu_j(\lambda)[\mathbf{f}_{XX}(\lambda)^{-1}\mathbf{f}_{XY}(\lambda)\mathbf{f}_{YY}(\lambda)^{-1/2}\mathbf{V}_j(\lambda)] \tag{10.3.24}$$

and

$$\mathbf{f}_{YY}(\lambda)^{-1}\mathbf{f}_{YX}(\lambda)\mathbf{f}_{XX}(\lambda)^{-1}\mathbf{f}_{XY}(\lambda)[\mathbf{f}_{YY}(\lambda)^{-1/2}\mathbf{V}_j(\lambda)] = \nu_j(\lambda)[\mathbf{f}_{YY}(\lambda)^{-1/2}\mathbf{V}_j(\lambda)] \tag{10.3.25}$$

allowing us to identify $\mu_j(\lambda)$ as $\nu_j(\lambda)$ and to take $\mathbf{A}_j(\lambda)$ and $\mathbf{B}_j(\lambda)$ proportional to $\mathbf{f}_{XX}(\lambda)^{-1}\mathbf{f}_{XY}(\lambda)\mathbf{f}_{YY}(\lambda)^{-1/2}\mathbf{V}_j(\lambda)$ and $\mathbf{f}_{YY}(\lambda)^{-1/2}\mathbf{V}_j(\lambda)$, respectively.

Theorem 10.3.2 has the advantage, over Corollary 10.3.1, of treating the series $\mathbf{X}(t)$ and $\mathbf{Y}(t)$ symmetrically. The pair $\zeta_j(t)$ and $\eta_j(t)$, $t = 0, \pm 1, \ldots$ of the theorem is called the ***j*th pair of canonical series.** Their coherence, $\mu_j(\lambda)$, is called the ***j*th canonical coherence.** They could also have been introduced through an analog of Theorem 10.2.4.

In the case that the autocovariance functions involved fall off rapidly as $|u| \to \infty$, the filter coefficients appearing will similarly fall off. Specifically we have

Theorem 10.3.3 Suppose the conditions of Theorem 10.3.1 are satisfied and in addition

$$\sum_u [1 + |u|^P]|\mathbf{c}_{XX}(u)| < \infty \tag{10.3.26}$$

and

$$\sum_u [1 + |u|^P]|\mathbf{c}_{XY}(u)| < \infty \tag{10.3.27}$$

for some $P \geqslant 0$ and suppose that the latent roots of $\mathbf{f}_{YX}(\lambda)\mathbf{f}_{XX}(\lambda)^{-1}\mathbf{f}_{XY}(\lambda)$ are distinct. Then the $\mathbf{b}(u)$, $\mathbf{c}(u)$ given in Theorem 10.3.1 satisfy

$$\sum_u [1 + |u|^P]|\mathbf{b}(u)| < \infty \tag{10.3.28}$$

and

$$\sum_u [1 + |u|^P]|\mathbf{c}(u)| < \infty. \tag{10.3.29}$$

Likewise the autocovariance function of the error series $\boldsymbol{\varepsilon}(t)$, $t = 0, \pm 1, \ldots$ satisfies

$$\sum_u [1 + |u|^P]|\mathbf{c}_{\varepsilon\varepsilon}(u)| < \infty. \tag{10.3.30}$$

The following theorem provides a related result sometimes useful in simplifying the structure of a series under consideration.

Theorem 10.3.4 Suppose the conditions of Theorem 10.3.1 are satisfied and in addition

$$\sum_u [1 + |u|^P]|\mathbf{c}_{XX}(u)| < \infty \tag{10.3.31}$$

and

$$\sum_u [1 + |u|^P]|\mathbf{c}_{YY}(u)| < \infty \tag{10.3.32}$$

for some $P \geqslant 0$ and suppose the latent roots $\mu_1(\lambda), \ldots, \mu_s(\lambda)$ of $\mathbf{f}_{YY}(\lambda)^{-1/2}$ $\mathbf{f}_{YX}(\lambda)\mathbf{f}_{XX}(\lambda)^{-1}\mathbf{f}_{XY}(\lambda)\mathbf{f}_{YY}(\lambda)^{-1/2}$ are distinct and nonzero. Then there exist $r \times r$ and $s \times s$ filters $\{\mathbf{a}(u)\}$ and $\{\mathbf{b}(u)\}$ satisfying

$$\sum_u [1 + |u|^P]|\mathbf{a}(u)| < \infty \tag{10.3.33}$$

and

$$\sum_u [1 + |u|^P]|\mathbf{b}(u)| < \infty \tag{10.3.34}$$

such that the series

$$\begin{bmatrix} \sum_u \mathbf{a}(t-u)\mathbf{X}(u) \\ \sum_u \mathbf{b}(t-u)\mathbf{Y}(u) \end{bmatrix} \tag{10.3.35}$$

has spectral density matrix

$$\begin{bmatrix}
1 & 0 & \cdot & \cdot & \cdot & 0 & \sqrt{\mu_1(\lambda)} & 0 & \cdot & \cdot & \cdot & 0 \\
0 & 1 & \cdot & \cdot & \cdot & 0 & 0 & \sqrt{\mu_2(\lambda)} & \cdot & \cdot & \cdot & 0 \\
0 & 0 & \cdot & \cdot & \cdot & 0 & 0 & & & & & \cdot \\
\cdot & & & & & \cdot & \cdot & & & & & \cdot \\
\cdot & & & & & \cdot & \cdot & & & & & \cdot \\
\cdot & & & & & 0 & 0 & & & & & \sqrt{\mu_s(\lambda)} \\
\cdot & & & & & \cdot & 0 & & & & & 0 \\
\cdot & & & & & \cdot & \cdot & & & & & \cdot \\
0 & \cdot & \cdot & \cdot & \cdot & 1 & 0 & \cdot & \cdot & \cdot & \cdot & 0 \\
\sqrt{\mu_1(\lambda)} & \cdot & \cdot & \cdot & \cdot & 0 & 1 & 0 & \cdot & \cdot & \cdot & 0 \\
0 & \sqrt{\mu_2(\lambda)} & \cdot & \cdot & \cdot & 0 & 0 & 1 & \cdot & \cdot & \cdot & 0 \\
\cdot & & & & & \cdot & \cdot & & & & & \cdot \\
\cdot & & & & & \cdot & \cdot & & & & & \cdot \\
\cdot & & & & & \cdot & \cdot & & & & & \cdot \\
0 & \cdot & \cdot & \sqrt{\mu_s(\lambda)} & \cdot & 0 & 0 & \cdot & \cdot & \cdot & \cdot & 1
\end{bmatrix}. \tag{10.3.36}$$

Pinsker (1964) indicates that we can filter a stationary series in order to obtain a spectral density matrix of the form of (10.3.36).

10.4 THE CONSTRUCTION OF ESTIMATES AND ASYMPTOTIC PROPERTIES

Suppose that we have a stretch

$$\begin{bmatrix} \mathbf{X}(t) \\ \mathbf{Y}(t) \end{bmatrix} \qquad t = 0, \ldots, T-1 \tag{10.4.1}$$

of the $(r+s)$ vector-valued stationary series

$$\begin{bmatrix} \mathbf{X}(t) \\ \mathbf{Y}(t) \end{bmatrix} \qquad t = 0, \pm 1, \ldots \tag{10.4.2}$$

with spectral density matrix

$$\begin{bmatrix} \mathbf{f}_{XX}(\lambda) & \mathbf{f}_{XY}(\lambda) \\ \mathbf{f}_{YX}(\lambda) & \mathbf{f}_{YY}(\lambda) \end{bmatrix} \tag{10.4.3}$$

and that we wish to construct estimates of the latent roots and transfer functions, $\mu_j(\lambda)$, $\mathbf{A}_j(\lambda)$, $\mathbf{B}_j(\lambda)$, $j = 1, 2, \ldots$, described in Theorem 10.3.2. An obvious means in which to proceed is to construct an estimate

$$\begin{bmatrix} \mathbf{f}_{XX}{}^{(T)}(\lambda) & \mathbf{f}_{XY}{}^{(T)}(\lambda) \\ \mathbf{f}_{YX}{}^{(T)}(\lambda) & \mathbf{f}_{YY}{}^{(T)}(\lambda) \end{bmatrix} \tag{10.4.4}$$

of the matrix (10.4.3) and then to determine estimates as solutions of the equations

$$\mathbf{f}_{XX}{}^{(T)}(\lambda)^{-1}\mathbf{f}_{XY}{}^{(T)}(\lambda)\mathbf{f}_{YY}{}^{(T)}(\lambda)^{-1}\mathbf{f}_{YX}{}^{(T)}(\lambda)\mathbf{A}_j{}^{(T)}(\lambda) = \mu_j{}^{(T)}(\lambda)\mathbf{A}_j{}^{(T)}(\lambda) \tag{10.4.5}$$

$$\mathbf{f}_{YY}{}^{(T)}(\lambda)^{-1}\mathbf{f}_{YX}{}^{(T)}(\lambda)\mathbf{f}_{XX}{}^{(T)}(\lambda)^{-1}\mathbf{f}_{XY}{}^{(T)}(\lambda)\mathbf{B}_j{}^{(T)}(\lambda) = \mu_j{}^{(T)}(\lambda)\mathbf{B}_j{}^{(T)}(\lambda) \tag{10.4.6}$$

with the standardizations

$$\overline{\mathbf{A}_j{}^{(T)}(\lambda)}^{\tau}\mathbf{A}_j{}^{(T)}(\lambda) \quad\text{and}\quad \overline{\mathbf{B}_j{}^{(T)}(\lambda)}^{\tau}\mathbf{B}_j{}^{(T)}(\lambda) = 1. \tag{10.4.7}$$

Now let us investigate the statistical properties of estimates constructed in this way.

Suppose we take

$$2\pi T^{-1}\sum_{s=1}^{T-1} W^{(T)}\left(\lambda - \frac{2\pi s}{T}\right)\begin{bmatrix} \mathbf{I}_{XX}{}^{(T)}\left(\frac{2\pi s}{T}\right) & \mathbf{I}_{XY}{}^{(T)}\left(\frac{2\pi s}{T}\right) \\ \mathbf{I}_{YX}{}^{(T)}\left(\frac{2\pi s}{T}\right) & \mathbf{I}_{YY}{}^{(T)}\left(\frac{2\pi s}{T}\right)\end{bmatrix} \tag{10.4.8}$$

as the estimate (10.4.4) where

$$\begin{bmatrix} \mathbf{I}_{XX}{}^{(T)}(\alpha) & \mathbf{I}_{XY}{}^{(T)}(\alpha) \\ \mathbf{I}_{YX}{}^{(T)}(\alpha) & \mathbf{I}_{YY}{}^{(T)}(\alpha)\end{bmatrix} = (2\pi T)^{-1}\begin{bmatrix} \sum \mathbf{X}(t)\exp\{-i\alpha t\} \\ \sum \mathbf{Y}(t)\exp\{-i\alpha t\}\end{bmatrix}\overline{\begin{bmatrix} \sum \mathbf{X}(t)\exp\{-i\alpha t\} \\ \sum \mathbf{Y}(t)\exp\{-i\alpha t\}\end{bmatrix}}^{\tau} \tag{10.4.9}$$

and

$$W^{(T)}(\alpha) = \sum_{j=-\infty}^{\infty} W(B_T{}^{-1}[\alpha + 2\pi j]) \tag{10.4.10}$$

for some weight function $W(\alpha)$. Then we have

Theorem 10.4.1 Let the $(r + s)$ vector-valued series (10.4.2) satisfy Assumption 2.6.2(1). Let $\nu_j{}^{(T)}(\lambda)$, $\mathbf{R}_j{}^{(T)}(\lambda)$, $\mathbf{S}_j{}^{(T)}(\lambda)$ be the solutions of the system of equations:

$$\mathbf{g}_{XX}{}^{(T)}(\lambda)^{-1}\mathbf{g}_{XY}{}^{(T)}(\lambda)\mathbf{g}_{YY}{}^{(T)}(\lambda)^{-1}\mathbf{g}_{YX}{}^{(T)}(\lambda)\mathbf{R}_j{}^{(T)}(\lambda) = \nu_j{}^{(T)}(\lambda)\mathbf{R}_j{}^{(T)}(\lambda) \tag{10.4.11}$$

and

$$\mathbf{g}_{YY}{}^{(T)}(\lambda)^{-1}\mathbf{g}_{YX}{}^{(T)}(\lambda)\mathbf{g}_{XX}{}^{(T)}(\lambda)^{-1}\mathbf{g}_{XY}{}^{(T)}(\lambda)\mathbf{S}_j{}^{(T)}(\lambda) = \nu_j{}^{(T)}(\lambda)\mathbf{S}_j{}^{(T)}(\lambda) \tag{10.4.12}$$

where

$$\begin{bmatrix} \mathbf{g}_{XX}^{(T)}(\lambda) & \mathbf{g}_{XY}^{(T)}(\lambda) \\ \mathbf{g}_{YX}^{(T)}(\lambda) & \mathbf{g}_{YY}^{(T)}(\lambda) \end{bmatrix} = \int_0^{2\pi} W^{(T)}(\lambda - \alpha) \begin{bmatrix} \mathbf{f}_{XX}(\alpha) & \mathbf{f}_{XY}(\alpha) \\ \mathbf{f}_{YX}(\alpha) & \mathbf{f}_{YY}(\alpha) \end{bmatrix} d\alpha. \tag{10.4.13}$$

Let (10.4.4) be given by (10.4.8) where $W(\alpha)$ satisfies Assumption 5.6.1. Let $\mu_j^{(T)}(\lambda)$, $\mathbf{A}_j^{(T)}(\lambda)$, $\mathbf{B}_j^{(T)}(\lambda)$ be given by (10.4.5) and (10.4.6). If $B_T T \to \infty$ as $T \to \infty$, then

$$E\mu_j^{(T)}(\lambda) = \nu_j^{(T)}(\lambda) + O(B_T^{-1/2}T^{-1/2}). \tag{10.4.14}$$

If, in addition, the latent roots of $\mathbf{f}_{YY}(\lambda)^{-1/2}\mathbf{f}_{YX}(\lambda)\mathbf{f}_{XX}(\lambda)^{-1}\mathbf{f}_{XY}(\lambda)\mathbf{f}_{YY}(\lambda)^{-1/2}$ are distinct, then

$$\vec{\text{ave}}\ \mu_j^{(T)}(\lambda) = \nu_j^{(T)}(\lambda) + O(B_T^{-1}T^{-1}), \tag{10.4.15}$$

$$\vec{\text{ave}}\ \mathbf{A}_j^{(T)}(\lambda) = \mathbf{R}_j^{(T)}(\lambda) + O(B_T^{-1}T^{-1}), \tag{10.4.16}$$

and

$$\vec{\text{ave}}\ \mathbf{B}_j^{(T)}(\lambda) = \mathbf{S}_j^{(T)}(\lambda) + O(B_T^{-1}T^{-1}), \tag{10.4.17}$$

for $j = 1, 2, \ldots$.

If $B_T \to 0$ as $T \to \infty$, then clearly

$$E\mu_j^{(T)}(\lambda),\ \vec{\text{ave}}\ \mu_j^{(T)}(\lambda) \to \mu_j(\lambda) \tag{10.4.18}$$

and

$$\begin{aligned} \vec{\text{ave}}\ \mathbf{A}_j^{(T)}(\lambda) &\to \mathbf{A}_j(\lambda) \\ \vec{\text{ave}}\ \mathbf{B}_j^{(T)}(\lambda) &\to \mathbf{B}_j(\lambda). \end{aligned} \tag{10.4.19}$$

Theorem 10.4.1 suggests the importance of prefiltering. The distributions of $\mu_j^{(T)}(\lambda)$, $\mathbf{A}_j^{(T)}(\lambda)$, $\mathbf{B}_j^{(T)}(\lambda)$ are centered at the solutions of the equations (10.4.11) and (10.4.12). These equations will be near the desired (10.3.21) and (10.3.22) only when the weighted average (10.4.13) is near (10.4.3). The latter is more likely to be the case when the series have been prefiltered adequately.

Turning to an investigation of the asymptotic distribution of $\mu_j^{(T)}(\lambda)$ and $\mathbf{A}_j^{(T)}(\lambda)$, $\mathbf{B}_j^{(T)}(\lambda)$ we have

Theorem 10.4.2 Under the conditions of Theorem 10.4.1 and if the $\mu_j(\lambda_m)$, $j = 1, \ldots, \min(r,s)$ are distinct for $m = 1, \ldots, M$, the variates $\mu_j^{(T)}(\lambda_m)$, $\mathbf{A}_j^{(T)}(\lambda_m)$, $\mathbf{B}_j^{(T)}(\lambda_m)$, $j = 1, 2, \ldots, m = 1, \ldots, M$ are asymptotically jointly normal with asymptotic covariance structure

$$\begin{aligned} \lim_{T\to\infty} B_T T\ \vec{\text{cov}}\ \{\mu_j^{(T)}(\lambda_m), \mu_k^{(T)}(\lambda_n)\} &= 2\pi \textstyle\int W(\alpha)^2 d\alpha\, [\eta\{\lambda_m - \lambda_n\} + \eta\{\lambda_m + \lambda_n\}] 2\mu_j(\lambda_m)[1 - \mu_j(\lambda_m)]^2 \\ &\qquad\qquad \text{if } j = k, \\ &= 0 \qquad\qquad \text{if } j \neq k, \end{aligned} \tag{10.4.20}$$

$$\lim_{T\to\infty} B_T T \,\vec{\text{cov}}\, \{\mu_j^{(T)}(\lambda_m), \mathbf{A}_k^{(T)}(\lambda_n)\} = \mathbf{0}, \tag{10.4.21}$$

$$\lim_{T\to\infty} B_T T \,\vec{\text{cov}}\, \{\mu_j^{(T)}(\lambda_m), \mathbf{B}_k^{(T)}(\lambda_n)\} = \mathbf{0}, \tag{10.4.22}$$

and suppressing the dependence of population parameters on λ_m,

$$\begin{aligned}
&\lim_{T\to\infty} B_T T \,\vec{\text{cov}}\, \{\mathbf{A}_j^{(T)}(\lambda_m), \mathbf{A}_k^{(T)}(\lambda_n)\} \\
&\quad = 2\pi \int W(\alpha)^2 d\alpha\, \eta\{\lambda_m - \lambda_n\} \overline{\mathbf{A}}_j^\tau \mathbf{f}_{XX} \mathbf{A}_j (1 - \mu_j) \sum_{l \neq j} (\mu_j - \mu_l)^{-2} \\
&\qquad \times (\mu_j + \mu_l - 2\mu_j\mu_l) \mathbf{A}_l \overline{\mathbf{A}}_l^\tau (\overline{\mathbf{A}}_l^\tau \mathbf{f}_{XX} \mathbf{A}_l)^{-1} \qquad \text{if } j = k, \\
&\quad = 2\pi \int W(\alpha)^2 d\alpha\, \eta\{\lambda_m + \lambda_n\} (\mu_j - \mu_k)^{-2} (-\mu_j^2\mu_k - \mu_j\mu_k^2 \\
&\qquad + 4\mu_j\mu_k + \mu_j^2 + \mu_k^2 - \mu_j - \mu_k) \mathbf{A}_k \mathbf{A}_j^\tau (\overline{\mathbf{A}}_j^\tau \mathbf{f}_{XX} \mathbf{A}_j)^{-1/2} (\overline{\mathbf{A}}_k^\tau \mathbf{f}_{XX} \mathbf{A}_k)^{-1/2} \\
&\qquad \text{if } j \neq k,
\end{aligned} \tag{10.4.23}$$

with analagous expressions for $\vec{\text{cov}}\, \{\mathbf{A}_j^{(T)}(\lambda_m), \mathbf{B}_k^{(T)}(\lambda_n)\}$, $\vec{\text{cov}}\, \{\mathbf{B}_j^{(T)}(\lambda_m), \mathbf{B}_k^{(T)}(\lambda_n)\}$ deducible from (10.2.84) to (10.2.87).

Expression (10.4.20) implies that

$$\begin{aligned}
\vec{\text{var}} \tanh^{-1} \sqrt{\mu_j^{(T)}(\lambda)} &\backsim B_T^{-1} T^{-1} \pi \int W(\alpha)^2 d\alpha \qquad \text{if } \lambda \not\equiv 0 \pmod{\pi} \\
&\backsim B_T^{-1} T^{-1} 2\pi \int W(\alpha)^2 d\alpha \qquad \text{if } \lambda \equiv 0 \pmod{\pi}
\end{aligned} \tag{10.4.24}$$

in addition to which $\tanh^{-1} \sqrt{\mu_j^{(T)}(\lambda)}$ will be asymptotically normal. These results may be used to construct approximate confidence limits for the canonical coherences.

An alternate form of limiting distribution results if we consider the spectral estimate (8.5.4) corresponding to a simple average of a fixed number of periodogram ordinates.

Theorem 10.4.3 Let the $(r + s)$ vector-valued series (10.4.2) satisfy Assumption 2.6.1 and have spectral density matrix (10.4.3). Let this matrix be estimated by (8.5.4) where m, $s(T)$ are integers with $2\pi s(T)/T \to \lambda$ as $T \to \infty$. Then let

$$\begin{bmatrix} \mathbf{W}_{XX} & \mathbf{W}_{XY} \\ \mathbf{W}_{YX} & \mathbf{W}_{YY} \end{bmatrix} \tag{10.4.25}$$

be distributed as $(2m + 1)^{-1} W_{r+s}^C(2m + 1, (10.4.3))$ if $\lambda \not\equiv 0 \pmod{\pi}$, and as $(2m)^{-1} W_{r+s}(2m, (10.4.3))$ if $\lambda \equiv 0 \pmod{\pi}$. Then, as $T \to \infty$, $\mu_j^{(T)}(\lambda)$, $\mathbf{A}_j^{(T)}(\lambda)$, $\mathbf{B}_j^{(T)}(\lambda)$ tend in distribution to the distribution of $\hat{\mu}_j$, $\hat{\mathbf{A}}_j$, $\hat{\mathbf{B}}_j$—the solutions of the equations

$$\mathbf{W}_{XX}^{-1} \mathbf{W}_{XY} \mathbf{W}_{YY}^{-1} \mathbf{W}_{YX} \hat{\mathbf{A}}_j = \hat{\mu}_j \hat{\mathbf{A}}_j \tag{10.4.26}$$

and

$$\mathbf{W}_{YY}^{-1} \mathbf{W}_{YX} \mathbf{W}_{XX}^{-1} \mathbf{W}_{XY} \hat{\mathbf{B}}_j = \hat{\mu}_j \hat{\mathbf{B}}_j. \tag{10.4.27}$$

Constantine (1963) and James (1964) give the distribution of the $\hat{\mu}_j$, $j = 1, 2, \ldots$.

The distributions obtained in Theorems 10.4.2 and 10.4.3 are not inconsistent. If, as in Section 5.7, we make the identification

$$2m + 1 \backsim \frac{1}{\sum_s \left[\frac{2\pi}{T} W^{(T)}\left(\lambda - \frac{2\pi s}{T}\right)\right]^2} \backsim \frac{B_T T}{2\pi \int W(\alpha)^2 d\alpha} \tag{10.4.28}$$

and m is large, then, as Theorems 10.2.3 and 10.2.6 imply, the $\mu_j^{(T)}(\lambda)$, $\mathbf{A}_j^{(T)}(\lambda)$ and $\mathbf{B}_j^{(T)}(\lambda)$ are asymptotically normal with the appropriate first- and second-order moment structure.

10.5 FURTHER ASPECTS OF CANONICAL VARIATES

We begin by interpreting the canonical series introduced in this chapter, in terms of the usual canonical variates of vector-valued variables with real-valued components. Let $\mathbf{X}(t,\lambda)$ and $\mathbf{Y}(t,\lambda)$ signify the components of frequency λ of the series $\mathbf{X}(t)$, $t = 0, \pm 1, \ldots$ and $\mathbf{Y}(t)$, $t = 0, \pm 1, \ldots$, respectively. Then (see Sections 4.6, 7.1) the $2(r + s)$ vector-valued variate

$$\begin{bmatrix} \mathbf{X}(t,\lambda) \\ \mathbf{X}(t,\lambda)^H \\ \mathbf{Y}(t,\lambda) \\ \mathbf{Y}(t,\lambda)^H \end{bmatrix} \tag{10.5.1}$$

has covariance matrix proportional to

$$\begin{bmatrix} \operatorname{Re} \mathbf{f}_{XX}(\lambda) & \operatorname{Im} \mathbf{f}_{XX}(\lambda) & \operatorname{Re} \mathbf{f}_{XY}(\lambda) & \operatorname{Im} \mathbf{f}_{XY}(\lambda) \\ -\operatorname{Im} \mathbf{f}_{XX}(\lambda) & \operatorname{Re} \mathbf{f}_{XX}(\lambda) & -\operatorname{Im} \mathbf{f}_{XY}(\lambda) & \operatorname{Re} \mathbf{f}_{XY}(\lambda) \\ \operatorname{Re} \mathbf{f}_{YX}(\lambda) & \operatorname{Im} \mathbf{f}_{YX}(\lambda) & \operatorname{Re} \mathbf{f}_{YY}(\lambda) & \operatorname{Im} \mathbf{f}_{YY}(\lambda) \\ -\operatorname{Im} \mathbf{f}_{YX}(\lambda) & \operatorname{Re} \mathbf{f}_{YX}(\lambda) & -\operatorname{Im} \mathbf{f}_{YY}(\lambda) & \operatorname{Re} \mathbf{f}_{YY}(\lambda) \end{bmatrix}$$

$$= \begin{bmatrix} \mathbf{f}_{XX}(\lambda)^R & \mathbf{f}_{XY}(\lambda)^R \\ \mathbf{f}_{YX}(\lambda)^R & \mathbf{f}_{YY}(\lambda)^R \end{bmatrix}. \tag{10.5.2}$$

A standard canonical correlation analysis of the variate (10.5.1) would thus lead us to consider latent roots and vectors based on (10.5.2), specifically the roots and vectors of

$$[\mathbf{f}_{YY}(\lambda)^R]^{-1/2}\mathbf{f}_{YX}(\lambda)^R[\mathbf{f}_{XX}(\lambda)^R]^{-1}\mathbf{f}_{XY}(\lambda)^R[\mathbf{f}_{YY}(\lambda)^R]^{-1/2}$$
$$= [\mathbf{f}_{YY}(\lambda)^{-1/2}\mathbf{f}_{YX}(\lambda)\mathbf{f}_{XX}(\lambda)^{-1}\mathbf{f}_{XY}(\lambda)\mathbf{f}_{YY}(\lambda)^{-1/2}]^R. \tag{10.5.3}$$

Following Lemma 3.7.1, these are essentially the roots and vectors of

$\mathbf{f}_{YY}^{-1/2}\mathbf{f}_{YX}\mathbf{f}_{XX}^{-1}\mathbf{f}_{XY}\mathbf{f}_{YY}^{-1/2}$. In summary, we see that a frequency domain canonical analysis of the series

$$\begin{bmatrix} \mathbf{X}(t) \\ \mathbf{Y}(t) \end{bmatrix} \qquad t = 0, \pm 1, \ldots \tag{10.5.4}$$

may be considered to be a standard canonical correlation analysis carried out on the individual frequency components of the series $\mathbf{X}(t)$ and $\mathbf{Y}(t)$ and their Hilbert transforms.

Alternately we can view the variates appearing in Theorem 10.4.3 as resulting from a canonical correlation analysis carried out on complex valued variates of the sort considered in Theorem 10.2.6. Specifically Theorem 4.4.1 suggests that for $s(T)$ an integer with $2\pi s(T)/T \backsim \lambda \not\equiv 0 \pmod{\pi}$, the values

$$\begin{bmatrix} \mathbf{d}_X^{(T)}\left(\dfrac{2\pi[s(T)+s]}{T}\right) \\ \mathbf{d}_Y^{(T)}\left(\dfrac{2\pi[s(T)+s]}{T}\right) \end{bmatrix} \qquad s = 0, \pm 1, \ldots, \pm m \tag{10.5.5}$$

are approximately a sample of size $(2m + 1)$ from

$$N_{r+s}^C\left(\mathbf{0}, 2\pi T\begin{bmatrix} \mathbf{f}_{XX}(\lambda) & \mathbf{f}_{XY}(\lambda) \\ \mathbf{f}_{YX}(\lambda) & \mathbf{f}_{YY}(\lambda) \end{bmatrix}\right). \tag{10.5.6}$$

The discussion preceding Theorem 10.2.6 now leads us to the calculation of variates of the sort considered in Theorem 10.4.3.

We remark that the student who has available a computer program for the canonical correlation analysis of real-valued quantities may make use of the real-valued correspondence discussed above in order to compute estimates of the coefficients, rather than writing a new program specific to the complex case.

Further statistics we may wish to calculate in the present context include:

$$\mathbf{a}_j^{(T)}(u) = (2\pi)^{-1}\int_0^{2\pi} \mathbf{A}_j^{(T)}(\alpha)\exp\{iu\alpha\}\,d\alpha \tag{10.5.7}$$

and

$$\mathbf{b}_j^{(T)}(u) = (2\pi)^{-1}\int_0^{2\pi} \mathbf{B}_j^{(T)}(\alpha)\exp\{iu\alpha\}\,d\alpha \tag{10.5.8}$$

for $u = 0, \pm 1, \ldots$, where $\mathbf{A}_j^{(T)}(\lambda)$, $\mathbf{B}_j^{(T)}(\lambda)$ are given by the solutions of (10.4.5) and (10.4.6). These statistics are estimates of the time domain coefficients of the canonical series.

By analogy with what is done in multivariate analysis, we may wish to form certain real-valued measures of the association of the series $\mathbf{X}(t)$ and $\mathbf{Y}(t)$ such as **Wilks' Λ statistic**

$$\prod_j (1 - \mu_j(\lambda)), \tag{10.5.9}$$

the **vector alienation coefficient**

$$\sum_j (1 - \mu_j(\lambda)), \tag{10.5.10}$$

or the **vector correlation coefficient**

$$\pm \prod_j \sqrt{\mu_j(\lambda)}. \tag{10.5.11}$$

Sample estimates of these coefficients would be of use in estimating the degree of association of the series $\mathbf{X}(t)$ and $\mathbf{Y}(t)$ at frequency λ.

Miyata (1970) includes an example of an empirical canonical analysis of some oceanographic series.

10.6 EXERCISES

10.6.1 Show that if in Theorem 10.2.1 we set $q = r$, then we obtain the multiple regression results of Theorem 8.2.1.

10.6.2 If $\mathbf{\Gamma}$ is taken to be $\mathbf{\Sigma}_{YY}$ in Theorem 10.2.1, show that the criterion (10.2.5) is invariant under nonsingular linear transformations of $\mathbf{Y}$.

10.6.3 Under the conditions of Theorem 10.3.1, prove that $\mu_1(\lambda) = |R_{YX}(\lambda)|^2$ if $s = 1$.

10.6.4 Under the conditions of Theorem 10.3.2, prove $|\mu_j(\lambda)| \leqslant 1$.

10.6.5 Under the conditions of Theorem 10.3.1, prove the canonical coherences $\mu_j(\lambda), j = 1, 2, \ldots$ are invariant under nonsingular filterings of the series $\mathbf{X}(t)$, $t = 0, \pm 1, \ldots$ or the series $\mathbf{Y}(t)$, $t = 0, \pm 1, \ldots$.

10.6.6 Suppose the conditions of Theorem 10.3.1 are satisfied. Also, $\mathbf{c}_{XX}(u)$, $\mathbf{c}_{XY}(u)$, $\mathbf{c}_{YY}(u) = \mathbf{0}$ for $u \neq 0$. Then $\{\mathbf{b}(u)\}$, $\{\mathbf{c}(u)\}$ given in the theorem satisfy $\mathbf{b}(u)$, $\mathbf{c}(u) = \mathbf{0}$ for $u \neq 0$.

10.6.7 Demonstrate that the coherence $|R_{YX}(\lambda)|^2$ can be interpreted as the largest squared canonical correlation of the variate $\{X(t,\lambda), X(t,\lambda)^H\}$ with the variate $\{Y(t,\lambda), Y(t,\lambda)^H\}$.

10.6.8 Prove that if in the estimate (8.5.4), used in Theorem 10.4.3, we smooth across the whole frequency domain, then the proposed analysis reduces to a standard canonical correlation analysis of the sample covariance matrix

$$\begin{bmatrix} \mathbf{c}_{XX}{}^{(T)}(0) & \mathbf{c}_{XY}{}^{(T)}(0) \\ \mathbf{c}_{YX}{}^{(T)}(0) & \mathbf{c}_{YY}{}^{(T)}(0) \end{bmatrix}.$$

10.6.9 Suppose the data is tapered with tapering function $h(t/T)$ prior to calculating the estimates of Theorem 10.4.2. Under the conditions of the theorem, prove the asymptotic covariances appearing become multiplied by $\int h(t)^4 dt / [\int h(t)^2 dt]^2$.

10.6.10 Suppose that there exist J groups of r vector-valued observations with K observations in each group. Suppose the vectors have complex entries and are denoted $\mathbf{Y}_{jk}; j = 1, \ldots, J; k = 1, \ldots, K$. Let

$$\mathbf{Y}_{j.} = K^{-1} \sum_{k=1}^{K} \mathbf{Y}_{jk}$$

$$\mathbf{Y}_{..} = J^{-1} \sum_{j=1}^{J} \mathbf{Y}_{j.}$$

$$\mathbf{S}_W = \sum_j \sum_k (\mathbf{Y}_{jk} - \mathbf{Y}_{j.})\overline{(\mathbf{Y}_{jk} - \mathbf{Y}_{j.})}^\tau$$

$$\mathbf{S}_B = K \sum_j (\mathbf{Y}_{j.} - \mathbf{Y}_{..})\overline{(\mathbf{Y}_{j.} - \mathbf{Y}_{..})}^\tau.$$

(a) Show that the **linear discriminant functions,** $\bar{\boldsymbol{\beta}}^\tau \mathbf{Y}$, providing the extrema of the ratio $\bar{\boldsymbol{\beta}}^\tau \mathbf{S}_B \boldsymbol{\beta} / \bar{\boldsymbol{\beta}}^\tau \mathbf{S}_W \boldsymbol{\beta}$ (of between to within group sums of squares) are solutions of the determinental equation

$$(\mathbf{S}_B - \nu \mathbf{S}_W)\boldsymbol{\beta} = \mathbf{0}$$

for some ν.

(b) Define a $J - 1$ vector-valued indicator variable $\mathbf{X} = [X_j]$ with $X_j = 1$ if $\mathbf{Y}$ is in the jth group and equal 0 otherwise, $j = 1, \ldots, J - 1$. Show that the analysis above is equivalent to a canonical correlation analysis of the values

$$\begin{bmatrix} X_{jk} \\ Y_{jk} \end{bmatrix}, \qquad j = 1, \ldots, J; k = 1, \ldots, K.$$

See Glahn (1968).

(c) Indicate extensions of these results to the case of stationary time series $\mathbf{Y}_{jk}(t)$, $t = 0, \pm 1, \ldots$.

PROOFS OF THEOREMS

PROOFS FOR CHAPTER 2

Proof of Theorem 2.3.1 Directly by identification of the indicated coefficient.

Proof of Lemma 2.3.1 First the "only if" part. If the partition is not indecomposable, then following (2.3.5), the $\phi(r_{ij_1}) - \phi(r_{ij_2})$; $1 \leqslant j_1 \neq j_2 \leqslant I_i$; $i = i_1, \ldots, i_0$ generate only the values $s_{m'} - s_{m''}$; $m', m'' = m_1, \ldots, m_N$. There is no way the values $s_{m'} - s_{m''}$; $m' = m_1, \ldots, m_N$ and $m'' \neq m_1, \ldots, m_N$ can be generated.

Next the "if" part. Suppose the $\phi(r_{ij_1}) - \phi(r_{ij_2})$; $1 \leqslant j_1 \neq j_2 \leqslant J_i$; $i = 1, \ldots, I$ generate the $s_{m_1} - s_{m_2}$; $1 \leqslant m_1 \neq m_2 \leqslant M$. It follows that each pair $P_{m'}, P_{m''}$ of the partition communicate otherwise $s_{m'} - s_{m''}$ would not have been generated. The indecomposability is thus shown.

We next demonstrate the alternate formulation. If the partition is not indecomposable, then following (2.3.5), the $\psi(r_{ij}) - \psi(r_{i'j'})$; $(i,j), (i',j') \in P_m$; $m = m_1, \ldots, m_N$ generate only the values $t_i - t_{i'}$; $i, i' = i_1, \ldots, i_0$. There is no way the values $t_i - t_{i'}$, $i = i_1, \ldots, i_0$; $i' \neq i_1, \ldots, i_0$ can be generated.

On the other hand, if the $\psi(r_{ij}) - \psi(r_{i'j'})$ generate all the $t_i - t_{i'}$, then there must be some sequence of the P_m beginning with an i and ending with an i' and so all sets communicate.

Proof of Theorem 2.3.2 We will proceed by induction on j. From Theorem 2.3.1 we see that

$$EY_1 \cdots Y_j = \sum_{\mu} D_{\mu_1} \cdots D_{\mu_p} \qquad (*)$$

where $D_\mu = \text{cum}(Y_{\alpha_1}, \ldots, Y_{\alpha_m})$ when $\mu = (\alpha_1, \ldots, \alpha_m)$ and the sum extends over all partitions $(\mu_1, \ldots, \mu_p)$ of $(1, \ldots, j)$. Also

$$EY_1 \cdots Y_j = E \prod_{m=1}^{j} \prod_{n=1}^{k_m} X_{mn} = \sum_\nu C_{\nu_1} \cdots C_{\nu_p} \qquad (**)$$

where $C_\nu = \text{cum}(X_{a_1}, \ldots, X_{a_m})$ when $\nu = (a_1, \ldots, a_m)$, (the a's being pairs of integers) and the sum extends over all partitions of the set $\{(m,n) \mid m = 1, \ldots, j \text{ and } n = 1, \ldots, k_m\}$.

From (*) and (**) we see

$$\text{cum}(Y_1, \ldots, Y_j) = \sum_\nu C_{\nu_1} \cdots C_{\nu_p} - \sum_\mu{}' D_{\mu_1} \cdots D_{\mu_p}$$

with $\sum_\mu{}'$ extending over partitions with $p \geqslant 2$. We see that the terms coming from decomposable partitions are subtracted out in the above expression yielding the stated result.

Proof of Theorem 2.5.1 We will write $\mathbf{A} \geqslant \mathbf{0}$ to mean that the matrix $\mathbf{A}$ is non-negative definite. The fact that $\mathbf{f}_{XX}(\lambda)$ is Hermitian follows from (2.5.7) and $\mathbf{c}_{XX}(u)^\tau = \mathbf{c}_{XX}(-u)$.

Next suppose $E\mathbf{X}(t) = \mathbf{0}$ as we may. Consider

$$\mathbf{I}_{XX}^{(T)}(\lambda) = (2\pi T)^{-1} \left[\sum_{t=0}^{T-1} \mathbf{X}(t) \exp\{-i\lambda t\}\right] \overline{\left[\sum_{t=0}^{T-1} \mathbf{X}(t) \exp\{-i\lambda t\}\right]}^\tau.$$

By construction $\mathbf{I}_{XX}^{(T)}(\lambda) \geqslant \mathbf{0}$ and so therefore $E\mathbf{I}_{XX}^{(T)}(\lambda) \geqslant \mathbf{0}$. We have

$$E\mathbf{I}_{XX}^{(T)}(\lambda) = (2\pi)^{-1} \sum_{u=-T+1}^{T-1} \left(1 - \frac{|u|}{T}\right) \exp\{-i\lambda u\} \mathbf{c}_{XX}(u),$$

that is, it is a Cesàro mean of the series for $\mathbf{f}_{XX}(\lambda)$. By assumption this series is convergent, and so Exercise 1.7.10 implies

$$\lim_{T \to \infty} E\mathbf{I}_{XX}^{(T)}(\lambda) = \mathbf{f}_{XX}(\lambda)$$

and the latter must therefore be $\geqslant \mathbf{0}$.

Proof of Theorem 2.5.2 Suppose $E\mathbf{X}(t) = \mathbf{0}$. Set

$$\mathbf{I}_{XX}^{(T)}(\lambda) = (2\pi T)^{-1} \left[\sum_{t=0}^{T-1} \mathbf{X}(t) \exp\{-i\lambda t\}\right] \overline{\left[\sum_{t=0}^{T-1} \mathbf{X}(t) \exp\{-i\lambda t\}\right]}^\tau$$

for $-\pi \leqslant \lambda < \pi$. By construction $\mathbf{I}_{XX}^{(T)}(\lambda) \geqslant \mathbf{0}$. Next set

$$\mathbf{J}_{XX}^{(T)}(\lambda) = E\mathbf{I}_{XX}^{(T)}(\lambda) = (2\pi)^{-1} \sum_{u=-T+1}^{T-1} \left(1 - \frac{|u|}{T}\right) \exp\{-i\lambda u\} \mathbf{c}_{XX}(u). \qquad (*)$$

It will also be $\geqslant 0$. From (*)

$$\left(1 - \frac{|u|}{T}\right) \mathbf{c}_{XX}(u) = \int_{-\pi}^{\pi} \exp\{i\lambda u\} \mathbf{J}_{XX}^{(T)}(\lambda) d\lambda \qquad (**)$$

for $u = 0, \pm 1, \ldots, \pm T$. Now the sequence of matrix-valued measures

$$\int_{-\pi}^{\lambda} \mathbf{J}_{XX}^{(T)}(\alpha)d\alpha \qquad -\pi \leqslant \lambda < \pi,$$

lies between $\mathbf{0}$ and $\mathbf{c}_{XX}(0)$. By an extension of Helly's selection theorem this sequence will contain a subsequence converging weakly to a matrix-valued measure at all continuity points of the limit. Suppose the limit of such a convergent subsequence, $\mathbf{J}_{XX}^{(T')}(\lambda)$, is $\mathbf{F}_{XX}(\lambda)$. By approximating the integrals involved by finite sums we can see that

$$\int_{-\pi}^{\pi} \exp\{i\lambda u\}\ \mathbf{J}_{XX}^{(T)}(\lambda)d\lambda \rightarrow \int_{-\pi}^{\pi} \exp\{i\lambda u\}\ d\mathbf{F}_{XX}(\lambda).$$

In addition from (**) it tends to $\mathbf{c}_{XX}(u)$. This gives (2.5.8).

Expression (2.5.9) follows from the usual inversion formula for Fourier-Stieltjes transforms. The increments of $\mathbf{F}_{XX}(\lambda)$ are $\geqslant \mathbf{0}$ by construction.

Proof of Lemma 2.7.1 We have

$$\begin{aligned}\mathcal{A}[T^u\mathbf{e}](t) &= \exp\{i\lambda u\}\ \mathcal{A}[\mathbf{e}](t)\\ &= \mathcal{A}[\mathbf{e}](t+u)\end{aligned}$$

using the properties of linearity and time invariance. Setting $t = 0$ it follows that

$$\mathcal{A}[\mathbf{e}](u) = \exp\{i\lambda u\}\ \mathcal{A}[\mathbf{e}](0)$$

and we have (2.7.6) with $\mathbf{A}(\lambda) = \mathcal{A}[\mathbf{e}](0)$.

Proof of Lemma 2.7.2 The properties indicated are standard results concerning the Fourier transforms of absolutely summable sequences; see Zygmund (1968), for example.

Proof of Lemma 2.7.3 We note that

$$\sum_u |\mathbf{a}(t-u)|\ E\,|\mathbf{X}(u)| \leqslant \sum_u |\mathbf{a}(u)|\ E\,|\mathbf{X}(0)| < \infty.$$

It follows that

$$E\,\Big|\sum_u \mathbf{a}(t-u)\mathbf{X}(u)\Big| < \infty$$

and so $\sum_u \mathbf{a}(t-u)\mathbf{X}(u)$ is finite with probability 1. The stationarity of $\mathbf{Y}(t)$ follows from the fact that the operation is time invariant.

Continuing, we have

$$\begin{aligned} E\,|Y_b(t)|^k &\leqslant \sum_{j_1}\cdots\sum_{j_k}\sum_{u_1}\cdots\sum_{u_k} |a_{bj_1}(t-u_1)\cdots a_{bj_k}(t-u_k)| \\ &\qquad \times E\,|X_{j_1}(u_1)\cdots X_{j_k}(u_k)| \\ &\leqslant \sup_j E\,|X_j(0)|^k \sum_{j_1}\cdots\sum_{j_k}\sum_{u_1}\cdots\sum_{u_k} |a_{bj_1}(u_1)\cdots a_{bj_k}(u_k)| < \infty, \end{aligned}$$

completing the proof.

Proof of Theorem 2.7.1 Set

$$\mathbf{Y}_T(t) = \sum_{u=-T}^{T} \mathbf{a}(u)\mathbf{X}(t-u)$$

for $T = 1, 2, \ldots$. Write $\sum'_u$ for a sum over terms with $T' < |u| \leqslant T$. Then for $T' < T$,

$$E[\mathbf{Y}_T(t) - \mathbf{Y}_{T'}(t)][\mathbf{Y}_T(t) - \mathbf{Y}_{T'}(t)]^\tau = \int_{-\pi}^{\pi} [\textstyle\sum' \mathbf{a}(u) \exp\{-i\lambda u\}]\mathbf{f}_{XX}(\lambda)$$
$$\times [\textstyle\sum' \mathbf{a}(u) \exp\{i\lambda u\}]d\lambda \leqslant K \int [\sum' \mathbf{a}(u) \exp\{-i\lambda u\}][\sum' \mathbf{a}(u) \exp\{-i\lambda u\}]d\lambda$$

for some finite K. The latter tends to 0 as $T, T' \to \infty$ in view of (2.7.23). The sequence $\mathbf{Y}_T(t)$, $T = 1, 2, \ldots$ is therefore Cauchy and so the limit (2.7.25) exists by completeness.

Proof of Lemma 2.7.4 Set

$$Y(t) = \exp\{i\lambda_0 t\}\, X(t)$$
$$W(t) = \sum_u a(t-u)Y(u)$$
$$V(t) = \exp\{-i\lambda_0 t\}\, W(t)$$
$$= \sum_u \exp\{-i\lambda_0(t-u)\}a(t-u)X(t),$$

showing that the operations are indeed linear and time invariant. Now

$$V_1(t) = \operatorname{Re} V(t)$$
$$V_2(t) = -\operatorname{Im} V(t)$$

from which (2.7.34) and (2.7.35) follow.

Proof of Theorem 2.8.1 The series $\mathbf{X}(t)$ is r vector-valued, strictly stationary, with cumulant spectra $f_{a_1,\ldots,a_k}(\lambda_1, \ldots, \lambda_k)$. $\mathbf{Y}(t) = \sum_{u=-\infty}^{\infty} \mathbf{a}(t-u)$ $\mathbf{X}(u)$; where the $\mathbf{a}(u)$ are the coefficients of an $s \times r$ filter and $\sum_{u=-\infty}^{\infty}$ $|a_{ij}(u)| < \infty$, $i = 1, \ldots, s$, $j = 1, \ldots, r$. From Lemma 2.7.3, $\mathbf{Y}(t)$ is also strictly stationary and its cumulant functions exist. Indicate these by $d_{b_1,\ldots,b_k}(v_1, \ldots, v_{k-1})$. We have

$$d_{b_1,\ldots,b_k}(v_1, \ldots, v_{k-1})$$
$$= \operatorname{cum}\{Y_{b_1}(v_1), \ldots, Y_{b_{k-1}}(v_{k-1}), Y_{b_k}(0)\}$$
$$= \operatorname{cum}\left\{\sum_{j_1=1}^{r} \sum_{u_1=-\infty}^{\infty} a_{b_1 j_1}(v_1 - u_1)X_{j_1}(u_1), \ldots,\right.$$
$$\sum_{j_{k-1}=1}^{r} \sum_{u_{k-1}=-\infty}^{\infty} a_{b_{k-1}j_{k-1}}(v_{k-1} - u_{k-1})X_{j_{k-1}}(a_{k-1}),$$
$$\left.\sum_{j_k=1}^{r} \sum_{u_k=-\infty}^{\infty} a_{b_k j_k}(-u_k)X_{j_k}(u_k)\right\} \quad (*)$$

$$= \sum_{j_1=1}^{r} \cdots \sum_{j_k=1}^{r} \sum_{u_1=-\infty}^{\infty} \cdots \sum_{u_k=-\infty}^{\infty} a_{b_1 j_1}(v_1 - u_1) \cdots$$
$$a_{b_{k-1} j_{k-1}}(v_{k-1} - u_{k-1}) a_{b_k j_k}(-u_k) c_{j_1 \ldots j_k}(u_1, \ldots, u_k)$$
$$= \sum_{j_1=1}^{r} \cdots \sum_{j_k=1}^{r} \sum_{u_1=-\infty}^{\infty} \cdots \sum_{u_{k-1}=-\infty}^{\infty} c_{j_1,\ldots,j_k}(u_1, \ldots, u_{k-1})$$
$$a_{j_1 \ldots j_k}(v_1 - u_1, \ldots, v_{k-1} - u_{k-1}) \quad (**)$$

where

$$a_{j_1,\ldots,j_k}(v_1 - u_1, \ldots, v_{k-1} - u_{k-1})$$
$$= \sum_{u=-\infty}^{\infty} a_{b_1 j_1}(v_1 - u_1 - u) \cdots a_{b_{k-1} j_{k-1}}(v_{k-1} - u_{k-1} - u) a_{b_k j_k}(-u).$$

In view of the absolute summability of the $a_{b_j}(u)$, $a_{j_1,\ldots,j_k}(u_1, \ldots, u_{k-1})$ is absolutely summable by Fubini's theorem. Therefore

$$\sum_{u_1} \cdots \sum_{u_{k-1}} |a_{j_1,\ldots,j_k}(v_1 - u_1, \ldots, v_{k-1} - u_{k-1}) c_{j_1,\ldots,j_k}(u_1, \ldots, u_{k-1})| < \infty$$

and the interchange of averaging and summation needed in going from (*) to (**) above is justified. Now, $d_{b_1,\ldots,b_k}(v_1, \ldots, v_{k-1})$ is a sum of convolutions of absolutely summable functions and is, therefore, absolutely summable. We see that $\mathbf{Y}(t)$ satisfies Assumption 2.6.1. Expression (2.8.1) follows on taking the Fourier transform of the cumulant function of $\mathbf{Y}(t)$ and noting that it is a sum of convolutions.

Proof of Lemma 2.9.1 If $X(t)$ is Markov and Gaussian, then $X(s + t) - E\{X(s + t) \mid X(s)\}$, $t > 0$, $s \geq 0$, is independent of $X(0)$. Therefore, $\operatorname{cov}\{X(0), [X(s + t) - E\{X(s + t) \mid X(s)\}]\} = 0$ and since $E\{X(s + t) \mid X(s)\} = K + X(s)c_{XX}(t)/c_{XX}(0)$, K a constant, we have

$$c_{XX}(s + t) = \frac{c_{XX}(t)c_{XX}(s)}{c_{XX}(0)} \qquad \text{for } s, t > 0.$$

From this we see that

$$c_{XX}(t) = c_{XX}(0)[c_{XX}(1)/c_{XX}(0)]^t \qquad \text{for } t > 0.$$

The proof is completed by noting that $c_{XX}(t) = c_{XX}(-t)$.

Proof of Theorem 2.9.1 We begin by noting that under the stated assumptions $Y(t)$ exists, with probability 1, since

$$\sum_{u_1} |a_1(t - u_1)E \mid X(u_1)| + \sum_{u_1} \sum_{u_2} |a_2(t - u_1, t - u_2)| E|X(u_1)X(u_2)| + \cdots$$
$$< E|X(t)| \sum_{u_1} |a_1(u_1)| + E|X(t)|^2 \sum_{u_1,u_2} |a_2(u_1,u_2)| + \cdots < \infty.$$

Consider

$$\operatorname{cum}\{Y(t_1),\ldots, Y(t_I)\} = \sum_{J_1=0}^{L} \cdots \sum_{J_I=0}^{L} \sum_{u_{ij}} a_{J_1}(u_{11},\ldots, u_{1J_1}) \cdots a_{J_I}(u_{I1},\ldots, u_{IJ_I})$$
$$\times \operatorname{cum}\{X(t_1 - u_{11}) \cdots X(t_1 - u_{1J_1}), \ldots, X(t_I - u_{I1}) \cdots X(t_I - u_{IJ_I})\}.$$

The cumulant involving the X's is, from Theorem 2.3.2, the sum over indecomposable partitions of products of joint cumulants in the $X(t_i - u_{ij})$, say

$$\sum_{M \geq 1} c(t_i - u_{ij};(i,j) \in P_1) \cdots c(t_i - u_{ij};(i,j) \in P_M).$$

Because the series is stationary, the cumulants will be functions of the differences $t_i - t_{i'} - u_{ij} + u_{i'j'}$. Following Lemma 2.3.1, $I - 1$ of the differences $t_i - t_{i'}$ will be independent. Suppose that these are $t_1 - t_I, \ldots, t_{I-1} - t_I$.

Setting $t_I = 0$ we now see that

$$\sum_{t_1} \cdots \sum_{t_{I-1}} |\mathrm{cum}\, \{Y(t_1), \ldots, Y(t_{I-1}), Y(0)\}|$$
$$\leqslant \sum_{t_1} \cdots \sum_{t_{I-1}} \sum_{J_1} \cdots \sum_{J_I} \sum_{u_{ij}} |a_{J_1}(u_{11}, \ldots, u_{1J_1}) \cdots a_{J_1}(u_{I1}, \ldots, u_{IJ_I})|$$
$$\times \sum_{M} g(t_1 - u_{1j} + u_{Ij'}, \ldots, t_{I-1} - u_{I-1j''} + u_{Ij'''}),$$

where g is absolutely summable as a function of its arguments. Making the change of variables

$$s_1 = t_1 - u_{1j} + u_{Ij'}$$
$$\vdots$$

we see that the cumulants of the series $Y(t)$ are absolutely summable.

Proof of Theorem 2.10.1 We anticipate our development somewhat in this proof. In Section 4.6 we will see that we can write

$$X(t) = \int_{-\pi}^{\pi} \exp\, \{i\lambda t\} dZ_X(\lambda)$$
$$Y(t) = \int_{-\pi}^{\pi} \exp\, \{i\lambda t\} dZ_Y(\lambda)$$

where

$$\mathrm{cum}\, \{dZ_X(\lambda_1), \ldots, dZ_X(\lambda_k)\}$$
$$= \eta(\lambda_1 + \cdots + \lambda_k) f_{X \ldots X}(\lambda_1, \ldots, \lambda_k) d\lambda_1 \cdots d\lambda_k$$
$$\mathrm{cum}\, \{dZ_Y(\lambda_1), \ldots, dZ_Y(\lambda_k)\}$$
$$= \eta(\lambda_1 + \cdots + \lambda_k) f_{Y \ldots Y}(\lambda_1, \ldots, \lambda_k) d\lambda_1 \cdots d\lambda_k.$$

Substituting into (2.9.15) shows that

$$dZ_Y(\lambda) = \sum_J \int \cdots \int \eta(\alpha_1 + \cdots + \alpha_J - \lambda) A_J(\alpha_1, \ldots, \alpha_J) dZ_X(\alpha_1) \cdots dZ_X(\alpha_J).$$

Now using Theorem 2.3.2 and the expressions set down above gives the desired expression (2.10.10).

PROOFS FOR CHAPTER 3

Proof of Theorem 3.3.1 We quickly see the first relation of (3.3.18). It may be rewritten as

$$\int_{-\infty}^{\infty} H(\alpha)A\left(\lambda - \frac{\alpha}{n}\right)d\alpha = \int_{-\infty}^{\infty} H(\alpha)\left[\sum_{p=0}^{P-1} \frac{\alpha^p}{p!n^p}\frac{d^pA(\lambda)}{d\lambda^p} + O(n^{-P})|\alpha|^P\right]d\alpha$$

from which the second part of (3.3.18) follows.

Proof of Lemma 3.4.1 We have

$$\begin{aligned}\mathbf{d}_Y^{(T)}(\lambda) &= \sum_{u=-\infty}^{\infty} \mathbf{a}(u)\left[\sum_{t=0}^{T-1} \mathbf{X}(t-u)\exp\{-i\lambda t\}\right]\\ &= \sum_{u=-\infty}^{0} \mathbf{a}(u)\exp\{-i\lambda u\}\left[\sum_{v=0}^{T-1} - \sum_{v=0}^{-u-1} + \sum_{v=T}^{T-1-u}\right][\mathbf{X}(v)\exp\{-i\lambda v\}]\\ &\quad + \sum_{u=1}^{\infty} \mathbf{a}(u)\exp\{-i\lambda u\}\left[\sum_{v=0}^{T-1} + \sum_{v=-u}^{-1} - \sum_{v=T-u}^{T-1}\right][\mathbf{X}(v)\exp\{-i\lambda v\}]\\ &= \mathbf{A}(\lambda)\mathbf{d}_X^{(T)}(\lambda) + \mathbf{e}^{(T)}(\lambda)\end{aligned}$$

where $|\mathbf{e}^{(T)}(\lambda)| < L\sum_u |\mathbf{a}(u)|\cdot|u|$ for some finite L because the components of $\mathbf{X}(t)$ are bounded.

Proof of Theorem 3.5.1 The theorem follows directly from the substitutions $j = j_1T_2 + j_2$, $t = t_1 + t_2T_1$ and the fact that $\exp\{-i2\pi k\} = 1$ for k an integer.

Proof of Theorem 3.5.2 See the proof of Theorem 3.5.3.

Proof of Theorem 3.5.3 We first note that the integers

$$\frac{t_1T}{T_1} + \cdots + \frac{t_kT}{T_k} \qquad 0 \leqslant t_j \leqslant T_j - 1, \tag{*}$$

when reduced mod T run through integers t, $0 \leqslant t \leqslant T - 1$. We see that there are $T_1 \cdots T_k = T$ possible values for (*), each of which is an integer. Suppose that two of these, when reduced mod T, are equal, that is

$$\frac{t_1T}{T_1} + \cdots + \frac{t_kT}{T_k} = \frac{\tau_1T}{T_1} + \cdots + \frac{\tau_kT}{T_k} + lT$$

for some integer l. This means

$$\frac{(t_1 - \tau_1)T}{T_1} = \frac{(\tau_2 - t_2)T}{T_2} + \cdots + \frac{(\tau_k - t_k)T}{T_k} + lT.$$

The left side of this equation is not divisible by T_1, whereas the right side is. We have a contradiction and so the values (*) are identical with the integers $t = 0, \ldots, T - 1$. The theorem now follows on substituting

$$t = \frac{t_1T}{T_1} + \cdots + \frac{t_kT}{T_k}$$

and reducing mod T.

Proof of Lemma 3.6.1 See the proof of Lemma 3.6.2.

Proof of Lemma 3.6.2 If we make the substitution

$$d_j^{(T)}(\lambda) = \sum_{t=0}^{T-1} X_j(t) \exp\{-i\lambda t\},$$

then expression (3.6.11) becomes

$$\sum_{t_1} \cdots \sum_{t_r} X_1(t_1) \cdots X_r(t_r) \eta\left\{\frac{2\pi(t_1 - t_r + u_1)}{S}\right\} \cdots \eta\left\{\frac{2\pi(t_{r-1} - t_r + u_{r-1})}{S}\right\}.$$

This last gives (3.6.5) in the case $r = 2$. It is also seen to give (3.6.10) in the case $|u_j| \leqslant S - T$.

Proof of Lemma 3.7.1 The results of this lemma follow directly once the correspondence (3.7.7) has been set up.

Proof of Theorem 3.7.1 See Bellman (1960).

Proof of Theorem 3.7.2 The matrix $\overline{\mathbf{Z}}^\tau\mathbf{Z}$ is non-negative definite and Hermitian. It therefore has latent values μ_j^2 for some $\mu_j \geqslant 0$. Take $\mathbf{V}$ to be the associated matrix of latent vectors satisfying $\overline{\mathbf{Z}}^\tau\mathbf{Z}\mathbf{V} = \mathbf{V}\mathbf{D}$ where $\mathbf{D} = \text{diag}\{\mu_j^2\}$. Let $\mathbf{M} = \text{diag}\{\mu_j\}$ be $s \times r$. Take $\mathbf{U}$ such that $\mathbf{U}\mathbf{M} = \mathbf{Z}\mathbf{V}$. We see that $\mathbf{U}$ is unitary and composed of the latent vectors of $\mathbf{Z}\,\overline{\mathbf{Z}}^\tau$. The proof is now complete.

Proof of Theorem 3.7.3 Let μ_k denote expression (3.7.11) and $\boldsymbol{\alpha}_k$ denote expression (3.7.12). We quickly check that $\mathbf{Z}\boldsymbol{\alpha}_k = \mu_k\boldsymbol{\alpha}_k$.

Proof of Theorem 3.7.4 Set $\mathbf{B} = \mathbf{Z} - \mathbf{A}$. By the Courant-Fischer theorem (see Bellman (1960) and Exercise 3.10.16),

$$\mu_j(\mathbf{B}\overline{\mathbf{B}}^\tau) = \inf_{\mathbf{D}} \sup_{\mathbf{D}\mathbf{x}=0} \frac{\overline{\mathbf{x}}^\tau\mathbf{B}\overline{\mathbf{B}}^\tau\mathbf{x}}{\overline{\mathbf{x}}^\tau\mathbf{x}}$$

where $\mathbf{D}$ is any $(j-1) \times J$ matrix and $\mathbf{x}$ is any J vector. Therefore

$$\begin{aligned}\mu_j(\mathbf{B}\overline{\mathbf{B}}^\tau) &\geqslant \sup_{\substack{\mathbf{D}\mathbf{x}=0\\ \overline{\mathbf{A}}^\tau\mathbf{x}=0}} \frac{\overline{\mathbf{x}}^\tau\mathbf{B}\overline{\mathbf{B}}^\tau\mathbf{x}}{\overline{\mathbf{x}}^\tau\mathbf{x}}\\ &\geqslant \sup_{\left[\begin{smallmatrix}\mathbf{D}\\ \overline{\mathbf{A}}^\tau\end{smallmatrix}\right]\mathbf{x}=0} \frac{\overline{\mathbf{x}}^\tau\mathbf{Z}\overline{\mathbf{Z}}^\tau\mathbf{x}}{\overline{\mathbf{x}}^\tau\mathbf{x}}\\ &\geqslant \mu_{j+L}(\mathbf{Z}\overline{\mathbf{Z}}^\tau)\end{aligned}$$

because the matrix

$$\begin{bmatrix}\mathbf{D}\\ \overline{\mathbf{A}}^\tau\end{bmatrix}$$

has rank at most $j + L - 1$. By inspection we see that this minimum is achieved by the matrix $\mathbf{A}$ of (3.7.19) completing the proof.

Proof of Theorem 3.8.1 See Bochner and Martin (1948) p. 39.

Proof of Theorem 3.8.2 The space $V_+(l)$ is a commutative normed ring; see Gelfand et al (1964). The space $\mathfrak{M}$ of maximal ideals of this ring is homomorphic with the strip $-\pi < \operatorname{Re} \lambda \leqslant \pi, \operatorname{Im} \lambda \geqslant 0$, in such a way that if $M \in \mathfrak{M}$ and λ are corresponding elements then

$$x(M) = \sum_{u=0}^{\infty} a(u) \exp \{-i\lambda u\}.$$

These are the functions of $V_+(l)$. The stated result now follows from Theorem 1, Gelfand et al (1964) p. 82. The result may be proved directly also.

Proof of Theorem 3.8.3 The space $V(l)$ is a commutative normed ring. Its space of maximum ideals is homomorphic with the interval $(-\pi,\pi]$ through the correspondence

$$x(M) = \sum_{u} a(u) \exp \{-i\lambda u\}$$

for M in the space of maximal ideals. The $x(M)$ are the elements of $V(l)$. The theorem now follows from Theorem 1, Gelfand et al (1964) p. 82.

Proof of Theorem 3.9.1 Consider the space consisting of finite linear combinations of $X_j(t + s)$, $j = 1, \ldots, r$ and $t = 0, \pm 1, \ldots$. An inner product may be introduced into this space by the definition

$$\langle Y_1,Y_2\rangle = \lim_{S\to\infty} (2S + 1)^{-1} \sum_{s=-S}^{S} Y_1(s)Y_2(s).$$

The space is then a pre-Hilbert space. It may be completed to obtain a Hilbert space H. There is a unitary operator, $\mathfrak{U}$, on H such that

$$\mathfrak{U}^t X_j(s) = X_j(t + s) \qquad j = 1, \ldots, r.$$

Following Stone's theorem (see Riesz and Nagy (1955)) the operator $\mathfrak{U}$ has a spectral representation

$$\mathfrak{U} = \int_{-\pi}^{\pi} \exp \{i\lambda\} dE(\lambda)$$

where $E(\lambda)$ is a spectral family of projection operators on H. This family has the properties; $E(\lambda)E(\mu) = E(\mu)E(\lambda) = E(\min\{\lambda, \mu\})$, $E(-\pi) = 0$, $E(\pi) = I$ and $E(\lambda)$ is continuous from the right. Also for $Y_1(t)$, $Y_2(t)$ in H, $\langle E(\lambda)Y_1,Y_2\rangle$ is of bounded variation and

$$\langle \mathfrak{U}Y_1,Y_2\rangle = \int_{-\pi}^{\pi} \exp \{i\lambda\} d\langle E(\lambda)Y_1,Y_2\rangle. \qquad (*)$$

If we define $Z_j(\lambda;s) = E(\lambda)X_j(s)$ we see that

$$\begin{aligned} X_j(t + s) &= \mathfrak{U}^t X_j(s) \\ &= \int_{-\pi}^{\pi} \exp \{i\lambda t\} dZ_j(\lambda;s) \end{aligned}$$

in the sense of (3.9.6). Also from (*) above

$$m_{jk}(u) = \int_{-\pi}^{\pi} \exp\{i\lambda u\} d\langle E(\lambda)X_j(s), X_k(s)\rangle$$
$$= \int_{-\pi}^{\pi} \exp\{i\lambda u\} d\langle Z_j(\lambda;s), Z_k(\lambda;s)\rangle.$$

Bochner's theorem now indicates that $G_{jk}(\lambda)$ defined by (3.9.4) is given by $\langle Z_j(\lambda;s), Z_k(\lambda;s)\rangle$. The remaining parts of the theorem follow from the properties of $E(\lambda)$.

Proof of Theorem 3.9.2 Set

$$\mathbf{Z}_X^{(T)}(\lambda) = \frac{\lambda}{2\pi}\mathbf{X}(0) + \frac{1}{2\pi}\sum_{1\leq|t|\leq T} \mathbf{X}(t)\exp\{-i\lambda t\}/(-it) \qquad -\pi < \lambda \leqslant \pi.$$

In view of (3.9.11), there exists $\mathbf{Z}_X(\lambda)$ such that

$$\int_{-\pi}^{\pi} \|\mathbf{Z}_X(\lambda)\|^2 d\lambda < \infty$$

and

$$\lim_{T\to\infty} \int_{-\pi}^{\pi} \|\mathbf{Z}_X(\lambda) - \mathbf{Z}_X^{(T)}(\lambda)\|^2 d\lambda = 0.$$

Now take an equivalent version of $\mathbf{Z}(\lambda)$ with the property $\mathbf{Z}_X(\pi) - \mathbf{Z}_X(-\pi) = \mathbf{X}(0)$. We see that

$$\int_{-\pi}^{\pi} \exp\{i\lambda t\}\mathbf{Z}_X(\lambda)d\lambda = 0 \qquad \text{if } t = 0$$
$$= \frac{\exp\{i\pi t\}}{it}\mathbf{X}(0) - \frac{1}{it}\mathbf{X}(t) \qquad \text{if } t \neq 0.$$

This gives (3.9.13). Turning to (3.9.15), we have

$$\lim_{\varepsilon\to 0}\frac{1}{2\varepsilon}\int_{-\pi}^{\pi} \exp\{iu\alpha\}[\mathbf{Z}_X(\alpha+\varepsilon) - \mathbf{Z}_X(\alpha-\varepsilon)]\overline{[\mathbf{Z}_X(\alpha+\varepsilon) - \mathbf{Z}_X(\alpha-\varepsilon)]}^{\tau} d\alpha$$
$$= \lim_{\varepsilon\to 0}\frac{1}{2\varepsilon}(2\pi)^{-2}\int_{-\pi}^{\pi} \exp\{iu\alpha\}\left[2\varepsilon\mathbf{X}(0) + \sum_s \mathbf{X}(s)\exp\{-is\alpha\}\frac{2\sin s\varepsilon}{s}\right]$$
$$\times\left[2\varepsilon\mathbf{X}(0) + \sum_t \mathbf{X}(t)\exp\{it\alpha\}\frac{2\sin t\varepsilon}{t}\right]^{\tau} d\alpha$$
$$= \lim_{\varepsilon\to 0}(2\varepsilon)^{-1}(2\pi)^{-1}\sum_t \mathbf{X}(u+t)\mathbf{X}(t)^{\tau}\frac{2\sin(u+t)\varepsilon}{u+t}\frac{2\sin t\varepsilon}{t}$$
$$= \mathbf{m}_{XX}(u) \qquad \text{for } u = 0, \pm 1, \ldots$$

where $\mathbf{m}_{XX}(u)$ is given by (3.9.3). Now because

$$\mathbf{m}_{XX}(u) = \int_{-\pi}^{\pi} \exp\{iu\alpha\} d\mathbf{G}_{XX}(\alpha)$$

the uniqueness theorem for Fourier-Stieltjes transforms gives (3.9.15).

PROOFS FOR CHAPTER 4

Before proving Theorems 4.3.1 and 4.3.2 we first set down some lemmas.

Lemma P4.1 If $h_a(u)$ satisfies Assumption 4.3.1 and if $h_a^{(T)}(t) = h_a(t/T)$ for $a = 1, \ldots, r$, then

$$\left|\sum_t h_{a_1}^{(T)}(t+u_1)\cdots h_{a_{k-1}}^{(T)}(t+u_{k-1})h_{a_k}^{(T)}(t)\exp\{-i\lambda t\} - H_{a_1\ldots a_k}^{(T)}(\lambda)\right|$$
$$\leqslant K(|u_1| + \cdots + |u_{k-1}|)$$

for some finite K.

Proof The expression in question is

$$\leqslant \sum_t |h_{a_1}^{(T)}(t+u_1)\cdots h_{a_{k-1}}^{(T)}(t+u_{k-1}) - h_{a_1}^{(T)}(t)\cdots h_{a_{k-1}}^{(T)}(t)|\,|h_{a_k}^{(T)}(t)|$$
$$\leqslant L\sum_{a=1}^{k-1}\sum_t |h_a^{(T)}(t+u_j) - h_a^{(T)}(t)|$$

for some finite L. Suppose for convenience $u_a > 0$. (The other cases are handled similarly.) The expression is now

$$\leqslant M\sum_{a=1}^{k-1}\sum_t\sum_{v=0}^{u_a-1}|h_a^{(T)}(t+v+1) - h_a^{(T)}(t+v)|$$
$$\leqslant M\sum_{a=1}^{k-1}\sum_v |\text{variation of } h_a|$$
$$\leqslant K(|u_1| + \cdots + |u_{k-1}|)$$

as desired.

Lemma P4.2 The cumulant of interest in Theorems 4.3.1 and 4.3.2 is given by

$$(2\pi)^{k-1}H_{a_1\ldots a_k}^{(T)}(\lambda_1 + \cdots + \lambda_k)\sum_{-S}^{S}\cdots\sum_{-S}^{S}\exp\{-i(\lambda_1 u_1 + \cdots + \lambda_{k-1}u_{k-1})\}$$
$$\times\, c_{a_1\ldots a_1}(u_1, \ldots, u_{k-1}) + \varepsilon_T$$

where $S = 2(T-1)$ and

$$|\varepsilon_T| \leqslant K\sum_{-S}^{S}\cdots\sum_{-S}^{S}(|u_1| + \cdots + |u_{k-1}|)|c_{a_1\ldots a_k}(u_1, \ldots, u_{k-1})|$$

for some finite K.

Proof The cumulant has the form

$$\sum_{t_1}\cdots\sum_{t_k} h_{a_1}^{(T)}(t_1)\cdots h_{a_k}^{(T)}(t_k)\exp\left\{-i\sum_1^k \lambda_j t_j\right\}c_{a_1\ldots a_k}(t_1 - t_k, \ldots, t_{k-1} - t_k)$$

$$= \sum_{u_j=-S}^{S} \cdots \sum \exp\left\{-i\sum_{1}^{k-1} \lambda_j u_j\right\} c_{a_1 \ldots a_k}(u_1, \ldots, u_{k-1}) \sum_t h_{a_1}{}^{(T)}(t + u_1)$$
$$\times \cdots h_{a_{k-1}}^{(T)}(t + u_{k-1}) h_{a_k}{}^{(T)}(t) \exp\left\{-i\sum_{1}^{k} \lambda_j t\right\}.$$

Using Lemma P4.1 this equals

$$\sum_{u_j=-S}^{S} \cdots \sum \exp\left\{-i\sum_{1}^{k-1} \lambda_j u_j\right\} c_{a_1 \ldots a_k}(u_1, \ldots, u_{k-1}) H_{a_1 \ldots a_k}^{(T)}(\lambda_1 + \cdots + \lambda_k) + \varepsilon_T$$

where ε_T has the indicated bound.

Lemma P4.3 Under the condition (4.3.6), $\varepsilon_T = o(T)$ as $T \to \infty$.

Proof

$$T^{-1}|\varepsilon_T| \leqslant K \sum_{-S}^{S} \cdots \sum T^{-1}(|u_1| + \cdots + |u_{k-1}|)|c_{a_1 \ldots a_k}(u_1, \ldots, u_{k-1})|.$$

Now $T^{-1}(|u_1| + \cdots + |u_{k-1}|) \to 0$ as $T \to \infty$. Because of (4.3.6) we may now use the dominated convergence theorem to see that $T^{-1}|\varepsilon_T| \to 0$ as $T \to \infty$.

Lemma P4.4 Under the condition (4.3.10), $\varepsilon_T = O(1)$.

Proof Immediate.

Proof of Theorem 4.3.1 Immediate from Lemmas P4.2, P4.3 and the fact that

$$f_{a_1 \ldots a_k}(\lambda_1, \ldots, \lambda_{k-1})$$
$$= (2\pi)^{k-1} \sum_{-S}^{S} \cdots \sum \exp\left\{-i\sum_{1}^{k-1} \lambda_j u_j\right\} c_{a_1 \ldots a_k}(u_1, \ldots, u_{k-1}) + o(1).$$

Proof of Theorem 4.3.2 Immediate from Lemmas P4.2, P4.4 and the fact that

$$f_{a_1 \ldots a_k}(\lambda_1, \ldots, \lambda_{k-1})$$
$$= (2\pi)^{k-1} \sum_{-S}^{S} \cdots \sum \exp\left\{-i\sum_{1}^{k-1} \lambda_j u_j\right\} c_{a_1 \ldots a_k}(u_1, \ldots, u_{k-1}) + O(T^{-1})$$

since (4.3.10) holds.

The following lemma will be needed in the course of the proof of Theorem 4.4.1.

Lemma P4.5 Let $\mathbf{Y}^{(T)}$, $T = 1, 2, \ldots$ be a sequence of r vector-valued random variables, with complex components, and such that all cumulants of the variate $[Y_1{}^{(T)}, \bar{Y}_1{}^{(T)}, \ldots, Y_r{}^{(T)}, \bar{Y}_r{}^{(T)}]$ exist and tend to the corresponding cumulants of a variate $[Y_1, \bar{Y}_1, \ldots, Y_r, \bar{Y}_r]$ that is determined by its moments. Then $\mathbf{Y}^{(T)}$ tends in distribution to a variate having components $Y_1, \ldots, Y_r$.

Proof All convergent subsequences of the sequence of cdf's of $\mathbf{Y}^{(T)}$ tend to cdf's with the given moments. By assumption there is only one cdf with these moments and we have the indicated result.

Proof of Theorem 4.4.1 We begin by noting that

$$
\begin{aligned}
Ed_a^{(T)}(\pm\lambda_j(T)) &= \Delta^{(T)}(\pm\lambda_j(T))EX_a(t) \\
&= 0 \qquad\qquad \text{if } \lambda_j(T) \not\equiv 0 \pmod{\pi} \\
&= TEX_a(t) \qquad \text{if } \lambda_j(T) \equiv 0 \pmod{2\pi} \\
&= 0 \text{ or } EX_a(t) \quad \text{if } \lambda_j(T) = \pm\pi, \pm 3\pi, \ldots
\end{aligned}
$$

We therefore see that the first cumulant of $\mathbf{d}_X^{(T)}(\lambda_j(T))$ behaves in the manner required by the theorem.

Next we note, from Theorem 4.3.1, that

$$
\begin{aligned}
&T^{-1} \operatorname{cov} \{d_a^{(T)}(\pm\lambda_j(T), d_b^{(T)}(\pm\lambda_k(T)\} \\
&\qquad = T^{-1}2\pi\Delta^{(T)}\left(\frac{2\pi[\pm s_j(T) \mp s_k(T)]}{T}\right) f_{ab}(\pm\lambda_j(T)) + o(1).
\end{aligned}
$$

The latter tends to 0 if $\lambda_j(T) \pm \lambda_k(T) \not\equiv 0 \pmod{2\pi}$. It tends to $2\pi f_{ab}(\pm\lambda_j)$ if $\pm\lambda_j(T) \equiv \pm\lambda_k(T) \pmod{2\pi}$. This indicates that the second-order cumulant behavior required by the theorem holds.

Finally, again from Theorem 4.3.1,

$$
\begin{aligned}
&T^{-k/2} \operatorname{cum} \{d_{a_1}^{(T)}(\pm\lambda_{j_1}(T)), \ldots, d_{a_k}^{(T)}(\pm\lambda_{j_k}(T))\} \\
&\quad = T^{-k/2}(2\pi)^{k-1}\Delta^{(T)}(\pm\lambda_{j_1}(T) \pm \cdots \pm \lambda_{j_k}(T)) f_{a_1\ldots a_k}(\pm\lambda_{j_1}(T), \ldots, \pm\lambda_{j_{k-1}}(T)) \\
&\qquad + o(T^{1-k/2})
\end{aligned}
$$

This last tends to 0 as $T \to \infty$ if $k > 2$ because $\Delta^{(T)}(\cdot)$ is $O(T)$.

Putting the above results together, we see that the cumulants of the variates at issue, and the conjugates of those variates, tend to the cumulants of a normal distribution. The conclusion of the theorem now follows from the lemma since the normal distribution is determined by its moments.

Before proving Theorem 4.4.2 we must state a lemma:

Lemma P4.6 Let $h_a(t)$ satisfy Assumption 4.3.1, $a = 1, \ldots, r$ and let $H_{a_1\ldots a_k}^{(T)}(\lambda)$ be given by (4.3.2). Then if $\lambda \not\equiv 0 \pmod{2\pi}$

$$
|H_{a_1\ldots a_k}^{(T)}(\lambda)| \leqslant \frac{K}{|\sin \lambda/2|}
$$

for some finite K.

Proof Suppose, for convenience, that $h(t)$ is nonzero only if $0 \leqslant t < T$. Using Exercise 1.7.13, we see that

$$
H_{a_1\ldots a_k}^{(T)}(\lambda) = -\sum_t \Delta^{(t)}(\lambda)\left[\prod_1^k h_{a_j}\left(\frac{t+1}{T}\right) - \prod_1^k h_{a_j}\left(\frac{t}{T}\right)\right].
$$

Now $|\Delta^{(t)}(\lambda)| \leqslant 1/|\sin \lambda/2|$, and so

$$|H_{a_1\ldots a_k}^{(T)}(\lambda)| \leqslant \frac{1}{|\sin \lambda/2|} \sum_t \left| \prod_1^k h_{a_j}\left(\frac{t+1}{T}\right) - \prod_1^k h_{a_j}\left(\frac{t}{T}\right) \right| \leqslant \frac{K}{|\sin \lambda/2|}$$

if we use the lemma required in the proof of Theorem 4.3.2.

Proof of Theorem 4.4.2 We proceed as in the proof of Theorem 4.4.1.

$$\begin{aligned} Ed_a^{(T)}(\pm\lambda_j) &= \sum_t h_a\left(\frac{t}{T}\right) \exp\{-i\lambda_j t\} c_a \\ &= 0(1) \qquad\qquad\qquad \text{if } \lambda_j \not\equiv 0 \pmod{2\pi} \\ &= TH_a(0)c_a + 0(1) \qquad \text{if } \lambda_j \equiv 0 \pmod{2\pi} \end{aligned}$$

using Lemma P4.6 and Lemma P4.1.

Next from Theorem 4.3.1,

$$T^{-1} \operatorname{cov}\{d_a^{(T)}(\pm\lambda_j), d_b^{(T)}(\pm\lambda_k)\} = T^{-1} 2\pi H_{ab}^{(T)}(\pm\lambda_j \mp \lambda_k) f_{ab}(\pm\lambda_j) + o(1).$$

This tends to 0 if $\lambda_j \pm \lambda_k \not\equiv 0 \pmod{2\pi}$ following Lemma P4.6. It tends to $2\pi\{\int h_a(t)h_b(t)dt\} f_{ab}(\pm\lambda_j) = 2\pi H_{ab}(0) f_{ab}(\pm\lambda_j)$ if $\pm\lambda_j \equiv \pm\lambda_k \pmod{2\pi}$.

Finally

$$\begin{aligned} &T^{-k/2} \operatorname{cum}\{d_{a_1}^{(T)}(\pm\lambda_{j_1}), \ldots, d_{a_k}^{(T)}(\pm\lambda_{j_k})\} \\ &\quad = T^{-k/2}(2\pi)^{k-1} H_{a_1\ldots a_k}^{(T)}(\pm\lambda_{j_1} \pm \cdots \pm \lambda_{j_k}) f_{a_1\ldots a_k}(\pm\lambda_{j_1}, \ldots, \pm\lambda_{j_{k-1}}) \\ &\qquad + o(T^{1-k/2}) \end{aligned}$$

This tends to 0 for $k > 2$ as $H_{a_1\ldots a_k}^{(T)}(\lambda) = O(T)$ and the proof of the theorem follows as before.

To prove Theorem 4.5.1, we proceed via a sequence of lemmas. Set

$$\begin{aligned} H_2 &= \int h(t)^2 dt, \\ \sigma_T &= \operatorname{var} \operatorname{Re} d_X^{(T)}(\lambda) = \tfrac{1}{4} \int |H^{(T)}(\lambda - \alpha) + H^{(T)}(-\lambda - \alpha)|^2 f_{XX}(\alpha) d\alpha. \end{aligned}$$

Lemma P4.7 Under the conditions of Theorem 4.5.1, for given λ, ε and α sufficiently small,

$$E \exp\{\alpha \operatorname{Re} d_X^{(T)}(\lambda)\} \leqslant \exp\{\alpha^2 \sigma_T (1 + \varepsilon)/2\}.$$

Proof From the first expression of the proof of Lemma P4.2, we see

$$|\operatorname{cum}\{d_X^{(T)}(\lambda_1), \ldots, d_X^{(T)}(\lambda_k)\}| \leqslant 2L^k T C_k \qquad (*)$$

where $L = \sup |h(u)|$ and C_k is given by (2.6.7). Therefore

$$|\log E \exp\{\alpha \operatorname{Re} d_X^{(T)}(\lambda)\} - \alpha^2 \sigma_T/2| \leqslant 2 \sum_3^\infty L^k T C_k |\alpha|^k / k!$$

The indicated expression now follows on taking α sufficiently small.

Corollary Under the conditions of Lemma P4.7.

$$E \exp \{\alpha \, |\mathrm{Re}\, d_X^{(T)}(\lambda)|\} \leqslant 2 \exp \{\alpha^2 \sigma_T (1 + \varepsilon)/2\}.$$

Lemma P4.8 Let $\lambda_r = 2\pi r/R$, $r = 0, \ldots, R - 1$ for some integer $R > 6\pi T$. Then

$$\sup_\lambda |\mathrm{Re}\, d_X^{(T)}(\lambda)| \leqslant \sup_r |\mathrm{Re}\, d_X^{(T)}(\lambda_r)|/(1 - 6\pi T R^{-1}).$$

Proof This follows immediately from Lemma 2.1 of Woodroofe and Van Ness (1967); see also Theorem 7.28, Zygmund (1968) Chap. 10.

Lemma P4.9 Under the conditions of Theorem 4.5.1

$$E \exp \{\alpha \sup_\lambda |\mathrm{Re}\, d_X^{(T)}(\lambda)|\}$$
$$\leqslant 2 \exp \{\log R + \alpha^2 2\pi T \textstyle\int h(u)^2 du\, (1 + \varepsilon) \sup_\lambda f_{XX}(\lambda)/[2(1 - 6\pi T R^{-1})^2].$$

Proof The indicated expected value is

$$\leqslant E \exp \{\alpha \sup_r |\mathrm{Re}\, d_X^{(T)}(\lambda_r)|/(1 - 6\pi T R^{-1})\}$$
$$\leqslant \sum_r E \exp \{\alpha \, |\mathrm{Re}\, d_X^{(T)}(\lambda_r)|/(1 - 6\pi T R^{-1})\}$$

giving the result because the sum runs over $R = \exp \{\log R\}$ points and

$$\sigma_T \leqslant \{\sup_\alpha f_{XX}(\alpha)\} \textstyle\int |H^{(T)}(\alpha)|^2 d\alpha = \{\sup f_{XX}(\alpha)\} 2\pi T \int h(u)^2 du.$$

Lemma P4.10 Given $\varepsilon, \delta > 0$, let $a^2 = 2\pi(1 + \varepsilon)(2 + \delta)T(\log T)H_2 \sup_\lambda f_{XX}(\lambda)$. Under the conditions of Theorem 4.5.1,

$$\mathrm{Prob}\, [\sup_\lambda |\mathrm{Re}\, d_X^{(T)}(\lambda)| \geqslant a] \leqslant K T^{-1-\delta}$$

for some K.

Proof The probability is

$$\leqslant \exp \{-\alpha a\}\, 2 \exp \{\log R$$
$$+ \alpha^2 2\pi T (1 + \varepsilon) H_2 \sup_\lambda f_{XX}(\lambda)/[2(1 - 6\pi T R^{-1})^2]\}.$$

Taking $R = T \log T$ and

$$\alpha = a(1 - 6\pi T R^{-1})^2/[2\pi T(1 + \varepsilon) H_2 \sup_\lambda f_{XX}(\lambda)]$$

we see it is

$$\leqslant 2 \exp \{-a^2 (1 - 6\pi T R^{-1})^2/[2\pi T (1 + \varepsilon) H_2 \sup_\lambda f_{XX}(\lambda)]\}$$
$$\times \exp \{\log T + \log \log T\}.$$

This last is $\leqslant K T^{-1-\delta}$ after the indicated choice of a.

Corollary Under the conditions of Theorem 4.5.1

$$\overline{\lim} \sup_{\lambda} |\text{Re } d_X^{(T)}(\lambda)|/(T \log T)^{1/2} \leqslant (2\pi \int h(t)^2 dt \sup_{\lambda} f_{XX}(\lambda))^{1/2}$$

with probability 1.

Proof From the Borel-Cantelli lemma (see Loève (1965)) and the fact that ε, δ above are arbitrary.

Proof of Theorem 4.5.1 We can develop a corollary, similar to the last one, for Im $d_X^{(T)}(\lambda)$. The theorem then follows from the fact that

$$|d_X^{(T)}(\lambda)| \leqslant |\text{Re } d_X^{(T)}(\lambda)| + |\text{Im } d_X^{(T)}(\lambda)|.$$

Proof of Theorem 4.5.2 We prove Theorem 4.5.3 below. The proof of Theorem 4.5.2 is similar with the key inequality of the first lemma below replaced by

$$|\text{cum } \{\zeta_{a_1}^{(T)}(\lambda_1), \ldots, \zeta_{a_k}^{(T)}(\lambda_k)\}| \leqslant M^k C_k.$$

To prove Theorem 4.5.3, we proceed via a sequence of lemmas.

Lemma P4.11 Suppose $h(u)$ has a uniformly bounded derivative and finite support. Let

$$\boldsymbol{\zeta}^{(T)}(\lambda) = \mathbf{d}_Y^{(T)}(\lambda) - \mathbf{A}(\lambda)\mathbf{d}_X^{(T)}(\lambda).$$

Then

$$|\text{cum } \{\zeta_{a_1}^{(T)}(\lambda_1), \ldots, \zeta_{a_k}^{(T)}(\lambda_k)\}| \leqslant T^{-k+1} M^k C_k$$

for some M.

Proof

$$\begin{aligned}
\mathbf{d}_Y^{(T)}(\lambda) &- \mathbf{A}(\lambda)\mathbf{d}_X^{(T)}(\lambda) \\
&= \sum_t h^{(T)}(t) \sum_v \mathbf{a}(t-v)\mathbf{X}(v) \exp\{-i\lambda t\} - \sum_u \mathbf{a}(u) \exp\{-i\lambda u\} \\
&\quad \times \sum_t h^{(T)}(t)\mathbf{X}(t) \exp\{-i\lambda t\} \\
&= \sum_{u,v} [h^{(T)}(u+v) - h^{(T)}(v)]\mathbf{a}(u)\mathbf{X}(v) \exp\{-i\lambda(u+v)\}.
\end{aligned}$$

Therefore

$$\begin{aligned}
\text{cum } &\{\zeta_{a_1}^{(T)}(\lambda_1), \ldots, \zeta_{a_k}^{(T)}(\lambda_k)\} \\
&= \sum_{u_1,v_1} \cdots \sum_{u_k,v_k} [h^{(T)}(u_1+v_1) - h^{(T)}(v_1)] \cdots [h^{(T)}(u_k+v_k) - h^{(T)}(v_k)] \\
&\quad \times \exp\{-i \sum_j \lambda_j(u_j+v_j)\} \sum_{b_1} \cdots \sum_{b_k} a_{a_1b_1}(u_1) \cdots a_{a_kb_k}(u_k) \\
&\quad \times c_{a_1 \ldots a_k}(v_1 - v_k, \ldots, v_{k-1} - v_k).
\end{aligned}$$

In absolute value this is

$$\leqslant \sum_{u_1,w_1} \cdots \sum_{u_{k-1},w_{k-1}} \sum_{u_k,v_k} |u_1|T^{-1}L \cdots |u_k|T^{-1}L \sum_{b_1} \cdots \sum_{b_k} |a_{a_1b_1}(u_1) \cdots a_{a_kb_k}(u_k)|$$
$$\times |c_{a_1 \ldots a_k}(w_1, \ldots, w_{k-1})|$$
$$\leqslant T^{-k+1}M^kC_k$$

for some finite M with L denoting a bound for the derivative of $h(u)$.

Lemma P4.12 For α sufficiently small there is a finite L such that

$$E \exp \{\alpha \operatorname{Re} \zeta_a^{(T)}(\lambda)\} \leqslant \exp \{\alpha^2 L/T\}.$$

Proof From the previous lemma

$$|\log E \exp \{\alpha \operatorname{Re} \zeta_a^{(T)}(\lambda)\}| \leqslant \sum_{2}^{\infty} T^{-k+1}M^kC_k \, |\alpha|^k/k!$$
$$\leqslant T^{-1} \, |\alpha|^2 L$$

for $|\alpha|$ sufficiently small and some finite L.

Lemma P4.13 Let $\lambda_r = 2\pi r/R$, $r = 0, \ldots, R-1$ for some integer $R > 12\pi T$, then there is a finite N such that

$$\sup_{\lambda} |\zeta_a^{(T)}(\lambda)| \leqslant \sup_{r} |\zeta_a^{(T)}(\lambda_r)|/(1 - 12\pi T R^{-1}) + NT^{-1} \sup |\mathbf{d}_X^{(T)}(\lambda)|.$$

Proof Let

$$\mathbf{A}^{(T)}(\lambda) = \sum_{u=-T}^{T} \mathbf{a}(u) \exp \{-i\lambda u\}.$$

In view of (4.5.8)

$$|\mathbf{A}(\lambda) - \mathbf{A}^{(T)}(\lambda)| \leqslant KT^{-1}$$

for some K. Now $\boldsymbol{\zeta}^{(T)}(\lambda)$ may be written

$$\{\mathbf{d}_Y^{(T)}(\lambda) - \mathbf{A}^{(T)}(\lambda)\mathbf{d}_X^{(T)}(\lambda)\} + \{\mathbf{A}^{(T)}(\lambda) - \mathbf{A}(\lambda)\}\mathbf{d}_X^{(T)}(\lambda). \quad (*)$$

The first term here is a trigonometric polynomial of order $2T$. From Lemma 2.1 of Woodroofe and Van Ness (1967) we therefore have

$$\sup_{\lambda} |[\mathbf{d}_Y^{(T)}(\lambda) - \mathbf{A}^{(T)}(\lambda)\mathbf{d}_X^{(T)}(\lambda)]_a|$$
$$\leqslant \sup_{r} |[\mathbf{d}_Y^{(T)}(\lambda_r) - \mathbf{A}^{(T)}(\lambda_r)\mathbf{d}_X^{(T)}(\lambda_r)]_a|/(1 - 12\pi T R^{-1})$$
$$\leqslant \sup_{r} |\zeta_a^{(T)}(\lambda_r)|/(1 - 12\pi T R^{-1}) + KT^{-1} \sup_{\lambda} |\mathbf{d}_X^{(T)}(\lambda)|/(1 - 12\pi T R^{-1}).$$

The latter and (*) now give the indicated inequality.

Lemma P4.14 Under the conditions of Theorem 4.5.3,

$$E \exp \{\alpha \sup_{r} |\zeta_a^{(T)}(\lambda_r)|/(1 - 12\pi T R^{-1})\}$$
$$\leqslant 2 \exp \{\log R + \alpha^2 L T^{-1}/(1 - 12\pi T R^{-1})^2\}.$$

Proof Immediate from Lemma P4.12 and the fact that the sup runs over R points.

Lemma P4.15 Given $\delta > 0$, let $a^2 = 4L(2 + \delta) \log T/T$, then under the conditions of Theorem 4.5.3,

$$\text{Prob}\,[\sup_r |\zeta_a^{(T)}(\lambda_r)|/(1 - 12\pi TR^{-1}) \geqslant a] \leqslant KT^{-1-\delta}$$

for some finite K.

Proof Set $R = T \log T$ and

$$\alpha = \frac{aT(1 - 12\pi TR^{-1})^2}{2L}.$$

The probability is then

$$\begin{aligned}&\leqslant 2 \exp \{\log T + \log\log T - a^2T(1 - 12\pi TR^{-1})^2/(4L)\}\\ &\leqslant K \exp \{\log T - (2 + \delta) \log T\}\\ &\leqslant KT^{-1-\delta}.\end{aligned}$$

Corollary Under the conditions of Theorem 4.5.3

$$\overline{\lim_{T\to\infty}} \sup_r |\zeta_a^{(T)}(\lambda_r)|/(1 - 12\pi TR^{-1})T^{-1/2}(\log T)^{1/2}] \leqslant K$$

for some finite K with probability 1.

Proof of Theorem 4.5.3 The result follows from Theorem 4.5.1, the previous corollary and Lemma P4.13.

Proof of Theorem 4.5.4 Exercise 3.10.34(b) gives

$$d_X^{(T)}(\lambda) = (2\pi)^{-1} \int_0^{2\pi} d_X^{(T)}(\alpha)\Delta^{(T)}(\lambda - \alpha)d\alpha.$$

Let k be a positive integer. Holder's inequality gives

$$\begin{aligned}&|d_X^{(T)}(\lambda)|\\ &\leqslant (2\pi)^{-1}\left[\int_0^{2\pi} |d_X^{(T)}(\alpha)|^{2k}d\alpha\right]^{1/2k}\left[\int_0^{2\pi} |\Delta^{(T)}(\lambda - \alpha)|^{2k/(2k-1)}d\alpha\right]^{(2k-1)/2k}\\ &\leqslant (2\pi)^{-1}\left[\int_0^{2\pi} d_X^{(T)}(\alpha)^k d_X^{(T)}(-\alpha)^k d\alpha\right]^{1/2k} KT^{1/2k}\end{aligned}$$

for some finite K following Exercise 3.10.28. It follows from Theorem 2.3.2 and (*) in the Proof of Lemma P4.7 that

$$E|d_X^{(T)}(\alpha)|^{2k} \leqslant MT^k$$

for some finite M and so

$$E[\sup_\lambda |d_X^{(T)}(\lambda)|]^{2k} \leqslant NT^{k+1}$$

for some finite N. This gives

$$E[T^{-1/2-\varepsilon} \sup_{\lambda} |d_X^{(T)}(\lambda)|]^{2k} \leqslant NT^{-(2k\varepsilon-1)}.$$

As $\Sigma\, T^{-(2k\varepsilon-1)} < \infty$ for k sufficiently large, we have the result of the theorem.

To prove Theorem 4.6.1, we first indicate a lemma.

Lemma P4.16 Suppose $\mathbf{X}(t)$, $t = 0, \pm 1, \ldots$ satisfies Assumption 2.6.1. Let

$$2\pi\, \mathbf{Z}_X^{(T)}(\lambda) = \sum_{t=-T}^{T} \mathbf{X}(t) \int_0^{\lambda} \exp\{-i\alpha t\}\, d\alpha$$

then

$$\lim_{T_j \to \infty} \operatorname{cum}\{Z_{a_1}^{(T_1)}(\lambda_1) \ldots, Z_{a_k}^{(T_k)}(\lambda_k)\}$$
$$= \int_0^{\lambda_1} \cdots \int_0^{\lambda_k} \eta\left(\sum_1^k \alpha_j\right) f_{a_1 \ldots a_k}(\alpha_1, \ldots, \alpha_{k-1})\, d\alpha_1 \cdots d\alpha_k$$

for $a_1, \ldots, a_k = 1, \ldots, r$; $k = 2, 3, \ldots$.

Proof The cumulant may be written as $(2\pi)^{-k}$ times

$$\int_0^{\lambda_1} \cdots \int_0^{\lambda_k} \sum_{t_1=-T_1}^{T_1} \cdots \sum_{t_k=-T_k}^{T_k} c_{a_1 \ldots a_k}(t_1 - t_k, \ldots, t_{k-1} - t_k)$$
$$\times \exp\left\{-i \sum_1^k \alpha_j t_j\right\} d\alpha_1 \cdots d\alpha_k$$
$$= \int_0^{\lambda_1} \cdots \int_0^{\lambda_k} \int_0^{2\pi} \cdots \int_0^{2\pi} f_{a_1 \ldots a_k}(\beta_1, \ldots, \beta_{k-1})$$
$$\times \sum_{t_1} \cdots \sum_{t_k} \exp\{it_1(\beta_1 - \alpha_1) + \cdots + it_{k-1}(\beta_{k-1} - \alpha_{k-1})$$
$$- it(\beta_1 + \cdots + \beta_k + \alpha_k)\}\, d\alpha_1 \cdots d\alpha_k d\beta_1 \cdots d\beta_{k-1}$$

if we substitute for the cumulant function.

The limit indicated in the lemma now results once we note that

$$\lim_{T \to \infty} (2\pi)^{-1} \sum_{t=-T}^{T} \exp\{i\gamma t\} = \eta(\gamma)$$

where $\eta(.)$ is the periodic extension of the Dirac delta function; see Exercise 2.13.33.

Corollary. If $\sum_1^k \lambda_j \equiv 0 \pmod{2\pi}$, then

$$\lim_{S,T \to \infty} \operatorname{cum}\{Z_{a_1}^{(S)}(\lambda_1) - Z_{a_1}^{(T)}(\lambda_1), \ldots, Z_{a_k}^{(S)}(\lambda_k) - Z_{a_k}^{(T)}(\lambda_k)\} = 0.$$

Proof The cumulant is a sum of terms of the form

$$\pm \operatorname{cum}\{Z_{a_1}^{(S \text{ or } T)}(\lambda_1), \ldots, Z_{a_k}^{(S \text{ or } T)}(\lambda_k)\}.$$

In view of the lemma above these all tend to $\pm$ the same limit. The sum therefore tends to 0.

Corollary

$$\lim_{S,T\to\infty} E|Z_a^{(S)}(\lambda) - Z_a^{(T)}(\lambda)|^{2k} = 0 \qquad \text{for } k = 1, 2, \ldots.$$

Proof The moment may be written as a sum of cumulants of the form of those appearing in the previous corollary. Each of these cumulants tend to 0 giving the result.

Proof of Theorem 4.6.1 From the last corollary above we see that the sequence $Z_a^{(T)}(\lambda)$, $T = 1, 2, \ldots$ is a Cauchy sequence in the space L_ν for any $\nu > 0$. Because this space is complete, the sequence has a limit $Z_a(\lambda)$ in the space.

To complete the proof we note that expression (4.6.7) follows from Lemma P4.16 above.

Proof of Theorem 4.6.2 Set

$$\mathbf{A}^{(N,T)}(t) = \frac{2\pi}{N}\sum_{n=0}^{N-1} \exp\{i2\pi nt/N\}\left[\mathbf{Z}_X^{(T)}\left(\frac{2\pi(n+1)}{N}\right) - \mathbf{Z}_X^{(T)}\left(\frac{2\pi n}{N}\right)\right]$$

then

$$\int_0^{2\pi} \exp\{i\lambda t\}d\mathbf{Z}_X(\lambda) = \underset{N\to\infty}{\text{l.i.m.}}\ \underset{T\to\infty}{\text{l.i.m.}}\ \mathbf{A}^{(N,T)}(t).$$

Also

$$\begin{aligned} E\{X_a(t)Z_a^{(T)}(\lambda)\} &= \sum_{u=-T}^{T} c_{aa}(t-u)[1 - \exp\{-i\lambda u\}]/(-iu)(2\pi) \\ &= \int_0^{\lambda}\int_0^{2\pi} f_{aa}(\beta) \sum_{u=-T}^{T} \exp\{i\beta(t-u) - i\alpha u\}d\alpha d\beta/(2\pi) \\ &\to \int_0^{\lambda} f_{aa}(-\alpha)\exp\{-i\alpha t\}d\alpha. \end{aligned}$$

From this we see that

$$\begin{aligned} &E\left\{X_a(t)\int_0^{2\pi} \exp\{i\lambda t\}dZ_a(\lambda)\right\} \\ &\quad = \underset{N\to\infty}{\text{l.i.m}}\ \underset{T\to\infty}{\text{l.i.m}}\ E\{X_a(t)A_a^{(N,T)}(t)\} \\ &\quad = \lim_{N\to\infty} \frac{2\pi}{N}\sum_{n=0}^{N-1} \exp\{i2\pi nt/N\}\int_{2\pi n/N}^{2\pi(n+1)/N} f_{aa}(-\alpha)\exp\{-i\alpha t\}d\alpha \\ &\quad = \int_0^{2\pi} f_{aa}(-\alpha)d\alpha \\ &\quad = EX_a(t)^2. \end{aligned}$$

In a similar manner we may show that

$$E\left[\int_0^{2\pi} \exp\{i\lambda t\} dZ_a(\lambda)\right]^2 = EX(t)^2.$$

From these last two we see

$$E\left[X_a(t) - \int_0^{2\pi} \exp\{i\lambda t\} dZ_a(\lambda)\right]^2 = 0$$

and so

$$X_a(t) = \int_0^{2\pi} \exp\{i\lambda t\} dZ_a(\lambda)$$

with probability 1, $t = 0, \pm 1, \ldots$ giving the desired representation.

Proof of Theorem 4.7.1 We may write

$$\begin{aligned} E\{(\mathbf{Y} - \mathbf{B} - \mathbf{C}\zeta)^\tau(\mathbf{Y} - \mathbf{B} - \mathbf{C}\zeta)\} &= \operatorname{tr}[\operatorname{cov}\{\mathbf{Y} - \mathbf{C}\zeta, \mathbf{Y} - \mathbf{C}\zeta\}] \\ &\quad + (E\mathbf{Y} - \mathbf{B} - \mathbf{C}E\zeta)^\tau(E\mathbf{Y} - \mathbf{B} - \mathbf{C}E\zeta). \end{aligned}$$

The latter is clearly minimized with respect to **B** by setting

$$\mathbf{B} = E\mathbf{Y} + \mathbf{C}E\zeta.$$

Now

$$\begin{aligned} \operatorname{tr}[\operatorname{cov}\{\mathbf{Y} - \mathbf{C}\zeta, \mathbf{Y} - \mathbf{C}\zeta\}] &= \operatorname{tr}[\mathbf{\Sigma}_{YY} - \mathbf{C}\mathbf{A}\mathbf{\Sigma}_{YY} - \mathbf{\Sigma}_{YY}\mathbf{A}^\tau\mathbf{C}^\tau + \mathbf{C}\mathbf{A}\mathbf{\Sigma}_{YY}\mathbf{A}^\tau\mathbf{C}^\tau] \\ &= \operatorname{tr}[(\mathbf{\Sigma}_{YY}^{1/2} - \mathbf{C}\mathbf{A}\mathbf{\Sigma}_{YY}^{1/2})(\mathbf{\Sigma}_{YY}^{1/2} - \mathbf{C}\mathbf{A}\mathbf{\Sigma}_{YY}^{1/2})^\tau]. \end{aligned}$$

Following Corollary 3.7.4 this is minimized by setting

$$\begin{aligned} \mathbf{C}\mathbf{A} &= \sum_{j=1}^{K} \mu_j^{1/2}\mathbf{U}_j\mathbf{U}_j^\tau\mathbf{\Sigma}_{YY}^{-1/2} \\ &= \sum_{j=1}^{K} \mathbf{U}_j\mathbf{U}_j^\tau \end{aligned}$$

in the notation of the theorem. This gives the required result.

PROOFS FOR CHAPTER 5

Proof of Theorem 5.2.1 We have

$$\begin{aligned} EI_{XX}^{(T)}(\lambda) &= (2\pi T)^{-1} \operatorname{cum}\{d_X^{(T)}(\lambda), d_X^{(T)}(-\lambda)\} + (2\pi T)^{-1}|Ed_X^{(T)}(\lambda)|^2 \\ &= (2\pi T)^{-1} \sum_{s,t=0}^{T-1} \exp\{-i\lambda(s-t)\}c_{XX}(s-t) \\ &\qquad + (2\pi T)^{-1}|\Delta^{(T)}(\lambda)|^2 c_X^2. \quad (*) \end{aligned}$$

Expression (5.2.6) now follows after the substitution

$$c_{XX}(u) = \int_{-\pi}^{\pi} \exp\{i\alpha u\} f_{XX}(\alpha) d\alpha.$$

Proof of Theorem 5.2.2 Proceeding as in the proof of Theorem 4.3.2, we see that (5.2.7) implies

$$\text{cum}\,\{d_X^{(T)}(\lambda), d_X^{(T)}(-\lambda)\} = 2\pi T f_{XX}(\lambda) + O(1)$$

and (5.2.8) follows from (*) immediately above.

Proof of Theorem 5.2.3 We begin by noting that

$$\int_{-\pi}^{\pi} |H^{(T)}(\alpha)|^2 d\alpha = 2\pi \sum_t h\left(\frac{t}{T}\right)^2.$$

Next we have

$$\begin{aligned}\text{cum}\,\{d_X^{(T)}(\lambda), d_X^{(T)}(-\lambda)] &= \sum_{s,t} \exp\,\{-i\lambda(s-t)\} h\left(\frac{s}{T}\right) h\left(\frac{t}{T}\right) c_{XX}(s-t)\\ &= \int_{-\pi}^{\pi} \sum_{s,t} \exp\,\{-i\lambda(s-t\} h\left(\frac{s}{T}\right) h\left(\frac{t}{T}\right) \exp\,\{i\alpha(s-t)\} f_{XX}(\alpha) d\alpha\\ &= \int_{-\pi}^{\pi} |H^{(T)}(\lambda-\alpha)|^2 f_{XX}(\alpha) d\alpha.\end{aligned}$$

Expression (5.2.17) now follows from the fact that

$$\begin{aligned}EI_{XX}^{(T)}(\lambda) = &\left(\int_{-\pi}^{\pi} |H^{(T)}(\alpha)|^2 d\alpha\right)^{-1} \text{cum}\,\{d_X^{(T)}(\lambda), d_X^{(T)}(-\lambda)\}\\ &+ \left(\int_{-\pi}^{\pi} |H^{(T)}(\alpha)|^2 d\alpha\right)^{-1} |H^{(T)}(\lambda)|^2 c_X^2.\end{aligned}$$

Proof of Theorem 5.2.4 See proof of Theorem 5.2.5 given immediately below.

Proof of Theorem 5.2.5 From Theorem 4.3.2 we have

$$\begin{aligned}&\text{cov}\,\{d_X^{(T)}(\lambda) d_X^{(T)}(-\lambda), d_X^{(T)}(\mu) d_X^{(T)}(-\mu)\}\\ &= (2\pi)^3 T f_{XXXX}(\lambda,-\lambda,\mu) + O(1)\\ &\quad + [\Delta^{(T)}(\lambda) f_X + O(1)][(2\pi)^2 \Delta^{(T)}(-\lambda) f_{XXX}(-\lambda,\mu) + O(1)]\\ &\quad + \text{three similar terms}\\ &\quad + [\Delta^{(T)}(\lambda) f_X + O(1)][\Delta^{(T)}(\mu) f_X + O(1)][2\pi\Delta^{(T)}(-\lambda-\mu) f_{XX}(-\lambda) + O(1)]\\ &\quad + \text{three similar terms}\\ &\quad + [2\pi\Delta^{(T)}(\lambda+\mu) f_{XX}(\lambda) + O(1)][2\pi\Delta^{(T)}(-\lambda-\mu) f_{XX}(-\lambda) + O(1)]\\ &\quad + [2\pi\Delta^{(T)}(\lambda-\mu) f_{XX}(\lambda) + O(1)][2\pi\Delta^{(T)}(-\lambda+\mu) f_{XX}(-\lambda) + O(1)]\end{aligned}$$

giving the indicated result.

We now set down a result that will be needed in the next proof and other proofs throughout this work.

Theorem P5.1 Let the sequence of r vector-valued random variables $\mathbf{X}_T$, $T = 1, 2, \ldots$ tend in distribution to the distribution of a random variable $\mathbf{X}$. Let $\mathbf{g} : R^r \rightarrow R^s$ be an s vector-valued measurable function whose discontinuities have $\mathbf{X}$ probability 0. Then the sequence of s vector-valued variables $\mathbf{g}(\mathbf{X}_T)$, $T = 1, 2, \ldots$ tends in distribution to the distribution of $\mathbf{g}(\mathbf{X})$.

Proof See Mann and Wald (1943a) and Theorem 5.1 of Billingsley (1968).

A related theorem that will also be needed later is

Theorem P5.2 Let the sequence of r vector-valued random variables $\sqrt{T}(\mathbf{Y}_T - \boldsymbol{\mu})$, $T = 1, 2, \ldots$ tend in distribution to $N_r(\mathbf{0},\boldsymbol{\Sigma})$. Let $\mathbf{g} : R^r \rightarrow R^s$ be an s vector-valued function differentiable in a neighborhood of $\boldsymbol{\mu}$ and having $s \times r$ Jacobian matrix $\mathbf{J}$ at $\boldsymbol{\mu}$. Then $\sqrt{T}(\mathbf{g}(\mathbf{Y}_T) - \mathbf{g}(\boldsymbol{\mu}))$ tends in distribution to $N_s(\mathbf{0},\mathbf{J}\boldsymbol{\Sigma}\mathbf{J}^\tau)$ as $T \rightarrow \infty$.

Proof See Mann and Wald (1943) and Rao (1965) p. 321.

Corollary P5.2 (*The Real-Valued Case*) Let $\sqrt{T}(Y_T - \mu)$, $T = 1, 2, \ldots$ tend in distribution to $N(0,\sigma^2)$. Let $g : R \rightarrow R$ have derivative g' in a neighborhood of μ. Then $\sqrt{T}(g(Y_T) - g(\mu)) \rightarrow N(0,[g'(\mu)]^2\sigma^2)$ as $T \rightarrow \infty$.

Proof of Theorem 5.2.6 Theorem 4.4.1 indicates that Re $d_X^{(T)}(\lambda_j(T))$, Im $d_X^{(T)}(\lambda_j(T))$ are asymptotically independent $N(0,\pi T f_{XX}(\lambda_j))$ variates. It follows from Theorem P5.1 that

$$I_{XX}^{(T)}(\lambda_j(T)) = (2\pi T)^{-1}\{[\text{Re}\, d_X^{(T)}(\lambda_j(T))]^2 + [\text{Im}\, d_X^{(T)}(\lambda_j(T))]^2\}$$

is asymptotically $f_{XX}(\lambda_j)\chi_2^2/2$. The asymptotic independence for different values of j follows in the same manner from the asymptotic independence of the $d_j^{(T)}(\lambda_j(T))$, $j = 1, \ldots, J$.

Proof of Theorem 5.2.7 This theorem follows from Theorem 4.4.2 as Theorem 5.2.6 followed from Theorem 4.4.1.

Proof of Theorem 5.2.8 From Theorem 4.3.2

$$\begin{aligned}
&\text{cov}\,\{d_X^{(T)}(\lambda)d_X^{(T)}(-\lambda),\, d_X^{(T)}(\mu)d_X^{(T)}(-\mu)\} \\
&\quad = [2\pi H_2^{(T)}(\lambda + \mu)f_{XX}(\lambda) + O(1)][2\pi H_2^{(T)}(-\lambda - \mu)f_{XX}(\lambda) + O(1)] \\
&\qquad + [2\pi H_2^{(T)}(\lambda - \mu)f_{XX}(\lambda) + O(1)][2\pi H_2^{(T)}(-\lambda - \mu)f_{XX}(\lambda) + O(1)] \\
&\qquad + (2\pi)^3 H_4^{(T)}(0) f_{XXXX}(\lambda,-\lambda,\mu) + O(1).
\end{aligned}$$

The indicated result follows as

$$H_2^{(T)}(0) = \sum h\left(\frac{t}{T}\right)^2 \backsim T \int h(\alpha)^2 d\alpha.$$

Proof of Theorem 5.3.1 This theorem is an immediate consequence of Exercise 4.8.23.

Proof of Theorem 5.3.2 Follows directly from Theorem 4.5.1 and the definition of $I_{XX}^{(T)}(\lambda)$.

Proof of Theorem 5.4.1 This theorem follows directly from expression (5.2.6) of Theorem 5.2.1 and the definitions of $A_T{}^m(\lambda)$, $B_T{}^m(\lambda)$, $C_T{}^m(\lambda)$.

The corollary follows from Theorem 5.2.2.

Proof of Theorem 5.4.2 This follows from Theorem 5.2.4.

Proof of Theorem 5.4.3 Follows from Theorem 5.2.6.

Proof of Theorem 5.5.1 This theorem follows from expression (5.2.6) of Theorem 5.2.1 and the definitions of $A_T(\lambda)$, $B_T(\lambda)$, $C_T(\lambda)$. The corollary follows from Theorem 5.2.2.

Proof of Theorem 5.5.2 From Theorem 5.2.4.

Proof of Theorem 5.5.3 From Theorem 5.2.6 and Theorem P5.1.

The following lemma will be required in the course of the proofs of several theorems.

Lemma P5.1 If a function $g(x)$ has finite total variation, V, on [0,1], then

$$\left| \int_0^1 g(x)dx - n^{-1} \sum_{k=1}^{n} g\left(\frac{k}{n}\right) \right| \leqslant \frac{V}{n}.$$

Proof See Polya and Szegö (1925) p. 37; a related reference is Cargo (1966). If g is differentiable, the right side may be replaced by $\int |g'(x)| dx/n$.

Further results are given as Exercises 1.7.14 and 5.13.28.

Proof of Theorem 5.6.1 The first expression in (5.6.7) follows directly from expression (5.2.8) and the definition (5.6.1).

If we use the lemma above to approximate the sum appearing by an integral, then we see that

$$\begin{aligned} Ef_{XX}^{(T)}(\lambda) &= \int_0^{2\pi} W^{(T)}(\lambda - \alpha) f_{XX}(\alpha) d\alpha + O(B_T^{-1}T^{-1}) \\ &= \int_{-\infty}^{\infty} B_T^{-1} W(B_T^{-1}[\lambda - \alpha]) f_{XX}(\alpha) d\alpha + O(B_T^{-1}T^{-1}) \end{aligned}$$

giving the final expression in (5.6.7).

Proof of Theorem 5.6.2 Using Theorem 5.2.5, the indicated covariance is given by

$$\left(\frac{2\pi}{T}\right)^2 \sum_{r=1}^{T-1} \sum_{s=1}^{T-1} W^{(T)}\left(\lambda - \frac{2\pi s}{T}\right) W^{(T)}\left(\mu - \frac{2\pi r}{T}\right) \\ \times \left[\eta\left\{\frac{2\pi}{T}(s + r)\right\} + \eta\left\{\frac{2\pi}{T}(s - r)\right\}\right] f_{XX}\left(\frac{2\pi s}{T}\right)^2 + O(T^{-1})$$

giving the indicated first expression. The second expression follows from this on replacing the sum by an integral making use of Lemma P5.1.

Proof of Theorem 5.6.3 See the proof of Theorem 7.4.4.

Proof of Corollary 5.6.3 This follows from Theorem 5.6.3 and Corollary P5.2.

Proof of Theorem 5.6.4 See the proof of Theorem 7.7.1.

Proof of Theorem 5.8.1

$$\int_{-\pi}^{\pi} K^{(T)}(\alpha) f_{XX}(\lambda - \alpha) d\alpha = (2\pi)^{-1} \sum_u k^{(T)}(u) c_{XX}(u) \exp\{-iu\lambda\}$$
$$= (2\pi)^{-1} \sum_{|u| \leq T} c_{XX}(u) \exp\{-iu\lambda\} + O(T^{-P})$$

in view of (5.8.7) and because $|k^{(T)}(u)| \leqslant 1$. This in turn equals

$$(2\pi)^{-1} \sum_{|u| \leq T} \left\{1 + k_1 \frac{u}{T} + \cdots + k_{P-1} \frac{u^{P-1}}{T^{P-1}} + O\left(\frac{|u|^P}{T^P}\right)\right\} c_{XX}(u) \exp\{-iu\lambda\}$$
$$+ O(T^{-P})$$

giving the desired result because

$$(2\pi)^{-1} \sum_{|u| \leq T} u^p c_{XX}(u) \exp\{-iu\lambda\} = i^p \frac{d^p f_{XX}(\lambda)}{d\lambda^p} + O(T^{p-P}).$$

Proof of Theorem 5.8.2 The first expression in (5.8.18) follows directly from the definition of $f_{XX}^{(T)}(\lambda)$ and expression (5.8.9). The second expression of (5.8.18) follows from the first, neglecting terms after the first, and Lemma P5.1.

Proof of Corollary 5.8.2 This follows after we substitute the Taylor expansion

$$f_{XX}(\lambda - \alpha) = f_{XX}(\lambda) + \sum_{p=1}^{P-1} \frac{(-\alpha)^p}{p!} f_{XX}^{\{p\}}(\lambda) + O(|\alpha|^P)$$

into the second expression of (5.8.18).

Proof of Theorem 5.9.1 We write X' for $X - c_X^{(T)}$ below. Now

$$\frac{2\pi}{T} \sum_{s=1}^{T-1} W^{(T)}\left(\lambda - \frac{2\pi s}{T}\right) I_{XX}^{(T)}\left(\frac{2\pi s}{T}\right) - \int_0^{2\pi} W^{(T)}(\lambda - \alpha) I_{X'X'}^{(T)}(\alpha) d\alpha$$
$$= \sum_{s=1}^{T-1} \int_{2\pi s/T}^{2\pi[s+1]/T} \left\{W^{(T)}\left(\lambda - \frac{2\pi s}{T}\right) I_{XX}^{(T)}\left(\frac{2\pi s}{T}\right) - W^{(T)}(\lambda - \alpha) I_{X'X'}^{(T)}(\alpha)\right\} d\alpha$$
$$+ \int_0^{2\pi/T} W^{(T)}(\lambda - \alpha) I_{X'X'}^{(T)}(\alpha) d\alpha$$
$$= \sum_s \int \left\{\left[W^{(T)}\left(\lambda - \frac{2\pi s}{T}\right) - W^{(T)}(\lambda - \alpha)\right] I_{XX}^{(T)}\left(\frac{2\pi s}{T}\right)\right.$$
$$\left. + W^{(T)}(\lambda - \alpha)\left[I_{XX}^{(T)}\left(\frac{2\pi s}{T}\right) - I_{X'X'}^{(T)}(\alpha)\right]\right\} d\alpha$$
$$+ \int_0^{2\pi/T} W^{(T)}(\lambda - \alpha) I_{X'X'}^{(T)}(\alpha) d\alpha.$$

The indicated result now follows as;

$$EI_{XX}^{(T)}\left(\frac{2\pi s}{T}\right) = f_{XX}\left(\frac{2\pi s}{T}\right) + O(T^{-1}),$$

$$\sum_{s=1}^{T-1} \int \sup_{\frac{2\pi s}{T} \leq \alpha \leq \frac{2\pi|s+1|}{T}} \left| W^{(T)}\left(\lambda - \frac{2\pi s}{T}\right) - W^{(T)}(\lambda - \alpha)\right| d\alpha = O(B_T^{-1})$$

$$E\left[I_{XX}^{(T)}\left(\frac{2\pi s}{T}\right) - I_{X'X'}^{(T)}(\alpha)\right] = O(T^{-1}) \qquad \text{for } \frac{2\pi s}{T} \leqslant \alpha \leqslant \frac{2\pi|s+1|}{T}$$

$$\text{var}\left[I_{XX}^{(T)}\left(\frac{2\pi s}{T}\right) - I_{X'X'}^{(T)}(\alpha)\right] = O(T^{-2}) \qquad \text{for } \frac{2\pi s}{T} \leqslant \alpha \leqslant \frac{2\pi|s+1|}{T}$$

giving

$$E\left|I_{XX}^{(T)}\left(\frac{2\pi s}{T}\right) - I_{X'X'}^{(T)}(\alpha)\right| = O(T^{-1}) \qquad \text{for } \frac{2\pi s}{T} \leqslant \alpha \leqslant \frac{2\pi|s+1|}{T}$$

and finally

$$EI_{X'X'}^{(T)}(\alpha) = O(T^{-1}) \qquad \text{for } 0 \leqslant \alpha \leqslant 2\pi s/T.$$

Proof of Theorem 5.9.2 Follows directly from Theorem 5.3.1.

Proof of Theorem 5.10.1 From Theorem 5.2.2

$$EI_{XX}^{(T)}\left(\frac{2\pi s}{T}\right) = f_{XX}\left(\frac{2\pi s}{T}\right) + O(T^{-1})$$

for $s \not\equiv 0 \pmod T$ and s an integer. This gives the first part of (5.10.12). The second part follows from Lemma P5.1.

Continuing, from Theorem 4.3.2

$$\begin{aligned}
&\text{cov}\,\{J^{(T)}(A_j), J^{(T)}(A_k)\} \\
&= \left(\frac{2\pi}{T}\right)^2 \sum_{r=1}^{T-1}\sum_{s=1}^{T-1} A_j\left(\frac{2\pi r}{T}\right)\overline{A_k\left(\frac{2\pi s}{T}\right)}(2\pi T)^{-2} \\
&\quad \times \text{cov}\left\{\left|d_X^{(T)}\left(\frac{2\pi r}{T}\right)\right|^2, \left|d_X^{(T)}\left(\frac{2\pi s}{T}\right)\right|^2\right\} \\
&= T^{-4}\sum_r\sum_s A_j\left(\frac{2\pi r}{T}\right)\overline{A_k\left(\frac{2\pi s}{T}\right)}\left\{\left[2\pi\Delta^{(T)}\left(\frac{2\pi[r+s]}{T}\right)f_{XX}\left(\frac{2\pi r}{T}\right) + O(1)\right]\right. \\
&\quad \times \left[2\pi\Delta^{(T)}\left(\frac{2\pi[-r-s]}{T}\right)f_{XX}\left(\frac{2\pi r}{T}\right) + O(1)\right] \\
&\quad + \left[2\pi\Delta^{(T)}\left(\frac{2\pi[r-s]}{T}\right)f_{XX}\left(\frac{2\pi r}{T}\right) + O(1)\right] \\
&\quad \times \left[2\pi\Delta^{(T)}\left(\frac{2\pi[-r+s]}{T}\right)f_{XX}\left(\frac{2\pi r}{T}\right) + O(1)\right] \\
&\quad \left. + (2\pi)^3 T f_{XXXX}\left(\frac{2\pi r}{T}, \frac{2\pi s}{T}, -\frac{2\pi r}{T}\right) + O(1)\right\}
\end{aligned}$$

$$= \left(\frac{2\pi}{T}\right)^2 \sum_{s=1}^{T-1} A_j\left(2\pi - \frac{2\pi s}{T}\right)\overline{A_k\left(\frac{2\pi s}{T}\right)} f_{XX}\left(\frac{2\pi s}{T}\right)^2$$

$$+ \left(\frac{2\pi}{T}\right)^2 \sum_{s=1}^{T-1} A_j\left(\frac{2\pi s}{T}\right)\overline{A_k\left(\frac{2\pi s}{T}\right)} f_{XX}\left(\frac{2\pi s}{T}\right)^2$$

$$+ \left(\frac{2\pi}{T}\right)^3 \sum_r \sum_s A_j\left(\frac{2\pi r}{T}\right)\overline{A_k\left(\frac{2\pi s}{T}\right)} f_{XXXX}\left(\frac{2\pi r}{T}, \frac{2\pi s}{T}, -\frac{2\pi r}{T}\right) + O(T^{-2})$$

giving expression (5.10.13).

Turning to the higher order cumulants we have

$$\operatorname{cum}\{J^{(T)}(A_{j_1}), \ldots, J^{(T)}(A_{j_L})\}$$

$$= \left(\frac{2\pi}{T}\right)^L \sum_{s_1=1}^{T-1} \cdots \sum_{s_L=1}^{T-1} A_{j_1}\left(\frac{2\pi s_1}{T}\right) \cdots A_{j_L}\left(\frac{2\pi s_L}{T}\right)(2\pi T)^{-L}$$

$$\times \operatorname{cum}\left\{\left|d_X^{(T)}\left(\frac{2\pi s_1}{T}\right)\right|^2, \ldots, \left|d_X^{(T)}\left(\frac{2\pi s_L}{T}\right)\right|^2\right\}$$

$$= T^{-2L} \sum_{s_1} \cdots \sum_{s_L} A_{j_1}\left(\frac{2\pi s_1}{T}\right) \cdots A_{j_L}\left(\frac{2\pi s_L}{T}\right) \sum_\nu \left[(2\pi)^{\nu_1 - 1}\Delta^{(T)}\left(\frac{2\pi}{T}\sum_{j\in\nu_1} \pm s_j\right)\right.$$

$$\left.\times f_{X\ldots X}\left(\frac{2\pi s_j}{T}; j \in \nu_1\right) + O(1)\right] \cdots \left[(2\pi)^{\nu_p - 1}\Delta^{(T)}\left(\frac{2\pi}{T}\sum_{j\in\nu_p} \pm s_j\right)\right.$$

$$\left.\times f_{X\ldots X}\left(\frac{2\pi s_j}{T}; j \in \nu_p\right) + O(1)\right]$$

the inner sum being over all indecomposable partitions of the table

$$\begin{matrix} d_X^{(T)}\left(\frac{2\pi s_1}{T}\right) & d_X^{(T)}\left(-\frac{2\pi s_1}{T}\right) \\ \vdots & \\ d_X^{(T)}\left(\frac{2\pi s_L}{T}\right) & d_X^{(T)}\left(-\frac{2\pi s_L}{T}\right). \end{matrix}$$

Taking note of the linear restrictions introduced by the $\Delta^{(T)}$ functions, we see that the dominant term in this cumulant is of order T^{-L+1}.

Now, when the variates $T^{1/2}J^{(T)}(A_j)$, $j = 1, \ldots, J$, are considered, we see that their joint cumulants of order greater than 2 all tend to 0. It follows that these variates are asymptotically normal.

Proof of Theorem 5.10.2 We proceed as in the proof of Theorem 5.9.1.

Proof of Theorem 5.11.1 In order to avoid cumbersome algebraic detail, we present a proof only in the case $J = 1$. The general J case follows in a similar manner.

The model is

$$X(t) = \theta\phi(t) + \varepsilon(t)$$

and the least squares estimate

$$\begin{aligned}\theta^{(T)} &= \sum_{t=0}^{T-1} X(t)\phi(t) / \sum_{t=0}^{T-1} \phi(t)^2 \\ &= \theta + \sum \varepsilon(t)\phi(t) / \sum \phi(t)^2.\end{aligned}$$

Because $E\varepsilon(t) = 0$, we see from the latter expression that $E\theta^{(T)} = \theta$. Also

$$\begin{aligned}\operatorname{var}\left\{\sum \varepsilon(t)\phi(t)\right\} &= \sum_{t_1}\sum_{t_2} c_{\varepsilon\varepsilon}(t_1 - t_2)\phi(t_1)\phi(t_2) \\ &= \sum_{u=-T+1}^{T-1} c_{\varepsilon\varepsilon}(u) \sum_{t=0}^{T-|u|} \phi(t+u)\phi(t).\end{aligned}$$

It follows by the bounded convergence criterion that

$$N_T^{-1} \operatorname{var}\left\{\sum \varepsilon(t)\phi(t)\right\} \rightarrow \sum_u c_{\varepsilon\varepsilon}(u) m_{\phi\phi}(u) = 2\pi \int f_{\varepsilon\varepsilon}(\alpha) dG_{\phi\phi}(\alpha).$$

At the same time

$$N_T^{-1} \sum_t \phi(t)^2 \rightarrow m_{\phi\phi}(0)$$

and so

$$N_T \operatorname{var} \theta^{(T)} \rightarrow \frac{2\pi \int f_{\varepsilon\varepsilon}(\alpha) dG_{\phi\phi}(\alpha)}{m_{\phi\phi}(0)^2}$$

as indicated in (5.11.20). In the case of higher order cumulants we see

$$\begin{aligned}\operatorname{cum}_L\left\{\sum \varepsilon(t)\phi(t)\right\} &= \sum_{t_1}\cdots\sum_{t_L} c_{\varepsilon\ldots\varepsilon}(t_1 - t_L, \ldots, t_{L-1} - t_L)\phi(t_1)\cdots\phi(t_L) \\ &= \sum_{u_1=-T+1}^{T-1} \cdots \sum_{u_{L-1}=-T+1}^{T-1} c_{\varepsilon\ldots\varepsilon}(u_1, \ldots, u_{L-1}) \\ &\quad \times \sum_t \phi(t+u_1)\cdots\phi(t+u_{L-1})\phi(t) = O(N_T)\end{aligned}$$

in view of the second condition of Assumption 5.11.1. It follows that

$$\begin{aligned}\operatorname{cum}_L\left\{N_T^{1/2}\theta^{(T)}\right\} &= N_T^{L/2} \operatorname{cum}\left\{\sum \varepsilon(t)\phi(t) / \sum \phi(t)^2\right\} \\ &= O(N_T^{L/2+1-L}) \rightarrow 0\end{aligned}$$

as $T \rightarrow \infty$ for $L > 2$ and so $\theta^{(T)}$ is asymptotically normal as indicated in the statement of the theorem.

We next consider the statistical behavior of $f_{ee}^{(T)}(\lambda)$. As

$$d_e^{(T)}(\lambda) = (\theta^{(T)} - \theta) d_\phi^{(T)}(\lambda) + d_\varepsilon^{(T)}(\lambda)$$

we have

$$\begin{aligned}f_{ee}^{(T)}(\lambda) = (\theta^{(T)} - \theta)^2 f_{\phi\phi}^{(T)}(\lambda) + (\theta^{(T)} - \theta) f_{\phi\varepsilon}^{(T)}(\lambda) \\ + (\theta^{(T)} - \theta) f_{\varepsilon\phi}^{(T)}(\lambda) + f_{\varepsilon\varepsilon}^{(T)}(\lambda).\end{aligned}$$

Now

$$\begin{aligned} |f_{\phi\phi}^{(T)}(\lambda)| &\leq MB_T^{-1}T^{-2}\sum_s \left|d_\phi^{(T)}\left(\frac{2\pi s}{T}\right)\right|^2 \\ &= MB_T^{-1}T^{-2}T\sum_t \phi(t)^2 \\ &\leq M'B_T^{-1}T^{-1}N_T \end{aligned}$$

for some finite M, M' while

$$\begin{aligned} E\,|f_{\phi\varepsilon}^{(T)}(\lambda)|^2 \leq T^{-4}\sum_r\sum_s &\left|W^{(T)}\left(\lambda - \frac{2\pi r}{T}\right)W^{(T)}\left(\lambda - \frac{2\pi s}{T}\right)d_\phi^{(T)}\left(\frac{2\pi r}{T}\right)d_\phi^{(T)}\left(\frac{2\pi s}{T}\right)\right| \\ &\left|2\pi T\eta\left\{\frac{2\pi(r-s)}{T}\right\} f_{\varepsilon\varepsilon}\left(\frac{2\pi r}{T}\right) + O(1)\right| \\ \leq NT^{-3}B_T^{-2}\sum_r &\left|d_\phi^{(T)}\left(\frac{2\pi r}{T}\right)\right|^2 \end{aligned}$$

for some finite N. Therefore

$$E\,|f_{\phi\varepsilon}^{(T)}(\lambda)| \leq N'T^{-1}B_T^{-1}N_T^{1/2}$$

for some finite N' and so

$$Ef_{ee}^{(T)}(\lambda) = Ef_{\varepsilon\varepsilon}^{(T)}(\lambda) + O(B_T^{-1}T^{-1})$$

giving (5.11.21) from Theorem 5.6.1. It also follows from these inequalities that

$$(B_TT)^{1/2}f_{ee}^{(T)}(\lambda) = (B_TT)^{1/2}f_{\varepsilon\varepsilon}^{(T)}(\lambda) + o_p(1)$$

showing that the asymptotic distribution of $f_{ee}^{(T)}(\lambda)$ is the same as that of $f_{\varepsilon\varepsilon}^{(T)}(\lambda)$ given in Theorem 5.6.3. ($o_p(1)$ denotes a variate tending to 0 in probability.)

The asymptotic independence of $\theta^{(T)}$ and $f_{ee}^{(T)}(\lambda_1), \ldots, f_{ee}^{(T)}(\lambda_K)$ follows from a consideration of joint asymptotic cumulants.

PROOFS FOR CHAPTER 6

Proofs of Theorems 6.2.1 and 6.2.2 These are classical results. Proofs may be found in Chapter 19, Kendall and Stuart (1961), for example.

Proofs of Theorems 6.2.3 and 6.2.4 These results follow from Theorems 6.2.1 and 6.2.2 when we rewrite (6.2.7) in the form

$$[\mathrm{Re}\,\mathbf{Y} \quad \mathrm{Im}\,\mathbf{Y}] = [\mathrm{Re}\,\mathbf{a} \quad \mathrm{Im}\,\mathbf{a}]\begin{bmatrix} \mathrm{Re}\,\mathbf{X} & \mathrm{Im}\,\mathbf{X} \\ -\mathrm{Im}\,\mathbf{X} & \mathrm{Re}\,\mathbf{X} \end{bmatrix} + [\mathrm{Re}\,\boldsymbol{\varepsilon} \quad \mathrm{Im}\,\boldsymbol{\varepsilon}].$$

This is a model of the form considered in those theorems.

Proof of Theorem 6.2.5 Follows directly from the properties of $\hat{\mathbf{a}}$ indicated in Theorem 6.2.4.

Proof of Lemma 6.3.1 We have

$$\begin{aligned} d_R^{(T)}(\beta) &= \sum_{u=-\infty}^{\infty} \mathbf{a}(u)\left[\sum_{t=0}^{T-1} \mathbf{X}(t-u)\exp\{-i\beta t\}\right] \\ &= \sum_{u=-\infty}^{0} \mathbf{a}(u)\exp\{-i\beta u\}\left[\sum_{v=0}^{T-1} - \sum_{v=0}^{-u-1} + \sum_{v=T}^{T-1-u}\right][\mathbf{X}(v)\exp\{-i\beta v\}] \\ &\quad + \sum_{u=1}^{\infty} \mathbf{a}(u)\exp\{-i\beta u\}\left[\sum_{v=0}^{T-1} + \sum_{v=-u}^{-1} - \sum_{v=T-u}^{T-1}\right][\mathbf{X}(v)\exp\{-i\beta v\}] \\ &= \mathbf{A}(\beta)\mathbf{d}_X^{(T)}(\beta) + e^{(T)}(\beta) \end{aligned}$$

where, because the components of $\mathbf{X}(t)$ are bounded $|e^{(T)}(\beta)| \leqslant 4m\sum_u |u\mathbf{a}(u)|$. The last part follows directly.

In the proofs below we will require the following lemma.

Lemma P6.1 Given a $1 \times M$ matrix $\mathbf{P}$ and an $r \times M$ matrix $\mathbf{Q}$ we have

$$\|\mathbf{P}\bar{\mathbf{Q}}^\tau(\mathbf{Q}\bar{\mathbf{Q}}^\tau)^{-1}\| \leqslant \|\mathbf{P}\bar{\mathbf{P}}^\tau\|^{1/2}\|(\mathbf{Q}\bar{\mathbf{Q}}^\tau)^{-1}\|^{1/2}.$$

Proof We begin by noting the matrix form of Schwarz's inequality

$$\mathbf{P}\bar{\mathbf{P}}^\tau \geqslant \mathbf{P}\bar{\mathbf{Q}}^\tau(\mathbf{Q}\bar{\mathbf{Q}}^\tau)^{-1}\mathbf{Q}\bar{\mathbf{P}}^\tau.$$

(This follows from the minimum achieved in Theorem 6.2.1.) This implies

$$\begin{aligned} \|\mathbf{P}\bar{\mathbf{P}}^\tau\| &\geqslant \|\mathbf{P}\bar{\mathbf{Q}}^\tau(\mathbf{Q}\bar{\mathbf{Q}}^\tau)^{-1}\mathbf{Q}\bar{\mathbf{P}}^\tau\| \\ &\geqslant \|\mathbf{P}\bar{\mathbf{Q}}^\tau(\mathbf{Q}\bar{\mathbf{Q}}^\tau)^{-1/2}\|^2. \end{aligned}$$

Now

$$\|\mathbf{P}\bar{\mathbf{Q}}^\tau(\mathbf{Q}\bar{\mathbf{Q}}^\tau)^{-1}\| \leqslant \|\mathbf{P}\bar{\mathbf{Q}}^\tau(\mathbf{Q}\bar{\mathbf{Q}}^\tau)^{-1/2}\| \cdot \|(\mathbf{Q}\bar{\mathbf{Q}}^\tau)^{-1/2}\|$$

and the result follows.

Proof of Theorem 6.4.1 Because $E\varepsilon(t) = 0$, we have $E\mathbf{A}^{(T)}(\lambda) = \mathbf{f}_{RX}^{(T)}(\lambda)\,\mathbf{f}_{XX}^{(T)}(\lambda)^{-1}$. By substitution

$$\mathbf{f}_{RX}^{(T)}(\lambda) = (2m+1)^{-1}\sum_{s=-m}^{m} \mathbf{A}\left(\frac{2\pi[s(T)+s]}{T}\right)\mathbf{I}_{XX}^{(T)}\left(\frac{2\pi[s(T)+s]}{T}\right) + \mathbf{G}^{(T)}$$

where following Lemma P6.1

$$\|\mathbf{G}^{(T)}\mathbf{f}_{XX}^{(T)}(\lambda)^{-1}\| \leqslant KT^{-1/2}\|\mathbf{f}_{XX}^{(T)}(\lambda)^{-1}\|^{1/2}$$

from which (6.4.9) follows.

Before proving Theorem 6.4.2 we state a lemma which is a slight extension of a result of Billingsley (1966).

Lemma P6.2 Let $\mathbf{Z}^{(T)}$ be a sequence of q vectors tending in distribution to $N_q^C(\mathbf{0},\mathbf{I})$ as $T \to \infty$. Let $\mathbf{U}^{(T)}$ be a sequence of $q \times q$ unitary matrices. Then $\mathbf{U}^{(T)}\mathbf{Z}^{(T)}$ also tends in distribution to $N_q^C(\mathbf{0},\mathbf{I})$.

Proof Consider any subsequence of $\mathbf{Z}^{(T)}$, say $\mathbf{Z}^{(T')}$. Because the group of unitary matrices is compact (see Weyl (1946)), $\mathbf{U}^{(T')}$ has a convergent subsequence, say $\mathbf{U}^{(T'')}$ tending to $\mathbf{U}$. Now, by Theorem P5.1 $\mathbf{U}^{(T'')}\mathbf{Z}^{(T'')}$ tends in distribution to $\mathbf{U}N_q^C(\mathbf{0},\mathbf{I}) = N_q^C(\mathbf{0},\mathbf{I})$. Therefore any subsequence of $\mathbf{U}^{(T)}\mathbf{Z}^{(T)}$ has a subsequence tending in distribution to $N_q^C(\mathbf{0},\mathbf{I})$ and so $\mathbf{U}^{(T)}\mathbf{Z}^{(T)}$ must tend to $N_q^C(\mathbf{0},\mathbf{I})$.

Proof of Theorem 6.4.2 Consider λ of Case A to begin. From Lemma 6.3.1

$$\begin{aligned} d_Y^{(T)}\left(\frac{2\pi[s(T)+s]}{T}\right) &= \mathbf{A}\left(\frac{2\pi[s(T)+s]}{T}\right)\mathbf{d}_X^{(T)}\left(\frac{2\pi[s(T)+s]}{T}\right) \\ &\quad + d_\varepsilon^{(T)}\left(\frac{2\pi[s(T)+s]}{T}\right) + O(1) \\ &= \mathbf{A}(\lambda)\mathbf{d}_X^{(T)}\left(\frac{2\pi[s(T)+s]}{T}\right) + d_\varepsilon^{(T)}\left(\frac{2\pi[s(T)+s]}{T}\right) \\ &\quad + O(1) \end{aligned}$$

$s = 0, \pm 1, \ldots, \pm m$ where the error term $O(1)$ is uniform in s because $\mathbf{A}(\lambda)$ has a uniformly bounded first derivative and $\|\mathbf{d}_X^{(T)}(\alpha)\| = O(T)$. (The equations above may be compared with (6.3.7).) Now let $\mathbf{D}_Y$ denote the $1 \times (2m+1)$ matrix whose columns are the values $(2\pi T)^{-1/2} d_Y^{(T)}(2\pi[s(T)+s]/T)$, $s = 0, \pm 1, \ldots, \pm m$ with a similar definition for $\mathbf{D}_X$ and $\mathbf{D}_\varepsilon$. The equations above now take the form

$$\mathbf{D}_Y = \mathbf{A}(\lambda)\mathbf{D}_X + \mathbf{D}_\varepsilon + O(T^{-1/2}).$$

Let $\mathbf{U}^{(T)}$ indicate a $(2m+1) \times (2m+1)$ unitary matrix whose first r columns are the matrix $\mathbf{U}_1^{(T)} = \overline{\mathbf{D}}_X^\tau[\mathbf{D}_X\overline{\mathbf{D}}_X^\tau]^{-1/2}$. Write $\mathbf{U}^{(T)} = [\mathbf{U}_1^{(T)}\mathbf{U}_2^{(T)}]$. Applying $\mathbf{U}^{(T)}$ to the matrix equation above gives

$$\mathbf{D}_Y\mathbf{U}^{(T)} = \mathbf{A}(\lambda)\mathbf{D}_X\mathbf{U}^{(T)} + \mathbf{D}_\varepsilon\mathbf{U}^{(T)} + O(T^{-1/2}).$$

The first r columns of the latter give

$$\{\mathbf{A}^{(T)}(\lambda) - \mathbf{A}(\lambda)\}\mathbf{f}_{XX}^{(T)}(\lambda)^{1/2}(2m+1)^{1/2} = \mathbf{D}_\varepsilon\mathbf{U}_1^{(T)} + O(T^{-1/2}).$$

The remaining give

$$\mathbf{D}_Y\mathbf{U}_2^{(T)} = \mathbf{D}_\varepsilon\mathbf{U}_2^{(T)} + O(T^{-1/2}).$$

Because $\mathbf{U}^{(T)}$ is unitary we have

$$\begin{aligned} (2m+1)\mathbf{f}_{YY}^{(T)}(\lambda) &= \mathbf{D}_Y\overline{\mathbf{D}}_Y^\tau \\ &= \mathbf{D}_Y\mathbf{U}_1^{(T)}\overline{\mathbf{U}}_1^{(T)\tau}\overline{\mathbf{D}}_Y^\tau + \mathbf{D}_Y\mathbf{U}_2^{(T)}\overline{\mathbf{U}}_2^{(T)\tau}\overline{\mathbf{D}}_Y^\tau \\ &= \mathbf{D}_Y\overline{\mathbf{D}}_X^\tau(\mathbf{D}_X\overline{\mathbf{D}}_X^\tau)^{-1}\mathbf{D}_X\overline{\mathbf{D}}_Y^\tau + \mathbf{D}_Y\mathbf{U}_2^{(T)}\overline{\mathbf{U}}_2^{(T)\tau}\overline{\mathbf{D}}_Y^\tau \end{aligned}$$

and so

$$g_{\varepsilon\varepsilon}^{(T)}(\lambda) = \mathbf{D}_Y\mathbf{U}_2^{(T)}\overline{\mathbf{U}}_2^{(T)\tau}\overline{\mathbf{D}}_Y^{\tau}$$
$$= \mathbf{D}_\varepsilon\mathbf{U}_2^{(T)}\overline{\mathbf{U}}_2^{(T)\tau}\overline{\mathbf{D}}_\varepsilon^{\tau} + O_p(T^{-1/2})$$

where $O_p(1)$ denotes a variate bounded in probability.

Now Theorem 4.4.1 applies indicating that because the series $\varepsilon(t)$, $t = 0, \pm 1, \ldots$ satisfies Assumption 2.6.1, $\mathbf{D}_\varepsilon^\tau$ tends to $N_{2m+1}^C(\mathbf{0}, f_{\varepsilon\varepsilon}(\lambda)\mathbf{I})$. Therefore $f_{\varepsilon\varepsilon}(\lambda)^{-1/2}\mathbf{D}_\varepsilon^\tau$ tends to $N_{2m+1}^C(\mathbf{0},\mathbf{I})$. Lemma P6.2 applies indicating that $f_{\varepsilon\varepsilon}(\lambda)^{-1/2}(\mathbf{D}_\varepsilon\mathbf{U}^{(T)})^\tau$ also tends to $N_{2m+1}^C(\mathbf{0},\mathbf{I})$ and so $(\mathbf{D}_\varepsilon\mathbf{U}^{(T)})^\tau$ tends to $N_{2m+1}^C(\mathbf{0}, f_{\varepsilon\varepsilon}(\lambda)\mathbf{I})$. The indicated asymptotic behavior of $\mathbf{A}^{(T)}(\lambda)$ and $g_{\varepsilon\varepsilon}^{(T)}(\lambda)$ now follows from the representations obtained for them above.

If λ is of Case B or Case C, then the above form of argument goes through with the unitary matrix replaced by an orthogonal one. The behavior of $\mu^{(T)}$ follows from its dependence on $\mathbf{A}^{(T)}(0)$.

We need the following lemma.

Lemma P6.3 (*Skorokhod* (*1956*)) Let $\mathbf{V}^{(T)}$, $T = 1, 2, \ldots$ be a sequence of vector-valued random variables tending in distribution to a random variable $\mathbf{V}$. Then, moving to an equivalent probability structure, we may write

$$\mathbf{V}^{(T)} = \mathbf{V} + o_{a.s.}(1).$$

This lemma provides us with another proof of Lemma P6.2. We may write

$$\mathbf{Z}^{(T)} = \mathbf{Z} + o_{a.s.}(1)$$

where $\mathbf{Z}$ is $N_q^C(\mathbf{0},\mathbf{I})$ and so

$$\mathbf{U}^{(T)}\mathbf{Z}^{(T)} = \mathbf{U}^{(T)}\mathbf{Z} + o_{a.s.}(1)$$

and $\mathbf{U}^{(T)}\mathbf{Z}$ is $N_q^C(\mathbf{0},\mathbf{I})$ for all T.

Proof of Theorem 6.4.3 The last lemma shows that we may write

$$d_\varepsilon^{(T)}\left(\frac{2\pi[s(T) + s]}{T}\right) = \zeta_s + o_{a.s.}(\sqrt{T}) \qquad s = 0, \pm 1, \ldots, \pm m,$$

where the ζ_s are independent $N_1^C(0, 2\pi T f_{\varepsilon\varepsilon}(\lambda))$ variates. Let $\lambda_s = 2\pi[s(T) + s]/T$. We may make the substitution $d_Y^{(T)}(\lambda_s) = \mathbf{A}(\lambda)\mathbf{d}_X^{(T)}(\lambda_s) + \zeta_s + o_{a.s.}(\sqrt{T})$. We have the sum of squares identity,

$$(2\pi T)^{-1}\sum_s |d_Y^{(T)}(\lambda_s)|^2 = (2\pi T)^{-1}\mathbf{A}^{(T)}(\lambda)\sum_s \mathbf{d}_X^{(T)}(\lambda_s)\overline{\mathbf{d}_X^{(T)}(\lambda_s)}^\tau\overline{\mathbf{A}^{(T)}(\lambda)}^\tau$$
$$+ (2\pi T)^{-1}\sum_s |d_Y^{(T)}(\lambda_s) - \mathbf{A}^{(T)}(\lambda)\mathbf{d}_X^{(T)}(\lambda_s)|^2.$$

The terms appearing are quadratic forms of ranks $2m + 1$, r, $2m + 1 - r$ respectively in the ζ_s, plus terms tending to 0 with probability 1. Exercise 4.8.7 applies to indicate that the first term on the right here may be written

$f_{\varepsilon\varepsilon}(\lambda)\chi'^2_{2r}(\mathbf{A}(\lambda)\mathbf{f}_{XX}{}^{(T)}(\lambda)\overline{\mathbf{A}(\lambda)}^{\tau}/f_{\varepsilon\varepsilon}(\lambda))/2 + o_{a.s.}(1)$, while the second term may be written $f_{\varepsilon\varepsilon}(\lambda)\chi^2_{2(2m+1-r)}/2 + o_{a.s.}(1)$ with the χ^2 variates independent. Expression (6.4.12) now follows by elementary algebra.

Proof of Theorem 6.5.1 Let $R(t) = \sum_u \mathbf{a}(t-u)\mathbf{X}(u)$. Now, because $E\varepsilon(t) = 0$, we have

$$E\mathbf{A}^{(T)}(\lambda) = \mathbf{f}_{RX}{}^{(T)}(\lambda)\mathbf{f}_{XX}{}^{(T)}(\lambda)^{-1}.$$

Let $d_R{}^{(T)}(\beta) = \mathbf{A}(\beta)\mathbf{d}_X{}^{(T)}(\beta) + e^{(T)}(\beta)$. From Lemma 6.3.1, $e^{(T)}(\beta)$ is uniformly bounded. By substitution we therefore have

$$\begin{aligned}\mathbf{f}_{RX}{}^{(T)}(\lambda) &= 2\pi T^{-1}\sum_{s=1}^{T-1} W^{(T)}\left(\lambda - \frac{2\pi s}{T}\right)(2\pi T)^{-1}\left[\mathbf{A}\left(\frac{2\pi s}{T}\right)\mathbf{d}_X{}^{(T)}\left(\frac{2\pi s}{T}\right)\right.\\ &\qquad \left. + e^{(T)}\left(\frac{2\pi s}{T}\right)\right]\overline{\mathbf{d}_X{}^{(T)}\left(\frac{2\pi s}{T}\right)}^{\tau}\\ &= 2\pi T^{-1}\sum_{s=1}^{T-1} W^{(T)}\left(\lambda - \frac{2\pi s}{T}\right)\mathbf{A}\left(\frac{2\pi s}{T}\right)\mathbf{I}_{XX}{}^{(T)}\left(\frac{2\pi s}{T}\right) + \mathbf{G}^{(T)}\end{aligned}$$

where, following Lemma P6.1,

$$\begin{aligned}\|\mathbf{G}^{(T)}\mathbf{f}_{XX}{}^{(T)}(\lambda)^{-1}\| &\leqslant \left\{\sum_{s=1}^{T-1} W^{(T)}\left(\lambda - \frac{2\pi s}{T}\right)(2\pi T)^{-1}\left|e^{(T)}\left(\frac{2\pi s}{T}\right)\right|^2\right\}^{1/2}\\ &\quad\times\left\{\left\|\left(\sum_{s=1}^{T-1} W^{(T)}\left(\lambda - \frac{2\pi s}{T}\right)\mathbf{I}_{XX}{}^{(T)}\left(\frac{2\pi s}{T}\right)\right)^{-1}\right\|^{1/2}\right\}\\ &\leqslant KT^{-1/2}\|\mathbf{f}_{XX}{}^{(T)}(\lambda)^{-1}\|^{1/2}\end{aligned}$$

for some finite K, where we use the facts that $e^{(T)}(\beta)$ is bounded and that $W(\beta)$ is non-negative. The first part of (6.5.14) now follows from Assumption 6.5.2. Turning to the second part: suppose $0 \leqslant \lambda < 2\pi$. The region in which $W^{(T)}$ is nonzero is $|\lambda - (2\pi s/T)| \leqslant B_T\pi$. In this region, $\mathbf{A}(2\pi s/T) = \mathbf{A}(\lambda) + O(B_T)$ because under the given assumptions $\mathbf{A}(\beta)$ has a uniformly bounded first derivative. The proof of the theorem is now completed by the substitution of this last into the first expression of (6.5.14).

Proof of Theorem 6.5.2 To begin we note that

$$\begin{aligned}E\,|\,|A_j{}^{(T)}(\lambda)| - |EA_j{}^{(T)}(\lambda)|\,| &\leqslant E|A_j{}^{(T)}(\lambda) - EA_j{}^{(T)}(\lambda)|\\ &\leqslant \{\mathrm{var}\, A_j{}^{(T)}(\lambda)\}^{1/2}\\ &= O(B_T^{-1/2}T^{-1/2})\end{aligned}$$

from (6.6.3), and so

$$|EG_j{}^{(T)}(\lambda) - |EA_j{}^{(T)}(\lambda)|\,| = O(B_T^{-1/2}T^{-1/2})$$

giving the first part of (6.5.19). The second part follows from (6.5.14) and the fact that $|\alpha + \varepsilon| = |\alpha| + 0(\varepsilon)$.

To prove the first parts of (6.5.20) and (6.5.21) we use the Taylor series expansions

$$\log|\zeta+\varepsilon| = \log|\zeta| + \frac{1}{2}\{\varepsilon/\zeta + \bar{\varepsilon}/\bar{\zeta}\} - \frac{1}{4}\{\varepsilon^2/\zeta^2 + \bar{\varepsilon}^2/\bar{\zeta}^2\} + \cdots$$

$$\arg[\zeta+\varepsilon] = \arg\zeta + \frac{1}{2i}\{\varepsilon/\zeta - \bar{\varepsilon}/\bar{\zeta}\} - \frac{1}{4i}\{\varepsilon^2/\zeta^2 - \bar{\varepsilon}^2/\bar{\zeta}^2\} + \cdots$$

taking $\zeta + \varepsilon = A_j^{(T)}(\lambda)$, $\zeta = EA_j^{(T)}(\lambda)$ and using (6.6.3). To prove the second parts, we again use these expansions; however, this time with $\zeta + \varepsilon = EA_j^{(T)}(\lambda)$, $\zeta = A_j(\lambda)$ and using (6.5.14).

Before developing the remaining proofs of this section we must first set down some notation and prove some lemmas. We define

$$d_\varepsilon^{(T)}(\lambda) = \sum_{t=0}^{T-1} \varepsilon(t)\exp\{-i\lambda t\}$$

$$f_{\varepsilon\varepsilon}^{(T)}(\lambda) = 2\pi T^{-1}\sum_{s=1}^{T-1} W^{(T)}\left(\lambda - \frac{2\pi s}{T}\right)(2\pi T)^{-1}\left|d_\varepsilon^{(T)}\left(\frac{2\pi s}{T}\right)\right|^2$$

$$\mathbf{f}_{X\varepsilon}^{(T)}(\lambda) = 2\pi T^{-1}\sum_{s=1}^{T-1} W^{(T)}\left(\lambda - \frac{2\pi s}{T}\right)(2\pi T)^{-1}\mathbf{d}_X^{(T)}\left(\frac{2\pi s}{T}\right)\overline{d_\varepsilon^{(T)}\left(\frac{2\pi s}{T}\right)}$$

$$g_{R\varepsilon}^{(T)}(\lambda) = f_{R\varepsilon}^{(T)}(\lambda) - \mathbf{A}(\lambda)\mathbf{f}_{X\varepsilon}^{(T)}(\lambda).$$

Because the series $\varepsilon(t)$, $t = 0, \pm 1, \ldots$ is unobservable, these variates are unobservable. However, we will see that the statistics of interest are elementary functions of these variates. Continuing to set up a notation, let $[\mathbf{d}_X^{(T)}(\lambda)]_k$ denote the kth entry of $\mathbf{d}_X^{(T)}(\lambda)$ with a similar notation for the entries of $\mathbf{f}_{X\varepsilon}^{(T)}(\lambda)$. We have

Lemma P6.4 If $\mathbf{f}_{XX}^{(T)}(\lambda)$ is uniformly bounded, then

$$\sum_{r=1}^{T-1} W^{(T)}\left(\lambda - \frac{2\pi r}{T}\right)\left[\mathbf{d}_X^{(T)}\left(\frac{2\pi r}{T}\right)\right]_k = O(T^{3/2}).$$

The error term is uniform in λ.

Proof We have

$$\left|2\pi T^{-1}\sum_r W^{(T)}\left(\lambda - \frac{2\pi r}{T}\right)\left[\mathbf{d}_X^{(T)}\left(\frac{2\pi r}{T}\right)\right]_k\right|$$
$$\leqslant \left\{2\pi T^{-1}\sum_r W^{(T)}\left(\lambda - \frac{2\pi r}{T}\right)\left|\left[\mathbf{d}_X^{(T)}\left(\frac{2\pi r}{T}\right)\right]_k\right|^2\right\}^{1/2}$$
$$\leqslant (2\pi T)^{1/2}\|\mathbf{f}_{XX}^{(T)}(\lambda)\|^{1/2}$$

from which the result follows.

Lemma P6.5 If $\mathbf{f}_{XX}^{(T)}(\lambda)$ is uniformly bounded, then

$$\sum_{r=1}^{T-1} W^{(T)}\left(\lambda - \frac{2\pi r}{T}\right)W^{(T)}\left(\mu - \frac{2\pi r}{T}\right)\left[d_X^{(T)}\left(\frac{2\pi r}{T}\right)\right]_k\left[d_X^{(T)}\left(\frac{2\pi(r+s)}{T}\right)\right]_l$$
$$= O(B_T^{-1}T^2).$$

The error term is uniform in λ, μ.

Proof By virtue of Schwarz's inequality the absolute value of the expression at issue is

$$\leqslant \left\{\sum_{r=1}^{T-1} W^{(T)}\left(\lambda - \frac{2\pi r}{T}\right)^2 \left|\left[\mathbf{d}_X{}^{(T)}\left(\frac{2\pi r}{T}\right)\right]_k\right|^2\right\}^{1/2}$$
$$\times \left\{\sum_{r=1}^{T-1} W^{(T)}\left(\mu - \frac{2\pi r}{T}\right)^2 \left|\left[\mathbf{d}_X{}^{(T)}\left(\frac{2\pi(r+s)}{T}\right)\right]_l\right|^2\right\}^{1/2}$$
$$\leqslant \{B_T{}^{-1}T^2\|\mathbf{f}_{XX}(\lambda)^{(T)}\|\}^{1/2}\left\{B_T{}^{-1}T^2\left\|\mathbf{f}_{XX}{}^{(T)}\left(\mu - \frac{2\pi s}{T}\right)\right\|\right\}^{1/2}$$

giving the desired result.

Lemma P6.6 Under the conditions of Theorem 6.5.1,

$$\begin{aligned}\text{cum } \{g_{R_\varepsilon}{}^{(T)}(\lambda_1), \ldots, g_{R_\varepsilon}{}^{(T)}(\lambda_L), [\mathbf{f}_{X_\varepsilon}{}^{(T)}(\mu_1)]_{m_1}, \ldots, [\mathbf{f}_{X_\varepsilon}{}^{(T)}(\mu_M)]_{m_M}, g_{\varepsilon\varepsilon}{}^{(T)}(\nu_1),\\ \ldots, g_{\varepsilon\varepsilon}{}^{(T)}(\nu_N)\} &= O(T^{-N+1}B_T^{-N+1}) && \text{if } L+M=0\\ &= O(T^{(1/2)-N}B_T^{-N+1}) && \text{if } L=0,\ M=1\\ &= O(T^{(1/2)-N}B_T^{-N+2}) && \text{if } L=1,\ M=0\\ &= O(T^{-(L/2)-(M/2)-N}B_T^{L-1}) && \text{if } L+M>1.\end{aligned}$$

Proof We begin by noting that

$$g_{R_\varepsilon}{}^{(T)}(\lambda) = 2\pi T^{-1}\sum_{q=1}^{T-1} W^{(T)}\left(\lambda - \frac{2\pi q}{T}\right)(2\pi T)^{-1}\left[\mathbf{A}\left(\frac{2\pi q}{T}\right)\mathbf{d}_X{}^{(T)}\left(\frac{2\pi q}{T}\right) + O(1)\right.$$
$$\left. - \mathbf{A}(\lambda)\mathbf{d}_X{}^{(T)}\left(\frac{2\pi q}{T}\right)\right]\overline{d_\varepsilon{}^{(T)}\left(\frac{2\pi q}{T}\right)}$$

if we use Lemma 6.3.1. Because

$$\mathbf{A}\left(\frac{2\pi q}{T}\right) = \mathbf{A}(\lambda) + O(B_T)$$

for $|\lambda - (2\pi q/T)| \leqslant B_T\pi$, we have

$$g_{R_\varepsilon}{}^{(T)}(\lambda) = 2\pi T^{-1}\sum_{q=1}^{T-1} W^{(T)}\left(\lambda - \frac{2\pi q}{T}\right)(2\pi T)^{-1}$$
$$\times\left[O(B_T)\mathbf{d}_X{}^{(T)}\left(\frac{2\pi q}{T}\right) + O(1)\right]\overline{d_\varepsilon{}^{(T)}\left(\frac{2\pi q}{T}\right)}.$$

The cumulant in question is given by

$$A = (2\pi T^{-1})^{L+M+N}\sum_{q_1}\cdots\sum_{q_L}\sum_{r_1}\cdots\sum_{r_M}\sum_{s_1}\cdots\sum_{s_N}\left[O(B_T)\mathbf{d}_X{}^{(T)}\left(\frac{2\pi q_1}{T}\right) + O(1)\right]$$
$$\times\left[O(B_T)\mathbf{d}_X{}^{(T)}\left(\frac{2\pi q_L}{T}\right) + O(1)\right]\left[\mathbf{d}_X{}^{(T)}\left(\frac{2\pi r_1}{T}\right)\right]_{m_1}\cdots\left[\mathbf{d}_X{}^{(T)}\left(\frac{2\pi r_M}{T}\right)\right]_{m_N}$$
$$\times W^{(T)}\left(\lambda_1 - \frac{2\pi q_1}{T}\right)\cdots W^{(T)}\left(\nu_N - \frac{2\pi s_N}{T}\right)(2\pi T)^{-L-M-N}$$

$$\times \operatorname{cum}\left\{d_\varepsilon^{(T)}\left(\frac{2\pi q_1}{T}\right), \ldots, d_\varepsilon^{(T)}\left(\frac{2\pi r_M}{T}\right), \left|d_\varepsilon^{(T)}\left(\frac{2\pi s_1}{T}\right)\right|^2, \ldots, \times \left|d_\varepsilon^{(T)}\left(\frac{2\pi s_N}{T}\right)\right|^2\right\}.$$

The cumulant appearing in the last expression has principal term

$$(2\pi)^{m+\rho_1-1}\Delta^{(T)}\left(\frac{2\pi}{T}\{q_1+\cdots+r_M+t_1+\cdots+t_{\rho_1}\}\right)f_{\varepsilon\ldots\varepsilon}\left(\frac{2\pi q_1}{T}, \ldots, \frac{2\pi t_{\rho_1}}{T}\right)$$
$$\times (2\pi)^{\rho_2-\rho_1-1}\Delta^{(T)}\left(\frac{2\pi}{T}\{t_{\rho_1+1}+\cdots+t_{\rho_2}\}\right)f_{\varepsilon\ldots\varepsilon}\left(\frac{2\pi t_{\rho_1+1}}{T}, \ldots, \frac{2\pi t_{\rho_2+1}}{T}\right)\cdots$$
$$(2\pi)^{\rho_p-\rho_{p-1}-1}\Delta^{(T)}\left(\frac{2\pi}{T}\{t_{\rho_{p-1}+1}+\cdots+t_{\rho_p}\}\right)f_{\varepsilon\ldots\varepsilon}\left(\frac{2\pi t_{\rho_{p-1}+1}}{T}, \ldots, \frac{2\pi t_{\rho_p}}{T}\right)$$

where $\rho_p = 2N$ and $t_1, \ldots, t_{2N}$ is a permutation of $(s_1, -s_1; \cdots; s_N, -s_N)$ corresponding to an indecomposable partition. We have $p \leqslant n$. We now use Lemmas P6.4, P6.5 to eliminate the summations on q and r and see that the principal term in A is

$$O\left(T^{-2L-2M-2N}B_T{}^L(T^{3/2})^{L+M-2}B_T{}^{-1}T^2T\sum_{s_1}\cdots\sum_{s_N}W^{(T)}\left(\nu_1-\frac{2\pi s_1}{T}\right)\right.$$
$$\cdots W^{(T)}\left(\nu_N-\frac{2\pi s_N}{T}\right)\left|\Delta^{(T)}\left(\frac{2\pi}{T}\{t_{\rho_1+1}+\cdots+t_{\rho_2}\}\right)\right.$$
$$\left.\left.\cdots\Delta^{(T)}\left(\frac{2\pi}{T}\{t_{\rho_{p-1}+1}+\cdots+t_{\rho_p}\}\right)\right|\right)$$
$$= O(T^{-2L-2M-2N}B_T^{L-1}(T^{3/2})^{L+M-2}T^3T^{p-1}T^{N-p+1})$$

giving the indicated result for $L + M > 1$. The other expressions follow in a similar manner.

These estimates of the order of the joint cumulants are sufficient for certain purposes; however, they are too crude in the second-order case. In greater detail in that case we have,

Lemma P6.7 Under the conditions of Theorem 6.5.1,

$$\operatorname{cov}\{[\mathbf{f}_{\varepsilon X}{}^{(T)}(\lambda)]_k, [\mathbf{f}_{\varepsilon X}{}^{(T)}(\mu)]_l\} = (2\pi T^{-1})^2\left\{\sum_s W^{(T)}\left(\lambda-\frac{2\pi s}{T}\right)W^{(T)}\left(\mu-\frac{2\pi s}{T}\right)\right.$$
$$\left.\times f_{\varepsilon\varepsilon}\left(\frac{2\pi s}{T}\right)\left[\mathbf{I}_{XX}{}^{(T)}\left(\frac{2\pi s}{T}\right)\right]_{kl}\right\} + O(T^{-1})$$
$$= f_{\varepsilon\varepsilon}(\lambda)(2\pi T^{-1})^2\left\{\sum_s W^{(T)}\left(\lambda-\frac{2\pi s}{T}\right)W^{(T)}\left(\mu-\frac{2\pi s}{T}\right)\right.$$
$$\left.\times\left[\mathbf{I}_{XX}{}^{(T)}\left(\frac{2\pi s}{T}\right)\right]_{kl}\right\} + O(T^{-1})$$
$$= O(B_T{}^{-1}T^{-1})$$

$$\operatorname{cov}\{[\mathbf{f}_{\varepsilon X}{}^{(T)}(\lambda)]_k, f_{\varepsilon\varepsilon}{}^{(T)}(\mu)\} = O(T^{-3/2})$$

$$\operatorname{cov}\{f_{\varepsilon\varepsilon}^{(T)}(\lambda), f_{\varepsilon\varepsilon}^{(T)}(\mu)\} = f_{\varepsilon\varepsilon}(\lambda)^2(2\pi T^{-1})^2 \sum_s \left\{W^{(T)}\left(\lambda - \frac{2\pi s}{T}\right)W^{(T)}\left(\mu - \frac{2\pi s}{T}\right)\right.$$
$$\left. + W^{(T)}\left(\lambda - \frac{2\pi s}{T}\right)W^{(T)}\left(\mu + \frac{2\pi s}{T}\right)\right\} + O(T^{-1})$$
$$= O(B_T^{-1}T^{-1}).$$

Proof Consider the second of these expressions. Following the first expression of the proof of Lemma P6.4, the required covariance is

$$T^{-4} \sum_{r\neq 0} \sum_s W^{(T)}\left(\lambda - \frac{2\pi r}{T}\right)W^{(T)}\left(\mu - \frac{2\pi s}{T}\right)\left[\mathbf{d}_X^{(T)}\left(\frac{2\pi r}{T}\right)\right]_k$$
$$\times \left\{(2\pi)^2\Delta^{(T)}\left(\frac{2\pi r}{T}\right)f_{\varepsilon\varepsilon\varepsilon}\left(\frac{2\pi r}{T}, \frac{2\pi s}{T}\right) + O(1)\right\} = O(T^{-4}T^{3/2}T).$$

The other covariances also follow from the expression of Lemma P6.4, the fact that $f_{\varepsilon\varepsilon}(\lambda)$ has a uniformly bounded derivative and the fact that the support of $W^{(T)}(\alpha)$ is $|\alpha| \leqslant B_T\pi$.

In the lemma below we let $C_T = B_T + T^{-1/2}$.

Lemma P6.8 Let $R(t) = \sum_u \mathbf{a}(t - u)\mathbf{X}(u)$. Under the assumptions of Theorem 6.5.1,

$$\mathbf{f}_{RX}^{(T)}(\lambda) = \mathbf{A}(\lambda)\,\mathbf{f}_{XX}^{(T)}(\lambda) + O(C_T)$$
$$\mathbf{f}_{YX}^{(T)}(\lambda) = \mathbf{f}_{RX}^{(T)}(\lambda) + \mathbf{f}_{\varepsilon X}^{(T)}(\lambda)$$
$$f_{RR}^{(T)}(\lambda) = \mathbf{A}(\lambda)\mathbf{f}_{XX}^{(T)}(\lambda)\overline{\mathbf{A}(\lambda)}^\tau + O(C_T)$$
$$\mathbf{f}_{\varepsilon X}^{(T)}(\lambda)\mathbf{f}_{XX}^{(T)}(\lambda)^{-1}\mathbf{f}_{X\varepsilon}^{(T)}(\lambda) = O\Big(\sum_k |[\mathbf{f}_{X\varepsilon}^{(T)}(\lambda)]_k|^2\Big)$$
$$g_{\varepsilon\varepsilon}^{(T)}(\lambda) = f_{\varepsilon\varepsilon}^{(T)}(\lambda) - \mathbf{f}_{\varepsilon X}^{(T)}(\lambda)\mathbf{f}_{XX}^{(T)}(\lambda)^{-1}\mathbf{f}_{X\varepsilon}^{(T)}(\lambda) + O(C_T)$$
$$+ g_{R\varepsilon}^{(T)}(\lambda) + g_{\varepsilon R}^{(T)}(\lambda)$$
$$= f_{\varepsilon\varepsilon}^{(T)}(\lambda) + O\Big(\sum_k |[\mathbf{f}_{X\varepsilon}^{(T)}(\lambda)]_k|^2\Big) + O(C_T)$$
$$+ g_{R\varepsilon}^{(T)}(\lambda) + g_{\varepsilon R}^{(T)}(\lambda).$$

Proof We derived the first expression in the course of the proof of (6.5.14). The second is immediate. For the third we note

$$f_{RR}^{(T)}(\lambda) = 2\pi T^{-1} \sum_s W^{(T)}\left(\lambda - \frac{2\pi s}{T}\right)(2\pi T)^{-1}\left\{\mathbf{A}\left(\frac{2\pi s}{T}\right)\mathbf{d}_X^{(T)}\left(\frac{2\pi s}{T}\right) + O(1)\right\}$$
$$\times \overline{\left\{\mathbf{d}_X^{(T)}\left(\frac{2\pi s}{T}\right)\mathbf{A}\left(\frac{2\pi s}{T}\right) + O(1)\right\}}^\tau$$

from which the indicated result follows. For the next

$$\|\mathbf{f}_{\varepsilon X}^{(T)}(\lambda)\mathbf{f}_{XX}^{(T)}(\lambda)^{-1}\mathbf{f}_{X\varepsilon}^{(T)}(\lambda)\| \leqslant K \operatorname{tr}\{\mathbf{f}_{X\varepsilon}^{(T)}(\lambda)\mathbf{f}_{\varepsilon X}^{(T)}(\lambda)\mathbf{f}_{XX}^{(T)}(\lambda)^{-1}\}$$
$$\leqslant L \operatorname{tr}\{\mathbf{f}_{X\varepsilon}^{(T)}(\lambda)\mathbf{f}_{\varepsilon X}^{(T)}(\lambda)\}$$

for finite K and L following Assumption 6.5.2. For the final statement we note that

$$g_{\varepsilon\varepsilon}^{(T)}(\lambda) = f_{RR}^{(T)}(\lambda) - f_{R\varepsilon}^{(T)}(\lambda) + f_{\varepsilon R}^{(T)}(\lambda) + f_{\varepsilon\varepsilon}^{(T)}(\lambda)$$
$$- \{\mathbf{f}_{RX}^{(T)}(\lambda) + \mathbf{f}_{\varepsilon X}^{(T)}(\lambda)\}\mathbf{f}_{XX}^{(T)}(\lambda)^{-1}\{\mathbf{f}_{XR}^{(T)}(\lambda) + \mathbf{f}_{X\varepsilon}^{(T)}(\lambda)\}$$

and the result follows from the earlier expressions of the lemma.

Proof of Theorem 6.5.3 From Lemma P6.8, we see that

$$Eg_{\varepsilon\varepsilon}^{(T)}(\lambda) = Ef_{\varepsilon\varepsilon}^{(T)}(\lambda) + O(\sum_k E|[\mathbf{f}_{\varepsilon X}^{(T)}(\lambda)]_k|^2) + O(C_T)$$
$$= Ef_{\varepsilon\varepsilon}^{(T)}(\lambda) + O(B_T^{-1}T^{-1}) + O(C_T)$$

from Lemma P6.7. From Theorem 5.6.1, we see that

$$Ef_{\varepsilon\varepsilon}^{(T)}(\lambda) = f_{\varepsilon\varepsilon}(\lambda) + O(B_T) + O(B_T^{-1}T^{-1})$$

and we have the indicated result.

Proof of Theorem 6.5.4 From (6.3.2) and Lemma 6.3.1 we see

$$Ed_Y^{(T)}(0) = T\mu + \mathbf{A}(0)\mathbf{d}_X^{(T)}(0) + O(1).$$

This gives

$$Ec_Y^{(T)} = \mu + \mathbf{A}(0)\mathbf{c}_X^{(T)} + O(T^{-1}).$$

Therefore

$$E\mu^{(T)} = E\{c_Y^{(T)} - \mathbf{A}^{(T)}(0)\mathbf{c}_X^{(T)}\}$$
$$= \mu + \mathbf{A}(0)\mathbf{c}_X^{(T)} - \{\mathbf{A}(0) + O(B_T) + O(T^{-1/2})\}\mathbf{c}_X^{(T)} + O(T^{-1})$$

using (6.5.14). The result now follows because under the indicated boundedness of $\mathbf{X}(t)$, $t = 0, \pm 1, \ldots$, $\mathbf{c}_X^{(T)}$ is uniformly bounded.

Proof of Theorem 6.6.1 Directly from the definition of $\mathbf{A}^{(T)}(\lambda)$, we see that

$$\text{cov}\,\{\mathbf{A}^{(T)}(\lambda)^\tau, \mathbf{A}^{(T)}(\mu)^\tau\} = \mathbf{f}_{XX}^{(T)}(\lambda)^{-1}\,\text{cov}\,\{\mathbf{f}_{X\varepsilon}^{(T)}(\lambda)\;\; \mathbf{f}_{X\varepsilon}^{(T)}(\mu)\}\mathbf{f}_{XX}^{(T)}(\mu)^{-1}$$

and (6.6.3) follows from the first expression of Lemma P6.7.

Proof of Theorem 6.6.2 As in the proof of Theorem 6.5.2, we have the Taylor expansions

$$\log G_j^{(T)}(\lambda) = \log |EA_j^{(T)}(\lambda)| + \frac{1}{2}\{[EA_j^{(T)}(\lambda)]^{-1}[A_j^{(T)}(\lambda) - EA_j^{(T)}(\lambda)]$$
$$+ [\overline{EA_j^{(T)}(\lambda)}]^{-1}[\overline{A_j^{(T)}(\lambda)} - \overline{EA_j^{(T)}(\lambda)}]\} + \cdots$$

$$\phi_j^{(T)}(\lambda) = \arg EA_j^{(T)}(\lambda) + \frac{1}{2i}\{[EA_j^{(T)}(\lambda)]^{-1}[A_j^{(T)}(\lambda) - EA_j^{(T)}(\lambda)]$$
$$- [\overline{EA_j^{(T)}(\lambda)}]^{-1}[\overline{A_j^{(T)}(\lambda)} - \overline{EA_j^{(T)}(\lambda)}]\} + \cdots.$$

The desired covariances now follow from these expansions and (6.6.3).

Proof of Theorem 6.6.3 From Lemma P6.8 we see

$$\operatorname{cov}\{g_{\varepsilon\varepsilon}^{(T)}(\lambda), g_{\varepsilon\varepsilon}^{(T)}(\mu)\} = \operatorname{cov}\{f_{\varepsilon\varepsilon}^{(T)}(\lambda), f_{\varepsilon\varepsilon}^{(T)}(\mu)\} + \text{a remainder.}$$

From Lemma P6.6 we see that the remainder term is

$$O(T^{-2}B_T^{-1} + T^{-1/2}B_T^{-1+1} + T^{-1}B_T^{-1+2}).$$

The indicated result now follows from (5.6.12).

Proof of Theorem 6.6.4 From (6.3.2) and Lemma 6.3.1 we see

$$c_Y^{(T)} = \mu + \mathbf{A}(0)\mathbf{c}_X^{(T)} + T^{-1}d_\varepsilon^{(T)}(0) + O(T^{-1}) = \mu^{(T)} + \mathbf{A}^{(T)}(0)\mathbf{c}_X^{(T)}.$$

Expression (6.6.13) now follows from Theorem 4.3.1.

Proof of Theorem 6.6.5 The first covariance required is

$$\mathbf{f}_{XX}^{(T)}(\lambda)^{-1} \operatorname{cov}\{\mathbf{f}_{X\varepsilon}^{(T)}(\lambda), g_{\varepsilon\varepsilon}^{(T)}(\mu)\} = O(T^{-3/2}B_T^{-1} + T^{-1})$$

if we use the representation of Lemma P6.8 and Lemma P6.6.

The second covariance follows from the representation of $\mu^{(T)}$ given in the Proof of Theorem 6.6.4 and from Lemmas P6.6 and P6.8. The final covariance follows likewise.

Proof of Theorem 6.7.1 We prove the first part of this theorem by evaluating joint cumulants of order greater than 2 of $\mathbf{A}^{(T)}(\mu)$, $g_{\varepsilon\varepsilon}^{(T)}(\nu)$ and proving that, when appropriately standardized, these joint cumulants tend to 0. From Lemma P6.6 we see that

$$\begin{aligned}(B_TT)^{(M+N)/2} \operatorname{cum}\{[\mathbf{A}^{(T)}(\mu_1)]_{m_1}, \ldots, [\mathbf{A}^{(T)}(\mu_M)]_{m_M}, g_{\varepsilon\varepsilon}^{(T)}(\nu_1), \ldots, g_{\varepsilon\varepsilon}^{(T)}(\nu_N)\}\\ &= (B_TT)^{N/2}O(T^{-N+1}B_T^{-N+1}) \quad \text{if } M = 0\\ &= (B_TT)^{(1+N)/2}O(T^{(1/2)-N}B_T^{-N+1}) \quad \text{if } M = 1\\ &= (B_TT)^{(M+N)/2}O(T^{-(M/2)-N}B_T^{-1}) \quad \text{if } M > 1\end{aligned}$$

and these each tend to 0 as $T \to \infty$. The second part of the theorem follows similarly by evaluating joint cumulants.

Proof of Theorem 6.8.1 From (6.8.2) we see that

$$\begin{aligned}E\mathbf{a}^{(T)}(\mu) &= P_T^{-1} \sum_{p=0}^{P_T-1} E\{\mathbf{A}^{(T)}(2\pi p/P_T)\} \exp\{i2\pi pu/P_T\}\\ &= P_T^{-1} \sum_{p=0}^{P_T-1} \{\mathbf{A}(2\pi p/P_T) + O(B_T) + O(T^{-1/2})\} \exp\{i2\pi pu/P_T\}\end{aligned}$$

where the error terms are uniform. This gives the first part of (6.8.4); the second part follows algebraically.

Proof of Theorem 6.8.2 We begin by examining (6.6.3) and noting that

$$\operatorname{cov}\{\mathbf{A}^{(T)}(2\pi p/P_T)^\tau, \mathbf{A}^{(T)}(2\pi q/P_T)^\tau\} = O(T^{-1})$$

for $p \neq q$, $1 \leqslant p, q \leqslant P_T^{-1}$ because $B_T \leqslant P_T^{-1}$ and so

$$W^{(T)}\left(\frac{2\pi p}{P_T} - \frac{2\pi s}{T}\right)W^{(T)}\left(\frac{2\pi q}{P_T} - \frac{2\pi s}{T}\right) = 0$$

in this case. Now from (6.8.2) we see that

$$\begin{aligned}
&\operatorname{cov}\{\mathbf{a}^{(T)}(u)^\tau, \mathbf{a}^{(T)}(v)^\tau\} \\
&= P_T^{-2} \sum_{p,q=0}^{P_T-1} \operatorname{cov}\{\mathbf{A}^{(T)}(2\pi p/P_T)^\tau, \mathbf{A}^{(T)}(2\pi q/P_T)^\tau\} \exp\{i2\pi(pu - qv)/P_T\} \\
&= P_T^{-2} \sum_{p=0}^{P_T-1} B_T^{-1}T^{-1}2\pi \int W(\alpha)^2 d\alpha\, \mathbf{\Psi}_{XX}{}^{(T)}(2\pi p/P_T) f_{\epsilon\epsilon}(2\pi p/P_T) \\
&\quad \times \exp\{i2\pi p(u - v)/P_T\} + O(T^{-1})
\end{aligned}$$

from (6.6.3) giving (6.8.7).

Proof of Theorem 6.8.3 This follows as did the proof of Theorem 6.8.2; however, we use (6.6.14) rather than (6.6.3).

Proof of Theorem 6.8.4 We prove that the standardized joint cumulants of order greater than 2 of the variates of the theorem tend to 0. We have

$$\begin{aligned}
&(P_TB_TT)^{(J+K)/2} \operatorname{cum}\{[\mathbf{a}^{(T)}(u_1)]_{j_1}, \ldots, [\mathbf{a}^{(T)}(u_J)]_{j_J}, g_{\epsilon\epsilon}{}^{(T)}(\lambda_1), \ldots, g_{\epsilon\epsilon}{}^{(T)}(\lambda_K)\} \\
&\quad = (P_TB_TT)^{(J+K)/2}P_T^{-J} \sum_{p_1}\cdots\sum_{p_J} \exp\{i2\pi(p_1u_1 + \cdots + p_Ju_J)/P_T\} \\
&\qquad \times \operatorname{cum}\{[\mathbf{A}^{(T)}(2\pi p_1/P_T)]_{j_1}, \ldots, \\
&\qquad\quad [\mathbf{A}^{(T)}(2\pi p_J/P_T)]_{j_J}, g_{\epsilon\epsilon}{}^{(T)}(\lambda_1), \ldots, g_{\epsilon\epsilon}{}^{(T)}(\lambda_K)\} \\
&\quad = (P_TB_TT)^{(J+K)/2}P_T^{-1}O(T^{-(J/2)-K-1})
\end{aligned}$$

where we use Lemma P6.6 and also the remark at the end of its proof to eliminate one of the summations on p. The cumulant is seen to tend to 0 because $P_TB_T \to 0$ as $T \to \infty$.

PROOFS FOR CHAPTER 7

Proof of Theorem 7.2.1

$$d_a{}^{(T)}(\lambda) = \sum_t h_a\left(\frac{t}{T}\right)X_a(t) \exp\{-i\lambda t\}$$

and so

$$Ed_a{}^{(T)}(\lambda) = c_a \sum_t h_a{}^{(T)}\left(\frac{t}{T}\right) \exp\{-i\lambda t\} = c_aH_a{}^{(T)}(0).$$

We also have

$$d_a{}^{(T)}(\lambda) = \int_{-\pi}^{\pi} H_a{}^{(T)}(\lambda - \alpha)dZ_a(\alpha)$$

from the Cramér representation. It follows from this that

$$\operatorname{cov}\{d_a^{(T)}(\lambda), d_b^{(T)}(\lambda)\} = \int_{-\pi}^{\pi} H_a^{(T)}(\lambda - \alpha)H_b^{(T)}(-\lambda + \alpha) f_{ab}(\alpha)d\alpha$$
$$= \int_{-\pi}^{\pi} H_a^{(T)}(\alpha)H_b^{(T)}(-\alpha) f_{ab}(\lambda - \alpha)d\alpha.$$

Finally from Parseval's formula

$$2\pi \sum_t h_a\left(\frac{t}{T}\right)h_b\left(\frac{t}{T}\right) = \int_{-\pi}^{\pi} H_a^{(T)}(\alpha)H_b^{(T)}(-\alpha)d\alpha$$

and we have (7.2.7).

Proof of Corollary 7.2.1 Suppose $h_a(u) = 0$ for $u < 0$. (The general case follows by writing h_a as a function vanishing for $u < 0$ plus a function vanishing for $u \geqslant 0$.) Now

$$H_a^{(T)}(\lambda) = \sum_t h_a\left(\frac{t}{T}\right) \exp\{-i\lambda t\}$$
$$= \sum_t \Delta^{(t)}(\lambda)\left[h_a\left(\frac{t+1}{T}\right) - h_a\left(\frac{t}{T}\right)\right]$$

using the Abel transformation of Exercise 1.7.13. If V_a denotes the variation of $h_a(u)$ we see that

$$|H_a^{(T)}(\lambda)| \leqslant V_a \sup_t |\Delta^{(t)}(\lambda)|$$
$$\leqslant V_a \{\sin \delta/2\}^{-1} \qquad \text{for } 0 < \delta \leqslant |\lambda| \leqslant \pi \qquad (*)$$

At the same time

$$\int_{-\pi}^{\pi} H_a^{(T)}(\alpha)H_b^{(T)}(-\alpha)d\alpha = 2\pi \sum_t h_a\left(\frac{t}{T}\right)h_b\left(\frac{t}{T}\right) \qquad (**)$$
$$\backsim 2\pi T \int h_a(t)h_b(t)dt$$

(*) and (**) show that the term in $c_a c_b$ tends to 0 as $T \to \infty$ if $\lambda \not\equiv 0 \pmod{2\pi}$ or if c_a or $c_b = 0$. Next consider

$$\left\{\int H_a^{(T)}(\alpha)H_b^{(T)}(-\alpha)d\alpha\right\}^{-1} \int_{-\pi}^{\pi} H_a^{(T)}(\alpha)H_b^{(T)}(-\alpha) f_{ab}(\lambda - \alpha)d\alpha - f_{ab}(\lambda)$$
$$= \left\{\int H_a^{(T)}(\alpha)H_b^{(T)}(-\alpha)d\alpha\right\}^{-1} \int_{-\pi}^{\pi} H_a^{(T)}(\alpha)H_b^{(T)}(-\alpha)[f_{ab}(\lambda - \alpha) - f_{ab}(\lambda)]d\alpha. \qquad (***)$$

We split the region of integration here into the regions $|\alpha| < \delta$ and $|\alpha| \geqslant \delta$. In the first region, $|f_{ab}(\lambda - \alpha) - f_{ab}(\lambda)|$ may be made arbitrarily small by choice of δ as f_{ab} is continuous. Also there

$$\int_{|\alpha|<\delta} |H_a^{(T)}(\alpha)|\,|H_b^{(T)}(-\alpha)|d\alpha \leqslant \left\{\int |H_a^{(T)}(\alpha)|^2 d\alpha \int |H_b^{(T)}(\alpha)|^2 d\alpha\right\}^{1/2} = O(T).$$

In the second region, $f_{ab}(\lambda - \alpha) - f_{ab}(\lambda)$ is bounded and

$$\int_{|\alpha| \geq \delta} |H_a^{(T)}(\alpha)|\, |H_b^{(T)}(-\alpha)| d\alpha = O(1)$$

from (*). It therefore follows from (**) that (***) tends to 0 as $T \to \infty$.

Proof of Theorem 7.2.2 From Theorem 4.3.2

$$\begin{aligned}
&\operatorname{cov}\{d_{a_1}^{(T)}(\lambda)d_{b_1}^{(T)}(-\lambda),\, d_{a_2}^{(T)}(\mu)d_{b_2}^{(T)}(-\mu)\} \\
&\quad = [H_{a_1}^{(T)}(\lambda)c_{a_1} + O(1)][H_{a_2}^{(T)}(-\mu)c_{a_2} + O(1)] \\
&\qquad \times [2\pi H_{b_1b_2}^{(T)}(-\lambda + \mu)f_{b_1b_2}(-\lambda) + O(1)] + \text{three similar terms} \\
&\qquad + [H_{a_1}^{(T)}(\lambda)c_{a_1} + O(1)][(2\pi)^2 H_{b_1a_2b_2}^{(T)}(-\lambda) f_{b_1a_2b_2}(\lambda, -\lambda, \mu) + O(1)] \\
&\qquad + \text{three similar terms} \\
&\qquad + [2\pi H_{a_1a_2}^{(T)}(\lambda - \mu)f_{a_1a_2}(\lambda) + O(1)][2\pi H_{b_1b_2}^{(T)}(-\lambda + \mu)f_{b_1b_2}(-\lambda) + O(1)] \\
&\qquad + [2\pi H_{a_1b_2}^{(T)}(\lambda + \mu)f_{a_1b_2}(\lambda) + O(1)][2\pi H_{b_1a_2}^{(T)}(-\lambda - \mu)f_{b_1a_2}(-\lambda) + O(1)] \\
&\qquad + [(2\pi)^3 H_{a_1b_1a_2b_2}^{(T)}(0) f_{a_1b_1a_2b_2}(\lambda, -\lambda, \mu) + O(1)].
\end{aligned}$$

This gives the required result once we note that $H_{ab}^{(T)}, H_{abcd}^{(T)} = O(T)$.

Proof of Corollary 7.2.2 We simply consider in turn the cases $\lambda \pm \mu \equiv 0$ (mod 2π) and $\lambda \pm \pi \not\equiv 0$ (mod 2π).

Proof of Theorem 7.2.3 Theorem 4.4.2 indicates that $\mathbf{d}_X^{(T)}(\lambda_1), \ldots, \mathbf{d}_X^{(T)}(\lambda_J)$ are asymptotically independent $N_r^C(\mathbf{0}, 2\pi T[H_{ab}(0)f_{ab}(\lambda)])$ variates. Theorem P5.1 now indicates that

$$[\{2\pi T H_{ab}(0)\}^{-1} d_a^{(T)}(\lambda_j)\overline{d_b^{(T)}(\lambda_j)}]$$

$j = 1, \ldots, J$ are asymptotically independent $W_r^C(1, \mathbf{f}_{XX}(\lambda_j))$ variates. The conclusions of the theorem now follow as

$$2\pi \sum_t h_a\left(\frac{t}{T}\right) h_b\left(\frac{t}{T}\right) \sim 2\pi T H_{ab}(0).$$

Proof of Theorem 7.2.4 This follows from Theorem 4.4.1 as Theorem 7.2.3 followed from Theorem 4.4.2.

Proof of Theorem 7.2.5 This follows directly from Exercise 4.8.23 and Theorem P5.1.

Proof of Theorem 7.3.1 From Exercise 7.10.21.

$$E\mathbf{I}_{XX}^{(T)}\left(\frac{2\pi r}{T}\right) = (2\pi T)^{-1} \int_{-\pi}^{\pi} \left|\Delta^{(T)}\left(\frac{2\pi r}{T} - \alpha\right)\right|^2 \mathbf{f}_{XX}(\alpha) d\alpha \qquad (*)$$

for r an integer $\not\equiv 0$ (mod T). If $\lambda \not\equiv 0$ (mod π), this gives

$$\begin{aligned}
&E\mathbf{f}_{XX}^{(T)}(\lambda) \\
&\quad = (2m+1)^{-1} \sum_{s=-m}^{m} (2\pi T)^{-1} \int_{-\pi}^{\pi} \left|\Delta^{(T)}\left(\frac{2\pi s(T)}{T} + \frac{2\pi s}{T} - \alpha\right)\right|^2 \mathbf{f}_{XX}(\alpha) d\alpha
\end{aligned}$$

giving (7.3.6). If $\lambda \equiv 0 \pmod{2\pi}$ or $\lambda = \pm\pi, \pm 3\pi, \ldots$ with T even, then

$$Ef_{XX}^{(T)}(\lambda) = (2m)^{-1} \sum_{s=1}^{m} (2\pi T)^{-1}$$
$$\times \int_{-\pi}^{\pi} \left\{ \left|\Delta^{(T)}\left(\lambda + \frac{2\pi s}{T} - \alpha\right)\right|^2 + \left|\Delta^{(T)}\left(\lambda - \frac{2\pi s}{T} - \alpha\right)\right|^2 \right\} \mathbf{f}_{XX}(\alpha) d\alpha$$

giving (7.3.7). If $\lambda = \pm\pi, \pm 3\pi, \ldots$ with T odd, then

$$Ef_{XX}^{(T)}(\lambda) = (2m)^{-1} \sum_{s=1}^{m} (2\pi T)^{-1} \int_{-\pi}^{\pi} \left\{ \left|\Delta^{(T)}\left(\lambda - \frac{\pi}{T} + \frac{2\pi s}{T} - \alpha\right)\right|^2 \right\}$$
$$+ \left|\Delta^{(T)}\left(\lambda + \frac{\pi}{T} - \frac{2\pi s}{T} - \alpha\right)\right|^2 \mathbf{f}_{XX}(\alpha) d\alpha$$

giving (7.3.8).

Proof of Corollary 7.3.1 As $\mathbf{f}_{XX}(\alpha)$ is a uniformly continuous function of α, expression (*), of the above proof, tends to $\mathbf{f}_{XX}(\lambda)$ as $T \to \infty$ if $2\pi r/T \to \lambda$. This gives the indicated result.

Proof of Theorem 7.3.2 If r, s are integers with $2\pi r/T, 2\pi s/T \not\equiv 0 \pmod{2\pi}$, Exercise 7.10.22(a) gives

$$\operatorname{cov}\left\{ I_{a_1b_1}^{(T)}\left(\frac{2\pi r}{T}\right), I_{a_2b_2}^{(T)}\left(\frac{2\pi s}{T}\right) \right\} = \eta\left\{\frac{2\pi(r-s)}{T}\right\} f_{a_1b_1}\left(\frac{2\pi r}{T}\right) f_{a_2b_2}\left(-\frac{2\pi r}{T}\right)$$
$$+ \eta\left\{\frac{2\pi(r+s)}{T}\right\} f_{a_1b_2}\left(\frac{2\pi r}{T}\right) f_{b_1a_2}\left(-\frac{2\pi r}{T}\right) + O(T^{-1}).$$

This together with the fact that

$$f_{ab}\left(\frac{2\pi r}{T}\right) = f_{ab}(\lambda) + O(T^{-1})$$

for $(2\pi r/T) - \lambda = O(T^{-1})$ gives (7.3.13) in the case $\lambda \not\equiv 0 \pmod{\pi}$ as $2m + 1$ terms of the estimates match up, while the other terms have covariance $O(T^{-1})$. Turning to the case $\lambda \equiv 0 \pmod{\pi}$, from Exercise 7.10.22(b) and the fact that m terms match up, the covariance is given by

$$\eta\{\lambda - \mu\}(f_{a_1a_2}(\lambda) f_{b_1b_2}(-\lambda) + f_{a_1b_2}(\lambda) f_{b_1a_2}(-\lambda))/(2m)$$

and we check that this can be written in the manner (7.3.13).

Proof of Theorem 7.3.3 This theorem follows directly from Theorem 7.2.4 and Theorem P5.1.

Proof of Theorem 7.3.4 This follows directly from Theorem 7.2.1 and its corollary.

Proof of Theorem 7.3.5 The pseudo tapers

$$\begin{aligned} h_a(u,l) &= h_a(u - l) && \text{for } l \leqslant u < l + 1 \\ &= 0 && \text{otherwise} \end{aligned}$$

have the property

$$\sum_t h_a\left(\frac{t}{V},l\right)h_b\left(\frac{t}{V},m\right)\exp\{-it\lambda\} = H_{ab}^{(V)}(\lambda) \qquad \text{if } l = m$$
$$= 0 \qquad \text{if } l \neq m$$

for $l, m = 0, \ldots, L-1$. The general expression of the proof of Theorem 7.2.2 with appropriate redefinition, now shows that

$$\begin{aligned}
\operatorname{cov}\{I_{a_1b_1}^{(V)}(\lambda,l), \overline{I_{a_2b_2}^{(V)}(\lambda,m)}\}
&= H_{a_1b_1}^{(V)}(0)^{-1}H_{a_2b_2}^{(V)}(0)^{-1}\{H_{a_1a_2}^{(V)}(\lambda-\mu)\overline{H_{b_1b_2}^{(V)}(\lambda-\mu)} \\
&\quad \times f_{a_1a_2}(\lambda)f_{b_1b_2}(-\lambda) \\
&\quad + H_{a_1b_2}^{(V)}(\lambda+\mu)\overline{H_{b_1a_2}^{(V)}(\lambda+\mu)} \\
&\quad \times f_{a_1b_2}(\lambda)f_{b_1a_2}(-\lambda)\} + O(T^{-1}) && \text{if } l = m \\
&= O(V^{-1}) && \text{if } l \neq m.
\end{aligned}$$

This now gives (7.3.18).

Proof of Theorem 7.3.6 Follows directly from Theorem 7.2.5 and Theorem P5.1.

Proof of Theorem 7.4.1 From Theorem 7.2.1

$$EI_{ab}^{(T)}\left(\frac{2\pi s}{T}\right) = (2\pi T)^{-1}\int_{-\pi}^{\pi}\left[\frac{\sin T\left(\frac{2\pi s}{T}-\alpha\right)/2}{\sin\left(\frac{2\pi s}{T}-\alpha\right)/2}\right]^2 f_{ab}(\alpha)d\alpha$$

for $s = 1, \ldots, T-1$. This gives the first part of (7.4.9). Beginning the proof of the second part, the right side of the above expression has the form

$$\sum_{u=-T+1}^{T-1}\left(1-\frac{|u|}{T}\right)c_{ab}(u)\exp\{-i2\pi su/T\} = f_{ab}\left(\frac{2\pi s}{T}\right) + o(1)$$

where $o(1)$ is uniform in s, from Exercise 1.7.10. Using this we see

$$Ef_{ab}^{(T)}(\lambda) = \sum_{s=1}^{T-1} W_{ab}^{(T)}\left(\lambda - \frac{2\pi s}{T}\right)f_{ab}\left(\frac{2\pi s}{T}\right)\left\{\sum_{s=1}^{T-1} W_{ab}^{(T)}\left(\lambda-\frac{2\pi s}{T}\right)\right\}^{-1} + o(1)$$
$$\sim f_{ab}(\lambda) + \frac{2\pi}{T}\sum_{s=1}^{T-1} W_{ab}^{(T)}\left(\lambda-\frac{2\pi s}{T}\right)\left\{f_{ab}\left(\frac{2\pi s}{T}\right) - f_{ab}(\lambda)\right\} + o(1).$$

The second term on the right side here may be made arbitrarily small by splitting the range of summation into a segment where $|(2\pi s/T) - \lambda| < \delta$ implies $|f_{ab}(2\pi s/T) - f_{ab}(\lambda)| < \varepsilon$ and a remainder where $\sum W^{(T)}(\lambda - (2\pi s/T))$ tends to 0 and $|f_{ab}(2\pi s/T) - f_{ab}(\lambda)|$ is bounded. This completes the proof of (7.4.9).

Turning to (7.4.11), from Theorem 4.3.2

$$EI_{ab}^{(T)}\left(\frac{2\pi s}{T}\right) = f_{ab}\left(\frac{2\pi s}{T}\right) + O(T^{-1}) \qquad \text{for } s = 1, \ldots, T-1$$

with the error term uniform in s. This gives the first part of (7.4.11). The second follows from Lemma P5.1.

Proof of Theorem 7.4.2 By Taylor series expansion of $f_{ab}(\lambda - B_T\alpha)$ as a function of α.

Proof of Theorem 7.4.3 From expression (7.2.14)

$$\begin{aligned}\operatorname{cov}&\left\{I_{a_1b_1}^{(T)}\left(\frac{2\pi s}{T}\right), I_{a_2b_2}^{(T)}\left(\frac{2\pi r}{T}\right)\right\}\\ &= \eta\left\{\frac{2\pi(s-r)}{T}\right\} f_{a_1a_2}\left(\frac{2\pi s}{T}\right) f_{b_1b_2}\left(-\frac{2\pi s}{T}\right)\\ &\quad + \eta\left\{\frac{2\pi(s+r)}{T}\right\} f_{a_1b_2}\left(\frac{2\pi s}{T}\right) f_{b_1a_2}\left(-\frac{2\pi s}{T}\right) + O(T^{-1})\end{aligned}$$

$r, s = 1, \ldots, T-1$ with the error term uniform in r, s. This gives

$$\begin{aligned}\operatorname{cov}&\left\{\sum W^{(T)}\left(\lambda - \frac{2\pi s}{T}\right) I_{a_1b_1}^{(T)}\left(\frac{2\pi s}{T}\right), \sum W^{(T)}\left(\mu - \frac{2\pi r}{T}\right) I_{a_2b_2}^{(T)}\left(\frac{2\pi r}{T}\right)\right\}\\ &= \left(\frac{2\pi}{T}\right)^2 \sum_s W_{a_1b_1}^{(T)}\left(\lambda - \frac{2\pi s}{T}\right) W_{a_2b_2}^{(T)}\left(\mu - \frac{2\pi s}{T}\right) f_{a_1a_2}\left(\frac{2\pi s}{T}\right) f_{b_1b_2}\left(-\frac{2\pi s}{T}\right)\\ &\quad + \left(\frac{2\pi}{T}\right)^2 \sum_s W_{a_1b_1}^{(T)}\left(\lambda - \frac{2\pi s}{T}\right) W_{a_2b_2}^{(T)}\left(\mu + \frac{2\pi s}{T}\right) f_{a_1b_2}\left(\frac{2\pi s}{T}\right)\\ &\quad \times f_{b_1a_2}\left(-\frac{2\pi s}{T}\right) + O(T^{-1})\end{aligned}$$

giving the first part of (7.4.15). The second part follows from Lemma P5.1.

Proof of Corollary 7.4.3 This follows directly from the final part of expression (7.4.15).

Proof of Theorem 7.4.4 We have already investigated the asymptotic first- and second-order moment structure of the estimates. We will complete the proof of asymptotic joint normality by showing that all standardized joint cumulants of order greater than 2 tend to 0 as $T \to \infty$ under the indicated conditions.

We have

$$\begin{aligned}&\operatorname{cum}\{\textstyle\sum W_{a_1b_1}^{(T)}(\lambda_1 - 2\pi s_1/T) I_{a_1b_1}^{(T)}(2\pi s_1/T), \ldots,\\ &\textstyle\sum W_{a_Kb_K}^{(T)}(\lambda_K - 2\pi s_K/T) I_{a_Kb_K}^{(T)}(2\pi s_K/T)\}\\ &= \sum \cdots \sum W_{a_1b_1}^{(T)}(\lambda_1 - 2\pi s_1/T) \cdots W_{a_Kb_K}^{(T)}(\lambda_K - 2\pi s_K/T)(2\pi T)^{-K}\\ &\quad \times \operatorname{cum}\{d_{a_1}^{(T)}(2\pi s_1/T) d_{b_1}^{(T)}(-2\pi s_1/T), \ldots, d_{a_K}^{(T)}(2\pi s_K/T)\\ &\quad \times d_{b_K}^{(T)}(-2\pi s_K/T)\}. \qquad (*)\end{aligned}$$

In the discussion below set $r_{k1} = s_k$, $r_{k2} = -s_k$, $k = 1, \ldots, K$. Also neglect the subscripts $a_1, \ldots, a_K, b_1, \ldots, b_K$ as they play no essential role. From Theorems 2.3.2 and 4.3.2 it follows that the cumulant in this last expression is given by

$$\sum_{\nu} [(2\pi)^{m_1-1}\Delta^{(T)}(2\pi \sum_{jk\in\nu_1} r_{jk}/T) f(2\pi r_{jk}/T; jk\in\nu_1) + o(T)] \cdots$$
$$[(2\pi)^{m_p-1}\Delta^{(T)}(2\pi \sum_{jk\in\nu_p} r_{jk}/T) f(2\pi r_{jk}/T; jk\in\nu_p) + o(T)]$$

where the summation extends over all indecomposable partitions $\nu = \{\nu_1, \ldots, \nu_p\}$ of the table

$$\begin{matrix} 11 & 12 \\ 21 & 22 \\ \cdot & \\ \cdot & \\ \cdot & \\ K1 & K2 \end{matrix}$$

and m_i denotes the number of elements in ν_i. The cumulant (*) therefore has the form

$$\sum \cdots \sum W^{(T)}(\lambda_1 - 2\pi s_1/T) \cdots W^{(T)}(\lambda_K - 2\pi s_K/T) \sum_{\nu} \Delta^{(T)}(2\pi \sum_{jk\in\nu_1} r_{jk}/T) \cdots$$
$$\Delta^{(T)}(2\pi \sum_{jk\in\nu_q} r_{jk}/T) o(T^{p-q}).$$

The effect of the $\Delta^{(T)}$ functions is to introduce q linear restraints if $q < K$ and $q - 1$ if $q = K$. We write this number as $q - [q/K]$. (Here [ ] denotes "integral part.") It follows that (*) is of order

$$\max_{q\le p\le K} (B_T{}^{-1})^{q-[q/K]} T^{K-q+[q/K]} T^q o(T^{p-q}) = B_T^{-K+1} T^{K+1}.$$

It follows that

$$\text{cum}\,\{(B_T T)^{1/2} f_{a_1b_1}^{(T)}(\pm\lambda_1), \ldots, (B_T T)^{1/2} f_{a_Kb_K}^{(T)}(\pm\lambda_K)\}$$

is of order $B_T^{-K/2+1} T^{-K/2+1}$ and so tends to 0 as $T \to \infty$ for $K > 2$. The desired result now follows from Lemma P4.5.

Proof of Theorem 7.6.1 From Theorem 4.3.1

$$EI_{ab}{}^{(T)}\left(\frac{2\pi s}{T}\right) = f_{ab}\left(\frac{2\pi s}{T}\right) + o(1) \qquad \text{for } s = 1, \ldots, T-1$$

with the error term uniform in s. This gives the first part of (7.6.6) directly. The second part follows from Lemma P5.1. Continuing, from Theorem 4.3.2

$$\operatorname{cov}\{J_{a_1b_1}^{(T)}(A_j), J_{a_2b_2}^{(T)}(A_k)\}$$

$$= \left(\frac{2\pi}{T}\right)^2 \sum_{r=1}^{T-1} \sum_{s=1}^{T-1} A_j\left(\frac{2\pi r}{T}\right)\overline{A_k\left(\frac{2\pi s}{T}\right)}(2\pi T)^{-2}$$

$$\times \operatorname{cov}\left\{d_{a_1}^{(T)}\left(\frac{2\pi r}{T}\right)d_{b_1}^{(T)}\left(-\frac{2\pi r}{T}\right), d_{a_2}^{(T)}\left(\frac{2\pi s}{T}\right)d_{b_2}^{(T)}\left(-\frac{2\pi s}{T}\right)\right\}$$

$$= T^{-4}\sum_r\sum_s A_j\left(\frac{2\pi r}{T}\right)\overline{A_k\left(\frac{2\pi s}{T}\right)}\left\{\left[2\pi\Delta^{(T)}\left(\frac{2\pi[r-s]}{T}\right)f_{a_1a_2}\left(\frac{2\pi r}{T}\right) + o(T)\right]\right.$$

$$\times\left[2\pi\Delta^{(T)}\left(\frac{2\pi[-r+s]}{T}\right)f_{b_1b_2}\left(-\frac{2\pi r}{T}\right) + o(T)\right]$$

$$+\left[2\pi\Delta^{(T)}\left(\frac{2\pi[r+s]}{T}\right)f_{a_1b_2}\left(\frac{2\pi r}{T}\right) + o(T)\right]$$

$$\times\left[2\pi\Delta^{(T)}\left(\frac{2\pi[-r-s]}{T}\right)f_{b_1a_2}\left(-\frac{2\pi r}{T}\right) + o(T)\right]$$

$$\left.+ (2\pi)^3 T f_{a_1b_1a_2b_2}\left(\frac{2\pi r}{T}, -\frac{2\pi r}{T}, -\frac{2\pi s}{T}\right) + o(T)\right\}$$

$$= \left(\frac{2\pi}{T}\right)^2 \sum_{r=1}^{T-1}\left\{A_j\left(\frac{2\pi r}{T}\right)\overline{A_k\left(\frac{2\pi r}{T}\right)}f_{a_1a_2}\left(\frac{2\pi r}{T}\right)f_{b_1b_2}\left(-\frac{2\pi r}{T}\right)\right.$$

$$\left.+ A_j\left(\frac{2\pi r}{T}\right)\overline{A_k\left(2\pi - \frac{2\pi r}{T}\right)}f_{a_1b_2}\left(\frac{2\pi r}{T}\right)f_{b_1a_2}\left(-\frac{2\pi r}{T}\right)\right\}$$

$$+ \left(\frac{2\pi}{T}\right)^3 \sum_r\sum_s A_j\left(\frac{2\pi r}{T}\right)\overline{A_k\left(\frac{2\pi s}{T}\right)}f_{a_1b_1a_2b_2}\left(\frac{2\pi r}{T}, -\frac{2\pi r}{T}, -\frac{2\pi s}{T}\right) + o(T^{-1})$$

giving expression (7.6.7).

Turning to the higher order cumulants we have, neglecting subscripts,

$$\operatorname{cum}\{J_{a_1b_1}^{(T)}(A_{j_1}), \ldots, J_{a_Lb_L}^{(T)}(A_{j_L})\}$$

$$= \left(\frac{2\pi}{T}\right)^L \sum\cdots\sum A(2\pi s_1/T)\cdots A(2\pi s_L/T)(2\pi T)^{-L}$$

$$\times \operatorname{cum}\{d^{(T)}(2\pi s_1/T)d^{(T)}(-2\pi s_1/T), \ldots, d^{(T)}(2\pi s_L/T)d^{(T)}(-2\pi s_L/T)\}$$

$$= T^{-2L}\sum\cdots\sum A(2\pi s_1/T)\cdots A(2\pi s_L/T)$$

$$\times \sum_\nu \left[(2\pi)^{m_1-1}\Delta^{(T)}\left(\frac{2\pi}{T}\sum_{j\in\nu_1} \pm s_j\right)f'(2\pi s_j/T; j\in\nu_1) + o(T)\right]\cdots$$

$$\left[(2\pi)^{m_p-1}\Delta^{(T)}\left(\frac{2\pi}{T}\sum_{j\in\nu_p} \pm s_j\right)f'(2\pi s_j/T; j\in\nu_p) + o(T)\right]$$

the inner sum being over all indecomposable partitions of the table

$$\begin{matrix} d^{(T)}(2\pi s_1/T) & d^{(T)}(-2\pi s_1/T) \\ \cdot & \\ \cdot & \\ \cdot & \\ d^{(T)}(2\pi s_L/T) & d^{(T)}(-2\pi s_L/T). \end{matrix}$$

Taking note of the linear restrictions introduced by the $\Delta^{(T)}$ functions we see that the dominant term in this cumulant is of order T^{-L+1}.

Now, when the variates $T^{1/2}J_{ab}^{(T)}(A_j)$, $j = 1, \ldots, J$, $a, b = 1, \ldots, r$ are considered, we see that their joint cumulants of order greater than 2 all tend to 0. It now follows from Lemma P4.5 that the variates are asymptotically normal as indicated.

Proof of Theorem 7.6.2 We use the Taylor series expansion

$$R_{ab}^{(T)}(\lambda) = R_{ab}(\lambda) + [\{f_{ab}^{(T)}(\lambda) - f_{ab}(\lambda)\} - \tfrac{1}{2}f_{aa}(\lambda)^{-1}\{f_{aa}^{(T)}(\lambda) - f_{aa}(\lambda)\} \\ - \tfrac{1}{2}f_{bb}(\lambda)^{-1}\{f_{bb}^{(T)}(\lambda) - f_{bb}(\lambda)\}]/[f_{aa}(\lambda)f_{bb}(\lambda)]^{1/2} + \cdots$$

to derive (7.6.15) and (7.6.16) from (7.4.13) and (7.4.17) using theorems of Brillinger and Tukey (1964). The indicated asymptotic normality follows from Theorem 7.4.4 and Theorem P5.2.

Proof of Theorem 7.6.3 We have already seen in Theorem 7.6.1 that the finite dimensional distributions converge as required. We also see that

$$EF_{ab}^{(T)}(\lambda) - F_{ab}(\lambda) = O(T^{-1})$$

uniformly in λ and so it is enough to consider the process $\mathbf{Y}^{(T)}(\lambda) = \sqrt{T}[\mathbf{F}_{XX}^{(T)}(\lambda) - E\mathbf{F}_{XX}^{(T)}(\lambda)]$; $0 \leqslant \lambda \leqslant \pi$. We have therefore to show that the sequence of probability measures is tight. It follows from Problem 6, p. 41 of Billingsley (1968) that we need show tightness only for the marginal probability distributions. Following Theorem 15.6 of Billingsley (1968) this will be the case if

$$E\{|Y_{ab}^{(T)}(\lambda) - Y_{ab}^{(T)}(\lambda_1)|^2|Y_{ab}^{(T)}(\lambda_2) - Y_{ab}^{(T)}(\lambda)|^2\} \leqslant K|\lambda_1 - \lambda_2|^2$$

for $0 \leqslant \lambda_1 \leqslant \lambda \leqslant \lambda_2 \leqslant \pi$ and some finite K. Now, for $\lambda \geqslant \mu$

$$Y_{ab}^{(T)}(\lambda) - Y_{ab}^{(T)}(\mu) = T\frac{2\pi}{T}\sum_{\mu<(2\pi s/T)\leq\lambda}\left(I_{ab}^{(T)}\left(\frac{2\pi s}{T}\right) - EI_{ab}^{(T)}\left(\frac{2\pi s}{T}\right)\right).$$

We see directly that

$$E\{Y_{ab}^{(T)}(\lambda) - Y_{ab}^{(T)}(\mu)\} = 0.$$

From the proof of Theorem 7.6.1 we see that all the second-order moments of the variates $Y_{ab}^{(T)}(\lambda) - Y_{ab}^{(T)}(\lambda_1)$, $Y_{ab}^{(T)}(\lambda_2) - Y_{ab}^{(T)}(\lambda)$ and their conjugates are $\leqslant L|\lambda_2 - \lambda_1|$ for some finite L. We have therefore only to consider

$$\begin{aligned}\mathrm{cum}_{2,2}\ &\{Y_{ab}^{(T)}(\lambda) - Y_{ab}^{(T)}(\lambda_1), Y_{ab}^{(T)}(\lambda_2) - Y_{ab}^{(T)}(\lambda)\} \\ &= T^2\left(\frac{2\pi}{T}\right)^4(2\pi T)^{-4}\sum_{r_1}\sum_{r_2}\sum_{s_1}\sum_{s_2}\mathrm{cum}\ \{d_a^{(T)}(2\pi r_1/T)d_b^{(T)}(-2\pi r_1/T), \\ &\quad d_a^{(T)}(2\pi r_2/T)d_b^{(T)}(-2\pi r_2/T), d_a^{(T)}(2\pi s_1/T)d_b^{(T)}(-2\pi s_1/T), \\ &\quad d_a^{(T)}(2\pi s_2/T)d_b^{(T)}(-2\pi s_2/T)\} \qquad (*)\end{aligned}$$

where $\lambda_1 < 2\pi r_1/T$, $2\pi r_2/T \leqslant \lambda$ and $\lambda < 2\pi s_1/T$, $2\pi s_2/T \leqslant \lambda_2$. Because these domains of summation are disjoint, the cumulants on the right side are of reduced order, in fact expression (*) is seen to be

$$T^2\left(\frac{2\pi}{T}\right)^4(2\pi T)^{-4}\sum\sum\sum\sum(\Delta^{(T)}(2\pi[r_1 - r_2]/T)O(T^2)$$
$$+ \Delta^{(T)}(2\pi[s_1 - s_2]/T)O(T^2) + O(T^2)) = |\lambda_2 - \lambda_1|^3O(1).$$

As

$$E\{U^2V^2\} = \text{cum}_{2,2}\{U, V\} + 2(E\{UV\})^2 + (EU^2)(EV^2)$$

when $EU, EV = 0$. This gives the desired result.

Before proving Theorem 7.7.1 we remark that as the estimate is translation invariant, we may act as if $E\mathbf{X}(t) = \mathbf{0}$. We set down some lemmas showing that mean correction has no asymptotic effect in the case that $E\mathbf{X}(t) = \mathbf{0}$.

Lemma P7.1 Let $\mathbf{X}(t)$, $t = 0, \pm 1, \ldots$ be an r vector-valued series satisfying Assumption 2.6.2(1) and having mean 0. Let $h_a(u)$, $-\infty < u < \infty$, satisfy Assumption 4.3.1, $a = 1, \ldots, r$. Let $c_{ab}^{(T)}(u)$ be given by (7.7.8) and

$$m_{ab}^{(T)}(u) = \{H_{ab}^{(T)}(0)\}^{-1}\sum_t h_a\left(\frac{[t+u]}{T}\right)h_b\left(\frac{t}{T}\right)X_a(t+u)X_b(t)$$
$$\text{for } u = 0, \pm 1, \ldots.$$

Then

$$E|c_{ab}^{(T)}(u) - m_{ab}^{(T)}(u)|^2 = O(T^{-2})$$

uniformly in u.

Proof Set $h_a^{(T)}(u) = h_a(u/T)$ below. Now

$$\sum_t h_a^{(T)}(t+u)h_b^{(T)}(t)[X_a(t+u) - c_a^{(T)}][X_b(t) - c_b^{(T)}]$$
$$- \sum_t h_a^{(T)}(t+u)h_b^{(T)}(t)X_a(t+u)X_b(t)$$
$$= -c_a^{(T)}\sum_t h_a^{(T)}(t+u)h_b^{(T)}(t)X_b(t)$$
$$- c_b^{(T)}\sum_t h_a^{(T)}(t+u)h_b^{(T)}(t)X_a(t+u)$$
$$+ c_a^{(T)}c_b^{(T)}\sum_t h_a^{(T)}(t+u)h_b^{(T)}(t). \qquad (*)$$

Now from Theorems 5.2.3 and 5.2.8 as $c_a = 0$

$$E(c_a^{(T)})^4 = E[2\pi\sum_t h_a^{(T)}(t)^2I_{aa}^{(T)}(0)/\{\sum_t h_a^{(T)}(t)\}^2]^2$$
$$= O(T^{-2}).$$

Also from the arguments of those theorems

$$E(\sum_t h_a^{(T)}(t+u)h_b^{(T)}(t)X_b(t))^4 = O(\sum_t h_a^{(T)}(t+u)^2h_b^{(T)}(t)^2)^2$$
$$= O(\sum_t h_a^{(T)}(t)^4\sum_t h_b^{(T)}(t)^4)$$
$$= O(T^2)$$

uniformly in u. It follows that

$$E|(*)|^2 = O(1)$$

giving the desired result.

Lemma P7.2 Suppose the conditions of the theorem are satisfied. Suppose $E\mathbf{X}(t) = \mathbf{0}$ and

$$g_{ab}^{(T)}(\lambda) = (2\pi)^{-1} \sum_u w_{ab}(B_T u) m_{ab}^{(T)}(u) \exp\{-i\lambda u\}$$

then

$$E|f_{ab}^{(T)}(\lambda) - g_{ab}^{(T)}(\lambda)|^2 = O(B_T^{-2}T^{-2})$$

uniformly in λ.

Proof This follows directly from Lemma P7.1 and the fact that

$$B_T \sum_u |w_{ab}(B_T u)| \backsim \int |w_{ab}(u)| du.$$

Proof of Theorem 7.7.1 Lemma P7.2 shows that the asymptotics of $f_{ab}^{(T)}(\lambda)$ are essentially the same as those of $g_{ab}^{(T)}(\lambda)$. We begin by considering $Eg_{ab}^{(T)}(\lambda)$. Now

$$g_{ab}^{(T)}(\lambda) = \int_0^{2\pi} W_{ab}^{(T)}(\lambda - \alpha) I_{ab}^{(T)}(\alpha) d\alpha$$

where

$$W_{ab}^{(T)}(\lambda) = \sum_u w_{ab}(B_T u) \exp\{-i\lambda u\}.$$

From Theorem 4.3.2

$$EI_{ab}^{(T)}(\alpha) = f_{ab}(\alpha) + O(T^{-1})$$

and so

$$\begin{aligned} Eg_{ab}^{(T)}(\lambda) &= \int_0^{2\pi} W_{ab}^{(T)}(\lambda - \alpha) f_{ab}(\alpha) d\alpha + O(T^{-1}) \\ &= \sum_u w_{ab}(B_T u) c_{ab}(u) \exp\{-i\lambda u\} + O(T^{-1}) \\ &= B_T^{-1} \int_{-\infty}^{\infty} W_{ab}(B_T^{-1}[\lambda - \alpha]) f_{ab}(\alpha) d\alpha + O(T^{-1}) \end{aligned}$$

giving (7.7.13).

Next, from Theorem 7.2.2

$$\begin{aligned} &\operatorname{cov}\{g_{a_1b_1}^{(T)}(\lambda), g_{a_2b_2}^{(T)}(\mu)\} \\ &= \int\int W_{a_1b_1}^{(T)}(\lambda - \alpha) W_{a_2b_2}^{(T)}(\mu - \beta) \operatorname{cov}\{I_{a_1b_1}^{(T)}(\alpha), I_{a_2b_2}^{(T)}(\beta)\} d\alpha\, d\beta \\ &= H_{a_1b_1}^{(T)}(0)^{-1} H_{a_2b_2}^{(T)}(0)^{-1} \int\int W_{a_1b_1}^{(T)}(\lambda - \alpha) \overline{W_{a_2b_2}^{(T)}(\mu - \beta)} \\ &\quad \times \{H_{a_1a_2}^{(T)}(\alpha - \beta) H_{b_1b_2}^{(T)}(\alpha - \beta) f_{a_1a_2}(\alpha) f_{b_1b_2}(-\alpha) \\ &\quad + H_{a_1b_2}^{(T)}(\alpha + \beta) H_{b_1a_2}^{(T)}(\alpha + \beta) f_{a_1b_2}(\alpha) f_{b_1a_2}(-\alpha)\} d\alpha\, d\beta + O(T^{-1}). \end{aligned}$$

We next show that

$$\int W_{a_2b_2}^{(T)}(\mu - \beta)H_{a_1a_2}^{(T)}(\alpha - \beta)\overline{H_{b_1b_2}^{(T)}(\alpha - \beta)}d\beta$$
$$= 2\pi W_{a_2b_2}^{(T)}(\mu - \alpha)\sum_t h_{a_1}{}^{(T)}(t)h_{a_2}{}^{(T)}(t)h_{b_1}{}^{(T)}(t)h_{b_2}{}^{(T)}(t) + O(B_T{}^{-2}) \quad (**)$$

uniformly in α. As

$$H_{ab}{}^{(T)}(\lambda) = \sum_t h_a{}^{(T)}(t)h_b{}^{(T)}(t) \exp\{-i\lambda t\}$$

we may write (**) as

$$\sum_{t_1}\sum_{t_2} h_{a_1}{}^{(T)}(t_1)h_{a_2}{}^{(T)}(t_1)h_{b_1}{}^{(T)}(t_2)h_{b_2}{}^{(T)}(t_2) \int W_{a_2b_2}^{(T)}(\mu - \beta)$$
$$\times \exp\{-i(\alpha - \beta)t_1 + i(\alpha - \beta)t_2\}d\beta$$
$$= \sum_{t_1}\sum_{t_2} h_{a_1}{}^{(T)}(t_1)h_{a_2}{}^{(T)}(t_1)h_{b_1}{}^{(T)}(t_2)h_{b_2}{}^{(T)}(t_2) \exp\{i(\mu - \alpha)(t_1 - t_2)\}$$
$$\times w(B_T[t_1 - t_2])$$
$$= \sum_u w(B_Tu) \exp\{i(\mu - \alpha)u\} \sum_t h_{a_1}{}^{(T)}(t + u)h_{a_2}{}^{(T)}(t + u)h_{b_1}{}^{(T)}(t)h_{b_2}{}^{(T)}(t)$$
$$= 2\pi W^{(T)}(\lambda - \alpha)\sum_t h_{a_1}{}^{(T)}(t)h_{a_2}{}^{(T)}(t)h_{b_1}{}^{(T)}(t)h_{b_2}{}^{(T)}(t) + R_T,$$

where from Lemma P4.1

$$|R_T| \leqslant H\sum_u |w(B_Tu)|\,|u| \backsim HB_T{}^{-2}\int |u|\,|w(u)|du$$

for some finite H. A similar result holds for the second term of the integral. The covariance being evaluated thus has the form

$$2\pi\{\sum_t h_{a_1}{}^{(T)}(t)h_{b_1}{}^{(T)}(t)\}^{-1}\{\sum_t h_{a_2}{}^{(T)}(t)h_{b_2}{}^{(T)}(t)\}^{-1}$$
$$\times \sum_t h_{a_1}{}^{(T)}(t)h_{a_2}{}^{(T)}(t)h_{b_1}{}^{(T)}(t)h_{b_2}{}^{(T)}(t)$$
$$\times \int \{W_{a_1b_1}^{(T)}(\lambda - \alpha)W_{a_2b_2}^{(T)}(\mu - \alpha)f_{a_1a_2}(\alpha)f_{b_1b_2}(\alpha)$$
$$+ W_{a_1b_1}^{(T)}(\lambda - \alpha)W_{a_2b_2}^{(T)}(\mu + \alpha)f_{a_1b_2}(\alpha)f_{b_1a_2}(-\alpha)\}d\alpha + O(B_T{}^{-2}T^{-2}) + O(T^{-1})$$

and the desired (7.7.14) follows.

Finally, we consider the magnitude of the joint cumulants of order K. We neglect the subscripts a, b henceforth. We have

$$\text{cum}\{g^{(T)}(\lambda_1), \ldots, g^{(T)}(\lambda_L)\}$$
$$= \{2\pi H_{ab}{}^{(T)}(0)\}^{-1}\cdots\{2\pi H_{ab}{}^{(T)}(0)\}^{-1}$$
$$\times \sum_{t_1}\cdots\sum_{t_{2L}} w(B_T[t_1 - t_2])\cdots w(B_T[t_{2L-1} - t_{2L}])$$
$$\times \exp\{-i\lambda_1[t_1 - t_2] - \cdots - i\lambda_L[t_{2L-1} - t_{2L}]\}h^{(T)}(t_1)\cdots h^{(T)}(t_{2L})$$
$$\times \text{cum}\{X(t_1)X(t_2), \ldots, X(t_{2L-1})X(t_{2L})\}. \quad (*)$$

Now

$$\text{cum}\{X(t_1)X(t_2), \ldots, X(t_{2L-1})X(t_{2L})\} = \sum_{\boldsymbol{\nu}} c_{X\ldots X}(t_j; j\in\nu_1)\cdots c_{X\ldots X}(t_j; j\in\nu_P)$$

where the summation is over all indecomposable partitions $\mathbf{v} = (\nu_1, \ldots, \nu_P)$ of the table

$$\begin{matrix} 1 & 2 \\ 3 & 4 \\ \cdot & \cdot \\ \cdot & \cdot \\ \cdot & \cdot \\ 2L-1 & 2L. \end{matrix}$$

As the partition is indecomposable, in each set ν_p of the partition we may find an element t_p^*, so that none of $t_j - t_p^*, j \in \nu_p, p = 1, \ldots, P$ is a $t_{2l-1} - t_{2l}, l = 1, 2, \ldots, L$. Define $2L - P$ new variables $u_1, \ldots, u_{2L-P}$ as the nonzero $t_j - t_p^*$. The cumulant (*) is now bounded by

$$M^L T^{-L} \sum_{\mathbf{v}} \sum_{t_1^*} \cdots \sum_{t_P^*} \sum_{u_1} \cdots \sum_{u_{2L-P}} |w(B_T[u_{\alpha_1} + t_{\beta_1}^* - u_{\alpha_2} - t_{\beta_2}^*]) \cdots$$
$$w(B_T[u_{\alpha_{2L-1}} + t_{\beta_{2L-1}}^* - u_{\alpha_{2L}} - t_{\beta_{2L}}^*])|$$
$$\times |h^{(T)}(t_1^*)| \, |c_{X\ldots X}(u_1, \ldots) \cdots c_{X\ldots X}(\ldots, u_{2L-P})|$$

for some finite M where $\alpha_1, \ldots, \alpha_{2L}$ are selected from $1, \ldots, 2L$ and $\beta_1, \ldots, \beta_{2L}$ are selected from $1, \ldots, P$. Defining $\phi(t_j) = t_p^*, j \in \nu_1$, we apply Lemma 2.3.1 to see that there are $P - 1$ linearly independent differences among the $t_{\beta_1}^* - t_{\beta_2}^*, \ldots, t_{\beta_{2L-1}}^* - t_{\beta_{2L}}^*$. For convenience suppose these are $t_{\beta_1}^* - t_{\beta_2}^*, \ldots, t_{\beta_{2P-3}}^* - t_{\beta_{2P-2}}^*$. Making a final change of variables

$$\begin{aligned} v_1 &= u_{\alpha_1} + t_{\beta_1}^* - u_{\alpha_2} - t_{\beta_2}^* \\ &\cdot \\ &\cdot \\ &\cdot \\ v_{P-1} &= u_{\alpha_{2P-3}} + t_{\beta_{2P-3}}^* - u_{\alpha_{2P-2}} - t_{\beta_{2P-2}}^* \end{aligned}$$

we see that the cumulant (*) is bounded by

$$M^L T^{-L} \sum_{\mathbf{v}} \sum_{t_1} \sum_{v_1} \cdots \sum_{v_{P-1}} \sum_{u_1} \cdots \sum_{u_{2L-P}} |w(B_T v_1) \cdots w(B_T v_{P-1})|$$
$$\times |h^{(T)}(t_1^*)| \, |c_{X\ldots X}(u_1, \ldots) \cdots c_{X\ldots X}(\ldots, u_{2L-P})|$$
$$\leqslant M^L T^{-L+1} \sum_{\mathbf{v}} B_T^{-P+1} C_{n_1} \cdots C_{n_P}$$
$$= O(T^{-L+1} B_T^{-L+1}) \qquad \text{as } P \leqslant L.$$

In the next to last expression, C_n is given by (2.6.7) and n_j denotes the number of elements in the jth set of the partition $\mathbf{v}$. We see that the standardized joint cumulant

$$\text{cum}\,\{(B_T T)^{1/2} g^{(T)}(\lambda_1), \ldots, (B_T T)^{1/2} g^{(T)}(\lambda_L)\}$$

for $L > 2$, tends to 0 as $Y \to \infty$. This means that the variates $g_{a_1 b_1}^{(T)}(\lambda_1), \ldots, g_{a_K b_K}^{(T)}(\lambda_K)$ are asymptotically normal with the moment structure of the theorem. From Lemma P7.2 the same is true of the $f^{(T)}$ and we have the theorem.

Proof of Corollary 7.7.1 Immediate from (7.7.13).

Proof of Theorem 7.7.2 Follows directly from Theorem 4.5.1.

Proof of Theorem 7.7.3 We prove this theorem by means of a sequence of lemmas paralleling those used in the proof of Theorem 4.5.1. Following Lemma P7.2 it is enough to consider $g_{ab}^{(T)}(\lambda)$ corresponding to the 0 mean case. In the lemmas below we use the notation

$$\hat{g}_{ab}^{(T)}(\lambda) = g_{ab}^{(T)}(\lambda) - Eg_{ab}^{(T)}(\lambda).$$

Lemma P7.3 Under the conditions of Theorem 7.7.3, for given λ, ε and α sufficiently small

$$E \exp \{\alpha \operatorname{Re} \hat{g}_{ab}^{(T)}(\lambda)\} \leqslant \exp \{\alpha^2 \operatorname{var} \operatorname{Re} g_{ab}^{(T)}(\lambda)(1 + \varepsilon)/2\}.$$

Proof In the course of the proof of Theorem 7.7.1 we saw that

$$|\operatorname{cum} \{\hat{g}^{(T)}(\lambda_1), \ldots, \hat{g}^{(T)}(\lambda_L)\}| \leqslant M^L T^{-L+1} B_T^{-L+1} \sum_{\mathbf{v}} C_{n_1} \cdots C_{n_P}$$

for some finite M. Therefore

$$|\log E \exp \{\alpha \operatorname{Re} \hat{g}_{ab}^{(T)}(\lambda)\} - \operatorname{var} \operatorname{Re} g_{ab}^{(T)}(\lambda)\alpha^2/2|$$
$$\leqslant T^{-1}B_T^{-1} \sum_{3}^{\infty} M^L \{\sum_{\mathbf{v}} C_{n_1} \cdots C_{n_P}\} |\alpha|^L/L!$$

The indicated expression now follows from (7.7.21) on taking $|\alpha|$ sufficiently small and the fact that, from (7.7.14),

$$\begin{aligned} \operatorname{var} \{\operatorname{Re} g_{ab}^{(T)}(\lambda)\} & \\ \sim B_T^{-1}T^{-1}2\pi & \int W_{ab}(\alpha)^2 d\alpha \{ \int h_a(t)h_b(t)dt\}^{-2} \\ & \times \{ \int h_a(t)^2 h_b(t)^2 dt\} [1 + \eta\{2\lambda\}][f_{aa}(\lambda) f_{bb}(\lambda) \\ & + \{\operatorname{Re} f_{ab}(\lambda)\}^2 - \{\operatorname{Im} f_{ab}(\lambda)\}^2]/2. \end{aligned}$$

In the discussion below let

$$\Phi = 2\pi \int W_{ab}(\alpha)^2 d\alpha \{ \int h_a(t)h_b(t)dt\}^{-2} \{ \int h_a(t)^2 h_b(t)^2 dt\}.$$

Corollary Under the conditions of Theorem 7.7.3, for given β

$$E \exp \{\alpha |\operatorname{Re} \hat{g}_{ab}^{(T)}(\lambda)|\} \leqslant 2 \exp \{\alpha^2 \operatorname{var} \operatorname{Re} g_{ab}^{(T)}(\lambda)(1 + \varepsilon)/2\}$$
$$\leqslant 2 \exp \{\alpha^2 B_T^{-1} T^{-1} \Phi \sup_\lambda f_{aa}(\lambda) \sup_\lambda f_{bb}(\lambda)(1 + \beta)\}$$

for T sufficiently large.

Lemma P7.4 Let $\lambda_r = 2\pi r/R$, $r = 0, \ldots, R - 1$ for some integer R, then

$$\sup_\lambda |\operatorname{Re} \hat{g}_{ab}^{(T)}(\lambda)| \leqslant \sup_r |\operatorname{Re} \hat{g}_{ab}^{(T)}(\lambda_r)|/(1 - KB_T^{-1}R^{-1})$$

for some finite K.

Proof We first note that because $w(u)$ is 0 for sufficiently large $|u|$, $g_{ab}^{(T)}(\lambda)$ is an entire function of order $\leqslant KB_T^{-1}$. The inequality of Lemma P7.4 now follows in the manner of Corollary 2.1 in Woodroofe and Van Ness (1967) using Bernstein's inequality for entire functions of finite order (see Timan (1963)).

Lemma P7.5 For T sufficiently large

$$E \exp \{\alpha \sup_{\lambda} |\text{Re } \hat{g}_{ab}^{(T)}(\lambda)| \leqslant 2 \exp \{\log R + \alpha^2 B_T^{-1} T^{-1} \Phi \sup_{\lambda} f_{aa}(\lambda) \sup_{\lambda} f_{bb}(\lambda)(1 + \beta)\}.$$

Lemma P7.6 Given η, $\beta > 0$, let $a^2 = (\log 1/B_T) B_T^{-1}T^{-1}\Phi \sup_\lambda f_{aa}(\lambda) \times \sup_\lambda f_{bb}(\lambda) 2(1 + \beta)(1 + \eta)$, then

$$\text{Prob} [\sup_{\lambda} |\text{Re } \hat{g}_{ab}^{(T)}(\lambda)| \geqslant a] \leqslant NB_T^{\eta}$$

for some finite N.

The proof of the theorem is now completed by developing similar lemmas for Im $\hat{g}_{ab}^{(T)}(\lambda)$ and applying the Borel-Cantelli lemma.

Proof of Theorem 7.7.4 Suppose $w_{ab}(u) = 0$ for $|u| > 1$. Then $f_{ab}^{(T)}(\lambda)$ is a trigonometric polynomial of degree $B_T^{-1} = n$. Exercise 3.10.35(b) gives

$$f_{ab}^{(T)}(\lambda) - Ef_{ab}^{(T)}(\lambda) = \int_0^{2\pi} [f_{ab}^{(T)}(\alpha) - Ef_{ab}^{(T)}(\alpha)] D_n(\lambda - \alpha) d\alpha.$$

Let k be a positive integer, then

$$|f_{ab}^{(T)}(\lambda) - Ef_{ab}^{(T)}(\lambda)| \leqslant \left[\int_0^{2\pi} |f_{ab}^{(T)}(\alpha) - Ef_{ab}^{(T)}(\alpha)|^{2k} d\alpha\right]^{1/2k} \times \left[\int_0^{2\pi} |D_n(\lambda - \alpha)|^{2k/(2k-1)} d\alpha\right]^{(2k-1)/2k}$$

From Exercise 3.10.28 the final integral here is $O(n^{1/2k})$. From the proof of Theorem 7.7.1, $E|f_{ab}^{(T)}(\alpha) - Ef_{ab}^{(T)}(\alpha)|^{2k} = O(B_T^{-k}T^{-k})$. This gives $E[(B_T T)^{1/2} B_T^{\varepsilon} \sup_\lambda |f_{ab}^{(T)}(\lambda) - Ef_{ab}^{(T)}(\lambda)|]^{2k} = O(B_T^{2k\varepsilon - 1})$. Taking k sufficiently large gives the two results of the theorem.

Proof of Theorem 7.7.5 We have

$$\sup_p \left| f_{ab}^{(T)}\left(\frac{2\pi p}{P}\right) - Ef_{ab}^{(T)}\left(\frac{2\pi p}{P}\right)\right|^{2k} \leqslant \sum_{p=0}^{P-1} \left| f_{ab}^{(T)}\left(\frac{2\pi p}{P}\right) - Ef_{ab}^{(T)}\left(\frac{2\pi p}{P}\right)\right|^{2k}$$

for positive integers k. From the proof of Theorem 7.4.4

$$E|f_{ab}^{(T)}(\lambda) - Ef_{ab}^{(T)}(\lambda)|^{2k} = O(B_T^{-k}T^{-k})$$

uniformly in λ. This gives

$$E\left[(B_T T)^{1/2} P_T^{-\epsilon} \sup \left| f_{ab}^{(T)}\left(\frac{2\pi p}{P}\right) - E f_{ab}^{(T)}\left(\frac{2\pi p}{P}\right)\right|\right]^{2k} = O(P_T^{1-2k\epsilon}).$$

Taking k sufficiently large gives the two results of the theorem.

Proof of Theorem 7.9.1 From Lemma P6.3 and Theorem 4.4.2 we may write

$$d_\alpha^{(T)}(\lambda) = \sqrt{2\pi T}\eta + o_{a.s.}(\sqrt{T})$$
$$d_{\beta_j}^{(T)}(\lambda) = \sqrt{2\pi T}\theta_j + o_{a.s.}(\sqrt{T})$$
$$d_{\varepsilon_{jk}}^{(T)}(\lambda) = \sqrt{2\pi T}\zeta_{jk} + o_{a.s.}(\sqrt{T})$$

where η is $N_1^C(0,f_{\alpha\alpha}(\lambda))$, $\theta_j, j=1,\ldots,J$ are independent $N_1^C(0,f_{\beta\beta}(\lambda))$ and ζ_{jk}, $j = 1,\ldots,J$, $k = 1,\ldots,K$ are independent $N_1^C(0,f_{\varepsilon\varepsilon}(\lambda))$. It follows that

$$(2\pi T)^{-1}\sum_{j=1}^{J}\sum_{k=1}^{K}|d_{X_{jk}}^{(T)}(\lambda) - d_{X_{j.}}^{(T)}(\lambda)|^2 = \sum_j\sum_k|\zeta_{jk} - \zeta_{j.}|^2 + o_{a.s.}(1)$$
$$(2\pi T)^{-1}K\sum_j|d_{X_{j.}}^{(T)} - d_{X_{..}}^{(T)}|^2 = K\sum_j|\zeta_{j.} - \zeta_{..} + \theta_j - \theta_.|^2 + o_{a.s.}(1)$$
$$JK|d_{X_{..}}^{(T)}|^2 = JK|\zeta_{..} + \theta_. + \eta|^2 + o_{a.s.}(1).$$

By evaluating covariances, we see that the $\zeta_{jk} - \zeta_{j.}$, the $\zeta_{j.} - \zeta_{..} + \theta_j - \theta_.$ and the $\zeta_{..} + \theta_. + \eta$ are statistically independent. This implies that the statistics of the theorem are asymptotically independent.

We have the identity

$$\sum_{j,k}|\zeta_{jk}|^2 = \sum_{j,k}|\zeta_{jk} - \zeta_{j.}|^2 + K\sum_j|\zeta_{j.}|^2.$$

Exercise 4.8.7 applies indicating that $\sum|\zeta_{jk} - \zeta_{j.}|^2$ is distributed as $f_{\varepsilon\varepsilon}(\lambda)\chi^2_{2J(K-1)}/2$. We also have the identity

$$\sum_j|\zeta_{j.} + \theta_j|^2 = \sum_j|\zeta_{j.} - \zeta_{..} + \theta_j - \theta_.|^2 + J|\zeta_{..} + \theta_.|^2$$

and Exercise 4.8.7 again applies to indicate that $\Sigma|\zeta_{j.} - \zeta_{..} + \theta_j - \theta_.|^2$ is distributed as $[f_{\beta\beta}(\lambda) + K^{-1}f_{\varepsilon\varepsilon}(\lambda)]\chi^2_{2(J-1)}/2$. Finally, $|\zeta_{..} + \theta_. + \eta|^2$ is distributed as $[f_{\alpha\alpha}(\lambda) + J^{-1}f_{\beta\beta}(\lambda) + J^{-1}K^{-1}f_{\varepsilon\varepsilon}(\lambda)]\chi_2^2/2$. This completes the proof of the theorem.

PROOFS FOR CHAPTER 8

Proof of Theorem 8.2.1 We may write

$$E\{[\mathbf{Y} - \boldsymbol{\mu} - \mathbf{a}\mathbf{X}][\mathbf{Y} - \boldsymbol{\mu} - \mathbf{a}\mathbf{X}]^\tau\} = [\boldsymbol{\mu}_Y - \boldsymbol{\mu} - \mathbf{a}\boldsymbol{\mu}_X][\boldsymbol{\mu}_Y - \boldsymbol{\mu} - \mathbf{a}\boldsymbol{\mu}_X]^\tau + \boldsymbol{\Sigma}_{YY} - \mathbf{a}\boldsymbol{\Sigma}_{XY} - \boldsymbol{\Sigma}_{YX}\mathbf{a}^\tau + \mathbf{a}\boldsymbol{\Sigma}_{XX}\mathbf{a}^\tau$$

$$
\begin{aligned}
&= [\boldsymbol{\mu}_Y - \boldsymbol{\mu} - \mathbf{a}\boldsymbol{\mu}_X][\boldsymbol{\mu}_Y - \boldsymbol{\mu} - \mathbf{a}\boldsymbol{\mu}_X]^\tau \\
&\quad + \boldsymbol{\Sigma}_{YY} - \boldsymbol{\Sigma}_{YX}\boldsymbol{\Sigma}_{XX}^{-1}\boldsymbol{\Sigma}_{XY} \\
&\quad + (\mathbf{a}\boldsymbol{\Sigma}_{XX} - \boldsymbol{\Sigma}_{YX})\boldsymbol{\Sigma}_{XX}^{-1}(\mathbf{a}\boldsymbol{\Sigma}_{XX} - \boldsymbol{\Sigma}_{YX})^\tau \\
&\geqslant \boldsymbol{\Sigma}_{YY} - \boldsymbol{\Sigma}_{YX}\boldsymbol{\Sigma}_{XX}^{-1}\boldsymbol{\Sigma}_{XY}
\end{aligned}
$$

with equality achieved by the choices (8.2.14) and (8.2.15).

Before proving Theorem 8.2.2, we state a lemma of independent interest.

Lemma P8.1 Suppose the conditions of Theorem 8.2.1 are satisfied. The s vector-valued function $\boldsymbol{\phi}(\mathbf{X})$, with $E\boldsymbol{\phi}(\mathbf{X})^\tau\boldsymbol{\phi}(\mathbf{X}) < \infty$ minimizing

$$E\{[\mathbf{Y} - \boldsymbol{\phi}(\mathbf{X})][\mathbf{Y} - \boldsymbol{\phi}(\mathbf{X})]^\tau\} \tag{*}$$

is given by the conditional expected value

$$\boldsymbol{\phi}(\mathbf{X}) = E\{\mathbf{Y} \mid \mathbf{X}\}.$$

Proof We may write (*) as

$$
\begin{aligned}
&E\{[\mathbf{Y} - E\{\mathbf{Y} \mid \mathbf{X}\}][\mathbf{Y} - E\{\mathbf{Y} \mid \mathbf{X}\}]^\tau\} \\
&\qquad + E\{[E\{\mathbf{Y} \mid \mathbf{X}\} - \boldsymbol{\phi}(\mathbf{X})][E\{\mathbf{Y} \mid \mathbf{X}\} - \boldsymbol{\phi}(\mathbf{X})]^\tau\} \\
&\qquad\qquad \geqslant E\{[\mathbf{Y} - E\{\mathbf{Y} \mid \mathbf{X}\}][\mathbf{Y} - E\{\mathbf{Y} \mid \mathbf{X}\}]^\tau\}
\end{aligned}
$$

with equality achieved by the indicated $\boldsymbol{\phi}(\mathbf{X})$.

Proof of Theorem 8.2.2 If the variate (8.2.10) is normal, it is a classical result, given in Anderson (1957) for example, that

$$E\{\mathbf{Y} \mid \mathbf{X}\} = \boldsymbol{\mu}_Y + \boldsymbol{\Sigma}_{YX}\boldsymbol{\Sigma}_{XX}^{-1}(\mathbf{X} - \boldsymbol{\mu}_X),$$

and the theorem follows from Lemma P8.1.

Proof of Theorem 8.2.3 We prove Theorem 8.2.5 below, this theorem follows in a similar manner.

Proof of Theorem 8.2.4 This follows directly as did the proof of Theorem 8.2.1.

Proof of Theorem 8.2.5 Let $\mathbf{x}$, $\mathbf{y}$ denote the matrices (8.2.25) and (8.2.26) respectively. We may write

$$\mathbf{y} = \mathbf{a}\mathbf{x} + \mathbf{e}$$

with $\mathbf{a} = \boldsymbol{\Sigma}_{YX}\boldsymbol{\Sigma}_{XX}^{-1}$ and $\mathbf{e} = \mathbf{y} - \mathbf{a}\mathbf{x}$. The columns of $\mathbf{e}$ are independent $N_s^C(\mathbf{0},\boldsymbol{\Sigma}_{\varepsilon\varepsilon})$ variates. Also $\mathbf{e}$ is independent of $\mathbf{x}$.

For fixed $\mathbf{x}$ it therefore follows from Exercise 6.12.20 that vec $(\hat{\mathbf{a}} - \mathbf{a})$ is distributed as $N_{rs}^C(\mathbf{0},\boldsymbol{\Sigma}_{\varepsilon\varepsilon} \otimes (\mathbf{x}\bar{\mathbf{x}}^\tau)^{-1})$ and $\hat{\boldsymbol{\Sigma}}_{\varepsilon\varepsilon}$ is independently $(n - r)^{-1}$ $W_s^C(n - r,\boldsymbol{\Sigma}_{\varepsilon\varepsilon})$. For fixed $\mathbf{x}$ the distribution of (8.2.53) is therefore $t_{2(n-r)}^C$. As this distribution does not depend on $\mathbf{x}$, it is also the unconditional distribution. Next $E\{\hat{\mathbf{a}} \mid \mathbf{x}\} = \mathbf{a}$ and so $E\hat{\mathbf{a}} = \mathbf{a}$ as desired. Also

$$\operatorname{cov}\{\text{vec } \hat{\mathbf{a}}, \text{vec } \hat{\mathbf{a}} \mid \mathbf{x}\} = \boldsymbol{\Sigma}_{\varepsilon\varepsilon} \otimes (\mathbf{x}\bar{\mathbf{x}}^\tau)^{-1}.$$

As $E(\mathbf{x}\bar{\mathbf{x}}^\tau)^{-1} = (n-r)^{-1}\mathbf{\Sigma}_{\varepsilon\varepsilon}^{-1}$ (see Exercise 8.16.47) and cov $\{E\hat{\mathbf{a}} \mid \mathbf{x}, E\hat{\mathbf{a}} \mid \mathbf{x}\} = \mathbf{0}$, we have (8.2.54). The asymptotic normality of $\hat{\mathbf{a}}$ follows from the joint asymptotic normality of the entries of $\mathbf{y}\bar{\mathbf{x}}^\tau$ and $\mathbf{x}\bar{\mathbf{x}}^\tau$ and the fact that $\hat{\mathbf{a}}$ is a differentiable function of those entries using Theorem P5.2.

It remains to demonstrate the independence of $\hat{\mathbf{a}}$ and $\hat{\mathbf{\Sigma}}_{\varepsilon\varepsilon}$. In terms of probability density functions we may write

$$p(\hat{\mathbf{a}},\hat{\mathbf{\Sigma}}_{\varepsilon\varepsilon} \mid \mathbf{x}) = p(\hat{\mathbf{a}} \mid \mathbf{x})p(\hat{\mathbf{\Sigma}}_{\varepsilon\varepsilon})$$

from the conditional independence indicated above. It follows that

$$p(\hat{\mathbf{a}},\hat{\mathbf{\Sigma}}_{\varepsilon\varepsilon},\mathbf{x}) = p(\hat{\mathbf{a}},\mathbf{x})p(\hat{\mathbf{\Sigma}}_{\varepsilon\varepsilon})$$

and the proof is completed.

Proof of Theorem 8.3.1 Let $\mathbf{A}(\lambda)$ be the transfer function of $\{\mathbf{a}(u)\}$. We shall see that it is well defined. We may write expression (8.3.2) in the form

$$\left[\mathbf{c}_Y - \boldsymbol{\mu} - \left(\sum_{-\infty}^{\infty} \mathbf{a}(u)\right)\mathbf{c}_X\right]\left[\mathbf{c}_Y - \boldsymbol{\mu} - \left(\sum_{-\infty}^{\infty} \mathbf{a}(u)\right)\mathbf{c}_X\right]^\tau$$
$$+ \int [\mathbf{f}_{YY}(\alpha) - \mathbf{f}_{YX}(\alpha)\mathbf{f}_{XX}(\alpha)^{-1}\mathbf{f}_{XY}(\alpha)]d\alpha$$
$$+ \int [\mathbf{A}(\alpha)\mathbf{f}_{XX}(\alpha) - \mathbf{f}_{YX}(\alpha)]\mathbf{f}_{XX}(\alpha)^{-1}\overline{[\mathbf{A}(\alpha)\mathbf{f}_{XX}(\alpha) - \mathbf{f}_{YX}(\alpha)]}^\tau d\alpha$$
$$\geqslant \int [\mathbf{f}_{YY}(\alpha) - \mathbf{f}_{YX}(\alpha)\mathbf{f}_{XX}(\alpha)^{-1}\mathbf{f}_{XY}(\alpha)]d\alpha$$

with equality under the choices (8.3.3) and (8.3.5).

The fact that $\mathbf{A}(\lambda)$ given by (8.3.5) is the Fourier transform of an absolutely summable function follows from Theorem 3.8.3 and the fact that $\mathbf{f}_{XX}(\lambda)$ is nonsingular $-\infty < \lambda < \infty$.

Proof of Theorem 8.3.2 We have seen that we can write

$$\mathbf{Y}(t) = \boldsymbol{\mu} + \sum_u \mathbf{a}(t-u)\mathbf{X}(u) + \boldsymbol{\varepsilon}(t)$$

with $\mathbf{E}\boldsymbol{\varepsilon}(t) = \mathbf{0}$, cov $\{\mathbf{X}(t+u), \boldsymbol{\varepsilon}(t)\} = \mathbf{0}$ for all u. Because the series are jointly normal this $\mathbf{0}$ covariance implies that $\mathbf{X}(t+u)$ and $\boldsymbol{\varepsilon}(t)$ are statistically independent for all u. We have, therefore,

$$E\{\boldsymbol{\varepsilon}(t) \mid \mathbf{X}(v), v = 0, \pm 1, \ldots\} = E\boldsymbol{\varepsilon}(t) = \mathbf{0}$$
$$\mathbf{Y}(t) - E\{\mathbf{Y}(t) \mid \mathbf{X}(v), v = 0, \pm 1, \ldots\} = \boldsymbol{\varepsilon}(t)$$

giving the required (8.3.21) and (8.3.22).

Proof of Theorem 8.5.1 This follows from Theorem 7.3.3 and Theorem P5.1.

Proof of Theorem 8.6.1 Under the indicated assumptions it follows from Theorem 7.4.1 that

$$E\mathbf{f}_{XX}^{(T)}(\lambda) = \int W^{(T)}(\lambda - \alpha)\mathbf{f}_{XX}(\alpha)d\alpha + O(B_T^{-1}T^{-1})$$
$$E\mathbf{f}_{XY}^{(T)}(\lambda) = \int W^{(T)}(\lambda - \alpha)\mathbf{f}_{XY}(\alpha)d\alpha + O(B_T^{-1}T^{-1})$$
$$E\mathbf{f}_{YY}^{(T)}(\lambda) = \int W^{(T)}(\lambda - \alpha)\mathbf{f}_{YY}(\alpha)d\alpha + O(B_T^{-1}T^{-1}).$$

The statistics $\mathbf{A}^{(T)}$, $\phi_{jk}{}^{(T)}$, $G_{jk}{}^{(T)}$, $R^{(T)}_{Y_jY_k.X}$, $|R_{YX}{}^{(T)}|^2$ are each differentiable functions of $\mathbf{f}_{XX}{}^{(T)}(\lambda)$, $\mathbf{f}_{XY}{}^{(T)}(\lambda)$, $\mathbf{f}_{YY}{}^{(T)}(\lambda)$. The indicated expressions now follow from a theorem of Brillinger and Tukey (1964).

Proof of Corollary 8.6.1 This follows directly from the expressions (8.6.11) to (8.6.15) and the convergence theorem of Exercise 1.7.4.

Proof of Theorem 8.7.1 $\mathbf{A}^{(T)}(\lambda)$, $\mathbf{g}_{\varepsilon\varepsilon}{}^{(T)}(\lambda)$, $R^{(T)}(\lambda)$, $|R^{(T)}(\lambda)|^2$ are all differentiable functions of the entries of $\mathbf{f}_{XX}{}^{(T)}(\lambda)$, $\mathbf{f}_{YX}{}^{(T)}(\lambda)$, $\mathbf{f}_{YY}{}^{(T)}(\lambda)$ and so perturbation expansions such as

$$\mathbf{A}^{(T)}(\lambda) = \mathbf{A}(\lambda) + \{\mathbf{f}_{YX}{}^{(T)}(\lambda) - \mathbf{f}_{YX}(\lambda)\}\mathbf{f}_{XX}(\lambda)^{-1} - \mathbf{A}(\lambda)\{\mathbf{f}_{XX}{}^{(T)}(\lambda) - \mathbf{f}_{XX}(\lambda)\}\mathbf{f}_{XX}(\lambda)^{-1} + \cdots$$

may be set down and used with Theorem 7.4.3 to deduce the indicated asymptotic covariances. In fact it is much more convenient to take advantage of the results of Section 8.2 to deduce the form of the covariances.

We begin by noting, from Corollary 7.4.3, that the covariances of variates at frequencies λ, μ are $o(B_T{}^{-1}T^{-1})$ unless $\lambda - \mu$ or $\lambda + \mu \equiv 0 \pmod{2\pi}$.

Suppose $\lambda - \mu \equiv 0 \pmod{2\pi}$ and $\lambda \not\equiv 0 \pmod{2\pi}$. The asymptotic covariance structure of

$$\begin{bmatrix} \mathbf{f}_{XX}{}^{(T)}(\lambda) & \mathbf{f}_{XY}{}^{(T)}(\lambda) \\ \mathbf{f}_{YX}{}^{(T)}(\lambda) & \mathbf{f}_{YY}{}^{(T)}(\lambda) \end{bmatrix}$$

is seen to be the same as that of

$$n^{-1} W^C_{r+s}\left(n, \begin{bmatrix} \mathbf{f}_{XX}(\lambda) & \mathbf{f}_{XY}(\lambda) \\ \mathbf{f}_{YX}(\lambda) & \mathbf{f}_{YY}(\lambda) \end{bmatrix}\right) \tag{*}$$

where

$$n = \frac{B_T T}{2\pi \int W(\alpha)^2 d\alpha}.$$

Covariances, to first asymptotic order, coming out of the perturbation expansions, will therefore be the same as those based on the variate (*). From Theorem 8.2.5 we can now say that

$$\overrightarrow{\text{cov}}\,\{\text{vec}\,\mathbf{A}^{(T)}(\lambda), \text{vec}\,\mathbf{A}^{(T)}(\lambda)\} \backsim n^{-1}\mathbf{f}_{\varepsilon\varepsilon}(\lambda) \otimes \mathbf{f}_{XX}(\lambda)^{-1}$$

here. From Theorem 7.4.3 we can say that

$$\overrightarrow{\text{cov}}\,\{g^{(T)}_{\varepsilon_j\varepsilon_k}(\lambda), g^{(T)}_{\varepsilon_l\varepsilon_m}(\lambda)\} \backsim n^{-1} f_{\varepsilon_j\varepsilon_l}(\lambda) f_{\varepsilon_k\varepsilon_m}(-\lambda)$$

here.

In the case that $\lambda + \mu \equiv 0 \pmod{2\pi}$ and $\lambda \not\equiv 0 \pmod{2\pi}$ we can say

$$\overrightarrow{\text{cov}}\,\{\text{vec}\,\mathbf{A}^{(T)}(\lambda), \text{vec}\,\mathbf{A}^{(T)}(-\lambda)\} = \text{cov}\,\{\text{vec}\,\mathbf{A}^{(T)}(\lambda), \text{vec}\,\overline{\mathbf{A}^{(T)}(\lambda)}\} = o(n^{-1})$$

as the limiting distribution of Theorem 8.2.5 is complex normal. Also here

$$\text{cov}\,\{g^{(T)}_{\varepsilon_j\varepsilon_k}(\lambda), g^{(T)}_{\varepsilon_l\varepsilon_m}(-\lambda)\} \backsim n^{-1} f_{\varepsilon_j\varepsilon_m}(\lambda) f_{\varepsilon_k\varepsilon_l}(-\lambda).$$

In the case that $\lambda, \mu \equiv 0 \pmod{2\pi}$, the statistics are real-valued and we must make use of Theorem 8.2.3 instead. We see that here

$$\operatorname{cov}\{\text{vec } \mathbf{A}^{(T)}(0), \text{vec } \mathbf{A}^{(T)}(0)\} \backsim n^{-1}\mathbf{f}_{\varepsilon\varepsilon}(0) \otimes \mathbf{f}_{XX}(0)^{-1}$$
$$\operatorname{cov}\{g_{\varepsilon_j\varepsilon_k}^{(T)}(0), g_{\varepsilon_l\varepsilon_m}^{(T)}(0)\} \backsim n^{-1}\{f_{\varepsilon_j\varepsilon_l}(0)f_{\varepsilon_k\varepsilon_m}(0) + f_{\varepsilon_j\varepsilon_m}(0)f_{\varepsilon_k\varepsilon_l}(0)\}.$$

This completes the development of expressions (8.7.1) and (8.7.2). Expressions (8.7.3) and (8.7.4) follow from Theorems 8.2.5 and 7.6.2.

Proof of Theorem 8.8.1 This follows from the remarks made at the beginning of the proof of Theorem 8.7.1, Theorem 7.4.4 and Theorem P5.2.

The asymptotic independence of $\mathbf{A}^{(T)}$ and $\mathbf{g}_{\varepsilon\varepsilon}^{(T)}$ follows from their negligible covariance indicated in Theorem 8.2.5.

Before proving Theorem 8.10.1 it will be convenient to set down some notation and a lemma. If $\lambda_p = 2\pi p/P_T$, $p = 0, \ldots, P_{T-1}$, we define

$$\mathbf{A}_p = E\mathbf{f}_{YX}^{(T)}(\lambda_p) \qquad \mathbf{A}_p + \boldsymbol{\alpha}_p = \mathbf{f}_{YX}^{(T)}(\lambda_p)$$
$$\mathbf{B}_p = E\mathbf{f}_{XX}^{(T)}(\lambda_p) \qquad \mathbf{B}_p + \boldsymbol{\beta}_p = \mathbf{f}_{XX}^{(T)}(\lambda_p).$$

We can now state

Lemma P8.1 Under the conditions of Theorem 8.10.1,

$$\mathbf{a}^{(T)}(u) = P_T^{-1}\sum_p \mathbf{A}_p\mathbf{B}_p^{-1}\exp\{i\lambda_p u\} + P_T^{-1}\sum_p \boldsymbol{\alpha}_p\mathbf{B}_p^{-1}\exp\{i\lambda_p u\}$$
$$- P_T^{-1}\sum_p \mathbf{A}_p\mathbf{B}_p^{-1}\boldsymbol{\beta}_p\mathbf{B}_p^{-1}\exp\{i\lambda_p u\} + o_p(P_T^{\delta}B_T^{-1}T^{-1})$$

for any $\delta > 0$.

Proof We have the identity

$$(\mathbf{A} + \boldsymbol{\alpha})(\mathbf{B} + \boldsymbol{\beta})^{-1} - \mathbf{A}\mathbf{B}^{-1} - \boldsymbol{\alpha}\mathbf{B}^{-1} + \mathbf{A}\mathbf{B}^{-1}\boldsymbol{\beta}\mathbf{B}^{-1}$$
$$= (\mathbf{A}\mathbf{B}^{-1}\boldsymbol{\beta}\mathbf{B}^{-1}\boldsymbol{\beta} - \boldsymbol{\alpha}\mathbf{B}^{-1}\boldsymbol{\beta})(\mathbf{B} + \boldsymbol{\beta})^{-1}.$$

The norm of the right side here is bounded by

$$|\boldsymbol{\beta}|\,|\mathbf{B}^{-1}|(|\mathbf{A}|\,|\mathbf{B}^{-1}|\,|\boldsymbol{\beta}| + |\boldsymbol{\alpha}|)|(\mathbf{B} - \boldsymbol{\gamma})^{-1}|$$

if $-\boldsymbol{\gamma} \leqslant \boldsymbol{\beta}$ with $\boldsymbol{\gamma} \geqslant \mathbf{0}$.

From Theorem 7.7.5

$$\sup_p |\boldsymbol{\alpha}_p|, \sup_p |\boldsymbol{\beta}_p| = o_p(P_T^{\varepsilon}B_T^{-1/2}T^{-1/2})$$

for any $\varepsilon > 0$. It follows that

$$|(\mathbf{A}_p\mathbf{B}_p^{-1}\boldsymbol{\beta}_p\mathbf{B}_p^{-1}\boldsymbol{\beta}_p - \boldsymbol{\alpha}_p\mathbf{B}_p^{-1}\boldsymbol{\beta}_p)(\mathbf{B}_p + \boldsymbol{\beta}_p)^{-1}|$$
$$\leqslant |\boldsymbol{\beta}_p|\,|\mathbf{B}_p^{-1}|(|\mathbf{A}_p|\,|\mathbf{B}_p^{-1}|\,|\boldsymbol{\beta}_p| + |\boldsymbol{\alpha}_p|)|(\mathbf{B}_p - \sup_q \boldsymbol{\beta}_q)^{-1}|$$

and $= o_p(P_T^{\varepsilon}B_T^{-1/2}T^{-1/2})$ uniformly in p. This gives the lemma.

Proof of Theorem 8.10.1 We must investigate the asymptotic behavior of the

$$\boldsymbol{\zeta}_p = \boldsymbol{\alpha}_p \mathbf{B}_p^{-1} - \mathbf{A}_p \mathbf{B}_p^{-1} \boldsymbol{\beta}_p \mathbf{B}_p^{-1}$$

$p = 0, \ldots, P_T - 1$. We have

$$\text{vec } \boldsymbol{\zeta}_p = [\mathbf{I} \otimes (\mathbf{B}_p^{-1})^\tau] \text{vec } \boldsymbol{\alpha}_p - [(\mathbf{A}_p \mathbf{B}_p^{-1}) \otimes (\mathbf{B}_p^{-1})^\tau] \text{ vec } \boldsymbol{\beta}_p.$$

We begin by noting that $E \text{ vec } \boldsymbol{\zeta}_p = \mathbf{0}$. Next, because $P_T B_T \leqslant 1$ and $W(\alpha)$ vanishes for $|\alpha| > \pi$, Exercise 7.10.41 takes the form

$$\begin{aligned} \operatorname{cov} \{ f_{a_1 b_1}^{(T)}(\lambda_p), f_{a_2 b_2}^{(T)}(\lambda_q) \} \\ = [\eta\{\lambda_p - \lambda_q\} f_{a_1 a_2}(\lambda_p) f_{b_1 b_2}(-\lambda_p) \\ + \eta\{\lambda_p + \lambda_q\} f_{a_1 b_2}(\lambda_p) f_{a_1 b_2}(-\lambda_p)] B_T^{-1} T^{-1} 2\pi \int W(\alpha)^2 d\alpha \\ + \frac{2\pi}{T} f_{a_1 b_1 a_2 b_2}(\lambda_p, -\lambda_p, -\lambda_q) + O(B_T^{-2} T^{-2}) + O(B_T T^{-1}). \end{aligned}$$

It follows that the covariance matrix of the variate

$$\begin{bmatrix} \text{vec } \boldsymbol{\alpha}_p \\ \text{vec } \boldsymbol{\beta}_p \end{bmatrix}$$

will be made up of two parts, a term involving only second-order spectra and a term involving fourth-order spectra.

From our investigation of $\mathbf{A}^{(T)}(\lambda)$, we can say that the contribution of the term in second-order spectra to cov $\{\text{vec } \boldsymbol{\zeta}_p, \text{ vec } \boldsymbol{\zeta}_q\}$ is asymptotically

$$\eta\{\lambda_p - \lambda_q\} [\mathbf{f}_{\epsilon\epsilon}(\lambda_p) \otimes \mathbf{f}_{XX}(\lambda_p)^{-1}] B_T^{-1} T^{-1} 2\pi \int W(\alpha)^2 d\alpha.$$

Suppose we denote the term in cov $\{\text{vec } \boldsymbol{\beta}_p, \text{ vec } \boldsymbol{\beta}_q\}$ that involves fourth-order spectra by $(2\pi/T)\mathbf{V}_{pq}$. Because of the model $\mathbf{Y}(t) = \boldsymbol{\mu} + \sum \mathbf{a}(t - u) \times \mathbf{X}(u) + \boldsymbol{\varepsilon}(t)$, with the series $\boldsymbol{\varepsilon}(t)$ independent of the series $\mathbf{X}(t)$, the corresponding terms in cov $\{\text{vec } \boldsymbol{\alpha}_p, \text{ vec } \boldsymbol{\beta}_q\}$ and cov $\{\text{vec } \boldsymbol{\alpha}_p, \text{ vec } \boldsymbol{\alpha}_q\}$ will be

$$\frac{2\pi}{T} [\mathbf{A}(\lambda_p) \otimes \mathbf{I}) \mathbf{V}_{pq}, \qquad \frac{2\pi}{T} [\mathbf{A}(\lambda_p) \otimes \mathbf{I}] \mathbf{V}_{pq} [\mathbf{A}(\lambda_q) \otimes \mathbf{I}].$$

It follows that their contribution to cov $\{\text{vec } \boldsymbol{\zeta}_p, \text{ vec } \boldsymbol{\zeta}_q\}$ will be

$$\begin{aligned} \frac{2\pi}{T} \{ [\mathbf{I} \otimes (\mathbf{B}_p^{-1})^\tau] [\mathbf{A}(\lambda_p) \otimes \mathbf{I}] \mathbf{V}_{pq} [\overline{\mathbf{A}(\lambda_q)}^\tau \otimes \mathbf{I}] [\mathbf{I} \otimes \overline{(\mathbf{B}_p^{-1})}] \} \\ - [\mathbf{I} \otimes (\mathbf{B}_p^{-1})^\tau] [\mathbf{A}(\lambda_p) \otimes \mathbf{I}] \mathbf{V}_{pq} [\overline{(\mathbf{A}_q \mathbf{B}_q^{-1})}^\tau \otimes \overline{(\mathbf{B}_p^{-1})}] \\ - [(\mathbf{A}_p \otimes \mathbf{B}_p^{-1}) \otimes (\mathbf{B}_p^{-1})^\tau] \mathbf{V}_{pq} [\overline{\mathbf{A}(\lambda_q)}^\tau \otimes \mathbf{I}] [\mathbf{I} \otimes \overline{(\mathbf{B}_q^{-1})}] \\ + [(\mathbf{A}_p \otimes \mathbf{B}_p^{-1}) \otimes (\mathbf{B}_p^{-1})^\tau] \mathbf{V}_{pq} \mid \overline{(\mathbf{A}_q \mathbf{B}_q^{-1})}^\tau \otimes \overline{(\mathbf{B}_p^{-1})}] \backsim \mathbf{0} \end{aligned}$$

as $\mathbf{A}(\lambda_p) \backsim \mathbf{A}_p \mathbf{B}_p^{-1}$.

We may deduce from all this that

$$\operatorname{cov}\{\operatorname{vec}\boldsymbol{\zeta}_p, \operatorname{vec}\boldsymbol{\zeta}_q\} \sim \eta\{\lambda_p - \lambda_q\}[\mathbf{f}_{\varepsilon\varepsilon}(\lambda_p) \otimes \mathbf{f}_{XX}(\lambda_p)^{-1}]B_T^{-1}T^{-1}2\pi \int W(\alpha)^2 d\alpha + O(B_T T^{-1}) + O(B_T^{-2}T^{-2})$$

and so

$$\operatorname{cov}\{P_T^{-1}\sum_p \exp\{i\lambda_p u\}\operatorname{vec}\boldsymbol{\zeta}_p, P_T^{-1}\sum_q \exp\{i\lambda_q v\}\operatorname{vec}\boldsymbol{\zeta}_q\}$$
$$= P_T^{-2}\sum_p [\mathbf{f}_{\varepsilon\varepsilon}(\lambda_p) \otimes \mathbf{f}_{XX}(\lambda_p)^{-1}]\exp\{i\lambda_p(u - v)\}B_T^{-1}T^{-1}2\pi \int W(\alpha)^2 d\alpha$$
$$+ O(B_T T^{-1}) + O(B_T^{-2}T^{-2}).$$

Exercise 7.10.42 may next be invoked to conclude that $P_T^{-1}\Sigma \exp\{i\lambda_p u\}$ vec $\boldsymbol{\zeta}_p$ is asymptotically normal. Putting this together we have the desired result.

Proof of Theorem 8.10.2 By substitution

$$\mathbf{c}_Y^{(T)} = \boldsymbol{\mu} + \mathbf{a}\mathbf{c}_X^{(T)} + \mathbf{c}_\varepsilon^{(T)}$$
$$\mathbf{c}_{YX}^{(T)}(0) = T\sum_t [\mathbf{Y}(t) - \mathbf{c}_Y^{(T)}][\mathbf{X}(t) - \mathbf{c}_X^{(T)}]^\tau$$
$$= \mathbf{a}\mathbf{c}_{XX}^{(T)}(0) + \mathbf{c}_{\varepsilon X}^{(T)}(0),$$

therefore

$$\boldsymbol{\mu}^{(T)} - \boldsymbol{\mu} = (\mathbf{a} - \mathbf{a}^{(T)})\mathbf{c}_X^{(T)} + \mathbf{c}_\varepsilon^{(T)}$$
$$\mathbf{a}^{(T)} - \mathbf{a} = \mathbf{c}_{\varepsilon X}^{(T)}(0)\mathbf{c}_{XX}^{(T)}(0)^{-1}. \qquad (*)$$

This last may be rewritten

$$\sqrt{T}\operatorname{vec}\{\mathbf{a}^{(T)} - \mathbf{a}\} = [\mathbf{I} \otimes \mathbf{c}_{XX}^{(T)}(0)^{-1}]\sqrt{T}\operatorname{vec}\mathbf{c}_{\varepsilon X}^{(T)}(0).$$

From Exercise 7.10.36, $\mathbf{c}_{\varepsilon X}^{(T)}(0)$ is asymptotically normal with mean $\mathbf{0}$ and

$$\operatorname{cov}\{c_{\varepsilon a_1 X b_1}^{(T)}(0), c_{\varepsilon a_2 X b_2}^{(T)}(0)\} \sim 2\pi T^{-1}\int f_{\varepsilon a_1 \varepsilon a_2}(\alpha) f_{X b_1 X b_2}(-\alpha)d\alpha,$$

and so

$$\operatorname{cov}\{\operatorname{vec}\mathbf{c}_{\varepsilon X}^{(T)}(0), \operatorname{vec}\mathbf{c}_{\varepsilon X}^{(T)}(0)\} \sim 2\pi T^{-1}\int \mathbf{f}_{\varepsilon\varepsilon}(\alpha) \otimes \mathbf{f}_{XX}(-\alpha)d\alpha.$$

This gives the indicated asymptotic distribution for vec $\mathbf{a}^{(T)}$ as $\mathbf{c}_{XX}^{(T)}(0)^{-1}$ tends to $\mathbf{c}_{XX}(0)^{-1}$ in probability.

Because $E\mathbf{X}(t) = \mathbf{0}$, $\mathbf{c}_X^{(T)} = o_p(1)$ and (*) shows that $\sqrt{T}(\boldsymbol{\mu}^{(T)} - \boldsymbol{\mu}) = \sqrt{T}\,\mathbf{c}_\varepsilon^{(T)} + o_p(1)$ giving the indicated limiting distribution for $\boldsymbol{\mu}^{(T)}$ from Theorem 4.4.1. The asymptotic independence of $\mathbf{a}^{(T)}$ and $\boldsymbol{\mu}^{(T)}$ follows from the asymptotic independence of $\mathbf{c}_\varepsilon^{(T)}$ and $\mathbf{c}_{\varepsilon X}^{(T)}(0)$.

Continuing

$$\mathbf{e}(t) = \mathbf{Y}(t) - \boldsymbol{\mu}^{(T)} - \mathbf{a}^{(T)}\mathbf{X}(t)$$
$$= \boldsymbol{\mu} - \boldsymbol{\mu}^{(T)} + (\mathbf{a} - \mathbf{a}^{(T)})\mathbf{X}(t) + \boldsymbol{\varepsilon}(t)$$
$$= (\mathbf{a} - \mathbf{a}^{(T)})(\mathbf{X}(t) - \mathbf{c}_X^{(T)}) + \boldsymbol{\varepsilon}(t) - \mathbf{c}_\varepsilon^{(T)}.$$

It follows that

$$\mathbf{f}_{ee}^{(T)}(\lambda) = (\mathbf{a} - \mathbf{a}^{(T)})\mathbf{f}_{XX}^{(T)}(\lambda)(\mathbf{a} - \mathbf{a}^{(T)})^\tau + (\mathbf{a} - \mathbf{a}^{(T)})\mathbf{f}_{X\varepsilon}^{(T)}(\lambda) + \mathbf{f}_{\varepsilon X}^{(T)}(\lambda)(\mathbf{a} - \mathbf{a}^{(T)}) + \mathbf{f}_{\varepsilon\varepsilon}^{(T)}(\lambda).$$

In view of the previously determined asymptotic distributions of $\mathbf{f}_{XX}^{(T)}(\lambda)$, $\mathbf{f}_{X\varepsilon}^{(T)}(\lambda)$ we have from the last expression

$$(B_T T)^{1/2}\mathbf{f}_{ee}^{(T)}(\lambda) = (B_T T)^{1/2}\mathbf{f}_{\varepsilon\varepsilon}^{(T)}(\lambda) + o_p(1)$$

giving the indicated asymptotic distribution for $\mathbf{f}_{ee}^{(T)}(\lambda)$.

Before proving Theorem 8.11.1 we first set down a lemma.

Lemma P8.2 Let $\mathbf{X}_T$, $T = 1, 2, \ldots$ be a sequence of vector-valued random variables, $\boldsymbol{\mu}$ a constant vector and α_T, $T = 1, 2, \ldots$ a sequence of constants tending to 0 with T. Suppose

$$\overline{\lim_{T\to\infty}} |\mathbf{X}_T - \boldsymbol{\mu}|/\alpha_T \leqslant 1$$

with probability 1. Let $\mathbf{f}(\mathbf{x})$ have a continuous first derivative in a neighborhood of $\boldsymbol{\mu}$ with $|\mathbf{f}'(\boldsymbol{\mu})| \neq 0$. Then

$$\overline{\lim_{T\to\infty}} |\mathbf{f}(\mathbf{X}_T) - \mathbf{f}(\boldsymbol{\mu})|/\alpha_T \leqslant |\mathbf{f}'(\boldsymbol{\mu})|$$

with probability 1.

Proof With probability 1, $\mathbf{X}_T$ will be in the indicated neighborhood of $\boldsymbol{\mu}$ for all large T. Take it as being there. Next because $\mathbf{f}(\mathbf{x})$ has a first derivative we have

$$\mathbf{f}(\mathbf{X}_T) = \mathbf{f}(\boldsymbol{\mu}) + \alpha_T\langle \mathbf{X}_T - \boldsymbol{\mu}, \mathbf{f}'(\boldsymbol{\zeta})\rangle$$

for some $\boldsymbol{\zeta}$ in the neighborhood. Because of the continuity of $\mathbf{f}'(\mathbf{x})$, $\mathbf{f}'(\boldsymbol{\zeta})$ becomes arbitrarily close to $\mathbf{f}'(\boldsymbol{\mu})$ as $T \to \infty$ and we have the indicated result.

Proof of Theorem 8.11.1 The theorem follows from Lemma P8.2, Theorem 7.7.3, and Theorem 7.4.2.

PROOFS FOR CHAPTER 9

Proof of Theorem 9.2.1 We prove Theorem 9.2.3 below; Theorem 9.2.1 follows in a similar manner.

Proof of Theorem 9.2.2 We prove Theorem 9.2.4 below; Theorem 9.2.2 follows in a similar manner.

Proof of Theorem 9.2.3 We prove that the jth latent root of (9.2.17) is $\geqslant \mu_{j+q}$ with equality achieved by the indicated $\boldsymbol{\mu}$, $\mathbf{B}$, $\mathbf{C}$.

From Theorem 8.2.4

$$E\{(\mathbf{X} - \boldsymbol{\mu} - \mathbf{CBX})\overline{(\mathbf{X} - \boldsymbol{\mu} - \mathbf{CBX})}^\tau\}$$
$$\geqslant \boldsymbol{\Sigma}_{XX} - \boldsymbol{\Sigma}_{XX}\overline{\mathbf{B}}^\tau\overline{\mathbf{C}}^\tau(\mathbf{CB}\boldsymbol{\Sigma}_{XX}\overline{\mathbf{B}}^\tau\overline{\mathbf{C}}^\tau)^{-1}\mathbf{CB}\boldsymbol{\Sigma}_{XX} \geqslant \boldsymbol{\Sigma}_{XX} - \mathbf{D},$$

where

$$\mathbf{D} = \boldsymbol{\Sigma}_{XX}\overline{\mathbf{B}}^\tau\overline{\mathbf{C}}^\tau(\mathbf{CB}\boldsymbol{\Sigma}_{XX}\overline{\mathbf{B}}^\tau\overline{\mathbf{C}}^\tau)^{-1}\mathbf{CB}\boldsymbol{\Sigma}_{XX}.$$

The matrix $\mathbf{D}$ has rank $\leqslant q$. Now

$$\mu_i(\boldsymbol{\Sigma}_{XX} - \mathbf{D}) = \sup_{\mathbf{L}\alpha=\mathbf{0}} \frac{\bar{\boldsymbol{\alpha}}^\tau(\boldsymbol{\Sigma}_{XX} - \mathbf{D})\boldsymbol{\alpha}}{\bar{\boldsymbol{\alpha}}^\tau\boldsymbol{\alpha}}$$

where $\mathbf{L}$ is $(i-1) \times r$. This is

$$\geqslant \sup_{\mathbf{L}\alpha,\mathbf{D}\alpha=\mathbf{0}} \frac{\bar{\boldsymbol{\alpha}}^\tau(\boldsymbol{\Sigma}_{XX} - \mathbf{D})\boldsymbol{\alpha}}{\bar{\boldsymbol{\alpha}}^\tau\boldsymbol{\alpha}}$$
$$\geqslant \sup_{\mathbf{L}\alpha,\mathbf{D}\alpha=\mathbf{0}} \frac{\bar{\boldsymbol{\alpha}}^\tau\boldsymbol{\Sigma}_{XX}\boldsymbol{\alpha}}{\bar{\boldsymbol{\alpha}}^\tau\boldsymbol{\alpha}}$$
$$\geqslant \mu_{q+i}(\boldsymbol{\Sigma}_{XX})$$

because the matrix $\begin{bmatrix}\mathbf{L}\\ \mathbf{D}\end{bmatrix}$ has rank $q+i-1$ at most. We quickly check that the indicated $\boldsymbol{\mu}$, $\mathbf{B}$, $\mathbf{C}$ lead to a matrix (9.2.17) of the form (9.2.21). Now equality in the above inequalities is achieved by the indicated choices because the ith latent root of (9.2.21) is μ_{q+i}.

We have here presented a complex version of the arguments of Okamoto and Kanazawa (1968).

Proof of Theorem 9.2.4 We have the Taylor series expansions

$$\hat{\mu}_j = \mu_j + \overline{\mathbf{V}}_j^\tau\{\hat{\boldsymbol{\Sigma}}_{XX} - \boldsymbol{\Sigma}_{XX}\}\mathbf{V}_j + \cdots \tag{*}$$
$$\hat{\mathbf{V}}_j = \mathbf{V}_j + \sum_{l \neq j} [\overline{\mathbf{V}}_l^\tau\{\hat{\boldsymbol{\Sigma}}_{XX} - \boldsymbol{\Sigma}_{XX}\}\mathbf{V}_j]\mathbf{V}_l/\{\mu_j - \mu_l\} + \cdots. \tag{**}$$

See Wilkinson (1965) p. 68.

We see that $\hat{\boldsymbol{\Sigma}}_{XX}$ is asymptotically normal with mean $\boldsymbol{\Sigma}_{XX}$ and

$$\operatorname{cov}\{\hat{\Sigma}_{a_1a_2}, \hat{\Sigma}_{b_1b_2}\} = \frac{\Sigma_{a_1b_1}\,\Sigma_{b_2a_2}}{n}.$$

This implies the useful result of Exercise 4.8.36(b)

$$\operatorname{cov}\{\bar{\boldsymbol{\alpha}}^\tau\hat{\boldsymbol{\Sigma}}\boldsymbol{\beta}, \bar{\boldsymbol{\gamma}}^\tau\hat{\boldsymbol{\Sigma}}\boldsymbol{\delta}\} = \Sigma_{a_1,a_2,b_1,b_2} \frac{\bar{\alpha}_{a_1}\beta_{a_2}\gamma_{b_1}\bar{\delta}_{b_2}\,\Sigma_{a_1b_1}\,\Sigma_{b_2a_2}}{n}$$
$$= \frac{(\bar{\boldsymbol{\alpha}}^\tau\boldsymbol{\Sigma}_{XX}\boldsymbol{\gamma})(\bar{\boldsymbol{\delta}}^\tau\boldsymbol{\Sigma}_{XX}\boldsymbol{\beta})}{n}$$

for r vectors $\boldsymbol{\alpha}$, $\boldsymbol{\beta}$, $\boldsymbol{\gamma}$, $\boldsymbol{\delta}$. The indicated asymptotic moments now follow directly from (*) and (**) using these expressions. For example,

$$\overrightarrow{\text{cov}}\,\{\hat{\mathbf{V}}_j, \hat{\mathbf{V}}_k\} = \sum_{l \neq j} \sum_{m \neq k} (\overline{\mathbf{V}}_l^\tau \mathbf{\Sigma}_{XX} \mathbf{V}_m)(\overline{\mathbf{V}}_k^\tau \mathbf{\Sigma}_{XX} \mathbf{V}_j) \mathbf{V}_l \overline{\mathbf{V}}_m^\tau / [(\mu_j - \mu_l)(\mu_k - \mu_m)]/n + \cdots$$

$$\overrightarrow{\text{cov}}\,\{\hat{\mathbf{V}}_j, \overline{\hat{\mathbf{V}}_k}\} = \sum_{l \neq j} \sum_{m \neq k} (\overline{\mathbf{V}}_l^\tau \mathbf{\Sigma}_{XX} \mathbf{V}_k)(\overline{\mathbf{V}}_m^\tau \mathbf{\Sigma}_{XX} \mathbf{V}_j) \mathbf{V}_l \mathbf{V}_m^\tau / [\mu_j - \mu_l)(\mu_k - \mu_m)]/n + \cdots$$

giving the indicated covariances because as the $\mathbf{V}_j$ are latent vectors

$$\begin{aligned} \overline{\mathbf{V}}_p^\tau \mathbf{\Sigma}_{XX} \mathbf{V}_q &= \mu_p \qquad \text{if } p = q \\ &= 0 \qquad \text{if } p \neq q. \end{aligned}$$

The asymptotic normality follows from the asymptotic normality of $\hat{\mathbf{\Sigma}}_{XX}$ and Theorem P5.2.

Proof of Theorem 9.3.1 We may write (9.3.3) as

$$\overline{\{[\mathbf{c}_X - \mathbf{\mu} - (\sum_u \mathbf{c}(u))\mathbf{c}_\zeta]^\tau}[\mathbf{c}_X - \mathbf{\mu} - (\sum_u \mathbf{c}(u))\mathbf{c}_\zeta]\} + \int \text{tr}\,\{[\mathbf{I} - \mathbf{A}(\alpha)]\mathbf{f}_{XX}(\alpha)\overline{[\mathbf{I} - \mathbf{A}(\alpha)]^\tau}\}\,d\alpha$$

where $\mathbf{A}(\alpha) = \mathbf{C}(\alpha)\mathbf{B}(\alpha)$. We may make the first term 0 by setting

$$\begin{aligned} \mathbf{\mu} &= \mathbf{c}_X - (\sum_u \mathbf{c}(u))\mathbf{c}_\zeta \\ &= \mathbf{c}_X - (\sum_u \mathbf{c}(u))(\sum_u \mathbf{b}(u))\mathbf{c}_X. \end{aligned}$$

We see that the second term is minimized if we minimize

$$\text{tr}\,\{[\mathbf{f}_{XX}(\alpha)^{1/2} - \mathbf{A}(\alpha)\mathbf{f}_{XX}(\alpha)^{1/2}]\overline{[\mathbf{f}_{XX}(\alpha)^{1/2} - \mathbf{A}(\alpha)\mathbf{f}_{XX}(\alpha)^{1/2}]^\tau}\}$$

for each α with $\mathbf{A}(\alpha)$ of rank $\leqslant q$. From Theorem 3.7.4 we see that we should take

$$\mathbf{A}(\lambda) = \sum_{j=1}^{q} \mathbf{V}_j(\lambda)\overline{\mathbf{V}_j(\lambda)}^\tau$$

where $\mathbf{V}_j(\lambda)$ is the jth latent vector of $\mathbf{f}_{XX}(\lambda)^{1/2}$ and *a fortiori* of $\mathbf{f}_{XX}(\lambda)$. The indicated $\mathbf{B}(\lambda)$, $\mathbf{C}(\lambda)$ are now seen to achieve the desired minimization.

Proof of Theorem 9.3.2 The cross-spectrum of $\zeta_j(t)$ with $\zeta_k(t)$ is given by

$$\begin{aligned} \overline{\mathbf{V}_j(\lambda)}^\tau \mathbf{f}_{XX}(\lambda)\mathbf{V}_k(\lambda) &= \mu_j(\lambda) \qquad \text{if } j = k \\ &= 0 \qquad \text{if } j \neq k \end{aligned}$$

giving the indicated results.

Proof of Theorem 9.3.3 Because the latent roots of $\mathbf{f}_{XX}(\lambda)$ are simple for all λ, its latent roots and vectors will be real holomorphic functions of its entries, see Exercises 3.10.19 to 3.10.21.

Expressions (9.3.29) and (9.3.30) now follow from Theorem 3.8.3. Expressions (9.3.31) and (9.3.32) follow directly from these and from expression (9.3.28).

Proof of Theorem 9.3.4 The desired $\mathbf{B}_j(\lambda)$ must be some linear combination of the $\overline{\mathbf{V}_k(\lambda)}^\tau$, $k = 1, \ldots,$ say

$$\mathbf{B}_j(\lambda) = G_{j1}(\lambda)\overline{\mathbf{V}_1(\lambda)}^\tau + \cdots + G_{jr}(\lambda)\overline{\mathbf{V}_r(\lambda)}^\tau.$$

The desired series is orthogonal to $\zeta_k(t)$, $k < j$ and so it must have $G_{jk}(\lambda) = 0$ for $k < j$. The variance of (9.3.33) may be written

$$\sum_{k \geq j} |G_{jk}(\lambda)|^2 \mu_k(\lambda)$$

with $\sum_k |G_{jk}(\lambda)|^2 = 1$. This variance is clearly maximized by taking

$$\begin{aligned} G_{jk}(\lambda) &= 1 \qquad \text{for } j = k \\ &= 0 \qquad \text{for } j \neq k \end{aligned}$$

and we have the result.

Proof of Theorem 9.3.5 The spectral density matrix of (9.3.35) is given by

$$[\mathbf{I} - \mathbf{A}(\lambda)]\mathbf{f}_{XX}(\lambda)\overline{[\mathbf{I} - \mathbf{A}(\lambda)]}^\tau$$

where $\mathbf{A}(\lambda) = \mathbf{C}(\lambda)\mathbf{B}(\lambda)$. We see from Theorem 9.2.3 that the latent roots of the latter are minimized by the indicated $\mathbf{B}(\lambda)$, $\mathbf{C}(\lambda)$.

Proof of Theorem 9.4.1 From the Wielandt-Hoffman theorem (see Wilkinson (1965))

$$\sum_{j=1}^{r} |\mu_j^{(T)}(\lambda) - \nu_j^{(T)}(\lambda)|^2 \leqslant \sum_{j,k=1}^{r} |f_{jk}^{(T)}(\lambda) - \int_0^{2\pi} W^{(T)}(\lambda - \alpha) f_{jk}(\alpha) d\alpha|^2.$$

Also from Theorems 7.4.1 and 7.4.3

$$E|f_{jk}^{(T)}(\lambda) - \int_0^{2\pi} W^{(T)}(\lambda - \alpha) f_{jk}(\alpha) d\alpha|^2 = O(B_T^{-1}T^{-1}).$$

As

$$\begin{aligned} |E[\mu_j^{(T)}(\lambda) - \nu_j^{(T)}(\lambda)]|^2 &\leqslant E[\mu_j^{(T)}(\lambda) - \nu_j^{(T)}(\lambda)]^2 \\ &\leqslant E\left[\sum_{k=1}^{r} |\mu_k^{(T)}(\lambda) - \nu_k^{(T)}(\lambda)|^2\right] \end{aligned}$$

expression (9.4.5) now follows.

Expressions (9.4.6) and (9.4.7) result from the following Taylor series expansions set down in the course of the proof of Theorem 9.2.4:

$$\begin{aligned} \mu_j^{(T)}(\lambda) &= \nu_j^{(T)}(\lambda) \\ &\quad + \overline{\mathbf{U}_j^{(T)}(\lambda)}^\tau \left\{ \mathbf{f}_{XX}^{(T)}(\lambda) - \int_0^{2\pi} W^{(T)}(\lambda - \alpha) \mathbf{f}_{XX}(\alpha) d\alpha \right\} \mathbf{U}_j^{(T)}(\lambda) + \cdots \end{aligned}$$

$$\mathbf{V}_j^{(T)}(\lambda) = \mathbf{U}_j^{(T)}(\lambda) + \sum_{l \neq j} \left[\overline{\mathbf{U}_l^{(T)}(\lambda)}^\tau \left\{ \mathbf{f}_{XX}^{(T)}(\lambda) - \int_0^{2\pi} W^{(T)}(\lambda - \alpha)\mathbf{f}_{XX}(\alpha)d\alpha \right\} \mathbf{U}_j^{(T)}(\lambda) \right] \times \mathbf{U}_l^{(T)}(\lambda)/\{\nu_j^{(T)}(\lambda) - \nu_l^{(T)}(\lambda)\} + \cdots.$$

Proof of Theorem 9.4.2 Expressions (9.4.13) and (9.4.14) follow from the following expressions given in the proof of Theorem 9.2.4:

$$\nu_j^{(T)}(\lambda) = \mu_j(\lambda) + \overline{\mathbf{V}_j(\lambda)}^\tau \left\{ \int_0^{2\pi} W^{(T)}(\lambda - \alpha)\mathbf{f}_{XX}(\alpha)d\alpha - \mathbf{f}_{XX}(\lambda) \right\} \mathbf{V}_j(\lambda) + \cdots$$

$$\mathbf{U}_j^{(T)}(\lambda) = \mathbf{V}_j(\lambda) + \sum_{l \neq j} \left[\overline{\mathbf{V}_j(\lambda)}^\tau \left\{ \int_0^{2\pi} W^{(T)}(\lambda - \alpha)\mathbf{f}_{XX}(\alpha)d\alpha - \mathbf{f}_{XX}(\lambda) \right\} \mathbf{V}_j(\lambda) \right] \times \mathbf{V}_l(\lambda)/\{\mu_j(\lambda) - \mu_l(\lambda)\} + \cdots$$

and the result (7.4.13)

$$\int_0^{2\pi} W^{(T)}(\lambda - \alpha)\mathbf{f}_{XX}(\alpha)d\alpha = \mathbf{f}_{XX}(\lambda) + \frac{1}{2} B_T^2 \frac{d^2\mathbf{f}_{XX}(\lambda)}{d\lambda^2} \int \alpha^2 W(\alpha)d\alpha + O(B_T^3)$$

under the given conditions.

Proof of Theorem 9.4.3 This follows from the expressions of the proof of Theorem 9.4.1 in the manner of the proof of Theorem 9.2.4.

Proof of Theorem 9.4.4 The latent roots and vectors of a matrix are continuous functions of its entries. This theorem consequently follows from Theorem 7.3.3 and Theorem P5.1.

PROOFS FOR CHAPTER 10

Proof of Theorem 10.2.1 Let $\mathbf{A} = \mathbf{CB}$, and write (10.2.5) as

$$\begin{aligned}
&[\boldsymbol{\mu}_Y - \boldsymbol{\mu} - \mathbf{A}\boldsymbol{\mu}_X]^\tau \boldsymbol{\Gamma}^{-1}[\boldsymbol{\mu}_Y - \boldsymbol{\mu} - \mathbf{A}\boldsymbol{\mu}_X] \\
&\quad + \operatorname{tr}\{\boldsymbol{\Gamma}^{-1}(\boldsymbol{\Sigma}_{YY} - \mathbf{A}\boldsymbol{\Sigma}_{XY} - \boldsymbol{\Sigma}_{YX}\mathbf{A}^\tau + \mathbf{A}\boldsymbol{\Sigma}_{XX}\mathbf{A}^\tau) \\
&\quad \geq \operatorname{tr}\{\boldsymbol{\Gamma}^{-1}(\boldsymbol{\Sigma}_{YY} - \boldsymbol{\Sigma}_{YX}\boldsymbol{\Sigma}_{XX}^{-1}\boldsymbol{\Sigma}_{XY})\} \\
&\quad + \operatorname{tr}\{\boldsymbol{\Gamma}^{-1/2}(\boldsymbol{\Sigma}_{YX} - \mathbf{A}\boldsymbol{\Sigma}_{XX})\boldsymbol{\Sigma}_{XX}^{-1}(\boldsymbol{\Sigma}_{YX} - \mathbf{A}\boldsymbol{\Sigma}_{XX})\boldsymbol{\Gamma}^{-1/2}\}.
\end{aligned}$$

From Theorem 3.7.4 this is minimized by setting

$$\boldsymbol{\Gamma}^{-1/2}\mathbf{A}\boldsymbol{\Sigma}_{XX}^{1/2} = \sum_{j=1}^{q} \mathbf{V}_j\mathbf{V}_j^\tau \boldsymbol{\Gamma}^{-1/2}\boldsymbol{\Sigma}_{YX}\boldsymbol{\Sigma}_{XX}^{-1/2}$$

or

$$\mathbf{A} = \boldsymbol{\Gamma}^{1/2} \sum_{j=1}^{q} \mathbf{V}_j\mathbf{V}_j^\tau \boldsymbol{\Gamma}^{-1/2}\boldsymbol{\Sigma}_{YX}\boldsymbol{\Sigma}_{XX}^{-1}.$$

The minimum achieved is seen to be as stated.

Proof of Theorem 10.2.2 First take $\mathbf{E}$ as fixed. Then Theorem 10.2.1 indicates that the minimum with respect to $\boldsymbol{\mu}$ and $\mathbf{D}$ is

$$\operatorname{tr}\{\mathbf{E}\boldsymbol{\Sigma}_{YY}\mathbf{E}^\tau\} - \sum_{j\leq q} \mu_j(\mathbf{E}\boldsymbol{\Sigma}_{YX}\boldsymbol{\Sigma}_{XX}{}^{-1}\boldsymbol{\Sigma}_{XY}\mathbf{E}^\tau).$$

Let $\mathbf{U} = \mathbf{E}\boldsymbol{\Sigma}_{YY}^{1/2}$, then write

$$\operatorname{tr}\{\mathbf{U}^\tau\mathbf{U}\} - \sum_{j\leq q} \mu_j(\mathbf{U}^\tau\boldsymbol{\Sigma}_{YY}^{-1/2}\boldsymbol{\Sigma}_{YX}\boldsymbol{\Sigma}_{XX}{}^{-1}\boldsymbol{\Sigma}_{XY}\boldsymbol{\Sigma}_{YY}^{-1/2}\mathbf{U})$$

with $\mathbf{U}^\tau\mathbf{U} = \mathbf{I}$. Now, the latent roots that appear are maximized by taking the columns of $\mathbf{U}$ to be the first q latent vectors of $\boldsymbol{\Sigma}_{YY}^{-1/2}\boldsymbol{\Sigma}_{YX}\boldsymbol{\Sigma}_{XX}{}^{-1}\boldsymbol{\Sigma}_{XY}\boldsymbol{\Sigma}_{YY}^{-1/2}$; see Bellman (1960) p. 117. The theorem follows directly.

Proof of Theorem 10.2.3 This follows as the proof of Theorem 10.2.6 given below.

Proof of Theorem 10.2.4 This follows as did the proof of Theorem 10.2.1.

Proof of Theorem 10.2.5 This follows as did the proof of Theorem 10.2.2.

Proof of Theorem 10.2.6 Let $\boldsymbol{\Delta}_{XX} = \hat{\boldsymbol{\Sigma}}_{XX} - \boldsymbol{\Sigma}_{XX}$ with a similar definition for $\boldsymbol{\Delta}_{XY}$, $\boldsymbol{\Delta}_{YY}$. Proceeding in the manner of Wilkinson (1965) p. 68 or Dempster (1966) we have the expansions

$$\hat{\mu}_j = \mu_j - \mu_j(\bar{\mathbf{a}}_j{}^\tau\boldsymbol{\Delta}_{XX}\mathbf{a}_j) - \mu_j(\bar{\mathbf{b}}_j{}^\tau\boldsymbol{\Delta}_{YY}\mathbf{b}_j) + (\bar{\mathbf{a}}_j{}^\tau\boldsymbol{\Delta}_{XY}\mathbf{b}_j) + (\bar{\mathbf{b}}_j{}^\tau\boldsymbol{\Delta}_{YX}\mathbf{a}_j) + \cdots,$$

$$\frac{\hat{\boldsymbol{\alpha}}_j}{(\bar{\boldsymbol{\alpha}}_j{}^\tau\boldsymbol{\Sigma}_{XX}\boldsymbol{\alpha}_j)^{1/2}} = \mathbf{a}_j + \sum_{l\neq j} g_{jl}\mathbf{a}_l + \cdots,$$

$$\frac{\hat{\boldsymbol{\beta}}_j}{(\bar{\boldsymbol{\beta}}_j{}^\tau\boldsymbol{\Sigma}_{YY}\boldsymbol{\beta}_j)^{1/2}} = \mathbf{b}_j + \sum_{l\neq j} h_{jl}\mathbf{b}_l + \cdots$$

where

$$g_{jl} = (\mu_j - \mu_l)^{-1}[-\rho_j{}^2(\bar{\mathbf{a}}_l{}^\tau\boldsymbol{\Delta}_{XX}\mathbf{a}_j) - \rho_j\rho_l(\bar{\mathbf{b}}_l{}^\tau\boldsymbol{\Delta}_{YY}\mathbf{b}_j) + \rho_j(\bar{\mathbf{a}}_l{}^\tau\boldsymbol{\Delta}_{XY}\mathbf{b}_j) + \rho_l(\bar{\mathbf{b}}_l{}^\tau\boldsymbol{\Delta}_{YX}\mathbf{a}_j)]$$

and

$$h_{jl} = (\mu_j - \mu_l)^{-1}[-\rho_j\rho_l(\bar{\mathbf{a}}_l{}^\tau\boldsymbol{\Delta}_{XX}\mathbf{a}_j) - \rho_j{}^2(\bar{\mathbf{b}}_l{}^\tau\boldsymbol{\Delta}_{YY}\mathbf{b}_j) + \rho_l(\bar{\mathbf{a}}_l{}^\tau\boldsymbol{\Delta}_{XY}\mathbf{b}_j) + \rho_j(\bar{\mathbf{b}}_l{}^\tau\boldsymbol{\Delta}_{YX}\mathbf{a}_j)].$$

Using the expression developed in the course of the proof of Theorem 9.2.4 we see that

$$\operatorname{cov}\{g_{jl}, g_{km}\} = (\mu_j - \mu_l)^{-2}(1 - \mu_j)(\mu_j + \mu_l - 2\mu_j\mu_l)$$

if $j = k$, $l = m$ and equals 0 otherwise. Similarly

$$\operatorname{cov}\{g_{jl}, \bar{g}_{km}\} = -(\mu_j - \mu_l)^{-2}(\mu_j{}^2\mu_l + \mu_j\mu_l{}^2 - 4\mu_j\mu_l - \mu_j{}^2 - \mu_l{}^2 + \mu_j + \mu_l)$$

if $j = m$, $l = k$ and equals 0 otherwise.

Continuing

$$\operatorname{cov}\{g_{jl}, h_{km}\} = (\mu_j - \mu_l)^{-2}(1 - \mu_j)(2 - \mu_j - \mu_l)$$

if $j = k$, $l = m$.

$$\operatorname{cov}\{g_{jl}, \bar{h}_{km}\} = -(\mu_j - \mu_l)^{-2}(\rho_j^5\rho_l + \rho_j^3\rho_l^3 - \rho_j^4 - \rho_j^2\rho_l^2 - 2\rho_j^3\rho_l - 2\rho_j\rho_l^3 + \rho_j^2 + \rho_l^2)$$

if $j = m$, $l = k$ and so on.

The expansions above and these moments now give the indicated first- and second-order asymptotic moments. The asymptotic normality follows from the asymptotic normality of the $\hat{\mathbf{\Sigma}}_{XX}$, $\hat{\mathbf{\Sigma}}_{XY}$, and $\hat{\mathbf{\Sigma}}_{YY}$ and the fact that the latent roots and vectors are differentiable functions of these matrices through Theorem P5.2.

Proof of Theorem 10.3.1 The expression (10.3.3) may be written

$$\int_0^{2\pi} \operatorname{tr}\{f_{Y-Y*,Y-Y*}(\alpha)\}d\alpha + \{E[\mathbf{Y}(t) - \mathbf{Y}^*(t)]\}^\tau\{E[\mathbf{Y}(t) - \mathbf{Y}^*(t)]\}$$

and we see that we should choose $\mathbf{\mu}$ so that $E\mathbf{Y}(t) = E\mathbf{Y}^*(t)$. Now

$$\begin{aligned} f_{Y-Y*,Y-Y*}(\alpha) &= \mathbf{f}_{YY}(\alpha) - \mathbf{C}(\alpha)\mathbf{B}(\alpha)\mathbf{f}_{XY}(\alpha) - \mathbf{f}_{YX}(\alpha)\overline{\mathbf{B}(\alpha)}^\tau\overline{\mathbf{C}(\alpha)}^\tau \\ &\quad + \mathbf{C}(\alpha)\mathbf{B}(\alpha)\mathbf{f}_{XX}(\alpha)\overline{\mathbf{B}(\alpha)}^\tau\overline{\mathbf{C}(\alpha)}^\tau \\ &= \mathbf{f}_{YY}(\alpha) - \mathbf{f}_{YX}(\alpha)\mathbf{f}_{XX}(\alpha)^{-1}\mathbf{f}_{XY}(\alpha) \\ &\quad + [\mathbf{f}_{YX}(\alpha)\mathbf{f}_{XX}(\alpha)^{-1/2} - \mathbf{C}(\alpha)\mathbf{B}(\alpha)\mathbf{f}_{XX}(\alpha)^{1/2}] \\ &\quad \times \overline{[\mathbf{f}_{YX}(\alpha)\mathbf{f}_{XX}(\alpha)^{-1/2} - \mathbf{C}(\alpha)\mathbf{B}(\alpha)\mathbf{f}_{XX}(\alpha)^{1/2}]}^\tau. \end{aligned}$$

It therefore follows from Corollary 3.7.4 that expression (10.3.3) is minimized by the indicated $\mathbf{B}(\alpha)$ and $\mathbf{C}(\alpha)$.

Proof of Corollary 10.3.1 This result follows from an application of Theorem 10.3.1 to the transformed variate

$$\mathbf{Y}'(t) = \int_0^{2\pi} \exp\{iu\alpha\}\mathbf{f}_{YY}(\alpha)^{-1/2}d\mathbf{Z}_Y(\alpha)$$

noting, for example, that

$$\mathbf{f}_{Y'X}(\lambda) = \mathbf{f}_{YY}(\lambda)^{-1/2}\mathbf{f}_{YX}(\lambda)$$

for this series.

Proof of Theorem 10.3.2 We are interested in the coherence

$$\frac{|\mathbf{A}(\lambda)\mathbf{f}_{XY}(\lambda)\overline{\mathbf{B}(\lambda)}^\tau|^2}{[\mathbf{A}(\lambda)\mathbf{f}_{XX}(\lambda)\overline{\mathbf{A}(\lambda)}^\tau][\mathbf{B}(\lambda)\mathbf{f}_{YY}(\lambda)\overline{\mathbf{B}(\lambda)}^\tau]} = \frac{|\mathbf{A}'(\lambda)\mathbf{f}_{XX}(\lambda)^{-1/2}\mathbf{f}_{XY}(\lambda)\mathbf{f}_{YY}(\lambda)^{-1/2}\overline{\mathbf{B}'(\lambda)}^\tau|^2}{[\mathbf{A}'(\lambda)\overline{\mathbf{A}'(\lambda)}^\tau][\mathbf{B}'(\lambda)\overline{\mathbf{B}'(\lambda)}^\tau]}$$

having defined

$$\mathbf{A}'(\lambda) = \mathbf{A}(\lambda)\mathbf{f}_{XX}(\lambda)^{1/2}$$
$$\mathbf{B}'(\lambda) = \mathbf{B}(\lambda)\mathbf{f}_{YY}(\lambda)^{1/2}.$$

By Schwarz's inequality the coherency is

$$\leqslant \frac{\mathbf{B}'(\lambda)\mathbf{f}_{YY}(\lambda)^{-1/2}\mathbf{f}_{YX}(\lambda)\mathbf{f}_{XX}(\lambda)^{-1}\mathbf{f}_{XY}(\lambda)\mathbf{f}_{YY}(\lambda)^{-1/2}\overline{\mathbf{B}'(\lambda)}^{\tau}}{\mathbf{B}'(\lambda)\overline{\mathbf{B}'(\lambda)}^{\tau}}$$
$$\leqslant \mu_j[\mathbf{f}_{YY}(\lambda)^{-1/2}\mathbf{f}_{YX}(\lambda)\mathbf{f}_{XX}(\lambda)^{-1}\mathbf{f}_{XY}(\lambda)\mathbf{f}_{YY}(\lambda)^{-1/2}]$$

for $\mathbf{B}'(\lambda)$ orthogonal to $\mathbf{V}_1(\lambda), \ldots, \mathbf{V}_{j-1}(\lambda)$ the first $j-1$ latent vectors of $\mathbf{f}_{YY}^{-1/2}\mathbf{f}_{YX}\mathbf{f}_{XX}^{-1}\mathbf{f}_{XY}\mathbf{f}_{YY}^{-1/2}$ by Exercise 3.10.26. Expression (10.3.25) indicates that $\mathbf{B}_j(\lambda)$ is as indicated in the theorem; that $\mathbf{A}_j(\lambda)$ achieves equality follows by inspection.

Proof of Theorem 10.3.3 Because the latent roots of $\mathbf{f}_{YX}\mathbf{f}_{XX}^{-1}\mathbf{f}_{XY}$ are simple for all λ, its latent roots and vectors are real holomorphic functions of the entries, see Exercises 3.10.19–21. Expressions (10.3.28) and (10.3.29) now follow from Theorem 3.8.3. Expression (10.3.30) follows from (10.3.26) to (10.3.29).

Proof of Theorem 10.3.4 Because the latent roots of $\mathbf{f}_{YY}^{-1/2}\mathbf{f}_{YX}\mathbf{f}_{XX}^{-1}\mathbf{f}_{XY}\mathbf{f}_{YY}^{-1/2}$ are simple for all λ, its latent roots and vectors are real holomorphic functions of its entries; see Exercises 3.10.19 to 3.10.21. Expressions (10.3.33) and (10.3.34) now follow from Theorem 3.8.3. That the spectral density is (10.3.36) either follows from Theorem 10.3.1 or by direct computation.

Proof of Theorem 10.4.1 This follows as did the proof of Theorem 9.4.1 with the exception that the perturbation expansions of the proof of Theorem 10.2.6 are now used.

Proof of Theorem 10.4.2 This follows from the above perturbation expansions in the manner of the proof of Theorem 10.2.6.

Proof of Theorem 10.4.3 The $\hat{\mu}_j$, $\hat{\mathbf{A}}_j$, and $\hat{\mathbf{B}}_j$ are continuous functions of the entries of (10.4.25). The theorem consequently follows from Theorem 7.3.3 and Theorem P5.1.

REFERENCES

ABELSON, R. (1953). *Spectral analysis and the study of individual differences.* Ph.D. Thesis, Princeton University.

ABRAMOWITZ, M., and STEGUN, I. A. (1964). *Handbook of Mathematical Functions.* Washington: National Bureau of Standards.

ACZÉL, J. (1969). *On Applications and Theory of Functional Equations.* Basel: Birkhäuser.

AITKEN, A. C. (1954). *Determinants and Matrices.* London: Oliver and Boyd.

AKAIKE, H. (1960). "Effect of timing-error on the power spectrum of sampled data." *Ann. Inst. Statist. Math.* **11**:145–165.

AKAIKE, H. (1962a). "Undamped oscillation of the sample autocovariance function and the effect of prewhitening operation." *Ann. Inst. Statist. Math.* **13**:127–144.

AKAIKE, H. (1962b). "On the design of lag windows for the estimation of spectra." *Ann. Inst. Statist. Math.* **14**:1–21.

AKAIKE, H. (1964). "Statistical measurement of frequency response function." *Ann. Inst. Statist. Math.*, Supp. III. **15**:5–17.

AKAIKE, H. (1965). "On the statistical estimation of the frequency response function of a system having multiple input." *Ann. Inst. Statist. Math.* **17**:185–210.

AKAIKE, H. (1966). "On the use of a non-Gaussian process in the identification of a linear dynamic system." *Ann. Inst. Statist. Math.* **18**:269–276.

AKAIKE, H. (1968a). "Low pass filter design." *Ann. Inst. Statist. Math.* **20**:271–298.

AKAIKE, H. (1968b). "On the use of an index of bias in the estimation of power spectra." *Ann. Inst. Statist. Math.* **20**:55–69.

AKAIKE, H. (1969a). "A method of statistical investigation of discrete time parameter linear systems." *Ann. Inst. Statist. Math.* **21**:225–242.

AKAIKE, H. (1969b). "Fitting autoregressive models for prediction." *Ann. Inst. Statist. Math.* **21**:243–247.

AKAIKE, H., and KANESHIGE, I. (1964). "An analysis of statistical response of backrash." *Ann. Inst. Statist. Math.*, Supp. III. **15**:99–102.

AKAIKE, H., and YAMANOUCHI, Y. (1962). "On the statistical estimation of frequency response function." *Ann. Inst. Statist. Math.* **14**:23–56.

AKCASU, A. Z. (1961). "Measurement of noise power spectra by Fourier analysis." *J. Appl. Physics.* **32**:565–568.

AKHIEZER, N. I. (1956). *Theory of Approximation.* New York: Ungar.

ALBERT, A. (1964). "On estimating the frequency of a sinusoid in the presence of noise." *Ann. Math. Statist.* **35**:1403.

ALBERTS, W. W., WRIGHT, L. E., and FEINSTEIN, B. (1965). "Physiological mechanisms of tremor and rigidity in Parkinsonism. *Confin. Neurol.* **26**:318–327.

ALEXANDER, M. J., and VOK, C. A. (1963). *Tables of the cumulative distribution of sample multiple coherence.* Res. Rep. 63–67. Rocketdyne Division, North American Aviation Inc.

AMOS, D. E., and KOOPMANS, L. H. (1962). *Tables of the distribution of the coefficient of coherence for stationary bivariate Gaussian processes.* Sandia Corporation Monograph SCR–483.

ANDERSON, G. A. (1965). "An asymptotic expansion for the distribution of the latent roots of the estimated covariance matrix." *Ann. Math. Statist.* **36**:1153–1173.

ANDERSON, T. W. (1957). *An Introduction to Multivariate Statistical Analysis.* New York: Wiley.

ANDERSON, T. W. (1963). "Asymptotic theory for principal component analysis." *Ann. Math. Statist.* **34**:122–148.

ANDERSON, T. W. (1971). *Statistical Analysis of Time Series.* New York: Wiley.

ANDERSON, T. W., and WALKER, A. M. (1964). "On the asymptotic distribution of the autocorrelations of a sample from a linear stochastic process." *Ann. Math. Statist.* **35**:1296–1303.

ARATO, M. (1961). "Sufficient statistics of stationary Gaussian processes." *Theory Prob. Appl.* **6**:199–201.

ARENS, R., and CALDERÓN, A. P. (1955). "Analytic functions of several Banach algebra elements." *Ann. Math.* **62**:204–216.

ASCHOFF, J. (1965). *Circadian Clocks.* Amsterdam: North Holland.

AUTONNE, L. (1915). "Sur les matrices hypohermitiennes et sur les matrices unitaires." *Ann. Univ. Lyon.* **38**:1–77.

BALAKRISHNAN, A. V. (1964). "A general theory of nonlinear estimation problems in control systems." *J. Math. Anal. App.* **8**:4–30.

BARLOW, J. S. (1967). "Correlation analysis of EEG-tremor relationships in man." In *Recent Advances in Clinical Neurophysiology, Electroenceph. Clin. Neurophysiol.*, Suppl. **25**:167–177.

BARTLETT, M. S. (1946). "On the theoretical specification of sampling properties of auto-correlated time series." *J. Roy. Statist. Soc.*, Suppl. **8**:27–41.

BARTLETT, M. S. (1948a). "A note on the statistical estimation of supply and demand relations from time series." *Econometrica.* **16**:323–329.

BARTLETT, M. S. (1948b). "Smoothing periodograms from time series with continuous spectra." *Nature.* **161**:686–687.

BARTLETT, M. S. (1950). "Periodogram analysis and continuous spectra." *Biometrika.* **37**:1–16.

BARTLETT, M. S. (1966). *An Introduction to Stochastic Processes*, 2nd ed. Cambridge: Cambridge Univ. Press.

BARTLETT, M. S. (1967). "Some remarks on the analysis of time series." *Biometrika.* **50**:25–38.

BASS, J. (1962a). "Transformées de Fourier des fonctions pseudo-aléatoires." *C. R. Acad. Sci.* **254**:3072.

BASS, J. (1962b). *Les Fonctions Pseudo-aléatoires.* Paris: Gauthier-Villars.

BATCHELOR, G. K. (1960). *The Theory of Homogeneous Turbulence.* Cambridge. Cambridge Univ. Press.

BAXTER, G. (1963). "A norm inequality for a finite section Weiner-Hopf equation." *Ill. J. Math.* **7**:97–103.

BELLMAN, R. (1960). *Introduction to Matrix Analysis.* New York: McGraw-Hill.

BENDAT, J. S., and PIERSOL, A. (1966). *Measurement and Analysis of Random Data.* New York: Wiley.

BERANEK, L. L. (1954). *Acoustics.* New York: McGraw-Hill.

BERGLAND, G. D. (1967). "The fast Fourier transform recursive equations for arbitrary length records." *Math. Comp.* **21**:236–238.

BERNSTEIN, S. (1938). "Equations differentielles stochastiques." *Act. Sci. Ind.* **738**:5–31.

BERTRAND, J., and LACAPE, R. S. (1943). *Théorie de l'Electro-encephalogram.* Paris: G. Doin.

BERTRANDIAS, J. B. (1960). "Sur le produit de deux fonctions pseudo-aléatoires." *C. R. Acad. Sci.* **250**:263

BERTRANDIAS, J. B. (1961). "Sur l'analyse harmonique généralisée des fonctions pseudo-aléatoires." *C. R. Acad. Sci.* **253**:2829.

BEVERIDGE, W. H. (1921). "Weather and harvest cycles." *Econ. J.* **31**:429.

BEVERIDGE, W. H. (1922). "Wheat prices and rainfall in Western Europe." *J. Roy. Statist. Soc.* **85**:412–459.

BILLINGSLEY, P. (1965). *Ergodic Theory and Information.* New York: Wiley.

BILLINGSLEY, P. (1966). "Convergence of types in *k*-space." *Zeit. Wahrschein.* **5**:175–179.

BILLINGSLEY, P. (1968). *Convergence of Probability Measures.* New York: Wiley.

BINGHAM, C., GODFREY, M. D., and TUKEY, J. W. (1967). "Modern techniques in power spectrum estimation." *IEEE Trans. Audio Electroacoust.* AU-**15**:56–66.

BLACKMAN, R. B. (1965). *Linear Data Smoothing and Prediction in Theory and Practice.* Reading, Mass.: Addison-Wesley.

BLACKMAN, R. B., and TUKEY, J. W. (1958). "The measurement of power spectra from the point of view of communications engineering." *Bell Syst. Tech. J.* **37**:183–282, 485–569.

BLANC-LAPIERRE, A., and FORTET, R. (1953). *Théorie des Fonctions Aléatoires.* Paris: Masson.

BLANC-LAPIERRE, A., and FORTET, R. (1965). *Theory of Random Functions.* New York: Gordon and Breach. Translation of 1953 French edition.

BOCHNER, S. (1936). "Summation of multiple Fourier series by spherical means." *Trans. Amer. Math. Soc.* **40**:175–207.

BOCHNER, S. (1959). *Lectures on Fourier Integrals.* Princeton: Princeton Univ. Press.

BOCHNER, S., and MARTIN, W. T. (1948). *Several Complex Variables.* Princeton: Princeton Univ. Press.

BODE, H. W. (1945). *Network Analysis and Feedback Amplifier Design.* New York: Van Nostrand.

BOHMAN, H. (1960). "Approximate Fourier analysis of distribution functions." *Ark. Mat.* **4**:99–157.

BORN, M., and WOLF, E. (1959). *Principles of Optics.* London: Pergamon.

BOWLEY, A. L. (1920). *Elements of Statistics.* London: King.

BOX, G. E. P. (1954). "Some theorems on quadratic forms applied in the study of analysis of variance problems." *Ann. Math. Statist.* **25**:290–302.

BOX, G. E. P. and JENKINS, G. M. (1970). *Time Series Analysis, Forecasting and Control.* San Francisco: Holden-Day.

BRACEWELL, R. (1965). *The Fourier Transform and its Applications.* New York: McGraw-Hill.

BRENNER, J. L. (1961). "Expanded matrices from matrices with complex elements." *SIAM Review.* **3**:165–166.

BRIGHAM, E. O., and MORROW, R. E. (1967). "The fast Fourier transform." *IEEE Spectrum.* **4**:63–70.

BRILLINGER, D. R. (1964a). "The generalization of the techniques of factor analysis, canonical correlation and principal components to stationary time series." Invited paper at Royal Statistical Society Conference in Cardiff, Wales. Sept. 29–Oct. 1.

BRILLINGER, D. R. (1964b). "A technique for estimating the spectral density matrix of two signals." *Proc. I.E.E.E.* **52**:103–104.

BRILLINGER, D. R. (1964c). "The asymptotic behavior of Tukey's general method of setting approximate confidence limits (the jackknife) when applied to maximum likelihood estimates." *Rev. Inter. Statis. Inst.* **32**:202–206.

BRILLINGER, D. R. (1965a). "A property of low-pass filters. "*SIAM Review.* **7**:65–67.

BRILLINGER, D. R. (1965b). "An introduction to polyspectra." *Ann. Math. Statist.* **36**:1351–1374.

BRILLINGER, D. R. (1966a). "An extremal property of the conditional expectation." *Biometrika.* **53**:594–595.

BRILLINGER, D. R. (1966b). "The application of the jackknife to the analysis of sample surveys." *Commentary.* **8**:74–80.

BRILLINGER, D. R. (1968). "Estimation of the cross-spectrum of a stationary bivariate Gaussian process from its zeros." *J. Roy. Statist. Soc.*, B. **30**:145–159.

BRILLINGER, D. R. (1969a). "A search for a relationship between monthly sunspot numbers and certain climatic series. "*Bull. ISI.* **43**:293–306.

BRILLINGER, D. R. (1969b). "The calculation of cumulants via conditioning." *Ann. Inst. Statist. Math.* **21**:215–218.

BRILLINGER, D. R. (1969c). "Asymptotic properties of spectral estimates of second-order." *Biometrika.* **56**:375–390.

BRILLINGER, D. R. (1969d). "The canonical analysis of stationary time series." In *Multivariate Analysis* — II, Ed. P. R. Krishnaiah, pp. 331–350. New York: Academic.

BRILLINGER, D. R. (1970a). "The identification of polynomial systems by means of higher order spectra." *J. Sound Vib.* **12**:301–313.

BRILLINGER, D. R. (1970b). "The frequency analysis of relations between stationary spatial series." *Proc. Twelfth Bien. Sem. Canadian Math. Congr*, Ed. R. Pyke, pp. 39–81. Montreal: Can. Math. Congr.

BRILLINGER, D. R. (1972). "The spectral analysis of stationary interval functions." In *Proc. Seventh Berkeley Symp. Prob. Statist.* Eds. L. LeCam, J. Neyman, and E. L. Scott, pp. 483–513. Berkeley: Univ. of California Press.

BRILLINGER, D. R. (1973). "The analysis of time series collected in an experimental design." *Multivariate Analysis* — III, Ed. P. R. Krishnaiah, pp. 241–256. New York: Academic.

BRILLINGER, D. R., and HATANAKA, M. (1969). "An harmonic analysis of nonstationary multivariate economic processes. "*Econometrica.* **35**:131–141.

BRILLINGER, D. R., and HATANAKA, M. (1970). "A permanent income hypothesis relating to the aggregate demand for money (an application of spectral and moving spectral analysis)." *Economic Studies Quart.* **21**:44–71.

BRILLINGER, D. R., and ROSENBLATT, M. (1967a). "Asymptotic theory of k-th order spectra." *Spectral Analysis of Time Series*, Ed. B. Harris, pp. 153–188. New York: Wiley.

BRILLINGER, D. R., and ROSENBLATT, M. (1967b). "Computation and interpretation of k-th order spectra." In *Spectral Analysis of Time Series*, Ed. B. Harris, pp. 189–232. New York: Wiley.

BRILLINGER, D. R., and TUKEY, J. W. (1964). *Asymptotic variances, moments, cumulants and other average values*. Unpublished manuscript.

BRYSON, R. A., and DUTTON, J. A. (1961). "Some aspects of the variance spectra of tree rings and varves." *Ann. New York Acad. Sci.* **95**:580–604.

BULLARD, E. (1966). "The detection of underground explosions." *Sci. Am.* **215**:19.

BUNIMOVITCH, V. I. (1949). The fluctuation process as a vibration with random amplitude and phase." *J. Tech. Phys.* (USSR) **19**:1237–1259.

BURGERS, J. M. (1948). "Spectral analysis of an irregular function." *Proc. Acad. Sci. Amsterdam.* **51**:1073.

BURKHARDT, H. (1904). "Trigonometrische Reihen und Integrale." *Enzykl. Math. Wiss.* **2**:825–1354.

BURLEY, S. P. (1969). "A spectral analysis of the Australian business cycle." *Austral. Econ. Papers.* **8**:193–128.

BUSINGER, P. A., and GOLUB, G. H. (1969). "Singular value decomposition of a complex matrix." *Comm. ACM.* **12**:564–565.

BUTZER, P. L., and NESSEL, R. J. (1971). *Fourier Analysis and Approximations*, Vol. 1. New York: Academic.

CAIRNS, T. W. (1971). "On the fast Fourier transform on a finite Abelian group." *IEEE Trans. Computers.* C–**20**:569–571.

CAPON, J. (1969). "High resolution frequency wavenumber spectral analysis." *Proc. I.E.E.E.* **57**:1408–1418.

CAPON, J. and GOODMAN, N. R. (1970). "Probability distributions for estimators of the frequency wavenumber spectrum." *Proc. I.E.E.E.* **58**:1785–1786.

CARGO, G. T. (1966). "Some extension of the integral test." *Amer. Math. Monthly.* **73**:521–525.

CARPENTER, E. W. (1965). "Explosions seismology." *Science.* **147**:363–373.

CARTWRIGHT, D. E. (1967). "Time series analysis of tides and similar motions of the sea surface." *J. Appl. Prob.* **4**:103–112.

CHAMBERS, J. M. (1966). *Some methods of asymptotic approximation in multivariate statistical analysis.* Ph.D. Thesis, Harvard University.

CHAMBERS, J. M. (1967). "On methods of asymptotic approximation for multivariate distributions." *Biometrika.* **54**:367–384.

CHANCE, B., PYE, K., and HIGGINS, J. (1967). "Waveform generation by enzymatic oscillators." *IEEE Spectrum.* **4**:79–86.

CHAPMAN, S., and BARTELS, J. (1951). *Geomagnetism*, Vol. 2. Oxford: Oxford Univ. Press.

CHERNOFF, H., and LIEBERMAN, G. J. (1954). "Use of normal probability paper." *J. Amer. Statist. Assoc.* **49**:778–785.

CHOKSI, J. R. (1966). "Unitary operators induced by measure preserving transformations." *J. Math. and Mech.* **16**:83–100.

CHOW, G. C. (1966). "A theorem on least squares and vector correlation in multivariate linear regression." *J. Amer. Statist. Assoc.* **61**:413–414.

CLEVENSON, M. L. (1970). *Asymptotically efficient estimates of the parameters of a moving average time series.* Ph.D. Thesis, Stanford University.

CONDIT, H. R., and GRUM, F. (1964). "Spectral energy distribution of daylight." *J. Optical Soc. Amer.* **54**:937–944.

CONSTANTINE, A. G. (1963). "Some noncentral distributions in multivariate analysis." *Ann. Math. Statist.* **34**:1270–1285.

COOLEY, J. W., LEWIS, P. A. W., and WELCH, P. D. (1967a). "Historical notes on the fast Fourier transform." *IEEE Trans. on Audio and Electroacoustics.* AU-**15**:76–79.

COOLEY, J. W., LEWIS, P. A. W., and WELCH, P. D. (1967b). *The fast Fourier transform algorithm and its applications.* IBM Memorandum RC 1743.

COOLEY, J. W., LEWIS, P. A. W., and WELCH, P. D. (1970). "The application of the Fast Fourier Transform Algorithm to the estimation of spectra and cross-spectra." *J. Sound Vib.* **12**:339–352.

COOLEY, J. W., and TUKEY, J. W. (1965). "An algorithm for the machine calculation of complex Fourier series." *Math. Comp.* **19**:297–301.

COOTNER, P. H. (1964). *The Random Character of Stock Market Prices.* Cambridge: MIT Press.

COVEYOU, R. R., and MACPHERSON, R. D. (1967). "Fourier analysis of uniform random number generators." *J. Assoc. Comp. Mach.* **14**:100–119.

CRADDOCK, J. M. (1965). "The analysis of meteorological time series for use in forecasting." *Statistician.* **15**:167–190.

CRADDOCK, J. M., and FLOOD, C. R. (1969). "Eigenvectors for representing the 500 mb geopotential surface over the Northern Hemisphere." *Quart. J. Roy. Met. Soc.* **95**:576–593.

CRAMÉR, H. (1939). "On the representation of functions by certain Fourier integrals." *Trans. Amer. Math. Soc.* **46**:191–201.

CRAMÉR, H. (1942). "On harmonic analysis in certain functional spaces." *Arkiv Math. Astr. Fysik.* **28**:1–7.

CRAMÉR, H., and LEADBETTER, M. R. (1967). *Stationary and Related Stochastic Processes.* New York: Wiley.

CRANDALL, I. B. (1958). *Random Vibration,* I. Cambridge: MIT Press.

CRANDALL, I. B. (1963). *Random Vibration,* II. Cambridge: MIT Press.

CRANDALL, I. B., and SACIA, C. F. (1924). "A dynamical study of the vowel sounds." *Bell Syst. Tech. J.* **3**:232–237.

DANIELL, P. J. (1946). "Discussion of paper by M. S. Bartlett," *J. Roy. Statist. Soc.*, Suppl. **8**:27.

DANIELS, H. E. (1962). "The estimation of spectral densities." *J. Roy. Statist. Soc.*, B. **24**:185–198.

DARROCH, J. N. (1965). "An optimal property of principal components." *Ann. Math. Statist.* **36**:1579–1582.

DARZELL, J. F., and PIERSON, W. J., Jr. (1960). *The apparent loss of coherency in vector Gaussian processes due to computational procedures with applications to ship motions and random seas.* Report of Dept. of Meteorology and Oceanography, New York University.

DAVIS, C., and KAHAN, W. M. (1969). "Some new bounds on perturbation of subspaces." *Bull. Amer. Math. Soc.* **75**:863–868.

DAVIS, R. C. (1953). "On the Fourier expansion of stationary random processes." *Proc. Amer. Math. Soc.* **24**:564–569.

DEEMER, W. L., and Olkin, I. (1951). "The Jacobians of certain matrix transformations." *Biometrika.* **38**:345–367.

DEMPSTER, A. P. (1966). "Estimation in multivariate analysis." In *Multivariate Analysis,* Ed. P. R. Krishmaiah, pp. 315–334. New York: Academic.

DEMPSTER, A. P. (1969). *Continuous Multivariate Analysis.* Reading: Addison-Wesley.

DEUTSCH, R. (1962). *Nonlinear Transformations of Random Processes.* Englewood Cliffs: Prentice-Hall.

DICKEY, J. M. (1967). "Matricvariate generalizations of the multivariate t distributions and the inverted multivariate t distribution." *Ann. Math. Statist.* **38**:511–519.

DOEBLIN, W. (1938). "Sur l'equation matricielle $A(t + s) = A(t)A(s)$ et ses applications aux probabilités en chaîne." *Bull. Sci. Math.* **62**:21–32.

DOOB, J. L. (1953). *Stochastic Processes.* New York: Wiley.

DRAPER, N. R., and SMITH, H. (1966). *Applied Regression Analysis.* New York: Wiley.

DRESSEL, P. L. (1940). "Semi-invariants and their estimates." *Ann. Math. Statist.* **11**:33–57.

DUGUNDJI, J. (1958). "Envelopes and pre-envelopes of real waveforms." *IRE Trans. Inf. Theory.* IT–**4**:53–57.

DUNCAN, D. B., and JONES, R. H. (1966). "Multiple regression with stationary errors." *J. Amer. Statist. Assoc.* **61**:917–928.

DUNFORD, N., and SCHWARTZ, J. T. (1963). *Linear Operators*, Part II. New York: Wiley, Interscience.

DUNNETT, C. W., and SOBEL, M. (1954). "A bivariate generalization of Student's *t*-distribution, with tables for certain special cases." *Biometrika*. **41**:153–169.

DURBIN, J. (1954). "Errors in variables." *Rev. Inter. Statist. Inst.* **22**:23–32.

DURBIN, J. (1960). "Estimation of parameters in time series regression models." *J. Roy. Statist. Soc.*, B. **22**:139–153.

DYNKIN, E. B. (1960). *Theory of Markov Processes*. London: Pergamon.

ECKART, C., and YOUNG, G. (1936). "On the approximation of one matrix by another of lower rank." *Psychometrika*. **1**:211–218.

ECONOMIC TRENDS (1968). No. 178. London, Central Statistical Office.

EDWARDS, R. E. (1967). *Fourier Series: A Modern Introduction*, Vols. I, II. New York: Holt, Rinehart and Winston.

EHRLICH, L. W. (1970). "Complex matrix inversion versus real." *Comm. A. C. M.* **13**:561–562.

ENOCHSON, L. D., and GOODMAN, N. R. (1965). *Gaussian approximations to the distribution of sample coherence*. Tech. Rep. AFFDL — TR — 65–57, Wright-Patterson Air Force Base.

EZEKIEL, M. A., and FOX, C. A. (1959). *Methods of Correlation and Regression Analysis*. New York: Wiley.

FEHR, U., and MCGAHAN, L. C. (1967). "Analog systems for analyzing infrasonic signals monitored in field experimentation." *J. Acoust. Soc. Amer.* **42**: 1001–1007.

FEJÉR, L. (1900). "Sur les fonctions bornées et integrables." *C. R. Acad. Sci.* (Paris) **131**:984–987.

FEJÉR, L. (1904). "Untersuchungen über Fouriersche Reihen." *Mat. Ann.* **58**:501–569.

FELLER, W. (1966). *Introduction to Probability Theory and its Applications*, Vol. 2. New York: Wiley.

FIELLER, E. C. (1954). "Some problems in interval estimation." *J. Roy. Statist. Soc.*, B. **16**:175–185.

FISHER, R. A. (1928). "The general sampling distribution of the multiple correlation coefficient." *Proc. Roy. Soc.* **121**:654–673.

FISHER, R. A. (1962). "The simultaneous distribution of correlation coefficients." *Sankhya A*. **24**:1–8.

FISHER, R. A., and MACKENZIE, W. A. (1922). "The correlation of weekly rainfall" (with discussion). *J. Roy. Met. Soc.* **48**:234–245.

FISHMAN, G. S. (1969). *Spectral Methods in Econometrics*. Cambridge: Harvard Univ. Press.

FISHMAN, G. S., and KIVIAT, P. J. (1967). "Spectral analysis of time series generated by simulation models. *Management Science*. **13**:525–557.

FOX, M. (1956). "Charts of the power of the *F*-test." *Ann. Math. Statist.* **27**:484–497.

FREIBERGER, W. (1963). "Approximate distributions of cross-spectral estimates for Gaussian processes." In *Time Series Analysis*, Ed. M. Rosenblatt, pp. 244–259. New York: Wiley.

FREIBERGER, W., and GRENANDER, U. (1959). "Approximate distributions of noise power measurements." *Quart. Appl. Math.* **17**:271–283.

FRIEDLANDER, S. K., and TOPPER, L. (1961). *Turbulence; Classic Papers on Statistical Theory*. New York: Wiley Interscience.

FRIEDMAN, B. (1961). "Eigenvalues of composite matrices." *Proc. Camb. Philos. Soc.* **57**:37–49.

GABOR, D. (1946). "Theory of communication." *J. Inst. Elec. Engrs.* **93**:429–457.

GAJJAR, A. V. (1967). "Limiting distributions of certain transformations of multiple correlation coefficient." *Metron.* **26**:189–193.

GAVURIN, M. K. (1957). "Approximate determination of eigenvalues and the theory of perturbations." *Uspehi Mat. Nauk.* **12**:173–175.

GELFAND, I., RAIKOV, D., and SHILOV, G. (1964). *Commutative Normed Rings*. New York: Chelsea.

GENTLEMAN, W. M., and SANDE, G. (1966). "Fast Fourier transforms — for fun and profit." *AFIPS*. 1966 Fall Joint Computer Conference. **28**:563–578. Washington: Spartan.

GERSCH, W. (1972). "Causality or driving in electrophysiological signal analysis." *J. Math. Bioscience.* **14**:177–196.

GIBBS, F. A., and GRASS, A. M. (1947). "Frequency analysis of electroencephalograms." *Science.* **105**:132–134.

GIKMAN, I. I., and SKOROKHOD, A. V. (1966). "On the densities of probability measures in function spaces." *Russian Math. Surveys.* **21**:83–156.

GINZBURG, J. P. (1964). "The factorization of analytic matrix functions." *Soviet Math.* **5**:1510–1514.

GIRI, N. (1965). "On the complex analogues of T^2 and R^2 tests." *Ann. Math. Statist.* **36**:664–670.

GIRSHICK, M. A. (1939). "On the sampling theory of roots of determinental equations." *Ann. Math. Statist.* **10**:203–224.

GLAHN, H. R. (1968). "Canonical correlation and its relationship to discriminant analysis and multiple regression." *J. Atmos. Sci.* **25**:23–31.

GODFREY, M. D. (1965). "An exploratory study of the bispectrum of an economic time series." *Applied Statistics.* **14**:48–69.

GODFREY, M. D., and KARREMAN, H. F. (1967). "A spectrum analysis of seasonal adjustment." In *Essays in Mathematical Economics*, Ed. M. Shubik, pp. 367–421. Princeton: Princeton Univ. Press.

GOLDBERGER, A. S. (1964). *Econometric Theory*. New York: Wiley.

GOLUB, G. H. (1969). "Matrix decompositions and statistical calculations." In *Statistical Computation*, Eds. R. C. Milton, J. A. Nelder, pp. 365–397. New York: Academic.

GOOD, I. J. (1950). "On the inversion of circulant matrices." *Biometrika.* **37**:185–186.

GOOD, I. J. (1958). "The interaction algorithm and practical Fourier series." *J. Roy. Stat. Soc.*, B. **20**:361–372. Addendum (1960), **22**:372–375.

GOOD, I. J. (1963). "Weighted covariance for detecting the direction of a Gaussian source. In *Time Series Analysis*, Ed. M. Rosenblatt, pp. 447–470. New York: Wiley.

GOOD, I. J. (1971). "The relationship between two fast Fourier transforms." *IEEE Trans. Computers*. C–**20**:310–317.

GOODMAN, N. R. (1957). *On the joint estimation of the spectra, cospectrum and quadrature spectrum of a two-dimensional stationary Gaussian process*. Ph.D. Thesis, Princeton University.

GOODMAN, N. R. (1960). "Measuring amplitude and phase." *J. Franklin Inst.* **270**:437–450.

GOODMAN, N. R. (1963). "Statistical analysis based upon a certain multivariate complex Gaussian distribution (an introduction)." *Ann. Math. Statist.* **34**:152–177.

GOODMAN, N. R. (1965). *Measurement of matrix frequency reponse functions and multiple coherence functions*. Research and Technology Division, AFSC, AFFDL TR 65–56, Wright-Patterson AFB, Ohio.

GOODMAN, N. R. (1967). *Eigenvalues and eigenvectors of spectral density matrices*. Seismic Data Lab. Report 179.

GOODMAN, N. R., and DUBMAN, M. R. (1969). "Theory of time-varying spectral analysis and complex Wishart matrix processes." In *Multivariate Analysis* II, Ed. P. R. Krishnaiah, pp. 351–366. New York: Academic.

GOODMAN, N. R., KATZ, S., KRAMER, B. H., and KUO, M. T. (1961). "Frequency response from stationary noise: two case histories." *Technometrics*. **3**:245–268.

GORMAN, D., and ZABORSZKY, J. (1966). "Functional expansion in state space and the *s* domain." *IEEE Trans. Aut. Control*. AC–**11**:498–505.

GRANGER, C. W. J. (1964). *Spectral Analysis of Economic Time Series*. Princeton: Princeton Univ. Press.

GRANGER, C. W. J., and ELLIOTT, C. M. (1968). "A fresh look at wheat prices and markets in the eighteenth century." *Economic History Review*. **20**:257–265.

GRANGER, C. W. J., and HUGHES, A. O. (1968). "Spectral analysis of short series — a simulation study." *J. Roy. Statist. Soc.*, A. **131**:83–99.

GRANGER, C. W. J., and MORGENSTERN, O. (1963). "Spectral analysis of stock market prices." *Kyklos*. **16**:1–27.

GRENANDER, U. (1950). "Stochastic processes and statistical inference." *Ark. Mat.* **1**:195–277.

GRENANDER, U. (1951a). "On empirical spectral analysis of stochastic processes." *Ark. Mat.* **1**:503–531.

GRENANDER, U. (1951b). "On Toeplitz forms and stationary processes." *Ark. Mat.* **1**:551–571.

GRENANDER, U. (1954). "On the estimation of regression coefficients in the case of an autocorrelated disturbance." *Ann. Math. Statist.* **25**:252–272.

GRENANDER, U., POLLAK, H. O., and SLEPIAN, D. (1959). "The distribution of quadratic forms in normal variates: a small sample theory with applications to spectral analysis." *J. Soc. Indust. Appl. Math.* **7**:374–401.

GRENANDER, U., and ROSENBLATT, M. (1953). "Statistical spectral analysis of time series arising from stochastic processes." *Ann. Math. Stat.* **24**:537–558.

GRENANDER, U., and ROSENBLATT, M. (1957). *Statistical Analysis of Stationary Time Series*. New York: Wiley.

GRENANDER, U., and SZEGÖ, G. (1958). *Toeplitz Forms and Their Applications*. Berkeley: Univ. of Cal. Press.

GROVES, G. W., and HANNAN, E. J. (1968). "Time series regression of sea level on weather." *Rev. Geophysics.* **6**:129–174.

GROVES, G. W., and ZETLER, B. D. (1964). "The cross-spectrum of sea level at San Francisco and Honolulu." *J. Marine Res.* **22**:269–275.

GUPTA, R. P. (1965). "Asymptotic theory for principal component analysis in the complex case." *J. Indian Statist. Assoc.* **3**:97–106.

GUPTA, S. S. (1963a). "Probability integrals of multivariate normal and multivariate *t*." *Ann. Math. Statist.* **34**:792–828.

GUPTA, S. S. (1963b). "Bibliography on the multivariate normal integrals and related topics." *Ann. Math. Statist.* **34**:829–838.

GURLAND, J. (1966). "Further consideration of the distribution of the multiple correlation coefficient." *Ann. Math. Statist.* **37**:1418.

GYIRES, B. (1961). "Über die Spuren der verallgemeinerten Toeplitzschen Matrize." *Publ. Math. Debrecen.* **8**:93–116.

HÁJEK, J. (1962). "On linear statistical problems in stochastic processes." *Czech. Math. J.* **12**:404–443.

HALL, P. (1927). "Multiple and partial correlation coefficients." *Biometrika.* **19**: 100–109.

HALMOS, P. R. (1956). *Lectures in Ergodic Theory*. Tokyo: Math. Soc. Japan.

HALPERIN, M. (1967). "A generalisation of Fieller's theorem to the ratio of complex parameters." *J. Roy. Statist. Soc.*, B. **29**:126–131.

HAMBURGER, H., and GRIMSHAW, M. E. (1951). *Linear Transformations in n-dimensional Vector Space*. Cambridge: Cambridge Univ. Press.

HAMMING, R. W. (1962). *Numerical Methods for Scientists and Engineers*. New York: McGraw-Hill.

HAMMING, R. W., and TUKEY, J. W. (1949). *Measuring noise color*. Bell Telephone Laboratories Memorandum.

HAMON, B. V., and HANNAN, E. J. (1963). "Estimating relations between time series." *J. Geophys. Res.* **68**:6033–6041.

HANNAN, E. J. (1960). *Time Series Analysis*. London: Methuen.

HANNAN, E. J. (1961a). "The general theory of canonical correlation and its relation to functional analysis." *J. Aust. Math. Soc.* **2**:229–242.

HANNAN, E. J. (1961b). "Testing for a jump in the spectral function." *J. Roy. Statist. Soc.*, B. **23**:394–404.

HANNAN, E. J. (1963a). "Regression for time series with errors of measurement." *Biometrika.* **50**:293–302.

HANNAN, E. J. (1963b). "Regression for time series." In *Time Series Analysis*, Ed. M. Rosenblatt, pp. 17–37. New York: Wiley.

HANNAN, E. J. (1965). "The estimation of relationships involving distributed lags." *Econometrica.* **33**:206–224.

HANNAN, E. J. (1967a). "The estimation of a lagged regression relation." *Biometrika.* **54**:409–418.

HANNAN, E. J. (1967b). "Fourier methods and random processes." *Bull. Inter. Statist. Inst.* **42**:475–494.

HANNAN, E. J. (1967c). "Canonical correlation and multiple equation systems in economics." *Econometrica.* **35**:123–138.

HANNAN, E. J. (1968). "Least squares efficiency for vector time series." *J. Roy. Statist. Soc.*, B. **30**:490–498.

HANNAN, E. J. (1970). *Multiple Time Series*. New York: Wiley.

HASSELMAN, K., MUNK, W., and MACDONALD, G. (1963). "Bispectrum of ocean waves." In *Time Series Analysis*, Ed. M. Rosenblatt, pp. 125–139. New York: Wiley.

HAUBRICH, R. A. (1965). "Earth noise, 5 to 500 millicycles per second. 1. Spectral stationarity, normality, nonlinearity." *J. Geophys. Res.* **70**:1415–1427.

HAUBRICH, R. A., and MACKENZIE, G. S. (1965). "Earth noise, 5 to 500 millicydes per second. 2. Reaction of the earth to ocean and atmosphere." *J. Geophys. Res.* **70**:1429–1440.

HENNINGER, J. (1970). "Functions of bounded mean square and generalized Fourier-Stieltjes transforms." *Can. J. Math.* **22**:1016–1034.

HERGLOTZ, G. (1911). "Über Potenzreihen mit positivem reellem Teil im Einheitskreis." *Sitzgsber. Sachs Akad. Wiss.* **63**:501–511.

HEWITT, E., and ROSS, K. A. (1963). *Abstract Harmonic Analysis*. Berlin: Springer.

HEXT, G. R. (1966). *A new approach to time series with mixed spectra*. Ph.D. Thesis, Stanford University.

HINICH, M. (1967). "Estimation of spectra after hard clipping of Gaussian processes." *Technometrics*. **9**:391–400.

HODGSON, V. (1968). "On the sampling distribution of the multiple correlation coefficient." *Ann. Math. Statist.* **39**:307.

HOFF, J. C. (1970). "Approximation with kernels of finite oscillations, I. Convergence." *J. Approx. Theory*. **3**:213–228.

HOOPER, J. W. (1958). "The sampling variance of correlation coefficients under assumptions of fixed and mixed variates." *Biometrika*. **45**:471–477.

HOOPER, J. W. (1959). "Simultaneous equations and canonical correlation theory." *Econometrica*. **27**:245–256.

HOPF, E. (1937). *Ergodentheorie*. Berlin: Springer.

HOPF, E. (1952). "Statistical hydromechanics and functional calculus." *J. Rat. Mech. Anal.* **1**:87–123.

HORST, P. (1965). *Factor Analysis of Data Matrices*. New York: Holt, Rinehart and Winston.

HOTELLING, H. (1933). "Analysis of a complex of statistical variables into principal components." *J. Educ. Psych.* **24**:417–441, 498–520.

HOTELLING, H. (1936). "Relations between two sets of variates." *Biometrika*. **28**:321–377.

HOWREY, E. P. (1968). "A spectrum analysis of the long-swing hypothesis." *Int. Econ. Rev.* **9**:228–252.

HOYT, R. S. (1947). "Probability functions for the modulus and angle of the normal complex variate." *Bell System Tech. J.* **26**:318–359.

HSU, P. L. (1941). "On the limiting distribution of canonical correlations." *Biometrika*. **33**:38–45.

HSU, P. L. (1949). "The limiting distribution of functions of sample means and application to testing hypotheses." In *Proc. Berkeley Symp. Math. Statist. Prob.*, Ed. J. Neyman, pp. 359–401. Berkeley: Univ. of Cal. Press.

HUA, L. K. (1963). *Harmonic Analysis of Functions of Several Variables in Classical Domains*. Providence: American Math. Society.

IBRAGIMOV, I. A. (1963). "On estimation of the spectral function of a stationary Gaussian process." *Theory Prob. Appl.* **8**:366–401.

IBRAGIMOV, I. A. (1967). "On maximum likelihood estimation of parameters of the spectral density of stationary time series." *Theory Prob. Appl.* **12**:115–119.

IOSIFESCU, M. (1968). "The law of the interated logarithm for a class of dependent random variables." *Theory Prob. Appl.* **13**:304–313.

IOSIFESCU, M., and THEODORESCU, R. (1969). *Random Processes and Learning*. Berlin: Springer.

ISSERLIS, L. (1918). "On a formula for the product moment coefficient of any order of a normal frequency distribution in any number of variables." *Biometrika.* **12**:134–139.

ITO, K., and NISIO, M. (1964). "On stationary solutions of a stochastic differential equation." *J. Math. Kyoto.* **4**:1–75.

IZENMAN, A. J. (1972). *Reduced rank regression for the multivariate linear model.* Ph.D. Thesis, University of California, Berkeley.

JAGERMAN, D. L. (1963). "The autocorrelation function of a sequence uniformly distributed modulo 1." *Ann. Math. Statist.* **34**:1243–1252.

JAMES, A. T. (1964). "Distributions of matrix variates and latent roots derived from normal samples." *Ann. Math. Statist.* **35**:475–501.

JAMES, A. T. (1966). "Inference on latent roots by calculation of hypergeometric functions of matrix argument." In *Multivariate Analysis*, Ed. P. R. Krishnaiah, pp. 209–235. New York: Academic.

JENKINS, G. M. (1961). "General considerations in the analysis of spectra." *Technometrics.* **3**:133–166.

JENKINS, G. M. (1963a). "Cross-spectral analysis and the estimation of linear open loop transfer functions." In *Time Series Analysis*, Ed. M. Rosenblatt, pp. 267–278. New York: Wiley.

JENKINS, G. M. (1963b). "An example of the estimation of a linear open-loop transfer function." *Technometrics.* **5**:227–245.

JENKINS, G. M., and WATTS, D. G. (1968). *Spectrum Analysis and Its Applications.* San Francisco: Holden-Day.

JENNISON, R. C. (1961). *Fourier Transforms and Convolutions for the Experimentalist.* London: Pergamon.

JONES, R. H. (1962a). "Spectral estimates and their distributions, II." *Skand. Aktuartidskr.* **45**:135–153.

JONES, R. H. (1962b). "Spectral analysis with regularly missed observations." *Ann. Math. Statist.* **33**:455–461.

JONES, R. H. (1965). "A reappraisal of the periodogram in spectral analysis." *Technometrics.* **7**:531–542.

JONES, R. H. (1969). "Phase free estimation of coherence." *Ann. Math. Statist.* **40**:540–548.

KABE, D. G. (1966). "Complex analogues of some classical non-central multivariate distributions." *Austral. J. Statist.* **8**:99–103.

KABE, D. G. (1968a). "On the distribution of the regression coefficient matrix of a normal distribution." *Austral. J. Statist.* **10**:21–23.

KABE, D. G. (1968b). "Some aspects of analysis of variance and covariance theory for a certain multivariate complex Gaussian distribution." *Metrika.* **13**:86–97.

KAHANE, J. (1968). *Some Random Series of Functions.* Lexington: Heath.

KAMPÉ de FÉRIET, J. (1954). "Introduction to the statistical theory of turbulence." *J. Soc. Ind. Appl. Math.* **2**:244–271.

KAMPÉ de FÉRIET, J. (1965). "Random integrals of differential equations." In *Lectures on Modern Mathematics,* Ed. T. L. Saaty, **3**:277–321. New York: Wiley.

KANESHIGE, I. (1964). "Frequency response of an automobile engine mounting." *Ann. Inst. Stat. Math.*, Suppl. **3**:49–58.

KAWASHIMA, R. (1964). "On the response function for the rolling motion of a fishing boat on ocean waves." *Ann. Inst. Stat. Math.*, Suppl. **3**:33–40.

KAWATA, T. (1959). "Some convergence theorems for stationary stochastic processes." *Ann. Math. Statist.* **30**:1192–1214.

KAWATA, T. (1960). "The Fourier series of some stochastic processes." *Japanese J. Math.* **29**:16–25.

KAWATA, T. (1965). "Sur la série de Fourier d'un processus stochastique stationaire." *C. R. Acad. Sci.* (Paris). **260**:5453–5455.

KAWATA, T. (1966). "On the Fourier series of a stationary stochastic process." *Zeit. Wahrschein.* **6**:224–245.

KEEN, C. G., MONTGOMERY, J., MOWAT, W. M. H., and PLATT, D. C. (1965). "British seismometer array recording systems." *J. Br. Instn. Radio Engrs.* **30**:279.

KENDALL, M. (1946). *Contributions to the Study of Oscillatory Time Series.* Cambridge: Cambridge Univ. Press.

KENDALL, M. G., and STUART, A. (1958). *The Advanced Theory of Statistics,* Vol. I. London: Griffin.

KENDALL, M. G., and STUART, A. (1961). *The Advanced Theory of Statistics,* Vol. II. London: Griffin.

KENDALL, M. G., and STUART, A. (1968). *The Advanced Theory of Statistics,* Vol. III. London: Griffin.

KHATRI, C. G. (1964). "Distribution of the 'generalised' multiple correlation matrix in the dual case." *Ann. Math. Statist.* **35**:1801–1806.

KHATRI, C. G. (1965a). "Classical statistical analysis based on a certain multivariate complex Gaussian distribution." *Ann. Math. Statist.* **36**:98–114.

KHATRI, C. G. (1965b). "A test for reality of a covariance matrix in a certain complex Gaussian distribution." *Ann. Math. Statist.* **36**:115–119.

KHATRI, C. G. (1967). "A theorem on least squares in multivariate linear regression." *J. Amer. Statist. Assoc.* **62**:1494–1495.

KHINTCHINE, A. (1934). "Korrelationstheorie der stationären Prozesse." *Math. Annalen.* **109**:604–615.

KINOSITA, K. (1964). "On the behaviour of tsunami in a tidal river." *Ann. Inst. Stat. Math.*, Suppl. **3**:78–88.

KIRCHENER, R. B. (1967). "An explicit formula for exp *At*." *Amer. Math. Monthly.* **74**:1200–1203.

KNOPP, K. (1948). *Theory and Application of Infinite Series.* New York: Hafner.

KOLMOGOROV, A. N. (1941a). "Interpolation und Extrapolation von stationären zufälligen Folgen." *Bull. Acad. Sci. de l'U.R.S.S.* **5**:3–14.

KOLMOGOROV, A. N. (1941b). "Stationary sequences in Hilbert space." (In Russian.) *Bull. Moscow State U. Math.* **2**:1–40. [Reprinted in Spanish in *Trab. Estad.* **4**:55–73, 243–270.]

KOOPMANS, L. H. (1964a). "On the coefficient of coherence for weakly stationary stochastic processes." *Ann. Math. Statist.* **35**:532–549.

KOOPMANS, L. H. (1964b). "On the multivariate analysis of weakly stationary stochastic processes." *Ann. Math. Statist.* **35**:1765–1780.

KOOPMANS, L. H. (1966). "A note on the estimation of amplitude spectra for stochastic processes with quasi-linear residuals." *J. Amer. Statist. Assoc.* **61**: 397–402.

KRAMER, H. P., and MATHEWS, M. V. (1956). "A linear coding for transmitting a set of correlated signals." *IRE Trans. Inf. Theo.* IT–**2**:41–46.

KRAMER, K. H. (1963). "Tables for constructing confidence limits on the multiple correlation coefficient." *J. Amer. Statist. Assoc.* **58**:1082–1085.

KRISHNAIAH, P. R., and WAIKAR, V. B. (1970). *Exact joint distributions of few roots of a class of random matrices.* Report ARL 70-0345. Aerospace Res. Labs.

KROMER, R. E. (1969). *Asymptotic properties of the autoregressive spectral estimator.* Ph.D. Thesis, Stanford University.

KSHIRSAGAR, A. M. (1961). "Some extensions of the multivariate *t*-distribution and the multivariate generalization of the distribution of the regression coefficient." *Proc. Camb. Philos. Soc.* **57**:80–85.

KSHIRSAGAR, A. M. (1971). "Goodness of fit of a discriminant function from the vector space of dummy variables." *J. Roy. Statist. Soc.*, B. **33**:111–116.

KUHN, H. G. (1962). *Atomic Spectra.* London: Longmans.

KUO, F. F., and KAISER, J. F. (1966). *System Analysis by Digital Computer.* New York: Wiley.

LABROUSTE, M. H. (1934). "L'analyse des séismogrammes." *Mémorial des Sciences Physiques*, Vol. 26. Paris: Gauthier-Villars.

LAMPERTI, J. (1962). "On covergence of stochastic processes." *Trans. Amer. Math. Soc.* **104**:430–435.

LANCASTER, H. O. (1966). "Kolmogorov's remark on the Hotelling canonical correlations." *Biometrika.* **53**:585–588.

LANCZOS, C. (1955). "Spectroscopic eigenvalue analysis." *J. Wash. Acad. Sci.* **45**:315–323.

LANCZOS, C. (1956). *Applied Analysis.* Englewood Cliffs: Prentice-Hall.

LATHAM, G., et al. (1970). "Seismic data from man-made impacts on the moon." *Science.* **170**:620–626.

LAUBSCHER, N. F. (1960). "Normalizing the noncentral *t* and *F* distributions." *Ann. Math. Statist.* **31**:1105–1112.

LAWLEY, D. N. (1959). "Tests of significance in canonical analysis." *Biometrika.* **46**:59–66.

LEE, Y. W. (1960). *Statistical Theory of Communication.* New York: Wiley.

LEE, Y. W., and WIESNER, J. B. (1950). "Correlation functions and communication applications." *Electronics.* **23**:86–92.

LEONOV, V. P. (1960). "The use of the characteristic functional and semi-invariants in the ergodic theory of stationary processes." *Soviet Math.* **1**:878–881.

LEONOV, V. P. (1964). *Some Applications of Higher-order Semi-invariants to the Theory of Stationary Random Processes* (in Russian). Moscow: Izdatilstvo, Nauka.

LEONOV, V. P., and SHIRYAEV, A. N. (1959). "On a method of calculation of semi-invariants." *Theor. Prob. Appl.* **4**:319–329.

LEONOV, V. P., and SHIRYAEV, A. N. (1960). "Some problems in the spectral theory of higher moments, II." *Theory Prob. Appl.* **5**:460–464.

LEPPINK, G. J. (1970). "Efficient estimators in spectral analysis. "*Proc. Twelfth Biennial Seminar Can. Math. Cong.*, Ed. R. Pyke, pp. 83–87. Montreal: Can. Math. Cong.

LÉVY, P. (1933). "Sur la convergence absolue des séries de Fourier." *C. R. Acad. Sci. Paris.* **196**:463–464.

LEWIS, F. A. (1939). "Problem 3824." *Amer. Math. Monthly.* **46**:304–305.

LIGHTHILL, M. J. (1958). *An Introduction to Fourier Analysis and Generalized Functions.* Cambridge: Cambridge Univ. Press.

LOÈVE, M. (1963). *Probability Theory.* Princeton: Van Nostrand.

LOMNICKI, Z. A., and ZAREMBA, S. K. (1957a). "On estimating the spectral density function of a stochastic process." *J. Roy. Statist. Soc.*, B. **19**:13–37.

LOMNICKI, Z. A., and ZAREMBA, S. K. (1957b). "On some moments and distributions occurring in the theory of linear stochastic processes, I." *Mh. Math.* **61**:318–358.

LOMNICKI, Z. A., and ZAREMBA, S. K. (1959). "On some moments and distributions occurring in the theory of linear stochastic processes, II." *Mh. Math.* **63**:128–168.

LOYNES, R. M. (1968). "On the concept of the spectrum for non-stationary processes." *J. Roy. Statist. Soc.*, B. **30**:1–30.

MACDONALD, N. J., and WARD, F. (1963). "The prediction of geomagnetic disturbance indices. 1. The elimination of internally predictable variations." *J. Geophys. Res.* **68**:3351–3373.

MACDUFFEE, C. C. (1946). *The Theory of Matrices.* New York: Chelsea.

MACNEIL, I. B. (1971). "Limit processes for co-spectral and quadrature spectral distribution functions." *Ann. Math. Statist.* **42**:81–96.

MADANSKY, A., and OLKIN, I. (1969). "Approximate confidence regions for constraint parameters." In *Multivariate Analysis* — II, Ed. P. R. Krishnaiah, pp. 261–286. New York: Academic.

MADDEN, T. (1964). "Spectral, cross-spectral and bispectral analysis of low frequency electromagnetic data." *Natural Electromagnetic Phenomena Below 30 kc/s*, Ed. D. F. Bleil, pp. 429–450. New York: Wiley.

MAJEWSKI, W., and HOLLIEN, H. (1967). "Formant frequency regions of Polish vowels." *J. Acoust. Soc. Amer.* **42**:1031–1037.

MALEVICH, T. L. (1964). "The asymptotic behavior of an estimate for the spectral function of a stationary Gaussian process." *Theory Prob. Appl.* **9**:350–353.

MALEVICH, T. L. (1965). "Some properties of the estimators of the spectrum of a stationary process." *Theory Prob. Appl.* **10**:447–465.

MALINVAUD, E. (1964). *Statistical Methods of Econometrics.* Amsterdam: North-Holland.

MALLOWS, C. L. (1961). "Latent vectors of random symmetric matrices." *Biometrika*. **48**:133–149.

MANN, H. B., and WALD, A. (1943a). "On stochastic limit and order relationships." *Ann. Math. Statist.* **14**:217–226.

MANN, H. B., and WALD, A. (1943b). "On the statistical treatment of linear stochastic difference equations." *Econometrica*. **11**:173–220.

MANWELL, T., and SIMON, M. (1966). "Spectral density of the possibly random fluctuations of 3 C 273." *Nature*. **212**:1224–1225.

MARUYAMA, G. (1949). "The harmonic analysis of stationary stochastic processes." *Mem. Fac. Sci. Kyusyu Univ.* Ser. A. **4**:45–106.

MATHEWS, M. V. (1963). "Signal detection models for human auditory perception." In *Time Series Analysis*, Ed. M. Rosenblatt, pp. 349–361. New York: Wiley.

MCGUCKEN, W. (1970). *Nineteenth Century Spectroscopy*. Baltimore: Johns Hopkins.

MCNEIL, D. R. (1967). "Estimating the covariance and spectral density functions from a clipped stationary time series." *J. Roy. Statist. Soc.*, B. **29**:180–195.

MCSHANE, E. J. (1963). "Integrals devised for special purposes." *Bull. Amer. Math. Soc.* **69**:597–627.

MEDGYESSY, P. (1961). *Decomposition of Superpositions of Distribution Functions*. Budapest: Hungar. Acad. Sci.

MEECHAM, W. C. (1969). "Stochastic representation of nearly-Gaussian nonlinear processes." *J. Statist. Physics*. **1**:25–40.

MEECHAM, W. C., and SIEGEL, A. (1964). "Wiener-Hermite expansion in model turbulence at large Reynolds numbers." *Physics Fluids*. **7**:1178–1190.

MEGGERS, W. F. (1946). "Spectroscopy, past, present and future." *J. Opt. Soc. Amer*. **36**:431–448.

MIDDLETON, D. (1960). *Statistical Communication Theory*. New York: McGraw-Hill.

MILLER, K. S. (1968). "Moments of complex Gaussian processes." *Proc. IEEE*. **56**:83–84.

MILLER, K. S. (1969). "Complex Gaussian processes." *SIAM Rev.* **11**:544–567.

MILLER, R. G. (1966). *Simultaneous Statistical Inference*. New York: McGraw-Hill.

MIYATA, M. (1970). "Complex generalization of canonical correlation and its application to sea level study." *J. Marine Res.* **28**:202–214.

MOORE, C. N. (1966). *Summable Series and Convergence Factors*. New York: Dover.

MORAN, J. M., et al. (1968). "The 18-cm flux of the unresolved component of 3 C 273." *Astrophysical J.* **151**:L99–L101.

MORRISON, D. F. (1967). *Multivariate Statistical Methods*. New York: McGraw-Hill.

MORTENSEN, R. E. (1969). "Mathematical problems of modeling stochastic non-linear dynamic systems." *J. Statist. Physics*. **1**:271–296.

MUNK, W. H., and CARTWRIGHT, D. E. (1966). "Tidal spectroscopy and prediction." *Phil. Trans.*, A. **259**:533–581.

MUNK, W. H., and MACDONALD, G. J. F. (1960). *The Rotation of the Earth.* Cambridge: Cambridge Univ. Press.

MUNK, W. H., and SNODGRASS, F. E. (1957). "Measurements of southern swell at Guadalupe Island." *Deep-Sea Research.* **4**:272–286.

MURTHY, V. K. (1963). "Estimation of the cross-spectrum." *Ann. Math. Statist.* **34**:1012–1021.

NAKAMURA, I. (1964). "Relation between superelevation and car rolling." *Ann. Inst. Stat. Math.*, Suppl. **3**:41–48.

NAKAMURA, H., and MURAKAMI, S. (1964). "Resonance characteristic of the hydraulic system of a water power plant." *Ann. Inst. Stat. Math.*, Suppl. **3**:65–70.

NAYLOR, T. H., WALLACE, W. H., and SASSER, W. E. (1967). "A computer simulation model of the textile industry." *J. Amer. Stat. Assoc.* **62**:1338–1364.

NERLOVE, M. (1964). "Spectral analysis of seasonal adjustment procedures." *Econometrica.* **32**:241–286.

NETTHEIM, N. (1966). *The estimation of coherence.* Technical Report, Statistics Department, Stanford University.

NEUDECKER, H. (1968). "The Kronecker matrix product and some of its applications in econometrics." *Statistica Neerlandica.* **22**:69–82.

NEWTON, H. W. (1958). *The Face of the Sun.* London: Penguin.

NICHOLLS, D. F. (1967). "Estimation of the spectral density function when testing for a jump in the spectrum." *Austral. J. Statist.* **9**:103–108.

NISIO, M. (1960). "On polynomial approximation for strictly stationary processes." *J. Math. Soc. Japan.* **12**:207–226.

NISIO, M. (1961). "Remarks on the canonical representation of strictly stationary processes." *J. Math. Kyoto.* **1**:129–146.

NISSEN, D. H. (1968). "A note on the variance of a matrix." *Econometrica.* **36**:603–604.

NOLL, A. M. (1964). "Short-time spectrum and 'cepstrum' techniques for vocal-pitch detection." *J. Acoust. Soc. Amer.* **36**:296–302.

OBUKHOV, A. M. (1938). "Normally correlated vectors." *Izv. Akad. Nauk SSR.* Section on Mathematics. **3**:339–370.

OBUKHOV, A. M. (1940). "Correlation theory of vectors." *Uchen. Zap. Moscow State Univ.* Mathematics Section. **45**:73–92.

OCEAN WAVE SPECTRA (1963). National Academy of Sciences. Englewood Cliffs: Prentice-Hall.

OKAMOTA, M. (1969). "Optimality of principal components." In *Multivariate Analysis* — II, Ed. P. R. Krisknaiah, pp. 673–686. New York: Academic.

OKAMOTO, M., and KANAZAWA, M. (1968). "Minimization of eigenvalues of a matrix and optimality of principal components." *Ann. Math. Statist.* **39**:859–863.

OLKIN, I., and PRATT, J. W. (1958). "Unbiased estimation of certain correlation coefficients." *Ann. Math. Statist.* **29**:201–210.

OLSHEN, R. A. (1967). "Asymptotic properties of the periodogram of a discrete stationary process." *J. Appl. Prob.* **4**:508–528.

OSWALD, J. R. V. (1956). "Theory of analytic bandlimited signals applied to carrier systems." *IRE Trans. Circuit Theory.* CT-**3**:244–251.

PANOFSKY, H. A. (1967). "Meteorological applications of cross-spectrum analysis." In *Advanced Seminar on Spectral Analysis of Time Series*, Ed. B. Harris, pp. 109–132. New York: Wiley.

PAPOULIS, A. (1962). *The Fourier Integral and its Applications*. New York: McGraw-Hill.

PARTHASARATHY, K. R. (1960). "On the estimation of the spectrum of a stationary stochastic process." *Ann. Math. Statist.* **31**:568–573.

PARTHASARATHY, K. R., and VARADAHN, S. R. S. (1964). "Extension of stationary stochastic processes." *Theory Prob. Appl.* **9**:65–71.

PARZEN, E. (1957). "On consistent estimates of the spectrum of a stationary time series." *Ann. Math. Statist.* **28**:329–348.

PARZEN, E. (1958). "On asymptotically efficient consistent estimates of the spectral density function of a stationary time series." *J. Roy. Statist. Soc.*, B. **20**:303–322.

PARZEN, E. (1961). "Mathematical considerations in the estimation of spectra." *Technometrics*. **3**:167–190.

PARZEN, E. (1963a). "On spectral analysis with missing observations and amplitude modulation." *Sankhya*. A. **25**:180–189.

PARZEN, E. (1963b). "Notes on Fourier analysis and spectral windows." Included in Parzen (1967a).

PARZEN, E. (1963c). "Probability density functionals and reproducing kernel Hilbert spaces." In *Times Series Analysis*, Ed. M. Rosenblatt, pp. 155–169. New York: Wiley.

PARZEN, E. (1964). "An approach to empirical-time series analysis." *Radio Science*. **68D**:937–951.

PARZEN, E. (1967a). *Time Series Analysis Papers*. San Francisco: Holden-Day.

PARZEN, E. (1967b). "Time series analysis for models of signals plus white noise." In *Advanced Seminar on Spectral Analysis of Time Series*, Ed. B. Harris, pp. 233–257. New York: Wiley.

PARZEN, E. (1967c). "On empirical multiple time series analysis." In *Proc. Fifth Berkeley Symp. Math. Statist. Prob.*, 1, Eds. L. Le Cam and J. Neyman, pp. 305–340. Berkeley: Univ. of Cal. Press.

PARZEN, E. (1969). "Multiple time series modelling." In *Multivariate Analysis — II*, Ed. P. R. Krishnaiah, pp. 389–409. New York: Academic.

PEARSON, E. S., and HARTLEY, H. O. (1951). "Charts of the power function for analysis of variance tests derived from the non-central *F* distribution." *Biometrika*. **38**:112–130.

PEARSON, K., and FILON, L. N. G. (1898). "Mathematical contributions to the theory of evolution. IV. On the probable errors of frequency constants and on the influence of random selection on variation and correlation." *Phil. Trans.*, A. **191**:229–311.

PEARSON, K., JEFFERY, G. B., and ELDERTON, E. M. (1929). "On the coefficient of the first product moment coefficient in samples drawn from an indefinitely large normal population." *Biometrika*. **21**:164–201.

PHILIPP, W. (1967). "Das Gesetz vom iterierten Logarithmus für stark mischende stationare Prozesse." *Zeit. Wahrschein.* **8**:204–209.

PHILIPP, W. (1969). "The central limit problem for mixing sequences of random variables." *Z. Wahrschein. verw. Gebiet*. **12**:155–171.

PICINBONO, B. (1959). "Tendence vers le caractère gaussien par filtrage selectif." *C. R. Acad. Sci. Paris.* **248**:2280.

PICKLANDS, J. (1970). "Spectral estimation with random truncation." *Ann. Math. Statist.* **41**:44–58.

PINSKER, M. S. (1964). *Information and Information Stability of Random Variables and Processes.* San Francisco: Holden-Day.

PISARENKO, V. F. (1970). "Statistical estimates of amplitude and phase corrections." *Geophys. J. Roy. Astron. Soc.* **20**:89–98.

PISARENKO, V. F. (1972). "On the estimation of spectra by means of non-linear functions of the covariance matrix." *Geophys. J. Roy Astron. Soc.* **28**:511–531.

PLAGEMANN, S. H., FELDMAN, V. A., and GRIBBIN, J. R. (1969). "Power spectrum analysis of the emmission-line redshift distribution of quasi-stellar and related objects." *Nature.* **224**:875–876.

POLYA, G., and SZEGÖ, G. (1925). *Aufgaben und Lehrsätze aus der Analysis I.* Berlin: Springer.

PORTMANN, W. O. (1960). "Hausdorff-analytic functions of matrices." *Proc. Amer. Math. Soc.* **11**:97–101.

POSNER, E. C. (1968). "Combinatorial structures in planetary reconnaissance." In *Error Correcting Codes*, Ed. H. B. Mann, pp. 15–47. New York: Wiley.

PRESS, H., and TUKEY, J. W. (1956). *Power spectral methods of analysis and their application to problems in airplane dynamics.* Bell Telephone System Monograph 2606.

PRIESTLEY, M. B. (1962a). "Basic considerations in the estimation of spectra." *Technometrics.* **4**:551–564.

PRIESTLEY, M. B. (1962b). "The analysis of stationary processes with mixed spectra." *J. Roy. Statist. Soc.*, B. **24**:511–529.

PRIESTLEY, M. B. (1964). "Estimation of the spectra density function in the presence of harmonic components." *J. Roy. Statist. Soc.*, B. **26**:123–132.

PRIESTLEY, M. B. (1965). "Evolutionary spectra and non-stationary processes." *J. Roy. Statist. Soc.*, B. **27**:204–237

PRIESTLEY, M. B. (1969). "Estimation of transfer functions in closed loop stochastic systems." *Automatica.* **5**:623–632.

PUPIN, M. I. (1894). "Resonance analysis of alternating and polyphase currents." *Trans. A.I.E.E.* **9**:523.

QUENOUILLE, M. H. (1957). *The Analysis of Multiple Time Series.* London: Griffin.

RAO, C. R. (1964). "The use and interpretation of principal component analysis in applied research." *Sankhya*, A. **26**:329–358.

RAO, C. R. (1965). *Linear Statistical Inference and Its Applications.* New York: Wiley.

RAO, M. M. (1960). "Estimation by periodogram." *Trabajos Estadistica.* **11**:123–137.

RAO, M. M. (1963). "Inference in stochastic processes. I." *Teor. Verojatnest. i Primemen.* **8**:282–298.

RAO, M. M. (1966). "Inference in stochastic processes, II." *Zeit. Wahrschein.* **5**:317–335.

RAO, S. T. (1967). "On the cross-periodogram of a stationary Gaussian vector process." *Ann. Math. Statist.* **38**:593–597.

RICHTER, C. P. (1967). "Biological clocks in medicine and psychiatry." *Proc. Nat. Acad. Sci.* **46**:1506–1530.

RICKER, N. (1940). The form and nature of seismic waves and the structure of seismograms." *Geophysics.* **5**:348–366.

RIESZ, F., and NAGY, B. Sz. (1955). *Lessons in Functional Analysis.* New York: Ungar.

ROBERTS, J. B., and BISHOP, R. E. D. (1965). "A simple illustration of spectral density analysis." *J. Sound Vib.* **2**:37–41.

ROBINSON, E. A. (1967a). *Multichannel Time Series Analysis with Digital Computer Programs.* San Francisco: Holden-Day.

ROBINSON, E. A. (1967b). *Statistical Communication and Detection with Special Reference to Digital Data Processing of Radar and Seismic Signals.* London: Griffin.

RODEMICH, E. R. (1966). "Spectral estimates using nonlinear functions." *Ann. Math. Statist.* **37**:1237–1256.

RODRIGUEZ-ITURBE, I., and YEVJEVICH, V. (1968). *The investigation of relationship between hydrologic time series and sunspot numbers.* Hydrology Paper. No. 26. Fort Collins: Colorado State University.

ROOT, W. L., and PITCHER, T. S. (1955). "On the Fourier expansion of random functions." *Ann. Math. Statist.* **26**:313–318.

ROSENBERG, M. (1964). "The square-integrability of matrix-valued functions with respect to a non-negative Hermitian measure." *Duke Math. J.* **31**:291–298.

ROSENBLATT, M. (1956a). "On estimation of regression coefficients of a vector-valued time series with a stationary disturbance." *Ann. Math. Statist.* **27**:99–121.

ROSENBLATT, M. (1956b). "On some regression problems in time series analysis." *Proc. Third Berkeley Symp. Math. Statist. Prob.*, Vol 1. Ed. J. Neyman, pp. 165–186. Berkeley: Univ. of Cal. Press.

ROSENBLATT, M. (1956c). "A central limit theorem and a strong mixing condition." *Proc. Nat. Acad. Sci.* (U.S.A.). **42**:43–47.

ROSENBLATT, M. (1959). "Statistical analysis of stochastic processes with stationary residuals." In *Probability and Statistics*, Ed. U. Grenander, pp. 246–275. New York: Wiley.

ROSENBLATT, M. (1960). "Asymptotic distribution of the eigenvalues of block Toeplitz matrices." *Bull. Amer. Math. Soc.* **66**:320–321.

ROSENBLATT, M. (1961). "Some comments on narrow band-pass filters." *Quart. Appl. Math.* **18**:387–393.

ROSENBLATT, M. (1962). "Asymptotic behavior of eigenvalues of Toeplitz forms." *J. Math. Mech.* **11**:941–950.

ROSENBLATT, M. (1964). "Some nonlinear problems arising in the study of random processes." *Radio Science.* **68D**:933–936.

ROSENBLATT, M., and VAN NESS, J. S. (1965). "Estimation of the bispectrum." *Ann. Math. Statist.* **36**:1120–1136.

ROZANOV, Yu. A. (1967). *Stationary Random Processes.* San Francisco: Holden-Day.

SALEM, R., and ZYGMUND, A. (1956). "A note on random trigonometric polynomials. In *Proc. Third Berkeley Symp. Math. Statist. Prob.*, Ed. J. Neyman, pp. 243–246. Berkeley: Univ. of Cal. Press.

SARGENT, T. J. (1968). "Interest rates in the nineteen-fifties." *Rev. Econ. Stat.* **50**:164–172.

SATO, H. (1964). "The measurement of transfer characteristic of ground-structure systems using micro tremor." *Ann. Inst. Stat. Math.*, Suppl. **3**:71–78.

SATTERTHWAITE, F. E. (1941). "Synthesis of variance." *Psychometrica.* **6**:309–316.

SAXENA, A. K. (1969). "Classification into two multivariate complex normal distributions with different covariance matrices." *J. Ind. Statist. Assoc.* **7**:158–161.

SCHEFFÉ, H. (1959). *The Analysis of Variance.* New York: Wiley.

SCHOENBERG, I. J. (1946). "Contributions to the problem of approximation of equidistant data by analytic functions." *Quart. Appl. Math.* **4**:45–87, 112–141.

SCHOENBERG, I. J. (1950). "The finite Fourier series and elementary geometry." *Amer. Math. Monthly.* **57**:390–404.

SCHUSTER, A. (1894). "On interference phenomena." *Phil. Mag.* **37**:509–545.

SCHUSTER, A. (1897). "On lunar and solar periodicities of earthquakes." *Proc. Roy. Soc.* **61**:455–465.

SCHUSTER, A. (1898). "On the investigation of hidden periodicities with application to a supposed 26 day period of meteorological phenomena." *Terr. Magn.* **3**:13–41.

SCHUSTER, A. (1900). "The periodogram of magnetic declination as obtained from the records of the Greenwich Observatory during the years 1871–1895." *Camb. Phil. Trans.* **18**:107–135.

SCHUSTER, A. (1904). *The Theory of Optics.* London: Cambridge Univ. Press.

SCHUSTER, A. (1906a). "The periodogram and its optical analogy." *Proc. Roy. Soc.* **77**:137–140.

SCHUSTER, A. (1906b). "On the periodicities of sunspots." *Philos. Trans. Roy. Soc.*, A. **206**:69–100.

SCHWARTZ, L. (1957). *Théorie des Distributions*, Vol. I. Paris: Hermann.

SCHWARTZ, L. (1959). *Théorie des Distributions*, Vol. II. Paris: Hermann.

SCHWERDTFEGER, H. (1960). "Direct proof of Lanczos's decomposition theorem." *Amer. Math. Mon.* **67**:856–860.

SEARS, F. W. (1949). *Optics.* Reading: Addison-Wesley.

SHAPIRO, H. S. (1969). *Smoothing and Approximation of Functions.* New York: Van Nostrand.

SHIRYAEV, A. N. (1960). "Some problems in the spectral theory of higher-order moments, I." *Theor. Prob. Appl.* **5**:265–284.

SHIRYAEV, A. N. (1963). "On conditions for ergodicity of stationary processes in terms of higher order moments." *Theory Prob. Appl.* **8**:436–439.

SHUMWAY, R. H. (1971). "On detecting a signal in N stationarily correlated noise series." *Technometrics.* **13**:499–519.

SIMPSON, S. M. (1966). *Time Series Computations in FORTRAN and FAP.* Reading: Addison-Wesley.

SINGLETON, R. C. (1969). "An algorithm for computing the mixed radix fast Fourier transform." *IEEE Trans. Audio Elec.* AU-**17**:93–103.

SINGLETON, R. C., and POULTER, T. C. (1967). "Spectral analysis of the call of the male killer whale." *IEEE Trans. on Audio and Electroacoustics.* AU–**15**: 104–113.

SIOTANI, M. (1967). "Some applications of Loewner's ordering of symmetric matrices." *Ann. Inst. Statist. Math.* **19**:245–259.

SKOROKHOD, A. V. (1956). "Limit theorems for stochastic processes." *Theory Prob. Appl.* **1**:261–290.

SLEPIAN, D. (1954). "Estimation of signal parameters in the presence of noise." *Trans. I.R.E.* PGIT–**3**:82–87.

SLEPIAN, D. (1958). "Fluctuations of random noise power." *Bell Syst. Tech. J.* **37**:163–184.

SLUTSKY, E. (1929). "Sur l'extension de la théorie de periodogrammes aux suites des quantités dépendentes." *Comptes Rendues.* **189**:722–733.

SLUTSKY, E. (1934). "Alcuni applicazioni di coefficienti di Fourier al analizo di sequenze eventuali coherenti stazionarii." *Giorn. d. Instituto Italiano degli Atuari.* **5**:435–482.

SMITH, E. J., HOLZER, R. E., MCLEOD, M. G., and RUSSELL, C. T. (1967). "Magnetic noise in the magnetosheath in the frequency range 3–300 Hz." *J. Geophys. Res.* **72**:4803–4813.

SOLODOVNIKOV, V. V. (1960). *Introduction to the Statistical Dynamics of Automatic Control Systems.* New York: Dover.

SRIVASTAVA, M. S. (1965). "On the complex Wishart distribution." *Ann. Math. Statist.* **36**:313–315.

STIGUM, B. P. (1967). "A decision theoretic approach to time series analysis." *Ann. Inst. Statist. Math.* **19**:207–243.

STOCKHAM, T. G., Jr., (1966). "High speed convolution and correlation." *Proc. Spring Joint Comput. Conf.* **28**:229–233.

STOKES, G. G. (1879). *Proc. Roy. Soc.* **122**:303.

STONE, R. (1947). "On the interdependence of blocks of transactions." *J. Roy. Statist. Soc.*, B. **9**:1–32.

STRIEBEL, C. (1959). "Densities for stochastic processes." *Ann. Math. Statist.* **30**:559–567.

STUMPFF, K. (1937). *Grundlagen und Methoden der Periodenforschung.* Berlin: Springer.

STUMPFF, K. (1939). *Tafeln und Aufgaben zur Harmonischen Analyse und Periodogrammrechnung.* Berlin: Springer.

SUGIYAMA, G. (1966). "On the distribution of the largest latent root and corresponding latent vector for principal component analysis." *Ann. Math. Statist.* **37**:995–1001.

SUHARA, K., and SUZUKI, H. (1964). "Some results of EEG analysis by analog type analyzers and finer examinations by a digital computer." *Ann. Inst. Statist. Math.*, Suppl. **3**:89–98.

TAKEDA, S. (1964). "Experimental studies on the airplane response to the side gusts." *Ann. Inst. Statist. Math.*, Suppl. **3**:59–64.

TATE, R. F. (1966). "Conditional-normal regression models." *J. Amer. Statist. Assoc.* **61**:477–489.

TICK, L. J. (1963). "Conditional spectra, linear systems and coherency." In *Time Series Analysis*, Ed. M. Rosenblatt, pp. 197–203. New York: Wiley.

TICK, L. J. (1966). "Letter to the Editor." *Technometrics*. **8**:559–561.

TICK, L. J. (1967). "Estimation of coherency." In *Advanced Seminar on Spectral Analysis of Time Series*, Ed. B. Harris, pp. 133–152. New York: Wiley.

TIMAN, M. F. (1962). "Some linear summation processes for the summation of Fourier series and best approximation." *Soviet Math*. **3**:1102–1105.

TIMAN, A. F. (1963). *Theory of Approximation of Functions of a Real Variable*. New York: Macmillan.

TUKEY, J. W. (1949). "The sampling theory of power spectrum estimates." *Proc. on Applications of Autocorrelation Analysis to Physical Problems*. NAVEXOS-P-735, pp. 47–67. Washington, D.C.: Office of Naval Research, Dept. of the Navy.

TUKEY, J. W. (1959a). "An introduction to the measurement of spectra." In *Probability and Statistics*, Ed. U. Grenander, pp. 300–330. New York: Wiley.

TUKEY, J. W. (1959b). "The estimation of power spectra and related quantities." In *On Numerical Approximation*, pp. 389–411. Madison: Univ. of Wisconsin Press.

TUKEY, J. W. (1959c). "Equalization and pulse shaping techniques applied to the determination of initial sense of Rayleigh waves." In *The Need of Fundamental Research in Seismology*, Appendix 9, pp. 60–129. Washington: U.S. Department of State.

TUKEY, J. W. (1961). "Discussion, emphasizing the connection between analysis of variance and spectrum analysis." *Technometrics*. **3**:1–29.

TUKEY, J. W. (1965a). "Uses of numerical spectrum analysis in geophysics." *Bull. I.S.I.* 35 Session. 267–307.

TUKEY, J. W. (1965b). "Data analysis and the frontiers of geophysics." *Science*. **148**:1283–1289.

TUKEY, J. W. (1967). "An introduction to the calculations of numerical spectrum analysis." In *Advanced Seminar on Spectral Analysis of Time Series*, Ed. B. Harris, pp. 25–46. New York: Wiley.

TUMURA, Y. (1965). "The distributions of latent roots and vectors." *TRU Mathematics*. **1**:1–16.

VAN DER POL, B. (1930). "Frequency modulation." *Proc. Inst. Radio. Eng*. **18**: 227.

VARIOUS AUTHORS (1966). "A discussion on recent advances in the technique of seismic recording and analysis." *Proc. Roy. Soc*. **290**:288–476.

VOLTERRA, V. (1959). *Theory of Functionals and of Integrals and Integro-differential Equations*. New York: Dover.

VON MISES, R. (1964). *Mathematical Theory of Probability and Statistics*. New York: Academic.

VON MISES, R., and DOOB, J. L. (1941). "Discussion of papers on probability theory." *Ann. Math. Statist*. **12**:215–217.

WAHBA, G. (1966). *Cross spectral distribution theory for mixed spectra and estimation of prediction filter coefficients*. Ph.D. Thesis, Stanford University.

WAHBA, G. (1968). "One the distribution of some statistics useful in the analysis of jointly stationary time series." *Ann. Math. Statist*. **39**:1849–1862.

WAHBA, G. (1969). "Estimation of the coefficients in a distributed lag model." *Econometrica*. **37**:398–407.

WALDMEIR, M. (1961). *The Sunspot Activity in the Years 1610–1960*. Zurich: Schulthess.

WALKER, A. M. (1954). "The asymptotic distribution of serial correlation coefficients for autoregressive processes with dependent residuals." *Proc. Camb. Philos. Soc.* **50**:60–64.

WALKER, A. M. (1965). "Some asymptotic results for the periodogram of a stationary time series." *J. Austral. Math. Soc.* **5**:107–128.

WALKER, A. M. (1971). "On the estimation of a harmonic component in a time series with stationary residuals." *Biometrika*. **58**:21–36.

WEDDERBURN, J. H. M. (1934). *Lectures on Matrices*. New York: Amer. Math. Soc.

WEGEL, R. L., and MOORE, C. R. (1924). "An electrical frequency analyzer." *Bell Syst. Tech. J.* **3**:299–323.

WELCH, P. D. (1961). "A direct digital method of power spectrum estimation." *IBM J. Res. Dev.* **5**:141–156.

WELCH, P. D. (1967). "The use of the fast Fourier transform for estimation of spectra: a method based on time averaging over short, modified periodograms." *IEEE Trans. Electr. Acoust.* AU–**15**:70.

WEYL, H. (1946). *Classical Groups*. Princeton: Princeton Univ. Fress.

WHITTAKER, E. T., and ROBINSON, G. (1944). *The Calculus of Observations*. Cambridge: Cambridge Univ. Press.

WHITTLE, P. (1951). *Hypothesis Testing in Time Series Analysis*. Uppsala: Almqvist.

WHITTLE, P. (1952a). "Some results in time series analysis." *Skand. Aktuar*. **35**: 48–60.

WHITTLE, P. (1952b). "The simultaneous estimation of a time series' harmonic and covariance structure." *Trab. Estad.* **3**:43–57.

WHITTLE, P. (1953). "The analysis of multiple stationary time series." *J. Roy. Statist. Soc.*, B. **15**:125–139.

WHITTLE, P. (1954). "A statistical investigation of sunspot observations with special reference to H. Alven's sunspot model." *Astrophys. J.* **120**:251–260.

WHITTLE, P. (1959). "Sur la distribution du maximim d'un polynome trigonométrique á coefficients aléatoires." *Colloques Internationaux du Centre National de la Recherche Scientifique*. **87**:173–184.

WHITTLE, P. (1961). "Gaussian estimation in stationary time series." *Bull. Int. Statist. Inst.* **39**:105–130.

WHITTLE, P. (1963a). *Prediction and Regulation*. London: English Universities Press.

WHITTLE, P. (1963b). "On the fitting of multivariate auto-regressions and the approximate canonical factorization of a spectral density matrix." *Biometrika*. **50**:129–134.

WIDOM, H. (1965). "Toeplitz matrices." In *Studies in Real and Complex Analysis*, Ed. I. I. Hirschman, Jr., pp. 179–209. Englewood Cliffs: Prentice-Hall.

WIENER, N. (1930). "Generalized harmonic analysis." *Acta. Math.* **55**:117–258.

WIENER, N. (1933). *The Fourier Integral and Certain of its Applications*. Cambridge: Cambridge Univ. Press.

WIENER, N. (1938). "The historical background of harmonic analysis." *Amer. Math. Soc. Semicentennial Pub.* **2**:56–68.

WIENER, N. (1949). *The Extrapolation, Interpolation and Smoothing of Stationary Time Series with Engineering Applications*. New York: Wiley.

WIENER, N. (1953). "Optics and the theory of stochastic processes." *J. Opt. Soc. Amer.* **43**:225–228.

WIENER, N. (1957). "Rhythms in physiology with particular reference to encephalography." *Proc. Rud. Virchow Med. Soc. in New York*. **16**:109–124.

WIENER, N. (1958). *Non-linear Problems in Random Theory*. Cambridge: MIT Press.

WIENER, N., SIEGEL, A., RANKIN, B., and MARTIN, W. T. (1967). *Differential Space, Quantum Systems and Prediction*. Cambridge: MIT Press.

WIENER, N., and WINTNER, A. (1941). "On the ergodic dynamics of almost periodic systems." *Amer. J. Math.* **63**:794–824.

WILK, M. B., GNANADESIKAN, R., and HUYETT, M. J. (1962). "Probability plots for the gamma distribution." *Technometrics*. **4**:1–20.

WILKINS, J. E. (1948). "A note on the general summability of functions." *Ann. Math.* **49**:189–199.

WILKINSON, J. H. (1965). *The Algebraic Eigenvalue Problem*. Oxford: Oxford Univ. Press.

WILLIAMS, E. J. (1967). "The analysis of association among many variates." *J. Roy. Statist. Soc.*, B. **29**:199–242.

WINTNER, A. (1932). "Remarks on the ergodic theorem of Birkhoff." *Proc. Nat. Acad. Sci.* (U.S.A.). **18**:248–251.

WISHART, J. (1931). "The mean and second moment coefficient of the multiple correlation coefficient in samples from a normal population." *Biometrika*. **22**:353–361.

WISHART, J., and BARTLETT, M. S. (1932). "The distribution of second order moment statistics in a normal system." *Proc. Camb. Philos. Soc.* **28**:455–459.

WOLD, H. O. A. (1948). "On prediction in stationary time series." *Ann. Math. Statist.* **19**:558–567.

WOLD, H. O. A. (1954). *A Study in the Analysis of Stationary Time Series*, 2nd ed. Uppsala: Almqvist and Wiksells.

WOLD, H. O. A. (1963). "Forecasting by the chain principle." In *Time Series Analysis*, Ed. M. Rosenblatt, pp. 471–497. New York: Wiley.

WOLD, H. O. A. (1965). *Bibliography on Time Series and Stochastic Processes*. London: Oliver and Boyd.

WONG, E. (1964). "The construction of a class of stationary Markov processes." *Proc. Symp. Applied Math.* **16**:264–276. Providence: Amer. Math. Soc.

WOOD, L. C. (1968). "A review of digital pass filtering." *Rev. Geophysics*. **6**:73–98.

WOODING, R. A. (1956). "The multivariate distribution of complex normal variates." *Biometrika*. **43**:212–215.

WOODROOFE, M. B., and VAN NESS, J. W. (1967). "The maximum deviation of sample spectral densities." *Ann. Math. Statist.* **38**:1558–1570.

WORLD WEATHER RECORDS. Smithsonian Miscellaneous Collections, Vol. 79 (1927), Vol. 90 (1934), Vol. 105 (1947). Smithsonian Inst. Washington.

WORLD WEATHER RECORDS. 1941–1950 (1959) and 1951–1960 (1965). U.S. Weather Bureau, Washington, D.C.

WRIGHT, W. D. (1906). *The Measurement of Colour*. New York: Macmillan.

YAGLOM, A. M. (1962). *An Introduction to the Theory of Stationary Random Functions*. Englewood Cliffs: Prentice-Hall.

YAGLOM, A. M. (1965). "Stationary Gaussian processes satisfying the strong mixing condition and best predictable functionals." In *Bernoulli, Bayes, Laplace*, Ed. J. Neyman and L. M. LeCam, pp. 241–252. New York: Springer.

YAMANOUCHI, Y. (1961). "On the analysis of the ship oscillations among waves—I, II, III." *J. Soc. Naval Arch.* (Japan). **109**:169–183; **110**:19–29; **111**: 103–115.

YULE, G. U. (1927). "On a method of investigating periodicities in disturbed series, with special reference to Wolfer's sunspot numbers." *Phil. Trans. Roy. Soc.*, A. **226**:267–298.

YUZURIHA, T. (1960). "The autocorrelation curves of schizophrenic brain waves and the power spectra." *Psych. Neurol. Jap.* **62**:911–924.

ZYGMUND, A. (1959). *Trigonometric Series*. Cambridge: Cambridge Univ. Press.

ZYGMUND, A. (1968). *Trigonometric Series*, Vols. I, II. Cambridge: Cambridge Univ. Press.

NOTATION INDEX

AUTHOR INDEX

SUBJECT INDEX

ADDENDUM

Fourier Analysis of Stationary Processes

***Abstract*—This paper begins with a description of some of the important procedures of the Fourier analysis of real-valued stationary discrete time series. These procedures include the estimation of the power spectrum, the fitting of finite parameter models, and the identification of linear time invariant systems. Among the results emphasized is the one that the large sample statistical properties of the Fourier transform are simpler than those of the series itself. The procedures are next generalized to apply to the cases of vector-valued series, multidimensional time series or spatial series, point processes, random measures, and finally to stationary random Schwartz distributions. It is seen that the relevant Fourier transforms are evaluated by different formulas in these further cases, but that the same constructions are carried out after their evaluation and the same statistical results hold. Such generalizations are of interest because of current work in the fields of picture processing and pulse-code modulation.**

I. INTRODUCTION

THE FOURIER analysis of data has a long history, dating back to Stokes [1] and Schuster [2], for example. It has been done by means of arithmetical formulas (Whittaker and Robinson [3], Cooley and Tukey [4]), by means of a mechanical device (Michelson [5]), and by means of real-time filters (Newton [6], Pupin [7]). It has been carried out on discrete data, such as monthly rainfall in the Ohio valley (Moore [8]), on continuous data, such as radiated light (Michelson [5]), on vector-valued data, such as vertical and horizontal components of wind speed (Panofsky and McCormick [9]), on spatial data, such as satellite photographs (Leese and Epstein [10]), on point processes, such as the times at which vehicles pass a position on a road (Bartlett [11]), and on

This invited paper is one of a series planned on topics of general interest—The Editor.

Manuscript received June 7, 1974; revised August 13, 1974. This paper was prepared while the author was a Miller Research Professor and was supported by NSF under Grant GP-31411.

The author is with the Department of Statistics, University of California, Berkeley, Calif. 94720.

point processes in space, such as the positions of pine trees in a field (Bartlett [12]). It has even been carried out on the logarithm of a Fourier transform (Oppenheim *et al.* [13]) and on the logarithm of a power spectrum estimate (Bogert *et al.* [14]).

The summary statistic examined has been: the Fourier transform itself (Stokes [1]), the modulus of the transform (Schuster [2]), the smoothed modulus squared (Bartlett [15]), the smoothed product of two transforms (Jones [16]), and the smoothed product of three transforms (Hasselman *et al.* [17]).

The summary statistics are evaluated in an attempt to measure population parameters of interest. Foremost among these parameters is the power spectrum. This parameter was initially defined for real-valued-time phenomena (Wiener [18]). In recent years it has been defined and shown useful for spatial series, point processes, and random measures as well. Our development in this paper is such that the definitions set down and mathematics employed are virtually the same for all of these cases.

Our method of approach to the topic is to present first an extensive discussion of the Fourier analysis of real-valued discrete-time series emphasizing those aspects that extend directly to the cases of vector-valued series, of continuous spatial series, of point processes, and finally of random distributions. We then present extensions to the processes just indicated. Throughout, we indicate aspects of the analysis that are peculiar to the particular process under consideration. We also mention higher order spectra and nonlinear systems. Wold [19] provides a bibliography of papers on time series analysis written prior to 1960. Brillinger [20] presents a detailed description of the Fourier analysis of vector-valued discrete-time series.

We now indicate several reasons that suggest why Fourier analysis has proved so useful in the analysis of time series.

II. Why the Fourier Transform?

Several arguments can be advanced as to why the Fourier transform has proved so useful in the analysis of empirical functions. For one thing, many experiments of interest have the property that their essential character is not changed by moderate translations in time or space. Random functions produced by such experiments are called *stationary*. (A definition of this term is given later.) Let us begin by looking for a class of functions that behave simply under translation. If, for example, we wish

$$f(t+u) = C_u f(t), \qquad t, u = 0, \pm 1, \pm 2, \cdots$$

with $C_1 \neq 0$, then by recursion

$$f(t) = C_1 f(t-1) = C_2 f(t-2) = \cdots = C_1^t f(0)$$

for $t \geqslant 0$ and so $f(t) = f(0) \exp\{\alpha t\}$ for $\alpha = \ln C_1$. If $f(t)$ is to be bounded, then $\alpha = i\lambda$, for $i = \sqrt{-1}$ and λ real. We have been led to the functions $\exp\{i\lambda t\}$. Fourier analysis is concerned with such functions and their linear combinations.

On the other hand, we might note that many of the operations we would like to apply to empirical functions are linear and translation invariant, that is such that; if $X_1(t) \to Y_1(t)$ and $X_2(t) \to Y_2(t)$ then $\alpha_1 X_1(t) + \alpha_2 X_2(t) \to \alpha_1 Y_1(t) + \alpha_2 Y_2(t)$ and if $X(t) \to Y(t)$ then $X(t-u) \to Y(t-u)$. Such operations are called *linear filters*. It follows from these conditions that if $X(t) = \exp\{i\lambda t\} \to Y_\lambda(t)$ then

$$X(t+u) = \exp\{i\lambda u\}\, X(t) \to \exp\{i\lambda t\}\, Y_\lambda(t) = Y(t+u).$$

Setting $u = t$, $t = 0$ gives $Y_\lambda(t) = \exp\{i\lambda t\}\; Y_\lambda(0)$. In summary, $\exp\{i\lambda t\}$ the complex exponential of frequency λ is carried over into a simple multiple of itself by a linear filter. $A(\lambda) = Y_\lambda(0)$ is called the *transfer function* of the filter. If the function $X(t)$ is a Fourier transform, $X(t) = \int \exp\{i\alpha t\}\, x(\alpha)\, d\alpha$, then from the linearity (and some continuity) $X(t) \to \int \exp i\alpha t\, A(\alpha)\, x(\alpha)\, d\alpha$. We see that the effect of a linear filter is easily described for a function that is a Fourier transform.

In the following sections, we will see another reason for dealing with the Fourier transforms of empirical functions, namely, in the case that the functions are realizations of a stationary process, the large sample statistical properties of the transforms are simpler than the properties of the functions themselves.

Finally, we mention that with the discovery of fast Fourier transform algorithms (Cooley and Tukey [4]), the transforms may often be computed exceedingly rapidly.

III. Stationary Real-Valued Discrete-Time Series

Suppose that we are interested in analyzing T real-valued measurements made at the equispaced times $t = 0, \cdots, T-1$. Suppose that we are prepared to model these measurements by the corresponding values of a realization of a stationary discrete-time series $X(t)$, $t = 0, \pm 1, \pm 2, \cdots$. Important parameters of such a series include its *mean*,

$$c_X = EX(t) \tag{1}$$

giving the average level about which the values of the series are distributed and its *autocovariance function*

$$c_{XX}(u) = \text{cov}\,\{X(t+u), X(t)\}$$
$$= E\{[X(t+u) - c_X]\,[X(t) - c_X]\}, \qquad u = 0, \pm 1, \cdots \tag{2}$$

providing a measure of the degree of dependence of values of the process $|u|$ time units apart. (These parameters do not depend on t because of the assumed stationarity of the series.) In many cases of interest the series is *mixing*, that is, such that values well separated in time are only weakly dependent in a formal statistical sense to be described later. Suppose, in particular, that $c_{XX}(u) \to 0$ sufficiently rapidly as $|u| \to \infty$ for

$$f_{XX}(\lambda) = (2\pi)^{-1} \sum_{u=-\infty}^{\infty} c_{XX}(u) \exp\{-i\lambda u\}, \qquad -\infty < \lambda < \infty \tag{3}$$

to be defined. The parameter $f_{XX}(\lambda)$ is called the *power spectrum* of the series $X(t)$ at frequency λ. It is symmetric about 0 and has period 2π. The definition (3) may be inverted to obtain the representation

$$c_{XX}(u) = \int_{-\pi}^{\pi} \exp\{i\alpha u\}\, f_{XX}(\alpha)\, d\alpha \tag{4}$$

of the autocovariance function in terms of the power spectrum.

If the series $X(t)$ is passed through the linear filter

$$X(t) \to Y(t) = \sum_u a(t-u)\, X(u)$$

with well-defined transfer function

$$A(\lambda) = \sum_u a(u) \exp\{-i\lambda u\}$$

then we can check that

$$c_{YY}(u) = \sum_u \sum_v a(u+v)\, a(w)\, c_{XX}(w-v) \tag{5}$$

and, by taking Fourier transforms, that

$$f_{YY}(\lambda) = |A(\lambda)|^2 f_{XX}(\lambda) \tag{6}$$

under some regularity conditions. Expression (6), the frequency domain description of linear filtering, is seen to be much nicer than (5), the time-domain description.

Expressions (4) and (6) may be combined to obtain an interpretation of the power spectrum at frequency λ. Suppose that we consider a narrow band-pass filter at frequency λ having transfer function

$$A(\alpha) \doteq \begin{cases} 1, & |\alpha \pm \lambda| \leqslant \Delta \\ 0, & \text{otherwise} \end{cases}$$

with Δ small. Then the variance of the output series $Y(t)$, of the filter, is given by

$$\begin{aligned} \operatorname{var} Y(t) &= c_{YY}(0) \\ &= \int f_{YY}(\alpha)\, d\alpha \\ &= \int |A(\alpha)|^2 f_{XX}(\alpha)\, d\alpha \\ &= 4\Delta f_{XX}(\lambda). \end{aligned} \tag{7}$$

In words, the power spectrum of the series $X(t)$ at frequency λ is proportional to the variance of the output of a narrow band-pass filter of frequency λ. In the case that $\lambda \neq 0, \pm 2\pi, \pm 4\pi, \cdots$ the mean of the output series is 0 and the variance of the output series is the same as its mean-squared value. Expression (7) shows incidentally that the power spectrum is nonnegative.

We mention, in connection with the representation (4), that Khintchine [21] shows that for $X(t)$ a stationary discrete time series with finite second order moments, we necessarily have

$$c_{XX}(u) = \int_{-\pi}^{\pi} \exp\{i\alpha u\}\, dF_{XX}(\alpha) \tag{8}$$

where $F_{XX}(\alpha)$ is a monotonic nondecreasing function. $F_{XX}(\lambda)$ is called the *spectral measure.* Its derivative is the power spectrum. Going along with (8), Cramér [22] demonstrated that the series itself has a Fourier representation

$$X(t) = \int_{-\pi}^{\pi} \exp\{i\alpha t\}\, dZ_X(\alpha), \qquad t = 0, \pm 1, \cdots \tag{9}$$

where $Z_X(\lambda)$ is a random function with the properties;

$$E\, dZ_X(\lambda) = \eta(\lambda)\, c_X\, d\lambda \tag{10}$$

$$\operatorname{cov}\{dZ_X(\lambda), dZ_X(\mu)\} = \eta(\lambda - \mu)\, dF_{XX}(\lambda)\, d\mu. \tag{11}$$

(In these last expressions, if $\delta(\lambda)$ is the Dirac delta function then $\eta(\lambda) = \Sigma\, \delta(\lambda - 2\pi j)$ is the Kronecker comb.) Also expression (11) concerns the covariance of two complex-varied variates. Such a covariance is defined by cov $\{X, Y\} = E\{(X - EX)\overline{(Y - EY)}\}$.) Expression (9) writes the series $X(t)$ as a Fourier transform. We can see that if the series $X(t)$ is passed through a linear filter with transfer function $A(\lambda)$, then the output series has Fourier representation

$$\int_{-\pi}^{\pi} \exp\{i\alpha t\} A(\alpha)\, dZ_X(\alpha), \qquad t = 0, \pm 1, \cdots.$$

In Section XV, we will see that the first and second-order relations (10), (11) may be extended to kth order relations with the definition of kth order spectra.

IV. The Finite Fourier Transform

Let the values of the series $X(t)$ be available for $t = 0, 1, 2, \cdots, T - 1$ where T is an integer. The *finite Fourier transform* of this stretch of series is defined to be

$$d_X^{(T)}(\lambda) = \sum_{t=0}^{T-1} X(t) \exp\{-i\lambda t\}, \qquad -\infty < \lambda < \infty. \tag{12}$$

A number of interpretations may be given for this variate. For example, suppose we take a linear filter with transfer function concentrated at the frequency λ, namely $A(\alpha) = \delta(\alpha - \lambda)$. The corresponding time domain coefficients of this filter are

$$\begin{aligned} a(u) &= (2\pi)^{-1} \int A(\alpha) \exp\{iu\alpha\}\, d\alpha \\ &= (2\pi)^{-1} \exp\{iu\lambda\}, \qquad u = 0, \pm 1, \cdots. \end{aligned}$$

The output of this filter is the series

$$(2\pi)^{-1} \sum_u X(u) \exp\{i\lambda(t - u)\} \doteq (2\pi)^{-1} \exp\{i\lambda t\}\, d_X^{(T)}(\lambda).$$

These remarks show that the finite Fourier transform may be interpreted as, essentially, the result of narrow band-pass filtering the series.

Before presenting a second interpretation, we first remark that the sample covariance of pairs of values $X(t)$, $Y(t)$, $t = 0, 1, \cdots, T - 1$ is given by $T^{-1} \Sigma X(t) Y(t)$, when the $Y(t)$ values have 0 mean. This quantity is a measure of the degree of linear relationship of the $X(t)$ and $Y(t)$ values. The finite Fourier transform is essentially, then, the sample covariance between the $X(t)$ values and the complex exponential of frequency λ. It provides some measure of the degree of linear relationship of the series $X(t)$ and phenomena of exact frequency λ.

In the case that $\lambda = 0$, the finite Fourier transform (12) is the sample sum. The central limit theorem indicates conditions under which a sum of random variables is asymptotically normal as the sample size grows to ∞. Likewise, there are theorems indicating that $d_X^{(T)}(\lambda)$ is asymptotically normal as $T \to \infty$. Before indicating some aspects of these theorems we set down a definition. A complex-valued variate w is called *complex normal* with mean 0 and variance σ^2 when its real and imaginary parts are independent normal variates with mean 0 and variance $\sigma^2/2$. The density function of w is proportional to $\exp\{-|w|^2/\sigma^2\}$. The variate $|w|^2$ is exponential with mean σ^2 in this case.

In the case that the series $X(t)$ is stationary, with finite second-order moments, and mixing (that is, well-separated values are only weakly dependent) the finite Fourier transform has the following useful asymptotic properties as $T \to \infty$:

a) $d_X^{(T)}(0) - Tc_X$ is asymptotically normal with mean 0 and variance $2\pi T f_{XX}(0)$;

b) for $\lambda \neq 0, \pm\pi, \pm 2\pi, \cdots, d_X^{(T)}(\lambda)$ is asymptotically complex normal with mean 0 and variance $2\pi T f_{XX}(\lambda)$;

c) for $s^j(T)$, $j = 1, \cdots, J$ integers with $\lambda^j(T) = 2\pi s^j(T)/T \to \lambda \neq 0, \pm\pi, \pm 2\pi, \cdots$ the variates $d_X^{(T)}(\lambda^1(T)), \cdots, d_X^{(T)}(\lambda^J(T))$ are asymptotically independent complex normals with mean 0 and variance $2\pi T f_{XX}(\lambda)$,

d) for $\lambda \neq 0, \pm\pi, \pm 2\pi, \cdots$ and $U = T/J$ and integer, the variates

$$d_X^{(U)}(\lambda, j) = \sum_{u=0}^{U-1} X(u + jU) \exp\{-i\lambda u\}, \qquad j = 0, \cdots, J-1$$

are asymptotically independent complex normals with mean 0 and variance $2\pi U f_{XX}(\lambda)$.

These results are developed in Brillinger [20]. Related results are given in Section XV and proved in the Appendix. Other references include: Leonov and Shiryaev [23], Picinbono [24], Rosenblatt [25], Brillinger [26], Hannan and Thomson [27]. We have seen that $\exp\{i\lambda t\}\, d_X^{(T)}(\lambda)$ may be interpreted as the result of narrow band-pass filtering the series $X(t)$. It follows that the preceding result b) is consistent with the "engineering folk" theorem to the effect that narrow band-pass noise is approximately Gaussian.

Result a) suggests estimating the mean c_X by

$$c_X^{(T)} = T^{-1} \sum_{t=0}^{T-1} X(t)$$

and approximating the distribution of this estimate by a normal distribution with mean 0 and variance $2\pi f_{XX}(0)/T$. Result b) suggests estimating the power spectrum $f_{XX}(\lambda)$ by the *periodogram*

$$I_{XX}^{(T)}(\lambda) = (2\pi T)^{-1} |d_X^{(T)}(\lambda)|^2 \tag{13}$$

in the case $\lambda \neq 0, \pm 2\pi, \cdots$. We will say more about this statistic later. It is interesting to note, from c) and d), that asymptotically independent statistics with mean 0 and variance proportional to the power spectrum at frequency λ may be obtained by either computing the Fourier transform at particular distinct frequencies near λ or by computing them at the frequency λ but based on different time domains. We warn the reader that the results a)–d) are asymptotic. They are to be evaluated in the sense that they might prove reasonable approximations in practice when the domain of observation is large and when values of the series well separated in the domain are only weakly dependent.

On a variety of occasions we will *taper* the data before computing its Fourier transform. This means that we take a data window $\phi^{(T)}(t)$ vanishing for $t < 0, t > T - 1$, and compute the transform

$$d_X^{(T)}(\lambda) = \sum_t \phi^{(T)}(t) \exp\{-i\lambda t\} X(t) \tag{14}$$

for selected values of λ. One intention of tapering is to reduce the interference of neighboring frequency components. If

$$\Phi^{(T)}(\lambda) = \sum_t \phi^{(T)}(t) \exp\{-i\lambda t\}$$

then the Cramér representation (9) shows that (14) may be written

$$d_X^{(T)}(\lambda) = \int \Phi^{(T)}(\lambda - \alpha)\, dZ_X(\alpha). \tag{15}$$

From what we have just said, we will want to choose $\phi^{(T)}(t)$ so that $\Phi^{(T)}(\alpha)$ is concentrated near $\alpha = 0, \pm 2\pi, \cdots$. (One convenient choice of $\phi^{(T)}(t)$ takes the form $\phi(t/T)$ where $\phi(u) = 0$ for $u < 0, u \geqslant 1$.) The asymptotic effect of tapering may be seen to be to replace the variance in b) by $2\pi \sum \phi^{(T)}(t)^2 f_{XX}(\lambda)$.

Hannan and Thomson [27] investigate the asymptotic distribution of the Fourier transform of tapered data in a case where $f_{XX}(\lambda)$ depends on T in a particular manner. The hope is to obtain better approximations to the distribution.

V. Estimation of the Power Spectrum

In the previous section, we mentioned the periodogram, $I_{XX}^{(T)}(\lambda)$, as a possible estimate of the power spectrum $f_{XX}(\lambda)$ in the case that $\lambda \neq 0, \pm 2\pi, \cdots$. If result b) holds true, then $I_{XX}^{(T)}(\lambda)$, being a continuous function of $d_X^{(T)}(\lambda)$, will be distributed asymptotically as $|w|^2$, where w is a complex normal variate with mean 0 and variance $f_{XX}(\lambda)$. That is $I_{XX}^{(T)}(\lambda)$ will be distributed asymptotically as an exponential variate with mean $f_{XX}(\lambda)$. From the practical standpoint this is interesting, but not satisfactory. It suggests that no matter how large the sample size T is, the variate $I_{XX}^{(T)}(\lambda)$ will tend to be distributed about $f_{XX}(\lambda)$ with an appreciable scatter. Luckily, results c) and d) suggest means around this difficulty. Following c), the variates $I_{XX}^{(T)}(\lambda^j(T)), j = 1, \cdots, J$ are distributed asymptotically as independent exponential variates with mean $f_{XX}(\lambda)$. Their average

$$f_{XX}^{(T)}(\lambda) = J^{-1} \sum_{j=1}^{J} I_{XX}^{(T)}(\lambda^j(T)) \tag{16}$$

will be distributed asymptotically as the average of J independent exponential variates having mean $f_{XX}(\lambda)$. That is, it will be distributed as

$$f_{XX}(\lambda)\chi_{2J}^2/2J \tag{17}$$

where χ_{2J}^2 denotes a chi-squared variate with $2J$ degrees of freedom. The variance of the variate (17) is

$$f_{XX}(\lambda)^2/J = f_{XX}(\lambda)^2\, U/T \tag{18}$$

if $U = T/J$. By choice of J the experimenter can seek to obtain an estimate of which the sampling fluctuations are small enough for his needs. From the standpoint of practice, it seems to be useful to compute the estimate (16) for a number of values of J. This allows us to tailor the choice of J to the situation at hand and even to use different values of J for different frequency ranges. Result d) suggests our consideration of the estimate

$$f_{XX}^{(T)}(\lambda) = J^{-1} \sum_{j=0}^{J-1} (2\pi U)^{-1} |d_X^{(U)}(\lambda, j)|^2. \tag{19}$$

It too will have the asymptotic distribution (17) with variance (18).

We must note that it is not sensible to take J in (16) and (19) arbitrarily large as the preceding arguments might have suggested. It may be seen from (15) that

$$E I_{XX}^{(T)}(\lambda) = \int_{-\pi}^{\pi} F_T(\lambda - \alpha) f_{XX}(\alpha)\, d\alpha + F_T(\lambda)\, c_X^2 \quad (20)$$

where

$$F_T(\lambda) = (2\pi T)^{-1} \left| \frac{\sin \frac{T\lambda}{2}}{\sin \frac{\lambda}{2}} \right|^2$$

is the Fejér kernel. This kernel, or frequency window, is non-negative, integrates to 1, and has most of its mass in the interval $(-2\pi/T, 2\pi/T)$. The term in c_X^2 may be neglected for $\lambda \neq 0, \pm 2\pi, \cdots$ and T large. From (16) and (20) we now see that

$$E f_{XX}^{(T)}(\lambda) \doteq \int_{-\pi}^{\pi} J^{-1} \sum_{j=1}^{J} F_T(\lambda^j(T) - \alpha) f_{XX}(\alpha)\, d\alpha. \quad (21)$$

If we are averaging J periodogram values at frequencies $2\pi/T$ apart and centered at λ, then the bandwidth of the kernel of (21) will be approximately $4\pi J/T$. If J is large and $f_{XX}(\alpha)$ varies substantially in the interval $-2\pi J/T < \alpha - \lambda < 2\pi J/T$, then the value of (21) can be very far from the desired $f_{XX}(\lambda)$. In practice we will seek to have J large so that the estimate is reasonably stable, but not so large that it has appreciable bias. This same remark applies to the estimate (19). Parzen [28] constructed a class of estimates such that $E f_{XX}^{(T)}(\lambda) \to f_{XX}(\lambda)$ and var $f_{XX}^{(T)}(\lambda) \to 0$. These estimates have an asymptotic distribution that is normal, rather than χ^2, Rosenblatt [29]. Using the notation preceding these estimates correspond to having J depend on T in such a way that $J_T \to \infty$, but $J_T/T \to 0$ as $T \to \infty$.

Estimates of the power spectrum have proved useful; i) as simple descriptive statistics, ii) in informal testing and discrimination, iii) in the estimation of unknown parameters, and iv) in the search for hidden periodicities. As an example of i), we mention their use in the description of the color of an object, Wright [30]. In connection with ii) we mention the estimation of the spectrum of the seismic record of an event in attempt to see if the event was an earthquake or a nuclear explosion, Carpenter [31], Lampert *et al.* [32]. In case iii), we mention that Munk and MacDonald [33] derived estimates of the fundamental parameters of the rotation of the Earth from the periodogram. Turning to iv), we remind the reader that the original problem that led to the definition of the power spectrum, was that of the search for hidden periodicities. As a modern example, we mention the examination of spectral estimates for the periods of the fundamental vibrations of the Earth, MacDonald and Ness [34].

VI. Other Estimates of the Power Spectrum

We begin by mentioning minor modifications that can be made to the estimates of Section V. The periodograms of (16) may be computed at frequencies other than those of the form $2\pi s/T$, s an integer, and they may be weighted unequally. The periodograms of the estimate (19) may be based on overlapping stretches of data. The asymptotic distributions are not so simple when these modifications are made, but the estimate is often improved. The estimate (19) has another interpretation. We saw in Section IV that $\exp\{i\lambda t\}\, d_X^{(U)}(\lambda, j)$ might be interpreted as the output of a narrow band-pass filter centered at λ. This suggests that (19) is essentially the first power spectral estimate widely employed in practice, the average of the squared output of a narrow band-pass filter (Wegel and Moore [35]). We next turn to a discussion of some spectral estimates of quite different character.

We saw in Section III that if the series $X(t)$ was passed through a linear filter with transfer function $A(\lambda)$, then the output series $Y(t)$ had power spectrum given by $f_{YY}(\lambda) = |A(\lambda)|^2 f_{XX}(\lambda)$. In Section V, we saw that the estimates (16), (19) could have substantial bias were there appreciable variation in the value of the population power spectrum. These remarks suggest a means of constructing an improved estimate, namely: we use our knowledge of the situation at hand to devise a filter, with transfer function $A(\lambda)$, such that the output series $Y(t)$ has spectrum nearer to being constant. We then estimate the power spectrum of the filtered series in the manner of Section V and take $|A(\lambda)|^{-2} f_{YY}^{(T)}(\lambda)$ as our estimate of $f_{XX}(\lambda)$. This procedure is called spectral estimation by prewhitening and is due to Tukey (see Panofsky and McCormick [9]). We mention that in many situations we will be content to just examine $f_{YY}^{(T)}(\lambda)$. This would be necessary were $A(\lambda) = 0$.

One useful means of determining an $A(\lambda)$ is to fit an autoregressive scheme to the data by least squares. That is, for some K, choose $\hat{a}(1), \cdots, \hat{a}(K)$ to minimize

$$\sum [X(t) + a(1)\, X(t-1) + \cdots + a(K)\, X(t-K)|^2$$

where the summation extends over the available data. In this case $\hat{A}(\lambda) = 1 + \hat{a}(1) \exp\{-i\lambda\} + \cdots + \hat{a}(K) \exp\{-i\lambda K\}$. An algorithm for efficient computation of the $\hat{a}(u)$ is given in Wiener [36, p. 136]. This procedure should prove especially effective when the series $X(t)$ is near to being an autoregressive scheme of order K. Related procedures are discussed in Grenander and Rosenblatt [37, p. 270], Parzen [38], Lacoss [39], and Burg [40]. Berk [41] discusses the asymptotic distribution of the estimate $|\hat{A}(\lambda)|^{-2}(2\pi T)^{-1} \Sigma\, [X(t) + \hat{a}(1)\, X(t-1) + \cdots + \hat{a}(K)\, X(t-K)|^2$. Its asymptotic variance is shown to be (18) with $U = 2K$.

Pisarenko [42] has proposed a broad class of estimates including the high resolution estimate of Capon [43] as a particular case. Suppose $\hat{\Sigma}$ is an estimate of the covariance matrix of the variate

$$\begin{bmatrix} X(1) \\ \vdots \\ X(U) \end{bmatrix}$$

determined from the sample values $X(0), \cdots, X(T-1)$. Suppose $\hat{\mu}_u$, $\hat{\alpha}_u$, $u = 1, \cdots, U$ are the latent roots and vectors of $\hat{\Sigma}$. Suppose $H(\mu)$, $0 < \mu < \infty$, is a strictly monotonic function with inverse $h(\cdot)$. Pisarenko proposed the estimate

$$h\left(\sum_{u=1}^{U} H(\hat{\mu}_u)(2\pi U)^{-1} \left|\sum_{j=1}^{U} \hat{\alpha}_{uj} \exp\{-i\lambda j\}\right|^2\right). \tag{22}$$

He presents an argument indicating that the asymptotic variance of this estimate is also (18). The hope is that it is less biased. Its character is that of a nonlinear average of periodogram values in contrast to the simple average of (16) and (19). The estimates (16) and (19) essentially correspond to the case $H(\mu) = \mu$. The high resolution estimate of Capon [43] corresponds to $H(\mu) = \mu^{-1}$.

The autoregressive estimate, the high-resolution estimate and the Pisarenko estimates are not likely to be better than an ordinary spectral estimate involving steps of prewhitening, tapering, naive spectral estimation and recoloring. They are probably better than a naive spectral estimate for a series that is a sum of sine waves and noise.

VII. Finite Parameter Models

Sometimes a situation arises in which we feel that the form of the power spectrum is known except for the value of a finite dimensional parameter θ. For example existing theory may suggest that the series $X(t)$ is generated by the mixed moving average autoregressive scheme

$$X(t) + a(1)X(t-1) + \cdots + a(K)X(t-K) = \epsilon(t) + b(1)\epsilon(t-1) + \cdots + b(L)\epsilon(t-L) \tag{23}$$

where U, V are nonnegative integers and $\epsilon(t)$ is a series of independent variates with mean 0 and variance σ^2. The power spectrum of this series is

$$f_{XX}(\lambda;\theta) = \frac{\sigma^2}{2\pi} \frac{|1 + b(1)\exp\{-i\lambda\} + \cdots + b(L)\exp\{-i\lambda L\}|^2}{|1 + a(1)\exp\{-i\lambda\} + \cdots + a(K)\exp\{-i\lambda K\}|^2} \tag{24}$$

with $\theta = \sigma^2, a(1), \cdots, a(K), b(1), \cdots, b(L)$. A number of procedures have been suggested for estimating the parameters of the model (23), see Hannan [44] and Anderson [45], for example.

The following procedure is useful in situations more general than the above. It is a slight modification of a procedure of Whittle [46]. Choose as an estimate of θ the value that maximizes

$$\prod_{0<s<T/2} f_{XX}\left(\frac{2\pi s}{T};\theta\right)^{-1} \exp\left\{-I_{XX}^{(T)}\left(\frac{2\pi s}{T}\right) f_{XX}\left(\frac{2\pi s}{T};\theta\right)^{-1}\right\}. \tag{25}$$

Expression (25) is the likelihood corresponding to the assumption that the periodogram values $I_{XX}^{(T)}(2\pi s/T)$, $0 < s < T/2$, are independent exponential variates with means $f_{XX}(2\pi s/T;\theta)$, $0 < s < T/2$, respectively. Under regularity conditions we can show that this estimate, $\hat{\theta}$, is asymptotically normal with mean θ and covariance matrix $2\pi T^{-1}A^{-1}(A + B)A^{-1}$ where; if $\nabla f_{XX}(\lambda;\theta)$ is the gradient vector with respect to θ and f_{XXXX} the 4th order cumulant spectrum (see Section XV)

$$A = \int_0^\pi \nabla f_{XX}(\alpha;\theta) \cdot \nabla f_{XX}(\alpha;\theta) f_{XX}(\alpha;\theta)^{-2}\, d\alpha$$

$$B = \int_0^\pi \int_0^\pi \nabla f_{XX}(\alpha;\theta) \cdot \nabla f_{XX}(\beta;\theta) f_{XX}(\alpha;\theta)^{-2} f_{XX}(\beta;\theta)^{-2} \cdot f_{XXXX}(\alpha, -\alpha, -\beta)\, d\alpha\, d\beta.$$

We may carry out the maximization of (25) by a number of computer algorithms, see the discussion in Chambers [47]. In [48], we used the method of scoring. Other papers investigating estimates of this type are Whittle [49], Walker [50], and Dzaparidze [51].

The power spectrum itself may now be estimated by $f_{XX}(\lambda;\hat{\theta})$. This estimate will be asymptotically normal with mean $f_{XX}(\lambda;\theta)$ and variance $2\pi T^{-1}\nabla f_{XX}(\lambda;\theta)^{\tau} A^{-1}(A + B) \cdot A^{-1}\nabla f_{XX}(\lambda;\theta)$ following the preceding asymptotic normal distribution for θ. In the case that we model the series by an autoregressive scheme and proceed in the same way, the estimate $f_{XX}(\lambda; \hat{\theta})$ has the character of the autoregressive estimate of the previous section.

VIII. Linear Models

In some circumstances we may find ourselves considering a linear time invariant model of the form

$$X(t) = \mu + \sum_{u=-\infty}^{\infty} a(t-u)S(u) + \epsilon(t) \tag{26}$$

where the values $X(t)$, $S(t)$, $t = 0, 1, \cdots, T-1$ are given, $\epsilon(t)$ is an unknown stationary error series with mean 0 and power spectrum $f_{\epsilon\epsilon}(\lambda)$, the $a(u)$ are unknown coefficients, μ is an unknown parameter, and $S(t)$ is a fixed function. For example, we might consider the linear trend model

$$X(t) = \mu + at + \epsilon(t)$$

with μ and a unknown, and be interested in estimating $f_{\epsilon\epsilon}(\lambda)$. Or we might have taken $S(t)$ to be the input series to a linear filter with unknown impulse-response function $a(u)$, $u = 0, \pm 1, \cdots$ in an attempt to identify the system, that is, to estimate the transfer function $A(\lambda) = \Sigma\, a(u) \exp\{-i\lambda u\}$ and the $a(u)$. The model (26) for the series $X(t)$ differs in an important way from the previous models of this paper. The series $X(t)$ is not generally stationary, because $EX(t) = \mu + \Sigma\, a(t-u)S(u)$.

Estimates of the preceding parameters may be constructed as follows: define

$$d_X^{(T)}(\lambda) = \sum_{t=0}^{T-1} X(t) \exp\{-i\lambda t\}$$

with similar definitions for $d_S^{(T)}(\lambda)$, $d_\epsilon^{(T)}(\lambda)$. Then (26) leads to the approximate relationship

$$d_X^{(T)}(\lambda) \doteq \mu \sum_{t=0}^{T-1} \exp\{-i\lambda t\} + A(\lambda) d_S^{(T)}(\lambda) + d_\epsilon^{(T)}(\lambda). \tag{27}$$

Suppose $\lambda^1(T), \cdots, \lambda^J(T) \doteq \lambda$ are as in Section IV. Then

$$d_X^{(T)}(\lambda^j(T)) \doteq A(\lambda) d_S^{(T)}(\lambda^j(T)) + d_\epsilon^{(T)}(\lambda^j(T)) \tag{28}$$

for $j = 1, \cdots, J$. Following b) of Section IV, the $d_\epsilon^{(T)}(\lambda^j(T))$ are, for large T, approximately independent complex normal variates with mean 0 and variance $2\pi T f_{\epsilon\epsilon}(\lambda)$. The approximate model (28) is seen to take the form of linear regression. The results of linear least-squares theory now suggest our consideration of the estimates,

$$A^{(T)}(\lambda) = f_{XS}^{(T)}(\lambda) f_{SS}^{(T)}(\lambda)^{-1} \tag{29}$$

and

$$f_{\epsilon\epsilon}^{(T)}(\lambda) = f_{XX}^{(T)}(\lambda) - f_{XS}^{(T)}(\lambda) f_{SS}^{(T)}(\lambda)^{-1} f_{SX}^{(T)}(\lambda)$$

where

$$f_{SX}^{(T)}(\lambda) = J^{-1} \sum_{j=1}^{J} (2\pi T)^{-1} d_S^{(T)}(\lambda^j(T)) \overline{d_X^{(T)}(\lambda^j(T))}$$

with similar definitions for $f_{XS}^{(T)}$, $f_{XX}^{(T)}$, $f_{SS}^{(T)}$. The impulse response could be estimated by an expression such as

$$a^{(T)}(u) = P^{-1} \sum_{p=0}^{P-1} A^{(T)}\left(\frac{2\pi p}{P}\right) \exp\left\{\frac{-i2\pi pu}{P}\right\}$$

for some integer P. In some circumstances it may be appropriate to taper the data prior to computing the Fourier transform. In others it might make sense to base the Fourier transforms on disjoint stretches of data in the manner of d) of Section IV.

Under regularity conditions the estimate $A^{(T)}(\lambda)$ may be shown to be asymptotically complex normal with mean $A(\lambda)$ and variance $J^{-1} f_{\epsilon\epsilon}(\lambda) f_{SS}^{(T)}(\lambda)^{-1}$ (see [20]). The degree of fit of the model (26) at frequency λ may be measured by the sample coherence function

$$|R_{XS}^{(T)}(\lambda)|^2 = |f_{XS}^{(T)}(\lambda)|^2 / [f_{SS}^{(T)}(\lambda) f_{XX}^{(T)}(\lambda)]$$

satisfying

$$f_{\epsilon\epsilon}^{(T)}(\lambda) = [1 - |R_{XS}^{(T)}(\lambda)|^2] f_{XX}^{(T)}(\lambda).$$

This function provides a time series analog of the squared coefficient of correlation of two variates (see Koopmans [52]).

The procedure of prefiltering is often essential in the estimation of the parameters of the model (26). Consider a common relationship in which the series $X(t)$ is essentially a delayed version of the series $S(t)$, namely

$$X(t) = \alpha S(t - v) + \epsilon(t)$$

for some v. In this case

$$A(\lambda) = \alpha \exp\{-i\lambda v\},$$

$$d_X^{(T)}(\lambda^j(T)) = \alpha \exp\{-i\lambda^j(T)v\} d_S^{(T)}(\lambda^j(T)) + d_\epsilon^{(T)}(\lambda^j(T))$$

and

$$f_{XS}^{(T)}(\lambda) = \alpha J^{-1} \sum_j \exp\{-i\lambda^j(T)v\} I_{SS}^{(T)}(\lambda^j(T)) + f_{\epsilon S}^{(T)}(\lambda). \tag{30}$$

If v is large, the complex exponential fluctuates rapidly about 0 as j changes and the first term on the right-hand side of (30) may be near 0 instead of the desired $\alpha \exp\{-i\lambda v\} f_{SS}^{(T)}(\lambda)$. A useful prefiltering for this situation is to estimate v by $\hat{v}$, the lag that maximizes the magnitude of the sample cross-covariance function, and then to carry out the spectral computations on the data $X(t)$, $S(t - \hat{v})$, see Akaike and Yamanouchi [53] and Tick [54]. In general, one should prefilter the $X(t)$ series or the $S(t)$ series or both, so that the relationship between the filtered series is as near to being instantaneous as is possible.

The most important use of the calculations we have described is in the identification of linear systems. It used to be the case that the transfer function of a linear system was estimated by probing the system with pure sine waves in a succession of experiments. Expression (29) shows, however, that we can estimate the transfer function, for all λ, by simply employing a single input series $S(t)$ such that $f_{SS}^{(T)}(\lambda) \neq 0$.

In some situations we may have reason to believe that the system (26) is *realizable* that is $a(u) = 0$ for $u < 0$. The factorization techniques of Wiener [36] may be paralleled on the data in order to obtain estimates of $A(\lambda)$, $a(u)$ appropriate to this case, see Bhansali [55]. In Section IX, we will discuss a model like (26), but for the case of stochastic $S(t)$.

Another useful linear model is

$$X(t) = \theta_1 \phi_1(t) + \cdots + \theta_K \phi_K(t) + \epsilon(t)$$

with $\phi_1(t), \cdots, \phi_K(t)$ given functions and $\theta_1, \cdots, \theta_K$ unknown. The estimation of these unknowns and $f_{\epsilon\epsilon}(\lambda)$ is considered in Hannan [44] and Anderson [45]. This model allows us to handle trends and seasonal effects.

Yet another useful model is

$$X(t) = \mu + \rho_1 \sin(\theta_1 t + \alpha_1) + \cdots + \rho_K \sin(\theta_K t + \alpha_K) + \epsilon(t)$$

with $\mu, \rho_1, \theta_1, \alpha_1, \cdots, \rho_K, \theta_K, \alpha_K$ unknown. The estimation of these unknowns and $f_{\epsilon\epsilon}(\lambda)$ is considered in Whittle [49]. It allows us to handle hidden periodicities.

IX. Vector-Valued Continuous Spatial Series

In this section we move on from a consideration of real-valued discrete time series to series with a more complicated domain, namely p-dimensional Euclidean space, and with a more complicated range, namely r-dimensional Euclidean space. This step will allow us to consider data such as: that received by an array of antennas or seismometers, picture or TV, holographic, turbulent field.

Provided we set down our notation judiciously, the changes

involved are not dramatic. The notation that we shall adopt includes the following: boldface letters such as $\boldsymbol{X}$, $\boldsymbol{a}$, $\boldsymbol{A}$ will denote vectors and matrices. $\boldsymbol{A}^T$ will denote the transpose of a matrix $\boldsymbol{A}$, tr $\boldsymbol{A}$ will denote its trace, det $\boldsymbol{A}$ will denote its determinant. $E\boldsymbol{X}$ will denote the vector whose entries are the expected values of the corresponding entries of the vector-valued variate $\boldsymbol{X}$. $\operatorname{cov}\{\boldsymbol{X}, \boldsymbol{Y}\} = E\{(\boldsymbol{X} - E\boldsymbol{X})\overline{(\boldsymbol{Y} - E\boldsymbol{Y})}^T\}$ will denote the covariance matrix of the two vector-valued variates $\boldsymbol{X}$, $\boldsymbol{Y}$ (that may have complex entries). t, u, λ will lie in p-dimensional Euclidean space, R^p, with

$$t = (t_1, \cdots, t_p) \qquad dt = dt_1 \cdots dt_p$$

$$u = (u_1, \cdots, u_p) \qquad du = du_1 \cdots du_p$$

$$\lambda = (\lambda_1, \cdots, \lambda_p) \qquad d\lambda = d\lambda_1 \cdots d\lambda_p$$

$$\langle \lambda, t \rangle = \lambda_1 t_1 + \cdots + \lambda_p t_p$$

$$\langle \lambda, u \rangle = \lambda_1 u_1 + \cdots + \lambda_p u_p$$

$$|u| = (u_1^2 + \cdots + u_p^2)^{1/2}$$

$$|\lambda| = (\lambda_1^2 + \cdots + \lambda_p^2)^{1/2}.$$

The limits of integrals will be from $-\infty$ to ∞, unless indicated otherwise.

We will proceed by paralleling the development of Sections III and IV. Suppose that we are interested in analyzing measurements made simultaneously on r series of interest at location t, for all locations in some subset of the hypercube $0 < t_1, \cdots, t_p < T$. Suppose that we are prepared to model the measurements by the corresponding values of a realization of an r vector-valued stationary continuous spatial series $\boldsymbol{X}(t)$, $t \in R^p$. We define the *mean*

$$\boldsymbol{c}_X = E\boldsymbol{X}(t)$$

the *autocovariance function*

$$\boldsymbol{c}_{XX}(u) = \operatorname{cov}\{\boldsymbol{X}(t+u), \boldsymbol{X}(t)\}$$

and the *spectral density matrix*

$$\boldsymbol{f}_{XX}(\lambda) = (2\pi)^{-p} \int \exp\{-i\langle\lambda, u\rangle\} \boldsymbol{c}_{XX}(u)\, du, \qquad \lambda \in R^p \quad (31)$$

in the case that the integral exists. (The integral will exist when well-separated values of the series are sufficiently weakly dependent.) The inverse of the relationship (31) is

$$\boldsymbol{c}_{XX}(u) = \int \exp\{i\langle\lambda, \alpha\rangle\} \boldsymbol{f}_{XX}(\alpha)\, d\alpha. \qquad (32)$$

Let

$$X(t) \rightarrow Y(t) = \int a(t - u) X(u)\, du$$

be a linear filter carrying the r vector-valued series $X(t)$ into the s vector-valued series $Y(t)$. Let

$$A(\lambda) = \int a(u) \exp\{-i\langle\lambda, u\rangle\}\, du$$

denote the transfer function of this filter. Then the spectral density matrix of the series $Y(t)$ may be seen to be

$$f_{YY}(\lambda) = A(\lambda) f_{XX}(\lambda) \overline{A(\lambda)}^{\tau}. \tag{33}$$

As in Section III, expressions (32) and (33) may be combined to see that the entry in row j, column k of the matrix $f_{XX}(\lambda)$ may be interpreted as the covariance of the series resulting from passing the jth and kth components of $X(t)$ through narrow band-pass filters with transfer functions $A(\alpha) = \delta(\alpha - \lambda)$.

The series has a Cramér representation

$$X(t) = \int \exp\{i\langle\alpha, t\rangle\}\, dZ_X(\alpha)$$

where $Z_X(\lambda)$ is an r vector-valued random function with the properties

$$E\, dZ_X(\lambda) = \delta(\lambda) c_X\, d\lambda$$

$$\operatorname{cov}\{dZ_X(\lambda), dZ_X(\mu)\} = \delta(\lambda - \mu) f_{XX}(\lambda)\, d\lambda\, d\mu.$$

If $Y(t)$ is the filtered version of $X(t)$, then it has Cramér representation

$$Y(t) = \int \exp\{i\langle\alpha, t\rangle\} A(\alpha)\, dZ_X(\alpha).$$

We turn to a discussion of useful computations when values of the series $X(t)$ are available for t in some subset of the hypercube $0 < t_1, \cdots, t_p < T$. Let $\phi^{(T)}(t)$ be a data window whose support (that is the region of locations where $\phi^{(T)}(t) \neq 0$) is the region of observation of $X(t)$. (We might take $\phi^{(T)}(t)$ of the form $\phi(t/T)$ where $\phi(t) = 0$ outside $0 < t_1, \cdots, t_p < 1$.) We consider the Fourier transform

$$d_X^{(T)}(\lambda) = \int X(t) \phi^{(T)}(t) \exp\{-i\langle\lambda, t\rangle\}\, dt$$

based on the observed sample values.

Before indicating an approximate large sample distribution for $d_X^{(T)}(\lambda)$, we must first define the complex multivariate normal distribution and the complex Wishart distribution. We say that a vector-valued variate X, with complex entries, is *multivariate complex normal* with mean $\mathbf{0}$ and covariance matrix Σ when it has probability density proportional to $\exp\{-\overline{X}^{\tau}\Sigma^{-1}X\}$. We shall say that a matrix-valued variate is *complex Wishart* with n degrees of freedom and parameter Σ when it has the form $X_1\overline{X}_1^{\tau} + \cdots + X_n\overline{X}_n^{\tau}$, where $X_1, \cdots, X_n$ are independent multivariate complex normal variates with mean $\mathbf{0}$ and covariance matrix Σ. In the one dimensional case, the complex Wishart with n degrees of freedom is a multiple of a chi-squared variate with $2n$ degrees of freedom.

In the case that well-separated values of the series $X(t)$ are only weakly dependent, the $d_X^{(T)}(\lambda)$ have useful asymptotic properties as $T \to \infty$. These include:

a′) $d_X^{(T)}(0)$ is asymptotically multivariate normal with mean $\int \phi^{(T)}(t)\, dt\, c_X$ and covariance matrix $(2\pi)^p \int \phi^{(T)}(t)^2\, dt f_{XX}(0)$;

b′) for $\lambda \neq 0$, $d_X^{(T)}(\lambda)$ is asymptotically multivariate complex normal with mean $\mathbf{0}$ and covariance matrix

$$(2\pi)^p \int \phi^{(T)}(t)^2\, dt\, f_{XX}(\lambda);$$

c′) for $\lambda^j(T) \to \lambda \neq 0$, with $\lambda^j(T) - \lambda^k(T)$ not tending to 0 too rapidly, $1 \leqslant j < k \leqslant J$, the variates $d_X^{(T)}(\lambda^1(T)), \cdots, d_X^{(T)}(\lambda^J(T))$ are asymptotically independent multivariate complex normal with mean $\mathbf{0}$ and covariance matrix

$$(2\pi)^p \int \phi^{(T)}(t)^2\, dt\, f_{XX}(\lambda);$$

d′) if $\phi_j^{(T)}(t)\phi_k^{(T)}(t) = 0$, for all t, $1 \leqslant j < k \leqslant J$, and if $\lambda \neq 0$ the variates

$$d_X^{(T)}(\lambda, j) = \int X(t)\phi_j^{(T)}(t) \exp\{-i\langle\lambda, t\rangle\}\, dt \qquad (34)$$

$j = 1, \cdots, J$ are asymptotically independent multivariate complex normal with mean $\mathbf{0}$ and respective covariance matrices $(2\pi)^p \int \phi^{(T)}(t, j)^2\, dt\, f_{XX}(\lambda), j = 1, \cdots, J$.

Specific conditions under which these results hold are given in Section XV. A proof is given in the Appendix.

Results a′), b′) are forms of the central limit theorem. In result d′) the Fourier transforms are based on values of $X(t)$ over disjoint domains. It is interesting to note, from c′) and d′) that asymptotically independent statistics may be obtained by either taking the Fourier transform at distinct frequencies or at the same frequency, but over disjoint domains.

Result a′) suggests estimating the mean c_X by

$$c_X^{(T)} = \frac{\int X(t)\phi^{(T)}(t)\,dt}{\int \phi^{(T)}(t)\,dt}. \tag{35}$$

Result b′) suggests the consideration of the *periodogram matrix*

$$I_{XX}^{(T)}(\lambda) = (2\pi)^{-p}\left(\int \phi^{(T)}(t)^2\,dt\right)^{-1} d_X^{(T)}(\lambda)\overline{d_X^{(T)}(\lambda)}^\tau \tag{36}$$

as an estimate of $f_{XX}(\lambda)$ when $\lambda \neq 0$. From b′) its asymptotic distribution is complex Wishart with 1 degree of freedom and parameter $f_{XX}(\lambda)$. This estimate is often inappropriate because of its instability and singularity. Result c′) suggests the consideration of the estimate

$$f_{XX}^{(T)}(\lambda) = J^{-1}\sum_{j=1}^{J} I_{XX}^{(T)}(\lambda^j(T)) \tag{37}$$

where J is chosen large enough to obtain acceptable stability, but not so large that the estimate becomes overly biased. From c′) the asymptotic distribution of the estimate (37) is complex Wishart with J degrees of freedom and parameter $f_{XX}(\lambda)$. In the case $J = 1$ this asymptotic distribution is that of $f_{XX}(\lambda)\chi_{2J}^2/2J$. Result d′) suggests the consideration of the periodogram matrices

$$I_{XX}^{(T)}(\lambda, j) = (2\pi)^{-p}\left(\int \phi_j^{(T)}(t)^2\,dt\right)^{-1} d_X^{(T)}(\lambda, j)\overline{d_X^{(T)}(\lambda, j)}^\tau \tag{38}$$

$j = 1, \cdots, J$ as estimates of $f_{XX}(\lambda)$, $\lambda \neq 0$. The estimate

$$f_{XX}^{(T)}(\lambda) = J^{-1}\sum_{j=1}^{J} I_{XX}^{(T)}(\lambda, j) \tag{39}$$

will have as asymptotic distribution J^{-1} times a complex Wishart with J degrees of freedom and parameter $f_{XX}(\lambda)$ following result d′). We could clearly modify the estimates (37), (39) by using a finer spacing of frequencies and by averaging periodograms based on data over nondisjoint domains. The exact asymptotic distributions will not be so simple in these cases.

The method of fitting finite parameter models, described in

Section VII, extends directly to this vector-valued situation. Result b′) suggests the replacement of the likelihood function (25) by

$$\prod_{0<s_j<S_j} \det f_{XX}\left(\frac{2\pi s}{T};\theta\right)^{-1} \cdot \exp\left\{-\operatorname{tr} I_{XX}^{(T)}\left(\frac{2\pi s}{T}\right) f_{XX}\left(\frac{2\pi s}{T};\theta\right)^{-1}\right\} \quad (40)$$

in this new case for some large values $S_1, \cdots, S_p$ such that there is little power left beyond the cutoff frequency $(2\pi S_1/T, \cdots, 2\pi S_p/T)$. Suppose that $\hat{\theta}$ is the value of θ leading to the maximum of (40). Under regularity conditions, we can show that $\hat{\theta}$ is asymptotically normal with mean θ and covariance matrix $2\pi T^{-1}A^{-1}(A+B)A^{-1}$ where if A_{jk}, B_{jk} are row j, column k of A, B

$$A_{jk} = \int_0^{2\pi S/T} \operatorname{tr} \frac{\partial f(\alpha)}{\partial \theta_j} f(\alpha)^{-1} \frac{\partial f(\alpha)}{\partial \theta_k} f(\alpha)^{-1}\, d\alpha$$

$$B_{jk} = \int_0^{2\pi S/T} \sum_a \sum_b \sum_c \sum_d C_{abj}(\alpha) C_{cdk}(\beta) f_{abcd}(\alpha, -\alpha, -\beta)\, d\alpha\, d\beta$$

with $C_{abj}(\alpha)$ the entry in row a column b of

$$f(\alpha)^{-1} \frac{\partial f(\alpha)}{\partial \theta_j} f(\alpha)^{-1}.$$

In a number of situations we find ourselves led to consider an $(r+s)$ vector-valued series,

$$\begin{bmatrix} S(t) \\ X(t) \end{bmatrix} \quad (41)$$

satisfying a linear model of the form

$$E\{X(t) \mid S(u), u \in R^p\} = \mu + \int a(t-u) S(u)\, du \quad (42)$$

for some s vector μ and some $s \times r$ matrix-valued function $a(u)$. The model says that the average level of the series $X(t)$ at position t, given the series $S(t)$, is a linear filtered version of the series $S(t)$. If (41) is a stationary series and if $A(\lambda)$ is the transfer function of the filter $a(u)$, then (42) implies

$$c_X = \mu + A(0) c_S \quad (43)$$

$$f_{XS}(\lambda) = A(\lambda) f_{SS}(\lambda). \tag{44}$$

If we define the error series $\epsilon(t)$ by

$$\epsilon(t) = X(t) - \mu - \int a(t-u) S(u)\, du$$

then the degree of fit of the model (42) may be measured by the error spectral density

$$f_{\epsilon\epsilon}(\lambda) = f_{XX}(\lambda) - f_{XS}(\lambda) f_{SS}(\lambda)^{-1} f_{SX}(\lambda). \tag{45}$$

The relationships (43)–(45) suggest the estimates

$$A^{(T)}(\lambda) = f_{XS}^{(T)}(\lambda) f_{SS}^{(T)}(\lambda)^{-1} \tag{46}$$

$$\mu^{(T)} = c_X^{(T)} - A^{(T)}(0) c_S^{(T)} \tag{47}$$

$$f_{\epsilon\epsilon}^{(T)}(\lambda) = f_{XX}^{(T)}(\lambda) - f_{XS}^{(T)}(\lambda) f_{SS}^{(T)}(\lambda)^{-1} f_{SX}^{(T)}(\lambda) \tag{48}$$

respectively. The asymptotic distributions of these statistics are given in [26].

If there is a possibility that the matrix $f_{SS}^{(T)}(\lambda)$ might become nearly singular, then we would be better off replacing the estimate (46) by a frequency domain analog of the ridge regression estimate (Hoerl and Kennard [56], Hunt [57]), such as

$$f_{XS}^{(T)}(\lambda)[f_{SS}^{(T)}(\lambda) + kI]^{-1} \tag{49}$$

for some $k > 0$ and I the identity matrix. This estimate introduces further bias, over what was already present, but it is hoped that its increased stability more than accounts for this. In some circumstances we might choose k to depend on λ and to be matrix-valued.

X. Additional Results in the Spatial Series Case

The results of the previous section have not taken any essential notice of the fact that the argument t of the random function under consideration is multidimensional. We now indicate some new results pertinent to the multidimensional character.

In some situations, we may be prepared to assume that the series $X(t)$, $t \in R^p$, is *isotropic*, that is the autocovariance function $c_{XX}(u) = \text{cov}\,\{X(t+u), X(t)\}$ is a function of $|u|$ only. In this case the spectral density matrix $f_{XX}(\lambda)$ is also rotationally symmetric, depending only on $|\lambda|$. In fact (see in Bochner and Chandrasekharan [58, p. 69])

$$f_{XX}(\lambda) = (2\pi)^{-p/2} |\lambda|^{(2-p)/2} \int_0^\infty |u|^{p/2} \cdot J_{(p-2)/2}(|\lambda||u|) c_{XX}(u)\, d|u| \tag{50}$$

where $J_k(t)$ is the Bessel function of the first kind of order k. The relationship (50) may be inverted as follows,

$$c_{XX}(u) = (2\pi)^{p/2} |u|^{(2-p)/2} \int_0^\infty |\lambda|^{p/2} \cdot J_{(p-2)/2}(|\lambda||u|) f_{XX}(\lambda)\, d|\lambda|.$$

The simplified character of $f_{XX}(\lambda)$ in the isotropic case makes its estimation and display much simpler. We can estimate it by an expression such as

$$J^{-1} \sum_{j=1}^{J} I_{XX}^{(T)}(\lambda^j(T)) \tag{51}$$

where the $\lambda^j(T)$ are distinct, but with $|\lambda^j(T)|$ near $|\lambda|$. There are many more $\lambda^j(T)$ with $|\lambda^j(T)|$ near $|\lambda|$ than there are $\lambda^j(T)$ with $\lambda^j(T)$ near λ. It follows that we generally obtain a much better estimate of the spectrum in this case over the estimate in the general case. Also the number of $\lambda^j(T)$ with $|\lambda^j(T)|$ near $|\lambda|$ increases as $|\lambda|$ increases. If follows that the estimate formed will generally be more stable for the frequencies with $|\lambda|$ large. Examples of power spectra estimated in this manner may be found in Mannos [59].

Another different thing that can occur in the general p dimensional case is the definition of marginal processes and marginal spectra. We are presently considering processes $X(t_1, \cdots, t_p)$. Suppose that for some n, $1 \leqslant n < p$, we are interested in the process with $t_{n+1}, \cdots, t_p$ fixed, say at $0, \cdots, 0$. By inspection we see that the marginal process $X(t_1, \cdots, t_n, 0, \cdots, 0)$ has autocovariance function $c_{XX}(u_1, \cdots, u_n, 0, \cdots, 0)$. The spectral density matrix of the marginal process is, therefore,

$$(2\pi)^{-n} \int \cdots \int c_{XX}(u_1, \cdots, u_n, 0, \cdots, 0) \cdot \exp\{-i(\lambda_1 u_1 + \cdots + \lambda_n u_n)\}\, du_1 \cdots du_n$$

$$= \int \cdots \int f_{XX}(\lambda_1, \cdots, \lambda_n, \lambda_{n+1}, \cdots, \lambda_p) \cdot d\lambda_{n+1} \cdots d\lambda_p.$$

We see that we obtain the spectral density of the marginal process by integrating the complete spectral density. The same remark applies to the Cramér representation for

$$X(t_1, \cdots, t_n, 0, \cdots, 0) = \int \cdots \int \exp\{i(t_1\lambda_1 + \cdots + t_n\lambda_n)\}$$

$$\cdot \int \cdots \int dZ_X(\lambda_1, \cdots, \lambda_p).$$

Vector-valued series with multidimensional domain are discussed in Hannan [44] and Brillinger [26].

XI. Additional Results in the Vector Case

In the case that the series $X(t)$ is r vector-valued with $r > 1$, we can describe analogs of the classical procedures of multivariate analysis including for example; i) partial correlation, ii) principal component analysis, iii) canonical correlation analysis, iv) cluster analysis, v) discriminant analysis, vi) multivariate analysis of variance, and vii) simultaneous equations. These analogs proceed from c′) or d′) of earlier section. The procedures listed are often developed for samples from multivariate normal distributions. We obtain the time series procedure by identifying the $d_X^{(T)}(\lambda^j(T)), j = 1, \cdots, J$ or $d_X^{(T)}(\lambda, j)$, $j = 0, \cdots, J - 1$ with independent multivariate normals having mean $\mathbf{0}$ and covariance matrix $(2\pi)^p \int \phi^{(T)}(t)^2\, dt\, f_{XX}(\lambda)$ and substituting into the formulas developed for the classical situation. For example, stationary time series analogs of correlation coefficients are provided by the

$$R_{jk}(\lambda) = f_{jk}(\lambda)/\sqrt{f_{jj}(\lambda) f_{kk}(\lambda)}$$

$$\sim \operatorname{cov}\{d_j^{(T)}(\lambda), d_k^{(T)}(\lambda)\} \Big/ \sqrt{\operatorname{var} d_j^{(T)}(\lambda) \operatorname{var} d_k^{(T)}(\lambda)}$$

the *coherency* at frequency λ of the jth component with the kth component of $X(t)$, where $f_{jk}(\lambda)$ is the entry in row j, column k of $f_{XX}(\lambda)$ and $d_j^{(T)}(\lambda)$ is the entry in row j of $d_X^{(T)}(\lambda)$ for $j, k = 1, \cdots, r$. The parameter $R_{jk}(\lambda)$ satisfies $0 \leqslant |R_{jk}(\lambda)| \leqslant 1$ and is seen to provide a measure of the degree of linear relationship of the series $X_j(t)$ with the series $X_k(t)$ at frequency λ. Its modulus squared, $|R_{jk}(\lambda)|^2$, is called the *coherence*. It may be estimated by

$$R_{jk}^{(T)}(\lambda) = f_{jk}^{(T)}(\lambda) \Big/ \sqrt{f_{jj}^{(T)}(\lambda) f_{kk}^{(T)}(\lambda)}$$

where $f_{jk}^{(T)}(\lambda)$ is an estimate of $f_{jk}(\lambda)$.

As time series papers on corresponding multivariate topics, we mention in case i) Tick [60], Granger [61], Goodman [62], Bendat and Piersol [63], Groves and Hannan [64], and Gersch [65]; in case ii) Goodman [66], Brillinger [67], [20], and Priestley *et al.* [68]; in case iii) Brillinger [67], [20],

Miyata [69], and Priestley *et al.* [68]; in case iv) Ligett [70]; in case v) Brillinger [20]; in case vi) Brillinger [71]; in case vii) Brillinger and Hatanaka [72], and Hannan and Terrell [73].

Instead of reviewing each of the time series analogs we content ourselves by indicating a form of discriminant analysis that can be carried out in the time series situation. Suppose that a segment of the r vector-valued series $X(t)$ is available and that its spectral density matrix may be any one of $f_i(\lambda)$, $i = 1, \cdots, I$. Suppose that we wish to construct a rule for assigning $X(t)$ to one of the $f_i(\lambda)$.

In the case of a variate U coming from one of I multivariate normal populations with mean $\mathbf{0}$ and covariance matrix Σ_i, $i = 1, \cdots, I$, a common discrimination procedure is to define a discriminant score

$$-\tfrac{1}{2} \log \det \Sigma_i - \tfrac{1}{2} U^T \Sigma_i^{-1} U$$

for the ith population and then to assign the observation U to the population for which the discriminant score has the highest value (see Rao [74, p. 488]). The discriminant score is essentially the logarithm of the probability density of the ith population.

Result 2) suggests a time series analog for this procedure. If the spectral density of the series $X(t)$ is $f_i(\lambda)$, the log density of $d_X^{(T)}(\lambda)$ is essentially

$$-\log \det f_i(\lambda) - \operatorname{tr} I_{XX}^{(T)}(\lambda) f_i(\lambda)^{-1}. \qquad (52)$$

This provides a discriminant score for each frequency λ. A more stable score would be provided by the smoothed version

$$-J^{-1} \log \det f_i(\lambda) - \operatorname{tr} f_{XX}^{(T)}(\lambda) f_i(\lambda)^{-1}$$

with $f_{XX}^{(T)}(\lambda)$ given by (37) or (39). These scores could be plotted against λ for $i = 1, \cdots, I$ in order to carry out the required discrimination. In the case that the $f_i(\lambda)$ are unknown, their values could be replaced by estimates in (52).

XII. Additional Results in the Continuous Case

In Section IX, we changed to a continuous domain in contrast to the discrete domain we began with in Section III. In many problems, we must deal with both sorts of domains, because while the phenomenon of interest may correspond to a continuous domain, observational and computational considerations may force us to deal with the values of the process for a discrete domain. This occurrence gives rise to the complication of *aliasing*. Let Z denote the set of integers, $Z = 0, \pm 1, \cdots$. Suppose $X(t)$, $t \in R^p$, is a stationary continuous spatial series with spectral density matrix $f_{XX}(\lambda)$ and Cramér representation

$$X(t) = \int \exp\{i\langle\alpha, t\rangle\}\, dZ_X(\alpha).$$

Suppose $X(t)$ is observable only for $t \in Z^p$. For these values of t

$$X(t) = \int_{(-\pi,\pi)^p} \exp\{i\langle\alpha, t\rangle\} \sum_{j\in Z^p} dZ_X(\alpha + 2\pi j).$$

This is the Cramér representation of a discrete series with spectral density matrix

$$\sum_{j\in Z^p} f_{XX}(\lambda + 2\pi j).$$

We see that if the series $X(t)$ is observable only for $t \in Z^p$, then there is no way of untangling the frequencies

$$\lambda + 2\pi j, \quad j \in Z^p.$$

These frequencies are called the aliases of the fundamental frequency λ.

XIII. Stationary Point Processes

A variety of problems, such as those of traffic systems, queues, nerve pulses, shot noise, impulse noise, and microscopic theory of gases lead us to data that has the character of times or positions in space at which certain events have occurred. We turn now to indicating how the formulas we have presented so far in this paper must be modified to apply to data of this new character.

Suppose that we are recording the positions in p-dimensional Euclidean space at which events of r distinct types occur. For $j = 1, \cdots, r$ let $X_j(t) = X_j(t_1, \cdots, t_p)$ denote the number of events of the jth type that occur in the hypercube $(0, t_1] \times \cdots \times (0, t_p]$. Let $dX_j(t)$ denote the number that occur in the small hypercube $(t_1, t_1 + dt_1] \times \cdots \times (t_p, t_p + dt_p]$. Suppose that joint distributions of variates such as $dX(t^1), \cdots, dX(t^k)$ are unaffected by simple translation of $t^1, \cdots t^k$, we then say that $X(t)$ is a stationary point process.

Stationary point process analogs of definitions set down previously include

$$c_X\, dt = E\, dX(t) \tag{53}$$

c_X is called the *mean intensity* of the process,

$$dC_{XX}(u)\, dt = \operatorname{cov}\{dX(t+u), dX(t)\} \tag{54}$$

$$f_{XX}(\lambda) = (2\pi)^{-p} \int \exp\{-i\langle\lambda, u\rangle\}\, dC_{XX}(u) \tag{55}$$

$$X(t) = \int \cdots \int \left[\frac{\exp\{i\lambda_1 t_1\} - 1}{i\lambda_1}\right] \cdots \left[\frac{\exp\{i\lambda_p t_p\} - 1}{i\lambda_p}\right] dZ_X(\lambda_1, \cdots, \lambda_p) \tag{56}$$

$$dX(t) = \int \exp\{i\langle\lambda, t\rangle\}\, dZ_X(\lambda)\, dt \tag{57}$$

$$E\{dX(t)|S(u), u \in R^p\} = \left[\mu + \int a(t-u)\, dS(u)\right] dt. \tag{58}$$

This last refers to an $(r + s)$ vector-valued point process. It says that the instantaneous intensity of the series $X(t)$ at position t, given the location of all the points of the process $S(u)$, is a linear translation invariant function of the process $S(u)$. The locations of the points of $X(t)$ are affected by where the points of $S(u)$ are located. We may define here a stationary random measure $d\epsilon(t)$ by

$$d\epsilon(t) = dX(t) - \left[\mu + \int a(t-u)\, dS(u)\right] dt. \tag{59}$$

We next indicate some statistics that it is useful to calculate when the process $X(t)$ has been observed over some region. The Fourier transform is now

$$d_X^{(T)}(\lambda) = \int \phi^{(T)}(t) \exp\{-i\langle\lambda, t\rangle\}\, dX(t) \tag{60}$$

for the data window $\phi^{(T)}(t)$ whose support corresponds to the domain of observation. If $r = 1$ and points occur at the positions $\tau_1, \tau_2, \cdots$, then this last has the form

$$\phi^{(T)}(\tau_1) \exp\{-i\langle\lambda, \tau_1\rangle\} + \phi^{(T)}(\tau_2) \exp\{-i\langle\lambda, \tau_2\rangle\} + \cdots.$$

We may compute Fourier transforms for different domains in which case we define

$$d_X^{(T)}(\lambda, j) = \int \phi_j^{(T)}(t) \exp\{-i\langle\lambda, t\rangle\}\, dX(t). \tag{61}$$

The change in going from the case of spatial series to the case of point processes is seen to be the replacement of $X(t)\, dt$ by $dX(t)$. In the case that well-separated increments of the process are only weakly dependent, the results a′)–d′) of Section IX hold without further redefinition.

References to the theory of stationary point processes include: Cox and Lewis [75], Brillinger [76], Daley and Vere-Jones [77], and Fisher [78]. We remark that the material of this section applies equally to the case in which $dX(t)$ is a general stationary random measure, for example with $p, r = 1$, we might take $dX(t)$ to be the amount of energy released by earthquakes in the time interval $(t, t + dt)$. In the next section we indicate some results that do take note of the specific character of a point process.

XIV. New Things in the Point Process Case

In the case of a point process, the parameters $\mathbf{c}_X$, $\mathbf{C}_{XX}(u)$ have interpretations further to their definitions (53), (54). Suppose that the process is *orderly*, that is the probability that a small region contains more than one point is very small. Then, for small dt

$$c_j\, dt = E\, dX_j(t) \doteq \Pr\,[\text{there is an event of type } j \text{ in } (t, t + dt]\,].$$

It follows that c_j may be interpreted as the intensity with which points of type j are occurring. Likewise, for $u \neq 0$

$$\begin{aligned} dC_{jk}(u)\, dt &= \operatorname{cov}\,\{dX_j(t+u), dX_k(t)\} \\ &\doteq \Pr\,[\text{there is an event of type } j \text{ in } (t+u, t+u+du] \text{ and an event of type } k \text{ in } (t, t+dt]] - c_j c_k\, dt\, du. \end{aligned}$$

It follows that

$$\frac{dC_{jk}(u) + c_j c_k\, du}{c_k} = \Pr\,[\text{event of sort } j \text{ in } (t+u, t+u+du] \text{ given an event of sort } k \text{ in } (t, t+dt]\,]. \tag{62}$$

In the case that the processes $X_j(t)$ and $X_k(t)$ are independent, expression (62) is equal to $c_j du$.

If the derivative $c_{jk}(u) = dC_{jk}(u)/du$ exists for $u \neq 0$ it is called the *cross-covariance density* of the two processes in the case $j \neq k$ and the *autocovariance density* in the case $j = k$. For many processes

$$dC_{jj}(u) = c_j \delta(u)\, du + c_{jj}(u)\, du$$

and so the power spectrum of the process $X_j(t)$ is given by

$$f_{jj}(\lambda) = (2\pi)^{-p} \left[c_j + \int \exp\,\{-i\langle\lambda, u\rangle\}\, c_{jj}(u)\, du \right].$$

For a Poisson process $c_{jj}(u) = 0$ and so $\mathbf{f}_{XX}(\lambda) = (2\pi)^{-p}\mathbf{c}_X$.

The parameter $(2\pi)^p f_{XX}(0)/c_X$ is useful in the classification of real-valued point processes. From 1)

$$\operatorname{var} X(T, \cdots, T) \sim (2\pi)^p T^p f_{XX}(0).$$

It follows that, for large T, $(2\pi)^p f_{XX}(0)/c_X$ is the ratio of the variance of the number of points in the hypercube $(0, T]^p$ for the process $X(t)$ to the variance of the number of points in the same hypercube for a Poisson process with the same intensity c_X. For this reason we say that the process $X(t)$ is *under-dispersed* or *clustered* if the ratio is greater than 1 and *over-dispersed* if the ratio is less than 1.

The estimation procedure described in Section XI for models with a finite number of parameters is especially useful in the point process case as, typically, convenient time domain estimation procedures do not exist at all. Results of applying such a procedure are indicated in [79].

XV. Stationary Random Schwartz Distributions

In this section, we present the theory of Schwartz distributions (or generalized functions) needed to develop properties of the Fourier transforms of random Schwartz distributions. These last are important as they contain the processes discussed so far in this paper as particular cases. In addition they contain other interesting processes as particular cases, such as processes whose components are a combination of the processes discussed so far and such as the processes with stationary increments that are useful in the study of turbulence, see Yaglom [80]. A further advantage of this abstract approach is that the assumptions needed to develop results are cut back to essentials. References to the theory of Schwartz distributions include Schwartz [81] and Papoulis [82].

Let $\mathfrak{D}$ denote the space of infinitely differentiable functions on R^p with compact support. Let $\mathcal{S}$ denote the space of infinitely differentiable functions on R^p with rapid decrease, that is such that if $\phi^{(q)}(t)$ denotes a derivative of order q then

$$\lim_{|t|\to\infty} (1 + |t|)^n \phi^{(q)}(t) \to 0 \quad \text{for all } n, q.$$

A continuous linear functional on $\mathfrak{D}$ is called a *Schwartz distribution* or *generalized function*. The Dirac delta function that we have been using throughout the paper is an example. A continuous linear functional on $\mathfrak{D}$ is called a *tempered distribution*.

Suppose now that a random experiment is being carried out, the possible results of which are continuous linear maps X from $\mathfrak{D}$ to $L^2(P)$, the space of square integrable functions for a probability measure P. Suppose that r of these maps are collected into an r vector, $\boldsymbol{X}(\phi)$. We call $\boldsymbol{X}(\phi)$ an r vector-valued *random Schwartz distribution*. It is possible to talk about things such as $E\,X(\phi)$, cov $\{X(\phi), X(\psi)\}$ in this case. An important family of transformations on $\mathfrak{D}$ consists of the shifts S^u defined by $S^u\phi(t) = \phi(t + u)$, $t, u \in R^p$. The random Schwartz distribution is called *wide-sense stationary* when

$$E\,X(S^u\phi) = E\,X(\phi)$$

$$\operatorname{cov}\{X(S^u\phi), X(S^u\psi)\} = \operatorname{cov}\{X(\phi), X(\psi)\}$$

for all $u \in R^p$ and $\phi, \psi \in \mathfrak{D}$. It is called *strictly stationary* when all the distributions of finite numbers of values are invariant under the shifts.

Let us denote the convolution of two functions $\phi, \psi \in \mathfrak{D}$ by

$$\phi * \psi(t) = \int \phi(t - u)\,\overline{\psi(u)}\,du$$

and the Fourier transform of a function in $\mathcal{S}$ by the corresponding capital letter

$$\Phi(\lambda) = \int \phi(u) \exp\{-i\langle\lambda, u\rangle\}\,du$$

then we can set down the following Theorem.

Theorem 1: (Ito [83], Yaglom [80].) If $X(\phi)$, $\phi \in \mathfrak{D}$ is a wide-sense stationary random Schwartz distribution, then

$$E\,X(\phi) = \boldsymbol{c}_X \int \phi(t)\,dt \tag{63}$$

$$\operatorname{cov}\{X(\phi), X(\psi)\} = \boldsymbol{c}_{XX}(\phi * \psi^{\neg}) \tag{64}$$

$$= \int \Phi(-\alpha)\,\overline{\Psi(-\alpha)}\,d\boldsymbol{F}_{XX}(\alpha) \tag{65}$$

and

$$X(\phi) = \int \Phi(-\alpha)\,dZ_X(\alpha) \tag{66}$$

where $\boldsymbol{c}_X$ is an r vector, $\boldsymbol{c}_{XX}(\cdot)$ is an $r \times r$ matrix of tempered distributions, $\boldsymbol{F}_{XX}(\lambda)$ is a nonnegative matrix-valued measure satisfying

$$\int (1 + |\alpha|)^{-k}\,d\boldsymbol{F}_{XX}(\alpha) < \infty \tag{67}$$

for some nonnegative integer k, and finally $Z_X(\lambda)$ is a random function satisfying

$$E\,dZ_X(\lambda) = \delta(\lambda)\,\boldsymbol{c}_X\,d\lambda \tag{68}$$

$$\operatorname{cov}\{dZ_X(\lambda), dZ_X(\mu)\} = \delta(\lambda - \mu)\, dF_{XX}(\lambda)\, d\mu. \quad (69)$$

The spatial series of Section IX is a random Schwartz distribution corresponding to the functional

$$X(\phi) = \int X(t)\phi(t)\, dt$$

for $\phi \in \mathfrak{D}$. The representations indicated in that section may be deduced from the results of Theorem 1. It may be shown that k of (67) may be taken to be 0 for this case.

The stationary point process of Section XII is likewise a random Schwartz distribution corresponding to the functional

$$X(\phi) = \int \phi(t)\, dX(t)$$

for $\phi \in \mathfrak{D}$. The representations of Section XII may be deduced from Theorem 1. It may be shown that k of (67) may be taken to be 2 for this case.

Gelfand and Vilenkin [84] is a general reference to the theory of random Schwartz distributions. Theorem 1 is proved there.

A linear model that extends those of (42) and (58) to the present situation is one in which the $(r + s)$ vector-valued stationary random Schwartz distribution

$$\begin{bmatrix} S(\phi) \\ X(\phi) \end{bmatrix}$$

satisfies

$$E\{X(\phi) \mid S(\psi), \psi \in \mathfrak{D}\} = \mu \int \phi(t)\, dt + S(\phi * a)$$

$$= \mu\Phi(0) + \int \Phi(-\alpha) A(\alpha)\, dZ_S(\alpha). \quad (70)$$

In the case that the spectral measure is differentiable this last implies that

$$f_{XS}(\lambda) = A(\lambda) f_{SS}(\lambda) \quad (71)$$

suggesting that the system may be identified if the spectral density may be estimated. We next set down a mixing assumption, before constructing such an estimate and determining its asymptotic properties.

Given k variates $X_1, \cdots, X_k$ let $\operatorname{cum}\{X_1, \cdots, X_k\}$ denote their joint cumulant or semi-invariant. Cumulants are defined and discussed in Kendall and Stuart [85] and Brillinger [20]. They are the elementary functions of the moments of the variates that vanish when the variates are independent. As such they provide measures of the degree of dependence of variates. We will make use of

Assumption 1. $X(\phi)$ is a stationary random Schwartz distribution with the property that for $\phi_1, \cdots, \phi_k \in \mathcal{S}$ and $a_1, \cdots, a_k = 1, \cdots, r; k = 2, 3, \cdots$,

$$\operatorname{cum}\{X_{a_1}(\phi_1), \cdots, X_{a_k}(\phi_k)\} = \int \cdots \int \Phi_1(-\alpha^1) \cdots \cdot \Phi_{k-1}(-\alpha^{k-1}) \Phi_k(\alpha^1 + \cdots + \alpha^{k-1}) \cdot f_{a_1 \cdots a_k}(\alpha^1, \cdots, \alpha^{k-1}) \, d\alpha^1 \cdots d\alpha^{k-1} \tag{72}$$

with

$$(1 + |\alpha^1|)^{-m_1} \cdots (1 + |\alpha^{k-1}|)^{-m_{k-1}} | f_{a_1 \cdots a_k} \cdot (\alpha^1, \cdots, \alpha^{k-1})| < L_k$$

for some finite $m_1, \cdots, m_{k-1}, L_k$.

In the case that the spectral measure $F_{XX}(\lambda)$ is differentiable, relation (65) corresponds to the case $k = 2$ of (72). The character of Assumption 1 is one of limiting the size of the cumulants of the functionals of the process $X(\phi)$. It will be shown that it is a form of weak dependence requirement, for functionals of the process that are far apart in t, in the Appendix. The function $f_{a_1 \cdots a_k}(\lambda^1, \cdots, \lambda^{k-1})$ appearing in (72) is called a *cumulant spectrum* of order k, see Brillinger [86] and the references therein. From (66) we see that it is also given by

$$\operatorname{cum}\{dZ_{a_1}(\lambda^1), \cdots, dZ_{a_k}(\lambda^k)\} = \delta(\lambda^1 + \cdots + \lambda^k) \cdot f_{a_1 \cdots a_k}(\lambda^1, \cdots, \lambda^{k-1}) \, d\lambda^1 \cdots d\lambda^k. \tag{73}$$

The fact that it only depends on $k - 1$ arguments results from the assumed stationarity of the process.

Let $\phi^{(T)}(t) = \phi(t/T)$ with $\phi \in \mathfrak{D}$. As an analog of the Fourier transforms of Sections IX and XII we now define

$$d_X^{(T)}(\lambda) = X(\exp\{-i\langle\lambda, \cdot\rangle\} \phi^{(T)}) \tag{74}$$

for the stationary random Schwartz distribution $X(\phi)$. We can now state the following theorem.

Theorem 2: If Assumption 1 is satisfied, if $d_X^{(T)}(\lambda)$ is given by (74) and if $T|\lambda^j(T) - \lambda^k(T)| \to \infty$, $1 \leqslant j < k \leqslant J$, then 1)–4) of Section IX hold.

This theorem is proved in the Appendix. It provides a justification for the estimation procedures suggested in the paper and for the large sample approximations suggested for the distributions of the estimates.

We end this section by mentioning that a point process with events at positions τ_k, $k = 1, \cdots$ may be represented by the generalized function

$$\sum_k \delta(t - \tau_k)$$

the sampled function of Section III may be represented by the generalized function

$$\sum_{j=-\infty}^{\infty} X(j)\delta(t - j)$$

and that a point process with associated variate S may be represented by

$$\sum_k S_k \delta(t - \tau_k)$$

see Beutler and Leneman [87]. Mathéron [92] discusses the use of random Schwartz distributions in the smoothing of maps.

XVI. Higher Order Spectra and Nonlinear Systems

In the previous section we have introduced the higher order cumulant spectra of stationary random Schwartz distributions. In this section we will briefly discuss the use of such spectra and how they may be estimated.

In the case that the process under consideration is Gaussian, the cumulant spectra of order greater than two are identically 0. In the non-Gaussian case, the higher order spectra provide us with important information concerning the distribution of the process. For example were the process real-valued Poisson on the line with intensity c_N, then the cumulant spectrum of order k would be constant equal to $c_N(2\pi)^{1-k}$. Were the process the result of passing a series of independent identically distributed variates through a filter with transfer function $A(\lambda)$, then the cumulant spectrum of order k would be proportional to

$$A(\lambda^1) \cdots A(\lambda^{k-1}) A(-\lambda^1 - \cdots - \lambda^{k-1}).$$

Such hypotheses might be checked by estimating higher cumulant spectra.

An important use of higher order spectra is in the identification of polynomial systems such as those discussed in Wiener [88] and Brillinger [86] and Halme [89]. Tick [90] shows that if $S(t)$ is a stationary real-valued Gaussian series, if $\epsilon(t)$ is an independent stationary series and if the series $X(t)$ is given by

$$X(t) = \mu + \int a(t-u)S(u)\,du + \iint b(t-u, t-v)S(u)S(v)\,du\,dv + \epsilon(t) \quad (75)$$

then

$$f_{SX}(\lambda) = A(-\lambda)f_{SS}(\lambda)$$
$$f_{SSX}(\lambda, \mu) = 2B(-\lambda, -\mu)f_{SS}(\lambda)f_{SS}(\mu)$$

where

$$A(\lambda) = \int a(u) \exp\{-i\lambda u\}\,du$$

$$B(\lambda, \mu) = \iint b(u, v) \exp\{-i(\lambda u + \mu v)\}\,du\,dv$$

and $f_{SSX}(\lambda, \mu)$ is a third-order cumulant spectrum. It follows that both the linear transfer function $A(\lambda)$ and the bitransfer function $B(\lambda, \mu)$ of the system may be estimated, from estimates of second- and third-order spectra, following the probing of the system by a single Gaussian series. References to the identification of systems of order greater than 2, and to the case of non-Gaussian $S(t)$ are given in [86].

We turn to the problem of constructing an estimate of a kth order cumulant spectrum. In the course of the proof of Theorem 2 given in the Appendix, we will see that

$$\text{cum}\,\{d_{a_1}^{(T)}(\lambda^1), \cdots, d_{a_k}^{(T)}(\lambda^k)\} \sim \begin{cases} (2\pi)^{p(k-1)} \int \phi^{(T)}(t)^k\,dt\, f_{a_1 \cdots a_k}(\lambda^1, \cdots, \lambda^{k-1}), & \text{if } \lambda^1 + \cdots + \lambda^{k-1} = 0 \\ 0, & \text{if } \lambda^1 + \cdots + \lambda^{k-1} \neq 0. \end{cases}$$

Suppose that no proper subset of $\lambda^1, \cdots, \lambda^k$ sums to 0. It then follows from the principal relation connecting moments and cumulants that

$$E\{d_{a_1}^{(T)}(\lambda^1)\cdots d_{a_k}^{(T)}(\lambda^k)\} \sim (2\pi)^{p(k-1)} \int \phi^{(T)}(t)^k \cdot dt\, f_{a_1\cdots a_k}(\lambda^1, \cdots, \lambda^{k-1})$$

provided $\lambda^1 + \cdots + \lambda^k = 0$. This last one suggests the use of kth order periodogram

$$I_{a_1\cdots a_k}^{(T)}(\lambda^1, \cdots, \lambda^{k-1}) = (2\pi)^{-p(k-1)} \left(\int \phi^{(T)}(t)^k dt\right)^{-1} \cdot d_{a_1}^{(T)}(\lambda^1)\cdots d_{a_{k-1}}^{(T)}(\lambda^{k-1}) \times d_{a_k}^{(T)}(-\lambda^1 - \cdots - \lambda^{k-1}) \qquad (76)$$

as a naive estimate of the spectrum $f_{a_1\cdots a_k}(\lambda^1, \cdots, \lambda^{k-1})$ provided that no proper subset of $\lambda^1, \cdots, \lambda^{k-1}$ sums to 0. From what we have seen in the case $k = 2$ this estimate will be unstable. It follows that we should in fact construct an estimate by smoothing the periodogram (76) over $(k - 1)$-tuples of frequencies in the neighborhood of $\lambda^1, \cdots, \lambda^{k-1}$, but such that no proper subset of the $(k - 1)$-tuple sums to 0. Details of this construction are given in Brillinger and Rosenblatt [91] for the discrete time case. We could equally well have constructed an estimate using the Fourier transforms $d_X^{(T)}(\lambda, j)$ based on disjoint domains.

Appendix

We begin by providing a motivation for Assumption 1 of Section XIV. Suppose that

$$\text{cum}\,\{X_{a_1}(\phi_1), \cdots, X_{a_k}(\phi_k)\}, \qquad \phi_1, \cdots, \phi_k \in \mathfrak{D}$$

is continuous in each of its arguments. Being a continuous multilinear functional it can be written

$$c_{a_1\cdots a_k}(\phi_1 \otimes \phi_2 \otimes \cdots \otimes \phi_k)$$

where $c_{a_1\cdots a_k}$ is a Schwartz distribution on $\mathfrak{D}(R^{pk})$, from the Schwartz nuclear theorem. If the process is stationary this distribution satisfies

$$c_{a_1\cdots a_k}(S^u\phi_1 \otimes S^u\phi_2 \otimes \cdots \otimes S^u\phi_k) = c_{a_1\cdots a_k} \cdot (\phi_1 \otimes \phi_2 \otimes \cdots \otimes \phi_k).$$

It follows that it has the form

$$\mathcal{C}\left(\int \phi(t + u^1, \cdots, t + u^{k-1}, t)\, dt\right)$$

for $\phi \in \mathfrak{D}(R^{pk})$ where $\mathcal{C}$ is a distribution on $\mathfrak{D}(R^{p(k-1)})$.

Now consider the case in which the process $X(\phi)$ has the property that

$$\text{cum}\,\{X_{a_1}(\phi_1), \cdots, X_{a_k}(\phi_k)\} = 0$$

when the supports of $\phi_1, \cdots, \phi_{k-1}$ are farther away from that of ϕ_k than some number ρ. This means that the distribution $\mathcal{C}$ has compact support. By the Schwartz–Paley–Wiener theorem, $\mathcal{C}$ is, therefore, the Fourier transform of a function of slow growth, say $f_{a_1 \cdots a_k}(\lambda^1, \cdots, \lambda^{k-1})$ and we may write the relation (72). In the case that values of the process $X(\phi)$ at a distance from each other are only weakly dependent, we can expect the cumulant to be small and for the representation (72) to hold with (73) satisfied.

Proof of Theorem 2: We see from (66) and (73)

$$\begin{aligned}
&\text{cum}\,\{d_{a_1}^{(T)}(\lambda^1), \cdots, d_{a_k}^{(T)}(\lambda^k)\} \\
&= \int \cdots \int \Phi_1^{(T)}(\alpha^1 - \lambda^1) \cdots \Phi_{k-1}^{(T)}(\alpha^{k-1} - \lambda^{k-1})\, \Phi_k^{(T)} \\
&\quad \cdot (-\alpha^1 - \cdots - \alpha^{k-1} - \lambda^k)\, f_{a_1 \cdots a_k}(\alpha^1, \cdots, \alpha^{k-1}) \\
&\quad \cdot d\alpha^1 \cdots d\alpha^{k-1} \\
&= T^p \int \cdots \int \Phi_1(\beta^1) \cdots \Phi_{k-1}(\beta^{k-1})\, \Phi_k \\
&\quad \cdot (-\beta^1 - \cdots - \beta^{k-1} - T(\lambda^1 + \cdots + \lambda^k))\, f_{a_1 \cdots a_k} \\
&\quad \cdot (\lambda^1 + T^{-1}\beta^1, \cdots, \lambda^{k-1} + T^{-1}\beta^{k-1})\, d\beta^1 \cdots d\beta^{k-1} \\
&\sim T^p \int \cdots \int \Phi_1(\beta^1) \cdots \Phi_{k-1}(\beta^{k-1})\, \Phi_k(-\beta^1 - \cdots - \beta^{k-1}) \\
&\quad \cdot d\beta^1 \cdots d\beta^{k-1} f_{a_1 \cdots a_k}(\lambda^1, \cdots, \lambda^{k-1}), \\
&\qquad\qquad\qquad\qquad \text{for } \lambda^1 + \cdots + \lambda^k = 0 \\
&= 0(T^p), \qquad\qquad\quad \text{for } \lambda^1 + \cdots + \lambda^k \neq 0.
\end{aligned}$$

It follows from this last that the standardized joint cumulants of order greater than 2 tend to 0 and so the Fourier transforms are asymptotically normal.

REFERENCES

[1] G. G. Stokes, "Note on searching for periodicities," *Proc. Roy. Soc.*, vol. 29, p. 122, 1879.

[2] A. Schuster, "The periodogram of magnetic declination," *Cambridge Phil. Soc.*, vol. 18, p. 18, 1899.

[3] E. T. Whittaker and A. Robinson, *The Calculus of Observations*. Cambridge, England: Cambridge Univ. Press, 1944.

[4] J. W. Cooley and J. W. Tukey, "An algorithm for the machine calculation of complex Fourier series," *Math. Comput.*, vol. 19, pp. 297-301, 1965.

[5] A. A. Michelson, *Light Waves and Their Uses*. Chicago, Ill.: Univ. Chicago Press, 1907.

[6] I. Newton, *Opticks*. London, England: W. Innys, 1730.

[7] M. I. Pupin, "Resonance analysis of alternating and polyphase currents," *Trans. AIEE*, vol. 9, p. 523, 1894.

[8] H. L. Moore, *Economic Cycles Their Law and Cause*. New York: Macmillan, 1914.

[9] H. A. Panofsky and R. A. McCormick, "Properties of spectra of atmospheric turbulence at 100 metres," *Quart. J. Roy. Meteorol. Soc.*, vol. 80, pp. 546-564, 1954.

[10] J. A. Leese and E. S. Epstein, "Application of two-dimensional spectral analysis to the quantification of satelite cloud photographs," *J. Appl. Meteorol.*, vol. 2, pp. 629-644, 1963.

[11] M. S. Bartlett, "The spectral analysis of point processes," *J. Roy. Stat. Soc.*, vol. B 25, pp. 264-296, 1963.

[12] —, "The spectral analysis of two dimensional point processes," *Biometrika*, vol. 51, pp. 299-311, 1964.

[13] A. V. Oppenheim, R. W. Schafer, and T. G. Stockham, Jr., "Nonlinear filtering of multiplied and convolved signals," *Proc. IEEE*, vol. 56, pp. 1264-1291, 1968.

[14] B. P. Bogert, M. J. Healey, and J. W. Tukey, "The quefrency alanysis of time series for echoes: cepstrum, pseudo-covariance, cross-cepstrum and saphe cracking," in *Time Series Analysis*, M. Rosenblatt, Ed. New York: Wiley, pp. 209-243, 1963.

[15] M. S. Bartlett, "Periodogram analysis and continuous spectra," *Biometrika*, vol. 37, pp. 1-16, 1950.

[16] R. H. Jones, "A reappraisal of the periodogram in spectral analysis," *Technometrics*, vol. 7, pp. 531-542, 1965.

[17] K. Hasselman, W. Munk, and G. J. F. MacDonald, "Bispectra of ocean waves," in *Time Series Analysis*, M. Rosenblatt, Ed. New York: Wiley, pp. 125-139, 1963.

[18] N. Wiener, "Generalized harmonic analysis," *Acta Math.*, vol. 55, pp. 117-258, 1930.

[19] H. Wold, *Bibliography on Time Series and Stochastic Processes*. London, England: Oliver and Boyd, 1965.

[20] D. R. Brillinger, *Time Series: Data Analysis and Theory*. New York: Holt, Rinehart and Winston, 1974.

[21] A. Ya. Khintchine, "Korrelationstheories der stationären stochastischen Prozesse," *Math. Ann.*, vol. 109, pp. 604-615, 1934.

[22] H. Cramér, "On harmonic analysis in certain functional spaces," *Ark. Mat. Astron. Fys.*, vol. 28B, pp. 1-7, 1942.

[23] V. P. Leonov and A. N. Shiryaev, "Some problems in the spectral theory of higher moments, II," *Theory Prob. Appl. (USSR)*, vol. 5, pp. 460-464, 1960.

[24] B. Picinbono, "Tendence vers le caractère Gaussien par filtrage seléctif," *C. R. Acad. Sci. Paris*, vol. 248, p. 2280, 1959.

[25] M. Rosenblatt, "Some comments on narrow band-pass filters," *Quart. Appl. Math.*, vol. 18, pp. 387-393, 1961.

[26] D. R. Brillinger, "The frequency analysis of relations between stationary spatial series," in *Proc. 12th Biennial Seminar of the Canadian Math. Congress*, R. Pyke, Ed. Montreal, P.Q., Canada: Can. Math. Congr., pp. 39-81, 1970.

[27] E. J. Hannan and P. J. Thomson, "Spectral inference over narrow

bands," *J. Appl. Prob.*, vol. 8, pp. 157–169, 1971.
[28] E. Parzen, "On consistent estimates of the spectrum of stationary time series," *Ann. Math. Statist.*, vol. 28, pp. 329–348, 1957.
[29] M. Rosenblatt, "Statistical analysis of stochastic processes with stationary residuals," in *Probability and Statistics*, U. Grenander, Ed. New York: Wiley, pp. 246–275, 1959.
[30] W. D. Wright, *The Measurement of Color*. New York: Macmillan. 1958.
[31] E. W. Carpenter, "Explosions seismology," *Science*, vol. 147, pp. 363–373, 1967.
[32] D. G. Lambert, E. A. Flinn, and C. B. Archambeau, "A comparative study of the elastic wave radiation from earthquakes and underground explosions," *Geophys. J. Roy. Astron. Soc.*, vol. 29, pp. 403–432, 1972.
[33] W. H. Munk and G. J. F. MacDonald, *Rotation of the Earth*. Cambridge, England: Cambridge Univ. Press, 1960.
[34] G. J. F. MacDonald and N. Ness, "A study of the free oscillations of the Earth," *J. Geophys. Res.*, vol. 66, pp. 1865–1911, 1961.
[35] R. L. Wegel and C. R. Moore, "An electrical frequency analyzer," *Bell Syst. Tech. J.*, vol. 3, pp. 299–323, 1924.
[36] N. Wiener, *Time Series*. Cambridge, Mass.: M.I.T. Press, 1964.
[37] U. Grenander and M. Rosenblatt, *Statistical Analysis of Stationary Time Series*. New York: Wiley, 1957.
[38] E. Parzen, "An approach to empirical time series analysis," *Radio Sci.*, vol. 68 D, pp. 551–565, 1964.
[39] R. T. Lacoss, "Data adaptive spectral analysis methods," *Geophysics*, vol. 36, pp. 661–675, 1971.
[40] J. P. Burg, "The relationship between maximum entropy spectra and maximum likelihood spectra," *Geophysics*, vol. 37, pp. 375–376, 1972.
[41] K. N. Berk, "Consistent autoregressive spectral estimates," *Ann. Stat.*, vol. 2, pp. 489–502, 1974.
[42] V. E. Pisarenko, "On the estimation of spectra by means of nonlinear functions of the covariance matrix," *Geophys. J. Roy. Astron. Soc.*, vol. 28, pp. 511–531, 1972.
[43] J. Capon, "Investigation of long-period noise at the large aperture seismic array," *J. Geophys. Res.*, vol. 74, pp. 3182–3194, 1969.
[44] E. J. Hannan, *Multiple Time Series*. New York: Wiley, 1970.
[45] T. W. Anderson, *The Statistical Analysis of Time Series*. New York: Wiley, 1971.
[46] P. Whittle, "Estimation and information in stationary time series," *Ark. Mat. Astron. Fys.*, vol. 2, pp. 423–434, 1953.
[47] J. M. Chambers, "Fitting nonlinear models: numerical techniques," *Biometrika*, vol. 60, pp. 1–14, 1973.
[48] D. R. Brillinger, "An empirical investigation of the Chandler wobble and two proposed excitation processes," *Bull. Int. Stat. Inst.*, vol. 39, pp. 413–434, 1973.
[49] P. Whittle, "Gaussian estimation in stationary time series," *Bull. Int. Stat. Inst.*, vol. 33, pp. 105–130, 1961.
[50] A. M. Walker, "Asymptotic properties of least-squares estimates of parameters of the spectrum of a stationary nondeterministic time-series," *J. Australian Math. Soc.*, vol. 4, pp. 363–384, 1964.
[51] K. O. Dzaparidze, "A new method in estimating spectrum parameters of a stationary regular time series," *Teor. Veroyat. Ee Primen.*, vol. 19, p. 130, 1974.
[52] L. H. Koopmans, "On the coefficient of coherence for weakly stationary stochastic processes," *Ann. Math. Stat.*, vol. 35, pp. 532–549, 1964.
[53] H. Akaike and Y. Yamanouchi, "On the statistical estimation of frequency response function," *Ann. Inst. Stat. Math.*, vol. 14, pp. 23–56, 1962.
[54] L. J. Tick, "Estimation of coherency," in *Advanced Seminar on Spectral Analysis of Time Series*, B. Harris, Ed. New York: Wiley, 1967, pp. 133–152.

[55] R. J. Bhansali, "Estimation of the Wiener filter," in *Contributed Papers 39th Session Int. Stat. Inst.*, vol. 1, pp. 82–88, 1973.
[56] A. E. Hoerl and R. W. Kennard, "Ridge regression: biased estimation for nonorthogonal problems," *Technometrics*, vol. 12, pp. 55–67, 1970.
[57] B. R. Hunt, "Biased estimation for nonparametric identification of linear systems," *Math. Biosci.*, vol. 10, pp. 215–237, 1971.
[58] S. Bochner and K. Chandrasekharan, *Fourier Transforms*. Princeton, N.J.: Princeton Univ. Press, 1949.
[59] J. Mannos, "A class of fidelity criteria for the encoding of visual images," Ph.D. dissertation, Univ. California, Berkeley, 1972.
[60] L. J. Tick, "Conditional spectra, linear systems and coherency," in *Time Series Analysis*, M. Rosenblatt, Ed. New York: Wiley, pp. 197–203, 1963.
[61] C. W. J. Granger, *Spectral Analysis of Economic Time Series*. Princeton, N.J.: Princeton Univ. Press, 1964.
[62] N. R. Goodman, "Measurement of matrix frequency response functions and multiple coherence functions," Air Force Dynamics Lab., Wright Patterson AFB, Ohio, Tech. Rep. AFFDL-TR-65-56, 1965.
[63] J. S. Bendat and A. Piersol, *Measurement and Analysis of Random Data*. New York: Wiley, 1966.
[64] G. W. Groves and E. J. Hannan, "Time series regression of sea level on weather," *Rev. Geophys.*, vol. 6, pp. 129–174, 1968.
[65] W. Gersch, "Causality or driving in electrophysiological signal analysis," *J. Math. Biosci.*, vol. 14, pp. 177–196, 1972.
[66] N. R. Goodman, "Eigenvalues and eigenvectors of spectral density matrices," Tech. Rep. 179, Seismic Data Lab., Teledyne, Inc., 1967.
[67] D. R. Brillinger, "The canonical analysis of time series," in *Multivariate Analysis–II*, P. R. Krishnaiah, Ed. New York: Academic, pp. 331–350, 1970.
[68] M. B. Priestley, T. Subba Rao, and H. Tong, "Identification of the structure of multivariable stochastic systems," in *Multivariate Analysis–III*, P. R. Krishnaiah, Ed. New York: Academic, pp. 351–368, 1973.
[69] M. Miyata, "Complex generalization of canonical correlation and its application to a sea-level study," *J. Marine Res.*, vol. 28, pp. 202–214, 1970.
[70] W. S. Ligett, Jr., "Passive sonar: Fitting models to multiple time series," paper presented at NATO Advanced Study Institute on Signal Processing, Loughborough, U. K., 1972.
[71] D. R. Brillinger, "The analysis of time series collected in an experimental design," *Multivariate Analysis–III*, P. R. Krishnaiah, Ed. New York: Academic, pp. 241–256, 1973.
[72] D. R. Brillinger and M. Hatanaka, "An harmonic analysis of nonstationary multivariate economic processes," *Econometrica*, vol. 35, pp. 131–141, 1969.
[73] E. J. Hannan and R. D. Terrell, "Multiple equation systems with stationary errors," *Econometrica*, vol. 41, pp. 299–320, 1973.
[74] C. R. Rao, *Linear Statistical Inference and its Applications*. New York: Wiley, 1965.
[75] D. R. Cox and P. A. W. Lewis, *The Statistical Analysis of Series of Events*. London, England: Methuen, 1966.
[76] D. R. Brillinger, "The spectral analysis of stationary interval functions," *Proc. 6th Berkeley Symp. Math. Stat. Prob. Vol. 1*, L. M. Le Cam, J. Neyman, and E. L. Scott, Eds. Berkeley, Calif.: Univ. California Press, pp. 483–513, 1972.
[77] D. J. Daley and D. Vere-Jones, "A summary of the theory of point processes," in *Stochastic Point Processes*, P. A. W. Lewis, Ed. New York: Wiley, pp. 299–383, 1972.
[78] L. Fisher, "A survey of the mathematical theory of multidimensional point processes," in *Stochastic Point Processes*, P. A. W. Lewis, Ed. New York: Wiley, pp. 468–513, 1972.

[79] A. G. Hawkes and L. Adamopoulos, "Cluster models for earthquakes—regional comparisons," *Bull. Int. Stat. Inst.*, vol. 39, pp. 454–460, 1973.
[80] A. M. Yaglom, "Some classes of random fields in *n*-dimensional space related to stationary random processes," *Theory Prob. Appl. (USSR)*, vol. 2, pp. 273–322, 1959.
[81] L. Schwartz, *Théorie des Distributions, Vols. 1, 2.* Paris, France: Hermann, 1957.
[82] A. Papoulis, *The Fourier Integral and Its Applications.* New York: McGraw-Hill, 1962
[83] K. Ito, "Stationary random distributions," *Mem. Col. Sci. Univ. Kyoto* A, vol. 28, pp. 209–223, 1954.
[84] I. M. Gelfand and N. Ya. Vilenkin, *Generalized Functions,* vol. 4. New York: Academic, 1964.
[85] M. G. Kendall and A. Stuart, *The Advanced Theory of Statistics,* vol. 1. London, England: Griffin, 1958.
[86] D. R. Brillinger, "The identification of polynomial systems by means of higher order spectra," *J. Sound Vibration*, vol. 12, pp. 301–313, 1970.
[87] F. J. Beutler and O. A. Z. Leneman, "On the statistics of random pulse processes, *Inform. Contr.*, vol. 18, pp. 326–341, 1971.
[88] N. Wiener, *Nonlinear Problems in Random Theory.* Cambridge, Mass.: M.I.T. Press, 1958.
[89] A. Halme, "Polynomial operators for nonlinear systems analysis," *Acta Polytech. Scandinavica*, no. 24, 1972.
[90] L. J. Tick, "The estimation of the transfer functions of quadratic systems," *Technometrics*, vol. 3, pp. 563–567, 1961.
[91] D. R. Brillinger and M. Rosenblatt, "Asymptotic theory of *k*th order spectra," in *Advanced Seminar on the Spectral Analysis of Time Series*, B. Harris, Ed. New York: Wiley, pp. 153–188, 1967.
[92] G. Matheron, "The intrinsic random functions and their applications," *Adv. Appl. Prob.*, vol. 5, pp. 439–468, 1973.